PASSAGE TIMES FOR MARKOV CHAINS

Studies in Probability, Optimization and Statistics

Editors
Richard W. Cottle, Sidney Resnick,
Roger J.-B. Wets, Willem R. van Zwet

Volume 1

Also published in this series
J.W. Cohen, Analysis of Random Walks

Passage Times for Markov Chains

by

R. Syski
University of Maryland
USA

IOS Press
1992
Amsterdam • Oxford • Washington • Tokyo

ISBN 90 5199 060 X

Library of Congress Cataloging–in–Publication Date

Syski, R., 1924-
Passage times for Markov chains / Ryszard Syski.
560 p. cm.
Includes bibliographical references and index.
ISBN 905199060X
1. Markov processes. I. Title.
QA274.7.S97 1992
519.2'33-dc20 91-75446
CIP

Publisher:

IOS Press
Van Diemenstraat 94
1013 CN Amsterdam
Netherlands

Sole distributor in the UK and Ireland:

IOS Press/Lavis Marketing
73 Lime Walk
Headington
Oxford OX3 7AD
England

Distributor in the USA and Canada:

IOS Press, Inc.
P.O. Box 10558
Burke, VA 22009-0558
USA

Distributor in Japan:

IOS Japan Dept.
Highway Development Co. Ltd.
1st Golden Bldg, 8-2-9 Ginza
104 Tokyo - Chuoku
Japan

PRINTED IN MALTA

PREFACE

The purpose of this book is to present a unifying treatment of passage times in Markov chains, written in the systematic manner and based on modern developments, and to describe basic results of theoretical and practical importance.

Passage times have been investigated since early days of probability theory and its applications. The best known example is the first entrance time to a set, which embraces waiting times, absorption problems, extinction phenomena, busy periods and other applications. Another example, which only recently gains prominence, is the last exit time from a set.

The book is a survey of work on passage times in stable Markov chains with a discrete state space and a continuous time parameter. Although from the theoretical point of view stable chains are rather trite, nevertheless their structure is sufficiently rich to illustrate even the most sophisticated ideas. Moreover, such chains constitute bulk of applications, and it is here where the systematic unifying presentation is most desirable.

The main theme of this book is that the appropriate unifying framework for discussion of passage times is provided by the probabilistic potential theory, and all results presented in the text are interpreted from this point of view. Concepts of potentials, excessive functions and measures, balayage and duality, are freely used here. In particular, the crucial role of the Dirichlet problem and the Poisson equation, and various decomposition theorems, is stressed.

The starting point is to show that study of passage times depends essentially on a topological duality between functions and sets, which leads to the probabilistic representation by the notion of taboo probability (corresponding to killing of a chain at the first entrance). Properties of passage times are then linked to those of the first entrance time, and can be expressed in terms of taboo probabilities. The taboo probability becomes the central notion for the purpose of this book, both from the point of view of theory and of convergence in computation.

The objective is to obtain equations, or explicit expressions, for joint distributions of passage times and associated passage positions. Such equations for the first entrance times have been used in literature (usually with *ad hoc* justification), but those for the last exit times only recently become available. Although passage times are in fact examples of stopping and co-stopping times, they enjoy important position in theoretical aspects and in practical applications.

As it is well known, there are two complementary approaches to study of Markov chains, namely the "analytic approach" (dealing with equations for transition probabilities and Markov semi-groups) and the "measure theoretic approach" (dealing with sample functions properties and random sets). In this book the interdependence of both approaches is stressed to obtain the best results. However, for convenience of presentation, the first half of the book is more analytic in character, and measure theoretic discussion using functionals and martingales is postponed to the later part of the text.

Besides taboo probabilities, there are alternative starting points which could be used for presentation. One could start with the balayage properties ("product theorem") and the Dynkin theorem for stopping times. It is also convenient to base presentation entirely on properties of Markovian functionals and use of martingales. In particular, the rigorous treatment of last exit times, excursions, and transformations of Markov chains is obtained in this manner. However, analysis is rather abstract, and the explicit determination again reduces to the use of taboo probabilities. Thus, although these methods are discussed in the text, taboo probabilities provide direct and convenient starting point. Concerning analytic tools, the book makes use of matrix formulation for the purpose of economy in notation, but matrix theory is not needed. Frequently, explicit expressions are provided to facilitate calculations.

The book consists of 4 chapters and an appendix. The first chapter ("Preliminaries") summarizes the background material on Markov chains and their potential theory, and also on taboo probabilities. Next, passage times are introduced and methodology used in the book is explained.

The second chapter ("Analytic theory") treats passage times in detail. The Dirichlet problem and the Poisson equation are discussed, and decomposition theorems and duality are examined.

The third chapter ("Measure theory") introduces functionals and martingales, and presents various transformation of Markov chains (like reversing, killing, birthing, and a random time change). Several extensions are indicated, including composite passage times. This chapter is more advanced than previous chapters, and some of the results obtained earlier are here re-derived from a different point of view.

The last chapter ("Applications") treats various practical systems, and discusses the notion of energy of a chain. One section presents rather meager introduction to boundary theory.

Illustrative examples are provided throughout the book, but Section 7 discusses some well-known examples (including classical absorption problems, busy periods and waiting times in Markovian queues, branching processes), whereas the last Section 13 presents more advanced examples (including perturbation, energy, random time change, martingales, and Revuz measure).

The presentation depends heavily on functional properties of Markov chains, like behavior of sample functions, strong Markov property, Kolmogorov equations, minimal process, and the jump chain. For the background material on Markov chains consult the monograph by K. L. Chung [1967], who was the first to stress importance of taboo probabilities. For the background material on potential theory reference is made here to the monograph of R. M. Blumenthal and R. K. Getoor [1968]. Numerous other references are given in the text.

Despite all this, the book is addressed to applied probabilists, and to those who are interested in applications of probabilistic methods. The level of presentation is that of a graduate text in applied stochastic processes. In view of these aims, clarity of presentation takes precedence over secondary mathematical details whenever no serious harm may be expected. Advanced concepts needed in the text gain nowadays growing acceptance in applied fields, and it is hoped that this book will serve as an useful introduction. Indeed, it is rather exciting that such a narrow topic like passage times illustrates so well advanced concepts from probabilistic theories.

The author wishes to express his gratitude to Professors J. Glover, H. Gzyl, J. Mitro and H. Kaspi for pre-prints of their papers which have been very helpful to him. The author kept constantly in mind the appeal of Professor J. Keilson for search of a practical interpretation behind abstract formulations. Also, the friendly comment by Dr. P. Le Gall that nature is not Markovian, has been well taken. Last but not least, special thanks are due to Professor J. W. Cohen for his advice and for numerous friendly discussions spread over 40 years which have been always source of inspiration.

However, the author wants to make it clear that these writers are in no way, explicit or implicit, responsible for any misstatements, errors or distortions which may occur in this book.

Finally, it is a pleasure to thank Dr. E. Fredriksson (Director, IOS Press) for his interest in the publication of this book, and to the editors of IOS Press for their great care taken in its production.

R. Syski

College Park
October 1991

To my wife Barbara in commemoration of
our marriage of 42 years (and counting)

CONTENTS

CHAPTER I

PRELIMINARIES

This chapter consists of two sections containing the background material. Section 1 (Markoviana) presents basic aspects of Markov chains and their potential theory, needed for discussion of passage times. This material is standard and is given here for convenient reference. This section includes also presentation of the main tool used in this book, namely the concept of taboo probability. For this reason, discussion of taboo probability is presented in some more detail.

Section 2 (Passage probabilities) introduces passage times and describes methodology to be used in Chapter II. For this reason, this section is essential for the rest of the book.

Markov chains with a discrete state space and a continuous time parameter possess special properties which justify their independent study. Theory of chains involves considerable amount of analysis and measure theory, and numerous applications to stochastic models make such chains an important practical tool. The language of probabilistic potential theory is very suitable for discussion of basic properties of chains, especially of those connected with passage times. Indeed, passage times illustrate well basic ideas deeply rooted in Markovian structure, and are indispensable in applications to all Markovian models.

Although further restrictions imposed here make rather drastic specialisation of general chains (known as Chung processes), advantages offered by simplicity compensate for regrettable loss of generality. Indeed, due to the simple structure of chains considered here highly technical aspects of general theory undergo considerable simplifications, and make basic ideas more transparent. Henceforth, presentation can proceed with sufficient rigor along the middle course between abstract and applied points of view.

SECTION 1

MARKOVIANA

This section summarizes basic properties of Markov chains with a discrete state space and a continuous parameter, and their potential theory, as needed in this book. Only preliminary material is listed here, additional material will be introduced later when needed (Appendix contains a general summary).

Both analytic and measure-theoretic aspects are discussed here, and the relationship between sample functions behavior and properties of Markov semi-group (defined by matrices of transition probabilities) is emphasized. As equations for passage distributions constitute the ultimate aim, the presentation in this section is mostly analytic but motivated by sample functions approach.

This section consists of three rather extensive sub-sections devoted to Markov chains, taboo probabilities and potential theory, respectively.

Sub-section 1.1 treats Markov chains, stressing equations for transition probabilities and their Laplace transforms (resolvent approach), the role of the generator (intensity matrix), the associated jump chain and its connection with the minimal semi-group, and finally the sample functions properties.

Sub-section 1.2 is devoted to presentation of basic properties of taboo probabilities. These properties resemble those of the ordinary transition probabilities (corresponding to the case when the taboo set is absent), but restriction to transitions avoiding the taboo set, introduce important modifications. The notion of taboo probability is fundamental for discussion of passage times because all passage probabilities discussed in this book can be expressed in terms of taboo probabilities.

Sub-section 1.3 treats some analytic aspects of potential theory, namely decomposition theorems and their ramifications, entrance and exit laws, duality etc. Potential theoretic notions lead to unification and clear formulation of many aspects of passage times.

The material presented here is classical and proofs are omitted; references should be consulted for complete discussion.

Presentation in Sub-section 1.1 is based primarily on the pioneering work by W. Feller [1960, 1966] and by K. L. Chung [1967, 1970, 1982]. See also books by E. Dynkin [1965], M. Loeve [1963], D. Williams [1979], S. N. Ethier and T. G. Kurtz [1986].

For complete account of results concerning the basic taboo probabilities (and its extension to "exit taboo") see K. L. Chung [1967] who was the first to stress importance of the notion of taboo; see also E. Archinard [1974].

Sub-section 1.3 is essentially based on the following books: R. M. Blumenthal and R. K. Getoor [1968], P. A. Meyer [1966], C. Dellacherie and P. A. Meyer

[1982] and books by Chung, Dynkin and Williams mentioned above. See also papers: G. E. Denzel *et al.* [1967], J. Neveu [1964], G. Giroux [1970], D. Williams [1969], M. W. Chamberlain [1974], A. O. Pittenger [1971]. The recent book by S. Asmussen [1987] presents very useful summary of Markov chain theory.

For complete discussion of examples mentioned briefly here, see books on applied stochastic processes, like S. Karlin and M. H. Taylor [1975], and books by Feller already mentioned.

1.1. MARKOV CHAINS

§ 1.1.1. Markov property. Let $X = (X_t,\ 0 \leqslant t < \infty)$ be a (time-homogeneous) Markov chain (M.C.) defined on a (complete) probability space $(\Omega, \mathcal{F}, \mathbb{P})$, with a discrete (countable) state space $\mathbb{I}$ and a continuous (time) parameter $0 \leqslant t < \infty$, and having a standard stochastic matrix of transition probabilities

$$\mathbf{P}(t) = \big(p_{ij}(t)\big)$$

with a conservative intensity matrix

$$\mathbf{Q} = (q_{ij}).$$

It is assumed that all states are stable ("totally stable chain") and (almost) all sample functions (paths) are right-continuous with left limits. Moreover the family (X_t) of random variables is adapted to the (right-continuous) increasing family $(\mathcal{F}_t)$ of the natural σ-fields generated by the chain ("natural filtration" or "history"):

$$\mathcal{F}_t = \sigma(X_s,\ s \leqslant t) \subset \mathcal{F}, \qquad 0 \leqslant t < \infty,$$

and the chain X has the strong Markov property. Furthermore, both the backward (B) and the forward (F) Kolmogorov equations for transition probabilities hold, and $\mathbf{P}(t)$ is their common minimal solution.

It will be convenient to call such a chain the *usual* M.C., or to say that a M.C. satisfies "the usual conditions" (any departure from this assumption will be explicitly mentioned). Basic properties of the usual M.C. to be used here are summarized in this section; see also Appendix A2 for further material.

A random variable (r.v.) X_t represents a state of the chain prevailing at time t, and by definition the event $(X_t = j)$ for each state $j \in \mathbb{I}$ belongs to $\mathcal{F}_t$ for each t (adaptation). The *Markov property* expresses the fact that in terms of conditional probability

$$\mathbb{P}\big(X_{t+s} = j \mid \mathcal{F}_s\big) = \mathbb{P}\big(X_{t+s} = j \mid X_s\big) \quad \text{a.e.}$$

for every $j \in \mathbb{I}$, $s, t \geqslant 0$. The value of the random variable $\mathbb{P}(X_{t+s} = j \mid X_s)$ at $X_s = i$ defines the *transition probability*

$$\mathbb{P}\big(X_{t+s} = j \mid X_s = i\big) = p_{ij}(t),$$

for i and j in $\mathbb{I}$, $t \geqslant 0$, independently of s (time homogeneity).

The following are the basic properties of transition probabilities. For every $i, j \in \mathbb{I}$ and $t, s \geqslant 0$:

1) $p_{ij}(t) \geqslant 0$ (positive);

2) $\sum_j p_{ij}(t) = 1$ (stochastic);

3) $\sum_k p_{ik}(s)\, p_{kj}(t) = p_{ij}(t+s)$ (Chapman–Kolmogorov);

4) $\lim_{t\to 0} p_{ij}(t) = p_{ij}(0) = \delta_{ij}$ (standard).

The associated intensities q_{ij}, with $q_{ii} = -q_i$, are defined by

5) $q_{ij} = \lim_{t\to 0} p_{ij}(t)/t, \quad i \neq j;$

6) $q_i = \lim_{t\to 0} \left[1 - p_{ii}(t)\right]/t;$

7) $q_i = \sum_{i\neq j} q_{ij}$ (conservative).

The Laplace transforms of transition probabilities are defined by

8) $u_{ij}^{\alpha} = \int_0^{\infty} e^{-\alpha t} p_{ij}(t)\, dt, \quad \alpha > 0.$

When $\alpha = 0$, this will be written as u_{ij}. Clearly,

9) $\alpha u_{ij}^{\alpha} \leqslant 1$ (for $\alpha > 0$), $\quad u_{ij} \leqslant \infty$.

The concise matrix notation will be used for convenience. Thus, the Chapman–Kolmogorov equation expresses the semi-group property

$$\mathbf{P}(s)\mathbf{P}(t) = \mathbf{P}(t+s), \qquad s, t \geqslant 0,$$

of the Markov semi-group $\mathbf{P}(t), t \geqslant 0$, of transition matrices, with $\mathbf{P}(0) = \mathbf{I} = (\delta_{ij})$ the identity matrix. Standardness of $\mathbf{P}(t)$ is expressed in matrix form as (point-wise limit)

$$\lim_{t\to 0} \mathbf{P}(t) = \mathbf{I} = \mathbf{P}(0).$$

The definition of the intensity matrix $\mathbf{Q}$ is in matrix form (point-wise limit)

$$\mathbf{Q} = \lim_{t\to 0} \left[\mathbf{P}(t) - \mathbf{I}\right]/t$$

indicating that $\mathbf{Q}$ is the (infinitesimal) generator of the Markov semi-group.

The matrix $\mathbf{U}^{\alpha} = (u_{ij}^{\alpha})$ of Laplace transforms

$$\mathbf{U}^{\alpha} = \int_0^{\infty} e^{-\alpha t} \mathbf{P}(t)\, dt, \qquad \alpha > 0,$$

is called the *resolvent*, or the α-potential, of the Markov semi-group. The semi-group property is then expressed by the resolvent equation

$$\mathbf{U}^{\alpha} - \mathbf{U}^{\beta} = (\beta - \alpha)\mathbf{U}^{\alpha}\mathbf{U}^{\beta}, \qquad \alpha, \beta > 0,$$

with $\mathbf{U}^\alpha\mathbf{U}^\beta = \mathbf{U}^\beta\mathbf{U}^\alpha$. Standardness of $\mathbf{P}(t)$ indicates that

$$\lim_{\alpha\to\infty} \alpha\mathbf{U}^\alpha = \mathbf{I}.$$

Writing $\mathbf{1}$ for the column vector of units, the facts that $\mathbf{P}(t)$ is stochastic and $\mathbf{Q}$ is conservative are expressed, respectively, by (matrix multiplication)

$$\mathbf{P}(t)\mathbf{1} = \mathbf{1} = \alpha\mathbf{U}^\alpha\mathbf{1}, \qquad t \geqslant 0, \quad \alpha > 0; \qquad \mathbf{Q1} = \mathbf{0}$$

(i.e., $\mathbf{Q}$ is a singular matrix with row sums zero).

Several explanatory remarks are in order (see references for details).

a) The Chapman–Kolmogorov equation is a consequence of the Markov property (by smoothing property of conditional expectation). The equation expresses the intuitive fact that a transition from $X_0 = i$ to $X_{t+s} = j$ is decomposed into disjoint transitions through intermediary position $X_s = k$, and that the probability of transition from k to j during the time interval of length t, does not depend on initial state i (by Markov property). It is important to stress that a transition probability is a much simpler object than a Markov chain. However, in analytic discussion of Markov chains, and especially in applications, one usually deals with transition probabilities, and the Chapman–Kolmogorov equation serves as a starting point of analysis. In this book, however, both the analytic and the measure-theoretic approaches will be used, because they complement each other.

b) Assumption that the transition matrix $\mathbf{P}(t)$ is standard is essential, and will never be violated. Assumptions that $\mathbf{P}(t)$ is stochastic and that $\mathbf{Q}$ is conservative are also essential. In some special situations (for example, when sub-matrices are considered) these two assumptions may be relaxed to sub-stochastic $\mathbf{P}(t)$ (i.e., $\mathbf{P}(t)\mathbf{1} \leqslant \mathbf{1}$) and to non-conservative $\mathbf{Q}$ (i.e., $\mathbf{Q1} \leqslant \mathbf{0}$).

c) Existence of limits q_{ij} is taken here as an assumption, although it can be proved. In applications, however, intensities q_{ij} are given as primary data. It is known that q_{ij} for $i \neq j$ is always finite, but q_i may be infinite. A state i is called *stable* when $q_i < \infty$ and *instantaneous* when $q_i = \infty$. Exclusion of instantaneous states is a serious restriction, imposed regrettably for simplicity reasons. Nevertheless, stable states still offer a rich area for theoretical and practical investigations, and may serve as an introduction to more advanced aspects of Markov chains.

A state i is called *absorbing* iff $p_{ii}(t) = 1$ for all $t > 0$, or equivalently, iff $q_i = 0$. Thus, for non-absorbing stable states $0 < q_i < \infty$. In many chains (for example, when state space $\mathbb{I}$ is finite), q_i are bounded; this would be the case iff standardness condition 4) holds uniformly in i. Boundness of q_i is too restrictive, and will not be assumed here.

When $\alpha = 0$, it will be dropped from notation, so the potential matrix is $\mathbf{U} = (u_{ij})$. However

$$u_{ij} \leqslant \infty.$$

d) The important generalisation of the Markov property extends it from fixed times to random times. A M.C. $X = (X_t)$ is said to have the *strong Markov property* when

$$\mathbb{P}(X_{t+S} = j \mid \mathcal{F}_S) = \mathbb{P}(X_{t+S} = j \mid X_S) \quad \text{a.e.}$$

for any stopping time S (relative to the natural σ-fields), i.e., $(S \leqslant t) \in \mathcal{F}_t$ for every $t \geqslant 0$. The compound random variable X_S (for finite S) represents the "stopping place" of the chain, and $\mathcal{F}_S$ is the "pre-S" σ-field of events determined by the chain up to time S (see Appendix A1 for precise definitions).

Under conditions imposed here, X_S is $\mathcal{F}_S$-measurable and the M.C. X is strongly Markov. It follows that

$$\mathbb{P}(X_{t+S} = j \mid X_S = i) = p_{ij}(t), \qquad t \geqslant 0, \quad i, j \in \mathbb{I},$$

and the "post-S" chain $(X_{t+S}, \mathcal{F}_{t+S}, t \geqslant 0)$ is also Markov with the same transition probabilities as the original chain X.

e) On some occasions extension of a (time) parameter to $0 \leqslant t \leqslant \infty$ may be needed (e.g., when stopping time S is infinite). Then, for the new chain $(X_t, \mathcal{F}_t, 0 \leqslant t \leqslant \infty)$ definitions of X_∞ and $\mathcal{F}_\infty$ are required (with X_∞ being $\mathcal{F}_\infty$-measurable). This would require augmentation of the state space. In some cases it suffices to enlarge the state space $\mathbb{I}$ by attaching an auxiliary "state" Δ, and require that X_∞ takes value Δ. If $\lim X_t$ exists (in a specified sense), it may be taken as X_∞; otherwise X_∞ can be defined arbitrarily. For $\mathcal{F}_\infty$ one takes usually the σ-field of events determined by the chain X

$$\mathcal{F}_\infty = \sigma(X_t, \ 0 \leqslant t < \infty),$$

i.e., the smallest σ-field with respect to which all random variables are measurable for $t < \infty$. (See also remarks on boundary theory in § 1.1.3.)

f) Customarily, if $\mathbf{P}(t)$ is sub-stochastic on $\mathbb{I}$, an auxiliary absorbing state Δ may be adjoined to the state space and $\mathbf{P}(t)$ is extended to a stochastic $\mathbf{P}^\Delta(t)$ on the enlarged state space $\mathbb{I} \cup \{\Delta\}$ in such a way that $\mathbf{P}(t)$ and $\mathbf{P}^\Delta(t)$ coincide on $\mathbb{I}$, and the probability of absorption in Δ is given by the deficiency

$$p_{i\Delta}(t) = 1 - \sum_j p_{ij}(t), \qquad i \in \mathbb{I}.$$

g) Denoting by π the initial distribution of X_0, the absolute distributions of the chain are determined by

$$\mathbb{P}(X_{t_1} = j_1, \ldots, X_{t_n} = j_n) = \sum_i \pi_i p_{ij_1}(t_1) p_{j_1 j_2}(t_2 - t_1) \ldots p_{j_{n-1} j_n}(t_n - t_{n-1}),$$

for any collection of instants $t_1 \leqslant t_2 \leqslant \ldots \leqslant t_n$. In particular, the absolute distribution of X_t is given by the row vector $\pi(t) = \pi \mathbf{P}(t)$.

Conversely, given a Markov semi-group $\mathbf{P}(t)$ and a distribution π, the extension theorem of Daniell–Kolmogorov assures existence of a M.C. X whose transition probabilities coincide with $\mathbf{P}(t)$.

§ 1.1.2. Kolmogorov equations. Difference-differential equations for transition probabilities form a basis for most of the analytic theory of Markov chains. These equations are deduced from the Chapman–Kolmogorov equation. Assumptions so far imposed assure the validity of the backward Kolmogorov equation (with "backward differentiation") — the (B)-equation for short — but are not sufficient for the forward equation (with "forward differentiation") — the (F)-equation for short. Indeed, for stable states, the (B)-equation holds iff the matrix $\mathbf{Q}$ is conservative. To assure validity of the (F)-equation further assumptions are needed.

The (B)- and (F)-equations have the following form

$$\frac{dp_{ij}(t)}{dt} = -q_i p_{ij}(t) + \sum_{k \neq i} q_{ik} p_{kj}(t), \tag{B}$$

$$\frac{dp_{ij}(t)}{dt} = -p_{ij}(t) q_j + \sum_{k \neq j} p_{ik}(t) q_{kj}, \tag{F}$$

each with the initial conditions $p_{ij}(0) = \delta_{ij}$.

In matrix notation, with the initial condition $\mathbf{P}(0) = \mathbf{I}$,

$$\frac{d\mathbf{P}(t)}{dt} = \mathbf{Q}\mathbf{P}(t), \tag{B}$$

$$\frac{d\mathbf{P}(t)}{dt} = \mathbf{P}(t)\mathbf{Q}. \tag{F}$$

Equivalently, in terms of Laplace transforms, the Kolmogorov equations are

$$(\alpha + q_i) u_{ij}^{\alpha} - \delta_{ij} = \sum_{k \neq i} q_{ik} u_{kj}^{\alpha}, \tag{B}$$

$$u_{ij}^{\alpha} (\alpha + q_j) - \delta_{ij} = \sum_{k \neq j} u_{ik}^{\alpha} q_{kj}, \tag{F}$$

or, in the matrix form,

$$\alpha \mathbf{U}^{\alpha} - \mathbf{I} = \mathbf{Q}\mathbf{U}^{\alpha}, \tag{B}$$

$$\alpha \mathbf{U}^{\alpha} - \mathbf{I} = \mathbf{U}^{\alpha}\mathbf{Q}, \qquad \alpha > 0. \tag{F}$$

Validity of Kolmogorov equations, as well as existence and uniqueness of solutions, are essentially tied up with behavior of sample functions of a M.C. (indicating that analytic and probabilistic aspects cannot be separated). These important problems received considerable attention in literature, but here only brief remarks can be made.

Introduce the following expressions defined for all i, j, and t

$$q_i v_{ij}(t) = \sum_{k \neq i} q_{ik} p_{kj}(t), \qquad w_{ij}(t) = \sum_{k \neq j} p_{ik}(t) q_{kj}.$$

Hence, Kolmogorov equations in the integrated form may now be written as

$$p_{ij}(t) = \delta_{ij} e^{-q_i t} + \int_0^t q_i e^{-q_i(t-s)} v_{ij}(s)\, ds, \tag{B}$$

$$p_{ij}(t) = \delta_{ij} e^{-q_j t} + \int_0^t w_{ij}(s)\, e^{-q_j(t-s)}\, ds, \tag{F}$$

Behavior of sample functions provides the following interpretation (see § 1.1.3 and § 1.1.5). The term $v_{ij}(t)$ is the probability that having left state i (through jump to intermediary state k) the M.C. will be at time t afterwards in state j. The term $w_{ij}(t)$ can be regarded as the renewal density of state j ($j \neq i$). Consequently, the (B)-equation holds iff initially $X_0 = i$ and the first discontinuity is a jump. On the other hand, the (F)-equation is valid iff for $X_j = j$ ($t > 0$) the last discontinuity is a jump. This explains why total stability and conservativeness are compatible with the (B)-equation, but not with the (F)-equation which would contain another term (eliminated by additional assumptions).

The interplay between these two dual equations is one of the highlights of the Markov chain theory. It is known that the (B)- and (F)-equations have a common minimal solution which in general is sub-stochastic. Then, there are infinitely many solutions of the (B)-equation (which need not satisfy the (F)-equation). If however the minimal solution is stochastic, then it is the unique solution of both the (B)- and (F)-equations. A sufficient condition for $\mathbf{P}(t)$ to be stochastic is that

$$\sum_j w_{ij}(t) = \sum_j p_{ij}(t) q_j < \infty.$$

In this book, $\mathbf{P}(t) = (p_{ij}(t))$ will denote the unique common solution (the minimal solution) of both Kolmogorov equations, assumed to be stochastic. This simplifying and rather restrictive assumption is nevertheless quite sufficient for study of basic properties of passage times. It will be seen from considerations in Chapter II that the (B)-equation leads to equations for distributions of passage times, and the (F)-equation provides explicit solutions.

§ 1.1.3. Sample functions. Behavior of sample functions is of fundamental importance in studies of M.C.'s, and interconnection between sample functions and analytic properties of Markov semi-group constitutes highlights of the theory. It is therefore useful to summarize briefly main properties of sample functions (under simplifying assumptions) needed in this book. As already mentioned, a M.C. is a much richer structure than a Markov semi-group, because of sample function properties.

Let $X = (X_t,\ 0 \leqslant t < \infty)$ be a M.C. on a probability space $(\Omega, \mathcal{F}, \mathbb{P})$, assumed to satisfy the usual conditions on $\mathbf{P}(t)$ and $\mathbf{Q}$. It is convenient to write $X(t)$ instead of X_t; the value of X at a point (t,ω) is denoted by $X(t,\omega)$. For a fixed ω, a function of t defined by $X(\cdot,\omega)$ is a *sample function* or a *path* of a M.C.

Standardness of the matrix $\mathbf{P}(t)$ implies that the M.C. X is continuous in probability:

$$\lim_{s\to t} \mathbb{P}(X_s \neq X_t) = 0 \qquad \text{for every} \quad t,$$

but this does not imply continuity of sample functions. Nevertheless, continuity in probability, together with conservative $\mathbf{Q}$ and total stability of $\mathbb{I}$, imply existence of a modified (separable and measurable) chain whose (almost) all sample functions are right-continuous with left limits ("càdlàg"). Thus, sample functions have simple discontinuities (jumps) but may have discontinuities of the second kind (limit points of jumps). This last contingency is eliminated by assumption that $\mathbf{P}(t)$ is stochastic (see § 1.1.5).

For each state $i \in \mathbb{I}$, the random set S_i on the time axis, defined by

$$S_i(\omega) = \big(t\colon\ X(t,\omega) = i\big)$$

is called a set of constancy, and

$$\big(\omega\colon\ t \in S_i(\omega)\big) = \big(\omega\colon\ X(t,\omega) = i\big).$$

For stable i and for (almost) all ω, $S_i(\omega)$ is a countable union of disjoint, left-closed and right-open intervals of positive length, and a number of such intervals is finite in any finite interval on the time axis (see Fig. 1.1). It is known (by Markov property) that for any $i \in \mathbb{I}$, $s, t \geqslant 0$:

$$\mathbb{P}(X_u = i,\ s < u < s+t \mid X_s = i) = e^{-q_i t}.$$

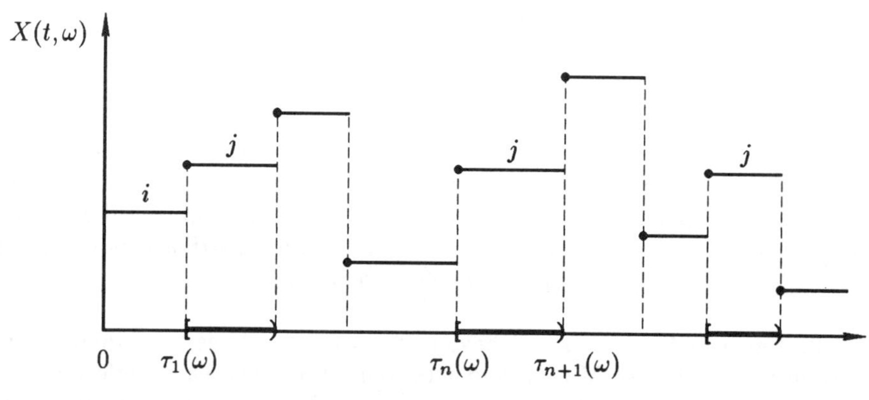

Figure 1.1. Realization of a sample function $X(t,\omega)$ and a random set $S_j(\omega)$.

Define now on the set $(X_0 = i)$ of positive probability, *the first exit time* from a state i by

$$\rho_i = \inf\,(t\colon\ t > 0,\ X_t \neq i)$$

with $\rho_i = \infty$ if the set in braces is empty. It follows from relation

$$(\rho_i > t,\ X_0 = i) = (X_u = i,\ 0 \leqslant u \leqslant t)$$

that ρ_i is a stopping time (relative to the natural σ-fields) with exponential distribution

$$\mathbb{P}(\rho_i > t \mid X_0 = i) = e^{-q_i t}, \qquad i \in \mathbb{I}, \quad t \geqslant 0.$$

Hence, when a M.C. enters a stable state i, it must stay in i for some positive time. The random time ρ_i is an example of a passage time of great importance in the sequel. The relationship between the above two (exponential) probabilities illustrates the general theory to be developed in Chapter II.

For non-absorbing stable states, the random variable $X(\rho_i)$ represents, by right-continuity of paths, the state entered at the instant of leaving a state i. One can now enumerate consecutive jumps of the chain until the first discontinuity that is not a jump. Let τ_n be the instant of occurrence of the n-th jump ($n \geqslant 0$). Then, $\tau_0 = 0$ and

$$\tau_n = \inf\big(s\colon\ s > \tau_{n-1},\ X(s) \neq X(s-)\big), \qquad n \geqslant 1,$$

with infimum taken to be ∞ when the set in braces is empty.

Clearly, each τ_n is a stopping time (relative to the natural σ-fields) and $\tau_n \leqslant \tau_{n+1}$ for all n. Hence

$$\lim_{n\to\infty} \tau_n = \tau$$

exists a.e. and represents the instant of occurrence of the first discontinuity, that is not a jump.

The stopping time τ is called the *life time* of a M.C., when the chain is considered only up to time τ. Note that for all sample functions (starting in $\mathbb{I}$) necessarily $\tau > 0$ from the right-continuity. Obviously, $\tau = \infty$ means that (almost) all sample functions have only jump discontinuities.

However, $X(\tau-)$ does not take values in the state space $\mathbb{I}$ and a "boundary point" Δ must be attached so that

$$\tau = \inf\big(t\colon\ t > 0,\ X(t-) = \Delta\big).$$

On the other hand, $X(\tau)$ takes values in $\mathbb{I}$ (by "return from the boundary"). The proper framework for discussion of these topics is the boundary theory for M.C.'s; see a brief discussion in Section 12 and also some comments in § 1.1.5. The fact that $\mathbf{Q}$ is conservative implies that there are no jumps to boundary (see § 1.1.4).

Furthermore, càdlàg property and continuity in probability imply that the M.C. X is continuous a.e. at each t, i.e.,

$$\mathbb{P}\big(\lim_{s\to t} X_s = X_t\big) = 1.$$

This means that the probability of a discontinuity at a fixed time t is zero. In other words, X has no "fixed times of discontinuity", but has "movable discontinuities" (depending on ω). More generally, with reference to stopping times, the chain X is known to be "quasi-left-continuous", and as a consequence all its jumps are "totally inaccessible" (see Appendix A1).

§ 1.1.4. Jump chain. The discrete parameter M.C. $X^* = (X_n,\ 0 \leqslant n < \infty)$ extracted from the original continuous parameter M.C. $X = (X_t,\ 0 \leqslant t < \infty)$ by restricting attention to instants τ_n at which transitions occur, is called the *jump chain*. Thus,

$$X_n = X(\tau_n), \qquad n = 0, 1, \ldots,$$

with σ-fields $\mathcal{F}_n = \sigma(X_k,\ k \leqslant n)$ for each n.

The jump chain has the same state space $\mathbb{I}$, and its one-step transition matrix is given by a *jump matrix* $\mathbf{R} = (r_{ij})$ associated with the intensity matrix $\mathbf{Q} = (q_{ij})$ by the following definition

$$r_{ij} = \begin{cases} q_{ij}/q_i & \text{for} \quad i \neq j, \\ 0 & \text{for} \quad i = j, \end{cases}$$

provided $q_i > 0$ (and by $r_{ij} = \delta_{ij}$ when $q_i = 0$).

Evidently, $\mathbf{R}$ is stochastic ($\mathbf{R1} = \mathbf{1}$). Denoting by $\mathbf{\Theta}$ the diagonal matrix of elements $q_i > 0$, the relation between $\mathbf{R}$ and $\mathbf{Q}$ takes the form

$$\mathbf{R} - \mathbf{I} = \mathbf{\Theta}^{-1}\mathbf{Q}.$$

The jump chain (X_n) is the important tool in the study of the original chain (X_t), because it is associated with the intensity matrix $\mathbf{Q}$. Hence, if attention is directed to changes of states in the original M.C., then properties depending only on such changes may be expressed in terms of the jump chain.

For this reason, it is useful to comment further on the jump chain and to provide probabilistic interpretation of the jump matrix $\mathbf{R}$. See references for details.

1) By the strong Markov property, the probability that a M.C. X will jump to state j, having just left state i, is

$$\mathbb{P}\big(X(\rho_i) = j \mid X(\rho_i -) = i\big) = r_{ij}, \qquad i, j \in \mathbb{I},$$

where ρ_i is the first exit time from state i.

This follows from the basic result for a strong Markov chain, which asserts that on the set $(X_0 = i)$ of positive probability, one has, for $t \geqslant \rho_i$,

$$\mathbb{P}\big(X_t = j \mid \rho_i = s\big) = v_{ij}(t - s), \qquad t \geqslant s,$$

where

$$v_{ij}(t) = \sum_k r_{ik} p_{kj}(t).$$

This relation indicates that a chain leaves the initial state i at time ρ_i by a jump to state k, and then goes to state j in the remaining time $t - s$ (see Fig. 1.2). In particular, for $\rho_i = t$, $v_{ij}(0) = r_{ij}$, which is the relation stated above.

Considering the post-ρ_i chain $X(t + \rho_i)$, $t \geqslant 0$, which originates at the instant of leaving a state i, one obtains from the above basic result that

$$\mathbb{P}\big(X(t + \rho_i) = j \mid \rho_i = s\big) = v_{ij}(t), \qquad t \geqslant 0.$$

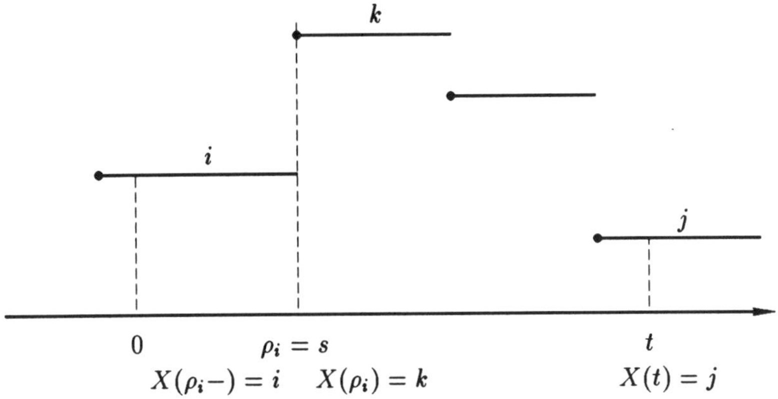

$$X(\rho_i-) = i,\ X(\rho_i) = k,\ X(t) = j,\ v_{ij}(0) = r_{ij}.$$

Figure 1.2. Interpretation of $v_{ij}(t-s)$ for $t \geqslant s$.

For $t = 0$, this yields again r_{ij} interpreted as a jump from i to j. Observe that the probability $v_{ij}(t)$ is the same as that mentioned in § 1.1.2, and the integrated form of the (B)-equation follows from partitioning of the event $(X_t = j)$ according to $\rho_i \leqslant t$ and $\rho_i > t$. Finally, the fact that $\mathbf{Q}$ is conservative (i.e, $\mathbf{R}$ stochastic) implies that there are no jumps to Δ,

$$\mathbb{P}\big(X(\rho_i) = \Delta \mid X(\rho_i-) = i\big) = 1 - \sum_j r_{ij} = 0.$$

2) The strong Markov property implies another result of importance

$$\begin{aligned}&\mathbb{P}\big(X_{n+1} = j,\ \tau_{n+1} - \tau_n > t \mid X_0, \dots, X_n,\ \tau_0, \dots, \tau_n\big)\\ &\quad = \mathbb{P}\big(X_{n+1} = j,\ \tau_{n+1} - \tau_n > t \mid X_n\big) = e^{-q_{X_n} t} r_{X_n j}, \qquad t \geqslant 0, \quad n \geqslant 0, \quad j \in \mathbb{I}.\end{aligned}$$

Taking $t = 0$, it indicates that $(X_n,\ n \geqslant 0)$ is indeed a M.C. For $n = 0$, one finds the joint distribution of ρ_i and $X(\rho_i)$ on the set $X_0 = i$,

$$\mathbb{P}\big(X(\rho_i) = j,\ t < \rho_i \mid X(\rho_i-) = i\big) = e^{-q_i t} r_{ij}.$$

Induction argument leads to

$$\begin{aligned}&\mathbb{P}\big(\tau_1 - \tau_0 > t_1, \dots, \tau_n - \tau_{n-1} > t_n \mid X_0 = i_0, \dots, X_n = i_n\big)\\ &\quad = e^{-q_{i_0} t_1} \dots e^{-q_{i_{n-1}} t_n}.\end{aligned}$$

This means that intervals $\tau_{n+1} - \tau_n$ between jumps are conditionally independent given the consecutive states being visited, and have exponential distributions.

3) The n-step transition matrix $\mathbf{R}^n = (r_{ij}^{(n)})$ is the n-th power of $\mathbf{R}$, and the semi-group property is

$$\mathbf{R}^{n+m} = \mathbf{R}^n \mathbf{R}^m, \qquad \mathbf{R}^0 = \mathbf{I},$$

or, explicitly,

$$r_{ij}^{(n+m)} = \sum_k r_{ik}^{(n)} r_{kj}^{(m)}, \qquad n, m \geqslant 0,$$

with $r_{ij}^{(1)} = r_{ij}$, $r_{ij}^{(0)} = \delta_{ij}$.

The matrix $\mathbf{G} = (g_{ij})$ defined by

$$\mathbf{G} = \sum_{n=0}^{\infty} \mathbf{R}^n, \qquad g_{ij} = \sum_{n=0}^{\infty} r_{ij}^{(n)} \leqslant \infty,$$

is called the (Green) potential for the jump matrix $\mathbf{R}$. One has

$$\mathbf{G}(\mathbf{I} - \mathbf{R}) = \mathbf{I} = (\mathbf{I} - \mathbf{R})\mathbf{G}.$$

Here g_{ij} represent the average number of visits to state j, conditional on the initial state i, for the jump chain. Moreover, denoting by s_{ij} the probability of infinitely many visits to state j (from the state i), the 0–1 law yields: $s_{ij} = 0$ for j transient and $s_{ij} = 1$ for j recurrent (see § 1.1.7).

§ 1.1.5. Minimal solution. The jump chain is used to construct the minimal solution of the Kolmogorov equations, providing clear illustration of interplay between sample functions and analytic aspects. Assume that all states are stable and non-absorbing ($0 < q_i < \infty$).

Let N_t be the number of jumps of a M.C. (X_t) in the interval $(0, t)$ or $[0, t]$, as the probability of a jump at a fixed point is zero. Then, write

$$p_{ij}^{(n)}(t) = \mathbb{P}^i(N_t = n,\ X_t = j) = \mathbb{P}^i(X_t = j,\ \tau_n \leqslant t < \tau_{n+1}),$$

$$p_{ij}(t) = \sum_{n=0}^{\infty} p_{ij}^{(n)}(t) = \mathbb{P}^i(N_t < \infty,\ X_t = j) = \mathbb{P}^i(X_t = j,\ \tau > t),$$

for $i, j \in \mathbb{I}$, $n \geqslant 0$, $t \geqslant 0$, where $\mathbb{P}^i$ denotes conditioning on $X_0 = i$.

Consideration of the first exit time from state i (exponential with parameter q_i) and subsequent jump to some state k (with probability r_{ik}) followed by n jumps in the remaining time interval, yields the recurrence relation (B)

$$p_{ij}^{(n+1)}(t) = \sum_{k \neq i} \int_0^t p_{kj}^{(n)}(s) q_{ik} e^{-q_i(t-s)}\, ds, \qquad n \geqslant 0, \quad t \geqslant 0,$$

with $p_{ij}^{(0)}(t) = \delta_{ij} e^{-q_i t}$.

Next, addition over all n yields the integrated form of the (B)-equation stated in § 1.1.2.

Alternatively, considering the last jump to state j before time t (from some state k, with probability r_{kj}), one can obtain the dual recurrence relation (F)

$$p_{ij}^{(n+1)}(t) = \sum_{k \neq j} \int_0^t p_{ik}^{(n)}(s)\, q_{kj} e^{-q_j(t-s)}\, ds, \qquad n \geqslant 0, \quad t \geqslant 0,$$

with the same initial condition at $n = 0$. This yields the integrated form of the (F)-equation stated in § 1.1.2.

It can be verified that both recurrence relations lead to the same $(p_{ij}^{(n)}(t))$, and that $(p_{ij}(t))$ obtained in this fashion does indeed satisfy the Chapman–Kolmogorov equation and yields the standard transition matrix.

Moreover, if $(p_{ij}^*(t))$ is any other solution of Kolmogorov equations with the same $\mathbf{Q}$ as $(p_{ij}(t))$, then

$$p_{ij}(t) \leqslant p_{ij}^*(t) \qquad \text{for all} \quad i,\ j,\ \text{and } t.$$

Thus, the solution obtained by the above construction is the *minimal solution*.

The following comment elucidates connection between uniqueness of solution of Kolmogorov equations and the fact that sample functions may have discontinuities which are not jumps. All M.C.'s with the same intensity matrix $\mathbf{Q}$ agree with the *minimal* M.C. till its life time τ, at which the minimal chain stops (at the "boundary"), but a non-minimal M.C. may continue beyond τ (by "return from the boundary" to the state space). Hence, the minimal solution is unique (and corresponds to a stopped chain), but general solutions are not unique.

If, however, the minimal solution is stochastic, then it is the unique solution of Kolmogorov equations, and (almost) all sample functions are step functions. Indeed, the above argument yields the distribution of the life time

$$\sum_j p_{ij}(t) = \mathbb{P}^i(N_t < \infty) = \mathbb{P}^i(\tau > t),$$

and this equals 1 for stochastic $\mathbf{P}(t)$, so then τ is infinite with probability one. Hence, only the minimal chain is present, as assumed in this book.

It is important to stress that the minimal solution is essentially associated with the jump chain. Although the minimal chain may be considered as a jump chain with randomized (exponential) durations of states, the theory of continuous parameter M.C.'s is much richer than that of associated discrete parameter chains. The time dependence introduces crucial modifications as it will seen later in this book.

It may be remarked that construction of recurrence relations used above, is an example of the "density argument" which traces the jumps of the chain. In particular, the (F)-relation actually uses renewal density.

§ 1.1.6. Feller relation. To establish connection between the resolvent of the minimal semi-group and the corresponding jump matrix, define the auxiliary α-transition matrix $\mathbf{P}^\alpha(t) = (p_{ij}^\alpha(t))$ of order $\alpha > 0$ by

$$p_{ij}^\alpha(t) = e^{-\alpha t} p_{ij}(t).$$

Obviously, $\mathbf{P}^\alpha(t) = e^{-\alpha t}\mathbf{P}(t)$, $t \geqslant 0$, form a sub-stochastic Markov semi-group. The M.C. X^α corresponding to $\mathbf{P}^\alpha(t)$, is called the α-chain of order $\alpha > 0$. This chain can be regarded as obtained from the original M.C. X by killing it at some random time independent of the chain and having exponential distribution. Evidently, the α-chain has only transient states, and the resolvent of the original chain X is now

$$\mathbf{U}^{\alpha} = \int_0^{\infty} \mathbf{P}^{\alpha}(t)\,dt.$$

The intensity matrix for the α-chain is

$$\mathbf{Q}^{\alpha} = (q_{ij}^{\alpha}) = \mathbf{Q} - \alpha\mathbf{I},$$

so $q_{ij}^{\alpha} = q_{ij}$ for $i \neq j$, $-q_{ii}^{\alpha} = q_i^{\alpha} = q_i + \alpha$. Although $\mathbf{Q}^{\alpha}$ is not conservative (obviously, $\mathbf{Q}^{\alpha}\mathbf{1} = -\alpha\mathbf{1}$), $\mathbf{P}^{\alpha}(t)$ satisfies Kolmogorov equations

$$\frac{d\mathbf{P}^{\alpha}(t)}{dt} = \begin{cases} \mathbf{Q}^{\alpha}\mathbf{P}^{\alpha}(t), & \text{(B)} \\ \mathbf{P}^{\alpha}(t)\mathbf{Q}^{\alpha}. & \text{(F)} \end{cases}$$

Define the jump matrix $\mathbf{R}^{\alpha} = (r_{ij}^{\alpha})$ by

$$r_{ij}^{\alpha} = q_i r_{ij}(q_i + \alpha)^{-1} \qquad \text{for} \quad i \neq j,$$

and $r_{ii}^{\alpha} = 0$ for $q_i > 0$, so

$$\mathbf{R}^{\alpha} - \mathbf{I} = (\mathbf{\Theta}^{\alpha})^{-1}\mathbf{Q}^{\alpha},$$

where $\mathbf{\Theta}^{\alpha} = \mathbf{\Theta} + \alpha\mathbf{I}$. Note that $\mathbf{R}^{\alpha}\mathbf{1} < \mathbf{1}$ for $\alpha > 0$.

Write

$$\mathbf{R}^{\alpha n} = \left(r_{ij}^{\alpha(n)}\right)$$

for the n-step transition matrix, and $\mathbf{G}^{\alpha} = (g_{ij}^{\alpha})$ for the corresponding (Green) potential,

$$\mathbf{G}^{\alpha} = \sum_{n=0}^{\infty} \mathbf{R}^{\alpha n} \qquad \text{with} \quad g_{ij}^{\alpha} = \sum_{n=0}^{\infty} r_{ij}^{\alpha(n)} < \infty, \quad \alpha > 0.$$

It should be remarked that when i is an absorbing state ($q_i = 0$), the definition may be modified to $r_{ij}^{\alpha} = \delta_{ij}$, so such a state remains absorbing for the α-chain. Note also that probabilities r_{ij}^{α} may be regarded as Laplace transform of expressions $q_i e^{-q_i t} r_{ij}$ from § 1.1.4.

The following important result which is used frequently in this book, will be called the *Feller relation*.

Let $\mathbf{U}^{\alpha} = (u_{ij}^{\alpha})$, $\alpha > 0$, be the resolvent of the minimal semi-group $\mathbf{P}(t)$, $t \geqslant 0$, and let $\mathbf{G}^{\alpha} = (g_{ij}^{\alpha})$ be the (Green) potential of the α-jump matrix $\mathbf{R}^{\alpha}$. Then

$$\mathbf{U}^{\alpha} = \mathbf{G}^{\alpha}(\mathbf{\Theta}^{\alpha})^{-1} \qquad \text{for} \quad \alpha > 0,$$

or, explicitly,

$$u_{ij}^{\alpha} = g_{ij}^{\alpha}(\alpha + q_j)^{-1}, \qquad i, j \in \mathbb{I}.$$

The result follows immediately by taking Laplace transforms of the (B)-recurrence relation or, equivalently, of the (F)-relation from construction of the minimal solution in § 1.1.5. Indeed, denoting by $\mathbf{U}^{\alpha(n)}$ the transform of the matrix $\mathbf{P}^{(n)}(t)$, the (B)-relation yields

$$\mathbf{U}^{\alpha(n)} = \mathbf{R}^{\alpha}\mathbf{U}^{\alpha(n-1)} = \ldots = (\mathbf{R}^{\alpha})^n\mathbf{U}^{\alpha(0)} = (\mathbf{R}^{\alpha})^n(\mathbf{\Theta}^{\alpha})^{-1}.$$

Dually, from the (F)-relation,

$$\mathbf{U}^{\alpha(n)} = \mathbf{U}^{\alpha(n-1)}\widehat{\mathbf{R}}^{\alpha} = \ldots = (\mathbf{\Theta}^{\alpha})^{-1}(\widehat{\mathbf{R}}^{\alpha})^n,$$

where

$$\widehat{\mathbf{R}}^{\alpha} = \mathbf{\Theta}^{\alpha}\mathbf{R}^{\alpha}(\mathbf{\Theta}^{\alpha})^{-1} \qquad \text{and} \qquad \widehat{\mathbf{G}}^{\alpha} = \mathbf{\Theta}^{\alpha}\mathbf{G}^{\alpha}(\mathbf{\Theta}^{\alpha})^{-1}.$$

Addition over all n completes the argument.

The meaning of the Feller relation is intuitively clear and indicates the renewal argument: for the α-chain, the average time spent in state j is equal to the average number of visits to j multiplied by the average duration of state j.

The Feller relation holds also for $\alpha = 0$, provided $\mathbf{U}$ is finite, so

$$\mathbf{G} = \mathbf{U}\mathbf{\Theta} \qquad \text{and} \qquad u_{ij}q_j = g_{ij}.$$

Thus, for the original M.C. X, u_{ij} is the average time spent in state j and g_{ij} is the average number of visits to j (both conditional on initial state i), and $1/q_j$ is the average duration of a state j.

§ 1.1.7. Passage relations. Passage times to a single state are well-known in the theory of M.C.'s, and their distributions lead to important classification of states. Equations given here are for the purpose of reference, and serve as the most familiar example of the general theory developed in Chapter II.

For a fixed state j, *the first entrance time* T_j to j is defined by

$$T_j = \inf(t\colon\ t > 0,\ X_t = j),$$

where as always the infimum of the empty set is taken to be ∞. By definition, T_j is identically zero when the initial state is j itself. As it is customary, write $F_{ij}(\cdot)$ for the distribution function of T_j, conditional on the initial state i (where $i \neq j$).

Assuming that a M.C. is initially in state i, *the first return time* T_ρ to a state i is defined as

$$T_\rho = \inf(t\colon\ t > \rho_i,\ X_t = i),$$

where ρ_i is the first exit time from state i (as defined in § 1.1.3). Note that this definition indicates that the chain leaves state i at instant ρ_i, and returns afterwards (excursion from i). Write $F_{ii}(\cdot)$ for the distribution function of T_ρ (conditional on initial state i). With this convention, $F_{ij}(\cdot)$ is defined for each i and j, although cases $i \neq j$ and $i = j$ are conceptually different.

It is known that $F_{ij}(\cdot)$ satisfies the "first entrance equation" for all i, j, and $t \geqslant 0$,

$$p_{ij}(t) = \delta_{ij}e^{-q_i t} + \int_{0-}^{t} dF_{ij}(s)\,p_{jj}(t-s),$$

where integration is taken over interval $[0,t]$. When j is stable and non-absorbing ($0 < q_j < \infty$), then $F_{ij}(\cdot)$ is absolutely continuous with density $f_{ij}(\cdot)$. The first entrance equation has intuitive meaning (path starting at i leads first to state j at time s, and then in remaining time $t - s$ goes from j to j), and will be obtained later in Section 3 from the general theory.

Define the Laplace–Stieltjes (L.S.) transform of $F_{ij}(t)$ by

$$f_{ij}^{\alpha} = \int_{0-}^{\infty} e^{-\alpha t}\, dF_{ij}(t), \qquad \alpha > 0,$$

with the integral taken over $[0,\infty)$. In terms of transforms, the first entrance equation takes a more convenient form

$$u_{ij}^{\alpha} = f_{ij}^{\alpha} u_{jj}^{\alpha}, \qquad i \neq j,$$

$$u_{ii}^{\alpha}(\alpha + q_i)(1 - f_{ii}^{\alpha}) = 1, \qquad i = j.$$

The expression

$$f_{ij} = F_{ij}(\infty) = \lim_{\alpha \to 0} f_{ij}^{\alpha}$$

represents the probability that the first entrance time to state j ($i \neq j$), or the first return time to state i ($i = j$), is finite. Similarly, denote by m_{ij} the average entrance time to j from state i ($i \neq j$), and by m_{ii} the average return time to i.

This leads to the basic classification of states. A state i is called *transient* when $f_{ii} < 1$ or *recurrent* when $f_{ii} = 1$. A recurrent state i is said to be *positive* if $m_{ii} < \infty$ or *null* if $m_{ii} = \infty$. A positive recurrent state is also called *ergodic*. A quantity $e_i = 1 - f_{ii}$ is called the *escape probability*, and $q_i e_i$ is the *escape rate*. It follows from the passage relations (as $\alpha \to 0$) that a state i is recurrent iff $u_{ii} = \infty$, and transient iff $u_{ii} < \infty$.

The case of absorbing states requires some comment. Clearly, an absorbing state is recurrent, because $u_{ii} = \infty$ and $f_{ii} = 1$. If j is an absorbing state, but i is not, then the entrance equations yield for the probability of absorption in j

$$\alpha u_{ij}^{\alpha} = f_{ij}^{\alpha} \qquad \text{or} \qquad p_{ij}(t) = F_{ij}(t) \qquad \text{for} \quad i \neq j.$$

It should be remarked that Feller relation implies that probabilities f_{ij} coincide with corresponding probabilities for the jump chain governed by the jump matrix $\mathbf{R}$. Consequently, the minimal chain and its jump chain have the same character of states, as it is intuitively obvious.

The dual passage relations, less known than the first entrance relations, concern the *last exit time* from a state i before time t, defined by

$$L_t = \sup\,(s\colon\ s \leqslant t,\ X_s = i),$$

where supremum of the empty set is taken as 0. Note that L_t is a co-stopping time (relative to the natural σ-fields) as its values are determined by the chain after time t. Also $0 \leqslant L_t \leqslant t$.

Analysis of sample functions yields the "last exit equation" for all $i \neq j,\ t \geqslant 0$,

$$p_{ij}(t) = \int_0^t p_{ii}(s)\, n_{ij}(t-s)\, ds,$$

where $n_{ij}(t)/q_i$ is the renewal density of the average time spent in a state j up to time t after the last exit from state i. Note that $n_{ij}(0) = q_{ij}$. The above equation has intuitive meaning, and it will be obtained later from the general theory (see Section 4).

Writing

$$n_i(t) = \sum_{i \neq j} n_{ij}(t) \qquad \text{with} \quad n_i(0) = q_i,$$

one can deduce that

$$1 - p_{ii}(t) = \int_0^t p_{ii}(s) n_i(t-s)\, ds.$$

In terms of Laplace transforms, the last exit relation become, for $\alpha > 0$,

$$u_{ij}^{\alpha} = u_{ii}^{\alpha} n_{ij}^{\alpha}, \qquad i \neq j,$$
$$u_{ii}^{\alpha}(1 + n_i^{\alpha}) = 1/\alpha.$$

Note that $\alpha n_i^{\alpha} \to q_i e_i$ as $\alpha \to 0$. The expression $n_i(t)$ is interpreted as the "Lévy measure" of the interval $(t, \infty]$, with an atom at ∞ of magnitude $q_i e_i$ and total mass q_i.

§ 1.1.8. Ergodic properties. The ergodic theorem for transition probabilities asserts that for every i and j the following limit exists

$$\lim_{t \to \infty} p_{ij}(t) = \lim_{\alpha \to 0} \alpha u_{ij}^{\alpha} = \varepsilon_{ij} \geqslant 0,$$

and

$$\varepsilon_{ij} = \sum_k p_{ik}(t) \varepsilon_{kj} = \sum_k \varepsilon_{ik} p_{kj}(t) = \sum_k \varepsilon_{ik} \varepsilon_{kj}$$

with

$$\sum_j \varepsilon_{ij} \leqslant 1.$$

Moreover, the (B)- and the (F)-equations give

$$\sum_k q_{ik} \varepsilon_{kj} = \sum_k \varepsilon_{ik} q_{kj} = 0.$$

In matrix notation, the limit matrix $\mathbf{E} = (\varepsilon_{ij})$ is,

sub-stochastic — $\mathbf{E1} \leqslant \mathbf{1}$;
idempotent — $\mathbf{EE} = \mathbf{E}$;
invariant — $\mathbf{EP}(t) = \mathbf{P}(t)\mathbf{E}, \qquad \mathbf{QE} = \mathbf{EQ} = \mathbf{0}$.

From passage equations (§ 1.1.7) it follows that, for a non-absorbing stable state j,

$$\varepsilon_{ij} = f_{ij}\varepsilon_{jj}, \qquad i \neq j,$$
$$\varepsilon_{jj} = (q_j m_{jj})^{-1},$$

whereas for an absorbing j one has $\varepsilon_{ij} = f_{ij}$ $(i \neq j)$, $\varepsilon_{jj} = 1$. Note that a recurrent state j is positive if $\varepsilon_{jj} > 0$ and null if $\varepsilon_{jj} = 0$. For a transient state j, necessarily $\varepsilon_{jj} = 0$.

A M.C., or its transition matrix $\mathbf{P}(t)$, is said to be irreducible if the state space $\mathbb{I}$ consists of a single communicating class (so all states have the same character). For the *ergodic* (i.e., positive recurrent, irreducible) chain, the limits $\varepsilon_{ij} = \varepsilon_j > 0$ are strictly positive, independent of an initial state i, and $\sum_j \varepsilon_j = 1$.

The probability measure $\varepsilon = (\varepsilon_j)$ is called the ergodic distribution of a M.C. It is unique, invariant, i.e.,

$$\varepsilon\, \mathbf{P}(t) = \varepsilon,$$

and the (F)-equation indicates that it can be found directly from the "balance equation"

$$\varepsilon\, \mathbf{Q} = 0$$

(or, alternatively with reference to the jump matrix $\mathbf{R}$, from the equation

$$\kappa \mathbf{R} = \kappa,$$

where $\kappa = \varepsilon\, \boldsymbol{\Theta}$, with $\kappa \bullet \mathbf{1} = 1$; the sign $\bullet$ stands for the inner product).

For the ergodic chain, the limit matrix $\mathbf{E}$ is stochastic, has all rows identical and equal to ε. Moreover, the matrices

$$\mathbf{Z}(t) = \mathbf{P}(t) - \mathbf{E}, \qquad t \geqslant 0,$$

form a semi-group with $\mathbf{Z}(0) = \mathbf{I} - \mathbf{E}$ and $\mathbf{Z}(t)\mathbf{1} = 0$. Its resolvent

$$\mathbf{Z}^{\alpha} = \mathbf{U}^{\alpha} - (1/\alpha)\mathbf{E}, \qquad \alpha > 0,$$

is called the ergodic α-potential. The limit $\mathbf{Z} = (z_{ij})$, where

$$z_{ij} = \int_0^{\infty} (p_{ij}(t) - \varepsilon_j)\, dt, \qquad i, j \in \mathbb{I},$$

is called an ergodic potential, or a fundamental matrix, of the M.C. In general, $\mathbf{Z}$ plays the same role for ergodic chains as $\mathbf{U}$ for transient chains.

§ 1.1.9. Examples. For illustrative purposes, it is convenient to mention here three examples which received considerable attention in a literature, and each of which embraces a large class of M.C.'s.

Example 1. *Birth-and-death process.* Its simple structure is responsible for interesting theoretical properties and significant applications, making this chain the most typical example.

The state space $\mathbb{I}$ is a set of non-negative integers (in particular, a finite set) forming a single communicating class (so all states are of the same character). The infinitesimal transitions are allowed only to the neighboring states, with intensities

$$\begin{aligned} q_{i,i-1} &= \mu_i > 0 && (\mu_0 = 0), \\ q_{i,i+1} &= \lambda_i > 0, && \\ q_{ij} &= 0 && \text{for } |i-j| > 1, \\ q_i &= \lambda_i + \mu_i, && i = 0, 1, \ldots, \end{aligned}$$

where λ_i and μ_i are known as the birth and the death coefficients, respectively.

The Kolmogorov equations, stated here in the resolvent form, reduce to the following simple expressions.

For fixed j, the (B)-equations are

$$\begin{aligned} (\alpha + \lambda_0)u^\alpha_{0j} - \delta_{0j} &= \lambda_0 u^\alpha_{1j}, && i = 0, \\ (\alpha + \lambda_i + \mu_i)u^\alpha_{ij} - \delta_{0j} &= \mu_i u^\alpha_{i-1,j} + \lambda_i u^\alpha_{i+1,j}, && i = 1, 2, \ldots, \end{aligned}$$

and, for fixed i, the (F)-equations are

$$\begin{aligned} u^\alpha_{i0}(\alpha + \lambda_0) - \delta_{i0} &= u^\alpha_{i1}\mu_1, && j = 0, \\ u^\alpha_{ij}(\alpha + \lambda_j + \mu_j) - \delta_{ij} &= u^\alpha_{i,j-1}\lambda_{j-1} + u^\alpha_{i,j+1}\mu_{j+1}, && j = 1, 2, \ldots . \end{aligned}$$

Except in a few well-known cases, the birth-and-death equations cannot be solved explicitly but various representations are available. Conditions for a solution to be minimal and stochastic, and for classification of states, can be expressed directly in terms of coefficients λ_i and μ_i. As it is well known, the (F)-equation yields the ergodic distribution

$$\varepsilon_j = \varepsilon_0 a_j \qquad \text{for} \quad j = 0, 1, \ldots,$$

with $1/\varepsilon_0 = \sum_{h=0}^\infty a_h$, provided the series converges. Here $a_0 = 1$ and, for $j > 0$,

$$a_j = (\lambda_0\lambda_1 \ldots \lambda_{j-1})/(\mu_1 \ldots \mu_j).$$

More generally, any invariant measure is a multiple of a_j. Note that a transient chain may posses an invariant measure, necessarily infinite.

Example 2. *Compound Poisson process.* Explicit form of its transition probabilities leads easily to explicit expressions for various quantities of interest in numerous applications (which include all finite chains).

Let $\mathbf{Q} = (q_{ij})$ be a bounded intensity matrix, and define a stochastic matrix $\mathbf{P} = (p_{ij})$ of a discrete parameter chain by

$$\begin{aligned} p_{ij} &= q_{ij}/\gamma \qquad \text{for} \quad i \neq j, \\ p_{ii} &= 1 - q_i/\gamma \end{aligned}$$

for any $\gamma \geqslant \sup_i q_i$. Next define $\mathbf{P}(t) = (p_{ij}(t))$ by

$$p_{ij}(t) = \sum_{n=0}^{\infty} p_{ij}^{(n)} \frac{(\gamma t)^n}{n!} e^{-\gamma t}, \qquad t \geqslant 0,$$

where $p_{ij}^{(n)}$ is the n-step transition probability, with $p_{ij}^{(1)} = p_{ij}$, $p_{ij}^{(0)} = \delta_{ij}$.

It is easy to verify that $\mathbf{P}(t)$ is the transition matrix satisfying Conditions 1–4 in § 1.1.1, independently of γ. Moreover, both the (B)- and (F)-equations are satisfied. Actually, the defining relation indicates that one has here Poisson randomization of the number of jumps in a discrete parameter chain.

In matrix notation

$$\mathbf{P}(t) = \sum_{n=0}^{\infty} \mathbf{P}^n \frac{(\gamma t)^n}{n!} e^{-\gamma t} = e^{\mathbf{Q}t},$$

$$\mathbf{U}^\alpha = \frac{1}{\gamma + \alpha} \sum_{n=0}^{\infty} \mathbf{P}^n \left(\frac{\gamma}{\gamma + \alpha}\right)^n = (\alpha \mathbf{I} - \mathbf{Q})^{-1},$$

with $\mathbf{Q} = (\mathbf{P} - \mathbf{I})\gamma$.

In the special case when q_i is constant, say γ, $\mathbf{P}$ coincides with the jump matrix $\mathbf{R}$. Hence, the Feller relation from § 1.1.6 reduces to

$$u_{ij}^\alpha (\alpha + \gamma) = g_{ij}^\alpha.$$

The limit matrix $\mathbf{E}$ may be found from the spectral representation of the bounded semi-group $\mathbf{P}(t)$, $t \geqslant 0$. Note that writing

$$\alpha u_{ij}^\alpha = \left(1 - \frac{\gamma}{\gamma + \alpha}\right) \sum_{n=0}^{\infty} \left(\frac{\gamma}{\gamma + \alpha}\right)^n r_{ij}^{(n)},$$

one finds from Abel equality and Cesaro summation that

$$\lim_{t \to \infty} p_{ij}(t) = \lim_{n \to \infty} \frac{1}{n + 1} \sum_{k=0}^{n} r_{ij}^{(k)}.$$

Finally, it is clear that now, for any invariant measure $\boldsymbol{\varepsilon}$,

$$\varepsilon \mathbf{Q} = 0 \qquad \text{iff} \qquad \varepsilon \mathbf{R} = \varepsilon.$$

Example 3. *Branching process.* This well-known chain has different structure from the previous examples. The state space is the set of all non-negative integers, and r.v. X_t represents the size of population of individuals whose life times are independent and exponentially distributed (with a parameter β). Individuals reproduce independently, and each individual produces j progeny with probability $b(j)$. Here b is a given distribution on non-negative integers such that

$$0 < b(0) < 1, \qquad b(1) = 0, \qquad \sum_{j=0}^{\infty} b(j) = 1, \qquad \text{with mean} \quad m.$$

Define intensities q_{ij} for $i, j = 0, 1, \ldots$, by

$$q_{ij} = \begin{cases} i\beta b(j-i+1) & \text{for} \quad j \geqslant i-1, \quad i \geqslant 1, \\ 0 & \text{for} \quad j < i-1, \end{cases}$$

and

$$q_i = i\beta \qquad \text{for} \quad i \geqslant 0.$$

Note that state 0 is absorbing, so $p_{i0}(t)$ is the probability of absorption (extinction) before time t, from initial population $i \geqslant 1$.

Introducing the generating functions for $|z| \leqslant 1$

$$B(z) = \sum_{i=0}^{\infty} b(i)\, z^i, \qquad \psi_i(t,z) = \sum_{j=0}^{\infty} p_{ij}(t) z^j,$$

the branching property is expressed by the relation

$$\psi_i(t,z) = [\psi_1(t,z)]^i, \qquad i \geqslant 0.$$

Furthermore, the (B)- and (F)-equations take the form

$$\frac{\partial \psi_i}{\partial t} = \beta\big[B(z) - z\big]\frac{\partial \psi_i}{\partial z}, \qquad i \geqslant 0, \tag{F}$$

$$\frac{\partial \psi_1}{\partial t} = \beta\big[B(\psi_1) - \psi_1\big], \qquad i = 1. \tag{B}$$

where $\psi_i = \psi_i(t,z)$. It is known that if m is finite, then $\psi_i(t,1) = 1$ (stochastic solution).

1.2. TABOO PROBABILITIES

§ **1.2.1. Taboo.** Let $X = (X_t)$ be an usual M.C. on a probability space $(\Omega, \mathcal{F}, \mathbb{P})$ having a standard stochastic transition matrix $\mathbf{P}(t) = (p_{ij}(t))$ and a conservative intensity matrix $\mathbf{Q} = (q_{ij})$, with all states in the state space $\mathbb{I}$ stable. Furthermore, let both the (B)- and (F)-equations hold, with $\mathbf{P}(t)$ being their common (minimal) solution.

Let $H \subset \mathbb{I}$ be an arbitrary set of states, to be kept fixed in the following discussion, and write for its complement $H^c = \mathbb{I} \setminus H$. It will be important to consider transitions of a M.C. X restricted to within the set H^c. For this reason the set H is called a *taboo set*. The following basic definition will be adopted — it refers to transition avoiding the taboo set H.

DEFINITION 1. The *taboo probability* on H^c, denoted by ${}_Hp_{ij}(t)$, is the transition probability from a state i to a state j at time t under the taboo H (i.e., avoiding set H), and is defined by

$${}_Hp_{ij}(t) = \mathbb{P}(X_s \in H^c,\ 0 < s < t,\ X_t = j \mid X_0 = i)$$

for both i and j in H^c and for $t \geqslant 0$.

Measurability of the defining set follows from separability of a M.C. Note that the probability of this set vanishes for $i \in H$ or for $j \in H$, because probability of a discontinuity at a fixed instant is zero. Hence, Definition 1 can be extended to

$$_Hp_{ij}(t) = 0 \qquad \text{if} \quad i \in H \quad \text{or} \quad j \in H.$$

For $t = 0$, write

$$_Hp_{ij}(0) = \begin{cases} \delta_{ij} & \text{for} \quad i, j \in H^c, \\ 0 & \text{otherwise.} \end{cases}$$

It is clear that for every i, j, and t,

$$_Hp_{ij}(t) \leqslant p_{ij}(t).$$

When H is empty, the taboo probability reduces to the ordinary transition probability.

It can be verified that $_Hp_{ij}(t)$ satisfy Conditions 1–4 from § 1.1.1, with the following modification caused by restriction to $i \in H^c$ and $j \in H^c$:

2) $\sum_{j \in H^c} {}_Hp_{ij}(t) \leqslant 1$ (sub-stochastic);

3) $\sum_{k \in H^c} {}_Hp_{ik}(s)\,{}_Hp_{kj}(t) = {}_Hp_{ij}(t+s)$ (Chapman—Kolmogorov).

The intensities $_Hq_{ij}$ are defined as in Items 5–6, § 1.1.1, and

$$_Hq_{ij} = \begin{cases} q_{ij} & \text{when} \quad i, j \in H^c, \quad i \neq j, \\ q_i & \text{when} \quad i, j \in H^c, \quad i = j, \\ 0 & \text{when} \quad i \in H \quad \text{or} \quad j \in H. \end{cases}$$

Consequently,

7) $\sum_{j \in H^c} {}_Hq_{ij} = -\bar{q}_i \leqslant 0, \qquad i \in H^c$ (non-conservative).

The quantity $\bar{q}_i$ (to be used frequently later on) is defined for all $i \in \mathbb{I}$ in terms of the original chain by

$$\bar{q}_i = \begin{cases} \sum_{j \in H} q_{ij} \geqslant 0, & i \in H^c, \\ \sum_{j \in H,\ j \neq i} q_{ij} \geqslant 0, & i \in H. \end{cases}$$

Clearly, one-step transition probability $_Hr_{ij}$ for the jump chain under taboo H coincide with the jump probabilities r_{ij} without taboo when i and j are in H^c, but vanish when $i \in H$ or $j \in H$. Note that $\bar{q}_i/q_i$ (for $i \in H^c$) is the probability of entry to H in a jump.

The Laplace transform $_Hu^\alpha_{ij}$ of the taboo probability $_Hp_{ij}(t)$ is defined in the usual way (see Item 8 in § 1.1.1). For $\alpha = 0$, write $_Hu_{ij}$ instead of $_Hu^0_{ij}$ and note that this is the average number of visits to state j (from state i) under taboo H.

THEOREM 2. *Suppose that a state $j \in H^c$ leads to H (i.e., there exists a state $k \in H$ which can be reached from j). Then*

$$ {}_H u_{ij} = \int_0^\infty {}_H p_{ij}(s)\, ds < \infty, \qquad i, j \in H^c. $$

Proof. See K. L. Chung [1967, p. 192]. □

§ 1.2.2. Matrices. When concise matrix notation is used, it would be helpful to partition matrices according to H^c and H. For any matrix indexed by $\mathbb{I}$, say $\mathbf{P}(t)$, the following notation for a partitioned form will be adopted:

$$ \mathbf{P}(t) = \begin{matrix} & H^c & H \\ H^c & \mathbf{P}_{11}(t) & \mathbf{P}_{12}(t) \\ H & \mathbf{P}_{21}(t) & \mathbf{P}_{22}(t) \end{matrix}. $$

Symbols H^c and H will be dropped subsequently. It is also convenient to introduce the partition of identity $\mathbf{I} = \mathbf{J}_1 + \mathbf{J}_2$, where

$$ \mathbf{J}_1 = \begin{pmatrix} \mathbf{I}_1 & \mathbf{0} \\ \mathbf{0} & \mathbf{0} \end{pmatrix}, \qquad \mathbf{J}_2 = \begin{pmatrix} \mathbf{0} & \mathbf{0} \\ \mathbf{0} & \mathbf{I}_2 \end{pmatrix}. $$

Similarly,

$$ \mathbf{\Theta} = \begin{pmatrix} \mathbf{\Theta}_1 & \mathbf{0} \\ \mathbf{0} & \mathbf{\Theta}_2 \end{pmatrix}. $$

Define now the *taboo matrix* ${}_H\mathbf{P}(t) = ({}_H p_{ij}(t))$ indexed by $\mathbb{I}$. Its partitioned form is

$$ {}_H\mathbf{P}(t) = \begin{pmatrix} {}_H\mathbf{P}_{11}(t) & \mathbf{0} \\ \mathbf{0} & \mathbf{0} \end{pmatrix}. $$

Clearly, ${}_H\mathbf{P}(t)$ is standard, sub-stochastic, dominated by $\mathbf{P}(t)$ and forms a semigroup for $t \geqslant 0$:

$$ \lim_{t\to 0} {}_H\mathbf{P}(t) = {}_H\mathbf{P}(0) = \mathbf{J}_1, \qquad {}_H\mathbf{P}(t)\mathbf{1} \leqslant \mathbf{1}, \qquad {}_H\mathbf{P}(t) \leqslant \mathbf{P}(t), $$
$$ {}_H\mathbf{P}(t+s) = {}_H\mathbf{P}(t)\, {}_H\mathbf{P}(s), \qquad s, t \geqslant 0. $$

The *taboo intensity matrix* ${}_H\mathbf{Q} = ({}_H q_{ij})$ is defined by

$$ {}_H\mathbf{Q} = \lim_{t\to 0} \left({}_H\mathbf{P}(t) - \mathbf{J}_1\right)/t $$

and equals

$$ {}_H\mathbf{Q} = \mathbf{J}_1\mathbf{Q}\mathbf{J}_1 = \begin{pmatrix} \mathbf{Q}_{11} & \mathbf{0} \\ \mathbf{0} & \mathbf{0} \end{pmatrix} \qquad \text{with} \quad {}_H\mathbf{Q}_{11} = \mathbf{Q}_{11}, $$

but it is not conservative:

$$ {}_H\mathbf{Q}\mathbf{1} \leqslant 0. $$

Hence, the corresponding jump matrix ${}_H\mathbf{R} = ({}_H r_{ij})$, defined by

$$ {}_H\mathbf{R} = \mathbf{J}_1\mathbf{R}\mathbf{J}_1 = \begin{pmatrix} {}_H\mathbf{R}_{11} & \mathbf{0} \\ \mathbf{0} & \mathbf{0} \end{pmatrix} \qquad \text{with} \quad {}_H\mathbf{R}_{11} = \mathbf{R}_{11}, $$

is sub-stochastic:

$$ {}_H\mathbf{R}\mathbf{1} \leqslant \mathbf{1}. $$

The *taboo resolvent matrix* ${}_H\mathbf{U}^\alpha = ({}_H u^\alpha_{ij})$ of Laplace transforms has the form

$$ {}_H\mathbf{U}^\alpha = \begin{pmatrix} {}_H\mathbf{U}^\alpha_{11} & \mathbf{0} \\ \mathbf{0} & \mathbf{0} \end{pmatrix}, \qquad \alpha > 0, $$

satisfies the resolvent equation, and

$$ {}_H\mathbf{U}^\alpha\mathbf{1} \leqslant \mathbf{1}. $$

Note that although ${}_H\mathbf{Q}_{11}$ and $\mathbf{Q}_{11}$ coincide, ${}_H\mathbf{P}_{11}(t)$ and $\mathbf{P}_{11}(t)$ differ.

§ 1.2.3. Equations. Under assumptions imposed here, the taboo probabilities satisfy the backward (B) and the forward (F) Kolmogorov equations of the form

$$ \frac{d\,{}_H p_{ij}(t)}{dt} = \begin{cases} -q_i\,{}_H p_{ij}(t) + \sum\limits_{k\in H^c,\; k\neq i} q_{ik}\,{}_H p_{kj}(t), & \text{(B)} \\ -{}_H p_{ij}(t)\,q_j + \sum\limits_{k\in H^c,\; k\neq j} {}_H p_{ik}(t) q_{kj}, & \text{(F)} \end{cases} $$

for $i, j \in H^c$, and with the initial conditions ${}_H p_{ij}(0) = \delta_{ij}$.

In matrix notation, with the initial condition ${}_H\mathbf{P}(0) = \mathbf{J}_1$,

$$ \frac{d_H\mathbf{P}(t)}{dt} = \begin{cases} {}_H\mathbf{Q}\,{}_H\mathbf{P}(t), & \text{(B)} \\ {}_H\mathbf{P}(t)\,{}_H\mathbf{Q}. & \text{(F)} \end{cases} $$

The only non-vanishing components are

$$ \frac{d_H\mathbf{P}_{11}(t)}{dt} = \begin{cases} \mathbf{Q}_{11}\,{}_H\mathbf{P}_{11}(t), & \text{(B)} \\ {}_H\mathbf{P}_{11}(t)\,\mathbf{Q}_{11}, & \text{(F)} \end{cases} $$

with ${}_H\mathbf{P}_{11}(0) = \mathbf{I}_1$.

Equivalently, in terms of Laplace transforms, the equations can be written for $i, j \in H^c$, as

$$ (\alpha + q_i)\,{}_H u^\alpha_{ij} - \delta_{ij} = \sum_{k\in H^c,\; k\neq i} q_{ik}\,{}_H u^\alpha_{kj}, \qquad \text{(B)} $$

$$ {}_H u^\alpha_{ij}(\alpha + q_j) - \delta_{ij} = \sum_{k\in H^c,\; k\neq j} {}_H u^\alpha_{ik}\,q_{kj}, \qquad \text{(F)} $$

or, in matrix notation, for non-vanishing components ($\alpha > 0$),

$$ \alpha\,{}_H\mathbf{U}^\alpha_{11} - \mathbf{I}_1 = \begin{cases} \mathbf{Q}_{11}\,{}_H\mathbf{U}^\alpha_{11}, & \text{(B)} \\ {}_H\mathbf{U}^\alpha_{11}\,\mathbf{Q}_{11}. & \text{(F)} \end{cases} $$

Introduce the following expressions defined for $i, j \in H^c$, and $t \geqslant 0$

$$q_i\,{}_H v_{ij}(t) = \sum_{k \in H^c,\ k \neq i} q_{ik}\,{}_H p_{kj}(t),$$

$${}_H w_{ij}(t) = \sum_{k \in H^c,\ k \neq j} {}_H p_{ik}(t)\, q_{kj},$$

and write ${}_H v_{ij}^{\alpha}$ and ${}_H w_{ij}^{\alpha}$ for their respective Laplace transforms. Then, the taboo equations can be written as

$$(\alpha + q_i)\,{}_H u_{ij}^{\alpha} - \delta_{ij} = q_i\,{}_H v_{ij}^{\alpha}, \tag{B}$$

$${}_H u_{ij}^{\alpha}(\alpha + q_j) - \delta_{ij} = {}_H w_{ij}^{\alpha}. \tag{F}$$

The term ${}_H v_{ij}(t)$ represents the probability that having left the state i, the M.C. will be at time t afterwards in state j, avoiding the taboo set H. The term ${}_H w_{ij}(t)$ may be regarded as a renewal density of the state j, under taboo H. See discussion in § 1.1.2.

It should be remarked that a special structure which a M.C. may possess (e.g., the birth-and-death process) may prohibit some transitions avoiding H. Hence, it may happen that for some i and j (and all $t > 0$) one has ${}_H p_{ij}(t) = 0$, even for $p_{ij}(t) \geqslant 0$. This fact is reflected in the matrix $\mathbf{Q}$, but as a consequence the equations for taboo probabilities may take reduced form in some subsets of H^c (see Ex. 3, § 1.2.6).

§ 1.2.4. Cut-off method. In the present situation ${}_H\mathbf{P}(t)$ is only sub-stochastic, but a more damaging fact is that ${}_H\mathbf{Q}$ is not conservative. Hence, the argument used to derive the (B)-equation (in § 1.1.2) does not apply as it stands. This deficiency is remedied easily by enlarging the state space by adjoining an absorbing state Δ.

As transitions on the taboo set H are not considered, the state space is taken to be $H^c \cup \{\Delta\}$. Define now the new transition matrix $\mathbf{P}^{\Delta}(t) = (p_{ij}^{\Delta}(t))$ by

$$\begin{aligned}
p_{ij}^{\Delta}(t) &= {}_H p_{ij}(t) && \text{for} \quad i, j \in H^c,\\
p_{i\Delta}^{\Delta}(t) &= 1 - \sum_{k \in H^c} {}_H p_{ik}(t) && \text{for} \quad i \in H^c,\\
p_{\Delta j}^{\Delta}(t) &= \delta_{\Delta j} && \text{for} \quad j \in H^c \cup \{\Delta\}.
\end{aligned}$$

The corresponding matrix $\mathbf{Q}^{\Delta}$ of intensities has the form

$$\begin{aligned}
q_{ij}^{\Delta} &= q_{ij} && \text{for} \quad i, j \in H^c,\\
q_{i\Delta}^{\Delta} &= \bar{q}_i && \text{for} \quad i \in H^c,\\
q_{\Delta j}^{\Delta} &= 0 && \text{for} \quad j \in H^c \cup \{\Delta\}.
\end{aligned}$$

Consequently, the matrix $\mathbf{P}^{\Delta}(t)$ is stochastic and the matrix $\mathbf{Q}^{\Delta}$ is conservative. Thus, results from § 1.1.2 apply, and the (B)- and (F)-equations hold for $\mathbf{P}^{\Delta}(t)$ on $H^c \cup \{\Delta\}$:

$$\frac{d\mathbf{P}^{\Delta}(t)}{dt} = \begin{cases} \mathbf{Q}^{\Delta}\mathbf{P}^{\Delta}(t), & \text{(B)}\\ \mathbf{P}^{\Delta}(t)\mathbf{Q}^{\Delta}, & \text{(F)} \end{cases}$$

with initial condition $\mathbf{P}^{\Delta}(0) = \mathbf{I}$. Recall that for validity of the (F)-equation additional conditions are required.

For i and j in H^c, the above equations reduce to the taboo equations already stated. On the other hand, for $i \in H^c$ and $j = \Delta$ one obtains the important relations (whose significance will be discussed later in Chapter II)

$$\frac{dp_{i\Delta}^{\Delta}(t)}{dt} = \begin{cases} \bar{q}_i + \sum_{k \in H^c} q_{ik} p_{k\Delta}^{\Delta}(t), & \text{(B)} \\ \sum_{k \in H^c} {}_H p_{ik}(t) \bar{q}_k. & \text{(F)} \end{cases}$$

Note that it follows from Definition 1, § 1.2.1, that

$$p_{i\Delta}^{\Delta}(t) = \mathbb{P}(X_s \in H \text{ for some } s,\ 0 < s < t \mid X_0 = i)$$

indicating "killing" of the chain at the entrance to H.

The above discussion gives the practical method of finding taboo probabilities, to be referred as the *"cut-off" method.* It says: Replace the taboo set H by a single absorbing state Δ (retaining unchanged intensities of transitions in and out of H^c), and write equations for transition probabilities for this modified chain with $\mathbf{P}^{\Delta}(t)$.

The detailed proof of argument presented here may be obtained by the adaptation of that used by K. L. Chung [1967, p. 135; 1970, p. 11] and based on entrance (and exit) laws for $\mathbf{P}^{\Delta}(t)$.

§ 1.2.5. Feller relation. Considerations of the jump chain restricted to transitions outside the taboo set H lead to the following definition of the n-step taboo probability

$${}_H r_{ij}^{(n)} = \mathbb{P}(X_1 \in H^c,\ \ldots,\ X_{n-1} \in H^c,\ X_n = j \mid X_0 = i)$$

for i and j in H^c and for $n \geqslant 1$. For $i \in H$ or $j \in H$ we write ${}_H r_{ij}^{(n)} = 0$.

This is the probability that starting from state i, the jump chain (X_n) will enter state j at the n-th step, under the restriction that none of the states in the taboo set H has been entered in-between, *inclusive* of both ends.

Much of the following analysis parallels that in § 1.1.4–1.1.6 (where the taboo set was absent), the main difference being that transitions are restricted to H^c. Thus, the taboo jump matrix ${}_H\mathbf{R}^n = ({}_H r_{ij}^{(n)})$ is sub-stochastic, with only non-zero term ${}_H\mathbf{R}_{11}^n$, and ${}_H\mathbf{R}^0 = \mathbf{J}_1$. The auxiliary α-jump chain under taboo H has one-step transition matrix ${}_H\mathbf{R}^{\alpha} = ({}_H r_{ij}^{\alpha})$ defined for $\alpha > 0$ by

$${}_H r_{ij}^{\alpha} = \begin{cases} q_i(\alpha + q_i)^{-1} {}_H r_{ij} & \text{when } i, j \in H^c,\ i \neq j, \\ 0 & \text{when } i, j \in H^c,\ i = j, \\ 0 & \text{when } i \in H \text{ or } j \in H. \end{cases}$$

This α-chain plays the same role under taboo as the α-chain with matrix $\mathbf{R}^{\alpha}$ when taboo is absent (see § 1.1.6).

Let ${}_H\mathbf{R}^{\alpha n} = ({}_H r_{ij}^{\alpha(n)})$ be the corresponding n-step transition matrix with ${}_H\mathbf{R}^{\alpha 1} = {}_H\mathbf{R}^{\alpha}$ and ${}_H\mathbf{R}^{\alpha 0} = \mathbf{J}_1$; its only non-zero component is ${}_H\mathbf{R}_{11}^{\alpha n}$. The associated (Green) potential for the α-chain under taboo is ${}_H\mathbf{G}^{\alpha} = ({}_H g_{ij}^{\alpha})$, given by

$$
{}_H\mathbf{G}^{\alpha} = \sum_{n=0}^{\infty} {}_H\mathbf{R}^{\alpha n}
$$

with only non-zero component ${}_H\mathbf{G}_{11}^{\alpha}$, where

$$
{}_H g_{ij}^{\alpha} = \sum_{n=0}^{\infty} {}_H r_{ij}^{\alpha(n)} \qquad \text{for} \quad i, j \in H^c.
$$

Observe that the (F)-equation yields

$$
{}_H g_{ij}^{\alpha} = \delta_{ij} + {}_H w_{ij}^{\alpha}.
$$

The primary use of the α-chain is to establish connection between the minimal semi-group ${}_H\mathbf{P}(t)$, $t \geqslant 0$, and its jump chain under taboo. This is provided by

Feller relation.

Let ${}_H\mathbf{U}^{\alpha} = ({}_H u_{ij}^{\alpha})$ be the resolvent of the minimal semi-group ${}_H\mathbf{P}(t)$, $t \geqslant 0$, and let ${}_H\mathbf{G}^{\alpha} = ({}_H g_{ij}^{\alpha})$ be the (Green) potential of the α-jump chain under taboo. Then, for $\alpha > 0$,

$$
{}_H\mathbf{U}^{\alpha} = {}_H\mathbf{G}^{\alpha} \, (\mathbf{\Theta} + \alpha \mathbf{I})^{-1}
$$

with the non-vanishing component

$$
{}_H\mathbf{U}_{11}^{\alpha} = {}_H\mathbf{G}_{11}^{\alpha} \, (\mathbf{\Theta}_1 + \alpha \mathbf{I}_1)^{-1},
$$

or, explicitly,

$$
{}_H u_{ij}^{\alpha} = {}_H g_{ij}^{\alpha} (\alpha + q_j)^{-1} \qquad \text{for} \quad i, j \in H^c.
$$

Proof of the above relations is analogous to that stated in § 1.1.6, but with taboo probabilities used. The interpretation is also the same. The average time spent in the state j (for α-chain) equals the product of the average number of visits to j (under taboo H) and the average duration of a visit (renewal argument). The Feller relation holds also for $\alpha = 0$, so

$$
{}_H u_{ij} q_j = {}_H g_{ij} \qquad \text{for} \quad i, j \in H^c.
$$

§ 1.2.6. Examples. The following well-known examples are given here for illustrative purpose. More involved examples will be discussed in later chapters.

In the case when the taboo set consists of a single state, say k, it is convenient to write ${}_kp_{ij}(t)$ for the taboo probability.

Example 1. Let k be a fixed state and take $H = \{k\}$. Then, using the cut-off method,

$$ {}_kp_{ij}(t) = \mathbb{P}\big(X_s \neq k,\ 0 < s < t,\ X_t = j \mid X_0 = i\big) \qquad \text{for} \quad i, j \neq k, $$

and

$$ p_{i\Delta}^{\Delta}(t) = \mathbb{P}\big(X_s = k \text{ for some } s,\ 0 < s < t \mid X_0 = i\big) = F_{ik}(t) \qquad \text{for} \quad i \neq k, $$

where $F_{ik}(t)$ is the passage distribution introduced in § 1.1.7. Note that now $\bar{q}_i = q_{ik}$ for $i \neq k$.

In particular, in the branching process (Ex. 3, § 1.1.9) with $k = 0$, the taboo ${}_0p_{ij}(t)$ coincides with $p_{ij}(t)$ for $i, j \geqslant 1$ (because state 0 is already absorbing), and $F_{i0}(t) = p_{i0}(t)$ is the distribution of time to extinction ($i \geqslant 1$).

Example 2. Let k be a fixed state and take $H = \{k\}^c$. Then

$$ {}_Hp_{kk}(t) = \mathbb{P}\big(X_s = k,\ 0 < s < t,\ X_t = k \mid X_0 = k\big) = e^{-q_k t} $$

and $p_{k\Delta}^{\Delta}(t)$ is the exponential distribution (with a parameter q_k) of the exit time ρ_k (see § 1.1.3).

Example 3. Let $\mathbb{I}$ consists of three states $-1, 0, 1$, and consider the birth-and-death process with coefficients

$$ \lambda_{-1} = \lambda, \quad \lambda_0 = \lambda, \qquad \mu_0 = \mu, \quad \mu_1 = \mu. $$

Take $H = \{0\}$ and $H^c = \{-1, 1\}$. It is clear that taboo probabilities ${}_0p_{-1,1}(t)$ and ${}_0p_{1,-1}(t)$ must vanish, because transitions between states -1 and 1 are impossible without passing through the state 0. The simple calculation yields

$$ {}_0p_{-1,-1}(t) = e^{-\lambda t}, \qquad {}_0p_{11}(t) = e^{-\mu t}. $$

Example 4. Consider a compound Poisson process with $q_i = \gamma > 0$, constant (see Ex. 2, § 1.1.9). It is evident that for any taboo set H, expressions for ${}_Hp_{ij}(t)$, ${}_Hu_{ij}^{\alpha}$ etc., are obtained by replacing $r_{ij}^{\alpha(n)}$ by ${}_Hr_{ij}^{\alpha(n)}$ ($i, j \in H^c$) in all defining expressions. In particular, the Feller relation under taboo becomes

$$ {}_Hu_{ij}^{\alpha}(\alpha + \gamma) = {}_Hg_{ij}^{\alpha}, \qquad i, j \in H^c. $$

1.3. POTENTIAL THEORY

§ 1.3.1. Operators. Matrices $\mathbf{P}(t)$, $\mathbf{Q}$, $\mathbf{U}^{\alpha}$, $\mathbf{R}$ etc. may be regarded as kernels defining transformations on functions and measures on the state space $\mathbb{I}$. For a function $f = (f(i))$ represented by a column vector and for a measure $\mu = (\mu(j))$ represented by a row vector, a function $\mathbf{P}(t)f$ and a measure $\mu\mathbf{P}(t)$ are defined on $\mathbb{I}$ component-wise by

$$\mathbf{P}(t)f(i) = \sum_j p_{ij}(t)f(j), \qquad \mu\mathbf{P}(t)(j) = \sum_i \mu(i)p_{ij}(t),$$

with summation taken over all states in $\mathbb{I}$. Similar notation will be used for other kernels, for example $\mathbf{Q}f$ and $\mu\mathbf{Q}$. For any measure μ and any function f on $\mathbb{I}$, their inner (scalar) product — denoted by $\mu \bullet f$ or simply by μf — is defined as

$$\mu \bullet f = \sum_k \mu(k)f(k).$$

Thus, $\mu \bullet \mathbf{1}$ is the total mass of a measure μ.

Let $\mathbf{P}(t)$ be a standard stochastic transition matrix with a conservative intensity matrix $\mathbf{Q}$. The operator $\mathbf{P}(t)$ induces two positive endomorphisms, both denoted by $\mathbf{P}(t)$, acting on the Banach space $\mathcal{B}$ of all bounded (measurable) functions on the state space $\mathbb{I}$ with supremum norm

$$\|f\| = \sup_i |f(i)|,$$

and on the Banach space $\mathcal{M}$ of all bounded (signed) measures on $\mathbb{I}$ with the total variation norm

$$\|\mu\| = \sum_i |\mu(i)|,$$

and defined by $\mathbf{P}(t)f$ and $\mu\mathbf{P}(t)$ for $f \in \mathcal{B}$ and $\mu \in \mathcal{M}$, respectively. The spaces $\mathcal{B}$ and $\mathcal{M}$ are connected by the inner product

$$\mu \bullet \mathbf{P}(t)f = \mu\mathbf{P}(t) \bullet f.$$

Furthermore, for any $f \in \mathcal{B}$ and any $\mu \in \mathcal{M}$,

$$\|\mathbf{P}(t)f\| \leqslant \|f\|, \qquad \|\mu\mathbf{P}(t)\| \leqslant \|\mu\|, \qquad |\mu \bullet f| \leqslant \|f\|\,\|\mu\|,$$

so $\mathbf{P}(t)$ is a contraction operator with norm

$$\|\mathbf{P}(t)\| = 1$$

(or $\leqslant 1$ in sub-stochastic case).

Similarly, the norm of the intensity matrix $\mathbf{Q}$ is

$$\|\mathbf{Q}\| = 2\sup_i q_i,$$

so in general $\mathbf{Q}$ is unbounded.

In this book, all limit operations on matrices and vectors will be taken point-wise, and operator topologies for semi-groups will not be used. The summary of basic properties follows.

1) For standard $\mathbf{P}(t)$ and conservative $\mathbf{Q}$, the point-wise limit

$$\lim_{t\to 0}\left[\mathbf{P}(t)f(i) - f(i)\right]/t = \mathbf{Q}f(i), \qquad i \in \mathbb{I},$$

exists for every $f \in \mathcal{B}$. Here $\mathbf{Q}$ is the generator defined in § 1.1.1, and it should be noted that the domain of $\mathbf{Q}$ coincides with the whole space.

Differentiability at $t = 0$ implies differentiability of $\mathbf{P}(t)f$ for all $t > 0$, for any bounded f, and leads to the (B)-equation (point-wise limit)

$$\frac{d\mathbf{P}(t)f(i)}{dt} = \mathbf{Q}\mathbf{P}(t)f(i) \qquad \text{for any} \quad i \in \mathbb{I}, \tag{B}$$

with the initial condition $\mathbf{P}(0)f = f$. Moreover, for $f \in \mathcal{B}$,

$$\mathbf{Q}\mathbf{P}(t) \cdot f = \mathbf{Q} \cdot \mathbf{P}(t)f.$$

The integrated form of the (B)-equation is

$$\mathbf{P}(t)f(i) - f(i) = \int_0^t \mathbf{Q}\mathbf{P}(s)f(i)\,ds, \qquad t \geqslant 0, \quad i \in \mathbb{I}. \tag{B}$$

For the (F)-equation, however, further conditions on $\mathbf{Q}$ must be imposed to justify the interchange of limits. Then, for any bounded f,

$$\frac{d\mathbf{P}(t)\,f(i)}{dt} = \mathbf{P}(t)\mathbf{Q}f(i) \qquad \text{for any} \quad i \in \mathbb{I} \tag{F}$$

(point-wise limit), or in the integrated form

$$\mathbf{P}(t)f(i) - f(i) = \int_0^t \mathbf{P}(s)\mathbf{Q}f(i)\,ds, \qquad t \geqslant 0, \quad i \in \mathbb{I}. \tag{F}$$

Under these conditions, associativity of matrix products holds:

$$\mathbf{P}(t) \cdot \mathbf{Q}f = \mathbf{P}(t)\mathbf{Q} \cdot f;$$

and

$$\mathbf{P}(t)\mathbf{Q}f(i) \longrightarrow \mathbf{Q}\,f(i) \qquad \text{as} \quad t \to 0.$$

The resolvent operator $\mathbf{U}^\alpha$ acts on the space $\mathcal{B}$, so, for any $f \in \mathcal{B}$,

$$\mathbf{U}^\alpha f(i) = \int_0^\alpha e^{-\alpha t}\mathbf{P}(t)f(i)\,dt = \sum_j u_{ij}^\alpha f(j), \qquad i \in \mathbb{I}, \quad \alpha > 0$$

(point-wise integral). Continuity and differentiability conditions are expressed, respectively, by

$$\lim_{\alpha\to\infty} \alpha\mathbf{U}^\alpha f(i) = f(i), \qquad \lim_{\alpha\to\infty} \alpha\big[\alpha\mathbf{U}^\alpha f(i) - f(i)\big] = \mathbf{Q}f(i), \qquad i \in \mathbb{I}.$$

The (B)- and (F)-equations in resolvent form are for $f \in \mathcal{B}$ and $i \in \mathbb{I}$

$$\alpha\mathbf{U}^\alpha f(i) - f(i) = \begin{cases} \mathbf{Q}\mathbf{U}^\alpha f(i), & \text{(B)} \\ \mathbf{U}^\alpha\mathbf{Q}f(i). & \text{(F)} \end{cases}$$

Moreover,

$$\|\mathbf{U}^\alpha\| = 1/\alpha, \qquad \alpha > 0.$$

As noted, the operator $\mathbf{Q}$ is unbounded in general, and so is $\mathbf{Q}\mathbf{P}(t) = \mathbf{P}(t)\mathbf{Q}$, but $\mathbf{Q}\mathbf{U}^\alpha = \mathbf{U}^\alpha\mathbf{Q}$ is bounded for $\alpha > 0$.

2) Much of the dual analysis for measures $\mu \in \mathcal{M}$ parallels that for functions. The continuity condition

$$\mu\mathbf{P}(t) \longrightarrow \mu \qquad \text{as} \quad t \to 0,$$

(point-wise limit) follows now from the dominated convergence theorem. The point-wise limit defining $\mu\mathbf{Q}$, as well as Kolmogorov equations, follow in the similar manner:

$$\frac{d\mu\mathbf{P}(t)}{dt} = \begin{cases} \mu\mathbf{Q}\mathbf{P}(t), & \text{(B)} \\ \mu\mathbf{P}(t)\mathbf{Q}, & \text{(F)} \end{cases}$$

or, in the resolvent form,

$$\alpha\mu\mathbf{U}^\alpha - \mu = \begin{cases} \mu\mathbf{Q}\mathbf{U}^\alpha, & \text{(B)} \\ \mu\mathbf{U}^\alpha\mathbf{Q}, & \text{(F)} \end{cases}$$

for $\mu \in \mathcal{M}$.

3) There is a need for an operator $\mathbf{P}(\infty)$ defined term-wise by

$$\mathbf{P}(\infty)f(i) = \lim_{t\to\infty} \mathbf{P}(t)f(i) = \lim_{\alpha\to 0} \alpha\mathbf{U}^\alpha f(i), \qquad i \in \mathbb{I},$$

for $f \in \mathcal{B}$, and similarly for $\mu\mathbf{P}(\infty)$ with $\mu \in \mathcal{M}$. This operator should not be in general identified with the limit matrix $\mathbf{E}$. Indeed, if a M.C. is transient and $\mathbf{P}(t)$ stochastic, then $\mathbf{P}(\infty)\mathbf{1} = \mathbf{1}$, but $\mathbf{E}\mathbf{1} = 0$. By the Helly–Bray theorem, one has $\mathbf{P}(\infty)f = \mathbf{E}f$ for $f \in \mathcal{B}$ iff $\mathbf{E}\mathbf{1} = \mathbf{1}$,. On the other hand, by the dominated convergence theorem always $\mu\mathbf{P}(\infty) = \mu\mathbf{E}$ for $\mu \in \mathcal{M}$. The analogue of $\mathbf{P}(\infty)$ for the jump matrix $\mathbf{R}$ will be denoted by $\mathbf{R}^\infty$.

4) It may be added that with the discrete topology on $\mathbb{I}$ (all sets open), the state space $\mathbb{I}$ is locally compact (but not compact in general), and all bounded functions on $\mathbb{I}$ are continuous. Thus, if f is continuous (i.e., a sequence) so is $\mathbf{P}(t)f$. Similarly, if $\mathcal{C}_0$ is a subspace of all bounded continuous functions "vanishing at infinity" (i.e., a space of sequences converging to zero), then $\mathcal{C}_0$ is invariant under

$\mathbf{P}(t)$. Hence, with the metric compatible with the discrete topology, the minimal stochastic Markov semi-group on $\mathcal{B}$ is Fellerian.

5) With reference to the taboo set H from Sub-section 1.2, functions (column vectors) and measures (row vectors) on $\mathbb{I}$ may be partitioned as

$$f = \begin{pmatrix} f_1 \\ f_2 \end{pmatrix}, \qquad \mu = (\mu_1, \mu_2).$$

Thus, for example, with the matrix kernel ${}_H\mathbf{P}(t)$, one has ${}_H\mathbf{P}(t)f$ and $\mu\,{}_H\mathbf{P}(t)$ in the partitioned form

$${}_H\mathbf{P}(t)f = \begin{pmatrix} {}_H\mathbf{P}_{11}(t)f_1 \\ \mathbf{0} \end{pmatrix}, \qquad \mu\,{}_H\mathbf{P}(t) = (\mu_1{}_H\mathbf{P}_{11}(t), \mathbf{0}).$$

§ 1.3.2. Definitions. For now on, all functions and measures defined on a state space $\mathbb{I}$ are assumed to be non-negative. To stress distinction between a continuous parameter M.C. with a transition matrix $\mathbf{P}(t)$, and its jump chain with a transition matrix $\mathbf{R}$, two sets of definitions are needed. Notation $\mathbf{P}(t)f$, $\mu\mathbf{P}(t)$ etc. to be used here has been already introduced in § 1.3.1.

DEFINITION 1. A non-negative function f on $\mathbb{I}$ is called:
a) excessive, when $\mathbf{P}(t)f \leqslant f$ for all $t \geqslant 0$;
b) invariant, when $\mathbf{P}(t)f = f$ for all $t \geqslant 0$;
c) purely excessive, when it is excessive and $\lim \mathbf{P}(t)f = 0$ as $t \to \infty$;
d) potential of a charge $a \geqslant 0$, when $f = \mathbf{U}a$.

DEFINITION 2. A non-negative function f on $\mathbb{I}$ is called:
a) superharmonic, when $\mathbf{R}f \leqslant f$;
b) harmonic, when $\mathbf{R}f = f$;
c) purely superharmonic, when it is superharmonic and $\lim_{n\to\infty} \mathbf{R}^n f = \mathbf{0}$;
d) potential of a charge $c \geqslant 0$, when $f = \mathbf{G}c$.

Analogues of Definition 1 and Definition 2 for non-negative measures are expressed by

$$\mu\mathbf{P}(t) \leqslant \mu, \qquad \mu\mathbf{R} \leqslant \mu,$$

with corresponding potentials of non-negative charges (measures) σ and γ:

$$\mu = \sigma\mathbf{U}, \qquad \mu = \gamma\mathbf{G}.$$

Observe that standardness of the matrix $\mathbf{P}(t)$ implies point-wise continuity at the origin for excessive functions and measures

$$\lim_{t\to 0} \mathbf{P}(t)f = f, \qquad \lim_{t\to 0} \mu\mathbf{P}(t) = \mu.$$

This continuity condition is a part of definition for general Markov processes, but for the usual M.C.'s it is automatically satisfied. Although the above definitions do not exclude infinite functions, the restriction to finite excessive functions will be imposed here. On the other hand, σ-finite measures will be considered.

In terms of a resolvent $\mathbf{U}^\alpha$, the following equivalent criterion is useful.

THEOREM 3. *A non-negative function f on $\mathbb{I}$ is:*

a) *excessive, iff* $\alpha \mathbf{U}^\alpha f \leqslant f$, $\alpha > 0$;

b) *invariant, iff* $\alpha \mathbf{U}^\alpha f = f$;

c) *purely excessive, iff it is excessive and* $\lim_{\alpha \to 0} \alpha \mathbf{U}^\alpha f = 0$.

Analogous conditions hold for excessive measures

$$\alpha \mu \mathbf{U}^\alpha \leqslant \mu.$$

Let $\mathbf{Q}$ be a generator defined point-wise as in § 1.3.1, and let $\mathbf{\Theta}$ be a diagonal matrix of elements q_i. From definitions it follows

PROPOSITION 4. i) *If f is excessive (invariant) function, then* $\mathbf{Q}f \leqslant 0$ ($\mathbf{Q}f = 0$). *If μ is excessive (invariant) measure, then* $\mu\mathbf{Q} \leqslant 0$ ($\mu\mathbf{Q} = 0$);

ii) *A function f is superharmonic (harmonic) iff* $\mathbf{Q}f \leqslant 0$ ($\mathbf{Q}f = 0$). *A measure* $\nu = \mu\mathbf{\Theta}$ *is superharmonic (harmonic) iff* $\mu\mathbf{Q} \leqslant 0$ ($\mu\mathbf{Q} = 0$).

The following properties are of interest (similar relations hold for measures).

THEOREM 5. i) *A function f is purely excessive (purely superharmonic) iff there exists no invariant (harmonic) function dominated by f and distinct from* 0.

ii) *A potential* $\mathbf{U}a$ *of a charge a is purely excessive. A function f is a potential* $\mathbf{G}c$ *of a charge c iff it is purely superharmonic.*

THEOREM 6. *Every purely excessive function p is the point-wise limit of a non-decreasing sequence of potentials of charges.*

Frequently, Definitions 1 and 2 may be restricted to subsets of a state space $\mathbb{I}$. On occasions when several transition matrices are considered, it is necessary to specify that a function (a measure) is excessive relative to some transition matrix. Note also that requirement that $\mathbf{P}(t)$ be stochastic (sub-stochastic) is the same as requirement that constant functions be invariant (excessive) on $\mathbb{I}$ relative to $\mathbf{P}(t)$; analogously for $\mathbf{R}$.

Remark. The following definitions are frequently used. Write $\mathbf{P}^\alpha(t) = e^{-\alpha t}\mathbf{P}(t)$ for $\alpha \geqslant 0$ (see § 1.1.6). A non-negative function f is said to be an α-supermean when

$$\mathbf{P}^\alpha(t) f \leqslant f \qquad \text{for every} \quad t \geqslant 0$$

(and α-invariant when equality holds). A function f is an α-excessive if it is α-supermean and

$$\lim_{t \to 0} \mathbf{P}^\alpha(t) f = f \qquad \text{(point-wise)}.$$

When $\alpha = 0$, it will be dropped from notation, and 0-excessive functions will simply be called excessive. The same comment applies to measures.

Furthermore, for a function $a \geqslant 0$ (a measure $\sigma \geqslant 0$), the function $\mathbf{U}^\alpha a$ (the measure $\sigma \mathbf{U}^\alpha$) is called an α-potential of a charge a (charge σ).

§ 1.3.3. Decompositions. i) The fundamental decomposition theorems named after F. Riesz, are stated here in two versions in order to stress distinction between continuous and discrete parameter M.C.'s. Moreover, dual versions for functions and measures differ slightly.

It is assumed that excessive functions, as well as superharmonic functions, are finite. Then, obviously, $\mathbf{P}(t)f$ and $\alpha\mathbf{U}^\alpha f$, and also $\mathbf{R}f$, are finite functions.

THEOREM 1 (Riesz decomposition I). *Every excessive function f on $\mathbb{I}$ has the unique decomposition*

$$f(i) = p(i) + h(i), \qquad i \in \mathbb{I},$$

where p is purely excessive and $h = \mathbf{P}(\infty)f$ is invariant.

THEOREM 2 (Riesz decomposition II). *Every superharmonic function f on $\mathbb{I}$ has the unique decomposition*

$$f(i) = u(i) + r(i), \qquad i \in \mathbb{I},$$

where $u = \mathbf{G}c$ is a potential of a charge $c = (\mathbf{I} - \mathbf{R})f$, and $r = \mathbf{R}^\infty f$ is harmonic.

Because of importance of Riesz decomposition I, a brief outline of its proof is given here. The fact that f is excessive implies that $\mathbf{P}(t)f$ decreases when t increases,

$$\mathbf{P}(t+s)f \leqslant \mathbf{P}(t)f \leqslant f < \infty,$$

so the point-wise limit

$$\lim_{t\to\infty} \mathbf{P}(t)f = h$$

exists and is finite. By the dominated convergence theorem one has, for each t,

$$\begin{aligned}\mathbf{P}(t)h = \mathbf{P}(t)\mathbf{P}(\infty)f &= \lim_{s\to\infty} \mathbf{P}(t)\mathbf{P}(s)f \\ &= \lim_{s\to\infty} \mathbf{P}(t+s)f = \mathbf{P}(\infty)f = h,\end{aligned}$$

so h is invariant. Next define $p = f - h \geqslant 0$ and note that p is excessive,

$$\mathbf{P}(t)p = \mathbf{P}(t)f - \mathbf{P}(t)h = \mathbf{P}(t)f - h \leqslant f - h = p.$$

Letting $t \to \infty$ in the above, one has $\mathbf{P}(\infty)p = 0$, so p is purely excessive.

Equivalent proof in terms of Laplace transforms follows the same pattern. Because this argument will be used frequently in the sequel, it is worthwhile to repeat the proof in terms of resolvent. As $\alpha\mathbf{U}^\alpha f$ increases as $\alpha \to 0$, it follows that the point-wise limit

$$\lim_{\alpha\to 0} \alpha\mathbf{U}^\alpha f = h$$

exists and is finite. Since $\|\alpha\mathbf{U}^\alpha\| \leqslant 1$, the dominated convergence theorem and the resolvent equation yield

$$\alpha\mathbf{U}^\alpha h = \alpha\mathbf{U}^\alpha \lim_{\beta\to 0} \beta\mathbf{U}^\beta f = \lim_{\beta\to 0} \alpha\beta\mathbf{U}^\alpha\mathbf{U}^\beta f = h, \qquad \alpha > 0,$$

so h is invariant. Letting $p = f - h$, one finds that $\alpha\mathbf{U}^\alpha p \leqslant p$ and $\lim_{\alpha\to 0} \alpha\mathbf{U}^\alpha p = 0$.

Due to frequent applications, it is of interest to associate the Riesz decomposition I with Kolmogorov equations. Consider first the (B)-equation for an excessive function f

$$f = \alpha \mathbf{U}^{\alpha} f + (-\mathbf{Q})\mathbf{U}^{\alpha} f, \qquad \alpha > 0. \tag{B}$$

The preceding resolvent argument indicates that the purely excessive component p is given by the point-wise limit

$$p = \lim_{\alpha \to 0} (-\mathbf{Q})\mathbf{U}^{\alpha} f.$$

However, p need not be a potential of a charge.

On the other hand, imposing a condition that a function $\mathbf{U}^{\alpha}\mathbf{\Theta} f$ is finite,

$$\sum_k u_{ik}^{\alpha} q_k f(k) < \infty \qquad \text{for all} \quad i \text{ and } \alpha > 0,$$

the associativity property holds (Fubini)

$$\mathbf{U}^{\alpha} \cdot \mathbf{Q} f = \mathbf{U}^{\alpha}\mathbf{Q} \cdot f, \qquad \alpha > 0,$$

(see § 1.1.9). Thus, assuming validity of the (F)-equation, one has

$$f = \alpha \mathbf{U}^{\alpha} f + \mathbf{U}^{\alpha} a, \qquad \alpha > 0, \tag{F}$$

where $a = (-\mathbf{Q})f \geqslant 0$, with $a(i) < \infty$ for all i. Consequently, in the limit another form of decomposition is obtained in which the purely excessive component p is a potential of a charge: $p = \mathbf{U}a$. It is of interest to contrast this result with the Riesz decomposition II (whose proof is similar, but simpler).

Dual statements apply to Riesz decomposition for measures. Assuming that excessive (and superharmonic) measures are σ-finite, one obtains

THEOREM 3 (Riesz decomposition III). *For any excessive measure μ on $\mathbb{I}$:*

$$\mu = \eta + \lambda,$$

where η is purely excessive and $\lambda = \mu\mathbf{P}(\infty) = \lim_{\alpha \to 0} \alpha\mu\mathbf{U}^{\alpha}$ is invariant.

THEOREM 4 (Riesz decomposition IV). *For any superharmonic measure μ on $\mathbb{I}$:*

$$\mu = \zeta + \kappa$$

where $\zeta = \gamma\mathbf{G}$ is a potential of a charge $\gamma = \mu(\mathbf{I} - \mathbf{R})$ and $\kappa = \mu\mathbf{R}^{\infty}$ is harmonic.

Proof of Riesz decomposition III proceeds in a manner similar to that of decomposition I, with similar approach through the (B)-equation for measures. Hence, the purely excessive component is

$$\eta = \lim_{\alpha \to 0} \sigma\mathbf{U}^{\alpha},$$

where $\sigma = \mu(-\mathbf{Q}) \geqslant 0$. Thus, in the limit $\eta = \sigma\mathbf{U}$ with $\mathbf{U}$ finite (transient) so the purely excessive measure η would always be a potential of a charge, and $\eta = 0$ when $\mathbf{U}$ has infinite terms (recurrent). Note that if σ were bounded ($\sigma \bullet \mathbf{1} < \infty$) then necessarily $\sigma = 0$, and μ would be invariant.

Remark. Potential theory for transient M.C.'s differs considerably from that for recurrent chains, and Riesz decomposition illustrates well this fact. For irreducible transient chains both components in the decomposition of functions are present, but for recurrent chains purely excessive (potential) component vanishes. In decomposition of measures, only one component is usually present. The well-known Blackwell theorem asserts that an irreducible M.C. is recurrent iff the only bounded (non-zero) $\mathbf{R}$-superharmonic functions are constant.

ii) It is rather remarkable that for the minimal M.C. classes of excessive functions and of superharmonic functions coincide. As a consequence, important simplifications are obtained.

THEOREM 5 (Watanabe). *A function f on $\mathbb{I}$ is excessive relative to the minimal $\mathbf{P}(t)$ iff it is superharmonic relative to the jump matrix $\mathbf{R}$.*

Proof (outline). a) f is $\mathbf{P}(t)$-excessive $\Rightarrow \mathbf{Q}f \leqslant 0 \Rightarrow \mathbf{R}f \leqslant f \Rightarrow f$ is superharmonic.

b) f is $\mathbf{R}$-superharmonic $\Rightarrow \mathbf{Q}f \leqslant 0 \Rightarrow \mathbf{Q}^\alpha f = (\mathbf{Q} - \alpha\mathbf{I})f \leqslant -\alpha f$.

Being $\mathbf{R}$-superharmonic, the function f is also $\mathbf{R}^\alpha$-superharmonic (see § 1.1.6), so its Riesz decomposition relative to $\mathbf{R}^\alpha$ is

$$f = \mathbf{G}^\alpha \cdot (\mathbf{I} - \mathbf{R}^\alpha)f + \mathbf{R}^{\alpha\infty} f.$$

Hence, using the Feller relation (§ 1.1.6),

$$f \geqslant \mathbf{G}^\alpha \cdot (\mathbf{I} - \mathbf{R}^\alpha)f = \mathbf{U}^\alpha \cdot (-\mathbf{Q}^\alpha)f \geqslant \alpha \mathbf{U}^\alpha f,$$

so f is $\mathbf{P}(t)$-excessive on $\mathbb{I}$. □

The following is the basic (minimal) decomposition theorem for functions satisfying the integrability condition $\mathbf{U}^\alpha \mathbf{\Theta} f < \infty$. The theorem indicates that for the usual M.C. distinction between invariant and harmonic, as well as between potential of charges and purely excessive functions, may be in most cases disregarded.

THEOREM 6 (Feller). *Let f be an excessive function on $\mathbb{I}$ relative to the minimal $\mathbf{P}(t)$, and let its decomposition with respect to $\mathbf{P}(t)$ be*

$$f = p + h,$$

where p is purely excessive and h is invariant, and let its decomposition with respect to $\mathbf{R}$ be

$$f = u + r,$$

where $u = \mathbf{G}c$ is a potential of a charge $c = (\mathbf{I} - \mathbf{R})f$, and $r = \mathbf{R}^\infty f$ is harmonic.

Then, f has further decomposition

$$f = \mathbf{U}a + h + w,$$

where $\mathbf{U}a = \mathbf{G}c = u$, with $a = \mathbf{\Theta}c = (-\mathbf{Q})f$, is a potential of a charge, h is invariant and w is a purely excessive component which is not a potential of a charge,

$$w = \lim_{\alpha \to 0} \mathbf{R}^{\alpha\infty} f.$$

Moreover,

$$p = u + w, \qquad r = h + w.$$

Proof (outline). By the preceding theorem, $\mathbf{P}(t)$-excessive f is also $\mathbf{R}$-superharmonic, and therefore $\mathbf{R}^\alpha$-superharmonic. Hence, as in the above proof, the Riesz decomposition of f relative to $\mathbf{R}^\alpha$ may be put in the following form (using the Feller relation)

$$\begin{aligned} f &= \mathbf{G}^\alpha \cdot (\mathbf{I} - \mathbf{R}^\alpha) f + \mathbf{R}^{\alpha\infty} f \\ &= \alpha \mathbf{U}^\alpha f + \mathbf{U}^\alpha \cdot (-\mathbf{Q}) f + \mathbf{R}^{\alpha\infty} f = h^\alpha + \mathbf{U}^\alpha a + w^\alpha, \end{aligned}$$

where

$$h^\alpha = \alpha \mathbf{U}^\alpha f, \qquad a = (-\mathbf{Q}) f = \Theta c, \qquad w^\alpha = \mathbf{R}^{\alpha\infty} f.$$

The limits

$$h^\alpha \longrightarrow h, \qquad \mathbf{U}^\alpha a \longrightarrow \mathbf{U} a, \qquad w^\alpha \longrightarrow w,$$

exist as $\alpha \to 0$, and the assertion follows by identification of components. □

The above theorem indicates that a harmonic function need not be invariant and that a purely excessive function need not be a potential of a charge. As noted in the proof, the component w, called the *boundary potential*, is either a purely excessive function which is not a potential of a charge, or a harmonic function which is not invariant.

iii) The boundary potential w is responsible for difference between both Riesz decomposition and elucidates also distinction between the (B)- and (F)-equation. For convenience write $\mathbf{W}^\alpha$ for the operator $\mathbf{R}^{\alpha\infty}$.

Recall that (under assumptions stated earlier) the (B)-equation for an excessive function f is

$$f = \alpha \mathbf{U}^\alpha f + (-\mathbf{Q}) \mathbf{U}^\alpha f, \qquad \alpha > 0, \tag{B}$$

whereas the proof of Theorem 6 yields

$$f = \alpha \mathbf{U}^\alpha f + \mathbf{U}^\alpha (-\mathbf{Q}) f + \mathbf{W}^\alpha f, \qquad \alpha > 0.$$

Comparison with the (F)-equation

$$f = \alpha \mathbf{U}^\alpha f + \mathbf{U}^\alpha (-\mathbf{Q}) f, \qquad \alpha > 0, \tag{F}$$

indicates that the term $\mathbf{W}^\alpha f$ is missing. Thus, $\mathbf{Q}$ and $\mathbf{U}^\alpha$ do not commute, in general, and

$$(-\mathbf{Q}) \mathbf{U}^\alpha f = \mathbf{U}^\alpha (-\mathbf{Q}) f + \mathbf{W}^\alpha f.$$

The validity of the (F)-equation requires vanishing of the boundary component. If it does not vanish, then the (F)-equation as it stands does not hold, and an extra term must be added to it (the "augmented (F)-equation").

The following theorem summarizes discussion (see Prop. 4, § 1.3.2).

THEOREM 7. *For the minimal* $\mathbf{P}(t)$:

i) *a function* f *is* $\mathbf{P}(t)$*-excessive iff* $\mathbf{Q} f \leqslant 0$;

ii) *a function* h *is* $\mathbf{P}(t)$*-invariant iff* $\mathbf{Q} h = 0$ *and* $\mathbf{W} h = 0$;

iii) *a function* p *is* $\mathbf{P}(t)$*-purely excessive iff it is a potential* $\mathbf{U} a$ *of a charge* a *and* $\mathbf{W} p = 0$.

Hence, when the (F)-equation is assumed to hold, invariant and harmonic functions, as well as purely excessive functions and potentials of charges, coincide respectively. For measures, the result simplifies because the boundary component vanishes (by dominated convergence).

iv) As an example, suppose that $\mathbf{P}(t)$ is sub-stochastic and consider the excessive function $f = \mathbf{1}$, with $\mathbf{Q1} = 0$. As noted in § 1.1.5,

$$\mathbf{P}(t)\mathbf{1}_i = \mathbb{P}^i(\tau > t), \qquad i \in \mathbb{I}.$$

The (B)-equation and the augmented (F)-equation give now

$$\mathbf{1} = \alpha \mathbf{U}^\alpha \mathbf{1} + (-\mathbf{Q})\mathbf{U}^\alpha \mathbf{1}, \tag{B}$$

$$\mathbf{1} = \alpha \mathbf{U}^\alpha \mathbf{1} + \mathbf{W}^\alpha \mathbf{1}. \tag{aug (F)}$$

It follows that the function $\mathbf{W}^\alpha \mathbf{1}$ is the L.S. transform of the distribution function of the life time τ (conditional on i). Letting $\alpha \to 0$, one has

$$\alpha \mathbf{U}^\alpha \mathbf{1}_i \longrightarrow \mathbb{P}^i(\tau = \infty), \qquad \mathbf{W}^\alpha \mathbf{1}_i \longrightarrow \mathbb{P}^i(\tau < \infty), \qquad i \in \mathbb{I}.$$

Hence, $\mathbf{P}(t)$ with conservative $\mathbf{Q}$ is stochastic iff $\mathbf{W}^\alpha \mathbf{1} = 0$. See § 12.2.3 for further discussion.

v) Explicit calculations of excessive functions for usual M.C.'s are rather involved, except in some special cases.

For a birth process, a bounded function $f \geqslant 0$ on $\mathbb{I} = \{0, 1, \dots\}$ is excessive whenever it is non-increasing:

$$a(i) = (-\mathbf{Q})f(i) = \lambda_i \big[f(i) - f(i+1)\big] \geqslant 0 \qquad \text{for all} \quad i \geqslant 0.$$

Under additional requirement that $f(i) \to 0$ as $i \to \infty$, one has $f = \mathbf{U}a$, so f is a potential of a charge $a \geqslant 0$, hence also purely excessive. However, constant functions are invariant provided $\mathbf{P}(t)$ is stochastic.

§ 1.3.4. Exit and entrance laws. The concept of exit and entrance laws is of importance in many aspects of analytic theory of M.C.'s. Its use greatly simplifies derivation of Kolmogorov equations, and is also essential in investigation of differentiability of various distributions. Here only basic properties are listed for purpose of reference.

Let $\mathbf{P}(t) = (p_{ij}(t))$ be a standard transition matrix which may be sub-stochastic. All functions and measures are defined on the state space $\mathbb{I}$.

DEFINITION 1. a) An *exit law* for $\mathbf{P}(t)$ is a family $f(\cdot) = \{f(t),\ t > 0\}$ of positive functions $f(t) = (f_i(t))_{i \in \mathbb{I}}$ such that

$$\mathbf{P}(t)f(s) = f(s+t) \qquad \text{for} \quad s > 0,\ t \geqslant 0.$$

The exit law $f(\cdot)$ is bounded if, for all $t > 0$,

$$\sup_i \int_0^t f_i(s)\, ds < \infty.$$

b) An *entrance law* for $\mathbf{P}(t)$ is a family $\mu(\cdot) = \{\mu(t),\ t > 0\}$ of positive measures $\mu(t) = (\mu_j(t))_{j\in\mathbb{I}}$ such that

$$\mu(s)\mathbf{P}(t) = \mu(s+t) \qquad \text{for} \quad s > 0,\ t \geqslant 0.$$

The entrance law $\mu(\cdot)$ is bounded if $\mu(t) \bullet \mathbf{1} < \infty$ for $t > 0$, and normalized when $\mu(t) \bullet \mathbf{1} = 1$ for all $t > 0$.

The natural examples are obtained by taking a bounded function $f \geqslant 0$ and a bounded measure $\mu \geqslant 0$, and defining for each $t > 0$,

$$f(t) = \mathbf{P}(t)f, \qquad \mu(t) = \mu\mathbf{P}(t).$$

In particular, $p_{\bullet j}(t)$ defines the exit law for fixed j, and $p_{i\bullet}(t)$ defines the entrance law for fixed i. With reference to § 1.1.2, it can be verified that for fixed j, $w_{\bullet j}(t)$ defines the exit law, and, for fixed i, $w_{i\bullet}(t)$ defines the (normalized) entrance law.

THEOREM 2. a) *Let $f(t)$ be an exit law for $\mathbf{P}(t)$. Then*

i) *$f_i(t)$ are continuous functions for $t > 0$ with finite limits $f_i(0+)$ at $t = 0$;*

ii) *$\mathbf{P}(t)f(0+) \leqslant f(t)$ for $t > 0$;*

iii) *if $f(t) = \mathbf{P}(t)f$ for some function $f \geqslant 0$, then $f(0+) = f$.*

b) *Dual expressions hold for an entrance law $\mu(t)$.*

THEOREM 3. a) *Let $F(t)$ be a positive function on $\mathbb{I}$ for $t > 0$ such that*

$$\mathbf{P}(t)F(s) = F(s+t) - F(t) \qquad \text{for} \quad s,t > 0.$$

Then, there exists an exit law $f(\cdot)$ such that

$$F(s+t) - F(t) = \int_t^{t+s} f(x)\,dx \qquad \text{for} \quad t > 0,\ s \geqslant 0,$$

and $F(\cdot)$ is continuously differentiable.

b) *Dual result holds for a positive measure $M(t)$ on $\mathbb{I}$ for $t > 0$ such that*

$$M(s)\mathbf{P}(t) = M(s+t) - M(t) \qquad \text{for} \quad s,t > 0.$$

THEOREM 4. a) *Let f be a purely excessive function on $\mathbb{I}$ relative to $\mathbf{P}(t)$. Then, there exists a (unique) exit law $f(t)$ such that*

$$f = \int_0^\infty f(t)\,dt$$

and

$$f(t) = -d\mathbf{P}(t)f/dt.$$

b) *Dual result holds for a purely excessive measure μ on $\mathbb{I}$.*

COROLLARY 5. a) *Let f be an excessive function on $\mathbb{I}$ relative to $\mathbf{P}(t)$ with Riesz decomposition $f = p + h$, where p is purely excessive and h is invariant. Then*

$$F(t) = f - \mathbf{P}(t)f = p - \mathbf{P}(t)p \geqslant 0, \qquad t \geqslant 0,$$

satisfies conditions of Theorem 3, so there exists an exit law $p(\cdot)$ such that

$$F(t) = \int_0^t p(s)\,ds \qquad \text{for} \quad 0 \leqslant t < \infty,$$

and

$$F(\infty) = \int_0^\infty p(s)\,ds = p.$$

b) *Dual result holds for an excessive measure μ on $\mathbb{I}$ relative to $\mathbf{P}(t)$.*

Observe that when the (B)-equation holds for $\mathbf{P}(t)f$ then, for $t > 0$,

$$p(t) = (-\mathbf{Q})\mathbf{P}(t)f = (-\mathbf{Q})\mathbf{P}(t)p,$$

and dually for $\mu\mathbf{P}(t)$.

Consider now an exit law $f(\cdot)$ and an entrance law $\mu(\cdot)$ relative to $\mathbf{P}(t)$, such that

$$f_k = \int_0^\infty f_k(t)\,dt, \qquad \mu_k = \int_0^\infty \mu_k(t)\,dt, \qquad k \in \mathbb{I}.$$

THEOREM 6. *The inner product*

$$\phi(t+s) = \sum_k \mu_k(s) f_k(t), \qquad s, t > 0,$$

depends only on $t + s$, and, for $t > 0$,

$$\sigma(t) := \int_t^\infty \phi(s)\,ds = \sum_k \mu_k f_k(t) = \sum_k \mu_k(t) f_k.$$

However, $\sigma(0)$ may be infinite.

Proof. The first assertion follows from observation that, for any $0 < h < t$,

$$\mu(s) \bullet f(t) = \mu(s)\mathbf{P}(t-h)f(h) = \mu(s+t-h) \bullet f(h). \qquad \square$$

Writing f_k^α, μ_k^α, ϕ^α, and σ^α for Laplace transforms of $f_k(t)$, $\mu_k(t)$, $\phi(t)$, and $\sigma(t)$, respectively, one has

COROLLARY 7. *For $\alpha > 0$:*

$$-d\phi^\alpha/d\alpha = \sum_k \mu_k^\alpha f_k^\alpha,$$

$$\sigma^\alpha = \sum_k \mu_k f_k^\alpha = \sum_k \mu_k^\alpha f_k, \qquad \sigma(0) = \lim_{\alpha\to\infty} \alpha\sigma^\alpha = \lim_{\alpha\to 0} \phi^\alpha.$$

As an example, taking $f_k(t) = p_{kj}(t)$ and $\mu_k(t) = p_{ik}(t)$ for fixed j and i, respectively, one has obviously $\phi(t+s) = p_{ij}(t+s)$, so $\sigma(0) = u_{ij}$. On the other hand, with $f_k = 1$ for all k, $\mathbf{P}(t)\mathbf{1} = \mathbf{1}$ implies that $\sigma(t) = \mu(t) \bullet \mathbf{1}$ is constant independent of t.

§ 1.3.5. Duality. Conceptually, the notion of duality is associated with reversing direction of time in a M.C. Here, however, only basic analytic properties of duality are summarized.

Let $\mathbf{P}(t) = (p_{ij}(t))$ be a standard stochastic transition matrix. Duality is defined relative to a fixed *reference measure* ξ, taken to be positive, σ-finite and $\mathbf{P}(t)$-excessive measure on $\mathbb{I}$. At least one such reference measure exists, namely the measure $\pi\mathbf{U}$.

DEFINITION 1. The dual to $\mathbf{P}(t)$, relative to the reference measure ξ, is a Markov semi-group $\widehat{\mathbf{P}}(t) = (\hat{p}_{ij}(t))$ defined by

$$\hat{p}_{ij}(t) = \xi(j)p_{ji}(t)/\xi(i), \qquad i, j \in \mathbb{I}.$$

Clearly, $\widehat{\mathbf{P}}(t)$ is only sub-stochastic. Adjoining an absorbing state Δ, one can extend $\widehat{\mathbf{P}}(t)$ to a stochastic matrix $\widehat{\mathbf{P}}^{\Delta}(t)$ on $\mathbb{I} \cup \{\Delta\}$ in the usual way, with

$$\hat{p}_{i\Delta}(t) = 1 - \widehat{\mathbf{P}}(t)\mathbf{1}_i, \qquad i \in \mathbb{I}.$$

It follows from Definition 1 that, relative to the reference measure ξ, the dual to a resolvent $\mathbf{U}^\alpha$ and the dual to a generator $\mathbf{Q}$ are, respectively, a resolvent $\widehat{\mathbf{U}}^\alpha = (\hat{u}_{ij}^\alpha)$ defined by

$$\hat{u}_{ij}^\alpha = \xi(j)u_{ji}^\alpha/\xi(i), \qquad i, j \in \mathbb{I},$$

with $\alpha\hat{u}_{i\Delta}^\alpha = 1 - \alpha\widehat{\mathbf{U}}^\alpha\mathbf{1}_i$, and a generator $\widehat{\mathbf{Q}} = (\hat{q}_{ij})$ defined by

$$\hat{q}_{ij} = \xi(j)q_{ji}/\xi(i), \qquad i, j \in \mathbb{I},$$

with $\hat{q}_{i\Delta} = -\widehat{\mathbf{Q}}\mathbf{1}_i$.

Note that always $\xi\widehat{\mathbf{P}}(t) = \xi$, $\xi\widehat{\mathbf{Q}} = 0$, and $\widehat{\mathbf{Q}}\mathbf{1} = 0$ iff $\xi\mathbf{Q} = 0$. It is easy to check that the dual of the (B)-equation for $\mathbf{P}(t)$ is the (F)-equation for $\widehat{\mathbf{P}}(t)$ on $\mathbb{I}$. Potential theoretic concepts relative to the dual will be designated with the prefix co-, when applied to the direct semi-group. Thus, a function f and a measure μ such that

$$\widehat{\mathbf{P}}(t)f \leqslant f, \qquad \mu\widehat{\mathbf{P}}(t) \leqslant \mu$$

are co-excessive with reference to $\mathbf{P}(t)$.

DEFINITION 2. A dual of a function f on $\mathbb{I}$ is a measure $\hat{f}$ on $\mathbb{I}$ defined by

$$\hat{f}(i) = f(i)\xi(i), \qquad i \in \mathbb{I}.$$

A dual of a measure μ on $\mathbb{I}$ is a function $\hat{\mu}$ on $\mathbb{I}$ defined by

$$\hat{\mu}(j) = \mu(j)/\xi(j), \qquad j \in \mathbb{I}.$$

It follows that a function f (a measure μ) is excessive for $\mathbf{P}(t)$ iff its dual is a co-excessive measure $\hat{f}$ (function $\hat{\mu}$), and

$$\widehat{\mathbf{P}(t)f} = \hat{f}\widehat{\mathbf{P}}(t), \qquad \widehat{\mu\mathbf{P}(t)} = \widehat{\mathbf{P}}(t)\hat{\mu}.$$

In particular, the reference measure ξ is both excessive and co-excessive, and its dual is a constant function $\mathbf{1}$. Note also that if $f(\cdot)$ is an exit law for $\mathbf{P}(t)$, then its dual is an entrance law for $\widehat{\mathbf{P}}(t)$; similarly, if $\mu(\cdot)$ is an entrance law for $\mathbf{P}(t)$, its dual is an exit law for $\widehat{\mathbf{P}}(t)$.

PROPOSITION 3. *Semi-groups* $\mathbf{P}(t)$ *and* $\widehat{\mathbf{P}}(t)$ *are in duality iff*

$$\mu\mathbf{P}(t)f = \hat{f}\widehat{\mathbf{P}}(t)\hat{\mu}$$

for positive functions f *and measures* μ *on* $\mathbb{I}$.

An equivalent approach to duality is based on the following definition.

DEFINITION 4. Two semi-groups $\mathbf{P}(t)$ and $\widehat{\mathbf{P}}(t)$, or their resolvents $\mathbf{U}^\alpha$ and $\widehat{\mathbf{U}}^\alpha$, are said to be in duality, relative to a reference measure ξ, if there exists a kernel $\mathbf{V}^\alpha = (v^\alpha(i,j))$ such that

$$u_{ij}^\alpha = v^\alpha(i,j)\,\xi(j), \qquad \hat{u}_{ji}^\alpha = \xi(i)\,v^\alpha(i,j).$$

The kernel $\mathbf{V}^\alpha$ is used to define new forms of potentials. Let μ be a (positive) measure on $\mathbb{I}$. Then, the α-potential of a measure μ is a function $\mathbf{U}^\alpha\mu$ defined by

$$\mathbf{U}^\alpha\mu(i) = \sum_j v^\alpha(i,j)\mu(j), \qquad i \in \mathbb{I},$$

and the α-co-potential of a measure μ is a function $\widehat{\mathbf{U}}^\alpha\mu$ defined by

$$\widehat{\mathbf{U}}^\alpha\mu(j) = \sum_i \mu(i)v^\alpha(i,j), \qquad j \in \mathbb{I}.$$

These functions should be distinguished from measures $\mu\mathbf{U}^\alpha$ and $\mu\widehat{\mathbf{U}}^\alpha$ defined as in § 1.3.1. Clearly,

$$\mu\mathbf{U}^\alpha(j) = \widehat{\mathbf{U}}^\alpha\mu(j)\,\xi(j), \qquad \mu\widehat{\mathbf{U}}^\alpha(i) = \xi(i)\mathbf{U}^\alpha\mu(i),$$

so functions $\mathbf{U}^\alpha\mu$ and $\widehat{\mathbf{U}}^\alpha\mu$ are densities, relative to the measure ξ, of measures $\mu\widehat{\mathbf{U}}^\alpha$ and $\mu\mathbf{U}^\alpha$, respectively.

Remark. Of special interest is the symmetric (self-dual) case for which

$$\xi(i)p_{ij}(t) = \xi(j)p_{ji}(t), \qquad i,j \in \mathbb{I}.$$

Then, $v^\alpha(i,j) = v^\alpha(j,i)$.

SECTION 2

PASSAGE PROBABILITIES

This section introduces passage times, a concept which includes both the first entrance time and the last exit time from a set, and their various extensions.

Both analytic and measure-theoretic aspects are discussed here, with stress on the role of the sample functions behaviour in determination of passage probabilities.

Sub-section 2.1 presents basic definitions of typical passage times, establishes their preliminary properties, and lists notation to be used later on. Some of the methods to be used in this book are introduced in sub-section 2.2. A notion of a delayed taboo probability (when taboo is imposed after some random time) is presented, with the important special case of "the exit taboo". Finally, a rather heuristic approach convenient in applications to usual Markov chains, known as "a density argument" (essentially linked to path analysis) is treated in some detail.

The material presented in this section has been known in Markov chain theory and applications since early days (see references in Section 1). The systematic presentation attempted here is based on R. Syski [1965, 1975]. Usefulness of the notion of the exit taboo was pointed out by K. L. Chung [1967].

2.1. DEFINITIONS

§ **2.1.1. Random sets.** Consider the usual Markov chain $X = (X_t) = (X_t, \mathcal{F}_t, 0 \leqslant t < \infty)$, introduced in § 1.1.1. Recall that this chain has a countable state space $\mathbb{I}$ with all states stable, and its sample functions are right-continuous with left limits; the chain is taken relative to the natural σ-fields $\mathcal{F}_t = \sigma(X_s,\ s \leqslant t)$.

Let H be a set in the state space, and write $H^c = \mathbb{I} \setminus H$. In the following discussion the set H will be kept fixed.

It is important in the considerations in this book to distinguish between time periods when the M.C. is in H or in H^c. In view of properties of sample functions, the transitions between H and H^c occur at jumps, and (by total stability) there are intervals of positive length when the chain remains in H or in H^c. More precisely, define a random set S_H called the *sojourn time* of H, by

$$S_H(\omega) = (t\colon\ X_t(\omega) \in H).$$

Its connected components are random intervals of the type $[\cdot, \cdot)$, by right-continuity of paths. For (almost) all ω, the number of such intervals is countable. In fact, S_H is a countable union of stable intervals, and S_H has no isolated points.

The complement of S_H is a union of countably many intervals of the type $[\cdot\,,\,\cdot\,)$; these intervals are said to be contiguous to S_H. In such intervals, the M.C. is in H^c. See Fig. 2.1.

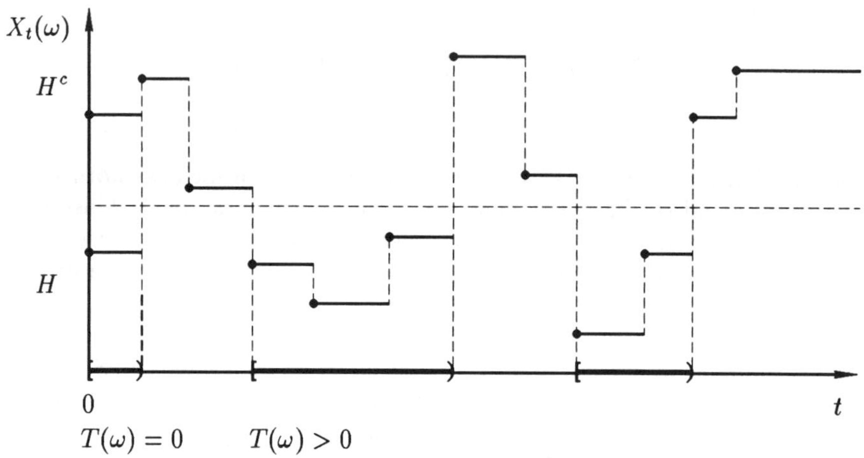

Figure 2.1. Realization of the random set S and the first entrance time T to the set H, for $X_0(\omega) \in H^c$ and $X_0(\omega) \in H$.

For each $t > 0$, define the following sets

$$K_t = (\omega\colon\ X_s(\omega) \in H \quad \text{for some } s,\ 0 < s \leqslant t),$$

$$K_{t-} = \bigcup_{0<s<t} K_s, \qquad K_{t+} = \bigcap_{s>t} K_s,$$

where the set operations may be taken over rationals. It is clear that (K_t) is an increasing family, and that the right-continuity of paths implies $K_t = K_{t+}$.

Shifting along time axis, define for a fixed $s > 0$

$$K_{s,s+t} = \theta_s^{-1} K_{0,t} = (\omega\colon\ X_u \in H \quad \text{for some } u,\ s < u \leqslant s+t),$$

where $K_{0,t} = K_t$, and θ_s is the shift operator defined below in $(*)$.

It will be of importance to consider a random set $W = W_H$ which consists of instants of departures from H (entrances to H^c). Clearly, W is at most countable.

As the initial state may be either in H or in H^c, it is convenient to restrict W to positive times. Observe that by right-continuity of paths

$$(s \in W\colon\ X_s \in H) = \varnothing, \qquad (s \in W\colon\ X_s \in H^c) = W.$$

Hence, W is the set of strictly positive left-end points of intervals contiguous to the sojourn set S_H.

Similarly, a random set $V = V_H$ consists of instants of entrances to H (departures from H^c), and is at most countable. By right-continuity of paths

$$(s \in V\colon\ X_s \in H^c) = \varnothing, \qquad (s \in V\colon\ X_s \in H) = V.$$

Hence, V is the set of strictly positive left-end points of intervals forming the sojourn set S_H.

Finally recall that the *shift* (translation) *operator* θ_t is a mapping $\Omega \to \Omega$ defined by

$$X_s(\theta_t\omega) = X_{s+t}(\omega) \tag{$*$}$$

or briefly, $X_s \circ \theta_t = X_{s+t}$ for all s, t. Hence, for sets:

$$\theta_t^{-1}(X_s = i) = (X_{s+t} = i), \qquad i \in \mathbb{I}.$$

If Y is an $\mathcal{F}_\infty$-measurable random variable of the form $Y = h(X_s,\ 0 \leqslant s < \infty)$, for some (Borel) function h, then

$$Y \circ \theta_t(\omega) = Y(\theta_t\omega),$$

so

$$Y \circ \theta_t = h(X_s,\ t \leqslant s < \infty).$$

Extension to random shift θ_S, where S is a stopping time, is defined by

$$\theta_S(\omega) = \theta_{S(\omega)}(\omega),$$

i.e., $\theta_S(\omega) = \theta_s(\omega)$ when $S(\omega) = s$. Hence,

$$X_t \circ \theta_S = X_{S+t}, \qquad X_S \circ \theta_t = X_{t+S\circ\theta_t}.$$

§ 2.1.2. First entrance times. a) The best known representative of passage times studied since the early days of Probability Theory is the first entrance time to a set, defined by the following

DEFINITION 1. On a probability space $(\Omega, \mathcal{F}, \mathbb{P})$ define $T = T_H$ by

$$T(\omega) = \inf\,(t\colon\ t > 0,\ X_t(\omega) \in H)$$

with infimum of the empty set taken to be ∞. T is called the *first entrance time* to a set H.

This name is most common in applications, although in theoretical works T is called the first hitting time of H. It is important to observe that when the chain starts in H (so $X_0 \in H$), then necessarily $T = 0$. The related concept of the first entry time T^0 to a set H is obtained by replacing $t > 0$ by $t \geqslant 0$ in Definition 1. However, for the usual M.C., stability and right-continuity imply that T and T^0 coincide.

The usual argument, using random sets K_t, yields the following useful result (see Fig. 2.2).

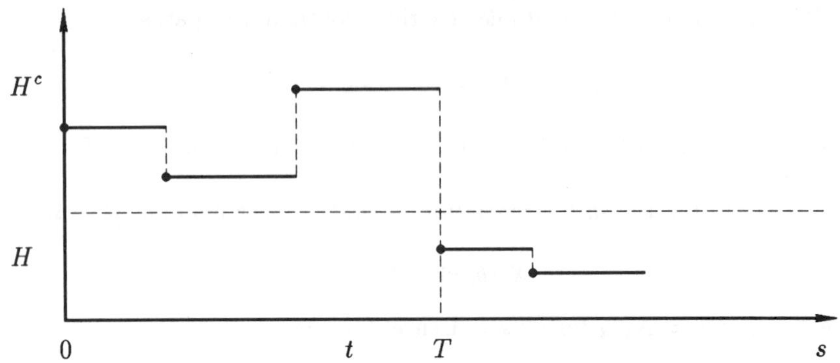

$$(T > t) = (X_s \in H^c,\ 0 < s \leqslant t)$$
$$(T \geqslant t) = (X_s \in H^c,\ 0 < s < t) \qquad X_T \in H$$

Figure 2.2. Comparison of two sets.

LEMMA 2. *For each* $t > 0$:

$$(T < t) = K_{t-} \subset K_t = K_{t+} = (T \leqslant t).$$

Consequently, T is a stopping time relative to the natural σ-fields $(\mathcal{F}_t)$. Note that this fact follows immediately from the "debut theorem", because T^0 coincides with the debut of the set

$$M = ((s,\omega)\colon\ X(s,\omega) \in H) \subset \mathbb{R}_+ \times \Omega,$$

(see Fig. 2.3). The set $[\![T]\!]$ defined by

$$[\![T]\!] = ((s,\omega)\colon\ T(\omega) = s)$$

is called a graph of T. Observe that Lemma 2 yields

$$(T < \infty) = (X_s \in H \text{ for some } s), \quad (T = \infty) = (X_s \in H^c \quad \text{for all } s).$$

The most common example is the first entrance time to a state j, obtained by taking $H = \{j\}$. Time to absorption, waiting times and busy periods are typical examples of the first entrance time. The first exit time ρ_i from a state i is the first entrance time to a set $H = \{i\}^c$. More substantial examples will be given later.

The random variable X_T represents the position or place of the first entrance (hit) to the set H. As T is a stopping time, X_T is $\mathcal{F}_T$-measurable, so

$$(X_T = j,\ T \leqslant t) \in \mathcal{F}_t \qquad \text{for each} \quad t.$$

Since all states are stable and sample functions are right-continuous, it follows that by definition X_T takes values in H only, when T is finite. Indeed, the M.C. stays in H^c up to time T, and at the instant T the jump to H occurs. Hence, for $T < \infty$,

$$X_{T-} \in H^c \quad \text{and} \quad X_T \in H.$$

b) Another important passage time is obtained by shifting (or delaying) the first entrance time by a fixed non-random amount.

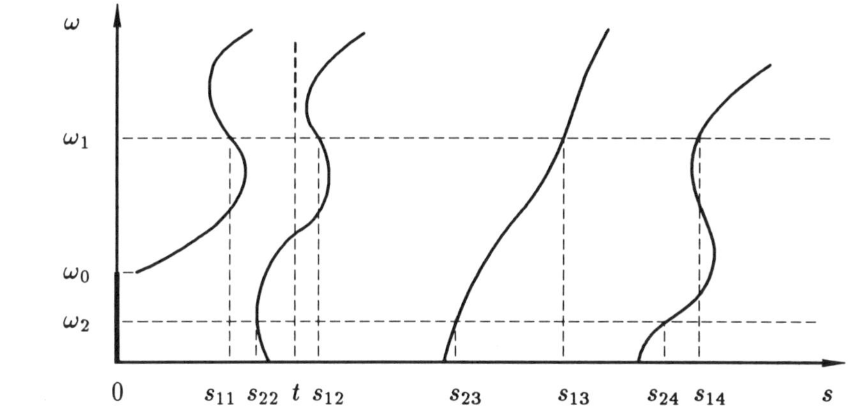

$$M = \{(s,\omega):\ X(s,\omega) \in H\} \qquad M_t = M \cap \{[0,1)\times\Omega\},$$

$$T^0 = \inf\{s: s \geqslant 0,\ (s,\omega) \in M\} \quad [T^0] = \{(s,\omega):\ T^0(\omega) = s\}$$

$$(T^0 < t) = \{X(s,\omega) \in H \text{ for some } s,\ 0 \leqslant s < t\} = \operatorname{proj}(M_t)$$

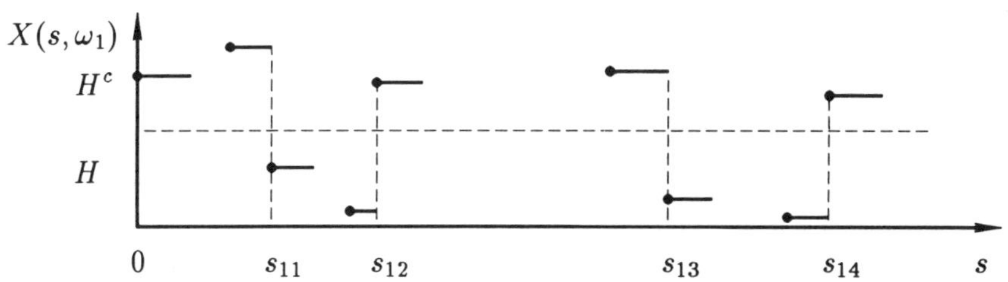

$T^0(\omega_1) = s_{11} > 0$

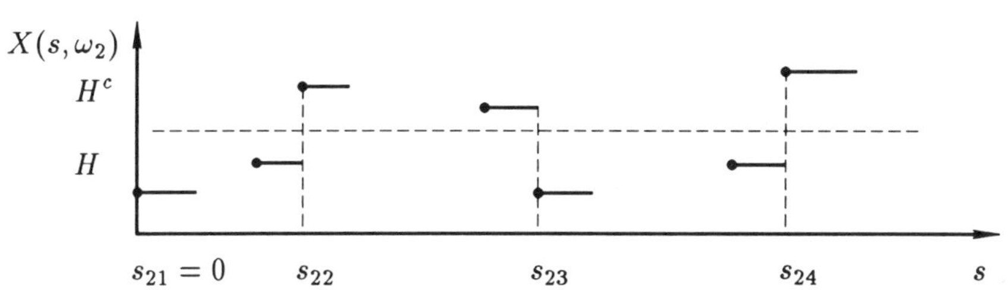

$T^0(\omega_2) = 0$

Note: $T^0(\omega_2) = 0,\ X(0,\omega_0) \in H \Rightarrow T^0(\omega_0) = 0,\ X(0,\omega_0) \in H^c \Rightarrow T^0(\omega_0) > 0$

Figure 2.3. Random set M and the debut theorem.

DEFINITION 3. On a probability space $(\Omega, \mathcal{F}, \mathbb{P})$ define $T_s = T_{s,H}$ for $s > 0$ by

$$T_s(\omega) = \inf(t\colon\ t > s,\ X_t(\omega) \in H)$$

with infimum of the empty set taken to be ∞. T_s is called the *first entrance time* to a set H *after* s.

Properties of T_s resemble those of T, subjected to a shift. As before, the first entry time T_s^0 obtained by modifying Definition 3 to have $t \geqslant s$, coincides with T_s for the usual M.C.

LEMMA 4. *For* $s \geqslant 0$:

$$T_s = s + T \circ \theta_s,$$

and $T \circ \theta_s = T - s$ *on the set* $(T > s)$.

Proof. In view of shift operator θ_s properties (see § 2.1.2), one has

$$T \circ \theta_s = \inf(u\colon\ u > 0,\ X_u \circ \theta_s \in H),$$

so

$$T_s = \inf(u + s\colon\ u + s > s,\ X_{u+s} \in H).$$

When $T > s$, it is clear that $T_s = T$ (this property is expressed by saying that the first entrance time T is a *terminal time*). □

It may be useful to remark that T_s represents time *of* occurrence, whereas $T \circ \theta_s$ represents time *to* occurrence of the first entrance after s (see Fig. 2.4 and Fig. 2.5).

It is clear that

$$\lim_{s \to 0} T_s = T,$$

and that $(T_s,\ 0 \leqslant s < \infty)$ is a right-continuous increasing process such that $T_0 = T$. The analogue of Lemma 2, extended under shift θ_s, holds. It then follows that T_s is a stopping time for each fixed s, relative to the natural σ-fields $(\mathcal{F}_t)$, but T_s need not be $\mathcal{F}_s$-measurable.

Moreover, T_s is not a terminal time for $s > 0$, as seen from the next Lemmas.

The behaviour under shift is expressed by

LEMMA 5. *For* $s, h \geqslant 0$:

$$T_{s+h} = h + T_s \circ \theta_h,$$
$$T_{h+T_s} = T_s + T_h \circ \theta_{T_s}.$$

Proof. By definition of shift $X_u \circ \theta_{T_s} = X_{u+T_s}$, so the assertion follows in the usual way. □

Define the following random set on the time axis

$$W = (s > 0\colon\ T_{s-} = s,\ T_s > s) = (s > 0\colon\ T \circ \theta_{s-} = 0,\ T \circ \theta_s > 0).$$

Clearly, this set coincides with the set W consisting of instants of departures from H (entrances to H^c), defined already in § 2.1.1 (see Fig. 2.4 and Fig. 2.5).

Also, if $s \in W$, then $T_s = s + T \circ \theta_s$ is the right-end point of the contiguous interval whose left-end point is s. Clearly,

$$(s\colon\ s > 0,\ T_{s-} = s) = S_H \setminus \{0\}.$$

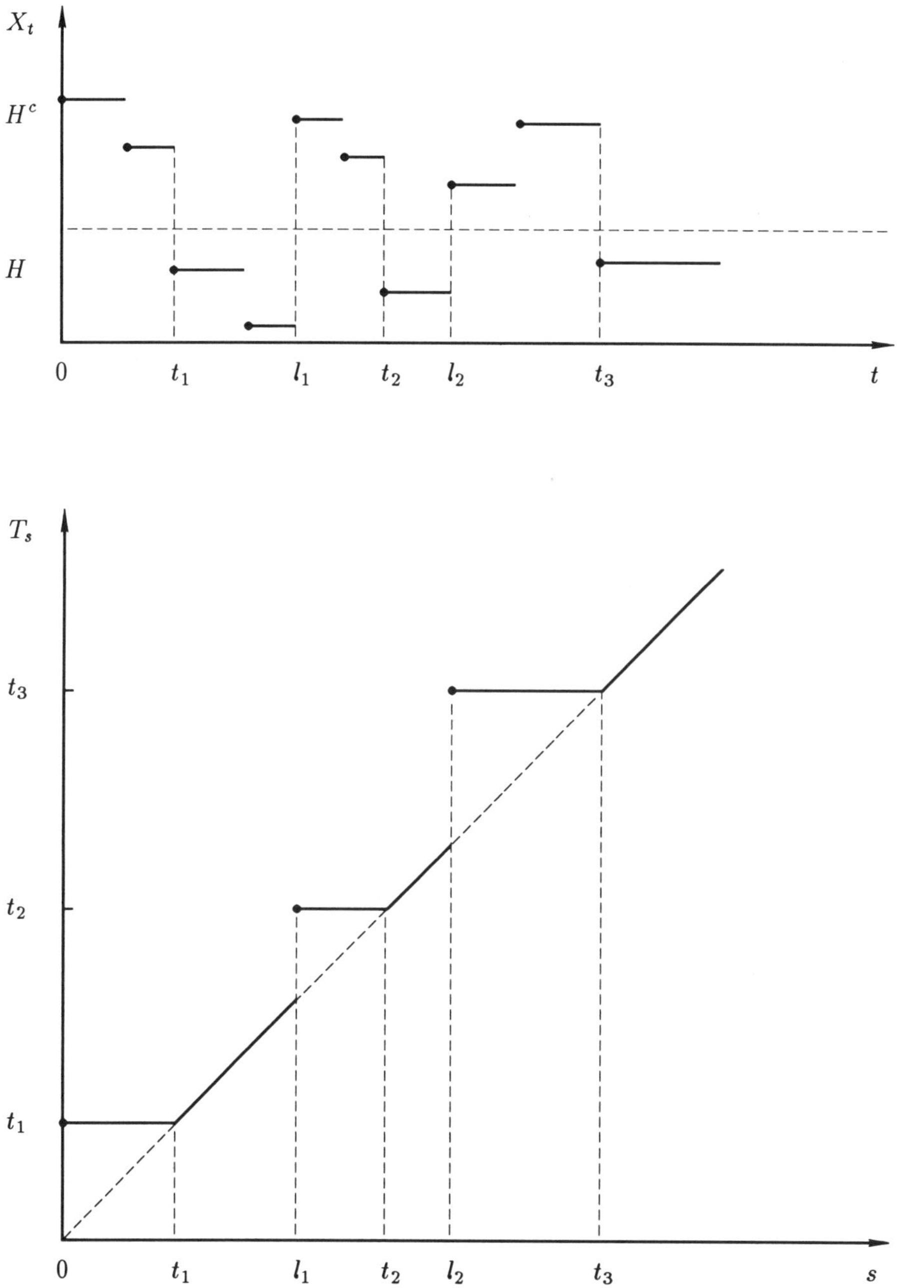

$$T_0 = T,\ T_s = \inf\{t\colon\ t > s,\ X_t \in H\} = s + T \circ \theta_s \geqslant s$$

Figure 2.4. The first entrance time T_s.

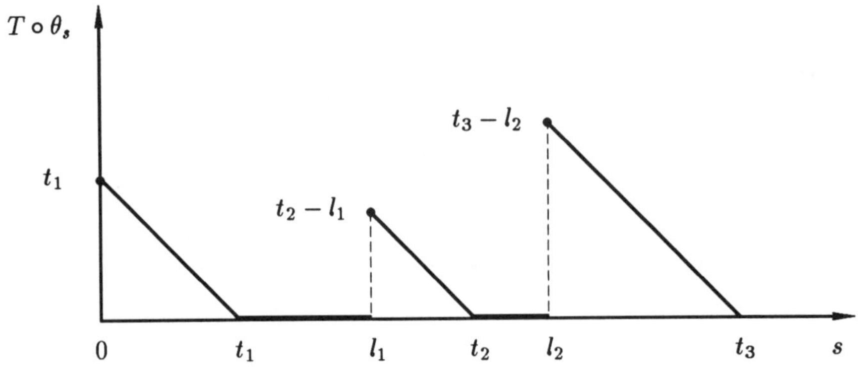

$$T\circ\theta_s = T_s - s$$

$$W = (s > 0\colon\ T\circ\theta_{s-} = 0,\ T\circ\theta_s > 0)$$

Figure 2.5. Time of occurrence of the first entrance after s.

The random variable $X(T_s)$ is the position or place of the first entrance (hit) to H after s. Since in the usual M.C. all states are stable and sample functions are right-continuous, $X(T_s)$ takes values in H only, when T_s is finite. Observe that

$$\theta_s^{-1}(X_T = j,\ T \leqslant t) = (X(T_s) = j,\ T_s \leqslant t + s).$$

§ **2.1.3. First return time.** In contrast with situation discussed in § 2.1.2, the first entrance time to H for a chain starting in H is now no longer taken as identically zero. Instead, when the M.C. is initially in a (stable) state $i \in H$, then it is assumed that the chain will leave H at the instant at which it ends its stay in i. Afterwards, the chain continues in H^c, and the subsequent first entrance to H is now regarded as the first return to H. On the other hand, if the initial position were in H^c, then the situation from § 2.1.2 would prevail.

It is clear that the new definition should depend on the first exit time ρ_i from a state i, already introduced in § 1.1.3.

DEFINITION 1. On a probability space $(\Omega, \mathcal{F}, \mathbb{P})$ define $T_\rho = T_{\rho_i, H}$ by

$$T_\rho(\omega) = \inf(t\colon\ t > \rho_i(\omega),\ X_t(\omega) \in H), \qquad i \in \mathbb{I},$$

with infimum of the empty set taken to be ∞. T_ρ is called the *first return time* to H.

The related concept T_ρ^0, defined by modified definition with $t \geqslant \rho_i$, coincides with T_ρ by the right-continuity of paths and stability of states. By taking $H = \{i\}$, one obtains a simple but important example of the first return time to a state i.

The following Lemma shows that T is in fact the first entrance time to H after ρ_i.

LEMMA 2. *We have*

$$T_\rho = \rho_i + T \circ \theta_{\rho_i}, \qquad i \in \mathbb{I},$$

and $T \circ \theta_{\rho_i} = T - \rho_i$ *on the set* $(T > \rho_i)$.

Proof. In view of properties of the random shift (see § 2.1.1) one has

$$T_\rho = \rho_i + \inf\big(s\colon\ s > 0,\ X(s + \rho_i) \in H\big) = \rho_i + T \circ \theta_{\rho_i}.$$

Furthermore, as T is the first entrance time after 0, so if $T > \rho_i$, then $T = T_\rho$. □

It can be verified in the manner analogous to that in § 2.1.2, that T_ρ is indeed a stopping time relative to the family of natural σ-fields. The following observation will be needed.

LEMMA 3. T_ρ *is a terminal time.*

Proof. Let $i \in H$. By definition, ρ_i is the time of a jump to H^c, so $t < T_\rho$ implies that *time to the first entrance* to H (after ρ_i) is $T_\rho \circ \theta_t = T_\rho - t$. □

It is important to stress dependence of T_ρ on the initial state $i \in \mathbb{I}$. By Lemma 2 (see also Fig. 2.6), if $i \in H^c$ then $\rho_i < T$, so $T = T_\rho$, and if $i \in H$ then $\rho_i > 0$, so $T_\rho \geqslant \rho_i$.

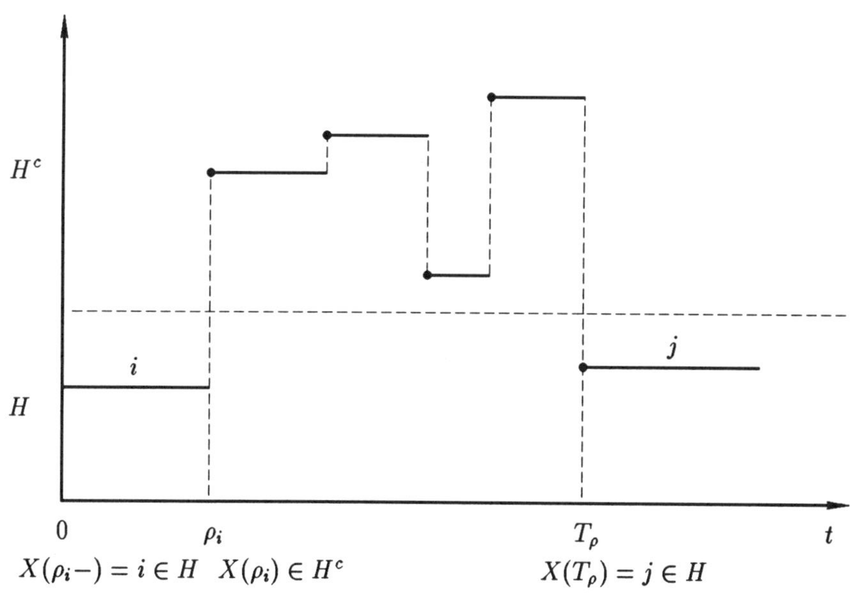

Figure 2.6. First return.

Although T_ρ is defined for all i in $\mathbb{I}$, a new situation prevails only when $i \in H$. Hence, for convenience, formulas are stated here for $i \in H$, on the understanding that extensions to $i \in H^c$ coincide with the corresponding formulas for T in § 2.1.2.

The random variable $X(T_\rho)$ is the position or place of the first return to H. As sample functions of the usual M.C. are right-continuous and all states are stable, $X(T_\rho)$ takes values in H only, when T_ρ is finite. Observe that

$$\theta_{\rho_i}^{-1}(X_T = j,\ T \leqslant t) = (X(T_\rho) = j,\ T_\rho \leqslant \rho_i + t).$$

§ **2.1.4. Last exit times.** a) Another well known representative of passage times is the last exit time L from a set H. Although less popular than its counterpart the first entrance time T, the last exit time receives increasing attention.

DEFINITION 1. On a probability space $(\Omega, \mathcal{F}, \mathbb{P})$ define $L = L_H$ by

$$L(\omega) = \sup\,(t\colon\ t > 0,\ X_t(\omega) \in H)$$

with supremum of the empty set taken to be 0. L is called the *last exit time* from H.

At time L, the M.C. leaves the set H never to return to it. Thus, the notion of transience is intrinsically connected with L, and it is clear that L depends on the future development of the process (in contrast to T which depends on the past) (see Fig. 2.7). The simplest example is provided by the last exit time from a single state.

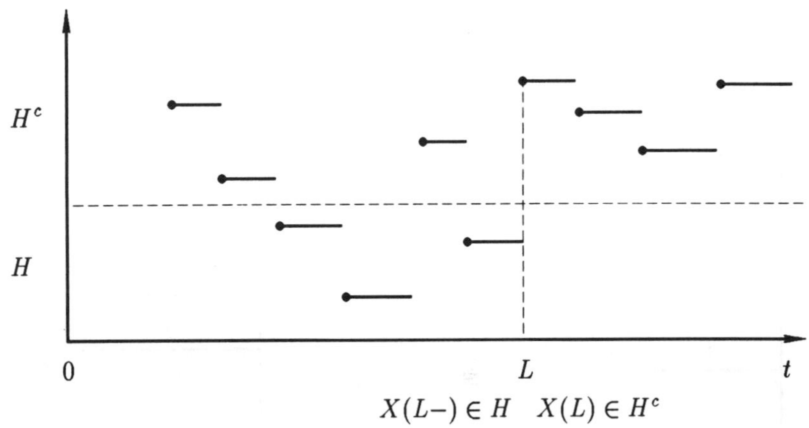

Figure 2.7. Last exit time.

The behaviour of L under shift is expressed by the following lemma, where $a^+ = \max(a, 0)$.

LEMMA 2. *For* $t \geqslant 0$:

$$L \circ \theta_t = (L - t)^+.$$

Proof.

$$L \circ \theta_t = \sup\,(s\colon\ s > 0,\ X_{s+t} \in H) = \sup\,(u\colon\ u > t,\ X_u \in H) - t,$$

and this is $L-t$ if $L>t$, and 0 if $L \leqslant t$. □

Examination of sample functions of the usual M.C. yields the basic relations

$$(L \leqslant t) = (L \circ \theta_t = 0) = (X_u \in H^c,\ u > t) = (T_t = \infty)$$

so L is a co-stopping time relative to the family $(\mathcal{G}_t)$, where $\mathcal{G}_t = \sigma(X_u,\ u \geqslant t)$, and, for $L > t$,

$$(L \circ \theta_t \leqslant s) = (L \leqslant t + s).$$

Moreover,

$$(L = 0) = (X_u \in H^c \text{ for all } u) = (T = \infty).$$

The random variable $X(L)$ represents the position or place of the last exit from H. As $X(L)$ is $\mathcal{G}_L$-measurable, so $(X(L) = j,\ L \geqslant t) \in \mathcal{G}_t$ for each t. By the right-continuity of paths and stability of states, $X(L)$ takes values in H^c only, provided $0 < L < \infty$. Strictly speaking, the last position from which the M.C. leaves the set H should obviously be in H, and is therefore given by $X(L-)$. The distinction between random variables $X(L)$ and $X(L-)$ is essential.

b) Another important passage time is obtained by restricting attention to those last exits which take place before a fixed time instant.

Definition 3. On a probability space $(\Omega, \mathcal{F}, \mathbb{P})$ define $L_s = L_{s,H}$, for $s > 0$, by

$$L_s(\omega) = \sup\,(t\colon\ 0 < t \leqslant s,\ X_t(\omega) \in H)$$

with supremum of the empty set taken to be 0. L_s is called the *last exit time* from *H before s*.

It should be remarked that after leaving the set H at time L_s and staying away from it, the M.C. may return to H after s. Taking $H = \{i\}$ is the most common special case. The passage time L_s represents the dual of its better known counterpart T_s, and receives increasing attention. However, analysis of L_s is much more delicate than that of T_s.

It is clear that $(L_s,\ 0 < s < \infty)$ is a right-continuous increasing process such that $L_s \leqslant s$ and $L_s \to L$ as $s \to \infty$. Moreover, discontinuity points of L_s are instants of consecutive entrances to the set H (see Fig. 2.8). Recalling definitions of random sets S_H (sojourn time) and W (discontinuity points of T_s), it is clear from properties of sample functions that

$$S_H = (s\colon\ T_s = s) = (s\colon\ L_s = s) \cap W^c = (s\colon\ T \circ \theta_s = 0),$$
$$(s\colon\ L_s = s) = S_H \cup W.$$

The behaviour of L_s under shift is expressed by the following lemma, where $a^+ = \max(a, 0)$.

Lemma 4. *For* $s > 0,\ t \geqslant 0$:

$$L_s \circ \theta_t = (L_{s+t} - t)^+.$$

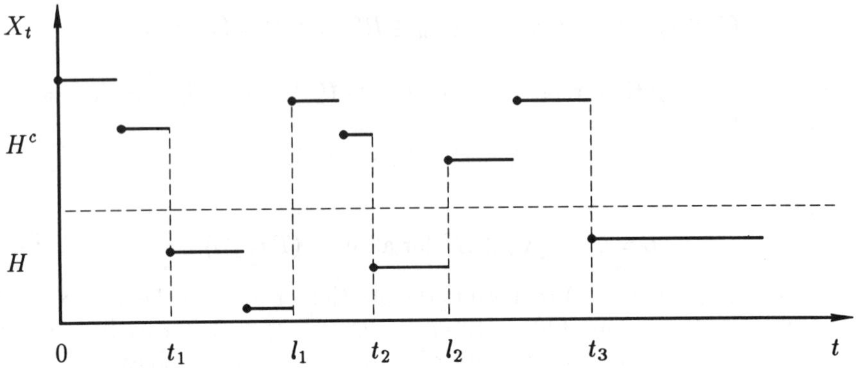

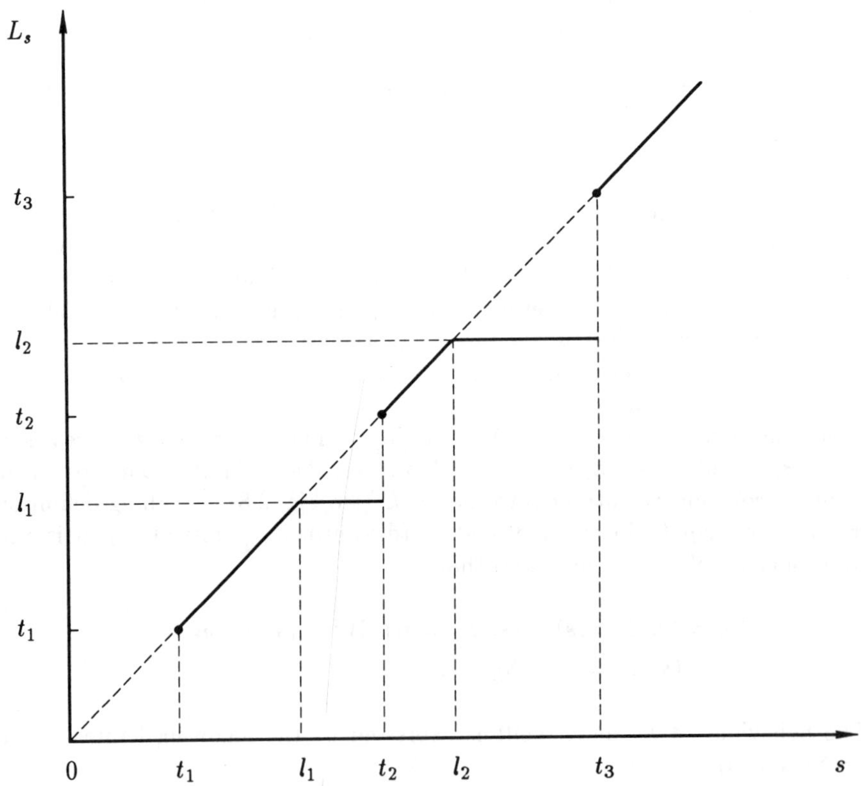

$$L_s = \inf(t\colon\ T_t > s) = \sup(t\colon\ t \leqslant s,\ X_t \in H) \leqslant s,\ (L_s) \text{ inverse of } (T_s).$$

Figure 2.8. Last exit time before s.

Proof.

$$L_s \circ \theta_t = \sup(u\colon\ u \leqslant s,\ X_{u+t} \in H) = \sup(v\colon\ v \leqslant s+t,\ X_v \in H) - t,$$

and this is $L_{s+t} - t$ if $L_{s+t} > t$, and 0 if $L_{s+t} \leqslant t$. □

It is shown later on that, for each s, L_s is a co-stopping time, relative to the decreasing family of σ-fields $\sigma(X_u,\ u \geqslant t)$.

The random variable $X(L_s)$ represents the position or place of the last exit from H before s. By the right-continuity of paths and stability of states, $X(L_s)$ takes values in H^c, only, provided $0 < L_s < s$. Strictly speaking, the last position from which the M.C. leaves the set H must obviously be in H, and is therefore given by $X(L_s-)$. The distinction between these two positions is essential.

c) It is more convenient to discuss right-continuous increasing processes (L_s) and (T_s) together, because Lemma 5 below indicates that they are the inverse of each other. It is also necessary to consider the left-continuous increasing processes (T_{s-}) and (L_{s-}), and by Lemma 5 they are also inverses of each other. Observe that $T_s = T_{s-}$ and $L_s = L_{s-}$ for all except countably many values of $s > 0$ (see Fig. 2.4 and Fig. 2.8).

From path properties of the usual M.C. one has

$$L_{s-} = \sup(t\colon\ t < s,\ X_t \in H), \qquad T_{s-} = \inf(t\colon\ t \geqslant s,\ X_{t-} \in H).$$

The following important relations are probabilistic versions of well known inversion formulae.

LEMMA 5. *For* $s > 0$:

$$\begin{aligned} \text{a)}\quad L_s &= \inf(t\colon\ T_t > s) = \sup(t\colon\ T_t \leqslant s),\\ T_s &= \inf(t\colon\ L_t > s) = \sup(t\colon\ L_t \leqslant s);\\ \text{b)}\quad L_{s-} &= \inf(t\colon\ T_t \geqslant s) = \sup(t\colon\ T_t < s),\\ T_{s-} &= \inf(t\colon\ L_t \geqslant s) = \sup(t\colon\ T_t < s). \end{aligned}$$

LEMMA 6. *For* $s > 0$:

$$\begin{aligned} (L_s < t) &= (T_{t-} > s), \qquad t > 0,\\ (L_{s-} \leqslant t) &= (T_t \geqslant s), \qquad t \geqslant 0. \end{aligned}$$

The following decomposition is the direct consequence of Lemma 6, and will be needed later.

LEMMA 7. *For* $s > 0$:

$$\begin{aligned} (L_s = s) &= (T_{s-} = s),\\ (0 < L_s < s) &= (T_{s-} > s,\ s > T),\\ (L_s = 0) &= (T_{s-} > s,\ s \leqslant T) = (s < T). \end{aligned}$$

Observe that $(L_s = s)$ is the set of all s which are either continuity points of T_s with $X_s \in H$, or are in W with $X_s \in H^c$. The set $(0 < L_s < s)$ contains those s which are continuity points of T_s with $X_s \in H^c$ (with $s > T$), whereas for all s in $(L_s = 0)$ obviously $X_s \in H^c$.

Furthermore, for $L_s = 0$, the initial position must be in H^c and the chain enters H at a time strictly larger than s. If the initial position is in H, then necessarily $L_s > 0$ (for $s > 0$).

The following useful result is the immediate consequence of Lemma 6.

LEMMA 8. *For* $0 \leqslant t < s$:

$$(L_{s-} \leqslant t) = (X_u \in H^c,\ t < u < s),$$

and $(L_s < t)$ *coincides with the above at continuity points* s.

It follows from Lemma 8 that L_s is indeed a co-stopping time for each $s > 0$.

With reference to the set M introduced in § 2.1.2, the stopping time T_s and the co-stopping time L_s are often called the beginning and the end of M, respectively.

The random interval (L_s, T_s) is called the *excursion interval* (from H) *straddling* s (for $s > 0$). The process $(X(L_s + h),\ 0 < h < T_s - L_s)$ is called the excursion process straddling s. By the right-continuity of the M.C., the excursion process takes values in H^c only, and $L_s < s < T_s$.

§ 2.1.5. Notation. When there is no danger of ambiguity, simplifying notation $\mathbb{P}^i$ and $\mathbb{E}^i$ will be used when conditioning on the event $X_0 = i$ is implicit. The following notation will be used.

a) For the joint distribution of X_T and T, conditional on the initial state $X_0 = i$, write

$$\mathbb{P}^i(X_T = j,\ T \leqslant t) = D_{ij}(t) \qquad \text{for} \quad i \in \mathbb{I},\ j \in H,\ t \geqslant 0.$$

The marginal distributions are

$$\mathbb{P}^i(T \leqslant t) = D_i(t) = \sum_{j \in H} D_{ij}(t), \qquad D_i^c(t) = \mathbb{P}^i(T > t),$$

$$\mathbb{P}^i(X_T = j,\ T < \infty) = D_{ij} = \lim_{t \to \infty} D_{ij}(t),$$

$$\mathbb{P}^i(T < \infty) = D_i = \sum_{j \in H} D_{ij}, \qquad D_i^c = \mathbb{P}^i(T = \infty).$$

For fixed i, (D_{ij}) is the *entrance* (or hitting) *distribution*, and D_i is the *first entrance probability*.

Introduce also

$$D_{ij}^c(t) = D_{ij} - D_{ij}(t) = \mathbb{P}^i(X_T = j,\ t < T < \infty),$$

so

$$\sum_{j \in H} D_{ij}^c(t) = D_i^c(t) - D_i^c.$$

For Laplace–Stieltjes (L.S.) transforms write (for $\alpha > 0$)

$$d_{ij}^\alpha = \int_{0-}^{\infty} e^{-\alpha t} dD_{ij}(t) = \mathbb{E}^i(e^{-\alpha T},\ T < \infty,\ X_T = j),$$

$$d_i^\alpha = \int_{0-}^{\infty} e^{-\alpha t} dD_i(t) = \mathbb{E}^i(e^{-\alpha T},\ T < \infty),$$

with integrals taken over $[0,\infty)$, so as $\alpha \to 0$,

$$d_{ij}^{\alpha} \longrightarrow D_{ij}, \qquad d_i^{\alpha} \longrightarrow D_i.$$

The mean first entrance time m_i restricted to finite T and conditional on initial state i is defined by

$$m_i = \mathbb{E}^i(T,\ T<\infty) = \int_{0-}^{\infty} t\,dD_i(t) = \int_0^{\infty} (D_i^c(t) - D_i^c)\,dt$$

and, similarly,

$$m_{ij} = \mathbb{E}^i(T,\ T<\infty,\ X_T = j) = \int_{0-}^{\infty} t\,dD_{ij}(t) = \int_0^{\infty} D_{ij}^c(t)\,dt.$$

The following relations are useful (with limits taken as $\alpha \to 0$):

$$\int_0^{\infty} e^{-\alpha t}(D_i^c(t) - D_i^c)\,dt = (D_i - d_i^{\alpha})/\alpha \longrightarrow m_i,$$

$$\int_0^{\infty} e^{-\alpha t} D_i^c(t)\,dt = (1 - d_i^{\alpha})/\alpha \longrightarrow \mathbb{E}^i T,$$

$$\int_0^{\infty} e^{-\alpha t} D_{ij}^c(t)\,dt = (D_{ij} - d_{ij}^{\alpha})/\alpha \longrightarrow m_{ij}.$$

b) Notation for the joint distribution of $X(T_s)$ and T_s will be the same as that for X_T and T, but with added superscript s. Thus, conditional on $X_0 = i$,

$$D_{ij}^s(t) = \mathbb{P}^i\big(X(T_s) = j,\ T_s \leqslant t\big) \qquad \text{for} \quad i \in \mathbb{I},\ j \in H,\ t \geqslant s,$$

with marginal distributions

$$D_i^s(t) = \mathbb{P}^i(T_s \leqslant t), \qquad D_{ij}^s = \mathbb{P}^i\big(X(T_s) = j,\ T_s < \infty\big).$$

For fixed i, (D_{ij}^s) is the *entrance* (or hitting) *distribution after* s, and D_i^s is the *first entrance probability after* s.

Introduce also

$$D_{ij}^{sc}(t) = D_{ij}^s - D_{ij}^s(t) = \mathbb{P}^i\big(X(T_s) = j,\ t < T_s < \infty\big).$$

Similarly, write $d_{ij}^{s\alpha}$, $d_i^{s\alpha}$ and m_{ij}^s, m_i^s for L.S. transforms and moments, respectively.

c) Notation for the joint distribution of $X(T_\rho)$ and T_ρ will be the same as that for X_T and T, but with added superscript ρ. Thus, conditional on $X_0 = i$,

$$D_{ij}^{\rho}(t) = \mathbb{P}^i\big(X(T_\rho) = j,\ T_\rho \leqslant t\big) \qquad \text{for} \quad i \in \mathbb{I},\ j \in H,$$

with marginal distributions

$$D_i^\rho(t) = \mathbb{P}^i(T_\rho \leqslant t), \qquad D_{ij}^\rho = \mathbb{P}^i\big(X(T_\rho) = j,\ T_\rho < \infty\big).$$

For fixed i, (D_{ij}^ρ) is the *return distribution*, and D_i^ρ is the probability of the first return to H, or simply the *return probability*.

Introduce also

$$D_{ij}^{\rho c}(t) = D_{ij}^\rho - D_{ij}^\rho(t) = \mathbb{P}^i\big(X(T_\rho) = j,\ t < T_\rho < \infty\big).$$

Similarly, write $d_{ij}^{\rho\alpha}, d_i^{\rho\alpha}$ and m_{ij}^ρ, m_i^ρ for L.S. transforms and moments, respectively.

All the above expressions are analogous to those in a), and are actually considered only for $i \in H$, because for $i \in H^c$ they coincide with the corresponding expressions from a).

d) The following notation will be used. For the joint distribution of $X(L)$ and L, conditional on the initial state $X_0 = i$, write

$$\Gamma_{ij}(t) = \mathbb{P}^i\big(X(L) = j,\ L \leqslant t\big) \qquad \text{for } i \in \mathbb{I},\ j \in H^c,\ 0 \leqslant t < \infty.$$

The marginal distributions are

$$\Gamma_i(t) = \mathbb{P}^i(L \leqslant t), \qquad \Gamma_{ij} = \mathbb{P}^i\big(X(L) = j,\ L < \infty\big),$$

with $\Gamma_i^c(t)$ denoting the complementary distribution of L. Moreover,

$$\Gamma_i = \mathbb{P}^i(L < \infty), \qquad \Gamma_i^c = \mathbb{P}^i(L = \infty).$$

For fixed i, (Γ_{ij}) is the *exit distribution* on H^c, and Γ_i is the *last exit probability*.

Introduce also

$$\Gamma_{ij}^c(t) = \Gamma_{ij} - \Gamma_{ij}(t) = \mathbb{P}^i\big(X(L) = j,\ t < L < \infty\big),$$

so

$$\sum_{j \in H^c} \Gamma_{ij}^c(t) = \Gamma_i^c(t) - \Gamma_i^c.$$

For L.S. transforms of $\Gamma_{ij}(t)$ and $\Gamma_i(t)$ write γ_{ij}^α and γ_i^α, respectively, where

$$\gamma_i^\alpha = \mathbb{E}^i(e^{-\alpha L},\ L < \infty) \quad \text{and} \quad \lim_{\alpha \to 0} \gamma_i^\alpha = \Gamma_i, \quad \lim_{\alpha \to \infty} \gamma_i^\alpha = \Gamma_i(0).$$

Note that

$$\int_0^\infty e^{-\alpha t}\big(\Gamma_i^c(t) - \Gamma_i^c\big)\, dt = \big(\Gamma_i - \gamma_i^\alpha\big)/\alpha, \qquad \alpha > 0.$$

The mean last exit time, restricted to finite L and conditional on the initial state i, is

$$\mathbb{E}^i(L,\ L < \infty) = \int_{0-}^\infty t\, d\Gamma_i(t) = \int_0^\infty \big(\Gamma_i^c(t) - \Gamma_i^c\big)\, dt.$$

It is important to consider the joint distribution of $X(L-)$ and L, written as

$$\Pi_{ij}(t) = \mathbb{P}^i\big(X(L-) = j,\ 0 < L \leqslant t\big) \qquad \text{for} \quad i \in \mathbb{I},\ j \in H,\ 0 < t < \infty,$$

with marginals

$$\Pi_i(t) = \mathbb{P}^i(0 < L \leqslant t), \qquad \Pi_{ij} = \mathbb{P}^i\big(X(L-) = j,\ 0 < L < \infty\big).$$

Observe that L is taken to be strictly positive. For fixed i, (Π_{ij}) is the *exit distribution* on H, and Π_i is the *last exit probability*.

One can also consider the L.S. transforms π^{α}_{ij} and π^{α}_{i}, and the probability

$$\Pi^c_{ij}(t) = \Pi_{ij} - \Pi_{ij}(t) = \mathbb{P}^i\big(X(L-) = j,\ t < L < \infty\big) \qquad \text{for} \quad t > 0.$$

e) Notation for the joint distribution of $X(L_s)$ and L_s will be the same as that for $X(L)$ and L, but with added superscript s. Thus, conditional on $X_0 = i$,

$$\Gamma^s_{ij}(t) = \mathbb{P}^i\big(X(L_s) = j,\ L_s \leqslant t\big) \qquad \text{for} \quad i \in \mathbb{I},\ j \in H^c,\ 0 \leqslant t < s.$$

The marginal distributions are

$$\Gamma^s_i(t) = \mathbb{P}^i(L_s \leqslant t), \qquad \Gamma^s_{ij} \equiv \Gamma^s_{ij}(s-) = \mathbb{P}^i\big(X(L_s) = j,\ L_s < s\big),$$

with

$$\Gamma^s_i \equiv \Gamma^s_i(s-) = \mathbb{P}^i(L_s < s) \leqslant \Gamma^s_i(s) = 1.$$

For fixed i, (Γ^s_{ij}) is the *exit distribution before s on* H^c, and Γ^s_i is the *last exit probability before s*. Introduce also

$$\Gamma^{sc}_{ij}(t) = \Gamma^s_{ij} - \Gamma^s_{ij}(t) = \mathbb{P}^i\big(X(L_s) = j,\ t < L_s < s\big).$$

It is important to consider also the joint distribution of $X(L_s-)$ and L_s

$$\Pi^s_{ij}(t) = \mathbb{P}^i\big(X(L_s-) = j,\ 0 < L_s \leqslant t\big) \qquad \text{for} \quad i \in \mathbb{I},\ j \in H,\ 0 < t < s,$$

with marginals

$$\Pi^s_i(t) = \mathbb{P}^i(0 < L_s \leqslant t), \qquad \Pi^s_{ij} = \mathbb{P}^i\big(X(L_s-) = j,\ 0 < L_s < s\big).$$

Observe that L_s is taken to be strictly positive. For fixed i, (Π^s_{ij}) is the *exit distribution before s on* H, and Π^s_i is the *last exit probability before s*. Introduce also

$$\Pi^{sc}_{ij}(t) = \Pi^s_{ij} - \Pi^s_{ij}(t) = \mathbb{P}^i\big(X(L_s-) = j,\ t < L_s < s\big),$$

with $\Pi^{sc}_{ij}(0) = \Pi^s_{ij}$.

2.2. METHODOLOGY

§ 2.2.1. Methods. It may be desirable to characterise briefly methods used in studies of passage times. Although analytic and measure-theoretic methods complement each other, it may be convenient to utilise some of their special properties.

In this book, the importance of taboo probabilities is emphasized by showing how they can be used to obtain all probabilities of interest. In this and in the next chapter, mainly analytic aspects are discussed, stressing the role of the (B)- and (F)-equations for distributions, as well as the inter-relation between the minimal chain and its jump chain. It is shown that distributions of individual passage times can be expressed directly in terms of taboo probabilities (either ordinary or extended). The same approach to the joint distributions of time and position must be supplemented by the "density argument", essentially the path analysis. Furthermore, the use of matrices greatly simplifies presentation and improves clarity.

Another approach, discussed in later sections, is based on the "product theorem" for stopping times, and on the "Dynkin formula". It is indispensable in discussion of more complex situations.

Still another approach, more general and essentially measure-theoretic, is based on the use of functionals of Markov chains and on associated martingales. This approach will be used later in the book and related to taboo argument. Indeed, the notion of taboo corresponds to a form of killing of a chain, done according to functionals preserving Markov property.

Each of these approaches can be taken as a starting point of discussion but taboo probabilities offer the most direct method.

This sub-section contains brief discussion of extended forms of taboo probabilities and explains a (rather heuristic) technique referred to as the "density argument".

§ 2.2.2. Delayed taboo. Let $X = (X_t)$ be the usual M.C., and let H be a taboo set. Let S be a stopping time relative to the natural σ-fields of X.

Suppose that up to time S the M.C. proceeds without restrictions and taboo is imposed after S. This leads to the following extension of the notion of taboo probability.

DEFINITION 1. The taboo probability *delayed by S*, denoted by ${}_Hp^S_{ij}(t)$, is given by

$${}_Hp^S_{ij}(t) = \mathbb{P}(X_u \in H^c, \ \min(t, S) < u < t, \ X_t = j \mid X_0 = i)$$

for $i, j \in \mathbb{I}$ and $t > 0$. For $t = 0$, ${}_Hp^S_{ij}(0) = \delta_{ij}$.

The measurability of the random set in Definition 1 follows from the separability of the chain. By definition, $X_u \in H^c$ for $S < u < t$ when $S < t$; otherwise no taboo is imposed. Thus,

$${}_Hp^S_{ij}(t) = \mathbb{P}^i(S \geqslant t, \ X_t = j) + \mathbb{P}^i(S < t, \ X_u \in H^c, \ S < u < t, \ X_t = j).$$

By right-continuity of paths and the fact that probability of a discontinuity at a fixed instant is zero, the second term on the right vanishes when $j \in H$, and also when $X(S)$ takes values in H. For $S = 0$, the delayed taboo probability reduces to the ordinary taboo probability ${}_Hp_{ij}(t)$. It is clear that, for all $i, j \in \mathbb{I}$,

$$ {}_Hp^S_{ij}(t) \leqslant p_{ij}(t), $$

and it is easy to verify that

$$ p_{ij}(t) - {}_Hp^S_{ij}(t) = \mathbb{P}^i\big(X_u \in H^c \text{ for some } u,\ S < u < t,\ X_t = j\big). $$

In contrast to the ordinary taboo probabilities, the delayed taboo matrices

$$ {}_H\mathbf{P}^S(t) = \big({}_Hp^S_{ij}(t)\big), \qquad t \geqslant 0, $$

do not form a semi-group, in general. Indeed, any taboo probability is obtained by killing the original chain at the first entry to the taboo set at some time after S, say, T_S. It is known, however, that a killed chain has the semi-group property provided the killing time T_S is a terminal time (see discussion in § 9.1.2). Brief discussion of this aspect would facilitate presentation of entrance times in Chapter II. The following definition embraces times T, T_s, and T_ρ, introduced in §§ 2.1.2 and 2.1.3.

DEFINITION 2. Let $T = T_H$ be the first entrance time to a set H, and let S be a stopping time. The *first entrance time to a set H after S*, denoted by T_S, is given by

$$ T_S = \inf(t\colon\ t > S,\ X_t \in H) $$

with infimum of the empty set taken to be ∞.

Using definition of the random shift (see § 2.1.1), one has

$$ T_S = S + T \circ \theta_S. $$

It can be verified that T_S is a stopping time relative to the natural σ-fields of the chain X. Evidently, $X(T_S)$ takes values in H only (for $T_S < \infty$).

Connection between delayed taboo probability and distribution of T_S is established by the following

LEMMA 3. *For $i, j \in \mathbb{I}$ and $t > 0$:*

$$ {}_Hp^S_{ij}(t) = \mathbb{P}^i(X_t = j,\ T_S > t). $$

Proof.

$$ \begin{aligned} \mathbb{P}^i(X_t = j,\ T_S > t) &= \mathbb{P}^i(S \geqslant t,\ T_S > t,\ X_t = j) + \mathbb{P}^i(S < t,\ T_S > t,\ X_t = j) \\ &= \mathbb{P}^i(S \geqslant t,\ X_t = j) + \mathbb{P}^i(S < t < T_S,\ X_t = j) \\ &= \mathbb{P}^i(S \geqslant t,\ X_t = j) + \mathbb{P}^i(S < t,\ X_u \in H^c,\ S < u < t,\ X_t = j) = {}_Hp^S_{ij}(t), \end{aligned} $$

where the last step follows from definition of T_S. □

Hence, one has immediately from Lemma 3 the following important result.

THEOREM 4. *For* $i \in \mathbb{I}$, $t > 0$:

$$\mathbb{P}^i(T_S > t) = \sum_j {}_Hp^S_{ij}(t).$$

Recall that T is a terminal time. Suppose now that S is also a terminal time, so by definition $S \circ \theta_t = S - t$ whenever $S > t$. The following proposition is basic.

PROPOSITION 5. *The first entrance time T_S to a set H after time S is a terminal time, i.e.,*

$$T_S \circ \theta_t = T_S - t \qquad \textit{on} \quad (T_S > t) \quad \textit{for all} \quad t \geqslant 0.$$

Proof. Using definitions of T_S, S_t, and properties of the shift operator, one has

$$\begin{aligned}(T_S)_t &= t + T_S \circ \theta_t = t + (S + T \circ \theta_S) \circ \theta_t = t + S \circ \theta_t + T \circ \theta_S \circ \theta_t \\ &= S_t + T_{t+S\circ\theta_t} = S_t + T \circ \theta_{S_t} = T_{S_t}.\end{aligned}$$

Therefore, if $t > S$, then $S_t = S$, so $(T_S)_t = T_S$.

On the other hand, up to an equivalence,

$$(S \leqslant t < T_S) = (X_u \in H^c,\ S \leqslant u \leqslant t),$$

so $(T_S)_t = T_S$. In other words, as $X_t \in H^c$, so time *to* the first entrance to H (after S) is

$$T_S \circ \theta_t = T_S - t.$$

Finally, the assertion follows by combining both cases. □

Lemma 3 and Proposition 5 indicate that delayed taboo probabilities are obtained by killing the original M.C. at the first entry time T_S to the taboo set H. Consequently, by the earlier remark the transition probabilities ${}_Hp^S_{ij}(t)$ of the killed chain form a Markov semi-group. Hence, one has the following basic result (which can also be verified directly by tedious calculation or by simple use of functionals as in § 9.1.2).

THEOREM 6. *Suppose that S is a terminal time and that no delay is imposed when the chain starts in H^c. Then, for $i, j \in \mathbb{I}$ and $t \geqslant 0$, $h \geqslant 0$,*

$${}_Hp^S_{ij}(t+h) = \sum_k {}_Hp^S_{ik}(t)\, {}_Hp^S_{kj}(h).$$

One can now apply Markov chain theory to transition matrix ${}_Hp^S(t)$.

Remark. When $S = 0$, Definition 2 reduces to T, and Lemma 3 interprets ordinary taboo probability in terms of killing at T,

$${}_Hp_{ij}(t) = \mathbb{P}^i(X_t = j,\ T > t).$$

Moreover, Theorem 6 holds, as noted in § 1.2.1.

§ 2.2.3. Constant delay. This special case will be used in Section 3. Suppose that $S = s > 0$, is a constant. Then, Definition 1, § 2.2.2, specializes to

DEFINITION 1. For fixed $s > 0$ and $i \in \mathbb{I}$:

$$ {}_Hp^s_{ij}(t) = \begin{cases} \mathbb{P}^i(X_u \in H^c,\ s < u < t,\ X_t = j) & \text{when } s \leqslant t,\ j \in H^c, \\ \mathbb{P}^i(X_t = j) & \text{when } s > t,\ j \in \mathbb{I}. \end{cases} $$

Definition 2, § 2.2.2, reduces to T_s, and Lemma 3, § 2.2.2, yields

$$ {}_Hp^s_{ij}(t) = \mathbb{P}^i(X_t = j,\ T_s > t). $$

However, Theorem 6, § 2.2.2, does not apply here. Nevertheless, Markov property yields immediately

THEOREM 2. *For $t \geqslant 0$ and $i \in \mathbb{I}$:*

$$ {}_Hp^s_{ij}(t) = \begin{cases} \sum\limits_{k \in H^c} p_{ik}(s)\, {}_Hp_{kj}(t-s) & \text{for } t \geqslant s,\ j \in H^c, \\ p_{ij}(t) & \text{for } t < s,\ j \in \mathbb{I}. \end{cases} $$

In matrix notation, one has

$$ {}_H\mathbf{P}^s(t) = \begin{cases} \mathbf{P}(s)\, {}_H\mathbf{P}(t-s), & t \geqslant s, \\ \mathbf{P}(t), & t < s. \end{cases} $$

Hence, obviously,

$$ {}_H\mathbf{P}^s(t+h) = {}_H\mathbf{P}^s(t)\, {}_H\mathbf{P}(h) \qquad \text{for} \quad t > s,\ h > 0, $$

and consequently, ${}_H\mathbf{P}^s(t)$, $t > s$, do not form a semi-group, as it may be verified directly.

Nevertheless, the formal differentiation yields the following modified form of Kolmogorov equations for $t > s$:

$$ \frac{d\, {}_H\mathbf{P}^s(t)}{dt} = \begin{cases} \mathbf{P}(s)\, {}_H\mathbf{Q}\, {}_H\mathbf{P}(t-s), & \text{(B)} \\ {}_H\mathbf{P}^s(t)\, {}_H\mathbf{Q}, & \text{(F)} \end{cases} $$

with the initial condition ${}_H\mathbf{P}^s(s) = \mathbf{P}(s)\mathbf{J}_1$ for $t = s$.

§ 2.2.4. Exit taboo. In some problems concerning the taboo set H, it is required to distinguish cases when the initial state of a M.C. is in H^c or in H. In the first case, the situation described in Sub-section 1.2 applies, but in the second case the taboo is imposed after the chain leaves the initial state, necessarily by a jump to H^c. This extended form of the taboo probability differs from that in Sub-section 1.2 only when the initial state is in H.

a) Let ρ_i be the first exit time from a state i (as defined in § 1.1.3). The following form of a delayed taboo probability refers to the above situation.

DEFINITION 1. The *exit taboo probability* is the transition probability from state i to state j at time t, under the taboo H, denoted by ${}_H p^\rho_{ij}(t)$, and is given by

$$ {}_H p^\rho_{ij}(t) = \delta_{ij} e^{-q_i t} + \mathbb{P}^i\big(\rho_i < t,\ X_u \in H^c,\ \rho_i < u < t,\ X_t = j \mid X_0 = i\big) $$

for $i, j \in \mathbb{I}$ and $t > 0$.

Observe that this is a special case of Definition 1, § 2.2.2, with $S = \rho_i$, and with requirement that $\rho_i = 0$ for $i \in H^c$. Thus, for $i \in H^c$ and $j \in \mathbb{I}$,

$$ {}_H p^\rho_{ij}(t) = {}_H p_{ij}(t), $$

whereas, for $i \in H$, the taboo is imposed after the exit from state i ($\rho_i > 0$). Thus, new probabilities are obtained for $i \in H$ and $j \in H^c$, but, for $i \in H$ and $j \in H$, one has

$$ {}_H p^\rho_{ij}(t) = \delta_{ij} e^{-q_i t}. $$

Thus, Definition 2, § 2.2.2, reduces to T_ρ, and Lemma 3, § 2.2.2, yields

$$ {}_H p^\rho_{ij}(t) = \mathbb{P}^i(X_t = j,\ T_\rho > t), $$

indicating killing at the time T_ρ.

The taboo matrix ${}_H\mathbf{P}^\rho(t) = ({}_H p^\rho_{ij}(t))$ has the partitioned form

$$ {}_H\mathbf{P}^\rho(t) = \begin{pmatrix} {}_H\mathbf{P}^\rho_{11}(t) & \mathbf{0} \\ {}_H\mathbf{P}^\rho_{21}(t) & {}_H\mathbf{P}^\rho_{22}(t) \end{pmatrix}, $$

where

$$ {}_H\mathbf{P}^\rho_{11}(t) = {}_H\mathbf{P}_{11}(t), \qquad {}_H\mathbf{P}^\rho_{22}(t) = (\delta_{ij} e^{-q_i t}). $$

Clearly, ${}_H\mathbf{P}^\rho(t)$ is standard: ${}_H\mathbf{P}^\rho(0) = \mathbf{I}$, sub-stochastic: ${}_H\mathbf{P}^\rho(t)\mathbf{1} \leqslant \mathbf{1}$, and it is dominated by the original matrix $\mathbf{P}(t)$. Moreover, it is important that the semi-group property holds:

$$ {}_H\mathbf{P}^\rho(t+h) = {}_H\mathbf{P}^\rho(t)\, {}_H\mathbf{P}^\rho(h), \qquad h, t \geqslant 0. $$

This is a consequence of Theorem 6, § 2.2.2, and the fact that ρ_i is a terminal time (Lemma 4, § 2.1.2).

The corresponding resolvent matrix ${}_H\mathbf{U}^{\rho\alpha} = ({}_H u^{\rho\alpha}_{ij})$ has the form

$$ {}_H\mathbf{U}^{\rho\alpha} = \begin{pmatrix} {}_H\mathbf{U}^{\rho\alpha}_{11} & \mathbf{0} \\ {}_H\mathbf{U}^{\rho\alpha}_{21} & {}_H\mathbf{U}^{\rho\alpha}_{22} \end{pmatrix} $$

where

$$ {}_H\mathbf{U}^{\rho\alpha}_{11} = {}_H\mathbf{U}^{\alpha}_{11}, \qquad {}_H\mathbf{U}^{\rho\alpha}_{22} = (\delta_{ij}(\alpha + q_i)^{-1}). $$

The intensity matrix ${}_H\mathbf{Q}^\rho = ({}_H q^\rho_{ij})$ defined in the usual way (as in § 1.1.1) has elements

$$ {}_H q^\rho_{ij} = \begin{cases} q_{ij} & \text{for } i \in \mathbb{I},\ j \in H^c, \\ 0 & \text{for } i \in \mathbb{I},\ j \in H, \end{cases} $$

for $i \neq j$, and $-{}_Hq^\rho_{ii} = {}_Hq^\rho_i$ for $i \in \mathbb{I}$. Consequently,

$$ {}_H\mathbf{Q}^\rho \mathbf{1}_i = -\bar{q}_i \leqslant 0 \qquad \text{for all} \quad i \in \mathbb{I}, $$

where $\bar{q}_i$ has been defined in § 1.2.1. Thus, the matrix ${}_H\mathbf{Q}^\rho$ is not conservative and has the partitioned form

$$ {}_H\mathbf{Q}^\rho = \begin{pmatrix} \mathbf{Q}_{11} & \mathbf{0} \\ \mathbf{Q}_{21} & -\mathbf{\Theta}_2 \end{pmatrix} = \mathbf{Q}\mathbf{J}_1 - \mathbf{\Theta}\mathbf{J}_2, $$

where $\mathbf{\Theta}_2 = (q_i \delta_{ij})$ is the diagonal sub-matrix of the matrix $\mathbf{\Theta}$ (defined in § 1.1.4).

b) The same argument as that in § 1.2.4, based on adjoining an absorbing state Δ (with state space extended to $\mathbb{I} \cup \{\Delta\}$) leads to the (B)- and (F)-equations for a stochastic matrix $\mathbf{P}^\Delta(t)$ with a conservative matrix $\mathbf{Q}^\Delta$. Hence, as before, these extended matrices coincide, respectively, with ${}_H\mathbf{P}^\rho(t)$ and ${}_H\mathbf{Q}^\rho$, when restricted to transitions within $\mathbb{I}$.

THEOREM 1. *For* $t \geqslant 0$:

$$ \frac{d\,{}_H\mathbf{P}^\rho(t)}{dt} = \begin{cases} {}_H\mathbf{Q}^\rho\,{}_H\mathbf{P}^\rho(t), & \text{(B)} \\ {}_H\mathbf{P}^\rho(t)\,{}_H\mathbf{Q}^\rho, & \text{(F)} \end{cases} $$

with ${}_H\mathbf{P}^\rho(0) = \mathbf{I}$, *or in terms of transforms, for* $\alpha > 0$,

$$ \alpha\,{}_H\mathbf{U}^{\rho\alpha} - \mathbf{I} = \begin{cases} {}_H\mathbf{Q}^\rho\,{}_H\mathbf{U}^{\rho\alpha}, & \text{(B)} \\ {}_H\mathbf{U}^{\rho\alpha}\,{}_H\mathbf{Q}^\rho. & \text{(F)} \end{cases} $$

Evidently, for the component ${}_H\mathbf{U}^\alpha_{11}$ (that is for i and j in H^c) the above equations are the same as those stated in § 1.2.3, whereas (for i and j in H)

$$ {}_H\mathbf{U}^{\rho\alpha}_{22} = (\alpha\mathbf{I}_2 + \mathbf{\Theta}_2)^{-1}. \tag{B)-(F} $$

The only new equations concern the component ${}_H\mathbf{U}^{\rho\alpha}_{21}$ (i.e., for $i \in H$, $j \in H^c$), and Theorem 1 gives

$$ (\alpha\mathbf{I}_2 + \mathbf{\Theta}_2)\,{}_H\mathbf{U}^{\rho\alpha}_{21} = \mathbf{Q}_{21}\,{}_H\mathbf{U}^\alpha_{11}, \tag{B} $$
$$ {}_H\mathbf{U}^{\rho\alpha}_{21}(\alpha\mathbf{I}_1 - \mathbf{Q}_{11}) = {}_H\mathbf{U}^{\rho\alpha}_{22}\mathbf{Q}_{21}. \tag{F} $$

In proof of Kolmogorov equations one considers again expressions ${}_Hv^\rho_{ij}(t)$ and ${}_Hw^\rho_{ij}(t)$ defined now for all i and j in $\mathbb{I}$; they vanish for $j \in H$. For i and j in H^c, these expressions agree, respectively, with ${}_Hv_{ij}(t)$ and ${}_Hw_{ij}(t)$ introduced in § 1.2.3. For $i \in H$ and $j \in H^c$, they are defined by

$$ q_i\,{}_Hv^\rho_{ij}(t) = \sum_{k \in H^c} q_{ik}\,{}_Hp_{kj}(t), \qquad {}_Hw^\rho_{ij}(t) = \sum_{k \neq j} {}_Hp^\rho_{ik}(t) q_{kj}. $$

It is important to observe that the (B)-equation provides the explicit expression for the exit taboo ${}_Hp^\rho_{ij}(t)$ in terms of the ordinary taboo ${}_Hp_{ij}(t)$.

THEOREM 2. *For $i \in H$ and $j \in H^c$:*

$$ {}_H p^{\rho}_{ij}(t) = \int_0^t q_i e^{-q_i(t-s)} {}_H v^{\rho}_{ij}(s)\, ds, $$

or, in terms of transforms,

$$ {}_H u^{\rho\alpha}_{ij} = \frac{q_i}{\alpha + q_i} {}_H v^{\rho\alpha}_{ij}. $$

The above result is intuitively clear. Indeed, the chain jumps to H^c on termination of its stay in an initial state i in H, and then proceeds avoiding H (see Lemma 1, § 3.4.1).

Finally, for the absorbing state Δ, one has

$$ p^{\Delta}_{i\Delta}(t) = 1 - \sum_j {}_H p^{\rho}_{ij}(t), \qquad q^{\Delta}_{i\Delta} = \bar{q}_i \qquad \text{for all} \quad i \in \mathbb{I}, $$

and the equations are obtained in the same manner as in § 1.2.4. It is of interest to give the explicit form of these equations.

THEOREM 3. *For $i \in H$:*

$$ \frac{dp^{\Delta}_{i\Delta}(t)}{dt} = \begin{cases} \bar{q}_i - q_i p^{\Delta}_{i\Delta}(t) + \sum_{k \in H^c} q_{ik} p^{\Delta}_{k\Delta}(t), & \text{(B)} \\ \bar{q}_i e^{-q_i t} + \sum_{k \in H^c} {}_H p^{\rho}_{ik}(t) \bar{q}_k, & \text{(F)} \end{cases} $$

whereas, for $i \in H^c$, the equations are the same as those stated in § 1.2.4.

c) For the exit taboo, the extended Feller relation provides inter-connection between the minimal solution of Kolmogorov equations and associated jump chain probabilities, under taboo imposed on the exit. The analysis is very similar to that in § 1.2.5.

The jump matrix ${}_H\mathbf{R}^{\rho}$, corresponding to ${}_H\mathbf{Q}^{\rho}$, is now ${}_H\mathbf{R}^{\rho} = \mathbf{I} + \Theta^{-1} {}_H\mathbf{Q}^{\rho}$, and the auxiliary α-jump matrix ${}_H\mathbf{R}^{\rho\alpha} = ({}_H r^{\rho\alpha}_{ij})$ is defined for $i \in \mathbb{I}$ by

$$ {}_H r^{\rho\alpha}_{ij} = \begin{cases} q_i(\alpha + q_i)^{-1} r_{ij} & \text{for} \quad j \in H^c,\ j \neq i, \\ 0 & \text{for} \quad j \in H^c,\ j = i, \\ 0 & \text{whenever} \quad j \in H. \end{cases} $$

Its n-th power $({}_H\mathbf{R}^{\rho\alpha})^n$ has only for $n \geqslant 1$ non-vanishing components $({}_H\mathbf{R}^{\alpha}_{11})^n$ and ${}_H\mathbf{R}^{\alpha}_{21}({}_H\mathbf{R}^{\alpha}_{11})^{n-1}$.

The corresponding (Green) α-potential is now

$$ {}_H\mathbf{G}^{\rho\alpha} = \sum_{n=0}^{\infty} ({}_H\mathbf{R}^{\rho\alpha})^n $$

whose non-zero components are

$$ {}_H\mathbf{G}^{\rho\alpha}_{11} = {}_H\mathbf{G}^{\alpha}_{11}, \qquad {}_H\mathbf{G}^{\rho\alpha}_{21} = {}_H\mathbf{R}^{\alpha}_{21}\,{}_H\mathbf{G}^{\alpha}_{11}, \qquad {}_H\mathbf{G}^{\rho\alpha}_{22} = \mathbf{I}_2, $$

where ${}_H\mathbf{G}^{\alpha}_{11}$ is the same as in § 1.2.5. Hence, the following *Feller relation* may be stated.

Let ${}_H\mathbf{U}^{\rho\alpha}$ be the resolvent of the minimal semi-group ${}_H\mathbf{P}^{\rho}(t)$, $t \geqslant 0$, and let ${}_H\mathbf{G}^{\rho\alpha}$ be the (Green) potential of the α-jump chain under exit taboo. Then, for $\alpha \geqslant 0$,

$$ {}_H\mathbf{U}^{\rho\alpha} = {}_H\mathbf{G}^{\rho\alpha}(\boldsymbol{\Theta} + \alpha\mathbf{I})^{-1}. $$

In terms of components,

$$ {}_H\mathbf{U}^{\alpha}_{11} = {}_H\mathbf{G}^{\alpha}_{11}(\boldsymbol{\Theta}_1 + \alpha\mathbf{I}_1)^{-1}, \qquad (\S\ 1.2.5) $$
$$ {}_H\mathbf{U}^{\rho\alpha}_{21} = {}_H\mathbf{G}^{\rho\alpha}_{21}(\boldsymbol{\Theta}_1 + \mathbf{I}_1)^{-1} = {}_H\mathbf{R}^{\alpha}_{21}\,{}_H\mathbf{U}^{\alpha}_{11}. \qquad (\text{Thm. } 2) $$

d) The following examples are continuation of those from § 1.2.6.

Example 1. Let k be a fixed state and take $H = \{k\}$. Then, for $j \neq k$,

$$ {}_kp^{\rho}_{kj}(t) = \mathbb{P}(\rho_k < t,\ X_u \neq k,\ \rho_k < u < t,\ X_t = j \mid X_0 = k) $$

and

$$ p^{\Delta}_{k\Delta}(t) = \mathbb{P}(\rho_k < t \text{ or } X_u = k \text{ for some } u,\ \rho_k < u < t \mid X_0 = k) = F_{kk}(t), $$

where $F_{kk}(t)$ is the return distribution introduced in § 1.1.7. In terms of transforms,

$$ f^{\alpha}_{kk} = \alpha u^{\Delta\alpha}_{k\Delta}. $$

Note also that $\bar{q}_k = 0$.

Example 2. Let k be a fixed state and take $H = \{k\}^c$. Then, for $i \neq k$,

$$ {}_Hp^{\rho}_{ik}(t) = \mathbb{P}(\rho_i < t,\ X_u = k,\ \rho_i < u < t,\ X_t = k \mid X_0 = i) $$

and therefore

$$ {}_Hu^{\rho\alpha}_{ik} = q_i(\alpha + q_i)^{-1} r_{ik}(\alpha + q_k)^{-1}, \qquad i \neq k. $$

Note that $\bar{q}_k = q_k$ and $\bar{q}_i = q_i - q_{ik}$ for $i \neq k$.

Example 3. Continuing Example 3 from § 1.2.6 with $H = \{0\}$, the easy calculation yields

$$ {}_0u^{\rho\alpha}_{0,-1} = \mu(\alpha + \lambda + \mu)^{-1}(\alpha + \lambda)^{-1}, $$
$$ {}_0u^{\rho\alpha}_{0,1} = \lambda(\alpha + \lambda + \mu)^{-1}(\alpha + \mu)^{-1}, $$

and

$$ \alpha u^{\Delta\alpha}_{0\Delta} = \lambda\,{}_0u^{\rho\alpha}_{0,-1} + \mu\,{}_0u^{\rho\alpha}_{0,1}. $$

Example 4. For compound Poisson chain, ${}_H\mathbf{U}^{\rho\alpha}$ is obtained as before from $\mathbf{U}^{\alpha}$ by replacing $\mathbf{R}$ in defining relation (see Ex. 2, § 1.1.9) by ${}_H\mathbf{R}^{\rho}$. In particular, the Feller relation gives

$$ {}_Hu^{\rho\alpha}_{ij}(\alpha + \gamma) = {}_Hg^{\rho\alpha}_{ij} \qquad \text{for } \ i \in H,\ j \in H^c. $$

§ 2.2.5. Density argument. This is a convenient method of finding joint distributions of passage times and positions, and is based on examination of sample functions of a M.C. The heuristic approach described here yields quick results, but proper justification requires delicate discussion of conditioning (see references for details).

a) Consider the first entrance time T_S to a set H after S. The basic result is

LEMMA 1. *For $i, j \in \mathbb{I}$, $k \in H^c$, $r \in H$, and $t > 0$, and for a stopping time S:*

$$\mathbb{P}^i(X(T_S-) = k,\ X(T_S) = r,\ S < T_S \leqslant t,\ X_t = j) = \int_0^t {}_H p_{ik}^S(\tau) q_{kr} p_{rj}(t-\tau)\, d\tau.$$

Proof (outline). By the strong Markov property (post-T_S chain), the probability on the LHS may be written in terms of conditional probabilities as follows (see Fig. 2.9):

$$\int_0^t \mathbb{P}^i(X(T_S-) = k,\ X(T_S) = r,\ S < T_S \in d\tau)\, \mathbb{P}(X_t = j \mid X(T_S) = r,\ T_S = \tau).$$

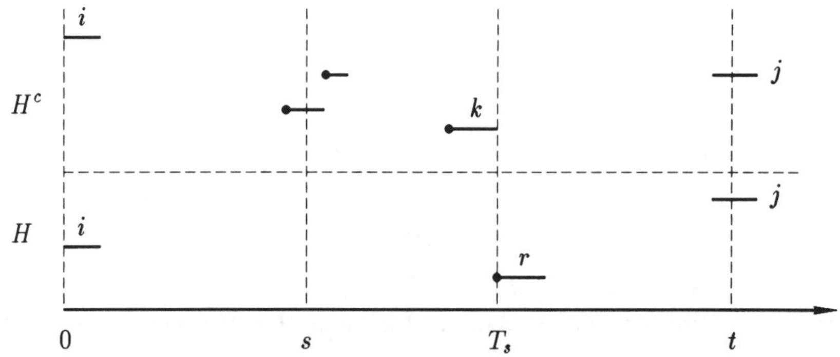

$$T_S = S + T \circ \theta_S, \qquad S < T_S.$$

Figure 2.9. Density argument – Lemma 1, § 2.2.5.

Analysis of paths allows to identify the jump at T_S (from k to r) with the jump at a stopping time ρ_k, so (recalling that a probability of a jump at fixed point is zero)

$$\begin{aligned}
&\mathbb{P}^i(X(T_S-) = k,\ X(T_S) = r,\ S < T_S \in d\tau) \\
&\quad = \mathbb{P}^i(X(T_S-) = k,\ S < T_S \in d\tau)\, \mathbb{P}(X(\rho_k) = r \mid X(\rho_k-) = k) \\
&\quad = {}_H p_{ik}^S(\tau) q_{kr}\, d\tau.
\end{aligned}$$

Again, the strong Markov property gives

$$\mathbb{P}\big(X_t = j \mid X(T_\rho) = r,\ T_S = \tau\big) = p_{rj}(t-\tau). \qquad \square$$

COROLLARY 2. *For a stopping time S, and for $i \in \mathbb{I}$, $k \in H^c$, $t > 0$:*

$$\mathbb{P}^i\big(X(T_S-) = k,\ S < T_S \leqslant t\big) = \int_0^t {}_H p_{ik}^S(\tau)\bar{q}_k\, d\tau,$$

$$\mathbb{P}^i(S < T_S \leqslant t) = \int_0^t d_i^S(\tau)\, d\tau,$$

where

$$d_i^S(\tau) = \sum_{k\in H^c} {}_H p_{ik}^S(\tau)\bar{q}_k,$$

with $\bar{q}_k$ defined in § 1.2.1.

Note that $S < T_S$ implies that X_S takes values in H^c. On the other hand, when $S = T_S$, then X_S takes values in H, but X_{S-} may be either in H or in H^c. Hence, the complementary term

$$\mathbb{P}^i\big(X(T_S) = r,\ S = T_S \leqslant t,\ X_t = j\big)$$

should be evaluated individually for a given S. Indeed, when T_S is a terminal time, then one has, in the analogous manner, the expression

$$\int_0^t \sum_{k\in H} {}_H p_{ik}^S(\tau) q_{kr} p_{rj}(t-\tau)\, d\tau$$

for the above probability, using the fact that now $X(S) \in H$.

It is desirable to express the joint distribution of T_S and $X(T_S)$ in terms of the delayed taboo probabilities. The expressions are more informative when it is assumed that T_S is a terminal time. The next lemma illustrates the kind of argument typical for Markov chains, to be used later on.

LEMMA 3. *Let T_S be a terminal time. Then, for $i \in \mathbb{I}$, $r \in H$, and $t, h > 0$:*

$$\mathbb{P}^i\big(t < T_S \leqslant t+h,\ X(T_S) = r\big) = \sum_v {}_H p_{iv}^S(t)\, \mathbb{P}^v\big(T_S \leqslant h,\ X(T_S) = r\big).$$

Proof. Consider

$$\begin{aligned} &\mathbb{P}^i(X_t = v,\ t < T_S \leqslant t+h,\ X(T_S) = r) \\ &\quad = \mathbb{P}^i(S \leqslant t,\ X_t = v,\ t < T_S \leqslant t+h,\ X(T_S) = r) \\ &\qquad + \mathbb{P}^i(S > t,\ X_t = v,\ t < T_S \leqslant t+h,\ X(T_S) = r). \end{aligned}$$

Using the fact that T_S is a terminal time ($T_S \circ \theta_t = T_S - t$), the first term on the right becomes

$$\begin{aligned}&\mathbb{P}^i(S \leqslant t,\ X_t = v,\ t < T_S,\ 0 < T_S \circ \theta_t \leqslant h,\ X(T_S) \circ \theta_t = r)\\ &\quad = \mathbb{P}^i(S \leqslant t,\ X_t = v,\ t < T_S)\ \mathbb{P}(0 < T_S \leqslant h,\ X(T_S) = r \mid X_0 = v).\end{aligned}$$

Next, $S > t$ implies that $T_S > t$, so the second term becomes

$$\begin{aligned}&\mathbb{P}^i(S > t,\ X_t = v,\ 0 < T_S \circ \theta_t \leqslant h,\ X(T_S) \circ \theta_t = r)\\ &\quad = \mathbb{P}^i(S > t,\ X_t = v)\ \mathbb{P}(0 < T_S \leqslant h,\ X(T_S) = r \mid X_0 = v).\end{aligned}$$

The assertion then follows from Lemma 3, § 2.2.2, and the fact that T_S has continuous distribution. □

In the case when T_S is not a terminal time, the joint distribution of T_S and $X(T_S)$ assumes rather cumbersome form when expressed in terms of taboo probabilities. Moreover, distribution of T_S need not be continuous.

b) Consider now the last exit time L_s from a set H before time $s > 0$ (see § 2.1.4). The basic result is

LEMMA 4. *For $i \in \mathbb{I}$, $r \in H$, $k, j, v \in H^c$, and $0 < t < s$:*

$$\begin{aligned}&\mathbb{P}^i(X(L_s-) = r,\ X(L_s) = k,\ 0 < L_s \leqslant t,\ X_t = j,\ X_s = v)\\ &\quad = \int_0^t p_{ir}(\sigma) q_{rk}\ {}_Hp_{kj}(t-\sigma)\, d\sigma\ {}_Hp_{jv}(s-t).\end{aligned}$$

Proof. The assertion follows by adaptation of the argument of K. L. Chung [1967, p. 201]; also by specialization of general results obtained by A. O. Pittenger and C. T. Shih [1973]. □

In view of importance of Lemma 4, the heuristic explanation may be useful. By definition (for $0 < L_s < s$), the M.C. is in H at L_s-, and at L_s a jump to H^c occurs, and then the chain stays in H^c till at least s, passing through state j at time $t < s$ (see Fig. 2.10).

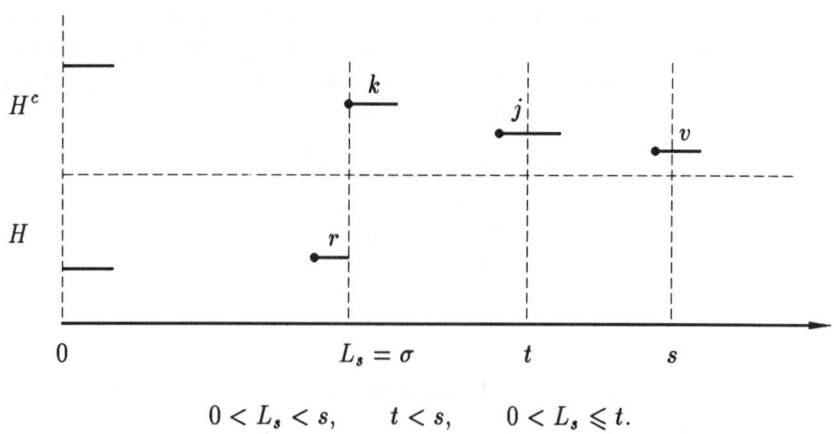

Figure 2.10. Density argument – Lemma 4, § 2.2.5.

It seems plausible to express the above probability in the conditional form for $0 < t < s$:

$$\int_0^t \mathbb{P}^i\big(X(L_s-) = r,\ L_s \in d\sigma\big)\, \mathbb{P}\big(X(L_s) = k \mid X(L_s-) = r,\ L_s = \sigma\big)$$

$$\times\, \mathbb{P}(X_u \in H^c,\ \sigma < u < t,\ X_t = j,\ t < u < s,$$
$$X_s = v \mid X(L_s-) = r,\ X(L_s) = k,\ L_s = \sigma)$$

$$= \int_0^t p_{ir}(\sigma) q_r\, d\sigma\, (q_{rk}/q_r)\, {}_H p_{kj}(t-\sigma)\, {}_H p_{jv}(s-t).$$

This requires the conditional independence of the past and the future, relative to the time L_s. As L_s is a co-stopping time, such an extension of the Markov property must be justified. The argument in Lemma 4 indicates that L_s indeed preserves the Markov property. It amounts basically to identification of L_s with the first exit time ρ_r (a stopping time) from a state $r \in H$, in the sense:

$$\mathbb{P}^i\big(X(L_s-) = r,\ X(L_s) = k,\, L_s \in d\sigma\big)$$
$$= \mathbb{P}^i\big(X(\rho_r-) = r,\ X(\rho_r) = k,\ \rho_r \in d\sigma\big) = p_{ir}(\sigma) q_{rk}\, d\sigma.$$

With the jump from r in H to k in H^c occurring with probability q_{rk}/q_r, as in the (F)-equation, one has the "density"

$$\mathbb{P}^i\big(X(L_s-) = r,\ L_s \in d\sigma\big) = p_{ir}(\sigma) q_r\, d\sigma.$$

For precise form of conditioning see § 9.2.5.

The following results are immediate consequence of Lemma 4.

COROLLARY 5. *For $i \in \mathbb{I}$, $r \in H$, $j, k \in H^c$ and $0 < t < s$:*

$$\mathbb{P}^i\big(X(L_s-) = r,\ X(L_s) = k,\ 0 < L_s \leqslant t,\ X_t = j\big)$$
$$= \int_0^t p_{ir}(\sigma) q_{rk}\, {}_H p_{kj}(t-\sigma)\, d\sigma D_j^c(s-t),$$

$$\mathbb{P}^i\big(X(L_s-) = r,\ X(L_s) = k,\ 0 < L_s \leqslant t\big) = \int_0^t p_{ir}(\sigma) q_{rk}\, D_k^c(s-\sigma)\, d\sigma.$$

In Corollary 5 use has been made of Theorem 4, § 2.2.2, applied to the first entrance time T. It is instructive to compare Corollary 5 with the following marginal distribution obtained from Lemma 4:

$$\mathbb{P}^i\big(X(L_s-) = r,\ X(L_s) = k,\ 0 < L_s \leqslant t,\ X_s = v\big)$$
$$= \int_0^t p_{ir}(\sigma) q_{rk}\, {}_H p_{kv}(s-\sigma)\, d\sigma.$$

Indeed, when $s \to \infty$, then $L_s \to L$, and one obtains results corresponding to L. Note that expressions involving X_s will vanish in the limit. On the other hand, Corollary 5 extends to

COROLLARY 6. *For* $i \in \mathbb{I}$, $r \in H$, $j,k \in H^c$ *and* $0 < t < \infty$:

$$\mathbb{P}^i(X(L-) = r,\ X(L) = k,\ 0 < L \leqslant t,\ X_t = j) = \int_0^t p_{ir}(\sigma) q_{rk}\, {}_Hp_{kj}(t-\sigma)\, D_j^c\, d\sigma,$$

$$\mathbb{P}^i(X(L-) = r,\ X(L) = k,\ 0 < L \leqslant t) = \int_0^t p_{ir}(\sigma) q_{rk} D_k^c\, d\sigma.$$

For further discussion of these joint distributions see Section 4.

c) It is of interest to consider times T_s and L_s together. Their joint distribution may be obtained directly by combination of arguments used in Lemma 1 and Lemma 4.

Full discussion of this important topic is presented later in Sub-section 10.3, and its application to analysis of excursions from a set H (see § 2.1.4) is treated in Sub-section 10.4.

§ 2.2.6. Notation. For easy reference, it is convenient to list in one place special symbols to be used later frequently.

$\bar{q}_i = \sum_{j \in H} q_{ij} \geqslant 0$	for $i \in H^c$,	
$\bar{q}_i = \sum_{j \neq i,\ j \in H} q_{ij} \geqslant 0$	for $i \in H$,	§ 1.2.1
$b_i = \sum_k q_{ik} D_k^c = q_i D_i^{\rho c}$,	escape rate,	§ 3.1.4, § 3.3.3
$h_{ij}(t) = \sum_{r \in H} q_{ir} p_{rj}(t)$,	$i \in H^c,\ j \in \mathbb{I}$,	§ 3.1.2
$c_{ij}(t) = \sum_{r \in H} p_{ir}(t) q_{rj}$,	$i \in \mathbb{I},\ j \in H^c$,	§ 4.1.2, § 10.3.1
$h_{ij}^{\rho}(t) = \sum_{r \neq i,\ r \in H} q_{ir} p_{rj}(t)$,	$i \in H,\ j \in \mathbb{I}$,	§ 3.3.3
$d_{ij}(t) = \sum_{k \in H^c} {}_Hp_{ik}(t) q_{kj}$,	$i \in H^c,\ j \in H$,	§ 3.1.2
$n_{ij}(t) = \sum_{k \in H^c} q_{ik}\, {}_Hp_{kj}(t)$,	$i \in H,\ j \in H^c$,	§ 3.4.1
$\mathbf{B}^{\alpha} = (d_{ij}^{\alpha})$,	α-balayage,	§ 3.1.5
$\mathbf{C}^{\alpha} = (n_{ij}^{\alpha})$,	α-co-balayage,	§ 3.4.4
$\widetilde{\mathbf{C}}^{\alpha} = (c_{ij}^{\alpha})$.		§ 6.2.1, § 6.3.1

As an immediate application of this notation, one can obtain convenient expressions for various marginal distributions from results in § 2.2.5, by summation over appropriate ranges.

CHAPTER II

ANALYTIC THEORY

This chapter constitutes the main part of the book, and is devoted to discussion of passage times using analytic methods. The chain is taken in its simplest form, namely a stable minimal Markov chain, referred to as the *usual chain*. Basic definitions of passage times relative to a fixed set H, have been already introduced in Section 2.

Attention here is concentrated on derivation of equations for distributions of passage times, and for joint distributions of times and positions. Interpretation of results in terms of potential theory is also presented. Hence, considerable attention is devoted to discussion of the Dirichlet problem, Poisson equation, decomposition theorems, balayage operators and duality.

The main feature of presentation is that required distributions are expressed in terms of taboo probabilities (convenient for calculations). The analytic approach makes use of Laplace transforms, differential equations and matrices. Approaches based on measure theory are postponed to later chapters.

The chapter comprises five sections. Section 3 treats first entrance times, including first entrance after a fixed time and also the first return time. It is shown that equations for distributions of time and position are related to the (B)-equations of a chain, but the explicit solutions make use of the (F)-equations. Moreover, the *density argument* is illustrated and the role of a jump chain is stressed. The individual sub-sections exhibit many similarities, and the use of taboo probabilities provides unifying treatment. The last sub-section treats interesting duality formulation.

Section 4 deals with the last exit times and positions. The material exhibits another kind of similarity, and is dual to that in the previous section. The analysis is more complex, however, and the *density argument* is frequently used. Because of the complexity of the problem, it is more convenient to treat both sub-sections independently. The matrix formulation is helpful.

Sections 5 and 6 are devoted to the fundamental problems of practical and theoretical importance. Section 5 discusses the Dirichlet problem and the Poisson equation, and illustrates interplay between potential theory and Markov chains. Section 6 treats decomposition theorems according to the first entrance and to the last exit, and stresses the role of duality. The last Section 7 provides several illustrative examples of explicit calculations.

The material in this chapter is rather standard, but attempt here is made to present it in a clear unifying fashion. Individual references are given in the text.

SECTION 3

FIRST ENTRANCE TIMES

This section presents results of fundamental importance for this book. Discussion concerns properties of the first entrance times to a fixed set H, and their distributions and related matters. Only analytic aspects are considered here, although the underlaying behavior of sample functions is indicated. The importance of taboo probabilities is emphasised by showing how they may be used to obtain all probabilities of interest. The role of the (B)- and (F)-equations for distributions is stressed.

Sub-section 3.1 considers the first entrance time T and the first entry position $X(T)$. Sub-section 3.2 treats the first entrance time T_s after an instant $s > 0$, and the first entry positions $X(T_s)$ and $X(T_s-)$, and is the natural extension of results from Sub-section 3.1. Sub-section 3.3 treats the first return time T_ρ, and parallels discussion in Sub-section 3.1 by extending its results to initial position in H.

Each of these sub-sections expresses the distribution of a first entrance time directly in terms of corresponding taboo probabilities. The same approach extended to the joint distributions of time and position must be supplemented by the "density argument", essentially the "path analysis". The alternative approach through "product theorem" is only briefly mentioned, but the approach through functionals will be discussed later. Each of these approaches can be taken as the starting point of discussion, but taboo probabilities offer the most direct method. The results are especially neat when summarised in convenient matrix form. Also matrices are helpful to illustrate connection between continuous time (minimal) chain and the randomized jump chain.

The last Sub-section 3.4 treats the situation dual to that in Sub-section 3.1. The discussion depends on T_ρ, because transition from H to H^c are involved. The results here illustrate general decomposition theorems to be discussed later.

The material presented in this section has been known in Markov chains theory and applications since early days, and belongs to the folklore of the subject. The results are scattered in literature, but their clear use may be found in K. L. Chung [1967] who pointed out the advantage of using T_ρ. The systematic presentation attempted here is based on R. Syski [1967, 1975].

3.1. FIRST ENTRANCE TIME

§ 3.1.1. Equations. a) Let T be the first entrance time to a set H for the usual M.C. $X = (X_t)$, as defined in § 2.1.2, and let $D_i(\cdot)$ be its distribution function, conditional on the initial state i. The following basic theorem links the distribution of the first entrance time with taboo probabilities.

THEOREM 1. *The complementary distribution function* $D_i^c(\cdot)$ *is a continuous function of* t, *and is given for* $t \geqslant 0$ *by*

$$D_i^c(t) = \begin{cases} \sum_{j \in H^c} {}_H p_{ij}(t), & i \in H^c, \\ 0, & i \in H. \end{cases}$$

Proof. Using a set K_{t-}^c from Lemma 2, § 2.1.2, one has

$$(T \geqslant t,\ X_t = j) = (X_s \in H^c,\ 0 < s < t,\ X_t = j).$$

Hence, for $i \in H^c$:

$$\mathbb{P}(T \geqslant t,\ X_t = j \mid X_0 = i) = \mathbb{P}(X_s \in H^c,\ 0 < s < t,\ X_t = j \mid X_0 = i)$$

and the RHS is the taboo probability ${}_H p_{ij}(t)$, vanishing for $j \in H$. As the chain is right-continuous, with stable states, and probability of a jump at fixed instant is zero, $T \geqslant t$ may be replaced by $T > t$ on the LHS. The first assertion follows then by addition over all $j \in H^c$.

On the other hand, for $i \in H$, $T = 0$ identically, so $D_i^c(t) \equiv 0$ in agreement with the fact that ${}_H p_{ij}(t) = 0$ for $i \in H$. □

The interpretation

$${}_H p_{ij}(t) = \mathbb{P}^i(X_t = j,\ T > t)$$

in the above proof indicates that taboo probabilities are in fact obtained from the original M.C. by observing it up to the first entrance to the taboo set H (i.e., killing the chain at T).

The following results are direct consequences of the defining relation from Theorem 1, but are stated here explicitly because of their importance.

COROLLARY 2. $D_i(0) = 0$ *for* $i \in H^c$, $D_i(0) = 1$ *for* $i \in H$.

This 0–1 law asserts in potential theoretic language that all points of H are *regular* for H, and all points of H^c are *irregular* for H. Furthermore, the following statement holds.

COROLLARY 3. *For* $i \in \mathbb{I}$:

$$D_i^c = \begin{cases} \lim_{t \to \infty} \sum_{j \in H^c} {}_H p_{ij}(t) \geqslant 0, & i \in H^c, \\ 0, & i \in H. \end{cases}$$

The first entrance probability $D_i = \mathbb{P}^i(T < \infty)$ is the important quantity to be studied in this sub-section. Note that when the set H^c is finite then, clearly, $D_i = 1$ for all $i \in \mathbb{I}$.

The semi-group property of taboo probabilities yields immediately Theorem 4, and letting $s \to \infty$, the Dominated Convergence Theorem gives Corollary 5.

THEOREM 4. *For $t, s \geqslant 0$, and $i \in H^c$:*

$$D_i^c(t+s) = \sum_{j \in H^c} {}_H p_{ij}(t) D_j^c(s),$$

$$D_i(t+s) - D_i(t) = \sum_{j \in H^c} {}_H p_{ij}(t) D_j(s).$$

COROLLARY 5. *For $t \geqslant 0$ and $i \in H^c$:*

$$D_i^c = \sum_{j \in H^c} {}_H p_{ij}(t) D_j^c,$$

$$D_i = D_i(t) + \sum_{j \in H^c} {}_H p_{ij}(t) D_j .$$

Probabilistic interpretation of Theorem 4 and Corollary 5 should be noted. Moreover, Corollary 5 indicates that relative to the taboo semi-group ${}_H\mathbf{P}(t)$, the function (D_i^c) is *invariant* and the function (D_i) is *purely excessive*:

$${}_H\mathbf{P}(t)\, D_i \leqslant D_i, \qquad \lim_{t \to \infty} {}_H\mathbf{P}(t) D_i = 0.$$

Hence, $\mathbf{1} = (D_i) + (D_i^c)$ is just the Riesz decomposition of the ${}_H\mathbf{P}(t)$-excessive function $\mathbf{1}$.

b) The basic equation for determination of the distribution of the first entrance time T is provided by the following (B)-equation.

THEOREM 6. *The d.f. of T satisfies the* (B)*-equation for $i \in H^c$ and $t > 0$:*

$$\frac{dD_i^c(t)}{dt} = -q_i D_i^c(t) + \sum_{k \in H^c,\ k \neq i} q_{ik} D_k^c(t),$$

with the initial condition $D_i^c(0) = 1$.

Proof. The result follows immediately from the (B)-equation for taboo probabilities and defining relation from Theorem 1. □

In contrast, the next result uses the (F)-equation for the taboo probabilities (see § 1.2.1).

THEOREM 7. *Under conditions of validity of the* (F)*-equation for the taboo probabilities:*

$$D_i(t) = \int_0^t \sum_{k \in H^c} {}_H p_{ik}(s)\, \bar{q}_k \, ds, \qquad i \in H^c, \quad t \geqslant 0,$$

$$D_i = \sum_{k \in H^c} {}_H u_{ik}\, \bar{q}_k, \qquad i \in H^c,$$

where

$$\bar{q}_i = \sum_{j \in H} q_{ij}, \qquad i \in H^c.$$

Note that the first equation in Theorem 4 indicates that $D^c(\cdot)$, as a function of t, is an exit law on H^c, relative to ${}_H\mathbf{P}(t)$. This leads in the usual way to equivalent derivation of Theorem 6.

In order to stress the role of density $d_i(t) = dD_i(t)/dt$, it is of interest to comment on the proof of Theorem 7. Since (D_i) is purely excessive on H^c, relative to ${}_H\mathbf{P}(t)$, the second equation in Theorem 4 implies that there exists a unique exit law $d(\cdot)$ such that

$$D_i(t) = \int_0^t d_i(s)\,ds \underset{t\to\infty}{\longrightarrow} \int_0^\infty d_i(s)\,ds = D_i, \qquad i \in H^c,$$

and

$$d_i(t+s) = \sum_{j\in H^c} {}_Hp_{ij}(t)\,d_j(s), \qquad t \geqslant 0, \quad s > 0,$$

(see § 1.3.4). When $s \downarrow 0$, one would only get inequality

$$d_i(t) \geqslant \sum_{j\in H^c} {}_Hp_{ij}(t) d_j(0), \qquad i \in H^c,$$

where $d_i(0) = \bar{q}_i$ (for $i \in H^c$) by Theorem 7. To obtain equality in the above expression for $d_i(t)$, stronger assumptions are needed, namely the validity of the (F)-equation.

Note that the (B)-equation is suitable for the direct solution for $D_i(t)$, whereas the (F)-equation yields the explicit expression, provided taboo probabilities are known.

Taking L.S. transforms of expressions in a) and of the (B)- and (F)-equations, with d_i^α denoting the transform of $D_i(t)$, one has the following useful results.

THEOREM 8. *For* $i \in H^c$, $\alpha > 0$:

$$\sum_{j\in H^c} {}_Hu_{ij}^\alpha = (1 - d_i^\alpha)/\alpha,$$

$$\sum_{j\in H^c} {}_Hu_{ij}^\alpha\, D_j = (D_i - d_i^\alpha)/\alpha.$$

THEOREM 9. *For* $i \in H^c$, $\alpha > 0$:

$$(\alpha + q_i) d_i^\alpha = \sum_{k\in H} q_{ik} + \sum_{k\in H^c,\ k\neq i} q_{ik} d_k^\alpha, \tag{B}$$

$$d_i^\alpha = \sum_{j\in H^c} {}_Hu_{ij}^\alpha \bar{q}_j\,. \tag{F}$$

THEOREM 10. *For* $i \in H^c$, *the conditional means satisfy*

$$m_i = \sum_{j\in H^c} {}_Hu_{ij}\, D_j, \qquad \mathbb{E}^i T = \sum_{j\in H^c} {}_Hu_{ij},$$

$$q_i m_i = D_i + \sum_{j\in H^c,\ j\neq i} q_{ij} m_j\,.$$

A rather curious fact may be noted from comparison of the above expression for m_i with that for D_i in Theorem 7.

§ 3.1.2. Joint distributions. a) The joint distribution $D_{ij}(t)$ of T and the first entrance position $X_T \equiv X(T)$ has been introduced in § 2.1.5 for $i \in \mathbb{I}$, $j \in H$, and $t \geqslant 0$. Recall that $D^c_{ij}(t) = D_{ij} - D_{ij}(t)$. The discussion presented here is the extension of that from § 3.1.1. The following basic result is the analogue of Theorem 1 and Corollary 5 in § 3.1.1.

THEOREM 1. *$D^c_{ij}(\cdot)$ is a continuous function of t, and is given for $t \geqslant 0$ and $j \in H$ by*

$$D^c_{ij}(t) = \begin{cases} \sum_{k \in H^c} {}_H p_{ij}(t) D_{kj}, & i \in H^c, \\ 0, & i \in H. \end{cases}$$

Proof. Recalling that $T_t = T$ for $t < T$ and using the Markov property, one has, for $i \in H^c$,

$$\begin{aligned} D^c_{ij}(t) &= \sum_{k \in H^c} \mathbb{P}^i\big(X_T = j,\, X_t = k,\ t < T < \infty\big) \\ &= \sum_{k \in H^c} \mathbb{P}^i\big(t < T,\, X_t = k,\ t < T_t < \infty,\, X(T_t) = j\big) \\ &= \sum_{k \in H^c} \mathbb{P}^i(t < T,\, X_t = k)\, \mathbb{P}\big(X(T_t) = j,\ t < T_t < \infty \mid X_t = k\big) \\ &= \sum_{k \in H^c} {}_H p_{ik}(t) \mathbb{P}^k\big(X_T = j,\ 0 < T < \infty\big). \end{aligned}$$

For $i \in H$, $T = 0$ identically, so $D^c_{ij}(t) = 0$. □

Note that for $t = 0$ and $t \to \infty$ Theorem 1 yields $D^c_{ij}(0) = D_{ij}$ and $D^c_{ij}(\infty) = 0$, respectively. The following result is the direct consequence of Theorem 1 and the semi-group property of taboo probabilities, but deserves to be stated explicitly.

THEOREM 2. *For $s, t \geqslant 0$ and $i \in H^c$, $j \in H$:*

$$D_{ij}(t+s) - D_{ij}(t) = \sum_{k \in H^c} {}_H p_{ik}(t) D_{kj}(s),$$

$$D^c_{ij}(t+s) = \sum_{k \in H^c} {}_H p_{ik}(t) D^c_{kj}(s).$$

Hence, the (B)-equation for $D^c_{ij}(t)$ follows from the corresponding equation for taboo probabilities.

THEOREM 3. *For $i \in H^c$, $j \in H$, and $t > 0$, with right-side derivative at $t = 0$:*

$$\frac{dD^c_{ij}(t)}{dt} = -q_i\, D^c_{ij}(t) + \sum_{k \in H^c,\ k \neq i} q_{ik}\, D^c_{kj}(t), \tag{B}$$

with the initial condition $D^c_{ij}(0) = D_{ij}$.

Similarly, the (F)-equation for taboo probabilities yields

THEOREM 4. *Under conditions of validity of the* (F)*-equation for taboo probabilities:*

$$D_{ij}(t) = \int_0^t \sum_{k\in H^c} {}_H p_{ik}(s) q_{kj}\, ds \qquad \text{for} \quad i \in H^c, \quad j \in H, \quad t \geqslant 0, \tag{F}$$

$$D_{ij} = \sum_{k\in H^c} {}_H u_{ik}\, q_{kj}.$$

To stress the role of the density $d_{ij}(t) = dD_{ij}(t)/dt$, some comments on Theorem 4 are needed. Note first that Theorem 1 indicates that, for fixed $j \in H$, $D_{\bullet j}$ is purely excessive relative to ${}_H\mathbf{P}(t)$ on H^c:

$$\sum_{k\in H^c} {}_H p_{ik}(t) D_{kj} \leqslant D_{ij}, \qquad i \in H^c,$$

$$\lim_{t\to\infty} \sum_{k\in H^c} {}_H p_{ik}(t) D_{kj} = 0, \qquad i \in H^c.$$

Hence, for fixed $j \in H$, there exists a unique exit law $d_{\bullet j}(\cdot)$, such that

$$d_{ij}(t+s) = \sum_{k\in H^c} {}_H p_{ik}(t) d_{kj}(s), \qquad t \geqslant 0, \quad s > 0, \quad i \in H^c.$$

Preservation of equality in the above expression, when $s \to 0$, is assured by validity of the (F)-equation; see the similar argument in § 3.1.1.

Nevertheless, these considerations do not determine $d_{ij}(0)$, so D_{ij} is also not determined. Therefore an independent argument is needed to determine $d_{ij}(t)$ explicitly. It is shown below in b) that

$$d_{ij}(0) = q_{ij}, \qquad i \in H^c, \quad j \in H,$$

so

$$d_{ij}(t) = \sum_{k\in H^c} {}_H p_{ik}(t)\, q_{kj}, \qquad t \geqslant 0,$$

which is the (F)-equation. Evidently, addition over all $j \in H$ reduce Theorems 1 and 2 to Theorems 1 and 4 in § 3.1.1, and Theorems 3 and 4 to Theorems 6 and 7 in § 3.1.1.

Similarly, taking L.S. transforms of expressions obtained above, with d_{ij}^{α} denoting the transform of $D_{ij}(t)$, one has the following results.

THEOREM 5. *For* $i \in H^c$, $j \in H$, *and* $\alpha > 0$:

$$\sum_{k\in H^c} {}_H u_{ik}^{\alpha} D_{kj} = (D_{ij} - d_{ij}^{\alpha})/\alpha.$$

THEOREM 6. *For $i \in H^c$, $j \in H$, and $\alpha > 0$:*

$$-\alpha d_{ij}^{\alpha} = \sum_{k \in H^c} q_{ik}\left(D_{kj} - d_{kj}^{\alpha}\right), \tag{B}$$

$$d_{ij}^{\alpha} = \sum_{k \in H^c} {}_H u_{ik}^{\alpha}\, q_{kj}. \tag{F}$$

THEOREM 7. *For $i \in H^c$, $j \in H$, and $\alpha > 0$:*

$$(\alpha + q_i)\, d_{ij}^{\alpha} = q_{ij} + \sum_{k \in H^c,\ k \neq i} q_{ik} d_{kj}^{\alpha}.$$

Proof. The result follows from the (B)-equation in Theorem 6 and from Theorem 3, § 3.1.3. □

THEOREM 8. *For $i \in H^c$, $j \in H$, the conditional means satisfy*

$$m_{ij} = \sum_{k \in H^c} {}_H u_{ik} D_{kj},$$

$$q_i m_{ij} = D_{ij} + \sum_{k \in H^c,\ k \neq i} q_{ik}\, m_{kj}.$$

A rather curious fact may be noted from comparison of the above expression for m_{ij} with that for $D_{ij} \leqslant 1$ in Theorem 4.

b) It is of interest to comment briefly on the role of the (B)- and (F)-equations. The equation for $D_{ij}^c(t)$ in Theorem 3 expresses the decomposition according to the first discontinuity (after the initial position). The M.C. starts at $i \in H^c$, and the first jump occurs either directly to $j \in H$ or to $k \in H^c$ and then the chain starts all over again.

In the explicit expression for $D_{ij}(t)$ in Theorem 4 the decomposition is according to the last discontinuity, namely the first entrance to H (before time t). Initially at $i \in H^c$, the M.C. stays in H^c till the instant s at which a jump to H occurs (from a state $k \in H^c$ to a state $j \in H$).

Thus, the (F)-equation considers explicitly the jump at T from a position $X(T-)$ in H^c to a position $X(T)$ in H. This leads to the alternative approach to study of joint distributions, known as the "density argument" (essentially path analysis).

Thus, Lemma 1, § 2.2.5, specializes to the following basic lemma (see Fig. 3.1).

LEMMA 9. *For $i, k \in H^c$, $r \in H$, $j \in \mathbb{I}$, and $t > 0$:*

$$\mathbb{P}^i\left(X(T-) = k,\ X(T) = r,\ T \leqslant t,\ X_t = j\right) = \int_0^t {}_H p_{ik}(s) q_{kr} p_{rj}(t-s)\, ds.$$

Lemma 9 yields immediately the expression for density $d_{ir}(t)$ used above, with $d_{ir}(0) = q_{ir}$. Similarly, expressions mentioned in § 3.1.3-b follow from Lemma 9.

Introduce now

$$h_{ij}(t) = \sum_{r \in H} q_{ir} p_{rj}(t), \qquad i \in H^c, \quad j \in \mathbb{I}.$$

Hence Lemma 9 yields the following joint distributions.

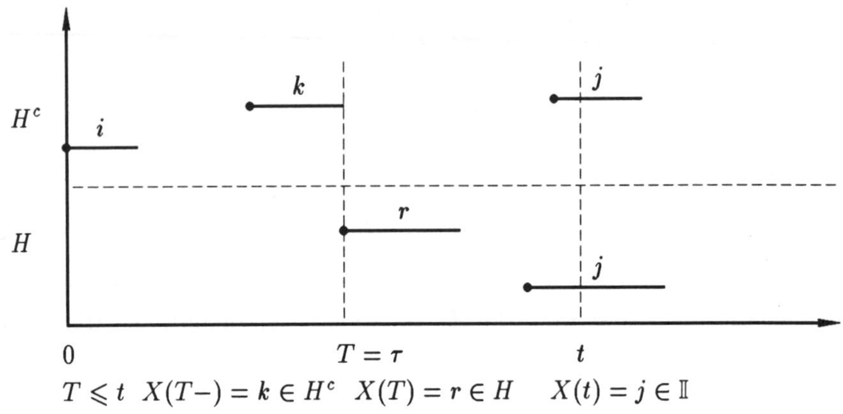

Figure 3.1. Illustration of Lemma 9, § 3.1.2.

THEOREM 10. *For $i, k \in H^c$, $r \in H$, $j \in \mathbb{I}$, and $t \geqslant 0$:*

$$\mathbb{P}^i\big(X(T) = r, T \leqslant t, X_t = j\big) = \int_0^t d_{ir}(s)\, p_{rj}(t-s)\, ds,$$

$$\mathbb{P}^i\big(X(T-) = k, T \leqslant t, X_t = j\big) = \int_0^t {}_Hp_{ik}(s)\, h_{kj}(t-s)\, ds.$$

COROLLARY 11. *For $i, k \in H^c$, and $t \geqslant 0$:*

$$\mathbb{P}^i\big(X(T-) = k, T \leqslant t\big) = \int_0^t {}_Hp_{ik}(s)\, \bar{q}_k\, ds,$$

$$\mathbb{P}^i\big(X(T-) = k, T < \infty\big) = {}_Hu_{ik}\, \bar{q}_k.$$

Corollary 11 interprets Theorem 7, § 3.1.1, and Theorem 4 here. The following conditional probability is of special interest.

COROLLARY 12. *For $k \in H^c$, $j \in \mathbb{I}$, and $t > s$:*

$$\mathbb{P}\big(X_t = j \mid X(T-) = k, T = s\big) = h_{kj}(t-s)/\bar{q}_k.$$

Remark. So far all discussion of X_T dealt with the case of finite T, and therefore occurrence of the event $(T < \infty)$ implies occurrence of the event $(X_T \in H)$. When $T = \infty$, extension is needed to define X_∞ (either arbitrarily or as a limit if it exists). Since $(T = \infty) = (X_t \in H^c$ for all $t)$, one may require that X_∞

should take values in H^c only. Alternatively, one can adjoin an isolated state Δ corresponding to X_T taking value Δ when $T = \infty$. Hence,

$$\mathbb{P}^i(X_T \in H) = \mathbb{P}^i(X_T \in H, T < \infty) = \mathbb{P}^i(T < \infty) = D_i,$$
$$\mathbb{P}^i(X_T = \Delta) = \mathbb{P}^i(X_T = \Delta, T = \infty) = \mathbb{P}^i(T = \infty) = D_i^c.$$

Analogous discussion may be carried out for $X(T-)$ with values in H^c for $T < \infty$, and in Δ for $T = \infty$:

$$\mathbb{P}^i(X(T-) \in H^c) = D_i, \qquad \mathbb{P}^i(X(T-) = \Delta) = D_i^c.$$

§ 3.1.3. Comments. a) The following results are the immediate consequences of Theorem 6, § 3.1.1, but because of their importance for passage times are stated here as theorems.

THEOREM 1. *The first entrance probability D_i satisfies the equations*

$$\sum_k q_{ik} D_k = 0, \qquad i \in H^c,$$
$$D_i = 1, \qquad i \in H.$$

THEOREM 2. *Define the function b on $\mathbb{I}$, to be called the escape rate, by*

$$b_i = -\sum_k q_{ik} D_k, \qquad i \in \mathbb{I}.$$

Then

$$b_i = \begin{cases} 0, & i \in H^c, \\ \sum_{k \in H^c} q_{ik} D_k^c \geqslant 0, & i \in H. \end{cases}$$

Theorem 1 represents the *Dirichlet problem* for the first entrance probability, relative to the intensity matrix $\mathbf{Q}$. It says that the function (D_i) is invariant (harmonic) on the set H^c and equal to the boundary function $\mathbf{1}$ on the "boundary" H. This interpretation will be examined at length in Sub-section 5.1. Similarly, Theorem 2 indicates that the function (D_i) is $\mathbf{P}(t)$-excessive on $\mathbb{I}$, and the function (D_i^c) is harmonic (invariant) on H^c, with transitions restricted to H^c. Theorem 2 represents the *Poisson equation* which will be discussed in Sub-section 5.2, and the function b will appear in the Poisson representation of the entrance probability D_i (see § 5.2.1).

b) It should be noted that the (B)-equation in Theorem 3, § 3.1.2, determines $d_{ij}(0)$ in terms of D_{ij} as

$$d_{ij}(0) = -\sum_{k \in H^c} q_{ik} D_{kj}, \qquad i \in H^c, \quad j \in H.$$

Equating this expression with that obtained from the (F)-equation in Theorem 4, § 3.1.2, one obtains the basic equation satisfied by the entrance distribution (D_{ij}).

THEOREM 3. *For $j \in H$:*

$$\sum_k q_{ik} D_{kj} = 0, \qquad i \in H^c,$$

$$D_{ij} = \delta_{ij}, \qquad i \in H.$$

It is clear that the following lemma holds.

LEMMA 4. *For a fixed $j \in H$, define a function $b_{\bullet j}$ on $\mathbb{I}$ by*

$$b_{ij} = -\sum_k q_{ik} D_{kj}, \qquad i \in \mathbb{I}.$$

Then

$$b_{ij} = 0 \quad \text{for} \quad i \in H^c,$$
$$b_{ij} \leqslant 0 \quad \text{for} \quad i \neq j,$$
$$b_{ii} \geqslant 0 \quad \text{for} \quad i \in H,$$

and the escape rate

$$b_i = \sum_{j \in H} b_{ij}, \qquad i \in \mathbb{I}.$$

The first entrance distribution (D_{ij}) is of fundamental importance in studies of passage times. It has been defined with $T < \infty$ for all i and j, but practically it is restricted to $i \in H^c$ and $j \in H$. Indeed, $D_{ij} = 0$ for $j \in H^c$, and $D_{ij} = \delta_{ij}$ for $i \in H$ and $j \in H$.

The explicit expression for D_{ij} has been found in Theorem 4, § 3.1.2. The discussion in § 3.1.2-b provides the natural interpretation, namely,

$$\mathbb{P}^i\left(X_{T-} = k,\, X_T = j,\, T < \infty\right) = {}_H u_{ik}\, q_{kj}, \qquad k \in H^c, \quad j \in H.$$

Then, the marginals are D_{ij} and the related entrance distribution (Cor. 11, § 3.1.2) are:

$$D_{ij} = \mathbb{P}^i\left(X_T = j,\, T < \infty\right), \qquad \mathbb{P}^i\left(X_{T-} = k,\, T < \infty\right) = {}_H u_{ik}\, \bar{q}_k.$$

c) Of great importance is the following *first entrance decomposition* whose full significance will be discussed in Section 6.

THEOREM 5. *For $i \in H^c$, $j \in \mathbb{I}$, and $t \geqslant 0$:*

$$p_{ij}(t) = {}_H p_{ij}(t) + \int_0^t \sum_{r \in H} d_{ir}(s) p_{rj}(t-s)\, ds$$
$$= {}_H p_{ij}(t) + \int_0^t \sum_{k \in H^c} {}_H p_{ik}(s) h_{kj}(t-s)\, ds.$$

Proof. Consider the decomposition

$$p_{ij}(t) = \mathbb{P}^i(X_t = j, t < T) + \mathbb{P}^i(X_t = j, T \leqslant t).$$

The first term is just ${}_Hp_{ij}(t)$, and the second follows from Theorem 10, § 3.1.2. □

It should be remarked that two forms of Theorem 5 differ by reference to $X(T)$ or to $X(T-)$. Equivalence of both forms has important consequences (see Subsection 6.1). Moreover, Theorem 5 is a special case of the well-known Dynkin formula (for stopping times).

§ 3.1.4. Examples. Illustrative examples given here are simple and well known. They have been already mentioned in § 1.2.6, but are deduced here directly from equations in this sub-section. More elaborate examples will be given later.

Example 1. Let k be a fixed state, and take $H = \{k\}$. Then, in the classical notation,

$$D_i(t) = D_{ik}(t) = F_{ik}(t), \qquad d_i^\alpha = d_{ik}^\alpha = f_{ik}^\alpha, \qquad D_i = D_{ik} = f_{ik},$$

for $i \neq k$. The situation covered in this example is the most common in applications.

Example 2. Let k be a fixed state and take $H^c = \{k\}$. Evidently, $T = \rho_k$ (see § 1.1.3) and $D_k(t)$ is the exponential distribution with the mean $1/q_k$ (see Ex. 2, § 1.2.6). The (B)-equation is now

$$\frac{dD_{kj}(t)}{dt} = q_{kj} - q_k D_{kj}(t), \qquad j \neq k,$$

so

$$D_{kj}(t) = r_{kj}\left(1 - e^{-q_k t}\right), \qquad j \neq k,$$

where $r_{kj} = q_{kj}/q_k$. Hence, for $j \neq k$:

$$d_{kj}(t) = q_{kj}e^{-q_k t}, \qquad d_{kj}^\alpha = \frac{q_{kj}}{(\alpha + q_k)},$$

$$D_{kj} = r_{kj} = {}_Hu_{kk}\, q_{kj}, \qquad {}_Hu_{kk} = \frac{1}{q_k}, \qquad m_{kj} = \frac{r_{kj}}{q_k}.$$

Addition over all $j \neq k$ yields

$$d_k(t) = q_k e^{-q_k t},$$

as noted above.

§ 3.1.5. Matrices. a) The concise matrix notation not only shortens formulae, but also provides additional insight into the structure of theorems in this section. The taboo matrix ${}_H\mathbf{P}(t)$, its intensity matrix ${}_H\mathbf{Q}$, and its resolvent matrix ${}_H\mathbf{U}^\alpha$ have been already defined in § 1.2.2. Here other important matrices will be introduced, and theorems proved earlier in this sub-section will be expressed in the matrix form. The partitioning of matrices according to H^c and H, to be used here, has been already explained in § 1.2.2.

The matrix $\mathbf{B}(t) = (D_{ij}(t))$ for $t \geqslant 0$, is called the time dependent balayage matrix and has the partitioned form

$$\mathbf{B}(t) = \begin{pmatrix} \mathbf{0} & \mathbf{B}_{12}(t) \\ \mathbf{0} & \mathbf{I}_2 \end{pmatrix}$$

with $\mathbf{B}(0) = \mathbf{J}_2$. The L.S. transform of $\mathbf{B}(t)$, denoted by $\mathbf{B}^\alpha = (d^\alpha_{ij})$ and called the α-balayage matrix, is

$$\mathbf{B}^\alpha = \begin{pmatrix} \mathbf{0} & \mathbf{B}^\alpha_{12} \\ \mathbf{0} & \mathbf{I}_2 \end{pmatrix}, \qquad \alpha > 0.$$

The limit matrix $\mathbf{B} = (D_{ij})$ defined by

$$\mathbf{B} = \lim_{t\to\infty} \mathbf{B}(t) = \lim_{\alpha\to 0} \mathbf{B}^\alpha = \begin{pmatrix} \mathbf{0} & \mathbf{B}_{12} \\ \mathbf{0} & \mathbf{I}_2 \end{pmatrix}$$

is called the balayage matrix. Note that $\mathbf{B}$ is sub-stochastic, idempotent and left-invariant:

$$\mathbf{B1} = D \leqslant 1, \qquad \mathbf{BB} = \mathbf{B}, \qquad \mathbf{BB}(t) = \mathbf{B}, \qquad \mathbf{B}(t)\mathbf{B} = \mathbf{B}(t).$$

Note that the matrix $\mathbf{B}^c(t) = \mathbf{B} - \mathbf{B}(t)$ has the only non-zero component $\mathbf{B}^c_{12}(t) = (D^c_{ij}(t))$. The defining relation (Thm. 1, § 3.1.2) is

$$\mathbf{B}^c(t) = {}_H\mathbf{P}(t)\mathbf{B}.$$

Hence, other matrix relations can be obtained directly. For example, Theorem 2, § 3.1.2, is

$$\mathbf{B}^c(t+s) = {}_H\mathbf{P}(t)\mathbf{B}^c(s),$$

$$\mathbf{B}(t+s) - \mathbf{B}(t) = {}_H\mathbf{P}(t)\mathbf{B}(s),$$

whose the only non-zero components are, respectively,

$$\mathbf{B}^c_{12}(t+s) = {}_H\mathbf{P}_{11}(t)\mathbf{B}^c_{12}(s),$$

$$\mathbf{B}_{12}(t+s) - \mathbf{B}_{12}(t) = {}_H\mathbf{P}_{11}(t)\,\mathbf{B}_{12}(s).$$

Similarly, one has for equations:

$$\frac{d\mathbf{B}^c_{12}(t)}{dt} = \mathbf{Q}_{11}\mathbf{B}^c_{12}(t) \qquad \text{(cf. Thm. 3, § 3.1.2)}, \qquad \text{(B)}$$

$$\frac{d\mathbf{B}_{12}(t)}{dt} = {}_H\mathbf{P}_{11}(t)\mathbf{Q}_{12} \qquad \text{(cf. Thm. 4, § 3.1.2)}, \qquad \text{(F)}$$

or in the transformed version (Thm. 6, § 3.1.2):

$$(\mathbf{Q}_{11} - \alpha \mathbf{I}_1)\mathbf{B}_{12}^{\alpha} = \mathbf{Q}_{11}\,\mathbf{B}_{12}, \tag{B}$$

$$\mathbf{B}_{12}^{\alpha} = {}_H\mathbf{U}_{11}^{\alpha}\,\mathbf{Q}_{12}. \tag{F}$$

Comparing expressions for $d\mathbf{B}_{12}(t)/dt$ at $t = 0$, obtained from the (B)- and (F)-equations, yields

$$\mathbf{Q}_{11}\mathbf{B}_{12} + \mathbf{Q}_{12} = \mathbf{0} \qquad \text{(cf. Thm. 3, § 3.1.3).}$$

Using this, the (B)-equation may be re-written as

$$\frac{d\mathbf{B}_{12}(t)}{dt} = \mathbf{Q}_{11}\mathbf{B}_{12}(t) + \mathbf{Q}_{12}$$

or in the transformed version:

$$\mathbf{Q}_{11}\mathbf{B}_{12}^{\alpha} + \mathbf{Q}_{12} = \alpha\mathbf{B}_{12}^{\alpha} \qquad \text{(cf. Thm. 7, § 3.1.2).}$$

b) Of great interest is the matrix $\mathbf{S}(t)$, $t \geqslant 0$, defined by

$$\mathbf{S}(t) = {}_H\mathbf{P}(t) + \mathbf{B}(t) = \begin{pmatrix} {}_H\mathbf{P}_{11}(t) & \mathbf{B}_{12}(t) \\ \mathbf{0} & \mathbf{I}_2 \end{pmatrix}.$$

It is important to observe that $\mathbf{S}(t)$ is a standard stochastic matrix with the semi-group property

$$\mathbf{S}(t+s) = \mathbf{S}(t)\mathbf{S}(s), \qquad t, s \geqslant 0,$$

and $\mathbf{S}(0) = \mathbf{I}$. Note also that all states in H are absorbing.

Direct differentiation yields the Kolmogorov equations

$$\frac{d\mathbf{S}(t)}{dt} = \begin{cases} \mathbf{Q}^a\mathbf{S}(t), & \text{(B)} \\ \mathbf{S}(t)\mathbf{Q}^a, & \text{(F)} \end{cases}$$

with the infinitesimal generator

$$\mathbf{Q}^a = \begin{pmatrix} \mathbf{Q}_{11} & \mathbf{Q}_{12} \\ \mathbf{0} & \mathbf{0} \end{pmatrix}.$$

The resolvent $\mathbf{S}^{\alpha}$ is now

$$\mathbf{S}^{\alpha} = \int_0^{\infty} e^{-\alpha t}\,\mathbf{S}(t)\,dt = {}_H\mathbf{U}^{\alpha} + \mathbf{B}^{\alpha}/\alpha, \qquad \alpha > 0,$$

and the Kolmogorov equations in the resolvent form are

$$\alpha\mathbf{S}^{\alpha} - \mathbf{I} = \mathbf{Q}^a\mathbf{S}^{\alpha}, \tag{B}$$

$$\alpha\mathbf{S}^{\alpha} - \mathbf{I} = \mathbf{S}^{\alpha}\mathbf{Q}^a. \tag{F}$$

It should be remarked here that the above (B)-equation yields the (B)-equation for the component $\mathbf{B}_{12}(t)$ in the form obtained at the end of a) (by identification of derivative at $t = 0$). Thus, another form of the (B)-equation is

$$\frac{d\,\mathbf{S}(t)}{dt} = {}_H\mathbf{Q}\,\mathbf{S}(t) - {}_H\mathbf{Q}\mathbf{B},$$

where ${}_H\mathbf{Q}\mathbf{B}$ has the only non-zero component $\mathbf{Q}_{11}\mathbf{B}_{12} = -\mathbf{Q}_{12}$.

Finally, note that

$$\mathbf{S}(t)\mathbf{B} = \mathbf{B}\,\mathbf{S}(t) = \mathbf{B}, \qquad \mathbf{Q}^a\mathbf{B} = \mathbf{B}\mathbf{Q}^a = \mathbf{0}.$$

c) The jump α-matrix $\mathbf{R}^{a\alpha}$ corresponding to the generator $\mathbf{Q}^a$ has the form

$$\mathbf{R}^{a\alpha} = \begin{pmatrix} \mathbf{R}_{11}^{\alpha} & \mathbf{R}_{12}^{\alpha} \\ \mathbf{0} & \mathbf{I}_2 \end{pmatrix}, \qquad \alpha \geqslant 0,$$

because states in H are absorbing. The elements $r_{ij}^{\alpha} = {}_H r_{ij}^{\alpha}$ have been defined in § 1.2.5 for $i \in H^c$, $j \in \mathbb{I}$, with $r_{ij}^{\alpha} = \delta_{ij}$ for $i \in H$. With this notation, Theorem 7, § 3.1.2, can be written as

$$d_{ij}^{\alpha} = \sum_k r_{ik}^{\alpha} d_{kj}^{\alpha}, \qquad i \in H^c, \quad j \in H,$$

or in the matrix form

$$\mathbf{B}_{12}^{\alpha} = \mathbf{R}_{12}^{\alpha} + \mathbf{R}_{11}^{\alpha}\mathbf{B}_{12}^{\alpha}, \qquad \alpha \geqslant 0.$$

Recalling definition of $\mathbf{B}^{\alpha}$, one can write now

$$\mathbf{B}^{\alpha} = \mathbf{R}^{a\alpha}\mathbf{B}^{\alpha}, \qquad \alpha \geqslant 0. \tag{PT}$$

The expression (PT) is the special form of the Product Theorem (see § 8.1.1), and may be taken as the starting point for study of the first entrance distribution. In particular, the Dirichlet problem for $D = \mathbf{B1}$ (see Thm. 1, § 3.1.3) requires that D be harmonic relative to the matrix jump $\mathbf{R}^a$: $D = \mathbf{R}^a D$.

As remarked earlier, the minimal chain $X = (X_t)$ is just a randomized jump chain $X^* = (X_n)$ with exponential duration of states (see § 1.1.4). From properties of a discrete chain with a matrix $\mathbf{R}^{a\alpha}$, it follows that the discrete analogue of $\mathbf{B}^{\alpha}$ is a discrete balayage matrix

$$\mathbf{B}^{*\alpha} = \begin{pmatrix} \mathbf{0} & {}_H\mathbf{G}_{11}^{\alpha}\mathbf{R}_{12}^{\alpha} \\ \mathbf{0} & \mathbf{I}_2 \end{pmatrix}, \qquad \alpha \geqslant 0,$$

where ${}_H\mathbf{G}_{11}^{\alpha}$ is the (Green) potential matrix defined in § 1.2.5.

The Feller relation for taboo probabilities (cf. § 1.2.5) and the (F)-equation stated in a) yield

$$\begin{aligned} \mathbf{B}_{12}^{\alpha} &= {}_H\mathbf{U}_{11}^{\alpha}\mathbf{Q}_{12} = {}_H\mathbf{U}_{11}^{\alpha}\,(\boldsymbol{\Theta}_1 + \alpha\,\mathbf{I}_1)(\boldsymbol{\Theta}_1 + \alpha\,\mathbf{I}_1)^{-1}\mathbf{Q}_{12} \\ &= {}_H\mathbf{G}_{11}^{\alpha}\,\mathbf{R}_{12}^{\alpha} = \mathbf{B}_{12}^{*\alpha}, \end{aligned}$$

so $\mathbf{B}^{\alpha} = \mathbf{B}^{*\alpha}$. This important fact may be expressed as

THEOREM 1. *The minimal chain X and its jump chain X^* have the same balayage matrix* $\mathbf{B}$.

The situation is intuitively clear, however. The first entrance to a set H occurs at a jump instant, and the jump chain is associated with the minimal chain (through the Feller relation). Explicitly, d_{ij}^{α} can now be expressed both in terms of the minimal chain and the jump chain:

$$d_{ij}^{\alpha} = \sum_{k \in H^c} {}_H u_{ik}^{\alpha}\, q_{kj} = \sum_{k \in H^c} {}_H g_{ik}^{\alpha}\, {}_H r_{kj}^{\alpha}, \qquad i \in H^c, \quad j \in H.$$

§ 3.1.6. Cut-off method. The principal role of taboo probabilities has been already stressed in this section. Indeed, the use of taboo probabilities (equivalent to killing of a chain at the first entrance time T) leads to explicit expressions for various passage probabilities. Such a use is facilitated, moreover, by the relative ease with which the taboo probabilities can be computed.

In order to find the taboo probabilities ${}_H\mathbf{P}(t) = ({}_Hp_{ij}(t))$ and the distribution $\mathbf{B}(t) = (D_{ij}(t))$, it is convenient in practice to use the "cut-off" method:

> make all states in H absorbing (retaining unchanged intensities of transitions in and out of H^c), and write equations for transition probabilities of this *modified* M.C. $\widetilde{X} = (\widetilde{X}_t)$.

Justification of the method follows immediately from observation that the transition matrix $\widetilde{\mathbf{P}}(t)$ for the modified chain $\widetilde{X}$ with all states in H absorbing ("the stopped chain") coincides with the matrix $\mathbf{S}(t)$ introduced in § 3.1.5. Indeed, the intensity matrix $\widetilde{\mathbf{Q}}$ for such a modified chain is just $\mathbf{Q}^a$ from § 3.1.5. Hence, $\widetilde{\mathbf{P}}(t)\mathbf{1} = \mathbf{1}$, $\widetilde{\mathbf{Q}}\,\mathbf{1} = 0$ and Kolmogorov equations for $\widetilde{\mathbf{P}}(t)$ are the same as those for $\mathbf{S}(t)$. In partitioned form

$$\widetilde{\mathbf{P}}_{11}(t) = {}_H\mathbf{P}_{11}(t), \qquad \widetilde{\mathbf{P}}_{12}(t) = \mathbf{B}_{12}(t).$$

In other words, transition probabilities $\tilde{p}_{ij}(t)$ for the modified chain yield taboo probabilities and the joint distributions for the original chain $X = (X_t)$:

$$\tilde{p}_{ij}(t) = \begin{cases} {}_Hp_{ij}(t) & \text{for} \quad i \in H^c,\ j \in H^c, \\ D_{ij}(t) & \text{for} \quad i \in H^c,\ j \in H, \end{cases}$$

and

$$\lim_{t \to \infty} \tilde{p}_{ij}(t) = D_{ij} \qquad \text{for} \quad i \in \mathbb{I},\ j \in H.$$

In terms of transforms

$$\widetilde{\mathbf{U}}_{11}^{\alpha} = {}_H\mathbf{U}_{11}^{\alpha}, \qquad \widetilde{\mathbf{U}}_{12}^{\alpha} = \mathbf{B}_{12}^{\alpha}/\alpha, \qquad \widetilde{\mathbf{U}}_{22}^{\alpha} = \mathbf{I}_2/\alpha,$$

so

$$\widetilde{\mathbf{U}}_{12}^{\alpha} = \mathbf{B}_{12}^{\alpha}\widetilde{\mathbf{U}}_{22}^{\alpha},$$

in agreement with the decomposition theorem (Thm. 5, § 3.1.3) applied to the matrix $\mathbf{P}(t)$.

Hence, for the original chain

$$D_i^c(t) = \sum_{j \in H^c} p_{ij}(t) \underset{t\to\infty}{\longrightarrow} D_i^c,$$

$$D_i(t) = \sum_{j \in H} p_{ij}(t) \underset{t\to\infty}{\longrightarrow} D_i, \qquad i \in H^c.$$

On the other hand, it is evident that D is the probability of absorption in H for the modified chain. The Dirichlet problem becomes now $\widetilde{\mathbf{Q}}\, D = 0$, so D is invariant relative to $\widetilde{\mathbf{P}}(t)$. Hence, the escape rate $\tilde{b}$ vanishes identically.

The cut-off method has been used extensively in applications since early days. Several interesting examples will be given later. Here only a few simple observations will be made.

Recall from § 1.2.4 that cut-off method restricted to H^c only yields taboo probabilities for the original chain. This corresponds to killing of a chain, which should be distinguished from stopping as discussed here. In particular, one finds here the entrance distribution (D_{ij}) for the original chain (i.e., absorption distribution for the modified chain). Further structure of the M.C. must be examined in order to determine, for example, when absorption is certain ($D = 1$).

It is also easy to verify that all states in H^c (leading to H) are transient for the modified chain $\widetilde{X}$ (namely, $\tilde{f}_{ii} < 1$ for $i \in H^c$).

3.2. FIRST ENTRANCE TIME AFTER A FIXED INSTANT

§ 3.2.1. Equations. a) Let T_s be the first entrance time to a set H after the fixed instant $s > 0$, as defined in § 2.1.2, for the usual M.C. $X = (X_t)$. It will be shown now that its distribution function $D_i^s(\cdot)$, conditional on the initial state i, can be expressed in terms of the distribution function of the first entrance time T (discussed in Sub-section 3.1) or, alternatively, in terms of the delayed taboo probabilities (introduced in § 2.2.3).

THEOREM 1. *For $i \in \mathbb{I}$, $t \geqslant 0$, the distribution function of T_s is given by*

$$D_i^s(t) = \begin{cases} \sum_{j \in H} p_{ij}(s) + \sum_{j \in H^c} p_{ij}(s) D_j(t-s), & t \geqslant s, \\ 0, & t < s, \end{cases}$$

and has a jump at $t = s$ of magnitude

$$D_i^s(s) = \sum_{j \in H} p_{ij}(s).$$

Proof. Recalling properties of shift, one has

$$\begin{aligned}\mathbb{P}^i(T_s \leqslant t) &= \mathbb{P}^i(T \circ \theta_s \leqslant t-s) = \sum_k \mathbb{P}^i(T \circ \theta_s \leqslant t-s,\, X_s = k)\\ &= \sum_k \mathbb{P}^i(X_s = k)\, \mathbb{P}(T \circ \theta_s \leqslant t-s \mid X_s = k)\\ &= \sum_k p_{ik}(s) D_k(t-s)\end{aligned}$$

and the assertion follows, noting that $D_k(t-s) = 1$ for all $k \in H$. □

The following theorem is the analogue of Theorem 1, § 3.1.1, with the delayed taboo probabilities.

THEOREM 2. *For $t \geqslant 0$, $i \in \mathbb{I}$, the complementary distribution function of T_s is given by*

$$D_i^{sc}(t) = \begin{cases} \sum_{j \in H^c} {}_H p_{ij}^s(t), & t \geqslant s,\\ 1, & t < s.\end{cases}$$

Proof. The result follows from Theorem 1 and from Theorem 2, § 2.2.3. □

It should be remarked that underlying the above argument is the fact that the right-continuity of paths and stability of states yield the natural interpretation of delayed taboo probabilities

$${}_H p_{ij}^s(t) = \mathbb{P}(X_t = j,\, T_s > t \mid X_0 = i), \qquad t > s, \quad i \in \mathbb{I}, \quad j \in H^c.$$

Recall, however, that T_s is not a terminal time and as a consequence ${}_H\mathbf{P}^s(t) = ({}_H p_{ij}^s(t))$ do not form a semi-group for a fixed $s > 0$. Nevertheless, one has the following analogues of Corollary 3, Theorem 4 and Corollary 5 from § 3.1.1, which are direct consequences of Theorem 1 above.

THEOREM 3. *For $i \in \mathbb{I}$, $t \geqslant s$, $h \geqslant 0$:*

$$D_i^s(t+h) - D_i^s(t) = \sum_{k \in H^c} {}_H p_{ik}^s(t) D_k(h),$$

$$D_i^{sc}(t+h) = \sum_{k \in H^c} {}_H p_{ik}^s(t) D_k^c(h).$$

COROLLARY 4. *For $i \in \mathbb{I}$:*

$$D_i^s = \sum_{j \in H} p_{ij}(s) + \sum_{j \in H^c} p_{ij}(s) D_j,$$

$$D_i^{sc} = \lim_{t \to \infty} \sum_{j \in H^c} {}_H p_{ij}^s(t).$$

COROLLARY 5. *For $i \in \mathbb{I}$, $t \geqslant s$:*

$$D_i^s - D_i^s(t) = \sum_{k \in H^c} {}_H p_{ik}^s(t) D_k,$$

$$D_i^{sc} = \sum_{k \in H^c} {}_H p_{ik}^s(t) D_k^c.$$

Observe that the expression for the probability D_i^s (Cor. 4) corresponds to the decomposition

$$D_i^s = \mathbb{P}^i(T_s = s) + \mathbb{P}^i(s < T_s < \infty).$$

Since ${}_H p_{ik}(t) \leqslant {}_H p_{ik}^s(t)$, comparison with Corollary 5, § 3.1.1, gives $D_i^s \leqslant D_i$. Hence, in the matrix notation,

$$D_i^s = \mathbf{P}(s) D_i \leqslant D_i, \qquad i \in \mathbb{I}.$$

Therefore, the function (D_i) is $\mathbf{P}(s)$-excessive on $\mathbb{I}$, as already noted in § 3.1.3.

b) Differentiability of $D_i^s(t)$ with respect to t for $t > s$ follows from differentiability of $D_i(t)$, and one has the representation of the following form.

THEOREM 6. *For $i \in \mathbb{I}$ and $t > s$:*

$$D_i^s(t) = D_i^s(s) + \int_s^t d_i^s(x)\,dx.$$

The specific expression for the density $d_i^s(t)$ may be obtained by taking a specific form of the density $d_k(t-s)$ from § 3.1.3. For further discussion see § 3.2.2.

Finally, direct verification shows that the mean m_i^s and the L.S. transform $d_i^{s\alpha}$ can be expressed in terms of m_i and d_i^α as follows.

THEOREM 7. *For $i \in \mathbb{I}$, $\alpha > 0$:*

$$d_i^{s\alpha} = e^{-\alpha s}\left[\sum_{j \in H} p_{ij}(s) + \sum_{j \in H^c} p_{ij}(s) d_j^\alpha\right].$$

THEOREM 8. *For $i \in \mathbb{I}$:*

$$m_i^s = s\,D_i^s + \sum_{j \in H^c} p_{ij}(s) m_j.$$

In the above discussion the analogy with the case $s = 0$ (Sub-sect. 3.1) is apparent.

c) It is of interest to consider D_i^s as a function of s, for $0 \leqslant s < \infty$. In matrix notation, one has

THEOREM 9. *For $i \in \mathbb{I}$ and $s, t \geqslant 0$:*

$$D_i^{s+t} = \mathbf{P}(s) D_i^t.$$

Proof. By Corollary 4,

$$D_i^{s+t} = \mathbf{P}(s+t)D_i,$$

and the result follows from the Chapman–Kolmogorov equation. □

Thus, Theorem 9 indicates that both $D^\bullet$ and $D^{\bullet c}$ are exit laws relative to the transition matrix $\mathbf{P}(s)$, hence continuous for $s > 0$ and right-continuous at $s = 0$. Hence, one obtains in the usual way the following equations, which also follow by direct differentiation of $D_i^s = \mathbf{P}(s)D_i$.

THEOREM 10. *For $i \in \mathbb{I}$ and $s > 0$:*

$$\frac{dD_i^s}{ds} = \begin{cases} \mathbf{Q}D_i^s, & \text{(B)} \\ \mathbf{P}(s)\mathbf{Q}D_i, & \text{(F)} \end{cases}$$

with $D_i^0 = D_i$, and the right-side derivative at $s = 0$.

Recall that the escape rate is $b_i = -\mathbf{Q}D_i$ (cf. § 3.1.3). Hence, the (F)-equation yields the explicit form of D_i^s.

COROLLARY 11. *For $i \in \mathbb{I}$ and $s \geqslant 0$:*

$$D_i^s = D_i - \int_0^s \mathbf{P}(x)b_i\,dx.$$

Note that D_i^s is a decreasing function of s, and it follows from Corollary 11 that the value of the limit is

$$D_i^\infty = \lim_{s\to\infty} \mathbf{P}(s)D_i = D_i - \mathbf{U}b_i$$

with finite potential $\mathbf{U}$. Moreover, Theorem 9 indicates that (D_i^∞) is $\mathbf{P}(s)$-invariant. The above is the Riesz decomposition of the excessive function (D_i) as a sum of an invariant component and a potential of a charge b.

§ 3.2.2. Joint distributions. a) The joint distribution $D_{ij}^s(t)$ of T_s and the position $X(T_s)$ has been introduced in § 2.1.5 for finite T_s. Recall that $s > 0$ is fixed.

LEMMA 1. *For $i \in \mathbb{I}$, $j \in H$, and $t \geqslant s$:*

$$\mathbb{P}^i\big(X_s = k,\ T_s \leqslant t,\ X(T_s) = j\big) = \begin{cases} p_{ik}(s)D_{kj}(t-s), & k \in H^c, \\ p_{ik}(s)\delta_{kj}, & k \in H, \end{cases}$$

and for $t < s$ this probability vanishes.

Proof. For $k \in H^c$, one has

$$\mathbb{P}^i\big(X_s = k,\ T\circ\theta_s \leqslant t-s,\ X_T\circ\theta_s = j\big) = \mathbb{P}^i\big(X_s = k\big)\,\mathbb{P}\big(T \leqslant t-s,\ X_T = j \mid X_0 = k\big),$$

and for $k \in H$, the only possibility is $T_s = s$ (cf. § 2.1.2). □

Addition over all $k \in \mathbb{I}$ yields the following basic result.

THEOREM 2. *For* $i \in \mathbb{I}$, $j \in H$:

$$D^s_{ij}(t) = \begin{cases} p_{ij}(s) + \sum_{k \in H^c} p_{ik}(s) D_{kj}(t-s), & t \geqslant s, \\ 0, & t < s, \end{cases}$$

with a jump at $t = s$ *of magnitude* $p_{ij}(s)$, *and a continuous component for* $t > s$.

The next theorems are analogues of theorems from § 3.1.2 with delayed taboo probabilities, and by addition over all $j \in H$ they reduce to corresponding theorems in § 3.2.1. Recall that $D^{sc}_{ij}(t) = D^s_{ij} - D^c_{ij}(t)$.

THEOREM 3. *For* $i \in \mathbb{I}$, $j \in H$, *and* $t \geqslant s$:

$$D^{sc}_{ij}(t) = \sum_{k \in H^c} {}_H p^s_{ik}(t) D_{kj} = \sum_{k \in H^c} p_{ik}(s) D^c_{kj}(t-s).$$

THEOREM 4. *For* $i \in \mathbb{I}$, $j \in H$, *and* $t \geqslant s$, $h \geqslant 0$:

$$D^s_{ij}(t+h) - D^s_{ij}(t) = \sum_{k \in H^c} {}_H p^s_{ik}(t) D_{kj}(h),$$

$$D^{sc}_{ij}(t+h) = \sum_{k \in H^c} {}_H p^s_{ik}(t) D^c_{kj}(h).$$

THEOREM 5. *For* $i \in \mathbb{I}$, $j \in H$:

$$D^s_{ij} = p_{ij}(s) + \sum_{k \in H^c} p_{ik}(s) D_{kj}.$$

Differentiability of $D^s_{ij}(t)$ with respect to t for $t > s$ follows from differentiability of $D_{ij}(t)$, and one has the representation of the following form.

THEOREM 6. *For* $i \in \mathbb{I}$, $j \in H$ *and* $t > s$:

$$D^s_{ij}(t) = p_{ij}(s) + \int_s^t d^s_{ij}(x)\, dx.$$

The specific expression for the density $d^s_{ij}(t)$ may be obtained by taking a specific form of density $d_{ij}(t)$ from § 3.1.2 — see Corollary 8 below.

COROLLARY 7. *For* $i \in \mathbb{I}$, $j \in H$, *and* $t > s$:

$$\frac{dD^{sc}_{ij}(t)}{dt} = \sum_{k \in H^c} p_{ik}(s) \sum_{v \in H^c} q_{kv}\, D^c_{vj}(t-s). \tag{B}$$

COROLLARY 8. *For* $i \in \mathbb{I}$, $j \in H$ *and* $t > s$:

$$d^s_{ij}(t) = \sum_{k \in H^c} {}_H p^s_{ik}(t)\, q_{kj} = \sum_{k \in H^c} p_{ik}(s) d_{kj}(t-s). \qquad \text{(F)}$$

It should be remarked that Theorem 4 leads to the relation

$$d^s_{ij}(t+h) = \sum_{k \in H^c} {}_H p^s_{ik}(t) d_{kj}(h), \qquad h \geqslant 0,$$

from which Corollary 8 follows by the argument similar to that in § 3.1.2. Note that $d_{kj}(0)$ is given by either one of two expressions stated in § 3.1.3.

Finally, taking the L.S. transform of Theorem 2 one has

THEOREM 9. *For* $i \in \mathbb{I}$, $j \in H$, *and* $\alpha > 0$:

$$d^{s\alpha}_{ij} = e^{-\alpha s}\left[p_{ij}(s) + \sum_{k \in H^c} p_{ik}(s) d^{\alpha}_{kj}\right].$$

b) As always, the (F)-equation considers the jump at T_s from a position $X(T_s-)$ in H^c to a position $X(T_s)$ in H. This leads to the "density argument" analogous to that of § 3.1.2. However, due to the discontinuity of distribution at $t = s$, consideration will be restricted to the case $s < T_s$. The following basic lemma is nothing else but Lemma 9, § 3.1.2, shifted by $s > 0$ (see Fig. 3.2); see also § 2.2.5.

LEMMA 10. *For* $i \in \mathbb{I}$, $k, n \in H^c$, $r \in H$, $j \in \mathbb{I}$, *and* $t > s$:

$$\mathbb{P}^i(X_s = n,\ X(T_s-) = k,\ X(T_s) = r,\ s < T_s \leqslant t,\ X_t = j)$$
$$= p_{in}(s) \int_s^t {}_H p_{nk}(x-s)\, q_{kr}\, p_{rj}(t-x)\, dx.$$

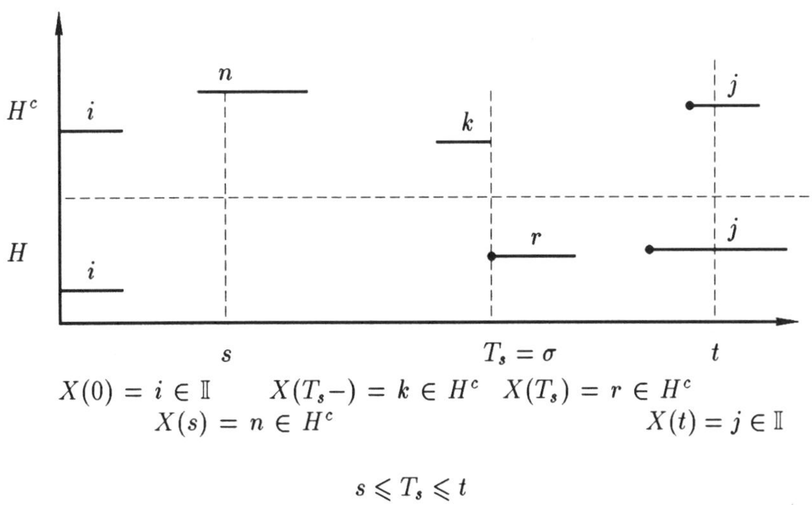

Figure 3.2. Illustration of Lemma 10, § 3.2.2.

THEOREM 11. *For $i \in \mathbb{I}$, $k \in H^c$, $r \in H$, and $t > s$:*

$$\mathbb{P}^i\left(X(T_s-) = k,\ X(T_s) = r,\ s < T_s \leqslant t\right) = \int_s^t {}_H p^s_{ik}(x)\, q_{kr}\, dx.$$

Theorem 11 yields the expression for density $d^s_{ij}(t)$ used in Theorem 6 and Corollary 8 above. Thus, we have

COROLLARY 12. *For $i \in \mathbb{I}$, $k \in H^c$, $r \in H$, and $t > s$:*

$$\mathbb{P}^i\left(X(T_s) = r,\ s < T_s \leqslant t\right) = \int_s^t d^s_{ir}(x)\, dx,$$

$$\mathbb{P}^i\left(X(T_s-) = k,\ s < T_s \leqslant t\right) = \int_s^t {}_H p^s_{ik}(x)\, \bar{q}_k\, dx.$$

Note analogies with expressions in § 3.1.2.

The delayed entrance distribution (D^s_{ij}) is of importance in the study of passage times. It has been defined with $T_s < \infty$ for all i and j, but vanishes for $j \in H^c$. The explicit expression for D^s_{ij} has been found in Theorem 5. The discussion here (with $t \to \infty$) provides the natural interpretation, namely

$$\mathbb{P}^i\left(X(T_s-) = k,\ X(T_s) = j,\ s < T_s < \infty\right) = {}_H u^s_{ik}\, q_{kj}, \qquad k \in H^c, \quad j \in H.$$

Then, the marginals yield Theorem 6 and the related entrance distribution (Cor. 12)

$$\mathbb{P}^i\left(X(T_s) = j,\ s < T_s < \infty\right) = D^s_{ij} - p_{ij}(s),$$
$$\mathbb{P}^i\left(X(T_s-) = k,\ s < T_s < \infty\right) = {}_H u^s_{ik}\, \bar{q}_k.$$

Recall that in the above formulas

$${}_H u^s_{ik} = \int_0^\infty {}_H p^s_{ik}(t)\, dt.$$

Remark. In the case T_s is infinite, one can adjoin an isolated state Δ. Then, as in Remark from § 3.1.2,

$$\mathbb{P}^i\left(X(T_s) \in H\right) = \mathbb{P}^i\left(X(T_s) \in H,\ T_s < \infty\right) = \mathbb{P}^i\left(T_s < \infty\right) = D^s_i,$$
$$\mathbb{P}^i\left(X(T_s) = \Delta\right) = \mathbb{P}^i\left(X(T_s) = \Delta,\ T_s = \infty\right) = \mathbb{P}^i\left(T_s = \infty\right) = D^{sc}_i.$$

Analogous discussion can be carried out for $X(T_s-)$.

§ 3.2.3. Comments. a) It is of interest to consider D^s_{ij} as a function of s, for $0 \leqslant s < \infty$. Proceeding as in § 3.2.1, the Chapman–Kolmogorov equation, and the (B)- and (F)-equations yield the following theorems.

THEOREM 1. *For $i \in \mathbb{I}$, $j \in H$, and $s, t \geqslant 0$:*

$$D^{s+t}_{ij} = \sum_k p_{ik}(s) D^t_{kj}.$$

THEOREM 2. *For $i \in \mathbb{I}$, $j \in H$, and $s > 0$:*

$$\frac{dD^s_{ij}}{ds} = \begin{cases} \sum_k q_{ik} D^s_{kj}, & \text{(B)} \\ \sum_k p_{ik}(s) b_{kj}, & \text{(F)} \end{cases}$$

with $D^0_{ij} = D_{ij}$, and the right-side derivative at $s = 0$.

COROLLARY 3. *For $i \in \mathbb{I}$, $j \in H$:*

$$D^s_{ij} = D_{ij} - \int_0^s \sum_k p_{ik}(t) b_{kj}\, dt.$$

Recall that quantity b_{ij} has been defined in Lemma 4, § 3.1.3. Letting $s \to \infty$, Corollary 3 yields

$$D_{ij} = D^\infty_{ij} + \sum_k u_{ik}\, b_{kj}$$

assuming the finiteness of potential (u_{ik}). Clearly, adding over all $j \in H$, the above results reduce to those for D^s_i in § 3.2.1.

b) The following first entrance decomposition (after s) generalizes Theorem 5, § 3.1.3, to the shift $s > 0$.

THEOREM 4. *For $i, j \in \mathbb{I}$, and $t \geqslant s$:*

$$\begin{aligned} p_{ij}(t) - {}_H p^s_{ij}(t) &= \int_s^t \sum_{r \in H} d^s_{ir}(x) p_{rj}(t-x)\, dx + a^s_{ij}(t) \\ &= \int_s^t \sum_{k \in H^c} {}_H p^s_{ik}(x) h_{kj}(t-x)\, dx + a^s_{ij}(t), \end{aligned}$$

where

$$a^s_{ij}(t) = \sum_{r \in H} p_{ir}(s) p_{rj}(t-s).$$

Proof. Consider the decomposition

$$p_{ij}(t) = \mathbb{P}^i(X_t = j,\, t < T_s) + \mathbb{P}^i(X_t = j,\, s < T_s \leqslant t) + \mathbb{P}^i(X_t = j,\, s = T_s \leqslant t).$$

The first term is just ${}_H p^s_{ij}(t)$. The second term follows from Lemma 10, § 3.2.2. The third term uses the fact that $s = T_s$ is equivalent to $X_s \in H$. □

It should be remarked that two forms of Theorem 4 differ only by reference to $X(T_s)$ or to $X(T_s-)$. Moreover, Theorem 4 is a special case of the well-known Dynkin formula (for stopping times).

§ 3.2.4. Examples. Examples given here are simple and serve for illustrative purposes. More elaborate examples will be given later.

Example 1. Take $H = \{k\}$, where k is a fixed state. Then $D_i^s(t) = D_{ik}^s(t)$ and

$$D_{ik}^s(t) = p_{ik}(s) + \sum_{v \neq k} p_{iv}(s) D_{vk}(t-s), \qquad t \geqslant s,$$

where $D_v(t-s) = D_{vk}(t-s) = F_{vk}(t-s)$ for $v \neq k$ (see Ex. 1, § 3.1.4).
Hence, $D_{ik}^s = D_i^s$ and

$$D_i^s = D_i - \int_0^s p_{ik}(x) b_k \, dx.$$

Example 2. Take $H^c = \{k\}$, where k is a fixed state. Then Example 2, § 3.1.4, yields

$$D_{ij}^s(t) = p_{ij}(s) + p_{ik}(s)\big(1 - e^{-q_k(t-s)}\big) r_{kj}, \qquad t \geqslant s, \quad j \neq k.$$

Hence,

$$D_i^s(t) = 1 - p_{ik}(s) e^{-q_k(t-s)}, \qquad t \geqslant s.$$

Moreover, b vanishes identically, and $D_i^s = 1$.

§ 3.2.5. Matrices. a) Matrices provide additional insight into structure of theorems in this sub-section. The matrix ${}_H\mathbf{P}^s(t) = ({}_Hp_{ij}^s(t))$ of delayed taboo probabilities has been defined in § 2.2.3 as

$${}_H\mathbf{P}^s(t) = \mathbf{P}(s)\, {}_H\mathbf{P}(t-s), \qquad t \geqslant s.$$

Introduce now the delayed balayage matrix $\mathbf{B}^s(t) = (D_{ij}^s(t))$ defined by

$$\mathbf{B}^s(t) = \mathbf{P}(s)\mathbf{B}(t-s), \qquad t \geqslant s,$$

where $\mathbf{B}(t-s)$ is the balayage matrix from § 3.1.5. Components of $\mathbf{B}^s(t)$ are from Theorem 2, § 3.2.2:

$$\mathbf{B}_{12}^s(t) = \mathbf{P}_{11}(s)\mathbf{B}_{12}(t-s) + \mathbf{P}_{12}(s),$$
$$\mathbf{B}_{22}^s(t) = \mathbf{P}_{21}(s)\mathbf{B}_{12}(t-s) + \mathbf{P}_{22}(s).$$

The L.S. transform of $\mathbf{B}^s(t)$, denoted by $\mathbf{B}^{s\alpha} = (d_{ij}^{s\alpha})$, is called the delayed α-balayage matrix. The limit matrix $\mathbf{B}^s = (D_{ij}^s)$ defined by

$$\mathbf{B}^s = \lim_{t \to \infty} \mathbf{B}^s(t) = \lim_{\alpha \to 0} \mathbf{B}^{s\alpha}$$

is called the delayed balayage matrix. Note that

$$\mathbf{B}^s\mathbf{1} = \mathbf{P}(s)D = D^s \leqslant 1, \qquad \mathbf{B}^s = \mathbf{P}(s)\mathbf{B}, \qquad \mathbf{B}^s\mathbf{B} = \mathbf{B}.$$

b) Define now

$$\mathbf{B}^{sc}(t) = \mathbf{B}^s - \mathbf{B}^s(t), \qquad t \geqslant s.$$

Then Theorem 3, § 3.2.2, becomes

$$\mathbf{B}^{sc}(t) = {}_H\mathbf{P}^s(t)\mathbf{B} = \mathbf{P}(s)\mathbf{B}^c(t-s), \qquad t \geqslant s.$$

Hence, using the (B)- and (F)-equations for ${}_H\mathbf{P}^s(t)$ from § 2.2.3, one obtains for $t > s$:

$$\frac{d\mathbf{B}^{sc}(t)}{dt} = \begin{cases} \mathbf{P}(s)_H\mathbf{Q}\mathbf{B}^c(t-s), & \text{(B)} \\ {}_H\mathbf{P}^s(t)_H\mathbf{Q}\mathbf{B}. & \text{(F)} \end{cases}$$

Note that for $t = s$, both equations yield the same value $\mathbf{P}(s)\mathbf{J}_1\,{}_H\mathbf{Q}\mathbf{B}$.

On the other hand, the (F)-equation obtained by the "density argument" (see § 3.2.2) yields, for $t > s$,

$$\frac{d\mathbf{B}^s(t)}{dt} = {}_H\mathbf{P}^s(t)_H\mathbf{Q},$$

whose value at $t = s$ is $\mathbf{P}(s)\mathbf{J}_1\,{}_H\mathbf{Q}$. Equating both values of the derivative at $t = s$, one recovers the relation (already used in § 3.1.5)

$$\mathbf{Q}_{11}\mathbf{B}_{12} + \mathbf{Q}_{12} = \mathbf{0}.$$

Hence, in terms of components the (B)-equation from Corollary 7, § 3.2.2, is

$$\frac{d\,\mathbf{B}^{sc}_{12}(t)}{dt} = \mathbf{P}_{11}(s)\,\mathbf{Q}_{11}\mathbf{B}^c_{12}(t-s),$$
$$\frac{d\,\mathbf{B}^{sc}_{22}(t)}{dt} = \mathbf{P}_{21}(s)\,\mathbf{Q}_{11}\mathbf{B}^c_{12}(t-s),$$

and the (F)-equation from Corollary 8, § 3.2.2, is

$$\frac{d\,\mathbf{B}^s_{12}(t)}{dt} = {}_H\mathbf{P}^s_{11}(t)\mathbf{Q}_{12},$$
$$\frac{d\mathbf{B}^s_{22}(t)}{dt} = {}_H\mathbf{P}^s_{21}(t)\mathbf{Q}_{12}.$$

Finally, the matrix form of the product theorem, applied to a constant stopping time s and to stopping time T (see § 8.1.1) is given by Theorem 9, § 3.2.2:

$$\mathbf{B}^{s\alpha} = e^{-\alpha s}\mathbf{P}(s)\mathbf{B}^{\alpha}, \qquad \alpha \geqslant 0. \tag{PT}$$

c) Consider now the matrix $\mathbf{S}^s(t)$ defined by

$$\mathbf{S}^s(t) = {}_H\mathbf{P}^s(t) + \mathbf{B}^s(t), \qquad t \geqslant s.$$

Observe that for $t \geqslant s$:

$$\mathbf{S}^s(t)\,\mathbf{1} = \mathbf{1}, \qquad \mathbf{S}^s(s) = \mathbf{P}(s), \qquad \mathbf{S}^s(t) = \mathbf{P}(s)\mathbf{S}(t-s), \qquad \mathbf{S}^s(t)\mathbf{B} = \mathbf{B}^s,$$

where the matrix $\mathbf{S}(t)$ has been defined in § 3.1.5. Although $\mathbf{S}(t)$, $t \geqslant 0$, form a semi-group, $\mathbf{S}^s(t)$, $t \geqslant s$, does not, and one has only

$$\mathbf{S}^s(t+h) = \mathbf{S}^s(t)\mathbf{S}(h).$$

Hence, formally,

$$\frac{d\,\mathbf{S}^s(t)}{dt} = \mathbf{S}^s(t)\mathbf{Q}^a,$$

in agreement with earlier results (§ 3.1.5).

d) The limit matrix $\mathbf{B}^s = \mathbf{P}(s)\mathbf{B}$ has the non-zero components

$$\mathbf{B}^s_{12} = \mathbf{P}_{11}(s)\mathbf{B}_{12} + \mathbf{P}_{12}(s), \qquad \mathbf{B}^s_{22} = \mathbf{P}_{21}(s)\mathbf{B}_{12} + \mathbf{P}_{22}(s).$$

The semi-group property of $\mathbf{P}(s)$, $s \geqslant 0$, implies that

$$\mathbf{B}^{s+t} = \mathbf{P}(s)\mathbf{B}^t \qquad \text{(Thm. 1, § 3.2.3).}$$

The (B)- and (F)-equations for $\mathbf{P}(s)$ yield the corresponding equations for $\mathbf{B}^s$:

$$\frac{d\mathbf{B}^s}{ds} = \begin{cases} \mathbf{Q}\mathbf{B}^s, & \text{(B)} \\ \mathbf{P}(s)\mathbf{Q}\mathbf{B}, & \text{(F)} \end{cases}$$

with the initial condition $\mathbf{B}^0 = \mathbf{B}$. This is the matrix form of Theorem 2, § 3.2.3.

Since $\mathbf{Q}^a\mathbf{B} = \mathbf{0}$ (cf. § 3.1.5), it is clear that $(-\mathbf{Q})\mathbf{B}$ has the only non-zero component of the form (see § 3.1.3)

$$(b_{ij}) = -(\mathbf{Q}_{21}\mathbf{B}_{12} + \mathbf{Q}_{22}).$$

Hence, Corollary 3, § 3.2.3, in the matrix form is

$$\mathbf{B}^s = \mathbf{B} - \int_0^s \mathbf{P}(t)\,(-\mathbf{Q})\mathbf{B}\,dt.$$

3.3. FIRST RETURN TIME

§ 3.3.1. Equations. a) Let T_ρ be the first return time to a set H for the usual M.C. $X = (X_t)$ as defined in § 2.1.3, and let $D_i^\rho(\cdot)$ be its distribution function, conditional on the initial state i. The fact that T_ρ is a terminal time (Lemma 3, § 2.1.3) allows to express distribution of T_ρ in terms of the exit taboo probabilities introduced in § 2.2.4. Much of the following discussion parallels that in Sub-section 3.1 extending it to the initial states in H.

LEMMA 1. *For $i, j \in \mathbb{I}$, and $t > 0$:*

$$\mathbb{P}^i\big(X_t = j,\ T_\rho > t\big) = {}_H p^\rho_{ij}(t),$$

where ${}_H p^\rho_{ij}(t)$ is the first exit taboo probability (Def. 1, § 2.2.4).

Proof (outline). From Lemma 2, § 2.1.2, one has

$$\big(\rho < t,\ T_\rho \geqslant t\big) = \big(X_u \in H^c \text{ for all } u \text{ with } \rho < u < t\big),$$

where $\rho \equiv \rho_i$. Thus,

$$\big(T_\rho \geqslant t\big) = \big(\rho \geqslant t\big) \cup \big(\rho < t,\ X_u \in H^c,\ \rho < u < t\big)$$

because $\rho \geqslant t$ implies $T_\rho \geqslant t$ (see Lemma 2, § 2.1.3). Hence, by right-continuity of paths and stability:

$$\mathbb{P}^i\big(X_t = j,\ T_\rho > t\big) = \mathbb{P}^i\big(\rho \geqslant t,\ X_t = j\big)$$
$$+\,\mathbb{P}^i\big(\rho < t,\ X_u \in H^c,\ \rho < u < t,\ X_t = j\big) = {}_H p^\rho_{ij}(t). \qquad \square$$

From Lemma 1, one has by summation over all $j \in \mathbb{I}$ the following

THEOREM 2. *The complementary distribution function $D^{\rho c}_i(\cdot)$ is a continuous function of $t \geqslant 0$, and for $i \in H$ it is given by*

$$D^{\rho c}_i(t) = e^{-q_i t} + \sum_{j \in H^c} {}_H p^\rho_{ij}(t),$$

with $D^{\rho c}_i(0) = 1$. For $i \in H^c$, it coincides with $D^c_i(t)$ of Theorem 1, § 3.1.1.

Observe that Theorem 2 is the analogue of Theorem 1, § 3.1.1, and extends it to the initial states in the taboo set H. Indeed, Lemma 1 indicates that the exit taboo probabilities are in fact obtained from the original M.C. by killing it at the instant of the first return to the taboo set. Consequently, arguments prevailing in Sub-section 3.1 apply here as well, the only difference being that summations of taboo probabilities are extended to all states in $\mathbb{I}$. Thus, for $i \in H^c$ the results coincide with those for $D_i(t)$, whereas new forms of $D^\rho_i(t)$ are obtained only for $i \in H$. For this reason, only the summary of results for $i \in H$ will be given here, with most of proofs omitted.

The following results are direct consequences of the defining relation in Theorem 2, but are stated explicitly because of their importance. Note first that now $D^\rho_i(0) = 0$ for all $i \in \mathbb{I}$.

COROLLARY 3. *For $i \in H$:*

$$D^{\rho c}_i = \lim_{t \to \infty} \sum_{j \in H^c} {}_H p^\rho_{ij}(t) \geqslant 0.$$

The semi-group property of exit taboo probabilities yields immediately Theorem 4, and Corollary 5 follows by the Dominated Convergence Theorem.

THEOREM 4. *For $i \in H$ and $s, t \geqslant 0$:*

$$D_i^{\rho c}(t+s) = e^{-q_i t} D_i^{\rho c}(s) + \sum_{j \in H^c} {}_H p_{ij}^{\rho}(t) D_j^c(s),$$

$$D_i^{\rho}(t+s) - D_i^{\rho}(t) = e^{-q_i t} D_i^{\rho}(s) + \sum_{j \in H^c} {}_H p_{ij}^{\rho}(t) D_j(s).$$

COROLLARY 5. *For $i \in H$ and $t \geqslant 0$:*

$$D_i^{\rho c} = \sum_{j \in H^c} {}_H p_{ij}^{\rho}(t) D_j^c + e^{-q_i t} D_i^{\rho c},$$

$$D_i^{\rho} = D_i^{\rho}(t) + \sum_{j \in H^c} {}_H p_{ij}^{\rho}(t) D_j + e^{-q_i t} D_i^{\rho}.$$

Observe that Theorem 4 and Corollary 5 are analogues of Theorem 4 and Corollary 5 in § 3.1.1 (extended to $i \in H$). Thus, Corollary 5 indicates that, relative to the taboo probabilities' matrix ${}_H\mathbf{P}^{\rho}(t)$, the function $(D_i^{\rho c})$ is invariant and the function (D_i^{ρ}) is purely excessive on the whole state space $\mathbb{I}$. Hence, $1 = D_i^{\rho} + D_i^{\rho c}$ is just the Riesz decomposition of the ${}_H\mathbf{P}^{\rho}(t)$-excessive function $\mathbf{1}$ on $\mathbb{I}$.

b) Theorems 6 and 7 below are direct consequences of Kolmogorov equations for the exit taboo probabilities (see § 2.2.4).

THEOREM 6. *The distribution function of T_ρ satisfies the* (B)*-equation for $i \in H$ and $t > 0$:*

$$\frac{dD_i^{\rho c}(t)}{dt} = -q_i D_i^{\rho c}(t) + \sum_{k \in H^c} q_{ik} D_k^c(t),$$

with the initial condition $D_i^{\rho c}(0) = 1$.

Observe that under summation sign the first entrance distribution $D_k^c(t)$ is taken as known, and Theorem 6 expresses $D_i^{\rho}(t)$ in terms of $D_i(t)$. Theorem 7 gives $D_i^{\rho}(t)$ directly in terms of taboo probabilities. Recall that the quantity $\bar{q}_i$ has been already defined in § 1.2.1.

THEOREM 7. *Under conditions of validity of the* (F)*-equation for the exit taboo probabilities, for $i \in H$ and $t \geqslant 0$:*

$$D_i^{\rho}(t) = (1 - e^{-q_i t}) \frac{\bar{q}_i}{q_i} + \int_0^t \sum_{k \in H^c} {}_H p_{ik}^{\rho}(s)\, \bar{q}_k \, ds,$$

$$D_i^{\rho} = \frac{\bar{q}_i}{q_i} + \sum_{k \in H^c} {}_H u_{ik}^{\rho}\, \bar{q}_k.$$

Concerning the (F)-equation, the situation is analogous to that in § 3.1.1. Indeed, the first equation in Theorem 4 indicates that $\mathbf{D}^{\rho c}(\cdot)$, as a function of t, is an exit law on $\mathbb{I}$, relative to the semi-group ${}_H\mathbf{P}^{\rho}(t)$. The second equation in Theorem 4 is now used (in the same manner as in § 3.1.1) to investigate the exit law $d_i^{\rho}(\cdot)$ giving the density $d_i^{\rho}(t) = d\, D_i^{\rho}(t)/dt$, with $d_i^{\rho}(0) = \bar{q}_i$ $(i \in \mathbb{I})$ by Theorem 6.

Taking L.S. transforms of expressions in a) and of the (B)- and (F)-equations, with $d_i^{\rho\alpha}$ denoting the transform of $D_i^{\rho}(t)$, one has the following theorems.

THEOREM 8. *For $i \in H$, $\alpha > 0$:*

$$(q_i + \alpha)^{-1} + \sum_{j \in H^c} {}_H u_{ij}^{\rho\alpha} = (1 - d_i^{\rho\alpha})/\alpha,$$

$$D_i^\rho (q_i + \alpha)^{-1} + \sum_{j \in H^c} {}_H u_{ij}^{\rho\alpha} D_j = (D_i^\rho - d_i^{\rho\alpha})/\alpha.$$

THEOREM 9. *For $i \in H$, $\alpha > 0$:*

$$(\alpha + q_i) d_i^{\rho\alpha} = \sum_{k \in H,\ k \neq 1} q_{ik} + \sum_{k \in H^c} q_{ik}\, d_k^{\alpha}, \tag{B}$$

$$d_i^{\rho\alpha} = \bar{q}_i\,(\alpha + q_i)^{-1} + \sum_{k \in H^c} {}_H u_{ik}^{\rho\alpha}\, \bar{q}_k. \tag{F}$$

THEOREM 10. *For $i \in H$, the conditional means satisfy:*

$$m_i^\rho = \frac{D_i^\rho}{q_i} + \sum_{j \in H^c} {}_H u_{ij}^\rho D_j, \qquad \mathbb{E}^i T_\rho = \frac{1}{q_i} + \sum_{j \in H^c} {}_H u_{ij}^\rho,$$

$$q_i m_i^\rho = D_i^\rho + \sum_{j \in H^c} q_{ij} m_j.$$

§ 3.3.2. Joint distributions. a) The joint distribution $D_{ij}^\rho(t)$ of T_ρ and the first return position $X(T_\rho)$ has been defined in § 2.1.5 for $i \in \mathbb{I}$, $j \in H$, and $t \geqslant 0$. Recall that $D_{ij}^{\rho c}(t) = D_{ij}^\rho - D_{ij}^\rho(t)$. The discussion presented here is the extension of that from § 3.3.1. The following theorem is the analogue of Theorem 2 and Corollary 5 in § 3.3.1.

THEOREM 1. *$D_{ij}^{\rho c}(\cdot)$ is a continuous function of t, and is given for $t \geqslant 0$ and $j \in H$ by*

$$D_{ij}^{\rho c}(t) = e^{-q_i t} D_{ij}^\rho + \sum_{k \in H^c} {}_H p_{ik}^\rho(t) D_{kj}$$

for $i \in H$. For $i \in H^c$ it coincides with $D_{ij}^c(t)$ of Theorem 1, § 3.1.2.

Proof. Recall that T_ρ is a terminal time. Hence, for $i \in H$,

$$\begin{aligned}
D_{ij}^{\rho c}(t) &= \sum_k \mathbb{P}^i\big(X(T_\rho) = j,\ X_t = k,\ t < T_\rho < \infty\big)\\
&= \sum_k \mathbb{P}^i\big(t < T_\rho,\ X_t = k,\ t < T_\rho \circ \theta_t + t < \infty,\ X(T_\rho \circ \theta_t + t) = j\big)\\
&= \sum_k \mathbb{P}^i\big(t < T_\rho,\ X_t = k\big)\, \mathbb{P}\big(X(T_\rho) \circ \theta_t = j,\ 0 < T_\rho \circ \theta_t < \infty \mid X_t = k\big)\\
&= \sum_k {}_H p_{ik}^\rho(t)\, \mathbb{P}^k\big(X(T_\rho) = j,\ 0 < T_\rho < \infty\big) = \sum_k {}_H p_{ik}^\rho(t) D_{kj}^\rho.
\end{aligned}$$

The result follows from definition of ${}_H p_{ik}^\rho(t)$ and the fact that $D_{kj}^\rho = D_{kj}$ for $k \in H^c$. □

Note that Theorem 1 yields $D_{ij}^{\rho c}(0) = D_{ij}^\rho$ and $D_{ij}^{\rho c}(\infty) = 0$. The following result is the direct consequence of Theorem 1, but deserves to be stated explicitly.

THEOREM 2. *For $s, t \geqslant 0$, and $i, j \in H$:*

$$D^{\rho}_{ij}(t+s) - D^{\rho}_{ij}(t) = e^{-q_i t} D^{\rho}_{ij}(s) + \sum_{k \in H^c} {}_H p^{\rho}_{ik}(t) D_{kj}(s),$$

$$D^{\rho c}_{ij}(t+s) = e^{-q_i t} D^{\rho c}_{ij}(s) + \sum_{k \in H^c} {}_H p^{\rho}_{ik}(t) D^{c}_{kj}(s),$$

whereas, for $i \in H^c$ and $j \in H$, Theorem 2, § 3.1.2, holds.

Observe that for fixed $j \in H$, the time dependent function $D^{c}_{\bullet j}(\cdot)$ is the exit law on $\mathbb{I}$ relative to the semi-group ${}_H\mathbf{P}^{\rho}(t)$. Finally, addition over all $j \in H$ reduces Theorems 1 and 2 to Theorems 2 and 4 in § 3.3.1, respectively.

The following equations are obtained in the same manner as their analogues in § 3.3.1. Observe that they express $D^{\rho}_{ij}(t)$ in terms of $D_{ij}(t)$. Thus, the (B)-equation follows from the corresponding equation for the exit taboo probabilities.

THEOREM 3. *For $i, j \in H$ and $t > 0$:*

$$\frac{dD^{\rho c}_{ij}(t)}{dt} = -q_i D^{\rho c}_{ij}(t) + \sum_{k \in H^c} q_{ik} D^{c}_{kj}(t) \tag{B}$$

with the initial condition $D^{\rho c}_{ij}(0) = D_{ij}$.

Concerning the (F)-equation, the situation is analogous to that described in §§ 3.1.2 and 3.3.1. Note first that Theorem 1 indicates that for fixed $j \in H$, $D^{\rho}_{\bullet j}$ is purely excessive relative to ${}_H\mathbf{P}^{\rho}(t)$ on $\mathbb{I}$:

$$\sum_k {}_H p^{\rho}_{ik}(t) D^{\rho}_{kj} \leqslant D^{\rho}_{ij}, \qquad \lim_{t \to \infty} \sum_k {}_H p^{\rho}_{ik}(t) D^{\rho}_{kj} = 0, \qquad i \in \mathbb{I}.$$

Hence, there exists a unique exit law $d^{\rho}_{ij}(\cdot)$ relative to ${}_H\mathbf{P}^{\rho}(t)$ such that, for fixed $j \in H$,

$$d^{\rho}_{ij}(t+s) = \sum_k {}_H p^{\rho}_{ik}(t) d^{\rho}_{kj}(s) \qquad \text{for} \quad t \geqslant 0,\ s > 0,\ i \in \mathbb{I}.$$

Preservation of equality in the above expression, when $s \to 0$, is assured by validity of the (F)-equation. Nevertheless, these considerations do not determine $d^{\rho}_{ij}(0)$, so D^{ρ}_{ij} is also not determined. Therefore, an independent argument is needed to determine $d^{\rho}_{ij}(t)$ explicitly. It is shown below in b) that

$$d^{\rho}_{ij}(0) = q_{ij}(1 - \delta_{ij}), \qquad i \in \mathbb{I}, \quad j \in H.$$

Hence,

$$d^{\rho}_{ij}(t) = d^{\rho}_{ij}(0) e^{-q_i t} + \sum_{k \in H^c} {}_H p^{\rho}_{ik}(t)\, q_{kj}, \qquad i, j \in H.$$

Consequently, one obtains the (F)-equation; recall that r_{ij} are jump probabilities.

THEOREM 4. *Under conditions of validity of the* (F)*-equation for exit taboo probabilities:*

$$D_{ij}^{\rho}(t) = r_{ij}\left(1 - e^{-q_i t}\right) + \int_0^t \sum_{k \in H^c} {}_H p_{ik}^{\rho}(s) q_{kj}\, ds, \qquad t \geqslant 0, \tag{F}$$

$$D_{ij}^{\rho} = r_{ij} + \sum_{k \in H^c} {}_H u_{ik}^{\rho} q_{kj} \qquad \text{for} \quad i, j \in H.$$

For $i \in H^c$, Theorem 4, § 3.1.2, holds.

Note that the first term represents direct transition from i to j, whereas the second term considers jumps from k to j. Indeed, Theorem 4 corresponds to the decomposition (see § 2.2.5)

$$\begin{aligned} \mathbb{P}^i\left(X(T_\rho) = j,\ T_\rho \leqslant t\right) &= \mathbb{P}^i\left(X(T_\rho) = j,\ T_\rho = \rho \leqslant t\right) \\ &+ \sum_{k \in H^c} \mathbb{P}^i\left(X(T_\rho -) = k,\ X(T_\rho) = j,\ \rho < T_\rho \leqslant t\right). \end{aligned}$$

Taking L.S. transforms of expressions obtained above, with $d_{ij}^{\rho\alpha}$ denoting the transform of $D_{ij}^{\rho}(t)$, one has the following results for $i \in H$. The corresponding results for $i \in H^c$ have been stated in § 3.1.2. It may be useful to recall Theorem 2, § 2.2.4, which expresses ${}_H u_{ij}^{\rho\alpha}$ in terms of known ${}_H u_{ij}^{\alpha}$.

THEOREM 5. *For $i, j \in H$ and $\alpha > 0$:*

$$D_{ij}^{\rho}\left(\alpha + q_i\right)^{-1} + \sum_{k \in H^c} {}_H u_{ik}^{\rho\alpha} D_{kj} = \left(D_{ij}^{\rho} - d_{ij}^{\rho\alpha}\right)/\alpha.$$

THEOREM 6. *For $i, j \in H$ and $\alpha > 0$:*

$$q_i D_{ij}^{\rho} - \left(\alpha + q_i\right) d_{ij}^{\rho\alpha} = \sum_{k \in H^c} q_{ik}\left(D_{kj} - d_{kj}^{\alpha}\right), \tag{B}$$

$$d_{ij}^{\rho\alpha} = \sum_{k \neq j} {}_H u_{ik}^{\rho\alpha} q_{kj}. \tag{F}$$

THEOREM 7. *For $i, j \in H$ and $\alpha > 0$:*

$$\left(\alpha + q_i\right) d_{ij}^{\rho\alpha} = q_{ij}\left(1 - \delta_{ij}\right) + \sum_{k \in H^c} q_{ik} d_{kj}^{\alpha}.$$

Proof. The result follows from the (B)-equation in Theorem 3 and from Theorem 2, § 3.3.3. □

THEOREM 8. *For $i, j \in H$, the conditional means satisfy*

$$m^{\rho}_{ij} = D^{\rho}_{ij}/q_i + \sum_{k\in H^c} {}_H u^{\rho}_{ik}\, D_{kj},$$

$$q_i m^{\rho}_{ij} = D^{\rho}_{ij} + \sum_{k\in H^c} q_{ik}\, m_{kj}.$$

Clearly, addition over all $j \in H$ yields the corresponding expressions from § 3.3.1.

b) Considerations of the (F)-equation and of jumps at T_ρ lead again to the "density argument". The following discussion parallel that in § 3.1.2, and proofs are omitted. However, by definition the taboo is now imposed after the initial state in H has been left, so possibility of a chain staying in that state in H is responsible for appearance of the additional term (corresponding to taboo on H). Recall that by right-continuity $X(T_\rho-)$ takes values in H^c, but $X(T_\rho)$ in H.

Thus, Lemma 1, § 2.2.5, specialises to the following basic lemma which extends Lemma 9, § 3.1.2, to $i \in H$ (see Fig. 3.3).

LEMMA 9. *For $i, r \in H$, $k \in H^c$, $j \in \mathbb{I}$ and $t \geqslant 0$:*

$$\mathbb{P}^i(X(T_\rho-) = k,\ X(T_\rho) = r,\ \rho < T_\rho \leqslant t,\ X_t = j) = \int_0^t {}_H p^{\rho}_{ik}(s) q_{kr}\, p_{rj}(t-s)\, ds,$$

$$\mathbb{P}^i(X(T_\rho) = r,\ T_\rho = \rho \leqslant t,\ X_t = j) = \int_0^t e^{-q_i s} q_{ir}\, p_{rj}(t-s)\, ds, \qquad i \neq r.$$

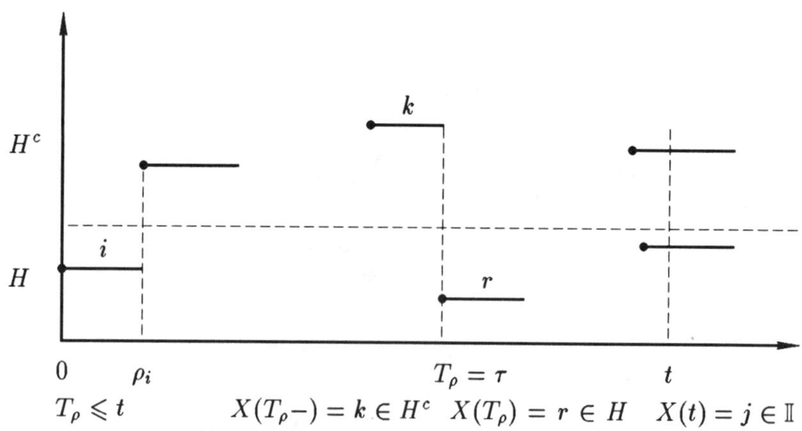

Figure 3.3. Illustration of Lemma 9, § 3.3.2.

THEOREM 10. *For $i, r \in H$, $k \in H^c$, and $t \geqslant 0$:*

$$\mathbb{P}^i\big(X(T_\rho-) = k,\ X(T_\rho) = r,\ \rho < T_\rho \leqslant t\big) = \int_0^t {}_Hp^\rho_{ik}(s) q_{kr}\, ds,$$

$$\mathbb{P}^i\big(X(T_\rho) = r,\ T_\rho = \rho \leqslant t\big) = \big(1 - e^{-q_i t}\big) q_{ir}/q_i, \qquad i \neq r.$$

It is immediate to obtain expressions for $D^\rho_{ir}(t)$ and for the density $d^\rho_{ir}(0)$, used above, and also the limiting form mentioned in § 3.3.3. Addition over all $r \in H$ yields expressions for $D^\rho_i(t)$, $d^\rho_i(0)$, and D^ρ_i corresponding to the (F)-equation (see § 3.3.1).

Introduce now

$$h^\rho_{ij}(t) = \sum_{r \in H,\ r \neq i} q_{ir}\, p_{rj}(t), \qquad i \in H, \quad j \in \mathbb{I}.$$

It is easy to deduce from Lemma 9 that the following analogue of Theorem 10, § 3.1.2, holds.

THEOREM 11. *For $i, r \in H$, $k \in H^c$, $j \in \mathbb{I}$, and $t \geqslant 0$:*

$$\mathbb{P}^i\big(X(T_\rho) = r,\ T_\rho \leqslant t,\ X_t = j\big) = \int_0^t d^{\rho'}_{ir}(s) p_{rj}(t-s)\, ds,$$

$$\mathbb{P}^i\big(X(T_\rho-) = k,\ \rho < T_\rho \leqslant t,\ X_t = j\big) = \int_0^t {}_Hp^\rho_{ik}(s) h^\rho_{kj}(t-s)\, ds.$$

Remark. In the case when T_ρ is infinite, one can adjoin an isolated state Δ. Then, as in Remark from § 3.1.2,

$$\mathbb{P}^i\big(X(T_\rho) \in H\big) = \mathbb{P}^i\big(X(T_\rho) \in H,\ T_\rho < \infty\big) = D^\rho_i,$$
$$\mathbb{P}^i\big(X(T_\rho) = \Delta\big) = \mathbb{P}^i\big(X(T_\rho) = \Delta,\ T_\rho = \infty\big) = D^{\rho c}_i.$$

Analogous discussion can be carried out for $X(T_\rho-)$.

§ 3.3.3. Comments. a) Introduce now the *escape probability* defined by $e_i = D^{\rho c}_i$ for $i \in H$, and $e_i = 0$ for $i \in H^c$, which represents probability that a chain starting in state i in H will leave H on termination of the initial state i, never to return.

The following result is the immediate consequence of Theorem 6, § 3.3.1, but because of its importance for passage times, it is stated here as a theorem.

THEOREM 1. *The first entrance probability D_i satisfies the equation*

$$\sum_k q_{ik}\, D_k = \begin{cases} 0, & i \in H^c, \\ -b_i, & i \in H, \end{cases}$$

where $b_i = q_i e_i$ is the escape rate (defined in § 3.1.3).

Proof. For $i \in H^c$ this is the Dirichlet problem (Thm. 1, § 3.1.3). For $i \in H$, the result follows from Theorem 2, § 3.1.3, by taking limit as $t \to \infty$, in Theorem 6, § 3.3.1. □

The above result is known as the *Poisson equation* for D, relative to the generator $\mathbf{Q}$. It expresses the fact that the function D is $\mathbf{P}(t)$-excessive on $\mathbb{I}$. Indeed, in matrix notation,

$$\mathbf{Q}\,D_i = \begin{cases} 0, & i \in H^c, \\ -b_i, & i \in H, \end{cases}$$

so $\mathbf{Q}D \leqslant 0$, and b is the charge supported on H. See § 3.1.3 and Sub-section 5.2.

b) It should be noted that the (B)-equation in Theorem 3, § 3.3.2, determines $d^{\rho}_{ij}(0)$ in terms of D^{ρ}_{ij} as

$$d^{\rho}_{ij}(0) = q_i D^{\rho}_{ij} - \sum_{k\in H^c} q_{ik}\, D_{kj}\,, \qquad i, j \in H.$$

Equating this expression with that obtained from the (F)-equation in Theorem 4, § 3.3.2, one obtains the basic equation satisfied by the first return distribution (D^{ρ}_{ij}).

THEOREM 2. *For $i, j \in H$:*

$$q_i D^{\rho}_{ij} = q_{ij}\,(1 - \delta_{ij}) + \sum_{k\in H^c} q_{ik}\, D_{kj}\,.$$

Recalling the expression for b_{ij} introduced in § 3.1.3, the above Theorem 2 and Theorem 3, § 3.1.3, yield the following important result.

THEOREM 3. *For $j \in H$:*

$$\sum_k q_{ik} D_{kj} = \begin{cases} 0, & i \in H^c, \\ -b_{ij}, & i \in H. \end{cases}$$

Observe that addition over all $j \in H$ yields the Poisson equation. The first return distribution (D^{ρ}_{ij}) is of importance in studies of passage times. It has been defined with $T_\rho < \infty$ for all i and j in $\mathbb{I}$. However, it vanishes when $j \in H^c$, and coincides with D_{ij} for $i \in H^c$. The explicit expression for D^{ρ}_{ij} has been found in Theorem 4, § 3.3.2. The discussion in § 3.3.2-b provides the natural interpretation, namely:

$$\mathbb{P}^i\big(X(T_\rho-) = k,\ X(T_\rho) = j,\ \rho < T_\rho < \infty\big) = {}_H u^{\rho}_{ik}\, q_{kj}\,, \qquad k \in H^c, \quad j \in H,$$
$$\mathbb{P}^i\big(X(T_\rho) = j,\ \rho = T_\rho < \infty\big) = r_{ij}$$

(see also Theorem 7, § 3.3.1).

c) Of great importance is the following first *return decomposition* theorem whose full significance will be discussed in Section 6.

THEOREM 4. *For $i \in H$, $j \in \mathbb{I}$, and $t \geqslant 0$:*

$$p_{ij}(t) = {}_H p^{\rho}_{ij}(t) + \int_0^t \sum_{r\in H} d^{\rho}_{ir}(s) p_{rj}(t-s)\,ds$$
$$= {}_H p^{\rho}_{ij}(t) + \int_0^t \sum_{k} {}_H p^{\rho}_{ik}(s) h^{\rho}_{kj}(t-s)\,ds.$$

Proof. Consider the decomposition

$$p_{ij}(t) = \mathbb{P}^i\big(X_t = j,\, t < T_\rho\big) + \mathbb{P}^i\big(X_t = j,\ T_\rho \leqslant t\big).$$

The first term is just ${}_H p^{\rho}_{ij}(t)$, whereas the second term follows from Lemma 9 and Theorem 11, § 3.3.2. □

It should be remarked that two forms of Theorem 4 differ only by reference to $X(T_\rho)$ and $X(T_\rho-)$. Equivalence of both forms has important consequences (see Sub-section 6.1). Moreover, Theorem 4 is a special case of the well-known Dynkin formula (for stopping times).

§ **3.3.4. Examples.** Illustrative examples given here are simple and well-known. They have been already mentioned in § 2.2.4, but are deduced here directly from equations in this sub-section. More elaborate examples will be given later.

Example 1. Let k be a fixed state, and take $H = \{k\}$. Then in classical notation

$$D^{\rho}_k(t) = F_{kk}(t), \qquad d^{\rho\alpha}_k = f^{\alpha}_{kk}, \qquad D^{\rho}_k = f_{kk} \leqslant 1,$$
$$D^{\rho}_{kk}(t) = D^{\rho}_k(t) = F_{kk}(t).$$

The situation covered by this example is common in applications. In literature the first entrance probability f_{ik} $(i \neq k)$ and the first return probability f_{kk} are usually considered together. There is, however, conceptual difference between cases $i \neq k$ and $i = k$ which corresponds, respectively, to $D_k = 1$ and $D^{\rho}_k \leqslant 1$. Observe that the escape probability is now $e_k = 1 - f_{kk}$.

Example 2. Let k be a fixed state and take $H^c = \{k\}$. Then (in agreement with Ex. 2, § 2.2.4)

$$D^{\rho}_{ij}(t) = \big(1 - e^{-q_i t}\big) r_{ij} + \Big(1 - \big(q_k - q_i\big)^{-1} q_k e^{-q_i t} + \big(q_k - q_i\big)^{-1} q_i e^{-q_k t}\Big) r_{ik} r_{kj}$$

for $i, j \neq k$. Therefore,

$$D^{\rho}_{ij} = r_{ij} + r_{ik}\, r_{kj}, \qquad i, j \neq k.$$

Summation over j yields

$$D^{\rho}_i(t) = 1 - e^{-q_i t} - q_{ik}\big(q_k - q_i\big)^{-1}\big(e^{-q_i t} - e^{-q_k t}\big), \qquad i \neq k,$$

with $\bar{q}_k = q_k$, and $\bar{q}_i = q_i - q_{ik}$ for $i \neq k$. Indeed, in this case the M.C. either stays in state i (with exponential distribution) or jumps from state i to state k and leaves k (after exponential sojourn in k), to return to H.

§ 3.3.5. Matrices. a) The concise matrix notation not only shortens formulae, but also provides additional insight into the structure of theorems proved in this section. The taboo matrix ${}_H\mathbf{P}^\rho(t)$, its intensity matrix ${}_H\mathbf{Q}^\rho$, and its resolvent matrix ${}_H\mathbf{U}^{\rho\alpha}$ have been already defined in § 2.2.4. Here other important matrices will be introduced, and theorems proved earlier in this section will be expressed in the matrix form. The partitioning of matrices according to H^c and H, to be used here, has been already explained in § 1.2.2.

The matrix $\mathbf{B}^\rho(t) = (D^\rho_{ij}(t))$ for $t \geqslant 0$ is called the time dependent balayage matrix, and has the partitioned form

$$\mathbf{B}^\rho(t) = \begin{pmatrix} \mathbf{0} & \mathbf{B}_{12}(t) \\ \mathbf{0} & \mathbf{B}^\rho_{22}(t) \end{pmatrix}$$

with $\mathbf{B}^\rho(0) = \mathbf{0}$; note that $\mathbf{B}^\rho_{12}(t) = \mathbf{B}_{12}(t) = (D_{ij}(t))$. The L.S. transform of $\mathbf{B}^\rho(t)$, denoted by $\mathbf{B}^{\rho\alpha} = (d^{\rho\alpha}_{ij})$ and called the α-balayage matrix, is

$$\mathbf{B}^{\rho\alpha} = \begin{pmatrix} \mathbf{0} & \mathbf{B}^\alpha_{12} \\ \mathbf{0} & \mathbf{B}^{\rho\alpha}_{22} \end{pmatrix}, \qquad \alpha > 0.$$

The limit matrix $\mathbf{B}^\rho = (D^\rho_{ij})$ defined by

$$\mathbf{B}^\rho = \lim_{t\to\infty} \mathbf{B}^\rho(t) = \lim_{\alpha\to 0} \mathbf{B}^{\rho\alpha} = \begin{pmatrix} \mathbf{0} & \mathbf{B}_{12} \\ \mathbf{0} & \mathbf{B}^\rho_{22} \end{pmatrix}$$

is called the balayage matrix. Note that $\mathbf{B}^\rho$ is sub-stochastic:

$$\mathbf{B}^\rho \mathbf{1} = D^\rho \leqslant 1.$$

Next, introduce the matrix

$$\mathbf{B}^{\rho c}(t) = \mathbf{B}^\rho - \mathbf{B}^\rho(t).$$

It is clear that defining relations (Thm. 1, § 3.1.2, and Thm. 1, § 3.3.2) can be written jointly as

$$\mathbf{B}^{\rho c}(t) = {}_H\mathbf{P}^\rho(t)\mathbf{B}^\rho, \qquad t \geqslant 0.$$

Hence, other matrix relations can be obtained directly. For example, Theorem 2, § 3.1.2, and Theorem 2, § 3.3.2, can be written jointly for $t, s \geqslant 0$, as

$$\mathbf{B}^{\rho c}(t+s) = {}_H\mathbf{P}^\rho(t)\mathbf{B}^{\rho c}(s), \qquad \mathbf{B}^\rho(t+s) - \mathbf{B}^\rho(t) = {}_H\mathbf{P}^\rho(t)\mathbf{B}^\rho(s).$$

Similarly, the (B)- and (F)-equations take the form

$$\frac{d\mathbf{B}^{\rho c}(t)}{dt} = {}_H\mathbf{Q}^\rho \mathbf{B}^{\rho c}(t), \tag{B}$$

$$\frac{d\mathbf{B}^{\rho}(t)}{dt} = {}_H\mathbf{P}^\rho(t)\,{}_B\mathbf{Q}^\rho, \tag{F}$$

where ${}_B\mathbf{Q}^\rho = (\mathbf{Q} + \mathbf{\Theta})\,\mathbf{J}_2 = \mathbf{Q} - {}_H\mathbf{Q}^\rho$.

Comparing expressions for $d\mathbf{B}^{\rho}(t)/dt$ at $t = 0$, obtained from the (B)- and (F)-equations, yields

$$\left(d_{ij}^{\rho}(0)\right) = {}_{B}\mathbf{Q}^{\rho} = -{}_{H}\mathbf{Q}^{\rho}\mathbf{B}^{\rho}$$

or, in terms of components,

$$\mathbf{Q}_{11}\,\mathbf{B}_{12} + \mathbf{Q}_{12} = \mathbf{0} \qquad \text{(Thm. 3, § 3.1.3)},$$

$$\mathbf{Q}_{21}\,\mathbf{B}_{12} + \mathbf{Q}_{22} = -\left(b_{ij}\right) = -\boldsymbol{\Theta}_2\left(\mathbf{I}_2 - \mathbf{B}_{22}^{\rho}\right) \qquad \text{(Thm. 2, § 3.3.3)},$$

which jointly give Theorem 3, § 3.3.3 (see also § 3.2.5). Using the above identity, the (B)-equation may be re-written in the form:

$$\frac{d\mathbf{B}^{\rho}(t)}{dt} = {}_{H}\mathbf{Q}^{\rho}\mathbf{B}^{\rho}(t) + {}_{B}\mathbf{Q}^{\rho} \qquad \text{(Thm. 7, § 3.3.2)}.$$

b) Consider now the matrix $\mathbf{S}^{\rho}(t)$, $t \geqslant 0$, defined by

$$\mathbf{S}^{\rho}(t) = {}_{H}\mathbf{P}^{\rho}(t) + \mathbf{B}^{\rho}(t).$$

Clearly, $\mathbf{S}^{\rho}(0) = \mathbf{I}$ and $\mathbf{S}^{\rho}(t)\mathbf{1} = \mathbf{1}$. However, $\mathbf{S}^{\rho}(t)$, $t \geqslant 0$, do not form a semigroup, as it may be easily verified. Nevertheless, direct differentiation of $\mathbf{S}^{\rho}(t)$ yields the equations

$$\frac{d\mathbf{S}^{\rho}(t)}{dt} = \begin{cases} {}_{H}\mathbf{Q}^{\rho}\mathbf{S}^{\rho}(t) - {}_{H}\mathbf{Q}^{\rho}\mathbf{B}^{\rho}, & \text{(B)} \\ {}_{H}\mathbf{P}^{\rho}(t)\mathbf{Q}. & \text{(F)} \end{cases}$$

Note that at $t = 0$, the value of each derivative is $\mathbf{Q}$.

c) Let $\mathbf{R}^{\alpha} = (r_{ij}^{\alpha})$ be the α-jump matrix (corresponding to the generator $\mathbf{Q}$) introduced in § 1.1.6 (see also § 2.2.4). Then Theorem 7, § 3.3.2, can be written as

$$d_{ij}^{\rho\alpha} = \sum_{k} r_{ik}^{\alpha} d_{kj}^{\alpha}, \qquad i, j \in H,$$

or, in the matrix form,

$$\mathbf{B}_{22}^{\rho\alpha} = \mathbf{R}_{22}^{\alpha} + \mathbf{R}_{21}^{\alpha}\mathbf{B}_{12}^{\alpha}, \qquad \alpha \geqslant 0.$$

This equation and its analogue for $i \in H^c$, already stated in § 3.1.5, can be combined in a single matrix equation

$$\mathbf{B}^{\rho\alpha} = \mathbf{R}^{\alpha}\mathbf{B}^{\alpha}, \qquad \alpha \geqslant 0. \qquad \text{(PT)}$$

The expression (PT) is the special form of the Product Theorem (see § 8.1.1), and may be taken as the starting point for the study of the first return distribution. In particular,

$$\mathbf{B}^{\rho} = \mathbf{R}\mathbf{B}, \qquad D^{\rho} = \mathbf{R}D$$

have intuitive interpretation. The transform $d_{ij}^{\rho\alpha}$ may be expressed equivalently in terms of the minimal chain or of the jump chain (using Feller relations). In particular,

$$\mathbf{B}_{12} = {}_{H}\mathbf{U}_{11}\mathbf{Q}_{12}, \qquad \mathbf{B}_{22}^{\rho} = \mathbf{R}_{22} + {}_{H}\mathbf{U}_{21}^{\rho}\mathbf{Q}_{12},$$

in agreement with Theorem 4, § 3.1.2, and Theorem 4, § 3.3.2. Also the analogue of Theorem 1, § 3.1.5, holds for $\mathbf{B}^{\rho}$.

3.4. DUALITY

§ 3.4.1. Equations. The first return time T and the fact that the initial position of a M.C. is in the set H, leads to duality between distributions $D_{ij}(t)$ from Subsection 3.1, and the functions $N_{ij}(t)$ defined below, which essentially interchanges the role of the (B)- and (F)-equations.

To begin with, recall that the (B)-equation in Theorem 2, § 2.2.4, yields the explicit expression for the exit taboo probability ${}_Hp^{\rho}_{ij}(t)$ in terms of functions ${}_Hv^{\rho}_{ij}(t)$, defined in § 2.2.4. The following lemma interprets this (B)-equation through "density argument" (sample functions, see § 2.2.5).

LEMMA 1. *For* $i \in H,\ j \in H^c$:

$$ {}_Hv^{\rho}_{ij}(t) = \mathbb{P}^i\left(X_{t+\rho} = j,\ t+\rho < T_\rho\right). $$

Proof. Here $\rho = \rho_i$ with jump from i to $k \in H^c$, so

$$ \begin{aligned} &\mathbb{P}^i\left(X(\rho) = k,\ X(t+\rho) = j, t+\rho < T_\rho\right) \\ &\quad = \mathbb{P}^i\left(X(\rho) = k\right) \mathbb{P}\left(X_t \circ \theta_\rho = j, t < T \circ \theta_\rho \mid X_\rho = k\right) \\ &\quad = \mathbb{P}^i\left(X_\rho = k\right) \mathbb{P}^k\left(X_t = j, t < T\right) = r_{ik}\, {}_Hp_{kj}(t). \end{aligned} $$ □

Define now for $i \in H,\ j \in H^c$, and $t \geqslant 0$:

$$ n_{ij}(t) = \sum_{k\in H^c} q_{ik}\, {}_Hp_{kj}(t) = q_i\, {}_Hv^{\rho}_{ij}(t), $$

$$ N_{ij}(t) = \int_0^t n_{ij}(s)\, ds. $$

Thus, $N_{ij}(t)/q_i$ is the average time spent in state $j \in H^c$ up to time t before the first return to H, when starting at $i \in H$ (with a jump to H^c when leaving the initial state i). Hence (by renewal argument), $N_{ij}(t)\, q_j/q_i$ is the average number of visits to state $j \in H^c$ under the same conditions.

Let $N_{ij} = N_{ij}(\infty)$ and write $N^c_{ij}(t) = N_{ij} - N_{ij}(t)$. The following result is the direct consequence of the semi-group property of taboo probabilities and the definition of $N_{ij}(t)$.

THEOREM 2. *For* $s,t \geqslant 0$ *and* $i \in H,\ j \in H^c$:

$$ \begin{aligned} N_{ij}(t+s) - N_{ij}(t) &= \sum_{k\in H^c} N_{ik}(s)\, {}_Hp_{kj}(t), \\ N^c_{ij}(t) &= \sum_{k\in H^c} N_{ik}\, {}_Hp_{kj}(t), \\ N^c_{ij}(t+s) &= \sum_{k\in H^c} N^c_{ik}(s)\, {}_Hp_{kj}(t). \end{aligned} $$

Observe that Theorem 2 is the dual of Theorem 1 and Theorem 2 from § 3.1.2. However, the expression for $N_{ij}(t)$ is known here, whereas the corresponding equations in § 3.1.2 do not determine $D_{ij}(t)$ explicitly, as already explained, and the analogue of Lemma 1 is needed.

The second formula in Theorem 2 indicates that for fixed $i \in H$, $N_{i\cdot}$ is a purely excessive measure on H^c, relative to ${}_H\mathbf{P}(t)$. Hence, there exists a unique entrance law $n_{ij}(\cdot)$, such that, for fixed $i \in H$

$$n_{ij}(t+s) = \sum_{k \in H^c} n_{ik}(s) {}_Hp_{kj}(t), \qquad t \geqslant 0, \quad s > 0,$$

with $N_{ij}(t) \uparrow N_{ij}$. Definition of $n_{ij}(t)$, based on Lemma 1, indicates that the above relation holds also for $s = 0$, with $n_{ij}(0) = q_{ij}$. Hence, the (B)-equation holds, being just definition of $N_{ij}(t)$.

On the other hand, the (F)-equation for $N_{ij}(t)$ is obtained from Theorem 2 by differentiation, provided the (F)-equation for taboo probabilities holds. This gives

THEOREM 3. *For $i \in H$, $j \in H^c$, and $t > 0$:*

$$\frac{dN_{ij}(t)}{dt} = \sum_{k \in H^c} q_{ik}\, {}_Hp_{kj}(t), \tag{B}$$

$$\frac{dN^c_{ij}(t)}{dt} = -N^c_{ij}(t) q_j + \sum_{k \in H^c,\ k \neq j} N^c_{ik}(t) q_{kj}, \tag{F}$$

with initial conditions $N_{ij}(0) = 0$.

It should be observed that the role of the (B)- and (F)-equations is reversed here, as compared with equations for $D_{ij}(t)$ in Theorems 3 and 4 in § 3.1.2. Indeed, the densities

(F): $$d_{ij}(t) = \sum_{k \in H^c} {}_Hp_{ik}(t)\, q_{kj}, \qquad i \in H^c, \quad j \in H,$$

(B): $$n_{ij}(t) = \sum_{k \in H^c} q_{ik}\, {}_Hp_{kj}(t), \qquad i \in H, \quad j \in H^c,$$

are dual in the sense that they determine the exit and the entrance laws, respectively.

Similarly, note that

(F): $$D_{ij} = \sum_{k \in H^c} {}_Hu_{ik}\, q_{kj}, \qquad i \in H^c, \quad j \in H,$$

(B): $$N_{ij} = \sum_{k \in H^c} q_{ik}\, {}_Hu_{kj}, \qquad i \in H, \quad j \in H^c,$$

but D_{ij} is a conditional probability, whereas N_{ij} is a conditional expectation. The following results are stated as theorems because of their importance, although their proofs are immediate consequences of results stated above. For example, Theorem 4 follows from comparison of two expressions for $n_{ij}(0)$ in Theorem 3.

THEOREM 4. *For* $i \in H,\ j \in H^c$:

$$q_{ij} + \sum_{k \in H^c} N_{ik}\, q_{kj} = 0.$$

THEOREM 5. *For* $i \in H,\ j \in H^c$:

$$N_{ij} = q_i\, {}_H u^{\rho}_{ij} < \infty.$$

THEOREM 6. *For* $i \in H$:

$$\sum_{j \in H^c} n_{ij}(t) D^c_j = b_i.$$

Define for $i \in H$ and $t \geqslant 0$:

$$n_i(t) = \sum_{j \in H^c} n_{ij}(t), \qquad N_i(t) = \int_0^t n_i(s)\, ds,$$

and note that $n_i(t) \to b_i$ as $t \to \infty$. Then, with $m_i = \mathbb{E}^i(T,\ T < \infty)$ the following result holds.

THEOREM 7. *For* $i \in H$:

$$\int_0^\infty \big(n_i(t) - b_i\big)\, dt = \sum_{k \in H^c} q_{ik} m_k.$$

Denote by n^α_{ij} the Laplace transform of $n_{ij}(t)$. Then one has immediately

THEOREM 8. *For* $i \in H,\ j \in H^c$, *and* $\alpha > 0$:

$$n^\alpha_{ij} = \sum_{k \in H^c} q_{ik}\, {}_H u^\alpha_{kj} = (\alpha + q_i)\, {}_H u^{\rho\alpha}_{ij},$$

$$n^\alpha_i = \sum_{j \in H^c} n^\alpha_{ij} = \sum_{k \in H^c} q_{ik}(1 - d^\alpha_k)\alpha^{-1} = (\alpha + q_i)(1 - d^{\rho\alpha}_i)\alpha^{-1} - 1,$$

$$n^\alpha_i - b_i \alpha^{-1} = \sum_{k \in H^c} q_{ik}(D_k - d^\alpha_k)\alpha^{-1},$$

$$\lim_{\alpha \to 0} \alpha n^\alpha_i = b_i, \qquad N_i = q_i \mathbb{E}^i T_\rho - 1.$$

§ 3.4.2. Relations. a) The following equation for n_{ij}^{α} is dual to the equation for d_{ij}^{α} (Thm. 7, § 3.1.2).

THEOREM 1. *For* $i \in H$, $j \in H^c$, *and* $\alpha > 0$:

$$n_{ij}^{\alpha}(\alpha + q_j) = q_{ij} + \sum_{k \in H^c,\ k \neq j} n_{ik}^{\alpha} q_{kj}.$$

Introduce now a new kernel (k_{ij}^{α}) defined for $i \in H$, $j \in H^c$, and $\alpha > 0$ by

$$k_{ij}^{\alpha} = (q_i + \alpha)^{-1} n_{ij}^{\alpha}(q_j + \alpha),$$

and write $K_{ij} = k_{ij}^{0}$. The equation in Theorem 1 takes now the form

$$k_{ij}^{\alpha} = r_{ij}^{\alpha} + \sum_{v \in H^c} k_{iv}^{\alpha} r_{vj}^{\alpha}, \qquad i \in H, \quad j \in H^c,$$

which is dual to the equation for d_{ij}^{α} stated in § 3.1.5. Moreover, recalling definition of ${}_H w_{ij}^{\rho\alpha}$ from § 2.2.4, one can show that $k_{ij}^{\alpha} = {}_H w_{ij}^{\rho\alpha}$.

It is clear that $K_{ij} = N_{ij} q_j / q_i$ represents the average number of visits to state $j \in H^c$, when starting at $i \in H$ and leaving H on the first jump. Duality of K_{ij} and D_{ij} is evident. It should be noted, however, that D_{ij} refers both to the jump chain and to the minimal chain, whereas N_{ij} and K_{ij} refer to the minimal chain and to the jump chain, respectively. See also § 8.2.3 for further comments.

b) The following theorem is of great importance in study of duality.

THEOREM 2. *For* $i \in H$ *and* $j \in H^c$:

$$\sum_{k \in H^c} n_{ik}^{\alpha} q_{kj} = \sum_{k \in H^c} q_{ik} d_{kj}^{\alpha}.$$

Proof. Direct consequence of the (F)-equation in Theorem 6, § 3.1.2, or the (B)-equation in Theorem 3, § 3.4.1. □

The connection with exit and entrance laws is another feature of duality. By Theorem 6, § 1.3.4, the inner product, for $i \in H$ and $j \in H$,

$$\phi_{ij}(t+s) = \sum_{k \in H^c} n_{ik}(s) d_{kj}(t)$$

depends only on $t + s$, so

$$\phi_{ij}(t) = \sum_{k \in H^c} n_{ik}(t) q_{kj} = \sum_{k \in H^c} q_{ik}\, d_{kj}(t).$$

Thus, Theorem 2 gives the expression for the Laplace transform ϕ_{ij}^{α}.

Next, define

$$\sigma_{ij}(t) = \int_t^{\infty} \phi_{ij}(s)\, ds, \qquad t > 0,$$

and note that

$$\sigma_{ij}(t) = \sum_{k\in H^c} N_{ik}\, d_{kj}(t) = \sum_{k\in H^c} n_{ik}(t) D_{kj}$$

with $\sigma_{ij}(0) < \infty$, and the Laplace transform

$$\sigma_{ij}^{\alpha} = \sum_{k\in H^c} N_{ik}\, d_{kj}^{\alpha} = \sum_{k\in H^c} n_{ik}^{\alpha}\, D_{kj}.$$

For further discussion see § 6.3.2.

§ 3.4.3. Examples. These examples are continuation of those in § 3.3.4. Calculations of n_{ij}^{α} are based on the formula stated in § 3.4.2.

Example 1. Let k be a fixed state, and take $H = \{k\}$. Then

$$n_{kj}^{\alpha} = (\alpha + q_k)\, {}_k u_{kj}^{\rho\alpha}, \qquad k_{kj}^{\alpha} = {}_k u_{kj}^{\rho\alpha}(q_j + \alpha), \qquad j \neq k,$$

where ${}_k u_{kj}^{\rho\alpha}$ has been found in Example 1, § 2.2.4.

Example 2. Let k be a fixed state and take $H^c = \{k\}$. Then

$$n_{ik}^{\alpha} = q_{ik}/(\alpha + q_k), \qquad k_{ik}^{\alpha} = r_{ik}^{\alpha}, \qquad i \neq k,$$

so $N_{ik} = q_{ik}/q_k$ and $K_{ik} = r_{ik}$.

See also equations for n_k^{α} in § 1.1.7.

§ 3.4.4. Dual matrices. a) The concise matrix notation not only shortens formulae, but also provides additional insight into the structure of theorems in this sub-section. Discussion here is dual to that in § 3.3.5.

The matrix $\mathbf{C}(t) = (N_{ij}(t))$, for $t \geqslant 0$, is called the time dependent co-balayage matrix, and has the partitioned form

$$\mathbf{C}(t) = \begin{pmatrix} \mathbf{0} & \mathbf{0} \\ \mathbf{C}_{21}(t) & \mathbf{I}_2 \end{pmatrix}$$

with $\mathbf{C}(0) = \mathbf{J}_2$. The L.S. transform of $\mathbf{C}(t)$, denoted by $\mathbf{C}^{\alpha} = (n_{ij}^{\alpha})$ and called the α-co-balayage matrix, is

$$\mathbf{C}^{\alpha} = \begin{pmatrix} \mathbf{0} & \mathbf{0} \\ \mathbf{C}_{21}^{\alpha} & \mathbf{I}_2 \end{pmatrix}, \qquad \alpha > 0.$$

The limit matrix $\mathbf{C} = (N_{ij})$ defined by

$$\mathbf{C} = \lim_{t\to\infty} \mathbf{C}(t) = \lim_{\alpha\to 0} \mathbf{C}^{\alpha} = \begin{pmatrix} \mathbf{0} & \mathbf{0} \\ \mathbf{C}_{21} & \mathbf{I}_2 \end{pmatrix}$$

is called the co-balayage matrix. Note that $\mathbf{C}$ is idempotent and right-invariant:

$$\mathbf{CC} = \mathbf{C}, \qquad \mathbf{C}(t)\mathbf{C} = \mathbf{C}, \qquad \mathbf{CC}(t) = \mathbf{C}(t).$$

Next, introduce the matrix

$$\mathbf{C}^c(t) = \mathbf{C} - \mathbf{C}(t),$$

whose the only non-vanishing component is $\mathbf{C}_{21}^c(t) = (N_{ij}^c(t))$. The basic relation (dual to that of Thm. 1, § 3.1.2) is now

$$\mathbf{C}^c(t) = \mathbf{C}\,{}_H\mathbf{P}(t), \qquad t \geqslant 0.$$

Hence, other matrix relations can be obtained directly. For example, Theorem 2, § 3.4.1, becomes

$$\mathbf{C}(t+s) - \mathbf{C}(t) = \mathbf{C}(s)\,{}_H\mathbf{P}(t),$$
$$\mathbf{C}^c(t+s) = \mathbf{C}^c(s)\,{}_H\mathbf{P}(t), \qquad t, s \geqslant 0,$$

whose only non-zero components are, respectively,

$$\mathbf{C}_{21}(t+s) = \mathbf{C}_{21}(t) + \mathbf{C}_{21}(s)\,{}_H\mathbf{P}_{11}(t),$$
$$\mathbf{C}_{21}^c(t+s) = \mathbf{C}_{21}^c(s)\,{}_H\mathbf{P}_{11}(t).$$

Similarly, Kolmogorov equations (Thm. 3, § 3.4.1) become

$$\frac{d\mathbf{C}_{21}(t)}{dt} = \mathbf{Q}_{21\,H}\mathbf{P}_{11}(t), \tag{B}$$
$$\frac{d\mathbf{C}_{21}^c(t)}{dt} = \mathbf{C}_{21}^c(t)\mathbf{Q}_{11}, \tag{F}$$

or, in terms of transforms,

$$\mathbf{C}_{21}^\alpha = \mathbf{Q}_{21\,H}\mathbf{U}_{11}^\alpha, \tag{B}$$
$$\mathbf{C}_{21}^\alpha(\mathbf{Q}_{11} - \alpha\mathbf{I}_1) = \mathbf{C}_{21}\mathbf{Q}_{11}. \tag{F}$$

Comparison of two expressions for $d\mathbf{C}_{21}(t)/dt$ at $t = 0$, obtained from the (B)- and (F)-equations, yields

$$\mathbf{Q}_{21} + \mathbf{C}_{21}\mathbf{Q}_{11} = \mathbf{0} \qquad \text{(Thm. 4, § 3.4.1)}.$$

Using this, the (F)-equation may be written as

$$\frac{d\mathbf{C}_{21}(t)}{dt} = \mathbf{Q}_{21} + \mathbf{C}_{21}(t)\mathbf{Q}_{11}$$

or, in the transformed form,

$$\mathbf{Q}_{21} + \mathbf{C}_{21}^\alpha\mathbf{Q}_{11} = \alpha\mathbf{C}_{21}^\alpha \qquad \text{(Thm. 1, § 3.4.2)}.$$

Finally, the matrix form of Theorem 2, § 3.4.2, is

$$\mathbf{C}_{21}^\alpha\mathbf{Q}_{12} = \mathbf{Q}_{21}\mathbf{B}_{12}^\alpha.$$

Hence, recalling the matrix $(-\mathbf{Q})\mathbf{B}$ from § 3.2.5, one has the duality relation

$$\mathbf{QB} = \mathbf{CQ},$$

whose the only non-zero term is that with index 22.

b) Consider now the matrix $\mathbf{W}(t)$, $t \geqslant 0$, defined by

$$\mathbf{W}(t) = \mathbf{C}(t) + {}_H\mathbf{P}(t) = \begin{pmatrix} {}_H\mathbf{P}_{11}(t) & \mathbf{0} \\ \mathbf{C}_{21}(t) & \mathbf{I}_2 \end{pmatrix}.$$

It should be noted that $\mathbf{W}(t)$ is a standard matrix with the semi-group property

$$\mathbf{W}(t+s) = \mathbf{W}(t)\mathbf{W}(s), \qquad t, s \geqslant 0,$$

with $\mathbf{W}(0) = \mathbf{I}$, but it is not a stochastic matrix, because sums along its rows equal $D_i^c(t)$ for $i \in H^c$, and $N_i(t) + 1$ for $i \in H$.

Direct differentiation yields the equations, dual to those for $\mathbf{S}(t)$ in § 3.1.5,

$$\frac{d\mathbf{W}(t)}{dt} = \begin{cases} \mathbf{Q}^d\mathbf{W}(t), & \text{(B)} \\ \mathbf{W}(t)\mathbf{Q}^d, & \text{(F)} \end{cases}$$

where in the (F)-equation the identification of derivative at $t = 0$ has been made, and where the (non-conservative) intensity matrix has the form

$$\mathbf{Q}^d = \begin{pmatrix} \mathbf{Q}_{11} & \mathbf{0} \\ \mathbf{Q}_{21} & \mathbf{0} \end{pmatrix}.$$

Finally, observe that $\mathbf{C}$ is invariant relative to $\mathbf{W}(t)$:

$$\mathbf{W}(t)\mathbf{C} = \mathbf{C}\mathbf{W}(t) = \mathbf{C}, \qquad \mathbf{C}\mathbf{Q}^d = \mathbf{Q}^d\mathbf{C} = \mathbf{0}.$$

c) Introduce now a new matrix

$$\mathbf{K}^\alpha = (\mathbf{\Theta} + \alpha\mathbf{I})^{-1}\mathbf{C}^\alpha(\mathbf{\Theta} + \alpha\mathbf{I}), \qquad \alpha > 0.$$

Tedious verification from the (B)- and (F)-equations (§ 3.4.1) and the Feller relation (§ 2.2.4) give

THEOREM 1. *The matrix* $\mathbf{K}^\alpha$ *has the form*

$$\mathbf{K}^\alpha = \begin{pmatrix} \mathbf{0} & \mathbf{0} \\ \mathbf{K}_{21}^\alpha & \mathbf{I}_2 \end{pmatrix},$$

where

$$\mathbf{K}_{21}^\alpha = \mathbf{R}_{21}^\alpha\, {}_H\mathbf{G}_{11}^\alpha = \mathbf{R}_{21}^\alpha + \mathbf{K}_{21}^\alpha\mathbf{R}_{11}^\alpha.$$

Evidently, $\mathbf{K}^\alpha = (k_{ij}^\alpha)$ and Theorem 1 is the matrix form of relation for k_{ij}^α stated in § 3.4.2. Indeed, with the jump α-matrix

$$\mathbf{R}^{d\alpha} = \begin{pmatrix} \mathbf{R}_{11}^\alpha & \mathbf{0} \\ \mathbf{R}_{21}^\alpha & \mathbf{I}_2 \end{pmatrix}$$

corresponding to the generator $\mathbf{Q}^d$, one can write Theorem 1 in the form

$$\mathbf{K}^\alpha \mathbf{R}^{d\alpha} = \mathbf{K}^\alpha, \qquad \alpha \geqslant 0. \tag{PT}$$

This is dual to the expression for $\mathbf{B}^\alpha$ stated in § 3.1.5. The matrix $\mathbf{K}^\alpha$ is called the α-co-balayage matrix for the jump chain, and expression (PT) is the special form of the dual Product Theorem (see § 8.1.1).

The limit

$$\mathbf{K} = \lim_{\alpha \to 0} \mathbf{K}^\alpha = \mathbf{\Theta}^{-1}\mathbf{C}\mathbf{\Theta}$$

is called the co-balayage matrix for the jump chain; its non-trivial component is

$$\mathbf{K}_{21} = \mathbf{\Theta}_2^{-1}\, \mathbf{C}_{21}\mathbf{\Theta}_1 = \mathbf{R}_{21\,H}\mathbf{G}_{11} = {}_H\mathbf{U}_{21}^\rho \mathbf{\Theta}_1.$$

It is important to stress that $\mathbf{K}$ and $\mathbf{C}$ are different, and refer to the co-balayage of the jump chain and of the minimal chain, respectively. In contrast, the balayage for both chains is the same, namely $\mathbf{B}$.

d) The components $\mathbf{K}_{22}^\alpha$ and $\mathbf{C}_{22}^\alpha$ of matrices $\mathbf{K}^\alpha$ and $\mathbf{C}^\alpha$, are both equal to $\mathbf{I}_2$. It is possible to extend the notion of α-co-balayage in the same fashion as $\mathbf{B}^\alpha$ has been extended to $\mathbf{B}^{\rho\alpha}$. Indeed, the situation is dual to that discussed in § 3.3.5. Proceeding formally, let $\mathbf{R}^\alpha$ be the full grown α-jump matrix (§ 1.1.6) and consider its dual

$$\widehat{\mathbf{R}}^\alpha = (\mathbf{\Theta} + \alpha\mathbf{I})\mathbf{R}^\alpha(\mathbf{\Theta} + \alpha\mathbf{I})^{-1}.$$

Using expression for $\mathbf{R}^\alpha$ (from § 1.1.6), one has $\widehat{\mathbf{R}}^\alpha - \mathbf{I} = \mathbf{Q}^\alpha\,(\mathbf{\Theta}^\alpha)^{-1}$, so elements of $\widehat{\mathbf{R}}^\alpha$ are

$$\hat{r}_{ij}^\alpha = q_{ij}\big(q_j + \alpha\big)^{-1} \qquad \text{for} \quad i \neq j,$$
$$\hat{r}_{ii}^\alpha = 0.$$

Define now

$$\mathbf{C}^{\rho\alpha} = \mathbf{C}^\alpha \widehat{\mathbf{R}}^\alpha, \qquad \alpha \geqslant 0. \tag{PT}$$

Then it is easy to see that

$$\mathbf{K}^{\rho\alpha} = \mathbf{K}^\alpha \mathbf{R}^\alpha, \qquad \alpha \geqslant 0, \tag{PT}$$

where $\mathbf{K}^{\rho\alpha}$ is the dual of $\mathbf{C}^{\rho\alpha}$:

$$\mathbf{K}^{\rho\alpha} = \big(\mathbf{\Theta} + \alpha\mathbf{I}\big)^{-1}\mathbf{C}^{\rho\alpha}\big(\mathbf{\Theta} + \alpha\,\mathbf{I}\big).$$

The above expressions (PT) are duals to $\mathbf{B}^{\rho\alpha} = \mathbf{R}^\alpha\, \mathbf{B}^\alpha$, and represent dual form of the Product Theorem (see § 8.1.1). An easy matrix algebra verifies that non-zero components of $\mathbf{K}^{\rho\alpha}$ and $\mathbf{C}^{\rho\alpha}$ are, respectively,

$$\mathbf{K}_{21}^{\rho\alpha} = \mathbf{K}_{21}^\alpha, \qquad \mathbf{K}_{22}^{\rho\alpha} = \mathbf{B}_{22}^{\rho\alpha},$$
$$\mathbf{C}_{21}^{\rho\alpha} = \mathbf{C}_{21}^\alpha, \qquad \mathbf{C}_{22}^{\rho\alpha} = \big(\mathbf{\Theta}_2 + \alpha\,\mathbf{I}_2\big)\mathbf{B}_{22}^{\rho\alpha}\big(\mathbf{\Theta}_2 + \alpha\mathbf{I}_2\big)^{-1}.$$

[illegible] to the generator $Q^{\dagger}$ one can write [illegible] in the form

$$K^{\dagger}R^{ab} = [illegible] \qquad \text{(FT)}$$

[illegible] to the expression for [illegible]. The [illegible] matrix for the [illegible] (FT) [illegible]

[illegible]

[illegible] the Baxter-like matrix for the [illegible] has non-trivial components

[illegible]

[illegible]

[illegible]

[illegible]

[illegible]

where [illegible] is the dual [illegible]

[illegible]

The above [illegible] (FT) [illegible] and [illegible] represent [illegible] (see § 8.1.2). [illegible] components [illegible] respectively

[illegible]

SECTION 4

LAST EXIT TIMES

This section presents further results of fundamental importance for this book. Discussion concerns properties of the last exit times from a fixed set H, and their distributions. Only analytic aspects are considered here, although the underlying behavior of sample functions is indicated. Again the importance of taboo probabilities and the role of the (B)- and (F)-equations are stressed. However, discussion is now more delicate because the concept of the last exit time is more difficult to handle than its counterpart the first entrance time.

Sub-section 4.1 treats the last exit time L and the last exit positions $X(L)$ and $X(L-)$. Sub-section 4.2 discusses the last exit time L_s before an instant s, and the last exit positions $X(L_s)$ and $X(L_s-)$. Although $L_s \to L$ as $s \to \infty$, it is preferable for clarity to treat both cases separately. In both sub-sections expressions for the distribution of the last exit time and its joint distribution with exit positions are obtained in terms of taboo probabilities. Distinction between expressions involving $X(L)$ and $X(L-)$, as well as $X(L_s)$ and $X(L_s-)$, is carefully observed, leading to a parallel treatment of both cases. Furthermore, duality between L_s and T_t is examined; also connection with the Poisson representation ("Chung equation") is discussed.

The approach is analytic and the "density argument" (path analysis) elucidates the formal presentation. The alternative approach through functionals is discussed later (in Sub-section 8.3). Again, matrices provide neat summary of results.

The material in this section is a modification of results scattered in literature. Recent work on Markov processes made the last exit times a useful and powerful tool. Its importance for Markov chains was demonstrated by K. L. Chung [1967], and further advantages were pointed out by R. K. Getoor and M. J. Sharpe [1973 a,b], R. K. Getoor [1979], and others.

4.1. LAST EXIT TIME

§ **4.1.1. Distribution function.** a) Let L be the last exit time from a set H for the usual M.C. $X = (X_t)$, as defined in § 2.1.4. It is of great importance that its distribution function $\Gamma_i(\cdot)$, conditional on the initial state i, can be expressed in terms of the first entrance probability, discussed in Sub-section 3.1.

THEOREM 1. *The distribution function of L is given for $i \in \mathbb{I}$ and $t \geqslant 0$ by*

$$\Gamma_i(t) = \sum_j p_{ij}(t) D_j^c,$$

with $\Gamma_i(0) = D_i^c$ *and* $\Gamma_i \leqslant 1$.

Proof. The result follows immediately from relations stated in § 2.1.4:

$$\mathbb{P}^i(L \leqslant t) = \sum_j \mathbb{P}^i(X_t = j,\ X_u \in H^c,\ t < u < \infty),$$
$$\mathbb{P}^i(L = 0) = \mathbb{P}^i(T = \infty).$$

Note that $\Gamma_i(t)$ has a discontinuity at $t = 0$, when $i \in H^c$. □

The duality relation (Lemma 6, § 2.1.4) indicates that

$$\Gamma_i(t) = D_i^{tc}$$

(in notation of § 3.2.1, but t is now a variable of a distribution function, and not a parameter). Hence, results of § 3.2.1 apply here without change, and one has in particular from Theorem 9, § 3.2.1, the following theorem.

THEOREM 2. *For* $t, s \geqslant 0$ *and* $i \in \mathbb{I}$:

$$\Gamma_i(t + s) = \sum_j p_{ij}(t)\Gamma_j(s)$$

where $\Gamma_i(\cdot)$ *may be replaced by* $\Gamma_i^c(\cdot)$.

The above theorem indicates that both $\Gamma(\cdot)$ and $\Gamma^c(\cdot)$ are exit laws relative to a stochastic matrix $\mathbf{P}(t)$. Therefore, one obtains the (B)-equation.

THEOREM 3. *For* $i \in \mathbb{I}$ *and* $t > 0$:

$$\frac{d\Gamma_i(t)}{dt} = \mathbf{Q}\Gamma_i(t)$$

with $\Gamma_i(0) = D_i^c$, *and the right-side derivative at* $t = 0$. *Here* $\Gamma_i(t)$ *may be replaced by* $\Gamma_i^c(t)$.

Again, the (F)-equation yields the explicit form of the distribution of L.

THEOREM 4. *For* $i \in \mathbb{I}$ *and* $t \geqslant 0$:

$$\Gamma_i(t) = D_i^c + \int_0^t \gamma_i(s)\, ds,$$

where

$$\gamma_i(t) = \sum_k p_{ik}(t) b_k.$$

Here b_i is the escape rate (vanishing on H^c). It is instructive to compare expressions for the density $\gamma_i(t)$ of the absolutely continuous component, obtained from Theorems 3 and 4. In particular, note that

$$\gamma_i(0) = \mathbf{Q}D_i^c = b_i.$$

Taking L.S. transforms, with γ_i^α denoting the transform of $\Gamma_i(t)$, one has

THEOREM 5. *For $i \in \mathbb{I}$ and $\alpha > 0$:*

$$\gamma_i^\alpha = \alpha \mathbf{U}^\alpha D_i^c,$$

$$(\alpha \mathbf{I} - \mathbf{Q})\gamma_i^\alpha = \alpha D_i^c, \tag{B}$$

$$\gamma_i^\alpha = D_i^c + \mathbf{U}^\alpha b_i. \tag{F}$$

Consider now the distribution of L restricted to positive values

$$\Pi_i(t) = \mathbb{P}^i(0 < L \leqslant t), \qquad t > 0.$$

Hence, by Theorem 4,

$$\Pi_i(t) = \int_0^t \gamma_i(s)\, ds, \qquad t > 0$$

and the following theorem is true.

THEOREM 6. *For $t, s > 0$ and $i \in \mathbb{I}$:*

$$\Pi_i(t+s) - \Pi_i(t) = \sum_j p_{ij}(t)\Pi_j(s).$$

This result should be compared with Theorem 2. Denoting by π_i^α the L.S. transform of $\Pi_i(t)$, it is clear that

$$\pi_i^\alpha = \mathbf{U}^\alpha b_i, \qquad i \in \mathbb{I}.$$

Finally, Corollary 11, § 3.2.1, yields

THEOREM 7. *For $i \in \mathbb{I}$ and $t > 0$:*

$$\Pi_i(t) = D_i - D_i^t.$$

b) It follows from Theorem 2 that the limit (Γ_i),

$$\Gamma_i = \lim_{t \to \infty} \sum_j p_{ij}(t) D_j^c,$$

is $\mathbf{P}(t)$-invariant; hence (Γ_i^c) is also $\mathbf{P}(t)$-invariant. On the other hand, Theorem 6 indicates that (Π_i) is purely excessive:

$$\Pi_i - \Pi_i(t) = \mathbf{P}(t)\Pi_i \longrightarrow 0 \qquad \text{as} \quad t \to \infty.$$

The next theorem indicates that (Π_i) is actually a potential of a charge. Indeed, for finite potential $\mathbf{U}$, the (F)-equation yields

THEOREM 8. *For* $i \in \mathbb{I}$:

$$\Gamma_i = D_i^c + \Pi_i, \qquad \text{where} \quad \Pi_i = \mathbf{U}b_i.$$

Note that this corresponds to the decomposition

$$\mathbb{P}^i(L < \infty) = \mathbb{P}^i(L = 0) + \mathbb{P}^i(0 < L < \infty),$$

which is fundamental for the Poisson equation (see § 3.3.3 and Sub-section 5.2).

Observe, however, that if $D_i = 1$ for all i, then Γ_i and b_i vanish identically. Thus, Γ_i^c equals 1 identically, so L is infinite with probability one. This is, for example, the case when set H^c is finite, because then $D_i^c = 0$ for all i.

The following theorem provides an important interpretation (here $\mathbf{B}$ is the balayage matrix).

THEOREM 9. *For* $i \in \mathbb{I}$:

$$\mathbf{B}\Gamma_i = \Pi_i, \qquad \mathbf{B}\Gamma_i^c = \Gamma_i^c, \qquad \mathbf{B}\Pi_i = \Pi_i.$$

Proof.

$$\Pi_i = \mathbb{P}^i(0 < L < \infty) = \sum_{k \in H} \mathbb{P}^i(X_T = k,\ T < \infty,\ L \circ \theta_T < \infty)$$
$$= \sum_{k \in H} D_{ik}\Gamma_k = \mathbf{B}\Gamma_i,$$

where use has been made of relation $(T < \infty) = (L > 0)$ and Lemma 6, § 2.1.4, applied to T, so $L \circ \theta_T = L - T > 0$. The second and the third relations follow from Theorem 8 and the fact that $\mathbf{B}D = D$. □

Using definition of the mean from § 2.1.5, one finds that

$$\mathbb{E}^i(L,\ L < \infty) = \int_0^\infty (\Gamma_i^c(t) - \Gamma_i^c)\, dt = \sum_j u_{ij}\Pi_j, \qquad i \in \mathbb{I}.$$

Evidently, when $\Gamma_i^c = 0$ for all i, then Π_j in the above expression can be replaced by D_j.

§ 4.1.2. Joint distributions. a) The joint distribution $\Gamma_{ij}(t)$ of L and $X(L)$ has been introduced in § 2.1.5 for $i \in \mathbb{I}$, $j \in H^c$, and $t \geqslant 0$. Recall that $X(L)$ takes values in H^c for $0 < L < \infty$. For $L = 0$, the M.C. cannot start in H, so

$$\Gamma_{ij}(0) = \delta_{ij} D_j^c.$$

However, it is more natural to consider the joint distribution $\Pi_{ij}(t)$ of L and $X(L-)$, for $i \in \mathbb{I}$, $j \in H$, and $t > 0$. In both cases, discussion presented here is the extension of that from § 4.1.1. As before, the case of $X(L)$ can be treated in two ways. Recall that $\Gamma_{ij}^c(t) = \Gamma_{ij} - \Gamma_{ij}(t)$ for $t \geqslant 0$.

LEMMA 1. *For $i \in \mathbb{I}$, $j \in H^c$, and $t \geqslant 0$:*

$$\Gamma^c_{ij}(t) = \sum_k p_{ik}(t)\,\Gamma^c_{kj}(0).$$

Proof.

$$\begin{aligned}\Gamma^c_{ij}(t) &= \sum_k \mathbb{P}^i\big(X_t = k,\ X(L) = j,\ t < L < \infty\big)\\ &= \sum_k \mathbb{P}^i\big(X_t = k,\ X(t + L\circ\theta_t) = j,\ t < t + L\circ\theta_t < \infty\big)\\ &= \sum_k p_{ik}(t)\,\mathbb{P}\big(X(L)\circ\theta_t = j,\ 0 < L\circ\theta_t < \infty \mid X_t = k\big)\\ &= \sum_k p_{ik}(t)\mathbb{P}^k\big(X(L) = j,\ 0 < L < \infty\big),\end{aligned}$$

where use has been made of Lemma 2, § 2.1.4. □

Lemma 1 yields immediately the following basic relations.

THEOREM 2. *For $i \in \mathbb{I}$, $j \in H^c$, and $t, s \geqslant 0$:*

$$\Gamma_{ij}(t+s) - \Gamma_{ij}(t) = \sum_k p_{ik}(t)\big(\Gamma_{kj}(s) - \Gamma_{kj}(0)\big),$$

$$\Gamma^c_{ij}(t+s) = \sum_k p_{ik}(t)\,\Gamma^c_{kj}(s).$$

THEOREM 3. *For $i \in \mathbb{I}$, $j \in H^c$, and $t > 0$, with right-side derivative at $t = 0$,*

$$\frac{d\Gamma^c_{ij}(t)}{dt} = \sum_k q_{ik}\Gamma^c_{kj}(t), \tag{B}$$

with the initial condition $\Gamma^c_{ij}(0) = \Gamma_{ij} - \Gamma_{ij}(0)$.

Furthermore, Lemma 1 indicates that $\Gamma^c_{\bullet j}(0)$, for fixed j, is purely excessive relative to $\mathbf{P}(t)$:

$$\Gamma^c_{ij}(0) \geqslant \mathbf{P}(t)\,\Gamma^c_{ij}(0) \longrightarrow 0 \qquad \text{as} \quad t \to \infty.$$

Hence, for fixed $j \in H^c$, there exists a unique exit law $\gamma_{\bullet j}(\cdots)$, such that

$$\gamma_{ij}(t+s) = \sum_k p_{ik}(t)\,\gamma_{kj}(s), \qquad t \geqslant 0, \quad s > 0.$$

Preservation of equality in the above expression when $s \to 0$, is assured by validity of the (F)-equation (see § 3.3.2 for the similar argument). Nevertheless, these considerations do not determine $\gamma_{ij}(0)$, so Γ_{ij} is also not determined. Therefore, an independent argument is needed to determine $\gamma_{ij}(t)$ explicitly. It is shown below in b) that

$$\gamma_{ij}(0) = \begin{cases} q_{ij}D^c_j, & i \in H,\\ 0, & i \in H^c,\end{cases}$$

so

$$\gamma_{ij}(t) = \sum_{k\in H} p_{ik}(t) q_{kj} D^c_j, \qquad i \in \mathbb{I}, \quad j \in H^c, \quad t \geqslant 0.$$

This leads to the (F)-equation.

THEOREM 4. *Under conditions of validity of the* (F)*-equation*:

$$\Gamma_{ij}(t) - \Gamma_{ij}(0) = \int_0^t \gamma_{ij}(s)\,ds \tag{F}$$

for $i \in \mathbb{I}$, $j \in H^c$, *and* $t \geqslant 0$.

Taking L.S. transforms, with γ_{ij}^α denoting the transform of $\Gamma_{ij}(t)$, one has

THEOREM 5. *For* $i \in \mathbb{I}$, $j \in H^c$, *and* $\alpha > 0$:

$$\gamma_{ij}^\alpha - \Gamma_{ij} = \alpha\Big(u_{ij}^\alpha D_j^c - \sum_k u_{ik}^\alpha \Gamma_{kj}\Big),$$

$$\alpha\big(\gamma_{ij}^\alpha - \Gamma_{ij}(0)\big) = \sum_k q_{ik}(\gamma_{kj}^\alpha - \Gamma_{kj}), \tag{B}$$

$$\gamma_{ij}^\alpha = \Gamma_{ij}(0) + \sum_{k\in H} u_{ik}^\alpha q_{kj} D_j^c. \tag{F}$$

Evidently, addition over all $j \in H^c$ reduces the above results to the corresponding expressions in § 4.1.1.

b) The density argument, essentially related to the (F)-equation, must be used to obtain the explicit expressions for $\Gamma_{ij}(t)$, $\Pi_{ij}(t)$, and related quantities. Thus, Corollary 6, § 2.2.5, yields immediately the following two basic results.

THEOREM 6. *For* $i \in \mathbb{I}$, $r \in H$, $k \in H^c$, *and* $0 < t < \infty$:

$$\mathbb{P}^i\big(X(L-) = r,\ X(L) = k,\ 0 < L \leqslant t\big) = \int_0^t p_{ir}(\sigma) q_{rk}\, D_k^c\, d\sigma,$$

and for $u_{ir} < \infty$:

$$\mathbb{P}^i\big(X(L-) = r,\ X(L) = k,\ 0 < L < \infty\big) = u_{ir}\, q_{rk}\, D_k^c.$$

THEOREM 7. *For* $i \in \mathbb{I}$, $r \in H$, $k, j \in H^c$, *and* $0 < t < \infty$:

$$\mathbb{P}^i\big(X(L) = k,\ 0 < L \leqslant t,\ X_t = j\big) = \int_0^t c_{ik}(\sigma)\, {}_H p_{kj}(t-\sigma) D_j^c\, d\sigma,$$

$$\mathbb{P}^i\big(X(L-) = r,\ 0 < L \leqslant t,\ X_t = j\big) = \int_0^t p_{ir}(\sigma) n_{rj}(t-\sigma) D_j^c\, d\sigma,$$

where functions $c_{ik}(t)$ *and* $n_{rj}(t)$ *have been defined in* §§ 2.2.6 *and* 3.4.1, *respectively.*

Addition over all j in Theorem 7 yields the following useful results, in view of Corollary 5, § 3.1.1, and Theorem 6, § 3.4.1, respectively.

COROLLARY 8. *For $i \in \mathbb{I}$, $k \in H^c$, and $t > 0$:*

$$\mathbb{P}^i\big(X(L) = k,\ 0 < L \leqslant t\big) = \int_0^t c_{ik}(\sigma) D_k^c \, d\sigma,$$

$$\mathbb{P}^i\big(X(L) = k,\ 0 < L < \infty\big) = C_{ik}\, D_k^c.$$

COROLLARY 9. *For $i \in \mathbb{I}$, $r \in H$, and $t > 0$:*

$$\mathbb{P}^i\big(X(L-) = r,\ 0 < L \leqslant t\big) = \int_0^t p_{ik}(\sigma) b_r \, d\sigma,$$

$$\mathbb{P}^i\big(X(L-) = r,\ 0 < L < \infty\big) = u_{ir} b_r,$$

where b_r is the escape rate (see § 3.1.3).

Hence, in agreement with Theorem 8, § 4.1.1,

$$\mathbb{P}^i(0 < L < \infty) = \mathbf{U} b_i = \Pi_i, \qquad i \in \mathbb{I}.$$

Furthermore, the explicit expressions for $\Gamma_{ik}(t) - \Gamma_{ik}(0)$ and $\Pi_{ik}(t)$ are direct consequences of Theorem 6. Indeed, Corollary 8 is just Theorem 4 repeated here for the purpose of comparison. The density takes now the form

$$\gamma_{ik}(t) = c_{ik}(t) D_k^c,$$

which gives the value of $\gamma_{ik}(0)$ used above in a).

Similarly one can obtain expressions for $\Pi_i(t) = \Gamma_i(t) - \Gamma_i(0)$.

Remark. Recall that X_L has been defined only for $L < \infty$, and takes values in H^c only (even when $L = 0$). When $L = \infty$, adjoin an isolated state Δ and write

$$\mathbb{P}^i(X_L \in H^c) = \mathbb{P}^i(X_L \in H^c,\ L < \infty) = \Gamma_i,$$

$$\mathbb{P}^i(X_L = \Delta) = \mathbb{P}^i(X_L = \Delta,\ L = \infty) = \Gamma_i^c.$$

Analogous discussion may be carried out for $X(L-)$ with values in H.

Remark. It is easy to verify that the jump chain will visit set H infinitely often with probability Γ_i^c, when $L = \infty$.

§ 4.1.3. Comments. a) Equating the expression for $\gamma_{ij}(0)$ from the (B)-equation in Theorem 3, § 4.1.2, with that from the (F)-equation in Theorem 4, § 4.1.2, one obtains the basic equation satisfied by the last exit distribution (Γ_{ij}).

THEOREM 1. *For $i \in \mathbb{I}$, $j \in H^c$:*

$$\sum_k q_{ik}\Gamma_{kj} = \begin{cases} q_{ij} D_j^c, & i \in H^c, \\ 0, & i \in H. \end{cases}$$

The last exit distribution (Γ_{ij}) is of importance in studies of passage times. It has been defined with $L < \infty$ for all i and j, but it vanishes for $j \in H$. The explicit expression for Γ_{ij} may be obtained from Theorem 4, § 4.1.2. The discussion in § 4.1.2-b provides the natural interpretation; see also § 4.1.5. Observe that addition over all $j \in H^c$ in Theorem 1 yields $\mathbf{Q}\Gamma_i = 0$, confirming that (Γ_i) is invariant.

Furthermore, Lemma 1, § 4.1.2, may be written as

$$\Gamma_{ij} - \sum_k p_{ik}(t)\Gamma_{kj} = \Gamma_{ij}(t) - p_{ij}(t)D_j^c.$$

Hence, addition over all $j \in H^c$ would give the expression for $\Gamma_i(t)$ from Theorem 1, § 4.1.1, if and only if (Γ_i) is invariant. However, it does not follow that $\Gamma_{\bullet j}$, for fixed j, is invariant nor that $\Gamma_{ij}(t)$ equals $p_{ij}(t)D_j^c$; this would contradict Theorem 1 above. Indeed,

$$\mathbb{P}^i(X_t = j,\ L \leqslant t) = p_{ij}(t)D_j^c = \mathbb{P}^i(X_t = j,\ T_t = \infty), \qquad j \in H^c,$$

(see Thm. 2 below).

Observe also that the (B)-equation from Theorem 3, § 4.1.2, yields the following expression for the density

$$\gamma_{ij}(t) = \sum_k q_{ik}\Gamma_{kj}(t) - \sum_k q_{ik}\Gamma_{kj}, \qquad t \geqslant 0,$$

which agrees with that found from the (F)-equation.

b) The following is the basic decomposition theorem.

THEOREM 2. *For $i \in \mathbb{I}$, $j \in H^c$, and $t > 0$:*

$$\mathbb{P}^i(0 < L \leqslant t,\ X_t = j) = p_{ij}(t)D_j^c - {}_Hp_{ij}(t)D_j^c$$
$$= \int_0^t \sum_{k \in H^c} c_{ik}(\sigma)\,{}_Hp_{kj}(t-\sigma)D_j^c\,d\sigma = \int_0^t \sum_{r \in H} p_{ir}(\sigma)n_{rj}(t-\sigma)D_j^c\,d\sigma.$$

Proof. From proof of Theorem 1, § 4.1.1, one has

$$\mathbb{P}^i(X_t = j,\ 0 \leqslant L \leqslant t) = p_{ij}(t)D_j^c, \qquad \mathbb{P}^i(X_t = j,\ L = 0) = {}_Hp_{ij}(t)D_j^c,$$

and the first assertion follows. The integral expressions follow immediately from Theorem 7, § 4.1.2. Alternatively, the result follows from tedious verification using properties of $n_{ij}(t)$. □

Theorem 2 may be re-written in the form exhibiting the last exit decomposition:

$$p_{ij}(t) = {}_Hp_{ij}(t)D_j^c + \mathbb{P}^i(X_t = j,\ 0 < L \leqslant t) + p_{ij}(t)D_j.$$

§ 4.1.4. Examples. These examples are of some interest. More elaborate examples are given later.

Example 1. Let $H = \{k\}$, where k is a fixed state. Then, for $j \neq k$,

$$D_j^c = 1 - f_{jk}, \qquad c_{ij}(t) = p_{ik}(t) q_{kj},$$
$$\gamma_{ij}(t) = c_{ij}(t) D_j^c, \qquad b_k = q_k(1 - f_{kk})$$

(see Ex. 1, § 3.3.4). Hence, for $j \neq k$,

$$\Gamma_{ij}(t) = \delta_{ij}(1 - f_{jk}) + \int_0^t p_{ik}(s)\, q_{kj}(1 - f_{jk})\, ds,$$

$$\Pi_{ik}(t) = \int_0^t p_{ik}(s) b_k\, ds,$$

and

$$\Gamma_i(t) = f_{ik} + \int_0^t p_{ik}(s) b_k\, ds, \qquad \Gamma_i = D_i^c + u_{ik} b_k, \qquad i \neq k.$$

Example 2. Let $H^c = \{k\}$, where k is a fixed state. Then, for all $i \in \mathbb{I}$, $D_i^c = 0$ and $b_i = 0$. Hence, $\Gamma_i(t) = 0$, and $\Pi_i(t) = 0$ for all i. Indeed, Γ_i vanishes identically, because H^c is finite (see also Ex. 2, § 3.2.4).

§ 4.1.5. Duality. a) As already noted, the last exit time L and the delayed first entrance time T_t are dual (see § 4.1.1). This duality is extended to pairs $(L, X(L-))$ and $(T_t, X(T_t))$, because of definitions imposed here. It is of great interest to examine this fact in some detail. In the present situation $X(L-)$ and $X(T_t)$ take values in H and $t \leqslant T_t < L$ (see Fig. 4.1).

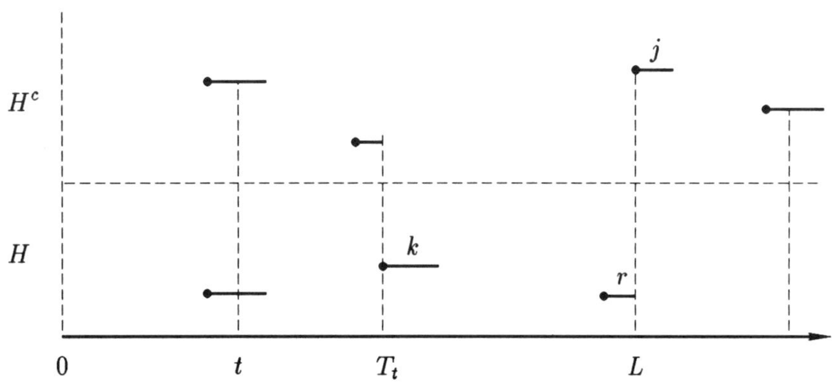

Figure 4.1. Duality.

Recall the definitions: for $i \in \mathbb{I}$, $j \in H$, and $t \geqslant 0$,

$$\mathbb{P}^i(X(T_t) = j,\ T_t < \infty) = D_{ij}^t, \qquad \mathbb{P}^i(X(L-) = j,\ t < L < \infty) = \Pi_{ij}^c(t),$$

where $\Pi_{ij}^c(t) = \Pi_{ij} - \Pi_{ij}(t)$ with $\Pi_{ij}^c(0) = \Pi_{ij}$, and $D_{ij}^0 = D_{ij}$. In particular, duality between D_{ij} and Π_{ij} is of special interest.

LEMMA 1. *For $i \in \mathbb{I}$, $j \in H$, and $t \geqslant 0$:*

$$\Pi_{ij}^c(t) = \sum_k p_{ik}(t)\Pi_{kj}.$$

Proof is the same as that of Lemma 1, § 4.1.2. From Lemma 1 the following basic relations are obtained in the usual way.

THEOREM 2. *For $i \in \mathbb{I}$, $j \in H$, and $t, s \geqslant 0$:*

$$\Pi_{ij}^c(t+s) = \sum_k p_{ik}(t)\Pi_{kj}^c(s).$$

THEOREM 3. *For $i \in \mathbb{I}$, $j \in H$, and $t > 0$, with right-side derivative at $t = 0$,*

$$\frac{d\Pi_{ij}^c(t)}{dt} = \sum_k q_{ik}\Pi_{kj}^c(t), \tag{B}$$

with the initial condition $\Pi_{ij}^c(0) = \Pi_{ij}$.

Observe that Theorems 2 and 3 are analogues of Theorems 1 and 2 in § 3.2.3.

By argument similar to that in § 4.1.2, for fixed j, $\Pi_{\cdot j}$ is purely excessive relative to $\mathbf{P}(t)$, and there exists a density $\pi_{ij}(t)$ which must take the form

$$\pi_{ij}(t) = p_{ij}(t)b_j,$$

provided the (F)-equation holds. Thus, one has Corollary 9, § 4.1.2, in the following form.

THEOREM 4. *Under conditions of validity of the* (F)*-equation,*

$$\Pi_{ij}^c(t) = \Pi_{ij} - \int_0^t p_{ij}(s)b_j\,ds, \tag{F}$$

for $i \in \mathbb{I}$, $j \in H$, and $t \geqslant 0$, with $\Pi_{ij}^c(t) \to 0$ as $t \to \infty$.

On the other hand, the density for D_{ij}^t has been found in Theorem 2, § 3.2.3, and one has

$$D_{ij}^t = D_{ij} - \int_0^t \sum_k p_{ik}(s)b_{kj}\,ds$$

for $i \in \mathbb{I}$, $j \in H$, and $t \geqslant 0$, with $D_{ij}^t \to D_{ij}^\infty$ as $t \to \infty$.

b) Comparing the expressions for $\pi_{ij}(0)$ in the (B)- and (F)-equations above, one obtains the basic equation satisfied by the last exit distribution (Π_{ij}).

THEOREM 5. *For* $i \in \mathbb{I}$, $j \in H$:

$$\sum_k q_{ik}\Pi_{kj} = \begin{cases} 0, & i \in H^c, \\ -\delta_{ij} b_j, & i \in H. \end{cases}$$

It is instructive to compare this with the corresponding equation for D_{ij} (Thm. 3, § 3.3.3). Adding over all $j \in H$, it follows that (Π_i) satisfies also the Poisson equation $\mathbf{Q}\Pi_i = -b_i$, confirming that (Π_i) is excessive.

§ 4.1.6. Matrices. a) Matrices provide additional insight into structure of theorems in this section. Write the matrix $\mathbf{\Lambda}(t) = (\Gamma_{ij}(t))$, $t \geqslant 0$, in the partitioned form,

$$\mathbf{\Lambda}(t) = \begin{pmatrix} \mathbf{\Lambda}_{11}(t) & \mathbf{0} \\ \mathbf{\Lambda}_{21}(t) & \mathbf{0} \end{pmatrix},$$

and $\mathbf{\Lambda}(0) = (\delta_{ij} D_j^c)$, $\mathbf{\Lambda} = (\Gamma_{ij})$, $\mathbf{\Lambda}^c(t) = \mathbf{\Lambda} - \mathbf{\Lambda}(t)$.

Theorem 2, § 4.1.2, is now

$$\mathbf{\Lambda}^c(t+s) = \mathbf{P}(t)\mathbf{\Lambda}^c(s), \qquad t, s \geqslant 0,$$

which for $s = 0$ yields the defining relation (Lemma 1, § 4.1.2)

$$\mathbf{\Lambda}^c(t) = \mathbf{P}(t)\,\mathbf{\Lambda}^c(0).$$

The formal differentiation yields for $t > 0$:

$$\frac{d\mathbf{\Lambda}^c(t)}{dt} = \mathbf{Q}\mathbf{\Lambda}^c(t) \qquad \text{(Thm. 3, § 4.1.2)}, \qquad \text{(B)}$$

$$\frac{d\mathbf{\Lambda}(t)}{dt} = \mathbf{P}(t)\mathbf{Q}(\mathbf{\Lambda}(0) - \mathbf{\Lambda}) \qquad \text{(Thm. 4, § 4.1.2)}, \qquad \text{(F)}$$

with Theorem 1, § 4.1.3, expressed as

$$\mathbf{Q}\mathbf{\Lambda} = (\mathbf{Q} - \mathbf{J}_2\mathbf{Q}\mathbf{J}_1)\mathbf{\Lambda}(0).$$

It should be noted that it is not true that $\mathbf{\Lambda}(t)$ equals $\mathbf{P}(t)\mathbf{\Lambda}(0)$. If this was true, then one would have $\mathbf{\Lambda} = \mathbf{P}(t)\mathbf{\Lambda}$ or $\mathbf{Q}\mathbf{\Lambda} = \mathbf{0}$, which is not possible.

b) Write the matrix $\mathbf{\Pi}(t) = (\Pi_{ij}(t))$, $t > 0$, in the partitioned form:

$$\mathbf{\Pi}(t) = \begin{pmatrix} \mathbf{0} & \mathbf{\Pi}_{12}(t) \\ \mathbf{0} & \mathbf{\Pi}_{22}(t) \end{pmatrix},$$

and $\mathbf{\Pi}(0) = \mathbf{0}$, $\mathbf{\Pi} = (\Pi_{ij})$, $\mathbf{\Pi}^c(t) = \mathbf{\Pi} - \mathbf{\Pi}(t)$.

The defining relation (Lemma 1, § 4.1.5) is

$$\mathbf{\Pi}^c(t) = \mathbf{P}(t)\mathbf{\Pi}.$$

The formal differentiation yields

$$\frac{d\mathbf{\Pi}^c(t)}{dt} = \mathbf{Q}\mathbf{\Pi}^c(t) \qquad \text{(Thm. 3, § 4.1.5)}, \tag{B}$$

$$\frac{d\mathbf{\Pi}(t)}{dt} = \mathbf{P}(t)(-\mathbf{Q})\mathbf{\Pi} \qquad \text{(Thm. 4, § 4.1.5)}, \tag{F}$$

where

$$(-\mathbf{Q})\mathbf{\Pi} = (\delta_{ij} b_j) \qquad \text{(Thm. 5, § 4.1.5)}.$$

Compare this with $(-\mathbf{Q})\mathbf{B} = (b_{ij})$ and the corresponding (F)-equation for $d\mathbf{B}^t/dt$ stated at the end of § 3.2.5. The analogies between matrices $\mathbf{B}$ and $\mathbf{\Pi}$ are now apparent.

4.2. LAST EXIT TIME BEFORE A FIXED INSTANT

§ 4.2.1. Distribution function. a) Let L_s be the last exit time from a set H before an instant s, for the usual M.C. $X = (X_t)$, as defined in § 2.1.4. It is of great importance that the distribution function $\Gamma_i^s(\cdot)$ of L_s, conditional on the initial state i, can be expressed in terms of the first entrance probability, discussed in Sub-section 3.1.

THEOREM 1. *For $i \in \mathbb{I}$ and $0 \leqslant t < s$, the distribution function of L_s is given by*

$$\Gamma_i^s(t) = \sum_{k \in H^c} p_{ik}(t) D_k^c(s-t)$$

and has two discontinuities at $t = 0$ and $t = s$ such that

$$\Gamma_i^s(0) = D_i^c(s), \qquad \Gamma_i^s \equiv \Gamma_i^s(s-) = \sum_{k \in H^c} p_{ik}(s), \qquad \Gamma_i^s(t) = 1 \quad \text{for} \quad t \geqslant s.$$

Proof. From Lemma 8, § 2.1.4,

$$\begin{aligned}\mathbb{P}^i(L_{s-} \leqslant t) &= \sum_{k \in H^c} \mathbb{P}^i(X_t = k,\ X_u \in H^c,\ t < u < s) \\ &= \sum_{k \in H^c} p_{ik}(t) D_k^c(s-t).\end{aligned}$$

Now, $L_{s-} = L_s$, except at a countable number of jump points (entrances to the set H), so events $(L_{s-} \leqslant t)$ and $(L_s \leqslant t)$ are $\mathbb{P}$-equivalent. It is clear that $\Gamma_i^s(t)$ is a continuous function of t, except at $t = 0$ and $t = s$. Moreover, by definition, $0 \leqslant L_s \leqslant s$. □

Comparison with Theorem 2, § 3.2.1, indicates that

$$\Gamma_i^s(t) = D_i^{tc}(s), \qquad t < s,$$

in agreement with the duality relation (Lemma 6, § 2.1.4). Hence, results of § 3.2.1 apply here without change. Thus, from Theorem 1 one has

THEOREM 2. *For* $0 \leqslant t+h < s$ *and* $i \in \mathbb{I}$:

$$\Gamma_i^s(t+h) = \sum_j p_{ij}(t)\,\Gamma_j^{s-t}(h),$$

where $\Gamma_i^s(\cdot)$ *may be replaced by* $\Gamma_i^{sc}(\cdot) = 1 - \Gamma_i^s(\cdot)$.

Consequently, the (B)-equation holds in the following form.

THEOREM 3. *For* $0 < t < s$ *and* $i \in \mathbb{I}$:

$$\frac{d\Gamma_i^s(t)}{dt} = \sum_k q_{ik}\,\Gamma_k^s(t) + \sum_{j \in H^c} p_{ij}(t)\,d_j(s-t), \tag{B}$$

with one sided derivatives at $t = 0$ *and* $t = s$.

In the same fashion, Theorem 2 yields in the limit (as $h \to t - s$)

THEOREM 4. *For* $i \in \mathbb{I}$ *and* $t < s$:

$$\Gamma_i^s = \sum_k p_{ik}(t)\Gamma_k^{s-t}.$$

Next, recall that for $0 < t < s$ and $i \in \mathbb{I}$:

$$\Pi_i^s(t) = \Gamma_i^s(t) - \Gamma_i^s(0).$$

Hence, Theorem 2 yields

THEOREM 5. *For* $0 < t+h < s$ *and* $i \in \mathbb{I}$:

$$\Pi_i^s(t+h) - \Pi_i^s(t) = \sum_j p_{ij}(t)\,\Pi_j^{s-t}(h).$$

Furthermore, with $\Pi_i^s = \Gamma_i^s - \Gamma_i^s(0)$, Theorem 5 yields in the limit (as $h \to t - s$)

THEOREM 6. *For* $i \in \mathbb{I}$ *and* $0 < t < s$:

$$\Pi_i^s - \Pi_i^s(t) = \sum_j p_{ij}(t)\Pi_j^{s-t}.$$

Finally observe that letting $s \to \infty$, the above results reduce to those in § 4.1.1.

b) Proceeding as in proof of Theorem 1, one obtains from Lemmas 7 and 8, § 2.1.4, the following expressions which provide additional insight into discussion in a).

LEMMA 7. *For* $i \in \mathbb{I}$, $k, j \in H^c$, *and* $t < s$:

$$\mathbb{P}^i(X_t = k,\ L_s = 0,\ X_s = j) = {}_Hp_{ik}(t)\,{}_Hp_{kj}(s-t),$$
$$\mathbb{P}^i(X_t = k,\ 0 \leqslant L_s \leqslant t,\ X_s = j) = p_{ik}(t)\,{}_Hp_{kj}(s-t),$$
$$\mathbb{P}^i(X_t = k,\ L_s = s,\ X_s = j) = 0.$$

PROPOSITION 8. *For $i \in \mathbb{I}$ and $j \in H^c$:*

$$\mathbb{P}^i(L_s = 0,\ X_s = j) = {}_Hp_{ij}(s),$$
$$\mathbb{P}^i(0 < L_s < s,\ X_s = j) = p_{ij}(s) - {}_Hp_{ij}(s),$$
$$\mathbb{P}^i(L_s = s,\ X_s = j) = 0.$$

Observe that the above proposition corresponds to the decomposition

$$\mathbb{P}^i(L_s = 0) + \mathbb{P}^i(0 < L_s < s) + \mathbb{P}^i(L_s = s) = 1$$

with

$$\mathbb{P}^i(L_s = 0) = \mathbb{P}^i(T > s) = D_i^c(s),$$
$$\mathbb{P}^i(L_s = s) = \mathbb{P}^i(T_s = s) = D_i^s(s),$$
$$\mathbb{P}^i(0 < L_s < s) = \sum_{j \in H^c} \left[p_{ij}(s) - {}_Hp_{ij}(s)\right].$$

c) It is shown in the next § 4.2.2, with the help of the density argument, essentially related to the (F)-equation, that the distribution function $\Gamma_i^s(\cdot)$ has an absolutely continuous component with density $\gamma_i^s(\cdot)$ defined by

$$\gamma_i^s(t) = \sum_{k \in H^c} c_{ik}(t) D_k^c(s-t) = \sum_{r \in H} p_{ir}(t) n_r(s-t), \qquad t < s,$$

with

$$\gamma_i^s(0) = \begin{cases} 0 & \text{when } i \in H^c, \\ n_i(s) & \text{when } i \in H. \end{cases}$$

Expressions $c_{ik}(t)$ and $n_r(t)$ have been defined in §§ 2.2.6 and 3.4.1, respectively. The following is the immediate corollary of Theorem 8, § 4.2.2.

THEOREM 9. *For $i \in \mathbb{I}$, $0 < t < s$:*

$$\Pi_i^s(t) = \mathbb{P}^i(0 < L_s \leqslant t) = \Gamma_i^s(t) - \Gamma_i^s(0) = \int_0^t \gamma_i^s(u)\, du.$$

It is also clear that for $i \in \mathbb{I}$, $0 < t + h < s$:

$$\gamma_i^s(t+h) = \sum_j p_{ij}(t)\, \gamma_j^{s-t}(h).$$

d) Letting $t \uparrow s$, one obtains several results of theoretical interest. The first one stated here is the analogue of the "Chung equation" from the boundary theory, which corresponds to the decomposition

$$\mathbb{P}^i(0 < L_s \leqslant s) = \mathbb{P}^i(L_s = s) + \mathbb{P}^i(0 < L_s < s).$$

THEOREM 10. *For $i \in \mathbb{I}$ and $s > 0$:*

$$D_i(s) = \Pi_i^s + \Gamma_i^{sc}.$$

Proof. This is just the re-statement of the relation $\Gamma_i^s - \Gamma_i^s(0) = \Pi_i^s$. Indeed, as noted earlier

$$\mathbb{P}^i(L_s = s) = \Gamma_i^{sc} = \sum_{r\in H} p_{ir}(s),$$

$$\mathbb{P}^i(0 < L_s < s) = \Pi_i^s = \int_0^s \gamma_i^s(\sigma)\, d\sigma. \qquad \square$$

Observe that Theorem 10 is actually a Poisson representation of $D_i(s)$ for finite s. Letting $s \to \infty$, one obtains Theorem 8, § 4.1.1.

Taking Laplace transform with respect to s, Theorem 10 assumes the following form.

COROLLARY 11. *For* $i \in \mathbb{I}$:

$$\sum_{r\in H} u_{ir}^\alpha (1 + n_r^\alpha) = d_i^\alpha / \alpha.$$

The proper Chung equation is obtained for $i \in H$, then $d_i^\alpha = 1$. Define a measure μ_r (for each $r \in H$) on the positive real line by

$$\mu_r(s, \infty] = n_r(s), \qquad 0 \leqslant s < \infty,$$

with $\mu_r[\infty] = b_r$. Then the Chung equation may be written as

$$\sum_{r\in H} u_{ir}^\alpha \left(\alpha + b_r + \int_{(0,\infty)} (1 - e^{-\alpha s})\, \mu_r(ds)\right) = 1, \qquad i \in H.$$

The second result of interest provides another illustration of duality between $D_i(s)$ and $N_i(s)$.

THEOREM 12. *For* $i \in \mathbb{I}$ *and* $s > 0$:

$$\sum_j q_{ij}\Pi_j^s + \sum_j q_{ij}\Gamma_j^{sc} = \begin{cases} d_i(s), & i \in H^c, \\ -n_i(s), & i \in H. \end{cases}$$

Proof. This is the immediate consequence of Theorem 10 and the fact that

$$\sum_{j\in H^c} q_{ij} D_j^c(s) = \begin{cases} -d_i(s) & \text{for} \quad i \in H^c \quad (\text{Thm. 6, § 3.1.1}), \\ n_i(s) & \text{for} \quad i \in H \quad (\text{§ 3.4.1}). \end{cases}$$

(Alternative longer proof may be obtained by differentiating Theorem 6 and equating the expression for $\gamma_i^s(0)$ with that stated above. The result then follows from Corollary 11.) $\square$

Letting $s \to \infty$, one obtains $\mathbf{Q}\Pi_i = -b_i$.

§ 4.2.2. Joint distributions. a) The joint distribution $\Gamma_{ij}^{s}(t)$ of L_s and $X(L_s)$ has been already introduced in § 2.1.5 for $i \in \mathbb{I}$, $j \in H^c$, and $0 \leqslant t < s$ (for fixed s). Recall that $X(L_s)$ takes values in H^c, provided $0 < L_s < s$. For $L_s = s$, the M.C. cannot start in H, so for $i \in \mathbb{I}$, $j \in H^c$:

$$\Gamma_{ij}^{s}(0) = \delta_{ij} D_j^c(s).$$

However, it is more natural to consider the joint distribution $\Pi_{ij}^{s}(t)$ of L_s and $X(L_s-)$, for $i \in \mathbb{I}$, $j \in H$, and $0 < t < s$.

As in previous discussions, the case of $X(L_s)$ can be treated in two ways. Recall from § 2.1.5 that $\Gamma_{ij}^{sc}(t) = \Gamma_{ij}^{s} - \Gamma_{ij}^{s}(t)$ for $0 \leqslant t < s$.

Lemma 1. *For $i \in \mathbb{I}$, $j \in H^c$, and $0 \leqslant t < s$:*

$$\Gamma_{ij}^{sc}(t) = \sum_k p_{ik}(t) \Gamma_{kj}^{(s-t)c}(0).$$

Proof.

$$\begin{aligned}
\Gamma_{ij}^{sc}(t) &= \mathbb{P}^i\left(X(L_s) = j,\ t < L_s < s\right) \\
&= \sum_k \mathbb{P}^i\left(X_t = k,\ X(L_{s-t}) \circ \theta_t = j,\ 0 < L_{s-t} \circ \theta_t < s - t\right) \\
&= \sum_k p_{ik}(t) \mathbb{P}^k\left(X(L_{s-t}) = j,\ 0 < L_{s-t} < s - t\right),
\end{aligned}$$

where use has been made of Lemma 4, § 2.1.4 (see Fig. 4.2). □

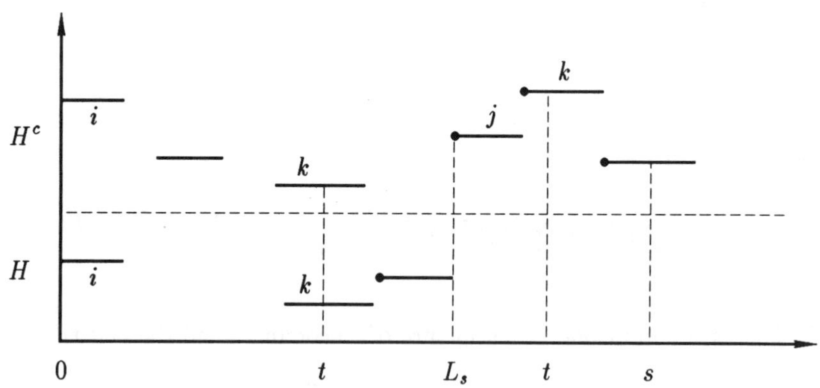

$i \in \mathbb{I}$, $k \in \mathbb{I}$, $j \in H^c$, $0 \leqslant t < s$

$$\begin{aligned}
&\mathbb{P}^i(X(L_s) = j,\ t < L_s < s) = \Gamma_{ij}^{s} - \Gamma_{ij}^{s}(t), \\
&\mathbb{P}^i(X_t = k,\ 0 \leqq L_s < t) = p_{ik}(t) D_k^c(s - t), \\
&\mathbb{P}^i(X(L_s) = j,\ 0 \leqslant L_s < t) = \Gamma_{ij}^{s}(t)
\end{aligned}$$

Figure 4.2. Illustration of Lemma 1, § 4.2.2.

Lemma 1 yields immediately the basic relations.

THEOREM 2. *For* $i \in \mathbb{I}$, $j \in H^c$, *and* $0 \leqslant t+h < s$:

$$\Gamma^s_{ij}(t+h) - \Gamma^s_{ij}(t) = \sum_k p_{ik}(t)\big(\Gamma^{s-t}_{kj}(h) - \Gamma^{s-t}_{kj}(0)\big),$$

$$\Gamma^{sc}_{ij}(t+h) = \sum_k p_{ik}(t)\Gamma^{(s-t)c}_{kj}(h).$$

Formally, one could obtain from Theorem 1 the (B)-equation, but its cumbersome form makes it inconvenient. As in the previous section, it is the (F)-equation which is more convenient here. It is shown below with the help of the density argument, essentially related to the (F)-equation, that distribution function $\Gamma^s_{ij}(\cdot)$ and $\Pi^s_{ij}(\cdot)$ have an absolutely continuous component with density $\gamma^s_{ij}(\cdot)$ and $\pi^s_{ij}(\cdot)$, respectively. Hence, Theorem 8 below yields the explicit expressions.

THEOREM 3. *Under conditions of validity of the* (F)*-equation, one has for* $i \in \mathbb{I}$, $0 < t < s$:

$$\mathbb{P}^i\big(X(L_s) = j,\ 0 < L_s \leqslant t\big) = \Gamma^s_{ij}(t) - \Gamma^s_{ij}(0) = \int_0^t \gamma^s_{ij}(u)\,du \qquad \text{for} \quad j \in H^c,$$

$$\mathbb{P}^i\big(X(L_s-) = r,\ 0 < L_s \leqslant t\big) = \Pi^s_{ir}(t) = \int_0^t \pi^s_{ir}(u)\,du \qquad \text{for} \quad r \in H,$$

where

$$\gamma^s_{ij}(t) = c_{ij}(t)D^c_j(s-t), \qquad \pi^s_{ir}(t) = p_{ir}(t)n_r(s-t) \qquad \text{for} \quad s > t.$$

Expressions $c_{ij}(t)$ and $n_r(t)$ have been defined in §§ 2.2.6 and 3.4.1, respectively. For convenience, the definitions are repeated here:

$$c_{ij}(t) = \sum_{r\in H} p_{ir}(t)q_{rj} \qquad \text{for} \quad j \in H^c,$$

$$n_r(t) = \sum_{j\in H^c} q_{rj}D^c_j(t) \qquad \text{for} \quad r \in H.$$

It follows from Theorem 3 that for $t+h < s$:

$$\gamma^s_{ij}(t+h) = \sum_k p_{ik}(t)\gamma^{s-t}_{kj}(h),$$

$$\pi^s_{ir}(t+h) = \sum_k p_{ik}(t)\pi^{s-t}_{kr}(h).$$

Note that adding all above expressions over $j \in H^c$ or $r \in H$, one recovers the corresponding expressions from § 4.2.1.

For the sake of completeness, the corresponding results for the joint distribution $\Pi^s_{ir}(t)$ of L_s and $X(L_s-)$ should be mentioned. Recall from § 2.1.5 that $\Pi^{sc}_{ir}(t) = \Pi^s_{ir} - \Pi^s_{ir}(t)$ with $\Pi^{sc}_{ir}(0) = \Pi^s_{ir}$, where $i \in \mathbb{I}$, $r \in H$, and $0 < t < s$. The argument analogous to that in Lemma 1 yields

LEMMA 4. *For $i \in \mathbb{I}$, $r \in H$, and $0 < t < s$:*

$$\Pi_{ir}^{sc}(t) = \sum_k p_{ik}(t)\Pi_{kr}^{s-t}.$$

From Lemma 4 the following basic relations are obtained in the usual way.

THEOREM 5. *For $i \in \mathbb{I}$, $r \in H$, and $0 < t + h < s$:*

$$\Pi_{ir}^{s}(t+h) - \Pi_{ir}^{s}(t) = \sum_k p_{ik}(t)\left(\Pi_{kr}^{s-t}(h) - \Pi_{kr}^{s-t}(0)\right),$$

$$\Pi_{ir}^{sc}(t+h) = \sum_k p_{ik}(t)\Pi_{kr}^{(s-t)c}(h).$$

Formally, one could obtain from Theorem 5 the (B)-equation, but it is the (F)-equation which is more convenient here. It has been given in Theorem 3 above. It should be stressed that $X(L_s-)$ takes values r in H.

b) The density argument, essentially related to the (F)-equation, must be used to obtain the explicit expressions for $\Gamma_{ij}^{s}(t)$, $\Pi_{ij}^{s}(t)$, and related quantities. One must consider here $X(L_s-)$ jointly with $X(L_s)$, noting that a jump from H to H^c occurs at instant L_s. Thus, Lemma 4, § 2.2.5, yields immediately the following fundamental result (see Fig. 4.3).

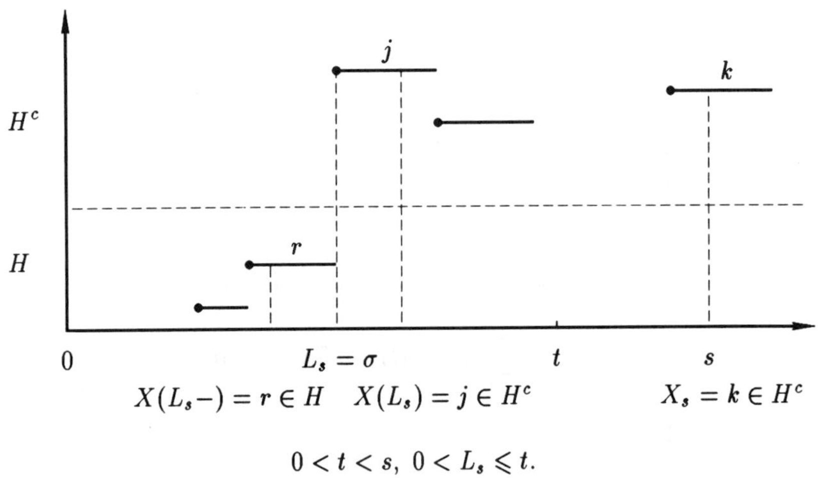

Figure 4.3. Illustration of Lemma 6, § 4.2.2.

LEMMA 6. *For $i \in \mathbb{I}$, $r \in H$, $j, k \in H^c$ and $0 < t < s$:*

$$\mathbb{P}^i\left(X(L_s-) = r,\ X(L_s) = j,\ 0 < L_s \leqslant t,\ X_s = k\right)$$

$$= \int_0^t p_{ir}(\sigma) q_{rj\,H}p_{jk}(s-\sigma)\,d\sigma.$$

Hence, one has immediate consequences of Lemma 6.

THEOREM 7. *For* $i \in \mathbb{I}$, $r \in H$, $j \in H^c$, *and* $0 < t < s$:

$$\mathbb{P}^i\big(X(L_s-) = r,\ X(L_s) = j,\ 0 < L_s \leqslant t\big) = \int_0^t p_{ir}(\sigma) q_{rj} D_j^c(s-\sigma)\, d\sigma.$$

THEOREM 8. *For* $i \in \mathbb{I}$, $r \in H$, $j,k \in H^c$, *and* $0 < t < s$:

$$\mathbb{P}^i\big(X(L_s) = j,\ 0 < L_s \leqslant t,\ X_s = k\big) = \int_0^t c_{ij}(\sigma)\, {}_H p_{jk}(s-\sigma)\, d\sigma,$$

$$\mathbb{P}^i\big(X(L_s-) = r,\ 0 < L_s \leqslant t,\ X_s = k\big) = \int_0^t p_{ir}(\sigma)\, n_{rk}(s-\sigma)\, d\sigma,$$

where functions $c_{ij}(t)$ *and* $n_{rk}(t)$ *have been defined in §§ 2.2.6 and 3.4.1, respectively.*

Note that at $t = 0$, there is a jump of magnitude

$$\mathbb{P}^i\big(X(L_s) = j,\ L_s = 0,\ X_s = k\big) = \delta_{ij}\, {}_H p_{jk}(s).$$

Hence, as the corollary to Theorem 8, one obtains expressions for $\Gamma^s_{ij}(t) - \Gamma^s_{ij}(0)$ and $\Pi^s_{ir}(t)$, together with expressions for $\gamma^s_{ij}(t)$ and $\pi^s_{ir}(t)$, all stated in Theorem 3.

The density argument from Lemma 6 is made precise by the conditional probabilities in the following way (see also discussion in §§ 2.2.5 and 9.2.5).

THEOREM 9. *For* $r \in H$, $j,k \in H^c$, *and* $0 < \sigma < s$:

$$\mathbb{P}(X_s = k \mid L_s = \sigma,\ X(L_s) = j) = {}_H p_{jk}(s-\sigma)/D_j^c(s-\sigma),$$
$$\mathbb{P}(X_s = k \mid L_s = \sigma,\ X(L_s-) = r) = n_{rk}(s-\sigma)/n_r(s-\sigma).$$

§ 4.2.3. Comments. a) As before, the (B)-equation does not determine $\Gamma^s_{ij}(t)$ or $\Pi^s_{ir}(t)$, because Γ^s_{ij} and Π^s_{ir} are unknown; they can be determined explicitly using the (F)-equation. On the other hand, determination of equations for last exit distributions (Γ^s_{ij}) and (Π^s_{ir}) by the method outlined in §§ 4.1.2 and 4.1.3, respectively, is too cumbersome now, and other more direct approach is more convenient.

Note that definition of $\gamma^s_{ij}(t)$ gives for $t = 0$ and $j \in H^c$:

$$\gamma^s_{ij}(0) = \begin{cases} 0 & \text{for } i \in H^c, \\ q_{ij} D_j^c(s) & \text{for } i \in H. \end{cases}$$

Equating the above expression with that which would be obtained from the (B)-equation for $t = 0$, one obtains rather cumbersome expression for $\sum q_{ik}\Gamma^s_{kj}$ as a function of s.

Similarly, from definition of $\pi^s_{ir}(t)$ for $t = 0$ and $r \in H$, one has

$$\pi^s_{ir}(0) = \begin{cases} 0 & \text{for } i \in H^c, \\ \delta_{ir} n_r(s) & \text{for } i \in H. \end{cases}$$

Equating the above expression with that which would be obtained from the (B)-equation for $t = 0$, the expression for $\sum q_{ik}\Pi^s_{kr}$ may be obtained. For simpler approach see Theorem 5 below.

Furthermore, Lemma 1, § 4.2.2, may be written (for $t < s$) as

$$\Gamma^s_{ij} - \Gamma^s_{ij}(t) = \sum_k p_{ik}(t)\Gamma^{s-t}_{kj} - p_{ij}(t)D^c_j(s-t).$$

Addition over all $j \in H^c$ would give the expression for $\Gamma^s_i(t)$ in Theorem 1, § 4.2.1, because of Theorem 4, § 4.2.1. However, it does not follow from the above relation that $\Gamma^s_{ij}(t)$ coincides with

$$p_{ij}(t)D^c_j(s-t) = \mathbb{P}^i(X_t = j,\ 0 \leqq L_s \leqslant t)$$

(see Lemma 7, § 4.2.1). See also the similar discussion in § 4.1.3.

Similarly, Lemma 4, § 4.2.1, may be written (for $0 < t < s$) as

$$\Pi^s_{ir} - \Pi^s_{ir}(t) = \sum_k p_{ik}(t)\Pi^{s-t}_{kr}.$$

Addition over all $r \in H$ would give only Theorem 6, § 4.2.1. It is important to stress here the difference between Π^s_{ir} and $D_{ir}(s)$.

b) Letting $t \uparrow s$ in the results obtained earlier in this sub-section, their extensions corresponding to $(L_s < s)$ are obtained. In essence, one obtains decomposition theorems according to events $(L_s = 0)$, $(0 < L_s < s)$, and $(L_s = s)$.

Thus, in particular, extension of Theorem 8, § 4.2.2, is

THEOREM 1. *For $i \in \mathbb{I}$, $r \in H$, $j, k \in H^c$:*

$$\mathbb{P}^i\left(X(L_s) = j,\ 0 < L_s < s,\ X_s = k\right) = \int_0^s c_{ij}(\sigma)\, {}_H p_{jk}(s-\sigma)\, d\sigma,$$

$$\mathbb{P}^i\left(X(L_s-) = r,\ 0 < L_s < s,\ X_s = k\right) = \int_0^s p_{ir}(\sigma) n_{rk}(s-\sigma)\, d\sigma.$$

Extension of Theorem 3, § 4.2.2, yields the explicit expressions for last exit distributions.

COROLLARY 2. *For $i \in \mathbb{I}$, and $j \in H^c$, $r \in H$:*

$$\mathbb{P}^i\left(X(L_s) = j,\ 0 < L_s < s\right) = \Gamma^s_{ij} - \Gamma^s_{ij}(0) = \int_0^s c_{ij}(\sigma) D^c_j(s-\sigma)\, d\sigma,$$

$$\mathbb{P}^i\left(X(L_s-) = r,\ 0 < L_s < s\right) = \Pi^s_{ir} = \int_0^s p_{ir}(\sigma) n_r(s-\sigma)\, d\sigma.$$

Of great importance is the following result, known as the *last exit decomposition*, whose full significance will be discussed in Section 6.

THEOREM 3. *For $i \in \mathbb{I}$, $k \in H^c$, and $s \geqslant 0$:*

$$p_{ik}(s) = {}_Hp_{ik}(s) + \int_0^s \sum_{j \in H^c} c_{ij}(\sigma)\, {}_Hp_{jk}(s-\sigma)\, d\sigma$$

$$= {}_Hp_{ik}(s) + \int_0^s \sum_{r \in H} p_{ir}(\sigma) n_{rk}(s-\sigma)\, d\sigma.$$

Proof. Consider the decomposition

$$\mathbb{P}^i(X_s = k) = \mathbb{P}^i(L_s = 0,\ X_s = k) + \mathbb{P}^i(0 < L_s < s,\ X_s = k) + \mathbb{P}^i(L_s = s,\ X_s = k).$$

The second term on the right-hand side follows from Theorem 1, by summation over $j \in H^c$ and $r \in H$, respectively. The first term is ${}_Hp_{ik}(s)$ and the third term vanishes (by Prop. 8, § 4.2.1). □

It should be noted that when added over all $k \in H^c$, Theorem 3 reduces to a version of the Poisson equation stated in Theorem 10, § 4.2.1.

c) Two methods used in this sub-section produce different expressions for $\Gamma^s_{ij}(s)$, (see Lemma 1 and Thm. 3 in § 4.2.2) and for $\Pi^s_{ir}(t)$ (see Thm. 3 and Lemma 4 in § 4.2.2), and for corresponding $\Gamma^s_i(t)$ and $\Pi^s_i(t)$ in § 4.2.1. It is not immediately clear that these expressions coincide, but the following proposition confirms their equivalence.

PROPOSITION 4. *For $i \in \mathbb{I}$, $k \in H^c$, and $t < s$:*

$$\sum_{m \in H^c} p_{im}(t)\, {}_Hp_{mk}(s-t) - {}_Hp_{ik}(s)$$

$$= \int_0^t \sum_{j \in H^c} c_{ij}(\sigma)\, {}_Hp_{jk}(s-\sigma)\, d\sigma = \int_0^t \sum_{r \in H} p_{ir}(\sigma) n_{rk}(s-\sigma)\, d\sigma.$$

Proof. Observe that these expressions represent $\mathbb{P}^i(0 < L_s \leqslant t,\ X_s = k)$. The first follows from Lemma 7, § 4.2.1, and the remaining two from Theorem 8, § 4.2.2. □

Note that addition over all $k \in H^c$ yields Theorems 1 and 9 in § 4.2.1.

d) Another result of interest is the following theorem which provides a preliminary link connecting Poisson representation, Chung equation, and the last exit distribution (Π^s_{ir}).

THEOREM 5. *For $i \in \mathbb{I}$, $r \in H$, and $s > 0$:*

$$\sum_k q_{ik} \Pi^s_{kr} = \int_0^s p'_{ir}(\sigma) n_r(s-\sigma)\, d\sigma$$

(where prime denotes differentiation), and

$$\sum_k q_{ik}(\Pi^s_{kr} + p_{kr}(s)) = \frac{d}{ds}(\Pi^s_{ir} + p_{ir}(s)) - \delta_{ir} n_r(s).$$

Furthermore,

$$\Pi^s_{ir} + p_{ir}(s) = \int_0^s g_{ir}(\sigma) D^{\rho c}_r(s-\sigma)\, d\sigma,$$

where $g_{ir}(\cdot)$ is the inverse transform from the Feller relation.

Proof. The first assertion is immediate from the (B)-equation for $p_{ir}(s)$ and definition of Π^s_{ir} from Corollary 2. Alternatively, it follows from differentiation of Lemma 4, § 4.2.2, and identification of two expressions for $\pi^s_{ir}(0)$, obtained from the (B)- and (F)-equations, respectively. The second assertion follows from the first by integration by parts.

To obtain the last assertion it suffices to consider the Laplace transform of the function $\Pi^s_{ir} + p_{ir}(s)$, namely

$$u^\alpha_{ir}(1+n^\alpha_r) = u^\alpha_{ir}(\alpha+q_r)(1-d^{\rho\alpha}_r)/\alpha = g^\alpha_{ir}(1-d^{\rho\alpha}_r)/\alpha,$$

where the second relation follows from § 3.4.1, and g^α_{ir} follows from the Feller relation (see § 1.1.6). □

It should be noted that when $s\to\infty$, the first two expressions reduce to Theorem 5, § 4.1.5, and the third expression yields $\Pi_{ir} = u_{ir}q_r D^{\rho c}_r = u_{ir}b_r$.

On the other hand, adding over $r\in H$, Theorem 5 reduces to Theorem 12, § 4.2.1. For further comments see e) below.

e) As defined in § 2.1.4, the event $(L_s < s)$ implies occurrence of the event $(X(L_s)\in H^c)$. Indeed, $X(L_s)$ has been defined only for $L_s < s$, and takes values only in H^c (even for $L_s = 0$). On the other hand, for $L_s = s$ clearly $X_s\in H$, except when $s\in W$; see Lemma 7, § 2.1.4. Actually, these events are $\mathbb{P}$-equivalent. Indeed, for $j\in H^c$ one has the decomposition

$$\mathbb{P}^i(X(L_s)=j) = \mathbb{P}^i(X(L_s)=j,\ 0\leqslant L_s<s) + \mathbb{P}^i(X(L_s)=j,\ L_s=s) = \Gamma^s_{ij},$$

where the second term in the middle equation vanishes (by Prop. 8, § 4.2.1). Hence, summing over all $j\in H^c$, one has

$$\mathbb{P}^i(X(L_s)\in H^c) = \mathbb{P}^i(X(L_s)\in H^c,\ L_s<s) = \mathbb{P}^i(L_s<s) = \Gamma^s_i.$$

On the other hand, one can adjoint an isolated state Δ corresponding to $X(L_s)$, taking values in H when $L_s = s$. Then

$$\mathbb{P}^i(X(L_s)=\Delta) = \mathbb{P}^i(X(L_s)=\Delta,\ L_s=s) = \mathbb{P}^i(L_s=s) = \Gamma^{sc}_i.$$

Similarly, discussion may be carried out for $X(L_s-)$ taking values in H. Thus, for $r\in H$,

$$\mathbb{P}^i(X(L_s-)=r,\ 0<L_s<s) = \Pi^s_{ir},$$
$$\mathbb{P}^i(X(L_s-)=r,\ L_s=s) = p_{ir}(s),$$
$$\mathbb{P}^i(X(L_s-)=r,\ L_s=0) = 0.$$

Hence,

$$\mathbb{P}^i(X(L_s-)=r) = \mathbb{P}^i(X(L_s-)=r,\ 0<L_s\leqslant s) = \Pi^s_{ir} + p_{ir}(s).$$

Summing over all $r \in H$, one has

$$\mathbb{P}^i\big(X(L_s-) \in H\big) = \mathbb{P}^i(0 < L_s \leqslant s) = \Pi_i^s + \Gamma_i^{sc} = D_i(s).$$

The above relations provide interpretation of Theorems 5 and 10, § 4.2.1.

On the other hand, one can adjoint an isolated state Δ corresponding to $X(L_s-)$, taking values in H^c when $L_s = 0$. Then,

$$\mathbb{P}^i\big(X(L_s-) = \Delta\big) = \mathbb{P}^i\big(X(L_s-) = \Delta,\ L_s = 0\big) = \mathbb{P}^i(L_s = 0) = D_i^c(s).$$

Remark. Finally observe that when $s \to \infty$, then results obtained in this Subsection reduce to those in sub-section 4.1, in agreement with that fact the $L_s \to L$. In particular,

$$\lim_{s\to\infty} \mathbb{P}^i(L_s = s) = \mathbb{P}^i(L = \infty).$$

§ 4.2.4. Examples. These illustrative examples are of some interest. More elaborate examples will be given later.

Example 1. Let $H = \{k\}$, where k is a fixed state. Then $D_i(t) = F_{ik}(t)$ (see Ex. 1, § 3.1.4). Hence

$$\Gamma_{ij}^s(t) = \big(1 - F_{ij}(s)\big)\delta_{ij} + \int_0^t \gamma_{ij}^s(\sigma)\,d\sigma, \qquad j \neq k, \quad t < s,$$

where

$$\gamma_{ij}^s(\sigma) = p_{ik}(\sigma)q_{kj}\big(1 - F_{jk}(s-\sigma)\big), \qquad j \neq k.$$

Consequently,

$$\Gamma_i^{sc}(t) = p_{ik}(t) + \sum_{j\neq k} p_{ij}(t)F_{jk}(s-t), \qquad t < s,$$

with $\Gamma_i^{sc} = p_{ik}(s)$ and

$$\Gamma_i^{sc}(0) = \begin{cases} F_{ik}(s) & \text{for } \ i \neq k, \\ 1 & \text{for } \ i = k. \end{cases}$$

On the other hand,

$$\Pi_{ik}^s(t) = \Pi_i^s(t) = \int_0^t p_{ik}(\sigma)n_k(s-\sigma)\,d\sigma,$$

where (see Ex. 1, § 3.4.3)

$$n_k(s-\sigma) = q_k - \sum_{j\neq k} q_{kj}F_{jk}(s-\sigma) = q_k\big(1 - F_{kk}(s-\sigma)\big) - f_{kk}(s-\sigma).$$

Hence again (as $t \to s$), the Chung equation takes the form

$$\Pi_{ik}^s + p_{ik}(s) = \begin{cases} F_{ik}(s) & \text{for } \ i \neq k, \\ 1 & \text{for } \ i = k, \end{cases}$$

or

$$\alpha u_{ik}^\alpha(1 + n_k^\alpha) = \begin{cases} f_{ik}^\alpha & \text{for } \ i \neq k, \\ 1 & \text{for } \ i = k. \end{cases}$$

Note that when $s \to \infty$, this example reduces to Example 1, § 4.1.4.

Example 2. Let $H = \{k\}^c$, where k is a fixed state. Then $D_k(t)$ is the exponential distribution (see Ex. 2, § 3.1.4) and, for $t < s$,

$$c_{ik}(t) = \sum_{j \neq k} p_{ij}(t) q_{jk},$$

$$\gamma_{ik}^s(t) = c_{ik}(t) e^{-q_k(s-t)},$$

$$\Gamma_{ik}^s(t) = \delta_{ik} e^{-q_k s} + \int_0^t \gamma_{ik}^s(\sigma)\, d\sigma.$$

Thus,

$$\Gamma_i^s(t) = p_{ik}(t) e^{-q_k(s-t)}, \qquad i \in \mathbb{I}, \quad t < s,$$

with

$$\Gamma_i^s = p_{ik}(s), \qquad \Gamma_i^s(0) = \delta_{ik} e^{-q_k s}.$$

On the other hand,

$$n_r(t) = q_{rk} e^{-q_k t}, \qquad r \neq k,$$

by Example 2, § 3.4.3, and

$$\Pi_{ir}^s(t) = \int_0^t p_{ir}(\sigma) q_{rk} e^{-q_k(s-\sigma)} d\sigma, \qquad r \neq k, \quad t < s.$$

From the above, the expression for $\Pi_{ir}^s + p_{ir}(s)$ may be obtained, and its Laplace transform is $u_{ir}^\alpha q_{rk}(\alpha + q_k)^{-1} + u_{ir}^\alpha$.

Furthermore,

$$\Pi_i^s(t) = \int_0^t \gamma_i^s(\sigma)\, d\sigma,$$

where

$$\gamma_i^s(t) = c_{ik}(t) e^{-q_k(s-t)}, \qquad t < s.$$

Hence, the Chung equation reduces to the integrated form of the (F)-equation for $p_{ik}(s)$ (see § 1.1.2) and

$$\Pi_i^s + (1 - p_{ik}(s)) = 1 - \delta_{ik} e^{-q_k s}, \qquad i \in \mathbb{I}.$$

As before, when $s \to \infty$, this example reduces to Example 2, § 4.1.4.

§ 4.2.5. Matrices. As always, matrices provide additional insight into theorems of this sub-section. Consider the matrix

$$\mathbf{\Lambda}^s(t) = \left(\Gamma^s_{ij}(t)\right), \qquad 0 \leqslant t < s,$$

whose non-zero components are (in partitioned form) $\mathbf{\Lambda}^s_{11}(t)$ and $\mathbf{\Lambda}^s_{21}(t)$. For $t = 0$ and $t \to s$ one has, respectively,

$$\mathbf{\Lambda}^s(0) = \left(\delta_{ij} D^c_j(s)\right), \qquad \mathbf{\Lambda}^s = (\Gamma^s_{ij}).$$

Put also

$$\mathbf{\Lambda}^{sc}(t) = \mathbf{\Lambda}^s - \mathbf{\Lambda}^s(t) \qquad \text{for} \quad t < s.$$

Similarly, write

$$\mathbf{\Pi}^s(t) = (\Pi^s_{ir}(t)), \qquad 0 < t < s,$$

whose non-zero components (in partitioned form) are $\mathbf{\Pi}^s_{12}(t)$ and $\mathbf{\Pi}^s_{22}(t)$.

Moreover, let

$$\mathbf{\Pi}^s(0) = \mathbf{0}, \qquad \mathbf{\Pi}^s = (\Pi^s_{ir}),$$
$$\mathbf{\Pi}^{sc}(t) = \mathbf{\Pi}^s - \mathbf{\Pi}^s(t) \qquad \text{for} \quad t < s.$$

Introduce also the matrix

$$\boldsymbol{\pi}^s(t) = \left(\pi^s_{ir}(t)\right), \qquad t < s.$$

The (F)-equation in Theorem 3, § 4.2.2, may be written as

$$\mathbf{\Pi}^s(t) = \int_0^t \boldsymbol{\pi}^s(\sigma)\, d\sigma, \qquad t < s,$$

where

$$\boldsymbol{\pi}^s(t) = \mathbf{P}(t)\boldsymbol{\pi}^s(0), \qquad \boldsymbol{\pi}^s(0) = (\delta_{ir} n_r(s)).$$

The argument based on identification of derivatives at the origin, briefly mentioned in § 4.2.2, becomes much simpler in matrix notation. Indeed, the formal differentiation of the defining relation (Lemma 4, § 4.2.2)

$$\mathbf{\Pi}^{sc}(t) = \mathbf{P}(t)\mathbf{\Pi}^{s-t}, \qquad t < s,$$

yields

$$\frac{d\mathbf{\Pi}^{sc}(t)}{dt} = \mathbf{Q}\mathbf{\Pi}^{sc}(t) - \mathbf{P}(t)\,\frac{d\mathbf{\Pi}^s}{ds}.$$

Hence, for $t = 0$, this gives Theorem 5, § 4.2.3:

$$\mathbf{Q}\mathbf{\Pi}^s = \frac{d\mathbf{\Pi}^s}{ds} - \boldsymbol{\pi}^s(0),$$

where $\boldsymbol{\pi}^s(0)$ is determined from the (F)-equation above.

Again, as $s \to \infty$, this reduces to the expression $\mathbf{Q\Pi}$ discussed in § 4.1.6.

SECTION 5

DIRICHLET PROBLEM AND POISSON EQUATION

This section discusses solutions of the Dirichlet problem and Poisson equation, establishing a link between probabilistic Potential Theory and Markov chains in terms of the notion of a balayage and of the first entrance probability D. In this sense, this section constitutes the main part of this book, all earlier section being preliminaries.

The Dirichlet problem and the Poisson equation play fundamental role in the study of passage times. In particular, it is shown that every first entrance time problem leads in fact to the Dirichlet problem. Similarly, the Poisson equation is associated with the last exit time L. Several examples are given at the end of the section, but more specific examples are discussed in Section 7.

It should be stressed that discussion in this section refers actually to the minimal chain, or what is equivalent, to the randomized jump chain. Although the method of presentation is essentially analytic, probabilistic aspects are also discussed. Probabilistic approach (through martingales and functionals) is postponed till Chapter III. The role of the balayage concept is stressed in Sub-section 5.3.

Although the Dirichlet problem and Poisson equation received considerable attention in literature, the case of M.C.'s seems to be rarely discussed. This is in very sharp contrast with numerous applications in diverse fields. It is hoped therefore that presentation given here, and based on author's papers [1984, 1986, 1988] will provide convincing argument for utility of this concept.

For general discussion of the Dirichlet problem and the Poisson equation see J. Doob [1983], E. B. Dynkin [1965], J. Bliedtner and W. Hansen [1986]; see also K. L. Chung [1982], D. Williams [1979], R. Durrett [1984] for Brownian motion, D. Revuz [1975], J. G. Kemeny et al. [1966], R. Syski [1973] for discrete parameter M.C.'s and K. Siegrist [1983] for revival processes. It should be added that the importance of the balayage notion for probabilistic applications was first stressed by G. A. Hunt [1957].

5.1. DIRICHLET PROBLEM

§ **5.1.1. Statement.** Let $X = (X_t)$ be the usual M.C., that is the minimal chain with a standard stochastic matrix $\mathbf{P}(t)$ of transition probabilities, a conservative intensity matrix $\mathbf{Q}$ (and a jump matrix $\mathbf{R}$), a resolvent $\mathbf{U}^\alpha$, and satisfying the (B)- and (F)-equations.

In this section it will be assumed that the chain is transient. Hence, the potential $\mathbf{U} = (u_{ij})$ is finite. Any departure from this assumption will be explicitly noted. Recall that expressions $\mathbf{Q}f$, $\mathbf{U}^\alpha f$ etc., for a function f on the state space $\mathbb{I}$, have been defined by

$$\mathbf{Q}f(i) = \sum_j q_{ij} f(j), \qquad \mathbf{U}^\alpha f(i) = \sum_j u_{ij}^\alpha f(j),$$

with summation taken over all states in $\mathbb{I}$.

The (exterior) *Dirichlet problem* for the usual chain X consists of finding a non-negative bounded function f on a state space $\mathbb{I}$ such that

$$\begin{aligned} \mathbf{Q}f(i) &= 0, \qquad & i \in H^c, \\ f(i) &= \varphi(i), \qquad & i \in H, \end{aligned}$$

where φ is a given (non-negative bounded) function. This is expressed by saying that the function f must be $\mathbf{Q}$-invariant (harmonic) on H^c and equal to the boundary function φ on the "boundary" H.

THEOREM 1. *A solution of the Dirichlet problem formulated above, has the unique representation*

$$f(i) = p(i) + h(i), \qquad i \in \mathbb{I},$$

where p is the purely excessive component which is a balayage of a function φ and equals also to a potential of a charge a

$$p(i) = \sum_{j \in H} D_{ij}\, \varphi(j) = \sum_{j \in H^c} {}_H u_{ij}\, a(j), \qquad \text{for} \quad i \in H^c,$$

$$p(i) = \varphi(i) \qquad \text{for} \quad i \in H,$$

where

$$a(j) = \sum_{k \in H} q_{jk}\, \varphi(k) \qquad j \in H^c;$$

and h is the invariant component of the form

$$h(i) = \begin{cases} \lim\limits_{t \to \infty} \sum\limits_{j \in H^c} {}_H p_{ij}(t) f(j), & i \in H^c, \\ 0, & i \in H. \end{cases}$$

Proof. For convenience, use partitioned form of matrices $\mathbf{Q}$ and ${}_H\mathbf{U}^\alpha$, and write f as a column vector with components f_1 and φ_2. Then, on H^c one has

$$\mathbf{Q}_{11} f_1 + \mathbf{Q}_{12}\, \varphi_2 = 0.$$

Hence f_1 is $\mathbf{Q}_{11}$-excessive on H^c. The (B)- and (F)-equations

$$f_1 = \begin{cases} \alpha\, {}_H\mathbf{U}_{11}^\alpha f_1 + (-\mathbf{Q}_{11})\, {}_H\mathbf{U}_{11}^\alpha f_1, & \text{(B)} \\ \alpha\, {}_H\mathbf{U}_{11}^\alpha f_1 + {}_H\mathbf{U}_{11}^\alpha (-\mathbf{Q}_{11}) f_1, & \text{(F)} \end{cases}$$

yield the Riesz decomposition on H^c:

$$f_1 = p_1 + h_1.$$

The $\mathbf{Q}_{11}$-invariant component is

$$h_1 = \lim_{\alpha\to 0} \alpha\, {}_H\mathbf{U}_{11}^{\alpha} f_1 = {}_H\mathbf{P}_{11}(\infty) f_1,$$

and the purely $\mathbf{Q}_{11}$-excessive component p_1 is obtained explicitly from the (F)-equation

$$p_1 = \lim_{\alpha\to 0} {}_H\mathbf{U}_{11}^{\alpha}(-\mathbf{Q}_{11}) f_1 = {}_H\mathbf{U}_{11}(-\mathbf{Q}_{11}) f_1 = {}_H\mathbf{U}_{11}\,\mathbf{Q}_{12}\,\varphi_2.$$

Recall that $\mathbf{B}_{12} = {}_H\mathbf{U}_{11}\,\mathbf{Q}_{12}$ and put $a_1 = \mathbf{Q}_{12}\,\varphi_2$. Hence, one can write

$$p_1 = \mathbf{B}_{12}\,\varphi_2 = {}_H\mathbf{U}_{11}\, a_1.$$

Recalling the definition of the balayage matrix $\mathbf{B}$, and extending φ to $\mathbb{I}$ by taking $\varphi = 0$ on H^c, one has now $p = \mathbf{B}\varphi$ on $\mathbb{I}$ with $p_2 = \varphi_2$ on H. Consequently, $h = {}_H\mathbf{P}(\infty) f$ with $h_2 = 0$ on H. □

As the component h depends on f, the component $p \leqslant f$ can be characterized as the *minimal solution* of the Dirichlet problem. This minimal solution can be expressed as the balayage of φ on H:

$$p(i) = \mathbf{B}\varphi(i) = \mathbb{E}^i\big(\varphi(X_T),\ T < \infty\big), \qquad i \in \mathbb{I},$$

where T is the first entrance time to H. For further comments see § 5.3.1. Proper uniqueness is assured by

COROLLARY 2. *Suppose that T is finite a.e. (i.e., $D \equiv 1$). Then, $f = \mathbf{B}\varphi$ on $\mathbb{I}$.*

Proof. Let $m = \sup_{i\in H^c} f(i)$. Then, for $i \in H^c$,

$$h(i) \leqslant \lim_{t\to\infty} \sum_{j\in H^c} {}_H p_{ij}(t)\, m = D_i^c m = 0. \qquad \square$$

Theorem 1 refers explicitly to the continuous parameter minimal M.C., but it also applies to the associated jump chain. By Watanabe theorem (Thm. 5, § 1.3.3), the function f_1 on H^c is ${}_H\mathbf{P}_{11}(t)$-excessive iff it is $\mathbf{R}_{11}$-superharmonic. Thus, the Dirichlet problem may be stated with reference to the jump chain as

$$\begin{aligned} \mathbf{R}f(i) &= f(i), \qquad i \in H^c,\\ f(i) &= \varphi(i), \qquad i \in H. \end{aligned}$$

Its solution is then expressed in the form

$$f(i) = u(i) + r(i), \qquad i \in H^c,$$

where $u_1 = {}_H\mathbf{G}_{11}c_1$ is the (Green) potential of a charge $c_1 = \mathbf{R}_{12}\varphi_2$, and $r_1 = \mathbf{R}_{11}^{\infty} f_1$ is harmonic. This can be proved directly (applying the Riesz decomposition to f_1 on H^c) or deduced from Theorem 1 using the Feller relation (§ 1.2.5):

$$p_1 = {}_H\mathbf{U}_{11}\, a_1 = {}_H\mathbf{U}_{11}\, \mathbf{\Theta}_1\, \mathbf{\Theta}_1^{-1} a_1 = {}_H\mathbf{G}_{11}\, c_1 = u_1$$

and

$$h_1 = f_1 - u_1 = r_1.$$

In this connection, it should be recalled (from Thm. 1, § 3.1.5) that the balayage matrix $\mathbf{B}$ is the same for the minimal chain and for the jump chain.

Remarks. The term "Dirichlet problem" comes from analogy with the Dirichlet problem in the classical Potential Theory. The problem requires to find a function f which is harmonic on an open set H^c (in a metric space, say), continuous and equal to the boundary function on the boundary ∂H^c of H^c. A harmonic function is defined by its mean value property which roughly means that the value of the function is the average of its values at the boundary.

In probabilistic approach, this definition is modified by requirement that the average is taken with respect to the distribution of the position of the first exit from H^c. For a large class of processes this requirement reduces to $\mathbf{Q}f = 0$ on H^c, where $\mathbf{Q}$ is now the Dynkin characteristic operator of the process. The classical case corresponds to the Brownian notion (the exit distribution from a ball is uniform on a sphere), where $\mathbf{Q}$ becomes the Laplacian Δ, and harmonic functions are solutions of partial differential equation $\Delta f = 0$ on H^c. For general discussion of probabilistic Dirichlet problem, see E. B. Dynkin [1965] and J. Doob [1983], where also historical remarks can be found.

In the case of a discontinuous process X considered here, the analogy is not so clearly noticeable as in the case of a continuous process. It suffices to consider the exit distribution from a single state, say i, and this is clearly $\mathbf{R} = (r_{ij})$. Hence, the definition of a function harmonic on H^c is that used here: $\mathbf{R}f = f$ on H^c.

Furthermore, the characteristic operator coincides with the generator, so the equivalent requirement is that $\mathbf{Q}f = 0$ on H^c. As already noted, these definitions refer to the jump chain. Moreover, for the usual M.C., harmonic and invariant functions coincide. This justifies form of Theorem 1 used here.

The mean value property of the minimal solution p of the Dirichlet problem may be expressed by $\mathbf{B}p = p$. This follows immediately from the fact that $\mathbf{B}$ is idempotent: $\mathbf{B}p = \mathbf{B}\mathbf{B}\varphi = \mathbf{B}\varphi = p$.

In practical applications the Dirichlet problem constitutes the basis for a large body of important results concerning passage times. Several illustrations are given in Sub-section 5.4, and explicit solutions are discussed in Section 7.

§ 5.1.2. Entrance probability. It has been already shown in Theorem 1, § 3.1.3, that the first entrance probability D satisfies the Dirichlet problem with the boundary function $\varphi \equiv 1$ on H. Thus, the following theorem is true.

THEOREM 1. *The first entrance probability D is the minimal (non-negative, bounded by one) solution of the Dirichlet problem on H^c, with the boundary function identically equal to 1 on H, and equals to the ${}_H\mathbf{U}_{11}$-potential on H^c:*

$$D_i = {}_H\mathbf{U}_{11}\, \bar{q}_i, \qquad i \in H^c,$$

of a charge $\bar{q}_1$ supported on H^c:

$$\bar{q}_i = \sum_{j \in H} q_{ij}, \qquad i \in H^c.$$

Proof. As noted earlier, $D = \mathbf{B1}$ is the minimal solution. Moreover,

$$D = {}_H\mathbf{U}_{11}\,\bar{q}_1, \qquad \text{where} \quad \bar{q}_1 = \mathbf{Q}_{12}\,\mathbf{1}_2,$$

in agreement with Theorem 7, § 3.1.1, which corresponds to the (F)-equation. □

Similarly, Corollary 5, § 3.1.1, indicates that D^c is $\mathbf{R}_{11}$-harmonic on H^c.

COROLLARY 2. *The probability D^c of no entry to H is the maximal (non-negative, bounded by one) solution of the harmonic equation on H^c:*

$$\mathbf{R}_{11}\, f(i) = f(i), \qquad i \in H^c.$$

The following result has been already obtained implicitly in Section 3. Nevertheless, it is instructive to illustrate its place within the framework of the Dirichlet problem.

THEOREM 3. *For the minimal* M.C., *one has on H^c:*

a) *D is purely excessive function relative to ${}_H\mathbf{P}_{11}(t)$ which is also an ${}_H\mathbf{U}_{11}$-potential of a charge $\bar{q}_1$ (or equivalently, an $\mathbf{R}_{11}$-potential of a charge $\Theta_1^{-1}\bar{q}_1$).*

b) *D^c is an invariant function relative to ${}_H\mathbf{P}_{11}(t)$ which is also $\mathbf{R}_{11}$-harmonic.*

Proof. The constant function $\mathbf{1}$ is ${}_H\mathbf{P}_{11}(t)$-excessive on H^c, so it is sufficient to identify components in its Riesz decomposition

$$p(i) = u(i) = D_i, \qquad h(i) = r(i) = D_i^c. \qquad \square$$

Equivalences discussed here are essential consequences of assumptions imposed on the usual M.C.

§ **5.1.3. Cut-off.** Another useful formulation of the Dirichlet problem may be obtained with reference to the cut-off method (see § 3.1.6). Let $\mathbf{Q}^a$ be the intensity matrix corresponding to all states in H made absorbing (see § 3.1.5). Then, the Dirichlet problem can be written in the matrix notation as $\mathbf{Q}^a f = 0$ on $\mathbb{I}$, where f is the column vector with components f_1 and φ_2.

As $\mathbf{Q}^a$ is the generator of a Markov semi-group $\mathbf{S}(t)$, $t \geqslant 0$, the above formulation indicates that now f is invariant on $\mathbb{I}$ relative to $\mathbf{S}(t)$

$$\mathbf{S}(t)f = f \quad \text{or} \quad \alpha\mathbf{S}^\alpha f = f \qquad \text{on} \quad \mathbb{I}.$$

Using the expression for $\mathbf{S}^\alpha$ from § 3.1.5, one has

$$\alpha\, {}_H\mathbf{U}^\alpha f + \mathbf{B}^\alpha f = f, \qquad \alpha > 0,$$

and, in the limit,

$$f = \mathbf{B}f + {}_H\mathbf{P}(\infty)f \qquad \text{on} \quad \mathbb{I}$$

in agreement with the solution stated in § 5.1.1.

Alternatively, with reference to the jump matrix $\mathbf{R}^a$ (see § 3.1.5), the Dirichlet problem may be expressed as $\mathbf{R}^a f = f$ on $\mathbb{I}$, so f is $\mathbf{R}^a$-harmonic on $\mathbb{I}$.

Remark. This formulation is convenient for a martingale method of solution of the Dirichlet problem (see § 12.1.1).

§ 5.1.4. Measure. It is of interest to consider the dual Dirichlet problem for measures. As the analysis is similar, only a brief outline is given here.

For the minimal chain, the Dirichlet problem consists of finding a (σ-finite) measure μ on $\mathbb{I}$ such that

$$\begin{aligned} \mu\mathbf{Q} &= 0 \qquad \text{on} \quad H^c, \\ \mu &= \varkappa \qquad \text{on} \quad H, \end{aligned}$$

where $\varkappa$ is a given boundary measure on H.

The solution on H^c, obtained from the (B)-equation on H^c, is

$$\mu_1 = \eta_1 + \lambda_1,$$

where

$$\eta_1 = \lim_{\alpha \to 0} \alpha \mu_1 \, {}_H\mathbf{U}_{11}^{\alpha}$$

is the invariant component, and

$$\lambda_1 = \sigma_1 \, {}_H\mathbf{U}_{11} = \varkappa_2 \, \mathbf{C}_{21}$$

is the purely excessive component which is a co-balayage of $\varkappa$ on H^c (see § 3.4.4) and also a potential of a charge

$$\sigma_1 = \mu_1(-\mathbf{Q}_{11}) = \varkappa_2 \mathbf{Q}_{21} \geqslant 0,$$

(recall that $\mathbf{C}_{21} = \mathbf{Q}_{21} \, {}_H\mathbf{U}_{11}$).

Observe that writing μ for the row vector $(\mu_1, \varkappa_2)$, the Dirichlet problem may be written in the form $\mu\mathbf{Q}^d = 0$ on $\mathbb{I}$, where $\mathbf{Q}^d$ is the generator of a semi-group $\mathbf{W}(t)$, $t \geqslant 0$, defined in § 3.4.4.

It should be remarked that in most situations only one component of the decomposition is present.

5.2. POISSON EQUATION

§ 5.2.1. Statement. Consider again the minimal chain $X = (X_t)$, and assume (unless explicitly stated otherwise) that the potential kernel $\mathbf{U}$ is finite.

As shown in Theorem 1, § 3.3.3, the first entrance probability D satisfies the *Poisson equation*

$$\mathbf{Q}D_i = \begin{cases} 0, & i \in H^c, \\ -b_i, & i \in H, \end{cases}$$

where b is the escape rate supported on H. Its solution has been already obtained in Theorem 8, § 4.1.1. Nevertheless, it is instructive to give the direct proof based on the (B)- and (F)-equations, as in the following theorem.

THEOREM 1 (Riesz decomposition). *The first entrance probability D can be expressed uniquely in the form*

$$D_i = \mathbf{U}b_i + \Gamma_i^c, \qquad i \in \mathbb{I},$$

where $\mathbf{U}b_i$ is the purely excessive component, which is also a potential of a charge (escape rate) with support H, and Γ_i^c is the invariant component.

Proof. The Poisson equation indicates that the function D is excessive on the state space $\mathbb{I}$ (see also § 3.1.3). Thus, it has the unique Riesz decomposition.

a) The (B)-equation applied to the function D yields for $i \in \mathbb{I}$ and $\alpha > 0$:

$$\begin{aligned} D_i &= \alpha\mathbf{U}^\alpha D_i + (-\mathbf{Q})\mathbf{U}^\alpha D_i \\ &= \alpha\mathbf{U}^\alpha D_i + (-\mathbf{Q})(1-\gamma_i^\alpha)/\alpha && \text{(Thm. 5, § 4.1.1)} \\ &= \alpha\mathbf{U}^\alpha D_i + \mathbf{Q}\gamma_i^\alpha/\alpha \\ &= \alpha\mathbf{U}^\alpha D_i + (\gamma_i^\alpha - D_i^c). && \text{((B)-equation)} \end{aligned}$$

Hence (as in the proof of Riesz decomposition),

$$\begin{aligned} &\lim_{\alpha\to 0} \alpha\mathbf{U}^\alpha D_i = \mathbf{P}(\infty)D_i = \Gamma_i^c && \text{(invariant)}, \\ &\lim_{\alpha\to 0} (\gamma_i^\alpha - D_i^c) = \Gamma_i - \Gamma_i(0) = \Pi_i && \text{(purely excessive)}. \end{aligned}$$

b) The explicit expression for the purely excessive component, as a potential of a charge, follows from the (F)-equation (for $i \in \mathbb{I}$, $\alpha > 0$):

$$D_i = \alpha\mathbf{U}^\alpha D_i + \mathbf{U}^\alpha(-\mathbf{Q})D_i = \alpha\mathbf{U}^\alpha D_i + \mathbf{U}^\alpha b_i.$$

Hence, as $\alpha \to 0$,

$$D_i = \Gamma_i^c + \mathbf{U}b_i. \qquad \square$$

In view of discussion of the last exit time L (see Sub-section 4.1), the above decomposition has the following probabilistic interpretation:

$$\begin{aligned} D_i &= \mathbb{P}^i(0 < L \leqslant \infty) = \mathbb{P}^i(T < \infty), \\ \mathbf{U}b_i &= \mathbb{P}^i(0 < L < \infty), \\ \Gamma_i^c &= \mathbb{P}^i(L = \infty). \end{aligned}$$

See also Theorem 8, § 4.1.1, and Corollaries 8 and 9 in § 4.1.2.

One should recall here connection with the Chung equation (see Thm. 10 and Cor. 11, § 4.2.1)

$$d_i^\alpha = \sum_{r\in H} u_{ir}^\alpha \alpha n_r^\alpha + \alpha \sum_{r\in H} u_{ir}^\alpha, \qquad i \in \mathbb{I}, \quad \alpha > 0.$$

Passage to the limit (as $\alpha \to 0$) yields assertion of Theorem 1

$$D_i = \sum_{r\in H} u_{ir} b_r + \lim_{s\to\infty} \sum_{r\in H} p_{ir}(s)$$

in view of relations in §§ 3.4.1, 3.4.2 and §§ 4.1.1, 4.2.1. See also comments in § 6.2.1.

The following re-formulation of the Poisson equation is of some interest. Recall that $\mathbf{Q}^\alpha = \mathbf{Q} - \alpha\mathbf{I}$ and introduce (for $\alpha > 0$)

$$b_i^\alpha = \begin{cases} \alpha(1+n_i^\alpha) & \text{for } i \in H, \\ 0 & \text{for } i \in H^c. \end{cases}$$

Note that $b_i^\alpha \to b_i$ as $\alpha \to 0$.

With this notation, taking transform of Theorem 12, § 4.2.1, one obtains immediately the transformed version of the Poisson equation

$$\mathbf{Q}^\alpha d_i^\alpha = \begin{cases} 0, & i \in H^c, \\ -b_i^\alpha, & i \in H, \end{cases}$$

for $\alpha > 0$ (it reduces to the form stated earlier when $\alpha \to 0$). The analogue of Theorem 1 is now

THEOREM 2. *The transformed Poisson equation has a unique solution given by*

$$d_i^\alpha = \mathbf{U}^\alpha b_i^\alpha, \qquad i \in \mathbb{I}, \quad \alpha > 0.$$

Proof. This is re-formulation of the above comment on the Chung equation. Alternatively, (d_i^α) is $\mathbf{Q}^\alpha$-excessive, but its invariant component vanishes and its purely excessive component is just a potential of a charge (b_i^α). □

Note that letting $\alpha \to 0$ in Theorem 2, one obtains the assertion of Theorem 1 (and not $\mathbf{U}b$ component alone).

§ 5.2.2. Extension. The solution of the Poisson equation so far considered is just the Riesz decomposition of the $\mathbf{Q}$-excessive function D on $\mathbb{I}$, with a charge supported on H. It is therefore natural to consider a generalization to any $\mathbf{Q}$-excessive function f on $\mathbb{I}$, with a charge $a \geqslant 0$ supported on H,

$$\mathbf{Q}f = \begin{cases} 0 & \text{on } H^c, \\ -a & \text{on } H. \end{cases}$$

It then follows from the Riesz decomposition, and from the (B)- and (F)-equations, that $f = p + h$, where

$$h = \lim_{\alpha \to 0} \alpha \mathbf{U}^\alpha f,$$
$$p = \lim_{\alpha \to 0} \mathbf{U}^\alpha a = \mathbf{U}a \qquad \text{with} \quad a = (-\mathbf{Q})f = (-\mathbf{Q})p \geqslant 0$$

(see discussion in § 1.3.3).

However, the argument in Theorem 1, § 5.2.1, uses special properties of the function D. It is therefore of interest to see how this argument can be extended to excessive functions.

LEMMA 1. *Let f be a $\mathbf{P}(t)$-excessive function on $\mathbb{I}$, with Riesz decomposition $f = p + h$. Suppose that g is $\mathbf{P}(t)$-invariant and such that $g \geqslant f$. Define*

$$G(t) = g - \mathbf{P}(t)f \geqslant 0, \qquad t \geqslant 0,$$

with $G(0) = g - f$, $G(\infty) = g - h$ (invariant), and

$$g^\alpha = \int_{0-}^{\infty} e^{-\alpha t}\, dG(t), \qquad \alpha > 0.$$

Then G is the exit law,

$$\mathbf{P}(t)G(s) = G(s+t), \qquad s, t \geqslant 0,$$

and, for $\alpha > 0$,

$$\alpha \mathbf{U}^\alpha f = g - g^\alpha, \qquad g^\alpha - G(0) = \mathbf{Q} g^\alpha / \alpha, \qquad \mathbf{Q}^\alpha g^\alpha = -\alpha(g - f).$$

Proof. The first two assertions follow from definition of $G(t)$, and the third uses the (B)-equation for $\mathbf{U}^\alpha f$; the last one is immediate. □

THEOREM 2. *The solution f of the Poisson equation with a charge $a \geqslant 0$ supported on H has the representation $f = p + h$, where the purely excessive component is $p = \mathbf{U}a = G(\infty) - G(0)$ with $a = (-\mathbf{Q})f$, and the invariant component is $h = f - G(\infty)$, where G is the exit law from Lemma 1.*

Proof. Lemma 1 applied to the (B)-equation for function f yields

$$\begin{aligned} f &= \alpha \mathbf{U}^\alpha f + (-\mathbf{Q})\mathbf{U}^\alpha f = \alpha \mathbf{U}^\alpha f + (-\mathbf{Q})(g - g^\alpha)/\alpha \\ &= \alpha \mathbf{U}^\alpha f + \mathbf{Q} g^\alpha / \alpha = \alpha \mathbf{U}^\alpha f + g^\alpha - G(0). \end{aligned} \tag{B}$$

Using the (F)-equation, one has

$$g^\alpha - G(0) = \mathbf{U}^\alpha \mathbf{Q}(g - f) = \mathbf{U}^\alpha a, \tag{F}$$

and the result follows by passage to the limit as $\alpha \to 0$. □

Remarks:

i) When f is bounded, then at least one such function g exists; take constant $\sup_i f(i)$.

ii) When $f = D$, then $g = 1$, $G(t) = \Gamma(t)$, and Theorem 2 reduces to Theorem 1, § 5.2.1.

iii) For further comments see Sub-section 6.3.

§ 5.2.3. Comments. Several comments on the Poisson equation are in order now.

a) Using the expression $\mathbf{R} - \mathbf{I} = \mathbf{\Theta}^{-1}\mathbf{Q}$, the Poisson equation from § 5.2.1 may be stated as an equation for the jump matrix $\mathbf{R}$

$$\mathbf{R}D_i - D_i = \begin{cases} 0, & i \in H^c, \\ -e_i, & i \in H, \end{cases}$$

where e is the escape probability supported on H (see § 3.3.3). Although both equations are equivalent, conceptually they refer to the minimal chain and to the jump chain, respectively. Consequently, each of these equations can be derived independently of the other, confirming the fact that D is the same for both chains (see Thm. 1, § 3.1.5).

Nevertheless, some comments on the Poisson equation for the jump chain may be of interest. The equation indicates that the function D is $\mathbf{R}$-superharmonic on $\mathbb{I}$, so it has the unique Riesz decomposition

$$D_i = u_i + s_i,$$

where $u = \mathbf{G}e$ in the (Green) potential of a charge e (with support H), and $s = \mathbf{R}^\infty D$ is the harmonic component. By Watanabe Theorem (Thm. 5, § 1.3.3) excessive functions relative to the minimal $\mathbf{P}(t)$ and superharmonic functions relative to the jump matrix $\mathbf{R}$, coincide. The Feller relation $\mathbf{G} = \mathbf{U\Theta}$ provides equality between components

$$u_i = \mathbf{G}e_i = \mathbf{U\Theta}e_i = \mathbf{U}b_i = \Pi_i, \qquad s_i = \Gamma_i^c.$$

In other words, potential component u and purely excessive component Π coincide, and so do the harmonic component s and the invariant component Γ^c (i.e., the term w in the Feller decomposition vanishes, see Theorem 6, § 1.3.3).

b) It is of interest to comment on the equality $s = \Gamma^c$ which analytically means $\mathbf{P}(\infty)D = \mathbf{R}^\infty D$. It is easy to verify that for the jump chain

$$s_i = \lim_{n\to\infty} \sum_j r_{ij}^{(n)} D_j = \mathbb{P}^i(X_n \in H \text{ i.o.}), \qquad i \in \mathbb{I}.$$

On the other hand, from Corollary 4, § 3.2.1, one has

$$D_i^\infty = \lim_{s\to\infty} \mathbf{P}(s)D_i = \lim_{s\to\infty} \mathbb{P}^i(T_s < \infty) = \mathbb{P}^i(L = \infty) = \Gamma_i^c, \qquad i \in \mathbb{I}.$$

Consequently, the events $(X_n \in H \text{ i.o.})$ and $(L = \infty)$ are $\mathbb{P}^i$-equivalent, as it is intuitively clear. In particular, note that (by Remarks in § 4.1.2)

$$\mathbb{P}^i(X_L = \Delta) = s_i, \qquad i \in \mathbb{I}.$$

These expressions should be compared with those based on the Chung equation:

$$\lim_{t\to\infty} \sum_{j\in H} p_{ij}(t) = \lim_{t\to\infty} \mathbb{P}^i(L_t = t) = \mathbb{P}^i(L = \infty) = s_i, \qquad i \in \mathbb{I}.$$

See also §§ 4.2.1, 4.2.2, and 5.2.1.

c) Note that if $e = 1$ identically on H, then $s = 0$ identically on $\mathbb{I}$. More generally, by Borel–Cantelli Lemma, the fact that

$$\sum_{n=0}^{\infty}\left(\sum_{j\in H} r_{ij}^{(n)}\right) = \sum_{j\in H} u_{ij}q_j < \infty$$

implies that $s_i = 0$.

On the other hand, as noted in § 4.1.1, when $D = 1$ identically on $\mathbb{I}$, then $s = 1$ identically (so H is visited infinitely often with probability one). In particular, this is the case when H^c is finite.

Theorem 9, § 4.1.1, provides further relations involving s. Of special interest is the fact that $\mathbf{B}s_i = s_i$, $i \in \mathbb{I}$.

d) Finally, another derivation of the Poisson equation can be obtained directly from the product theorem $\mathbf{B}^\rho = \mathbf{RB}$ by re-writting it as $(\mathbf{I} - \mathbf{R})\mathbf{B} = \mathbf{B} - \mathbf{B}^\rho$ and noting that $\mathbf{B1} = D$, $(\mathbf{B} - \mathbf{B}^\rho)\mathbf{1} = e$. As noted in § 3.3.5, $\mathbf{B}^\rho$ is the common balayage matrix for the minimal chain and for the jump chain.

Remark. The Poisson equation is also an analogue of the Poisson equation in Potential Theory. For its solution by martingale method see § 12.1.2.

§ 5.2.4. Equilibrium. Observe that if $D = 1$ identically on $\mathbb{I}$, then Theorem 1, § 4.1.1, implies that $s = 1$ identically on $\mathbb{I}$. In other words, if a set H is visited at least once from every state with probability one, then it is visited infinitely often with probability one (recurrence theorem). The set H such that $D_i = 1$ for all $i \in \mathbb{I}$ is therefore called *recurrent*.

On the other hand, a set H is called (*totally*) *transient* or an *equilibrium set* if $s_i = 0$ for all $i \in H$ (as s is invariant, its vanishing on H implies its vanishing on $\mathbb{I}$). Thus, if H is transient, there will be only a finite number of visits to H. Clearly, a transient state is a transient set.

The following important result is the specialization of a theorem for general Markov processes due to G. A. Hunt. For the usual M.C.'s, this is the immediate consequence of the Poisson equation. Note that D in the following theorem is necessarily purely excessive on $\mathbb{I}$.

THEOREM 1. *For an equilibrium set H:*

$$\mathbf{B1} = D = \mathbf{U}b = \Pi,$$

where b is the escape rate (supported on H).

Observe that notions of transient and recurrent sets are exclusive, but there are sets which do not belong to either category. Indeed, a recurrent set cannot be transient. Conversely, if H is transient, then $D = \mathbf{U}b$ is purely excessive and cannot be invariant (in particular, be a constant) unless it vanishes. Hence, D cannot be identically 1 on H^c.

Note that a set H is transient whenever $\mathbf{U}b_i = 1$ for all $i \in H$, i.e., when starting in H, the last exit time L from H is finite, but not zero, with probability one. Indeed, the expression $\mathbf{U}b_i = 1 - s_i$ for $i \in H$, may be called the *probability of escape* from a set H. This should not be confused with escape probability e_i (§ 3.3.3).

Recall finally from § 5.2.3 that

$$s_i = \mathbb{P}^i(L = \infty) = \mathbb{P}^i(T_\infty < \infty) = \mathbb{P}^i(X_n \in H \text{ i.o.}), \qquad i \in \mathbb{I},$$
$$\Pi_i = \mathbb{P}^i(0 < L < \infty) = \mathbf{U}b_i.$$

For further discussion see Sub-section 11.4.

5.3. BALAYAGE

§ 5.3.1. Remark. The importance of the concept of a balayage is now apparent. The name comes from the probabilistic interpretation, due to G. A. Hunt, of an analytic concept introduced by H. Poincaré in the study of the Dirichlet problem.

The notation of a balayage can be treated in several ways. Namely, in the analytic manner as the operator $\mathbf{B}^\alpha$ (introduced in § 3.1.5) acting on functions, or in the equivalent probabilistic fashion as the operator $\mathbf{P}^\alpha(T)$ with $T = T_H$ (see § 8.5.1). Still another analytic way is through "reduced functions" defined in this sub-section.

Let T be the first entrance time to a set H for the usual M.C. $X = (X_t)$. For any bounded function f on the state space $\mathbb{I}$ write

$$\mathbf{B}^\alpha f(i) = \mathbb{E}^i(e^{-\alpha T} f(X_T),\, T < \infty) = \sum_j d_{ij}^\alpha f(j), \qquad i \in \mathbb{I}, \quad \alpha \geqslant 0,$$

where $\mathbf{B}^\alpha = (d_{ij}^\alpha)$. Note that $\mathbf{B}^\alpha f(i) = f(i)$ for $i \in H$.

The function $\mathbf{B}^\alpha f$ is called the α-balayage of f on H. When $\alpha = 0$, α will be dropped from notation, and the word balayage will be used instead of 0-balayage.

The measure-theoretic approach to balayage will be discussed in Sub-section 8.5, where also references are given. Here only "reduced functions" are discussed; see also further comments in Sub-section 6.4.

§ 5.3.2. Reduced function. Let f be an $\mathbf{P}(t)$-excessive function on $\mathbb{I}$ (see § 1.3.2 for definition) and let $\mathbf{B}$ be the balayage operator. Write $\bar{f} = \mathbf{B}f$ for the balayage of f on H. In particular, if $f \equiv 1$, then $\bar{f} = D$.

The following propositions illustrate some analytic properties of the balayage. In particular, Proposition 2 is also used as the alternative definition of the balayage $\bar{f}$, called the *reduced function* of f.

Assume now that the potential kernel $\mathbf{U}$ is finite and impose the integrability condition $|\mathbf{Q}f(i)| < \infty$ for all $i \in \mathbb{I}$.

PROPOSITION 1. *For an excessive function f, its balayage $\bar{f}$ satisfies:*

a) *$\bar{f} \leqslant f$ on $\mathbb{I}$ and $\bar{f} = f$ on H,*

b) *$\bar{f}$ is excessive on $\mathbb{I}$, and invariant on H^c.*

Proof. For the first assertion, write in the partitioned form:

$$\bar{f}_1 = \mathbf{B}_{12} f_2, \qquad \bar{f}_2 = f_2.$$

The function f is excessive, so $\mathbf{Q}f \leqslant 0$, and in particular $\mathbf{Q}_{11} f_1 + \mathbf{Q}_{12} f_2 \leqslant 0$. Hence, using expressions from § 3.1.5,

$$\mathbf{B}_{12} f_2 = {}_H\mathbf{U}_{11}\, \mathbf{Q}_{12} f_2 \leqslant -{}_H\mathbf{U}_{11}\, \mathbf{Q}_{11} f_1 = f_1.$$

For the second assertion, evaluate

$$(-\mathbf{Q})\bar{f} = (-\mathbf{Q})\mathbf{B}f,$$

where the matrix $(-\mathbf{Q})\mathbf{B}$ has been found in § 3.2.5. The component on H^c vanishes, so $\bar{f}$ is invariant on H^c. On the other hand, on H one has

$$-(\mathbf{Q}_{21}\, \mathbf{B}_{12} f_2 + \mathbf{Q}_{22} f_2) = -(\mathbf{Q}_{21} \bar{f}_1 + \mathbf{Q}_{22} f_2) \geqslant -(\mathbf{Q}_{21} f_1 + \mathbf{Q}_{22} f_2) \geqslant 0,$$

because $-\mathbf{Q}f \geqslant 0$. Consequently, $-\mathbf{Q}\bar{f} \geqslant 0$. □

Observe that one has now (see the end of § 3.2.5)

$$\mathbf{P}(t)\bar{f} = \mathbf{P}(t)\mathbf{B}f = \mathbf{B}^t f \leqslant \mathbf{B}f = \bar{f}.$$

PROPOSITION 2. *The balayage $\bar{f}$ of an excessive function f on $\mathbb{I}$ is the smallest excessive function that dominates f on H.*

Proof. Let $\mathcal{G}$ be a class of excessive functions g such that $g \geqslant f$ on H. Hence,

$$g \geqslant \mathbf{B}g \geqslant \mathbf{B}f = \bar{f} \quad \text{and} \quad \bar{f} = \inf(g\colon\ g \in \mathcal{G}). \qquad \square$$

The next two propositions are special cases of what is known as the *maximum principle*.

PROPOSITION 3. *Let $u = \mathbf{B}\varphi$ be the minimal solution of the Dirichlet problem with a boundary function φ. Then*

$$\sup_{i\in\mathbb{I}} u(i) = \sup_{i\in H} u(i).$$

Proof. Recall that u is excessive and that $\mathbf{B}u = u$. Write M for $\sup \varphi(i)$ over $i \in H$. Therefore,

$$u(i) = \sum_{j\in H} D_{ij}\, \varphi(j) \leqslant D_i M \leqslant M \qquad \text{for all} \quad i \in \mathbb{I},$$

so $\sup_i u(i) \leqslant M$. Conversely,

$$M = \sup_{i\in H} u(i) \leqslant \sup_i u(i). \qquad \square$$

PROPOSITION 4. *Let h be an invariant function on $\mathbb{I}$ such that $D \geqslant h$ on $\mathbb{I}$. This is only possible provided $s \geqslant h$, where s is the invariant component of D.*

Proof. From the Poisson equation:

$$D = s + \mathbf{U}b \geqslant h \Longrightarrow s + \mathbf{P}(t)\mathbf{U}b \geqslant h.$$

As $\mathbf{U}b$ is purely excessive, letting $t \to \infty$, one finds $s \geqslant h$. □

§ 5.3.3. Balayage principles. For further discussion of balayage from the point of view of projection properties, see Sub-section 6.4. The following propositions are direct consequences of projection theorems, but are stated here because of their affinity with the preceeding propositions.

PROPOSITION 1 (balayage principle). *Let $p = \mathbf{U}a$ be a potential of a charge a. Then, there exists a unique potential $\bar{p} = \mathbf{U}\bar{a}$ (with support on H), with $\bar{a} = \mathbf{C}a$, and such that $\bar{p} \leqslant p$ on $\mathbb{I}$ and $\bar{p} = p$ on H.*

Proof. The assertion follows immediately from Theorem 3, § 6.4.1. □

PROPOSITION 2. *The balayage potential $\bar{p} = \mathbf{B}p$ (i.e., the balayage of a potential p of a charge) is:*

a) *the infimum of all potentials which dominate p on H,*

b) *the supremum of all potentials with support on H which are dominated by p on H.*

Proof. The first assertion follows from Proposition 2, § 5.3.2. For the second assertion, let v be any potential with the stated property. Then $\bar{p} = \mathbf{B}p \geqslant \mathbf{B}v = v$, where the last equality follows from Projection Theorem (Thm. 2, § 6.4.1). □

PROPOSITION 3. *Suppose that $u = \mathbf{B}\varphi$ is the unique solution of the Dirichlet problem with a boundary function φ. Suppose also that all states in H^c communicate (under taboo H) and lead to H. Then, u cannot assume maximum on H^c, unless it is constant everywhere.*

5.4. APPLICATIONS

§ 5.4.1. Special sets. The Dirichlet problem and the Poisson equation are fundamental in study of passage times, especially in determination of the entrance probability as well as the exit probabilities Γ and Π. Examples given here illustrate typical situations encountered in applications. More specific examples are given in Section 7.

Example 1. Let H^c be a finite set. Then

$$D_i^c = \lim_{t\to\infty} \sum_{j\in H^c} {}_H p_{ij}(t) = 0, \qquad i \in H^c.$$

Thus, $D_i = 1$ for all $i \in \mathbb{I}$, so H is a recurrent set. Furthermore, $\Gamma_i(t) = 0$ and $s_i = 1$ for all $i \in \mathbb{I}$. Evidently, $b_i = 0$, so $\Pi_i = 0$ for all $i \in \mathbb{I}$. In other words, with probability one, the first entrance time T is finite, whereas the last exit time L is infinite. See remarks in §§ 4.1.1 and 5.2.3.

Example 2. Let H be a finite set. Then, it can be only said that $D \leqslant 1$, and further structure of the M.C. must be considered in order to determine D. For example, in a transient chain a finite set will be visited only finitely often, so $s = 0$; hence, $D = \mathbf{U}b$.

Example 3. Let $H = \{k\}$, where k is a fixed state. Then, according to Example 1, § 3.1.4, and Example 1, § 3.3.4,

$$D_i = f_{ik} \quad (i \neq k), \qquad D_k = f_{kk}, \qquad b_k = (1 - f_{kk})q_k.$$

Suppose that k is a transient state ($f_{kk} < 1$). Then, passage relations (§ 1.1.7)

$$u_{ik} = f_{ik}u_{kk}, \qquad u_{kk}(1 - f_{kk})q_k = 1$$

imply that

$$\mathbf{U}b_i = \begin{cases} u_{ik}b_k = D_i u_{kk} b_k = D_i, & i \neq k, \\ u_{kk}b_k = 1, & i = k. \end{cases}$$

Thus, $s_i = 0$ for all i, or equivalently, $u_{kk}b_k = 1$; furthermore, D_i cannot be identically 1 for $i \neq k$.

On the other hand, when k is recurrent ($f_{kk} = 1$), then necessarily $D_i = s_i = 1$ for all i.

Example 4. Suppose that the set H consists of absorbing states only. Then, the Dirichlet problem and the Poisson equation for the absorption probability D reduce to a single equation $\mathbf{Q}D = 0$ (see § 5.1.3). Obviously now $b = 0$, so $D = s$. This situation has been examined in detail in § 3.1.6 (cut-off method).

Example 5. Suppose that the set H is partitioned into two sets Z and $H \setminus Z$, and suppose that the boundary function φ vanishes on $H \setminus Z$. Then, the minimal solution of the Dirichlet problem is

$$u(i) = \mathbf{B}\varphi(i) = \sum_{j \in Z} D_{ij}\, \varphi(j), \qquad i \in H^c.$$

In particular, if φ is the indicator of the set Z, $\varphi = I_Z$, then

$$\mathbf{B}I_Z(i) = \mathbb{P}^i(X_T \in Z,\ T < \infty),$$

and this is interpreted as probability of the first entry to Z, before the entry to $H \setminus Z$.

Example 6. Suppose that the set H^c contains a closed set S (a set which a M.C. cannot leave):

$$q_{ij} = 0 \qquad \text{for} \quad i \in S, \quad j \in S^c.$$

It is evident that now $D_i = 0$ for $i \in S$. Consequently, the equation for the Dirichlet problem reduces to

$$\sum_{j \in S^c} q_{ij} D_j = 0 \qquad \text{for} \quad i \in H^c \setminus S,$$

with $D_i = 1$ for all $i \in H$. The equation for $i \in S$ vanishes identically. Observe that now necessarily $D_i < 1$, even for the finite chain. Furthermore, $s_i = 0$ for $i \in S$.

§ 5.4.2. Special systems. It is of interest to consider some special forms of matrices $\mathbf{P}(t)$ which occur in many applications. Write in the partitioned form, relative to H and H^c,

$$\mathbf{P}(t) = \begin{pmatrix} \mathbf{P}_{11}(t) & \mathbf{P}_{12}(t) \\ \mathbf{P}_{21}(t) & \mathbf{P}_{22}(t) \end{pmatrix}, \qquad \mathbf{Q} = \begin{pmatrix} \mathbf{Q}_{11} & \mathbf{Q}_{12} \\ \mathbf{Q}_{21} & \mathbf{Q}_{22} \end{pmatrix},$$

and similarly for other matrices $\mathbf{U}^\alpha$, $\mathbf{B}^\alpha$, $\mathbf{C}^\alpha$ etc. Recall that a subset S of the state space $\mathbb{I}$ is said to be closed if transitions from S to outside of S are not possible, i.e., if

$$\sum_{j\in S} p_{ij}(t) = 1 \qquad \text{for every} \quad i \in S.$$

Obviously, in the cut-off method (§ 3.1.6) the taboo set H with all its states absorbing is trivially closed. See also Ex. 6 above in § 5.4.1.

Example 1. Suppose that the taboo set H is closed, so $\mathbf{P}_{21}(t) = \mathbf{0}$ and $\mathbf{Q}_{21} = \mathbf{0}$. Then, $\mathbf{P}_{11}(t)$ and $\mathbf{P}_{22}(t)$ satisfy separate equations:

$$\alpha\mathbf{U}_{11}^\alpha - \mathbf{I}_1 = \begin{cases} \mathbf{Q}_{11}\,\mathbf{U}_{11}^\alpha, & \text{(B)} \\ \mathbf{U}_{11}^\alpha\,\mathbf{Q}_{11}, & \text{(F)} \end{cases}$$

$$\alpha\mathbf{U}_{22}^\alpha - \mathbf{I}_2 = \begin{cases} \mathbf{Q}_{22}\,\mathbf{U}_{22}^\alpha, & \text{(B)} \\ \mathbf{U}_{22}^\alpha\,\mathbf{Q}_{22}. & \text{(F)} \end{cases}$$

The equation for $\mathbf{P}_{12}(t)$, however, has the form

$$\alpha\mathbf{U}_{12}^\alpha = \begin{cases} \mathbf{Q}_{11}\,\mathbf{U}_{12}^\alpha + \mathbf{Q}_{12}\,\mathbf{U}_{22}^\alpha, & \text{(B)} \\ \mathbf{U}_{11}^\alpha\,\mathbf{Q}_{12} + \mathbf{U}_{12}^\alpha\,\mathbf{Q}_{22}. & \text{(F)} \end{cases}$$

Combining these equations, one obtains

$$\mathbf{U}_{12}^\alpha = \mathbf{U}_{11}^\alpha\,\mathbf{Q}_{12}\,\mathbf{U}_{22}^\alpha. \qquad \text{(B)-(F)}$$

Comparison with equations for taboo probabilities (§ 1.2.3) indicates that $\mathbf{U}_{11}^\alpha = {}_H\mathbf{U}_{11}^\alpha$. Hence, recalling the expression for balayage $\mathbf{B}_{12}^\alpha$ from § 3.1.5, one obtains the following useful relation

$$\mathbf{U}_{12}^\alpha = \mathbf{B}_{12}^\alpha\,\mathbf{U}_{22}^\alpha, \qquad \alpha > 0.$$

To interpret this formula, observe that the matrix $\mathbf{B}^\alpha\mathbf{U}^\alpha$ appearing in the first entrance decomposition (Thm. 1, § 6.1.1) has only two non-zero terms, namely $\mathbf{U}_{12}^\alpha$ and $\mathbf{U}_{22}^\alpha$ related as above. When a M.C. enters the taboo set H at time T, with the first entrance distribution $\mathbf{B}_{12}^\alpha$, it continues in H without leaving it. By the strong Markov property, the transition probabilities remain the same, namely $\mathbf{U}_{22}^\alpha$. This is so-called post-T chain (see § 9.2.2).

Finally, it is clear that now $\mathbf{Q}D = 0$, so the Dirichlet problem and the Poisson equation yield

$$\mathbf{Q}_{11}\,D_1 + \mathbf{Q}_{12}\,\mathbf{1}_2 = 0, \qquad b = 0.$$

As H^c is not a closed set, so its states are transient (see Thm. 2, § 1.2.1), and passage to the limit yields, for the limit matrix $\mathbf{E}$,

$$\mathbf{E}_{12} = \mathbf{B}_{12}\,\mathbf{E}_{22}.$$

In particular, when all states in H are absorbing ($\mathbf{Q}_{22} = \mathbf{0}$), then $\mathbf{E}_{12} = \mathbf{B}_{12}$ (cut-off as in § 3.1.6).

Example 2. Dually, suppose that the set H^c is closed, so $\mathbf{P}_{12}(t) = \mathbf{0}$ and $\mathbf{Q}_{12} = \mathbf{0}$. Again, equations for $\mathbf{P}_{11}(t)$ and $\mathbf{P}_{22}(t)$ are separate, and by similar argument one finds that

$$\mathbf{U}_{21}^{\alpha} = \mathbf{U}_{22}^{\alpha}\,\mathbf{Q}_{21}\,\mathbf{U}_{11}^{\alpha}. \qquad \text{(B)-(F)}$$

Hence, recalling the expression the for co-balayage $\mathbf{C}_{21}^{\alpha}$ from § 3.4.4, one obtains the following useful relation

$$\mathbf{U}_{21}^{\alpha} = \mathbf{U}_{22}^{\alpha}\,\mathbf{C}_{21}^{\alpha}, \qquad \alpha > 0.$$

To interpret this formula, observe that the M.C. cannot leave the set H^c, so clearly $\mathbf{B}_{12}^{\alpha} = \mathbf{0}$. However, when in H, the chain may enter H^c on its last exit time, which is obviously the only exit time L from H. Thus, the matrix $\mathbf{U}^{\alpha}\mathbf{C}^{\alpha}$ in the last exit decomposition (Thm. 1, § 6.2.1) has only two non-zero terms, namely $\mathbf{U}_{21}^{\alpha}$ and $\mathbf{U}_{22}^{\alpha}$, related as above. Indeed, the situation here refers to so-called post-L chain (see § 9.2.3).

Evidently, $D_i = 0$ for $i \in H^c$, and the escape rate is now $b_i = q_i - \bar{q}_i$, $i \in H$. As H is not closed, so all its states are transient and, therefore, in the limit:

$$\begin{aligned} \mathbf{U}_{21} &= \mathbf{U}_{22}\,\mathbf{C}_{21} && \text{(transient on } H^c\text{)},\\ \mathbf{E}_{21} &= \mathbf{U}_{22}\,\mathbf{Q}_{21}\,\mathbf{E}_{11} && \text{(recurrent on } H^c\text{)}, \end{aligned}$$

where

$$\alpha\mathbf{C}_{21}^{\alpha} = \mathbf{Q}_{21}\,\alpha\mathbf{U}_{11}^{\alpha} \longrightarrow \mathbf{Q}_{21}\,\mathbf{E}_{11}.$$

In both cases, however, $\alpha\mathbf{C}_{21}^{\alpha}\mathbf{1}_1 = b_2$ is constant.

Indeed, recalling properties of n_{ij}^{α} from § 3.4.1, one has in present situation

$$n_{ij}^{\alpha} = \sum_{k \in H^c} q_{ik}u_{kj}^{\alpha} \longrightarrow N_{ij}, \qquad i \in H, \quad j \in H^c.$$

Hence,

$$\alpha n_i^{\alpha} = b_i, \qquad \bar{q}_i = (\alpha + q_i)d_i^{\rho\alpha} = q_i D_i^{\rho}, \qquad i \in H, \quad \alpha > 0,$$

so $N_i = \infty$ and $\mathbb{E}^i T_{\rho} = \infty$, $i \in H$.

Example 3. Suppose that the intensity matrix $\mathbf{Q}$ has the component $\mathbf{Q}_{11}$ of the form

$$\mathbf{Q}_{11} = -\mathbf{\Theta}_1 = -(\delta_{ij}q_j).$$

This indicates that there are no transitions on the set H^c. Indeed, the chain can enter a state in H^c only by a jump from H, and then must return to H at the end of its stay in that state in H^c.

Evidently, one finds from the Dirichlet problem and the Poisson equation that $D = 1$ identically on $\mathbb{I}$, and $b = 0$.

Direct calculation yields for elements of ${}_H\mathbf{P}_{11}(t)$ and $\mathbf{B}_{12}(t)$, respectively,

$$\begin{aligned} {}_Hp_{ij}(t) &= \delta_{ij}e^{-q_i t}, && i \in H^c, \quad j \in H^c,\\ D_{ij}(t) &= (1 - e^{-q_i t})r_{ij}, && i \in H^c, \quad j \in H. \end{aligned}$$

Here $\mathbf{R} = (r_{ij})$ is the jump matrix. Evidently, $\mathbf{R}_{11} = \mathbf{0}$, and the balayage matrix $\mathbf{B}$ has the form

$$\mathbf{B}_{12} = \Theta_1^{-1}\mathbf{Q}_{12} = \mathbf{R}_{12}.$$

Recall from § 3.3.5 that $\mathbf{B}^{\rho} = \mathbf{B} + \Theta^{-1}\mathbf{Q}\mathbf{B}$, so

$$\mathbf{B}_{22}^{\rho} = \mathbf{R}_{21}\,\mathbf{B}_{12} + \mathbf{R}_{22}, \qquad \Theta_2(\mathbf{B}_{22}^{\rho} - \mathbf{I}_2) = \mathbf{Q}_{22} + \mathbf{Q}_{21}\,\mathbf{B}_{12}.$$

It may be remarked that generators of the form $\mathbf{Q}_{21}\mathbf{B}_{12} + \mathbf{Q}_{22}$ correspond to a *trace chain* on a set H (see § 9.5.4 and Ex. 6, § 9.6.7).

SECTION 6

DECOMPOSITION THEOREMS

This section deals with decomposition theorems for transition probabilities. Such theorems form the backbone of Markov chain studies and have been used since early days. They are now of growing importance in theoretical studies, and of great practical interest in applications.

The best known is the first entrance decomposition (i.e., according to the first entrance to a set). Basically a special case of the Dynkin formula, this decomposition has provided a standard technique in Markov chains theory. Its dual, the last exit decomposition (i.e., according to the last exit from a set) is more delicate and only recently received considerable attention.

Although implicit in earlier sections, decomposition theorems are examined here directly to stress their different character. The approach is analytic and emphasises connections with taboo probabilities, but sample function analysis is also mentioned. Approach through the Product Theorem is briefly mentioned, but the use of functionals is postponed to Chapter III.

The role of duality is examined in detail, and notions of a balayage and a cobalayage are used extensively. Finally, regarded as operators on functions and measures the decomposition theorems represent projection theorems for potentials.

Although this section is the specialization of general results of R. K. Getoor and M. J. Sharpe [1973], K. L. Chung [1967, 1970, 1974], A. O. Pittenger [1970], P. W. Millar [1977], and others, the simple structure of the usual M.C.'s allows simple direct proofs.

6.1. FIRST ENTRANCE DECOMPOSITION

§ 6.1.1. Main theorem. Let $\mathbf{P}(t) = (p_{ij}(t))$ be a standard stochastic transition matrix, with a conservative intensity matrix $\mathbf{Q} = (q_{ij})$, and let ${}_H\mathbf{P}(t) = ({}_Hp_{ij}(t))$ be a taboo probability matrix, with a taboo set H (see § 1.2.1). Furthermore, let $T = T_H$ be the first entrance time to a set H, and $D_{ij}(t)$ be the joint distribution of T and X_T, discussed in Sub-section 3.1. Recall that $\mathbf{B}(t) = (D_{ij}(t))$ is the time dependent balayage matrix.

It is of great theoretical and practical interest to express transition probabilities $p_{ij}(t)$ in terms of quantities related to the first entrance to a set H. The following decomposition theorem represents $p_{ij}(t)$ as the sum of two terms, according to the situation before and after the first entrance to H. In the first case, transitions from state i to state j at t are considered avoiding the taboo set H. In the second case, the M.C. after leaving state i, enters H for the first time at T, and afterwards unrestricted transitions lead to state j during the remaining time interval.

The following basic theorem has been already proved in Theorem 5, § 3.1.3, by decomposition according to $T > t$ and $T \leqslant t$, relative to $X_T \in H$.

THEOREM 1. *For* $i \in H^c$, $j \in \mathbb{I}$, *and* $t \geqslant 0$:

$$p_{ij}(t) = {}_H p_{ij}(t) + \int_0^t \sum_{k \in H} d_{ik}(s) p_{kj}(t-s)\, ds$$

or, in terms of Laplace transforms, for $\alpha > 0$:

$$u_{ij}^\alpha = {}_H u_{ij}^\alpha + \sum_{k \in H} d_{ik}^\alpha\, u_{kj}^\alpha .$$

These expressions hold trivially for $i \in H$, $j \in \mathbb{I}$.

Theorem 1 may be expressed in the matrix notation as

$$\mathbf{P}(t) = {}_H\mathbf{P}(t) + \int_0^t d\mathbf{B}(s)\, \mathbf{P}(t-s), \qquad t \geqslant 0, \tag{1}$$

or, alternatively, in the transformed form:

$$\mathbf{U}^\alpha = {}_H\mathbf{U}^\alpha + \mathbf{B}^\alpha \mathbf{U}^\alpha, \qquad \alpha > 0, \tag{2}$$

where $\mathbf{B}^\alpha$ is the α-balayage matrix, namely the L.S. transform of $\mathbf{B}(s)$ (see § 3.1.5). The non-trivial components are

$$\mathbf{U}_{11}^\alpha = {}_H\mathbf{U}_{11}^\alpha + \mathbf{B}_{12}^\alpha\, \mathbf{U}_{21}^\alpha, \qquad \mathbf{U}_{12}^\alpha = \mathbf{B}_{12}^\alpha\, \mathbf{U}_{22}^\alpha .$$

In the special case when all states are transient, passage to the limit yields

COROLLARY 2. *For* $i \in H^c$, $j \in \mathbb{I}$, *and* $u_{ij} < \infty$:

$$u_{ij} = {}_H u_{ij} + \sum_{k \in H} D_{ik}\, u_{kj},$$

with trivial extension for $i \in H$, $j \in \mathbb{I}$.

Hence, in the matrix notation:

$$\mathbf{U} = {}_H\mathbf{U} + \mathbf{B}\mathbf{U}, \tag{3}$$

where $\mathbf{B}$ is the balayage matrix.

The second version of Theorem 5, § 3.1.3, yields decomposition according to the value $X(T-) \in H^c$.

THEOREM 3. *For $i \in H^c$, $j \in \mathbb{I}$, and $t \geqslant 0$:*

$$p_{ij}(t) = {}_Hp_{ij}(t) + \int_0^t \sum_{k \in H^c} {}_Hp_{ik}(s) h_{kj}(t-s)\, ds$$

or, in terms of Laplace transform, for $\alpha > 0$:

$$u_{ij}^\alpha = {}_Hu_{ij}^\alpha + \sum_{k \in H^c} {}_Hu_{ik}^\alpha\, h_{kj}^\alpha,$$

where h_{kj}^α is the transform of $h_{kj}(t)$ defined in § 3.1.2.

Recalling from § 3.1.5 that

$$\mathbf{B}_{12}^\alpha = {}_H\mathbf{U}_{11}^\alpha\, \mathbf{Q}_{12},$$

the non-trivial components can be written in the equivalent form as

$$\mathbf{U}_{11}^\alpha = {}_H\mathbf{U}_{11}^\alpha + {}_H\mathbf{U}_{11}^\alpha\, \mathbf{Q}_{12}\, \mathbf{U}_{21}^\alpha, \qquad \mathbf{U}_{12}^\alpha = {}_H\mathbf{U}_{11}^\alpha\, \mathbf{Q}_{12}\, \mathbf{U}_{22}^\alpha.$$

Because of importance of the first entrance decomposition, it is worthwhile to examine in some detail equations stated above.

i) Summation over all $j \in \mathbb{I}$ in Theorem 1 yields the obvious result

$$\mathbf{1} = \alpha\, {}_H\mathbf{U}^\alpha \mathbf{1} + \mathbf{B}^\alpha \mathbf{1}, \qquad \alpha > 0,$$

in agreement with Theorem 8, § 3.1.1. Letting $t \to 0$ in equation (1), one obtains the partition $\mathbf{I} = \mathbf{J}_1 + \mathbf{J}_2$ (see § 1.2.2 for notation). Similarly, equation (1) or equation (2) give the relation between generators

$$\mathbf{Q} = {}_H\mathbf{Q} + \big(\mathbf{J}_2\mathbf{Q} + \mathbf{J}_1\mathbf{Q}\mathbf{J}_2\big).$$

ii) It is of interest to verify directly that equation (1) preserves the semi-group property of $\mathbf{P}(t)$. For convenience, write (1) in the form

$$\mathbf{P}(t) = {}_H\mathbf{P}(t) + \mathbf{W}(t)$$

and note that

$$\begin{aligned} \mathbf{P}(t)\,\mathbf{P}(h) &= \big[{}_H\mathbf{P}(t) + \mathbf{W}(t)\big]\big[{}_H\mathbf{P}(h) + \mathbf{W}(h)\big] \\ &= {}_H\mathbf{P}(t+h) + {}_H\mathbf{P}(t)\,\mathbf{W}(h) + \mathbf{W}(t)\,\mathbf{P}(h) \\ &= {}_H\mathbf{P}(t+h) + \mathbf{W}(t+h) = \mathbf{P}(t+h). \end{aligned}$$

iii) Using (F)-equations for $\mathbf{U}^\alpha$ and for ${}_H\mathbf{U}^\alpha$, it is easy to deduce from equation (2) that

$$\mathbf{B}^\alpha - \mathbf{J}_2 = {}_H\mathbf{U}^\alpha\big(\mathbf{Q} - {}_H\mathbf{Q}\big) = {}_H\mathbf{U}^\alpha \mathbf{Q}\mathbf{J}_2,$$

where $\mathbf{B}(0) = \mathbf{J}_2$. Substituting this expression for $\mathbf{B}^\alpha$ into equation (2), one obtains the following form of Theorem 3

$$\mathbf{J}_1\mathbf{U}^\alpha = {}_H\mathbf{U}^\alpha + {}_H\mathbf{U}^\alpha\big(\mathbf{Q} - {}_H\mathbf{Q}\big)\mathbf{U}^\alpha = {}_H\mathbf{U}^\alpha + {}_H\mathbf{U}^\alpha \mathbf{Q}\mathbf{J}_2\, \mathbf{U}^\alpha. \tag{4}$$

iv) Observe that equations (1) and (2) have the form of a renewal equation for $\mathbf{U}^\alpha$, when ${}_H\mathbf{U}^\alpha$ and $\mathbf{B}^\alpha$ are given. However, the facts that $\mathbf{B}^\alpha$ is idempotent and that $\mathbf{B}^\alpha {}_H\mathbf{U}^\alpha = \mathbf{0}$, imply that the equation is indeterminate. Indeed, components $\mathbf{U}_{21}^\alpha$ and $\mathbf{U}_{22}^\alpha$ are arbitrary.

§ 6.1.2. Exit theorem. The analogous decomposition is obtained for the first return time T_ρ to a set H, discussed in Section 3. When the initial state i is in H^c, the decomposition coincides with that discussed in the preceding § 6.1.1. The main difference occurs when the initial state i is in H; the M.C. then leaves i by a jump to H^c.

Let ${}_H\mathbf{P}^\rho(t) = ({}_Hp^\rho_{ij}(t))$ be an exit taboo matrix defined in § 3.3.5 and let $D^\rho_{ij}(t)$ be the joint distribution of $X(T_\rho)$ and T_ρ, discussed in Sub-section 3.3. Recall that $\mathbf{B}^\rho(t) = (D^\rho_{ij}(t))$ is the time dependent balayage matrix.

The following basic relation has been already proved in Theorem 4, § 3.3.3, by decomposition according to $T_\rho > t$ and $T_\rho \leqslant t$, relative to $X(T_\rho) \in H$.

THEOREM 1. *For $i, j \in \mathbb{I}$ and $t \geqslant 0$:*

$$p_{ij}(t) = {}_Hp^\rho_{ij}(t) + \int_0^t \sum_{k\in H} d^\rho_{ik}(s) p_{kj}(t-s)\, ds$$

or, in terms of Laplace transforms, for $\alpha > 0$:

$$u^\alpha_{ij} = {}_Hu^{\rho\alpha}_{ij} + \sum_{k\in H} d^{\rho\alpha}_{ik}\, u^\alpha_{kj}.$$

Theorem 1 may be expressed in the matrix notation as

$$\mathbf{P}(t) = {}_H\mathbf{P}^\rho(t) + \int_0^t d\mathbf{B}^\rho(s)\,\mathbf{P}(t-s), \qquad t \geqslant 0, \tag{1}$$

or, alternatively, in the transformed form

$$\mathbf{U}^\alpha = {}_H\mathbf{U}^{\rho\alpha} + \mathbf{B}^{\rho\alpha}\,\mathbf{U}^\alpha, \qquad \alpha > 0, \tag{2}$$

where $\mathbf{B}^{\rho\alpha}$ is the α-balayage matrix, namely the L.S. transform of $\mathbf{B}^\rho(s)$ (see § 3.3.5). The components $\mathbf{U}^\alpha_{11}$ and $\mathbf{U}^\alpha_{12}$ are the same as in Theorem 1, § 6.1.1, and the new components are

$$\mathbf{U}^\alpha_{21} = {}_H\mathbf{U}^{\rho\alpha}_{21} + \mathbf{B}^{\rho\alpha}_{22}\,\mathbf{U}^\alpha_{21}, \qquad \mathbf{U}^\alpha_{22} = {}_H\mathbf{U}^{\rho\alpha}_{22} + \mathbf{B}^{\rho\alpha}_{22}\,\mathbf{U}^\alpha_{22}.$$

In the special case when all states are transient, passage to the limit yields

COROLLARY 2. *For $i, j \in \mathbb{I}$ and $u_{ij} < \infty$:*

$$u_{ij} = {}_Hu^\rho_{ij} + \sum_{k\in H} D^\rho_{ik}\, u_{kj}.$$

Hence, in the matrix notation:

$$\mathbf{U} = {}_H\mathbf{U}^\rho + \mathbf{B}^\rho\mathbf{U}, \tag{3}$$

where $\mathbf{B}^\rho$ is the balayage matrix.

The second version of Theorem 4, § 3.3.3, yields decomposition according to the value $X(T_\rho-) \in H^c$.

THEOREM 3. *For $i, j \in \mathbb{I}$ and $t \geqslant 0$:*

$$p_{ij}(t) = {}_H p_{ij}^{\rho}(t) + \int_0^t \sum_k {}_H p_{ik}^{\rho}(s) h_{kj}^{\rho}(t-s)\, ds,$$

or, in terms of Laplace transform, for $\alpha > 0$:

$$u_{ij}^{\alpha} = {}_H u_{ij}^{\rho\alpha} + \sum_k {}_H u_{ik}^{\rho\alpha} h_{kj}^{\rho\alpha},$$

where $h_{kj}^{\rho\alpha}$ is the transform of $h_{kj}^{\rho}(t)$ defined in § 3.3.2.

Theorem 3 may be expressed in the matrix form as

$$\mathbf{U}^{\alpha} = {}_H\mathbf{U}^{\rho\alpha} + {}_H\mathbf{U}^{\rho\alpha}{}_B\mathbf{Q}^{\rho}\,\mathbf{U}^{\alpha}, \tag{4}$$

where ${}_B\mathbf{Q}^{\rho} = (\mathbf{Q}+\boldsymbol{\Theta})\mathbf{J}_2$. The components are of course the same as in Theorem 1, but they can be written in equivalent form with

$$\mathbf{B}_{22}^{\rho\alpha}\,\mathbf{U}_{21}^{\alpha} = {}_H\mathbf{U}_{21}^{\rho\alpha}\,\mathbf{Q}_{12}\,\mathbf{U}_{21}^{\alpha} + {}_H\mathbf{U}_{22}^{\rho\alpha}(\mathbf{Q}_{22}+\boldsymbol{\Theta}_2)\mathbf{U}_{21}^{\alpha},$$
$$\mathbf{B}_{22}^{\rho\alpha}\,\mathbf{U}_{22}^{\alpha} = {}_H\mathbf{U}_{21}^{\rho\alpha}\,\mathbf{Q}_{12}\,\mathbf{U}_{22}^{\alpha} + {}_H\mathbf{U}_{22}^{\rho\alpha}(\mathbf{Q}_{22}+\boldsymbol{\Theta}_2)\mathbf{U}_{21}^{\alpha}.$$

These relations can be verified by recalling expressions for ${}_H\mathbf{U}_{21}^{\rho\alpha}$ and ${}_H\mathbf{U}_{22}^{\rho\alpha}$ from § 2.2.4 and the expression for $\mathbf{B}_{22}^{\rho\alpha}$ from § 3.3.5.

Because of importance of the first return decomposition, it is worthwhile to examine in some detail equations stated above.

i) Summation over all $j \in \mathbb{I}$ in Theorem 1 yields the obvious result

$$\mathbf{1} = \alpha\, {}_H\mathbf{U}^{\rho\alpha}\mathbf{1} + \mathbf{B}^{\rho\alpha}\mathbf{1}, \qquad \alpha > 0,$$

in agreement with Theorem 8, § 3.3.1. Letting $t \to 0$ in equation (1), one obtains identity $\mathbf{I} = \mathbf{I}$, because ${}_H\mathbf{P}^{\rho}(0) = \mathbf{I}$ and $\mathbf{B}^{\rho}(0) = \mathbf{0}$ (see §§ 2.2.4 and 3.3.5, respectively). Equation (1) or equation (2) give the relation between generators

$$\mathbf{Q} = {}_H\mathbf{Q}^{\rho} + {}_B\mathbf{Q}^{\rho} \qquad \text{with} \quad {}_B\mathbf{Q}^{\rho} = (\boldsymbol{\Theta}+\mathbf{Q})\mathbf{J}_2,$$

as already noted in § 3.3.5.

ii) To verify directly that equation (1) preserves the semi-group property of $\mathbf{P}(t)$, one can proceed in the same fashion as in § 6.1.1-ii.

iii) Using (F)-equations for $\mathbf{U}^{\alpha}$ and for ${}_H\mathbf{U}^{\rho\alpha}$, it is easy to deduce from equation (2) that

$$\mathbf{B}^{\rho\alpha} = {}_H\mathbf{U}^{\rho\alpha}(\mathbf{Q} - {}_H\mathbf{Q}^{\rho}).$$

Substituting this expression for $\mathbf{B}^{\rho\alpha}$ into equation (2), one obtains the following form of Theorem 3

$$\begin{aligned}\mathbf{U}^{\alpha} &= {}_H\mathbf{U}^{\rho\alpha} + {}_H\mathbf{U}^{\rho\alpha}(\mathbf{Q} - {}_H\mathbf{Q}^{\rho})\mathbf{U}^{\alpha}\\ &= {}_H\mathbf{U}^{\rho\alpha} + {}_H\mathbf{U}^{\rho\alpha}(\boldsymbol{\Theta}+\mathbf{Q})\mathbf{J}_2\mathbf{U}^{\alpha}.\end{aligned} \tag{5}$$

iv) Equation (2), regarded as a renewal equation for $\mathbf{U}^{\alpha}$, can be solved with the formal representation of solution

$$\mathbf{U}^{\alpha} = {}_H\mathbf{U}^{\rho\alpha} + \sum_{n=1}^{\infty} (\mathbf{B}^{\rho\alpha})^n\, {}_H\mathbf{U}^{\rho\alpha},$$

where

$$(\mathbf{B}^{\rho\alpha})^n = \begin{pmatrix} \mathbf{0} & \mathbf{B}_{12}^{\alpha}(\mathbf{B}_{22}^{\rho\alpha})^{n-1} \\ \mathbf{0} & (\mathbf{B}_{22}^{\rho\alpha})^n \end{pmatrix}, \qquad n \geqslant 1.$$

§ 6.1.3. Discrete version. It is of interest to examine the decomposition theorems in relation to the jump matrix $\mathbf{R}^\alpha$. Let $\mathbf{G}^\alpha$ be the (Green) potential of the α-jump chain, and let ${}_H\mathbf{G}^\alpha$ and ${}_H\mathbf{G}^{\rho\alpha}$ be the (Green) potentials corresponding to taboo probabilities. Recall also the corresponding Feller relations for $\mathbf{U}^\alpha$, ${}_H\mathbf{U}^\alpha$, and ${}_H\mathbf{U}^{\rho\alpha}$ (see §§ 1.1.6, 1.2.5, and 2.2.4), and the fact that $\mathbf{B}^\alpha$ and $\mathbf{B}^{\rho\alpha}$ refer both to the minimal chain and to the jump chain.

Hence, Theorem 1, § 6.1.1, and Theorem 1, § 6.1.2, lead respectively to the following theorems.

THEOREM 1. *For* $\alpha > 0$:

$$\mathbf{G}^\alpha = {}_H\mathbf{G}^\alpha + \mathbf{B}^\alpha\,\mathbf{G}^\alpha.$$

THEOREM 2. *For* $\alpha > 0$:

$$\mathbf{G}^\alpha = {}_H\mathbf{G}^{\rho\alpha} + \mathbf{B}^{\rho\alpha}\,\mathbf{G}^\alpha.$$

It is also of interest to deduce Theorem 2 directly from Theorem 1, using the Product Theorem $\mathbf{B}^{\rho\alpha} = \mathbf{R}^\alpha\mathbf{B}^\alpha$ (§ 3.3.4).

To illustrate the method, one may deduce equation (2) in § 6.1.2 from equation (2) in § 6.1.1, by multiplying it by $\mathbf{R}^\alpha$ from left:

$$\mathbf{R}^\alpha\mathbf{U}^\alpha = \mathbf{R}^\alpha{}_H\mathbf{U}^\alpha + \mathbf{R}^\alpha\mathbf{B}^\alpha\mathbf{U}^\alpha.$$

Using Feller relations for $\mathbf{U}^\alpha$ (§ 1.1.6), ${}_H\mathbf{U}^\alpha$ (§ 1.2.5), and ${}_H\mathbf{U}^{\rho\alpha}$ (§ 2.2.4), one has

$$\mathbf{R}^\alpha\mathbf{U}^\alpha = \mathbf{R}^\alpha\mathbf{G}^\alpha(\mathbf{\Theta}^\alpha)^{-1} = (\mathbf{G}^\alpha - \mathbf{I})(\mathbf{\Theta}^\alpha)^{-1} = \mathbf{U}^\alpha - (\mathbf{\Theta}^\alpha)^{-1},$$
$$\mathbf{R}^\alpha{}_H\mathbf{U}^\alpha = \mathbf{R}^\alpha{}_H\mathbf{G}^\alpha(\mathbf{\Theta}^\alpha)^{-1} = ({}_H\mathbf{G}^{\rho\alpha} - \mathbf{J}_2)(\mathbf{\Theta}^\alpha)^{-1} = {}_H\mathbf{U}^{\rho\alpha} - (\mathbf{\Theta}^\alpha)^{-1}.$$

Addition yields the result

$$\mathbf{U}^\alpha = {}_H\mathbf{U}^{\rho\alpha} + \mathbf{B}^{\rho\alpha}\,\mathbf{U}^\alpha.$$

§ 6.1.4. Examples. The following classical examples are easily deduced from results of this sub-section. More elaborate examples will be given later.

Example 1. Take $H = \{k\}$, where k is a fixed state. Then, in the classical notation,

$$D_{ik}(t) = F_{ik}(t) \quad \text{for} \quad i \neq k, \qquad D_{kk}(t) = F_{kk}(t)$$

(see Ex. 1, § 3.1.4, and Ex. 1, § 3.3.4). Now, Theorem 1, § 6.1.1, and Theorem 1, § 6.1.2, provide proof of the classic expressions for passage probabilities, already stated in § 1.1.7,

$$u_{ik}^\alpha = \begin{cases} f_{ik}^\alpha\, u_{kk}^\alpha, & i \neq k, \\ (\alpha + q_k)^{-1} + f_{kk}^\alpha\, u_{kk}^\alpha, & i = k, \end{cases}$$

and, for their extensions for $j \neq k$,

$$u_{ij}^\alpha = \begin{cases} {}_k u_{ij}^\alpha + f_{ik}^\alpha\, u_{kj}^\alpha, & i \neq k, \\ {}_k u_{kj}^{\rho\alpha} + f_{kk}^\alpha\, u_{kj}^\alpha, & i = k. \end{cases}$$

The case when k is an absorbing state ($q_k = 0$) requires a special comment. Clearly, $\alpha u_{kk}^{\alpha} = 1$, so from the first relation above one has

$$\alpha u_{ik}^{\alpha} = f_{ik}^{\alpha} \qquad \text{for} \quad i \neq k.$$

However, when k is absorbing, then ρ_k and consequently T_ρ, are both infinite and, therefore, the second term in decomposition vanishes, i.e.,

$$\mathbb{P}^k(X_t = k,\ T_\rho \leqslant t) = 0,$$

and the formula reduces to $u_{kk}^{\alpha} = 1/\alpha$, $\alpha > 0$. Hence, interpreting f_{kk} as the probability $\mathbb{P}^k(T_\rho \leqslant \infty) = 1$, the absorbing state k satisfies definition of recurrent state, with escape probability $e_k = 1 - f_{kk} = 0$. See comments in § 1.1.7.

Example 2. Take $H^c = \{k\}$, where k is a fixed state. Then, Theorem 1, § 6.1.1, yields the (B)-equation

$$u_{kj}^{\alpha} = \delta_{kj}(\alpha + q_k)^{-1} + \sum_{v \neq k} r_{kv}^{\alpha}\, u_{vj}^{\alpha},$$

using Example 2, § 1.2.6, and Example 2, § 3.1.4.

On the other hand, Theorem 1, § 6.1.2, yields

$$u_{ij}^{\alpha} = {}_H u_{ij}^{\rho\alpha} + \sum_{v \neq k} d_{iv}^{\rho\alpha}\, u_{vj}^{\alpha}, \qquad i \neq k,$$

where ${}_H u_{ij}^{\rho\alpha}$ and $d_{iv}^{\rho\alpha}$ are given in Example 2, § 2.2.4, and Example 2, § 3.3.4, respectively.

6.2. LAST EXIT DECOMPOSITION

§ 6.2.1. Main theorem. The dual problem to that discussed in the previous Sub-section 6.1 concerns the decomposition of transition probabilities according to the last exit from a set H. However, the analysis is more complex than that for the first entrance.

As before, let $\mathbf{P}(t) = (p_{ij}(t))$ be a standard stochastic transition matrix with a conservative intensity matrix $\mathbf{Q} = (q_{ij})$, and let ${}_H\mathbf{P}(t) = ({}_Hp_{ij}(t))$ be a taboo probability matrix with a taboo set H (see § 1.2.1). Furthermore, let L_t be the last exit time from a set H before an instant t, discussed at length in Section 4. Recall that $\mathbf{C}(t) = (N_{ij}(t))$ is the time dependent co-balayage matrix introduced in § 3.4.4.

The following decomposition theorem represents $p_{ij}(t)$ as the sum of two terms, according to whether the last exit occurred initially or not. In the first case, transitions avoid the taboo set till at least time t. In the second case, the chain may visit all states in $\mathbb{I}$ till the last exit from H at a positive time, and afterwards the chain remains outside H up to time t.

The following basic relation has been already proved in Theorem 3, § 4.2.3, by decomposition according to $L_t = 0$ and $L_t > 0$, relative to the last exit position $X(L_t-) \in H$.

THEOREM 1. *For $i \in \mathbb{I}$, $j \in H^c$, and $t \geqslant 0$:*

$$p_{ij}(t) = {}_Hp_{ij}(t) + \int_0^t \sum_{k \in H} {}_Hp_{ik}(s) n_{kj}(t-s)\, ds$$

or, in terms of Laplace transforms, for $\alpha > 0$:

$$u_{ij}^\alpha = {}_Hu_{ij}^\alpha + \sum_{k \in H} u_{ik}^\alpha\, n_{kj}^\alpha.$$

These expressions hold trivially for $i \in \mathbb{I}$, $j \in H$.

Theorem 1 may be expressed in matrix notation as

$$\mathbf{P}(t) = {}_H\mathbf{P}(t) + \int_0^t \mathbf{P}(s)\, d\mathbf{C}(t-s), \qquad t \geqslant 0, \tag{1}$$

or, alternatively, in the transformed form

$$\mathbf{U}^\alpha = {}_H\mathbf{U}^\alpha + \mathbf{U}^\alpha \mathbf{C}^\alpha, \qquad \alpha > 0, \tag{2}$$

where $\mathbf{C}^\alpha$ is the α-co-balayage matrix, namely the L.S. transform of $\mathbf{C}(t)$ (see § 3.4.4). The non-trivial components are

$$\mathbf{U}_{11}^\alpha = {}_H\mathbf{U}_{11}^\alpha + \mathbf{U}_{12}^\alpha\, \mathbf{C}_{21}^\alpha, \qquad \mathbf{U}_{21}^\alpha = \mathbf{U}_{22}^\alpha\, \mathbf{C}_{21}^\alpha.$$

In the special case when all states are transient, passage to the limit yields

COROLLARY 2. *For $i \in \mathbb{I}$, $j \in H^c$, and $u_{ij} < \infty$:*

$$u_{ij} = {}_Hu_{ij} + \sum_{k \in H} u_{ik} N_{kj},$$

with trivial extension for $i \in \mathbb{I}$, $j \in H$.

Hence, in the matrix notation, where $\mathbf{C}$ is the co-balayage matrix

$$\mathbf{U} = {}_H\mathbf{U} + \mathbf{UC}. \tag{3}$$

The second version of Theorem 3, § 4.2.3, yields decomposition according to the value $X(L_t) \in H^c$.

THEOREM 3. *For $i \in \mathbb{I}$, $j \in H^c$, and $t \geqslant 0$:*

$$p_{ij}(t) = {}_Hp_{ij}(t) + \int_0^t \sum_{k \in H^c} c_{ik}(s)\, {}_Hp_{kj}(t-s)\, ds$$

or, in terms of transforms, for $\alpha > 0$:

$$u_{ij}^\alpha = {}_Hu_{ij}^\alpha + \sum_{k \in H^c} c_{ik}^\alpha\, {}_Hu_{kj}^\alpha,$$

where c_{ik}^{α} is the transform of $c_{ik}(t)$ defined in § 2.2.6.

Hence, in the matrix notation, the non-trivial components are

$$\mathbf{U}_{11}^{\alpha} = {}_{H}\mathbf{U}_{11}^{\alpha} + \widetilde{\mathbf{C}}_{11}^{\alpha}\,{}_{H}U_{11}^{\alpha}, \qquad \mathbf{U}_{21}^{\alpha} = \widetilde{\mathbf{C}}_{21}^{\alpha}\,{}_{H}\mathbf{U}_{11}^{\alpha},$$

where

$$\widetilde{\mathbf{C}}_{11}^{\alpha} = \mathbf{U}_{12}^{\alpha}\,\mathbf{Q}_{21}, \qquad \widetilde{\mathbf{C}}_{21}^{\alpha} = \mathbf{U}_{22}^{\alpha}\,\mathbf{Q}_{21},$$

are components of the matrix $\widetilde{\mathbf{C}}^{\alpha} = (c_{ik}^{\alpha})$.

Recalling from § 3.4.4 the definition of the co-balayage matrix $\mathbf{C}^{\alpha}$ (not to be confused with $\widetilde{\mathbf{C}}^{\alpha}$), one has $\mathbf{C}_{21}^{\alpha} = \mathbf{Q}_{21}\,{}_{H}\mathbf{U}_{11}^{\alpha}$, so it is clear that expressions for $\mathbf{U}_{11}^{\alpha}$ and $\mathbf{U}_{21}^{\alpha}$ given here coincide with those in Theorem 1.

Because of importance of the last exit decomposition, it is worthwhile to examine in some detail equations stated above.

i) Summation over all $j \in \mathbb{I}$ in Theorem 1 yields

$$(1/\alpha)\mathbf{1} = {}_{H}\mathbf{U}^{\alpha}\,\mathbf{1} + \mathbf{U}^{\alpha}\,\mathbf{C}^{\alpha}\,\mathbf{1},$$

which is the Chung equation (Cor. 11, § 4.2.1). Passage to the limit (as $\alpha \to 0$) gives the Poisson relation (Thm. 8, § 4.1.1). See also comments in § 5.2.1.

ii) Direct verification that equation (1) preserves the semi-group property of $\mathbf{P}(t)$ follows the same pattern as in § 6.1.1-ii.

iii) Using (B)-equations for $\mathbf{U}^{\alpha}$ and for ${}_{H}\mathbf{U}^{\alpha}$, it is easy to deduce from equation (2) that

$$\mathbf{C}^{\alpha} - \mathbf{J}_2 = (\mathbf{Q} - {}_{H}\mathbf{Q})\,{}_{H}\mathbf{U}^{\alpha} = \mathbf{J}_2\,\mathbf{Q}\,{}_{H}\mathbf{U}^{\alpha},$$

where $\mathbf{C}(0) = \mathbf{J}_2$. Substituting this expression for $\mathbf{C}^{\alpha}$ into equation (2), one obtains the following form of Theorem 3:

$$\mathbf{U}^{\alpha}\mathbf{J}_1 = {}_{H}\mathbf{U}^{\alpha} + \mathbf{U}^{\alpha}(\mathbf{Q} - {}_{H}\mathbf{Q})\,{}_{H}\mathbf{U}^{\alpha} = {}_{H}\mathbf{U}^{\alpha} + \mathbf{U}^{\alpha}\,\mathbf{J}_2\mathbf{Q}\,{}_{H}\mathbf{U}^{\alpha}, \tag{4}$$

with $\mathbf{U}^{\alpha}\mathbf{J}_2\,\mathbf{Q} = \widetilde{\mathbf{C}}^{\alpha}$.

iv) Although (1) and (2) are renewal equations for $\mathbf{U}^{\alpha}$, the facts that $\mathbf{C}^{\alpha}$ is idempotent and that ${}_{H}\mathbf{U}^{\alpha}\mathbf{C}^{\alpha} = \mathbf{0}$, imply that the equations are indeterminate. Indeed, components $\mathbf{U}_{12}^{\alpha}$ and $\mathbf{U}_{22}^{\alpha}$ are arbitrary.

§ **6.2.2. Exit theorem.** The α-co-balayage matrix $\mathbf{C}^{\alpha}$ has the component $\mathbf{C}_{22}^{\alpha}$ equal to $\mathbf{I}_2$. It is possible to extend $\mathbf{C}^{\alpha}$ in a fashion similar to extension of α-balayage $\mathbf{B}^{\alpha}$ to $\mathbf{B}^{\rho\alpha}$, in such a way that the new matrix $\mathbf{C}^{\rho\alpha}$ has components $\mathbf{C}_{21}^{\rho\alpha} = \mathbf{C}_{21}^{\alpha}$, but $\mathbf{C}_{22}^{\rho\alpha}$ differs from $\mathbf{I}_2$. This has been already done in § 3.4.4, introducing the matrix

$$\mathbf{C}^{\rho\alpha} = \mathbf{C}^{\alpha}\widehat{\mathbf{R}}^{\alpha},$$

which is dual to matrix $\mathbf{B}^{\rho\alpha} = \mathbf{R}^{\alpha}\mathbf{B}^{\alpha}$.

Indeed, this would correspond to the requirement that the M.C. after leaving the set H (for the last time) must return to H in the next step and stay there. The situation is dual to that of the first return time T_{ρ} discussed in Section 3. One could introduce the "last return time" which would differ from L_s only when the last position is in H. This approach, however, will not be used here, as it is more convenient to proceed in the formal way.

The following theorem is dual to the decomposition (referring to T_{ρ}) as given by (4) in § 6.1.2. Note that in contrast with decomposition involving $\mathbf{C}^{\alpha}$ (Thm. 1, § 6.2.1), the components $\mathbf{U}_{12}^{\alpha}$ and $\mathbf{U}_{22}^{\alpha}$ have now non-trivial representation.

THEOREM 1. *In matrix notation:*

$$\mathbf{U}^\alpha = {}_H\widetilde{\mathbf{U}}_{\rho\alpha} + \mathbf{U}^\alpha\,\mathbf{C}^{\rho\alpha}, \qquad \alpha > 0, \tag{1}$$

where $\mathbf{C}^{\rho\alpha}$ *is as above, and*

$${}_H\widetilde{\mathbf{U}}_{\rho\alpha} = {}_H\mathbf{U}^\alpha + \mathbf{B}^\alpha(\boldsymbol{\Theta} + \alpha\mathbf{I})^{-1}.$$

Proof. Using notation from § 3.4.4, multiplication of equation (2), § 6.2.1, from right by $\widehat{\mathbf{R}}^\alpha$ yields

$$\mathbf{U}^\alpha\,\widehat{\mathbf{R}}^\alpha = {}_H\mathbf{U}^\alpha\,\widehat{\mathbf{R}}^\alpha + \mathbf{U}^\alpha\,\mathbf{C}^{\rho\alpha}.$$

By Feller relation for $\mathbf{U}^\alpha$ (§ 1.1.6):

$$\mathbf{U}^\alpha\,\widehat{\mathbf{R}}^\alpha = \mathbf{G}^\alpha\,\mathbf{R}^\alpha(\boldsymbol{\Theta}^\alpha)^{-1} = (\mathbf{G}^\alpha - \mathbf{I})(\boldsymbol{\Theta}^\alpha)^{-1} = \mathbf{U}^\alpha - (\boldsymbol{\Theta}^\alpha)^{-1}.$$

It follows from the discussion in § 1.2.2 and § 3.1.5, and also from § 3.4.4, that

$${}_H\mathbf{U}^\alpha\,\mathbf{Q}^\alpha = \mathbf{B}^\alpha - \mathbf{I}, \qquad \widehat{\mathbf{R}}^\alpha - \mathbf{I} = \mathbf{Q}^\alpha(\boldsymbol{\Theta}^\alpha)^{-1}.$$

Hence

$${}_H\mathbf{U}^\alpha\,\widehat{\mathbf{R}}^\alpha = {}_H\mathbf{U}^\alpha\,\mathbf{Q}^\alpha(\boldsymbol{\Theta}^\alpha)^{-1} + {}_H\mathbf{U}^\alpha = (\mathbf{B}^\alpha - \mathbf{I})(\boldsymbol{\Theta}^\alpha)^{-1} + {}_H\mathbf{U}^\alpha.$$

Addition yields the final result

$$\mathbf{U}^\alpha = {}_H\mathbf{U}^\alpha + \mathbf{B}^\alpha(\boldsymbol{\Theta}^\alpha)^{-1} + \mathbf{U}^\alpha\mathbf{C}^{\rho\alpha}.$$ □

§ 6.2.3. Discrete version. With reference to the jump chain, it should be recalled that the α-co-balayage $\mathbf{C}^\alpha$ for the minimal chain and the α-co-balayage $\mathbf{K}^\alpha$ for the jump chain are different, and are related by

$$(\boldsymbol{\Theta} + \alpha\mathbf{I})\mathbf{K}^\alpha = \mathbf{C}^\alpha(\boldsymbol{\Theta} + \alpha\mathbf{I}),$$

as noted in § 3.4.4.

Hence, the analogue of Theorem 1, § 6.2.1, is (by Feller relations)

THEOREM 1. *For* $\alpha > 0$:

$$\mathbf{G}^\alpha = {}_H\mathbf{G}^\alpha + \mathbf{G}^\alpha\,\mathbf{K}^\alpha.$$

In the same fashion, α-co-balayages $\mathbf{C}^{\rho\alpha}$ and $\mathbf{K}^{\rho\alpha}$ are related by

$$(\boldsymbol{\Theta} + \alpha\mathbf{I})\mathbf{K}^{\rho\alpha} = \mathbf{C}^{\rho\alpha}(\boldsymbol{\Theta} + \alpha\mathbf{I})$$

as noted in § 3.4.4.

Hence, the analogue of Theorem 1, § 6.2.2, is (by Feller relations)

THEOREM 2. *For* $\alpha > 0$:

$$\mathbf{G}^\alpha = ({}_H\mathbf{G}^\alpha + \mathbf{B}^\alpha) + \mathbf{G}^\alpha\,\mathbf{K}^{\rho\alpha}.$$

Recall that duals to these theorems have been stated in § 6.1.3.

§ 6.2.4. Examples. The following classical examples are easily deduced from results of this sub-section. More elaborate examples will be given later.

Example 1. Take $H = \{k\}$, where k is a fixed state. Then Theorem 1, § 6.2.1, yields for $j \neq k$ and $\alpha > 0$:

$$u_{ij}^{\alpha} = \begin{cases} {}_k u_{ij}^{\alpha} + u_{ik}^{\alpha}\, n_{kj}^{\alpha}, & i \neq k, \\ u_{kk}^{\alpha}\, n_{kj}^{\alpha}, & i = k. \end{cases}$$

These formulas are dual to the well-known first entrance formulas stated in Example 1, § 6.1.4. Observe that the second equation has been already stated without proof in § 1.1.7, with comments. One can also re-write it in the form (for $j \neq k$)

$$p_{kj}(t) = \int_0^t \gamma_k^t(s) n_{kj}(t-s)/n_k(t-s)\, ds$$

in agreement with Theorems 8 and 9 in § 4.2.2, with $\gamma_k^t(s) = p_{kk}(s) n_k(t-s)$. See also Example 1, § 4.2.4, and Example 1, § 3.4.3.

Example 2. Take $H^c = \{k\}$, where k is a fixed state. Recall from Example 2, § 1.2.6, and Example 2, § 3.4.3, that

$${}_H u_{kk}^{\alpha} = (\alpha + q_k)^{-1}, \qquad n_{ik}^{\alpha} = q_{ik}(\alpha + q_k)^{-1} \qquad \text{for} \quad i \neq k.$$

Then, Theorem 1, § 6.2.1, yields the (F)-equation

$$u_{ik}^{\alpha}(\alpha + q_k) = \delta_{ik} + \sum_{v \neq k} u_{iv}^{\alpha}\, q_{vk}.$$

Obviously, this is the dual to the (B)-equation from Example 2, § 6.1.4.

6.3. DUALITY

§ 6.3.1. Summary. It is evident that decompositions according to the first entrance or the last exit are dual to each other. This duality is perhaps best seen in the form of matrices. Thus, Theorem 1, § 6.1.1, and Theorem 1, § 6.2.1, yield

$$\mathbf{B}^{\alpha}\mathbf{U}^{\alpha} = \mathbf{U}^{\alpha}\mathbf{C}^{\alpha},$$

which essentially reduces to

$$\mathbf{B}_{12}^{\alpha}\mathbf{U}_{21}^{\alpha} = \mathbf{U}_{12}^{\alpha}\mathbf{C}_{21}^{\alpha}, \qquad \mathbf{U}_{21}^{\alpha} = \mathbf{U}_{22}^{\alpha}\mathbf{C}_{21}^{\alpha}, \qquad \mathbf{U}_{12}^{\alpha} = \mathbf{B}_{12}^{\alpha}\mathbf{U}_{22}^{\alpha}. \tag{1}$$

These symmetric relations reflect duality of the α-balayage $\mathbf{B}^{\alpha}$ and the α-co-balayage $\mathbf{C}^{\alpha}$. They hold also for $\alpha = 0$ for finite $\mathbf{U}$.

Formal analogies between $\mathbf{B}_{12}^{\alpha} = (d_{ij}^{\alpha})$ and $\mathbf{C}_{21}^{\alpha} = (n_{ij}^{\alpha})$ are implicit in their definitions, and various similarities have been pointed out in Sub-section 3.4 and § 4.1.5.

In particular, the (B)- and (F)-equations have been reversed:

$$\mathbf{B}_{12}^{\alpha} = {}_H\mathbf{U}_{11}^{\alpha}\,\mathbf{Q}_{12}, \tag{F}$$

$$\mathbf{C}_{21}^{\alpha} = \mathbf{Q}_{21}\,{}_H\mathbf{U}_{11}^{\alpha}. \tag{B}$$

Substitution of these equations into (1), yields duality relations

$$\begin{aligned} &{}_H\mathbf{U}_{11}^{\alpha}\,\mathbf{Q}_{12}\,\mathbf{U}_{21}^{\alpha} = \mathbf{U}_{12}^{\alpha}\,\mathbf{Q}_{21}\,{}_H\mathbf{U}_{11}^{\alpha},\\ &\mathbf{U}_{21}^{\alpha} = \mathbf{U}_{22}^{\alpha}\,\mathbf{Q}_{21}\,{}_H\mathbf{U}_{11}^{\alpha}, \qquad \mathbf{U}_{12}^{\alpha} = {}_H\mathbf{U}_{11}^{\alpha}\,\mathbf{Q}_{12}\,\mathbf{U}_{22}^{\alpha}. \end{aligned} \tag{2}$$

Another form of reciprocal relations is obtained with the help of functions $h_{ij}(t)$ and $c_{ij}(t)$ defined in §§ 3.1.2 and 2.2.6, respectively. In terms of their Laplace transforms h_{ij}^{α} and c_{ij}^{α}, the matrices $\overline{\mathbf{B}}^{\alpha} = (h_{ij}^{\alpha})$ and $\widetilde{\mathbf{C}}^{\alpha} = (c_{ij}^{\alpha})$ are defined by

$$\overline{\mathbf{B}}^{\alpha} = \mathbf{Q}\,\mathbf{J}_2\,\mathbf{U}^{\alpha}, \qquad \widetilde{\mathbf{C}}^{\alpha} = \mathbf{U}^{\alpha}\,\mathbf{J}_2\,\mathbf{Q}.$$

It has been seen, however, that only parts of these matrices were needed, namely rows $\mathbf{J}_1\,\overline{\mathbf{B}}^{\alpha}$ and columns $\widetilde{\mathbf{C}}^{\alpha}\,\mathbf{J}_1$. Thus, the non-zero components are

$$\overline{\mathbf{B}}_{11}^{\alpha} = \mathbf{Q}_{12}\,\mathbf{U}_{21}^{\alpha}, \qquad \overline{\mathbf{B}}_{12}^{\alpha} = \mathbf{Q}_{12}\,\mathbf{U}_{22}^{\alpha},$$

$$\widetilde{\mathbf{C}}_{11}^{\alpha} = \mathbf{U}_{12}^{\alpha}\,\mathbf{Q}_{21}, \qquad \widetilde{\mathbf{C}}_{21}^{\alpha} = \mathbf{U}_{22}^{\alpha}\,\mathbf{Q}_{21}.$$

Substitution of these relations into (2) yields equivalent form of duality, corresponding to (1):

$$\begin{aligned} &{}_H\mathbf{U}_{11}^{\alpha}\,\overline{\mathbf{B}}_{11}^{\alpha} = \widetilde{\mathbf{C}}_{11}^{\alpha}\,{}_H\mathbf{U}_{11}^{\alpha},\\ &\mathbf{U}_{21}^{\alpha} = \widetilde{\mathbf{C}}_{21}^{\alpha}\,{}_H\mathbf{U}_{11}^{\alpha}, \qquad \mathbf{U}_{12}^{\alpha} = {}_H\mathbf{U}_{11}^{\alpha}\,\overline{\mathbf{B}}_{12}^{\alpha}. \end{aligned} \tag{3}$$

It should be noted that comparison of equation (2) with equation (4) in § 6.1.1, as well as equations (2) and (4) in § 6.2.1, produces reciprocity relations

$$\begin{aligned} &\mathbf{B}_{12}^{\alpha}\,\mathbf{U}_{21}^{\alpha} = {}_H\mathbf{U}_{11}^{\alpha}\,\overline{\mathbf{B}}_{11}^{\alpha}, \qquad \mathbf{B}_{12}^{\alpha}\,\mathbf{U}_{22}^{\alpha} = {}_H\mathbf{U}_{11}^{\alpha}\,\overline{\mathbf{B}}_{12}^{\alpha},\\ &\mathbf{U}_{12}^{\alpha}\,\mathbf{C}_{21}^{\alpha} = \widehat{\mathbf{C}}_{11}^{\alpha}\,{}_H\mathbf{U}_{11}^{\alpha}, \qquad \mathbf{U}_{22}^{\alpha}\,\mathbf{C}_{21}^{\alpha} = \widehat{\mathbf{C}}_{21}^{\alpha}\,{}_H\mathbf{U}_{11}^{\alpha}. \end{aligned} \tag{4}$$

Basically, duality goes back to the duality between T_t and L_t, as expressed in Lemma 7, § 2.1.4, which in the present case leads to the relation

$$\mathbb{P}^i(X_t = j,\ T \leqslant t) = \mathbb{P}^i(X_t = j,\ 0 < L_t < t).$$

Note that the LHS is non-trivial for $i \in H^c$, $j \in \mathbb{I}$, and the RHS is non-trivial for $i \in \mathbb{I}$, $j \in H^c$. For an application of these ideas see Sub-sections 6.4 and 11.2.

Duality involving T_ρ leads to rather cumbersome analytic expressions, as it may be seen from comparison of Theorem 1, § 6.1.2, with Theorem 1, § 6.2.2. See also Sub-sections 3.3 and 4.1.

Finally, it may be of interest to list the following relations which are the immediate consequences of decomposition theorem from Sub-sections 6.1 and 6.2:

$$\mathbf{B}^{\alpha} - \mathbf{I} = {}_H\mathbf{U}^{\alpha}\,\mathbf{Q}^{\alpha}, \qquad \mathbf{C}^{\alpha} - \mathbf{I} = \mathbf{Q}^{\alpha}\,{}_H\mathbf{U}^{\alpha},$$

$$\mathbf{B}^{\rho\alpha} - \mathbf{I} = {}_H\mathbf{U}^{\rho\alpha}\,\mathbf{Q}^{\alpha}, \qquad \mathbf{C}^{\rho\alpha} - \mathbf{I} = \mathbf{Q}^{\alpha}\,{}_H\widetilde{\mathbf{U}}_{\rho\alpha},$$

where $\mathbf{Q}^{\alpha} = \mathbf{Q} - \alpha\mathbf{I}$ (see § 1.1.6).

§ 6.3.2. Theorems. a) Continuing discussion in § 3.4.4, duality between $\mathbf{B}^\alpha$ and $\mathbf{C}^\alpha$ may be expressed by the following theorem.

THEOREM 1. *For* $\alpha > 0$:

$$\mathbf{Q}^\alpha \mathbf{B}^\alpha = \mathbf{C}^\alpha \mathbf{Q}^\alpha,$$

which reduces to

$$\mathbf{Q}_{21} \mathbf{B}^\alpha_{12} = \mathbf{C}^\alpha_{21} \mathbf{Q}_{12}.$$

Proof. Using the (B)- and (F)-equations

$$\mathbf{Q}^\alpha \mathbf{U}^\alpha = -\mathbf{I} = \mathbf{U}^\alpha \mathbf{Q}^\alpha,$$

one has

$$\mathbf{Q}^\alpha \mathbf{B}^\alpha = -\mathbf{Q}^\alpha \mathbf{B}^\alpha \mathbf{U}^\alpha \mathbf{Q}^\alpha = -\mathbf{Q}^\alpha \mathbf{U}^\alpha \mathbf{C}^\alpha \mathbf{Q}^\alpha = \mathbf{C}^\alpha \mathbf{Q}^\alpha$$

(see also the expressions for $\mathbf{B}^\alpha$ and for $\mathbf{C}^\alpha$ stated above in § 6.3.1). It is immediate to verify that the only non-zero component is that indexed by 22 (see Thm. 2, § 3.4.2, for the explicit form). □

Recall that

$$d^\alpha = \mathbf{B}^\alpha \mathbf{1}, \qquad \flat^\alpha = \alpha \mathbf{C}^\alpha \mathbf{1},$$

so Theorem 1 yields the transformed version of the Poisson equation, already stated in § 5.2.1,

$$\mathbf{Q}^\alpha d^\alpha = -\flat^\alpha, \qquad \alpha > 0,$$

with solution

$$d^\alpha = \mathbf{U}^\alpha \flat^\alpha$$

(by Thm. 2, § 5.2.1).

Theorem 1 holds also for $\alpha = 0$, and one can write $(-\mathbf{Q})\mathbf{B} = \mathbf{C}(-\mathbf{Q})$ as mentioned in § 3.4.4. Note, however, that passage to the limit as $\alpha \to 0$, yields

$$\mathbf{C}(-\mathbf{Q}) \cdot \mathbf{1} = \flat \neq \mathbf{C} \cdot (-\mathbf{Q})\mathbf{1} = 0.$$

b) The next theorem illustrates the link between quantities γ^α_{ij} (introduced in § 4.1.2) and n^α_{ij} (introduced in § 3.4.1) as well as connection with the Poisson equation.

Using notation from § 4.1.6, consider the matrix $\mathbf{\Lambda}(t) = (\Gamma_{ij}(t))$, and write $\mathbf{\Lambda}^\alpha = (\gamma^\alpha_{ij})$ for the L.S. transform of $\mathbf{\Lambda}(t)$, as mentioned in § 4.1.2. Recall that $\mathbf{\Lambda}^\alpha \mathbf{1}_i = \gamma^\alpha_i$ for all $i \in \mathbb{I}$.

Then, one has immediately

THEOREM 2. *For* $\alpha > 0$ *and* $i \in \mathbb{I}$:

$$\flat_i = (-\mathbf{Q}^\alpha)\mathbf{B}^\alpha \gamma^\alpha_i = \mathbf{C}^\alpha(-\mathbf{Q}^\alpha)\gamma^\alpha_i,$$

where the escape rate $\flat$ *(see § 3.1.3) is independent of* α *(with* $\flat_i = 0$ *for* $i \in H^c$*).*

Proof. The result follows from Theorem 1, using Theorem 5, § 4.1.1, for γ_i^α and Theorem 6, § 3.4.1, for b_i. □

Observe that Theorem 2 preserves associativity for $\alpha > 0$, but for $\alpha = 0$:

$$b = (-\mathbf{Q})\mathbf{B} \cdot \Gamma = \mathbf{C}(-\mathbf{Q}) \cdot \Gamma \neq \mathbf{C} \cdot (-\mathbf{Q})\Gamma = 0,$$

because $\mathbf{Q}\Gamma = 0$ (as Γ is invariant).

c) Still another form of duality is that between L and T_t and their joint distributions, as discussed in § 4.1.5. In matrix form, using notation from § 3.2.5 and § 4.1.6, one has

$$\mathbf{B}^t = \mathbf{B} - \int_0^t \mathbf{P}(s)(-\mathbf{Q})\mathbf{B}\, ds, \tag{F}$$

$$\mathbf{\Pi}(t) = \int_0^t \mathbf{P}(s)(-\mathbf{Q})\mathbf{\Pi}\, ds, \tag{F}$$

and the (B)-equations have the similar form

$$\frac{d\mathbf{B}^t}{dt} = \mathbf{Q}\mathbf{B}^t, \qquad \frac{d\mathbf{\Pi}^c(t)}{dt} = \mathbf{Q}\mathbf{\Pi}^c(t), \tag{B}$$

with $\mathbf{B}^0 = \mathbf{B}$ and $\mathbf{\Pi}^c(0) = \mathbf{\Pi}$.

Recall that

$$(-\mathbf{Q})\mathbf{B} = (b_{ij}), \qquad (-\mathbf{Q})\mathbf{\Pi} = (\delta_{ij} b_j).$$

§ 6.3.3. Dual balayage. The dual of a resolvent $\mathbf{U}^\alpha = (u_{ij}^\alpha)$ relative to a reference measure ξ has been defined as a resolvent $\widehat{\mathbf{U}}^\alpha = (\hat{u}_{ij}^\alpha)$ such that

$$\hat{u}_{ij}^\alpha = \xi(j) u_{ji}^\alpha / \xi(i), \qquad i, j \in \mathbb{I},$$

with similar expressions for the dual generator $\widehat{\mathbf{Q}}$ and the dual transition matrix $\widehat{\mathbf{P}}(t)$; see § 1.3.5. The corresponding M.C. $\widehat{X} = (\widehat{X}_t)$ is said to be dual to the (direct) M.C. $X = (X_t)$; the term "reverse chain" is also used (see § 9.4.5 for interpretation).

To minimize confusion, symbols refered to the dual chain will be always marked with $\widehat{\ }$. Thus, in particular, T is the first entrance time and L is the last exit time for the direct chain X with

$$\mathbf{B} = (D_{ij}) \quad \text{and} \quad \mathbf{C} = (N_{ij}),$$

whereas $\widehat{T}$ is the first entrance time and $\widehat{L}$ is the last exit time for the dual chain $\widehat{X}$ with

$$\widehat{\mathbf{B}} = (\widehat{D}_{ij}) \quad \text{and} \quad \widehat{\mathbf{C}} = (\widehat{N}_{ij}).$$

The following lemma determines duals of $\mathbf{B}$ and $\mathbf{C}$. Unfortunately, it is also responsible for notational malice, because the dual of $\mathbf{B}$ is not $\widehat{\mathbf{B}}$ but $\widehat{\mathbf{C}}$; also the dual of $\mathbf{C}$ is $\widehat{\mathbf{B}}$ and not $\widehat{\mathbf{C}}$.

LEMMA 1. *The dual of* $\mathbf{B}^\alpha$ *is* $\widehat{\mathbf{C}}^\alpha$ *with elements*

$$\hat{n}_{ij}^\alpha = \xi(j) d_{ji}^\alpha / \xi(i), \qquad \alpha \geqslant 0,$$

which are non-trivial for $i \in H$, $j \in H^c$. *The dual of* $\mathbf{C}^\alpha$ *is* $\widehat{\mathbf{B}}^\alpha$ *with elements*

$$\hat{d}_{ij}^\alpha = \xi(j) n_{ji}^\alpha / \xi(i), \qquad \alpha \geqslant 0,$$

which are non-trivial for $i \in H^c$, $j \in H$.

Proof. By definition of duality

$$\begin{aligned} \operatorname{dual}\left(d_{ji}^\alpha\right) = \xi(j) d_{ji}^\alpha / \xi(i) &= \sum_{k \in H^c} \left(\xi(j)_H u_{jk}^\alpha / \xi(k)\right) \cdot \xi(k) q_{ki} / \xi(i) \\ &= \sum_{k \in H^c} \hat{q}_{ik\,H} \hat{u}_{kj}^\alpha = \hat{n}_{ij}^\alpha, \qquad i \in H, \quad j \in H^c, \end{aligned}$$

using Theorem 6, § 3.1.2, for d_{ji}^α and definition of $\hat{n}_{ij}^\alpha$ from § 3.4.1 (applied to the dual chain). The above formula holds trivially for i and j in other domains. The expression for $\hat{d}_{ij}^\alpha$ is obtained in the analogous manner. □

It is convenient to express Lemma 1 in the diagram whose arrows indicate duality:

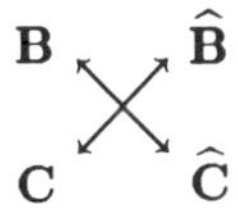

Thus, the dual of the balayage $\mathbf{B}$ for the direct chain is the co-balayage $\widehat{\mathbf{C}}$ for the reverse chain, and the dual of the co-balayage $\mathbf{C}$ of the direct chain is the balayage $\widehat{\mathbf{B}}$ for the reverse chain.

The following is the main result.

THEOREM 2. *The dual of the first entrance decomposition for the direct chain is the last exit decomposition for the reverse chain.*

Proof. It suffices to verify the assertions

$$\operatorname{dual}\left(\mathbf{B}^\alpha \mathbf{U}^\alpha\right) = \widehat{\mathbf{U}}^\alpha \widehat{\mathbf{C}}^\alpha, \qquad \operatorname{dual}\left(\mathbf{U}^\alpha \mathbf{C}^\alpha\right) = \widehat{\mathbf{B}}^\alpha \widehat{\mathbf{U}}^\alpha .$$

This follows from

$$\xi(j) \sum_{k \in H} d_{jk}^\alpha u_{ki}^\alpha / \xi(i) = \sum_{k \in H} \left(\xi(j) d_{jk}^\alpha / \xi(k)\right) \cdot \xi(k) u_{ki}^\alpha / \xi(i) = \sum_{k \in H} \hat{u}_{ik}^\alpha \hat{n}_{kj}^\alpha$$

for $i \in H$, $j \in H^c$ (non-trivial domain). The second assertion follows in the analogous manner. □

Recall that the dual of a function f is a measure $\hat{f}$ defined by $\xi(i) f(i) = \hat{f}(i)$. Thus, from relations $\mathbf{B1} = D$, $\widehat{\mathbf{B}}\mathbf{1} = \widehat{D}$ one obtains

$$\begin{aligned} \operatorname{dual}(D) &= \hat{\mathbf{1}} \cdot \operatorname{dual}(\mathbf{B}) = \xi \widehat{\mathbf{C}}, \\ \operatorname{dual}(\widehat{D}) &= \hat{\mathbf{1}} \cdot \operatorname{dual}(\widehat{\mathbf{B}}) = \xi \mathbf{C}, \end{aligned}$$

so

$$\xi(i) D_i = \sum_j \xi(j) \widehat{N}_{ji},$$

$$\xi(i) \widehat{D}_i = \sum_j \xi(j) N_{ji}.$$

§ 6.3.4. Other duals. In the similar manner one can obtain duals of other expressions discussed earlier in this section. For example, let $\Pi_{ij}^{\alpha} = u_{ij}^{\alpha} b_j$ be the L.S. transform of the joint distribution $\Pi_{ij}(t)$; see § 4.1.2 and § 4.1.5. Then

$$\operatorname{dual}\left(u_{ji}^{\alpha} b_i\right) = b_i \hat{u}_{ij}^{\alpha}, \qquad i \in H, \quad j \in \mathbb{I},$$

because the diagonal matrix $(\delta_{ij} b_j)$ is self-dual.

In another example, take $\xi(i) = (q_i + \alpha)^{-1}$. Then, using the Feller relations, one find that

$$\hat{g}_{ij}^{\alpha} = \hat{u}_{ij}^{\alpha}(q_j + \alpha) = \hat{u}_{ji}^{\alpha}(q_i + \alpha) = g_{ji}^{\alpha}$$

and

$$\operatorname{dual}\left(n_{ji}^{\alpha}\right) = \left(q_j + \alpha\right)^{-1} n_{ji}^{\alpha}(q_i + \alpha) \equiv k_{ji}^{\alpha}, \qquad i \in H^c, \quad j \in H,$$
$$\operatorname{dual}\left(d_{ji}^{\alpha}\right) = \left(q_j + \alpha\right)^{-1} d_{ji}^{\alpha}(q_i + \alpha), \qquad i \in H, \quad j \in H^c,$$

see Sub-section 3.4.

6.4. PROJECTION THEOREMS

§ 6.4.1. Functions. Decomposition theorems discussed earlier in this section, when applied to functions or measures, are collectively known as *projection theorems*, because of their involvement with the concept of a balayage.

Let H be a fixed subset of the state space $\mathbb{I}$, and consider the usual matrix operators: resolvent (α-potential) $\mathbf{U}^{\alpha}$, taboo potential ${}_H\mathbf{U}^{\alpha}$, α-balayage $\mathbf{B}^{\alpha}$, α-co-balayage $\mathbf{C}^{\alpha}$, and so on. For a function f and a measure μ, a function $\mathbf{U}^{\alpha} f$ and a measure $\mu \mathbf{U}^{\alpha}$ (similarily for other kernels) have been already introduced in § 1.3.1.

Let $a \geqslant 0$ be a non-negative charge (function) defined on $\mathbb{I}$, and let $p^{\alpha} = \mathbf{U}^{\alpha} a$ be its α-potential (see § 1.3.2 for definition). Note that conversely, $(-\mathbf{Q}^{\alpha}) p^{\alpha} = a$. It is not assumed that a vanishes on H^c; in fact, such an assumption would reduce projection theorems to a trivial special case.

DEFINITION 1. The α-balayage of an α-potential p^{α} is given by

$$\bar{p}^{\alpha} = \mathbf{B}^{\alpha} p^{\alpha},$$

or, explicitely,

$$\bar{p}^{\alpha}(i) = \begin{cases} \sum\limits_{j \in H} d_{ij}^{\alpha}\, p^{\alpha}(j), & i \in H^c, \\ p^{\alpha}(i), & i \in H. \end{cases}$$

THEOREM 2 (projection theorem). *Let $a \geqslant 0$ be a charge on $\mathbb{I}$, and let $p^{\alpha} = \mathbf{U}^{\alpha} a$ be its α-potential ($\alpha > 0$). Then, the α-balayage of p^{α} on H is given by*

$$\mathbf{B}^{\alpha} p^{\alpha}(i) = \begin{cases} p^{\alpha}(i) - {}_H\mathbf{U}^{\alpha} a(i), & i \in H^c, \\ p^{\alpha}(i), & i \in H. \end{cases}$$

Proof. This is the immediate consequence of the first entrance decomposition (Thm. 1, § 6.1.1):

$$p^\alpha = \mathbf{U}^\alpha a = {}_H\mathbf{U}^\alpha a + \mathbf{B}^\alpha \mathbf{U}^\alpha a = {}_H\mathbf{U}^\alpha a + \mathbf{B}^\alpha p^\alpha. \qquad \square$$

It is now clear that $\bar{p}^\alpha$ is the projection of p^α on H with

$$\mathbf{B}^\alpha \bar{p}^\alpha = \bar{p}^\alpha, \qquad \bar{p}^\alpha \leqslant p^\alpha \quad \text{and} \quad \bar{p}^\alpha = p^\alpha \quad \text{on } H.$$

If the charge a were supported on H, then $\bar{p}^\alpha$ and p^α would coincide. Theorem 2 holds also for $\alpha = 0$, provided $\mathbf{U}$ is finite.

Although p^α, for fixed α, is purely excessive relative to $\mathbf{P}(t)$, it is not immediately clear whether $\bar{p}^\alpha$ is even excessive. When dealing with the minimal chain, it may be expected that $\bar{p}^\alpha$ is also an α-potential of some charge (hence purely excessive). This important fact is verified by the next theorem which is the specialization of the general result due to G. A. Hunt; cf. R. M. Blumenthal and R. K. Getoor [1968, p. 136].

THEOREM 3. *Let $p^\alpha = \mathbf{U}^\alpha a$ be an α-potential of a charge $a \geqslant 0$ on $\mathbb{I}$ $(\alpha > 0)$, and let $\bar{p}^\alpha = \mathbf{B}^\alpha p^\alpha$ be its α-balayage on H. Then, there exists the unique charge $\bar{a}^\alpha$, supported by H, given by $\bar{a}^\alpha = \mathbf{C}^\alpha a$, such that $\bar{p}^\alpha$ is the α-potential of this charge: $\bar{p}^\alpha = \mathbf{U}^\alpha \bar{a}^\alpha$.*

Proof. Using duality (§ 6.3.1),

$$\bar{p}^\alpha = \mathbf{B}^\alpha p^\alpha = \mathbf{B}^\alpha\, \mathbf{U}^\alpha a = \mathbf{U}^\alpha\, \mathbf{C}^\alpha a = \mathbf{U}^\alpha\, \bar{a}^\alpha \qquad \text{with} \quad \bar{a}^\alpha = \mathbf{C}^\alpha a,$$

and it is clear that $\bar{a}^\alpha$ vanishes on H^c. Moreover, from the (B)-equation we have

$$(\alpha \mathbf{I} - \mathbf{Q})\bar{p}^\alpha = \bar{a}^\alpha. \qquad \square$$

Observe that the new charge is given explicitely by

$$\bar{a}^\alpha(i) = \begin{cases} a(i) + \sum\limits_{j \in H^c} n_{ij}^\alpha\, a(j), & i \in H, \\ 0, & i \in H^c. \end{cases}$$

Theorem 3 holds also for $\alpha = 0$, provided potentials are finite.

§ **6.4.2. Measures.** The operator $\mathbf{C}^\alpha$ acts on measures in the same manner as $\mathbf{B}^\alpha$ acts on functions. The following theorems are therefore dual to those from § 6.4.1. Let $\sigma \geqslant 0$ be a charge (measure) on $\mathbb{I}$; its α-potential has been defined as $\eta^\alpha = \sigma \mathbf{U}^\alpha$.

DEFINITION 1. The α-co-balayage on H of an α-potential η^α is given by

$$\bar{\eta}^\alpha = \eta^\alpha \mathbf{C}^\alpha.$$

THEOREM 2 (projection theorem). *The α-co-balayage on H of an α-potential $\eta^\alpha = \sigma \mathbf{U}^\alpha$ of a charge $\sigma \geqslant 0$ is given by*

$$\eta^\alpha \mathbf{C}^\alpha = \begin{cases} \eta^\alpha - \sigma\, {}_H\mathbf{U}^\alpha & \text{on} \quad H^c, \\ \eta^\alpha & \text{on} \quad H. \end{cases}$$

Proof. This is the immediate consequence of the last exit decomposition (Thm. 1, § 6.2.1), and the fact that $\mathbf{C}^\alpha$ is idempotent. □

Thus, $\bar{\eta}^\alpha$ is the projection of η^α on H with $\bar{\eta}^\alpha \leqslant \eta^\alpha$ and $\bar{\eta}^\alpha = \eta^\alpha$ on H. If the charge σ were supported on H, then $\bar{\eta}^\alpha$ and η^α would coincide on $\mathbb{I}$. Theorem 2 holds also for $\alpha = 0$, provided $\mathbf{U}$ is finite.

THEOREM 3. *Let $\bar{\eta}^\alpha = \eta^\alpha \mathbf{C}^\alpha$ be an α-co-balayage on H of an α-potential $\eta^\alpha = \sigma \mathbf{U}^\alpha$ of a charge σ. Then there exists the unique charge $\bar{\sigma}^\alpha$, given by $\bar{\sigma}^\alpha = \sigma \mathbf{B}^\alpha$ and supported by H, such that $\bar{\eta}^\alpha$ is the α-potential of this charge: $\bar{\eta}^\alpha = \bar{\sigma}^\alpha \mathbf{U}^\alpha$.*

Proof. The result follows from the duality and the (F)-equation. □

Theorem 3 holds also for $\alpha = 0$, provided potentials are finite. Observe that

$$\bar{\sigma}^\alpha = (-\mathbf{Q}^\alpha)\bar{\eta}^\alpha$$

and that $\eta^\alpha \mathbf{C}^\alpha \leqslant \eta^\alpha$ is equivalent to $\varpi^\alpha \mathbf{K}^\alpha \leqslant \varpi^\alpha$, where $\varpi^\alpha = \eta^\alpha \boldsymbol{\Theta}^\alpha$ (cf. § 3.4.4, and Prop. 4, § 1.3.3).

§ 6.4.3. Remark. As already mentioned in discussion of balayage $\mathbf{B}$ in § 5.3.3, the balayage principles discussed there, made use of projection theorems from § 6.4.1. For further comments on projection theorems see § 8.5.2.

The following argument makes interesting use of decomposition theorems (and has been already employed in Section 5). Theorem 1, § 5.1.1, (see also Thm. 5, § 3.1.3) is, in matrix notation,

$$\mathbf{U}^\alpha = {}_H\mathbf{U}^\alpha + \mathbf{B}^\alpha\, \mathbf{U}^\alpha, \qquad \alpha > 0.$$

Using the (F)-equation $\mathbf{U}^\alpha \mathbf{Q}^\alpha = -\mathbf{I}$, one can deduce from the above that

$$\mathbf{I} - \mathbf{B}^\alpha = \alpha\, {}_H\mathbf{U}^\alpha - {}_H\mathbf{U}^\alpha\, \mathbf{Q}, \qquad \alpha > 0.$$

Let f be an excessive function with respect to $\mathbf{U}^\alpha$, having the Riesz decomposition $f = p + h$ (see Thm. 1, § 1.3.3), and note that

$$\alpha\, {}_H\mathbf{U}^\alpha f \leqslant \alpha\, \mathbf{U}^\alpha f \leqslant f.$$

Hence, for the invariant component h, one has

$$h - \mathbf{B}^\alpha h = \alpha\, {}_H\mathbf{U}^\alpha h, \qquad \alpha > 0.$$

Suppose in addition that the invariant component h is also purely excessive relative to ${}_H\mathbf{U}^\alpha$. Then letting $\alpha \to 0$, one obtains $h = \mathbf{B}h$. As an example, take $h \equiv 1$, so one recovers

$$\mathbf{1} - d^\alpha = \alpha\, {}_H\mathbf{U}^\alpha \mathbf{1},$$

and taking the limit as $\alpha \to 0$, the second condition yields $D \equiv 1$.

On the other hand, the purely excessive component p is also purely excessive relative to ${}_H\mathbf{U}^\alpha$, so in the limit as $\alpha \to 0$

$$p - \mathbf{B}p = {}_H\mathbf{U}a, \qquad \text{where} \quad a = (-\mathbf{Q})p,$$

which should be compared with $p = \mathbf{U}a$.

SECTION 7

EXAMPLES

The classical examples presented in this section serve well as illustration of the main theme of this book. They have been selected both for their analytic properties and for general interest. Most of them have long history and result from work of a long series of researchers. Brevity of presentation should not detract from importance of these examples. Although they are different in content and complexity, yet they exhibit the unity of structure deduced from the general theory of passage times developed in this book.

The material is grouped in three sub-sections according to type of Markov chains. Analytic details are omitted (they can be found in references) with attention directed to passage times properties, in particular the Dirichlet problem and the Poisson equation for the first entrance probability to a set.

The Sub-section 7.1 treats briefly very simple situations with explicit calculation of taboo probabilities and balayage. The Sub-section 7.2 is devoted to the well-known birth-and-death process. Explicit expression for solution of the Dirichlet problem is stated for the most typical situation. Examples include the busy period in the M/M/1 queue, absorption time in the linear process, and others. The last Sub-section 7.3 treats some problems (not of the birth-and-death type), like a branching process, random walk, and "pole climbing" which leads to simple explicit expressions.

Examples presented here received considerable attention in literature, and are discussed, with various degree of completeness, in almost every book on M.C.'s and their applications. There is also enormous literature devoted to birth-and-death processes, in particular the early work of S. Karlin and J. McGregor. Applications to queueing problems (like waiting times, output) were discussed in R. Syski [1986]. The recent book by D. Aldous [1989] on "Poisson heuristic" approach contains many interesting examples.

Further examples involving martingales and functionals are presented later.

7.1. SIMPLE ILLUSTRATIONS

Examples collected here are rather intuitive, have very simple structure and do not require extensive calculations. Yet, they are very instructive illustrations of passage times properties.

§ **7.1.1. Elementary examples.** See Fig. 7.1.1 for details.

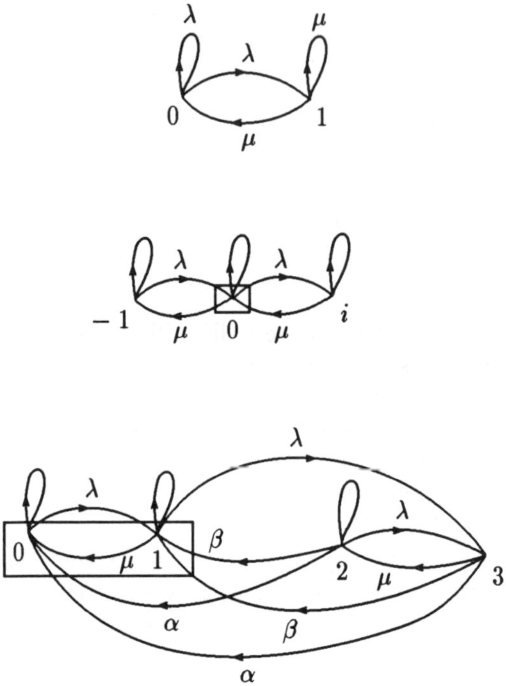

Ex. 1. $H = \{0\}$

Ex. 2. $H = \{0\}$

Ex. 3. $H = \{0,1\}, H^c = \{2,3\}$

Figure 7.1.1. Examples.

Example 1. Two-state system. This classical birth-and-death system has only two states 0 and 1 (idle-busy, on-off etc.) with the birth coefficient λ and the death coefficient μ. The common solution of the (B)- and (F)-equations in the transformed form is:

$$u_{00}^{\alpha} = \frac{\alpha+\mu}{\alpha(\alpha+\lambda+\mu)}, \qquad u_{01}^{\alpha} = \frac{\lambda}{\alpha(\alpha+\lambda+\mu)},$$

$$u_{10}^{\alpha} = \frac{\mu}{\alpha(\alpha+\lambda+\mu)}, \qquad u_{11}^{\alpha} = \frac{\alpha+\lambda}{\alpha(\alpha+\lambda+\mu)}.$$

The passage distributions are therefore exponential or convolutions of exponentials:

$$f_{00}^{\alpha} = \frac{\lambda}{\alpha+\lambda}\cdot\frac{\mu}{\alpha+\mu}, \qquad f_{01}^{\alpha} = \frac{\lambda}{\alpha+\lambda},$$

$$f_{10}^{\alpha} = \frac{\mu}{\alpha+\mu}, \qquad f_{11}^{\alpha} = \frac{\lambda}{\alpha+\lambda}\cdot\frac{\mu}{\alpha+\mu};$$

and the renewal densities are

$$n_{01}^{\alpha} = \frac{\lambda}{\alpha+\mu}, \qquad n_{10}^{\alpha} = \frac{\mu}{\alpha+\lambda}.$$

Obviously, the chain is ergodic, and the steady-state distribution is

$$\varepsilon_0 = \mu/(\lambda+\mu), \qquad \varepsilon_1 = \lambda/(\lambda+\mu).$$

Let T be the first entrance time to the state 0. It is obvious that now

$$D_1(t) = 1 - e^{-\mu t}, \qquad t \geqslant 0,$$

and $D_1 = 1$. The Dirichlet equation reduces now to triviality $D_0 = D_1$.

Let L be the last exit from state 0. It is obvious that $L = \infty$, so $s = 1$.

Example 2. Consider the system discussed in Example 3, § 1.2.6, consisting of three states -1, 0 and 1. Take now $H = \{0\}$ and $H^c = \{-1, 1\}$. It is clear that

$$D^c_{-1}(t) = e^{-\mu t}, \qquad D^c_1(t) = e^{-\lambda t}.$$

Example 3. Consider a finite chain with states 0, 1, 2, and 3, with the intensity matrix of the form

$$\begin{array}{llll} q_{01} = \lambda = q_0, & & & \\ q_{10} = \mu, & q_{13} = \lambda, & q_1 = \lambda + \mu, & \\ q_{20} = \alpha, & q_{21} = \beta, & q_{23} = \lambda, & q_2 = \lambda + \alpha + \beta, \\ q_{30} = \alpha, & q_{31} = \beta, & q_{32} = \mu, & q_3 = \mu + \alpha + \beta. \end{array}$$

Take $H = \{0, 1\}$ and $H^c = \{2, 3\}$. Then, Theorem 3, § 3.1.3, yields equations for D_{ij}:

$$\begin{aligned} &\alpha - (\lambda + \alpha + \beta)D_{20} + \lambda D_{30} = 0, \\ &\beta - (\lambda + \alpha + \beta)D_{21} + \lambda D_{31} = 0, \\ &\alpha + \mu D_{20} - (\mu + \alpha + \beta)D_{30} = 0, \\ &\beta + \mu D_{21} - (\mu + \alpha + \beta)D_{31} = 0. \end{aligned}$$

The solution is

$$D_{20} = D_{30} = \alpha/(\alpha + \beta), \qquad D_{21} = D_{31} = \beta/(\alpha + \beta),$$

and obviously $D_2 = D_3 = 1$.

§ **7.1.2. Gambler's ruin problem.** This classical birth-and-death process has a finite number of states $0, 1, \ldots, m$, with states 0 and m absorbing, and with constant coefficients λ and μ for $i = 1, \ldots, m-1$. The random variable X_t represents now gambler's capital at time t. It is required to find probability of absorption in 0 ("ruin") prior to absorption in m.

In order to cast the problem in the framework of the Dirichlet problem, it is natural to take

$$H = \{0, m\}, \qquad H^c = \{1, \ldots, m-1\}.$$

Let T be the first entrance time to H. The required probability of gambler's ruin is therefore

$$D_{i0} = \mathbb{P}^i(X_T = 0,\ T < \infty), \qquad i = 1, \ldots, m-1.$$

The situation is that corresponding to Example 5, § 5.4.1, and the equations reduce to

$$\begin{aligned} \mu - (\lambda+\mu)D_{10} + \lambda D_{20} &= 0, && i=1,\\ \mu D_{i-1,0} - (\lambda+\mu)D_{i0} + \lambda D_{i+1,0} &= 0, && i = 2, \dots, m-2,\\ \mu D_{m-2,0} - (\lambda+\mu)D_{m-1,0} &= 0, && i = m-1, \end{aligned}$$

with the boundary conditions $D_{00} = 1$, $D_{m0} = 0$.

The solution is found to be

$$D_{i0} = \begin{cases} \dfrac{1-(\lambda/\mu)^{m-i}}{1-(\lambda/\mu)^m} & \text{for } \lambda \neq \mu,\\ 1 - i/m & \text{for } \lambda = \mu, \end{cases}$$

for $i = 0, 1, \dots, m$.

Observe that here $D_i = D_{i0} + D_{im} = 1$ for $i \in H^c$, so the Dirichlet problem equations and the Poisson equations for D_i coincide (as required by Example 4, § 5.4.1).

For another approach, see the next sub-section.

7.2. BIRTH-AND-DEATH PROCESS

This well-known M.C. $X = (X_t)$ offers interesting properties and received considerable attention in literature. It is assumed here that the state space $\mathbb{I}$ is the set of all non-negative integers, and that the intensity matrix $\mathbf{Q} = (q_{ij})$ is determined by the birth and death coefficients λ_i and μ_i. See Example 1, § 1.1.9, for definitions.

§ 7.2.1. Single state taboo. This classical case considers the first entrance time T to the state 0. Thus, $H = \{0\}$ and $H^c = \{1, 2, \dots\}$, and D_i is the probability that T is finite, conditional on the initial state i. As typical applications represent time to absorption at 0 like busy periods and waiting times in Markovian queues, it is often convenient to refer to T as the absorption time (in agreement with the cut-off method).

It has been already shown in this chapter that determination of absorption probability D_i is equivalent to finding the minimal solution of the Dirichlet problem. The corresponding equations become now

$$\mu_i D_{i-1} - (\mu_i + \lambda_i) D_i + \lambda_i D_{i+1} = 0, \qquad i \geqslant 1,$$

with $D_0 = 1$.

The explicit solution found by standard methods is given by the following theorem; for the proof see S. Karlin and J. McGregor [1957], S. Karlin and M. H. Taylor [1975].

THEOREM 1. *In the birth-and-death process, probability D_i of the first entrance time to state* 0*, from an initial state $i \geqslant 1$, is given by*

$$D_i = \begin{cases} A_i / A_\infty & \text{when } A_\infty < \infty,\\ 1 & \text{when } A_\infty = \infty, \end{cases}$$

where

$$A_i = \sum_{j=1}^{\infty} \varrho_j^{-1},$$

and

$$\varrho_j = \prod_{k=1}^{j} (\lambda_k/\mu_k) \qquad \text{for} \quad j \geqslant 1, \qquad \varrho_0 = 1.$$

The proof uses the following lemma which depends essentially on the fact that the state space $\mathbb{I}$ is the set of all non-negative integers.

LEMMA 2. *Suppose that $D_1 < 1$. Then $D_i \to 0$ as $i \to \infty$.*

Proof. Assume the contrary, i.e., that $D_i \geqslant \varepsilon > 0$ for all i. Then, by § 4.1.1 and § 5.2.3, the probability s_{i0} of infinitely many visits to state 0 (starting at the initial state i) is

$$s_{i0} = \lim_{t \to \infty} \mathbf{P}(t) D_i \geqslant \varepsilon > 0, \qquad i \geqslant 0.$$

Recall that $s_{i0} = f_{i0} s_{00}$ (see Thm. 9, § 4.1.1, or K. L. Chung [1967, p. 19]). Hence, also $s_{00} \geqslant \varepsilon > 0$, and therefore by the 0–1 law, $s_{00} = 1$. Thus, the state 0 is recurrent and so (by the class property) is the whole chain. Then, clearly $D_i = f_{i0} = 1$ for all i and this is the contradiction. Hence, $D_\infty = 0$. □

As an immediate consequence of Lemma 2, one has

COROLLARY 3. *The birth-and-death process is recurrent iff $D_i = 1$ for all i (and then $D_i = s_{i0}$, $p_i = 0$), transient iff $D_i < 1$ for all i (and then $D_i = p_i$, $s_{i0} = 0$); the Riesz decomposition being $D_i = p_i + s_{i0}$.*

With reference to the Poisson equation, the escape rate b_i is found to be

$$b_i = 0 \qquad \text{for} \quad i \geqslant 1, \qquad b_0 = \lambda_0 D_1^c,$$

so $D_1^c = D_0^{\rho c}$. Furthermore, for the transient chain

$$p_i = \mathbf{U} b_i = u_{i0} b_0, \qquad i \geqslant 0,$$

and $p_i = 0$ for recurrent chain.

On the other hand, $\bar{q}_1 = \mu_1$, $\bar{q}_j = 0$ for $j > 1$, so using taboo probabilities, the solution of the Dirichlet problem may be written as

$$D_i = {}_0u_{i1} \mu_1, \qquad i \geqslant 1.$$

With reference to the passage probabilities from § 1.1.7, it is clear that

$$D_i = f_{i0} \qquad \text{for} \quad i \geqslant 1, \qquad D_0^\rho = f_{00}.$$

More generally,

$$d_i^\alpha = f_{i0}^\alpha \qquad \text{for} \quad i \geqslant 1, \qquad d_0^{\rho\alpha} = f_{00}^\alpha,$$

and

$$f_{i0}^\alpha = u_{i0}^\alpha / u_{00}^\alpha \qquad \text{for} \quad i \geqslant 1, \qquad f_{00}^\alpha = f_{10}^\alpha(\lambda_0/(\alpha + \lambda_0)).$$

Dually, one has from § 3.4.1 that for $\alpha > 0$:

$$n_{0j}^{\alpha} = \lambda_0 {}_0u_{1j}^{\alpha} = (\alpha + \lambda_0) {}_0u_{0j}^{\rho\alpha}, \qquad j \geqslant 1,$$

$$\alpha n_0^{\alpha} = \lambda_0 (1 - d_1^{\alpha}) = (\alpha + \lambda_0)(1 - d_0^{\rho\alpha}),$$

and

$$n_{0j}^{\alpha} = u_{0j}^{\alpha} / u_{00}^{\alpha} \qquad \text{for} \quad j \geqslant 1, \qquad \alpha u_{00}^{\alpha}(1 + n_0^{\alpha}) = 1.$$

The decomposition theorems take now the form $(i, j \geqslant 1)$

$$u_{ij}^{\alpha} = \begin{cases} {}_0u_{ij}^{\alpha} + d_{i0}^{\alpha}\, u_{0j}^{\alpha}, \\ {}_0u_{ij}^{\alpha} + u_{i0}^{\alpha}\, n_{0j}^{\alpha}. \end{cases}$$

Finally, the conditional mean $m_i = \mathbb{E}^i(T,\ T < \infty)$ satisfies the equation

$$\mu_i m_{i-1} - (\lambda_i + \mu_i) m_i + \lambda_i m_{i+1} = -D_i$$

for $i \geqslant 1$, with $m_0 = 0$. An explicit solution for m_i (when $D_i = 1$) is available; see references. Furthermore, the mean return time is

$$m_0^{\rho} = m_1 + D_0^{\rho} / \lambda_0.$$

Consider now the last exit time L from state 0. By Theorem 1, § 4.1.1, the complementary distribution function of L, conditional on initial state i, is given by

$$\Gamma_i^c(t) = \sum_{j=0}^{\infty} p_{ij}(t) D_j, \qquad i \geqslant 0, \quad t \geqslant 0.$$

On the other hand, Example 1, § 4.1.4, yields

$$\Gamma_i(t) = f_{i0} + \int_0^t p_{i0}(s) b_0\, ds,$$

with $\Gamma_i^c = s_{i0}$, as already noted.

Remark. Similar analysis may be applied to the case when the taboo set consists of a single state different from 0, say $H = \{k\}$. The structure of the birth-and-death chain indicates that state k can only be entered through its neighbouring states $k-1$ and $k+1$ (in infinitesimal time). The equations for the Dirichlet problem are now

$$\mu_i D_{i-1} - (\mu_i + \lambda_i) D_i + \lambda_i D_{i+1} = 0 \qquad \text{for} \quad i \neq k,$$

with $D_{-1} = 0$ and $D_k = 1$. It follows that

$$D_i = 1 \qquad \text{for} \quad i \leqslant k, \qquad D_i \leqslant 1 \qquad \text{for} \quad i > k.$$

§ 7.2.2. Stopped process. A birth-and-death process stopped at the first entrance to state 0 is obtained by modifying the original chain by making state 0 an absorbing state ($\lambda_0 = 0$). According to the cut-off method (see § 1.2.4) the transition probabilities $\tilde{p}_{ij}(t)$ of the stopped process coincide with the taboo probabilities of the original chain:

$$\tilde{p}_{ij}(t) = {}_0p_{ij}(t) \qquad \text{for} \quad i,j \geqslant 1,$$
$$\tilde{p}_{i0}(t) = D_{i0}(t) = D_i(t) \qquad \text{for} \quad i \geqslant 1, \quad j = 0.$$

As already explained, the first entrance time T to state 0 for the original chain coincides now with the absorption time for the stopped process.

The (F)-equations for α-potentials

$$\tilde{u}^{\alpha}_{ij} = {}_0u^{\alpha}_{ij}, \qquad i,j \geqslant 1,$$

for the stopped process are (for $i \geqslant 1$):

$$\begin{aligned} \alpha\tilde{u}^{\alpha}_{i0} &= \tilde{u}^{\alpha}_{i1}\,\mu_1, & j &= 0,\\ \tilde{u}^{\alpha}_{i1}(\alpha + \lambda_1 + \mu_1) - \delta_{i1} &= \tilde{u}^{\alpha}_{i2}\,\mu_2, & j &= 1,\\ \tilde{u}^{\alpha}_{ij}(\alpha + \lambda_j + \mu_j) - \delta_{ij} &= \tilde{u}^{\alpha}_{i,j-1}\lambda_{j-1} + \tilde{u}^{\alpha}_{i,j+1}\mu_{j+1}, & j &\geqslant 2. \end{aligned}$$

The (F)-equation for $D_i(t)$ reduces then to

$$D_i(t) = \int_0^t {}_0p_{i1}(s)\mu_1\,ds, \qquad i \geqslant 1,$$

with $D_i = {}_0u_{i1}\,\mu_1$, as already noted.

Next, the first passage relations (from § 1.1.7) give

$$\alpha\tilde{u}^{\alpha}_{i0} = \tilde{f}^{\alpha}_{i0} = \tilde{d}^{\alpha}_i \qquad \text{for} \quad i \geqslant 1,$$

because for the absorbing state $\tilde{u}^{\alpha}_{00} = 1/\alpha$. Hence, in the limit,

$$\varepsilon_{i0} = \tilde{f}_{i0} = D_i \qquad \text{for} \quad i \geqslant 1.$$

Obviously, for the stopped process, $\tilde{f}_{00} = 1$, $\tilde{f}_{ii} < 1$ for $i \geqslant 1$.

It should be pointed out that the function D_i is now invariant relative to the transition matrix $\tilde{\mathbf{P}}(t)$ of the stopped process (see § 5.1.3).

§ 7.2.3. Excessive functions. Let $\mathbf{P}(t)$ be a minimal matrix of the birth-and-death chain, with a conservative intensity matrix $\mathbf{Q}$. Assume that both equations (B) and (F) hold. For a function $f \geqslant 0$ on $\mathbb{I}$ define a charge by $a = (-\mathbf{Q})f$, so

$$a(0) = \lambda_0 f(0) - \lambda_0 f(1),$$
$$a(i) = -\mu_i f(i-1) + (\lambda_i + \mu_i) f(i) - \lambda_i f(i+1).$$

By Theorem 7, § 1.3.3, a finite function $f \geqslant 0$ on $\mathbb{I}$ is $\mathbf{P}(t)$-excessive iff $a(i) \geqslant 0$ for all i, and it is $\mathbf{P}(t)$-invariant iff all $a(i)$ vanish (the boundary potential w vanishes, because of the (F)-equation). It is easy to see that the only invariant functions are constant.

By Theorem 6, § 1.3.3, $\mathbf{P}(t)$-excessive function f has the unique decomposition $f = p + h$, where p is purely excessive and h is invariant. In the present case, one has also $p = \mathbf{U}a$, so p is a potential of a charge and $w = 0$, as noted. By the Blackwell theorem (§ 1.3.3) for recurrent chain $\mathbf{U}a = 0$ and all excessive functions are constant, but for the transient chain $\mathbf{U}a$ is finite. Evidently, $(-\mathbf{Q})f = (-\mathbf{Q})p$.

Although expressions for excessive functions can be found in terms of a charge $a \geqslant 0$, they are rather complicated. Instead, it is more instructive to consider the Poisson equation for f, with $H = \{0\}$. It turns out that $f(i)$ is a multiple of D_i; moreover, the effect of $f(\infty)$ is examined.

THEOREM 1. *Let $\mathbf{P}(t)$ be a stochastic matrix of the birth-and-death chain. Then the only solution of the Poisson equation for f:*

$$\mathbf{Q}f(i) = \begin{cases} 0, & i = 1, 2, \dots, \\ -a(0), & i = 0, \end{cases}$$

is given by $f(i) = D_i f(0)$.

Proof. Write $g(i) = f(i) - f(i+1)$ for all $i \geqslant 0$. Then, one has from the above equations for $i \geqslant 1$ that

$$g(i) = (\mu_i/\lambda_i)\, g(i-1) = (1/\varrho_i)\, g(0),$$

(see notation in § 7.2.1). Addition over $0, 1, \dots, i-1$ yields

$$f(0) - f(i) = A_i g(0), \qquad i \geqslant 1.$$

Clearly, $f(i)$ is decreasing, so, in the limit,

$$f(0) - f(\infty) = A_\infty g(0).$$

Hence, by elimination,

$$f(i) = (1 - A_i/A_\infty) f(0) + (A_i/A_\infty) f(\infty) = D_i f(i) + D_i^c f(\infty).$$

Note that for a recurrent chain $f(i) = f(0)$ for all i.

Next, using $f(1)$ from the above expression, one finds that

$$a(0) = \lambda_0 [f(0) - f(1)] = \lambda_0 D_1^c [f(0) - f(\infty)] = b_0 [f(0) - f(\infty)].$$

Suppose now that f has decomposition $f = h + \mathbf{U}a + w$, where h is invariant and w is a boundary potential. The potential of a charge $a \geqslant 0$ would then be

$$\mathbf{U}a(i) = u_{i0} b_0 [f(0) - f(\infty)].$$

Then from the decomposition one would have $w(i) = u_{i0}\, b_0 f(\infty)$. Thus, vanishing of w implies that $f(\infty) = 0$, so $\mathbf{U}a(i) = u_{i0}\, b_0 f(0)$. □

Remark. Consider an invariant function h (necessarily constant), and take its balayage $\bar{h} = \mathbf{B}h$, where $\mathbf{B} = (D_{i0})$ is the balayage operator. By Proposition 1, § 5.3.2, $\bar{h}$ is excessive on $\mathbb{I}$ and invariant on H^c, so

$$\mathbf{Q}\bar{h}(i) = \begin{cases} 0, & i \geqslant 0, \\ -\bar{a}(0), & i = 0. \end{cases}$$

Recall that $\mathbf{Q}\bar{h} = \mathbf{QB}h$, so (using Thm. 2, § 3.3.3)

$$\bar{a}(0) = (-\mathbf{Q})\bar{h}(0) = b_0 h(0).$$

Note that for $h = 1$, one has situation considered in Lemma 2, § 7.2.1.

§ 7.2.4. Examples. The following examples are well known, but are presented here for illustrative purpose; see references for details.

Example 1. M/M/1 *queue*. This classical example offers many features of theoretical interest, in addition to practical applications. The chain $X = (X_t)$ is of the birth-and-death type, with coefficients $\lambda_i = \lambda$ for $i \geqslant 0$, $\mu_i = \mu$ for $i \geqslant 1$ ($\mu_0 = 0$), and X_t represents the number of customers present at time t in the single server queue with infinite waiting room.

Although the equilibrium solution can be found in most of elementary textbooks, the time dependent solution for transition probabilities $p_{ij}(t)$ requires more advanced methods. The attempts to obtain the explicit solution have long history, and interest in this problem continues up to the present time. For full discussion see J. W. Cohen [1982] and R. Syski [1986], and for recent development P. R. Parthasarathy [1987] and R. Syski [1988]; see also series of papers by J. Abate and W. Whitt [1988]. The solution is expressed in terms of Bessel functions, but it will not be quoted here.

The expression $\varrho = \lambda/\mu$ is called a traffic intensity, and it is known that the chain is ergodic when $\varrho < 1$, recurrent null when $\varrho = 1$, and transient when $\varrho > 1$. In the ergodic case, the steady state solution has geometric form

$$\varepsilon_j = (1 - \varrho)\varrho^j, \qquad j = 0, 1, \ldots$$

In the transient case, the explicit expression for the potential $\mathbf{U} = (u_{ij})$ is

$$u_{ij} = \begin{cases} \dfrac{1}{\lambda - \mu}(\mu/\lambda)^{i-j}, & j \leqslant i, \\ \dfrac{1}{\lambda - \mu}, & j > i. \end{cases}$$

Take now $H = \{0\}$. Then, the first entrance time T to state 0 represents the busy period in the queue, starting at state i (the proper busy period corresponds to $i = 1$). Finding probability D_i that T is finite is equivalent to solving of the Dirichlet problem, and the equations for D_i are

$$\mu D_{i-1} - (\lambda + \mu) D_i + \lambda D_{i+1} = 0, \qquad i \geqslant 1,$$

with $D_0 = 1$. The solution is found from Theorem 1, § 7.2.1,

$$D_i = \begin{cases} 1 & \text{for } \varrho \leqslant 1, \\ (\mu/\lambda)^i & \text{for } \varrho > 1. \end{cases}$$

The escape rate obtained from the Poisson equation is

$$b_0 = \begin{cases} \lambda - \mu > 0 & \text{for transient case,} \\ 0 & \text{for recurrent case,} \end{cases}$$

so $D_0^\rho = D_1$.

Writing $D_i(t)$ for distribution of T, it follows that density $d_i(t)$ of the busy period has the form

$$d_i(t) = (\mu/\lambda)^{i/2}(i/t)\, e^{-(\lambda+\mu)t} I_i(2\sqrt{\lambda u t}), \qquad t > 0, \quad i \geqslant 1,$$

(which integrates to D_i) where $I_i(x)$ is the modified Bessel function of the first kind of order i. It then follows that $m_i = \mathbb{E}^i(T,\ T < \infty)$ is

$$m_i = \begin{cases} \dfrac{i}{\lambda - \mu}\,(\mu/\lambda)^i & \text{for } \lambda > \mu, \\[2ex] \dfrac{i}{\mu - \lambda} & \text{for } \lambda < \mu, \end{cases}$$

and $m_i = \infty$ when $\lambda = \mu$. Note however, that $\mathbb{E}^i T = m_i$ for $\varrho < 1$, but $\mathbb{E}^i T = \infty$ for $\varrho \geqslant 1$.

Next, using distribution of the last exit time L from state 0, stated in § 7.2.1, one finds for the transient case ($\lambda > \mu$) that

$$\Gamma_i^c(t) = \varphi_i(t, \mu/\lambda), \qquad t \geqslant 0,$$

where $\varphi_i(t,z)$ is the generating function of $p_{ij}(t)$. Thus, $\Gamma_i^c = 0$. For the recurrent case ($\lambda \leqslant \mu$), $\Gamma_i^c(t)$ is 1 identically, in agreement with notion of recurrence.

Consider now the reverse problem of finding an excessive function f given a charge $a \geqslant 0$. As all invariant functions are constant, it suffices to consider only a potential of a charge $p = \mathbf{U}a$. Assuming transience of the chain, take for simplicity $a(i) = (\mu/\lambda)^i$ for all i. Then

$$p(i) = \sum_{j=0}^{\infty} u_{ij}\, a(j) = \frac{\lambda + i(\lambda - \mu)}{(\lambda - \mu)^2} \left(\frac{\mu}{\lambda}\right)^i, \qquad i \geqslant 0.$$

Example 2. M/M/∞ *queue.* This is the system with infinitely many servers, described by the birth-and-death coefficients of the form $\lambda_i = \lambda$, $\mu_i = i\mu$ for all $i \geqslant 0$. As it is well known, the chain is always ergodic, and the equilibrium distribution is Poisson with mean λ/μ. Furthermore, the explicit expression for $p_{ij}(t)$ is the convolution of the Poisson and binomial distributions with generating function of the form

$$\sum_{j=0}^{\infty} p_{ij}(t) z^j = \left(1 - (1 - z)e^{-\mu t}\right)^i e^{-(1-z)a(t)},$$

where

$$a(t) = (\lambda/\mu)\big(1 - e^{-\mu t}\big),$$

with the mean

$$\mathbb{E}\big(X_t \mid X_0 = i\big) = a(t) + ie^{-\mu t}, \qquad t \geqslant 0.$$

Taking $H = \{0\}$, $D_i(t)$ represents distribution of the busy period (starting at i), and obviously $D_i = 1$ for all $i \geqslant 0$. Again $d_{i0}^{\alpha} = f_{i0}^{\alpha}$, and the passage relations give

$$d_{i0}^{\alpha} = u_{i0}^{\alpha}/u_{00}^{\alpha}, \qquad i \geqslant 1.$$

The explicit expressions are rather complicated. Indeed, it is known that

$$u_{i0}^{\alpha} = \int_0^{\infty} e^{-\alpha t}\big(1 - e^{-\mu t}\big)^i e^{-a(t)}\, dt.$$

Introducing the ergodic potential $\mathbf{Z} = (z_{ij})$ defined by

$$z_{ij} = \lim_{\alpha \to 0} \big(u_{ij}^{\alpha} - \varepsilon_j/\alpha\big),$$

one can find that the mean busy period is

$$m_i = (z_{00} - z_{i0})e^{\lambda/\mu}, \qquad i \geqslant 0.$$

For details see R. Syski [1978].

The first passage relation yields the mean return time

$$m_0^{\rho} = (1/\lambda)e^{\lambda/\mu}.$$

Finally, note that for the taboo probabilities

$${}_0u_{i0} = 1/\mu, \qquad i \geqslant 1.$$

Evidently, by ergodicity of the chain, the distribution $\Gamma_i(t) = 0$ identically.

Example 3. *Linear birth-and-death process.* This process received considerable attention in literature because of its interesting properties. Its coefficients are $\lambda_i = i\lambda$, $\mu_i = i\mu$ for all $i \geqslant 0$, so state 0 is absorbing by construction. It follows that the solution of Dirichlet problem is, by Theorem 1, § 7.2.1, for $i \geqslant 0$:

$$D_i = \begin{cases} (\mu/\lambda)^i & \text{if} \quad \lambda > \mu, \\ 1 & \text{if} \quad \lambda \leqslant \mu. \end{cases}$$

The escape rate b_i vanishes for all i, so $D_i = s_i$, for $i \geqslant 0$ (D being invariant in this case). Observe that now $p_{ij}(t)$ coincide with taboo probabilities ${}_0p_{ij}(t)$, by the cut-off method (see § 7.2.2), so $D_i(t) = p_{i0}(t)$.

It is known that

$$D_i(t) = \begin{cases} \left(\frac{\mu}{\lambda}\right)^i \left(\frac{e^{(\lambda-\mu)t}-1}{e^{(\lambda-\mu)t}-\mu/\lambda}\right)^i, & \lambda \neq \mu, \\ \left(\frac{\lambda t}{1+\lambda t}\right)^i, & \lambda = \mu, \end{cases}$$

for $t \geqslant 0,\ i \geqslant 1$. The mean for $i = 1$ is found to be

$$m_1 = \frac{1}{\lambda}\sum_{j=1}^{\infty}\frac{1}{j}\frac{\lambda}{\mu} = -\frac{1}{\lambda}\log\left(1-\frac{\lambda}{\mu}\right) \qquad \text{for} \quad \lambda < \mu.$$

Moreover, the generating function for transition probabilities is given by

$$\sum_{j=0}^{\infty} p_{ij}(t) z^j = \left(\frac{\lambda z - \mu + \mu(1-z)e^{(\lambda-\mu)t}}{\lambda z - \mu + \lambda(1-z)e^{(\lambda-\mu)t}}\right)^i, \qquad i \geqslant 1, \quad \lambda \neq \mu,$$

with the mean

$$\mathbb{E}(X_t \mid X_0 = i) = ie^{(\lambda-\mu)t}, \qquad i \geqslant 1.$$

Evaluating the above generating function for $z = \mu/\lambda$, one finds from § 7.2.1, that

$$\Gamma_i^c(t) = (\mu/\lambda)^i, \qquad i \geqslant 0,$$

for $\mu < \lambda$, independently of t. This agrees with the obvious fact that

$$\Gamma_i^c = D_i = \Gamma_i^c(0) \qquad \text{for} \quad \mu < \lambda.$$

For $\mu \geqslant \lambda$, D_i and $\Gamma_i^c(t)$ are both identically 1.

Finally, $\bar{q}_i = \mu_1$ for $i = 1$, and 0 otherwise. Thus,

$$D_i = {}_0u_{i1}\,\mu_1 \qquad \text{for} \quad i \geqslant 1.$$

§ 7.2.5. Birth process. Although a birth process is a special case of a birth-and-death process corresponding to the upward transitions only ($\mu_i = 0$ for all i), its simple structure admits explicit solutions. The results given here are summarized for purpose of reference; see standard texts for details.

a) The state space $\mathbb{I}$ is again the set of all non-negative integers, and the intensity matrix $\mathbf{Q} = (q_{ij})$ is of the form

$$q_{i,i+1} = \lambda_i > 0, \qquad q_i = \lambda_i \qquad \text{for all} \quad i \geqslant 0, \qquad q_{ij} = 0 \qquad \text{otherwise}.$$

By definition, transition probabilities are of interest only for $j \geqslant i$, and $p_{ij}(t) = 0$ for $i > j$. The (B)- and (F)-equations are easily written down, and their common solution (in transformed form) is

$$u_{ij}^{\alpha} = \frac{\lambda_i}{\alpha+\lambda_i} \cdots \frac{\lambda_{j-1}}{\alpha+\lambda_{j-1}} \frac{1}{\alpha+\lambda_j}, \qquad i \leqslant j,$$

with $u_{ij}^\alpha = 0$ for $i > j$. It is clear that in particular

$$p_{ii}(t) = e^{-\lambda_i t}$$

and that for $j \geqslant i$:

$$\lim_{t\to\infty} p_{ij}(t) = \lim_{\alpha\to 0} \alpha u_{ij}^\alpha = 0, \qquad u_{ij} = 1/\lambda_j,$$

so all states are transient.

The first passage relation $u_{ij}^\alpha = f_{ij}^\alpha\, u_{jj}^\alpha$ $(i \neq j)$ yields

$$f_{ij}^\alpha = u_{i,j-1}^\alpha\, \lambda_{j-1}, \qquad j > i,$$

which corresponds to convolution of exponentials with means $1/\lambda_i, \dots, 1/\lambda_{j-1}$. Similarly,

$$\left(1 - f_{ii}^\alpha\right)^{-1} = u_{ii}^\alpha\,(\alpha + \lambda_i) = 1$$

implies that for the first return to i one has $f_{ii}^\alpha = 0$, as it is obvious.

b) The birth process provides classical examples of sub-stochastic transition matrices $\mathbf{P}(t)$, as noted in § 1.1.5. Using notation introduced there, let τ be the life time of a birth process, and let $L_i(t)$ and l_i^α be its distribution function and its L.S. transform, respectively $(i \geqslant 0)$. Then, with $p_{ij}(t)$ denoting the minimal solution,

$$p_{i\mathrm{I\!I}}(t) := \sum_{j=0}^{\infty} p_{ij}(t) = \mathbb{P}^i(\tau > t),$$
$$L_i(t) = 1 - p_{i\mathrm{I\!I}}(t), \qquad t \geqslant 0.$$

Therefore (see § 1.3.3),

$$\alpha u_{i\mathrm{I\!I}}^\alpha + l_i^\alpha = 1, \qquad \alpha > 0, \quad i \geqslant 0.$$

It is known that

$$l_i^\alpha = \prod_{j=1}^{\infty} \frac{\lambda_j}{\lambda_j + \alpha}, \qquad \alpha > 0,$$

indicating that $L_i(t)$ is the infinite convolution of exponential distributions corresponding to life times of states $j \geqslant i$, and

$$\mathbb{E}^i(\tau,\ \tau < \infty) = \sum_{j=i}^{\infty} \frac{1}{\lambda_j}.$$

The famous 0–1 law (Feller–Lundberg) asserts then that $L_i = L_i(\infty)$ is either 0 or 1, according to whether the above sum diverges or converges.

Addition over all j of the (B)- and (F)-equations yields for the excessive function $\mathbf{1}$ with $\mathbf{Q1} = 0$:

$$\alpha u^{\alpha}_{i\mathbb{I}} + (-\mathbf{Q})u^{\alpha}_{i\mathbb{I}} = 1, \tag{B}$$

$$\alpha u^{\alpha}_{i\mathbb{I}} + w^{\alpha}_i = 1, \tag{aug (F)}$$

so $w^{\alpha}_i = (-\mathbf{Q})u^{\alpha}_{i\mathbb{I}} = l^{\alpha}_i$. Evidently, when $\mathbf{P}(t)$ is stochastic, then $w^{\alpha}_i = 0$ and $\alpha u^{\alpha}_{i\mathbb{I}} = 1$ (see comment at the end of § 1.3.3).

c) Let $\mathbf{P}(t)$ be a sub-stochastic minimal matrix of the birth process with a conservative intensity matrix $\mathbf{Q}$. By Theorem 7, § 1.3.3, a finite function $f \geqslant 0$ on $\mathbb{I}$ is $\mathbf{P}(t)$-excessive iff

$$a(i) = (-\mathbf{Q})f(i) = \lambda_i[f(i) - f(i+1)] \geqslant 0 \qquad \text{for all} \quad i \geqslant 0,$$

or

$$f(i) \geqslant f(i+1) \qquad \text{for all} \quad i \geqslant 0.$$

Note that constant functions are invariant, provided $\mathbf{P}(t)$ is stochastic. Indeed, a constant function f is only excessive, and has a representation $f = fL^c_i + fL_i$, where fL^c_i is invariant and $fL_i = w_i$ is the boundary potential.

The potential of a charge $a \geqslant 0$ now is

$$\mathbf{U}a(i) = \sum_{j=i}^{\infty} u_{ij}\, a(j) = \sum_{j=i}^{\infty} a(j)/\lambda_j = \sum_{j=i}^{\infty} [f(j) - f(j+1)] = f(i) - f(\infty).$$

Hence, one has a decomposition of an excessive function f

$$f(i) = \mathbf{U}a(i) + f(\infty)L_i + f(\infty)L^c_i, \qquad i \geqslant 0,$$

where $\mathbf{U}a(i) + f(\infty)L_i$ is a purely excessive component. Thus, when $f(\infty) = 0$, then f is necessarily a potential of a charge.

On the other hand, a measure $\xi \geqslant 0$ on $\mathbb{I}$ is excessive iff

$$\sigma(j) = \xi(-\mathbf{Q})(j) = \begin{cases} \xi(j)\lambda_j - \xi(j-1)\lambda_{j-1} \geqslant 0, & j \geqslant 1, \\ \xi(0)\lambda_0, & j = 0, \end{cases}$$

so all invariant ξ must vanish. Moreover,

$$\sum_{i=0}^{j} \sigma(i) = \xi(j)\lambda_j \uparrow \xi(\infty)\lambda_\infty = \sigma\bullet\mathbf{1},$$

so a non-zero excessive measure ξ must be infinite. Also, the boundary component vanishes (see remark after Thm. 7, § 1.3.3), so a non-vanishing excessive ξ is necessarily purely excessive and a potential of a charge $\sigma \geqslant 0$: $\xi = \sigma\mathbf{U}$.

d) Take now $H = \{k\}$, where k is a fixed state. Then, clearly, $D_i = 1$ for $i \leqslant k$, and $D_i = 0$ for $i > k$. The same result applies for $H = \{0, \ldots, k\}$. On the other hand, for $H = \{k+1, \ldots\}$, $D_i = 1$ for all i.

From Theorem 1, § 4.1.1, for distribution of the last exit from H one has

$$\Gamma_i(t) = \sum_{j=i}^{\infty} p_{ij}(t) \leqslant 1 \qquad \text{for} \quad i > k,$$

$$\Gamma_i(t) = \sum_{j=k+1}^{\infty} p_{ij}(t) \qquad \text{for} \quad i \leqslant k,$$

in the first two cases, and $\Gamma_i(t) = 0$ for all i in the third case.

e) Examples given here are well known, yet they are of great interest.

Example 1. *Poisson process.* Here $\lambda_i = \lambda$ for all $i \geqslant 0$, and the common solution of both equations is

$$p_{ij}(t) = \frac{(\lambda t)^{j-i}}{(j-i)!}\, e^{-\lambda t} \qquad \text{for} \quad j \geqslant i,$$

with the Laplace transform

$$u_{ij}^{\alpha} = \Big(\frac{\lambda}{\lambda + \alpha}\Big)^{j-i} \frac{1}{\lambda + \alpha}, \qquad j \geqslant i,$$

and $u_{ij} = 1/\lambda$ for $j \geqslant i$. Evidently,

$$f_{ij}^{\alpha} = \Big(\frac{\lambda}{\lambda + \alpha}\Big)^{j-i} \qquad \text{for} \quad j > i,$$

and $f_{ii}^{\alpha} = 0$, corresponding to the Erlang distribution and the impossibility of return.

Evidently, all excessive functions are decreasing, and invariant functions are constant. Moreover, every excessive function has decomposition $f = \mathbf{U}a + f(\infty)$, and f is purely excessive iff it is a potential of a charge a. On the other hand, the only invariant measure is the zero measure, and every excessive measure is increasing (hence infinite). Thus, every non-zero excessive measure is a potential of a charge, so also purely excessive.

Suppose that $H = \{0, 1, \dots, k\}$ for a fixed state k. As $\bar{q}_i = 0$ for $i \in H^c$, it follows that $D_i = 1$ for $i \leqslant k$, and $D_i = 0$ for $i \geqslant k+1$. Consequently, from the Poisson equation, the escape rate is $b_i = 0$ for $i = 0, 1, \dots, k-1$, and $b_k = \lambda$. Hence, the potential is $\mathbf{U}b_i = u_{ik}\lambda = 1$ for $i \leqslant k$, and 0 for $i \geqslant k+1$. Moreover, $s_i = 0$ for all i.

It is of interest to remark here that obviously the process will escape from a finite set H, but the escape probability $e_i = 0$ for $i < k$, yet $e_k = 1$. Indeed, the probability e_i refers to one step, as defined. On the other hand, the probability of escape from H is $\Gamma_i = 1$ for all i in H.

Example 2. *Linear growth* (*Yule–Furry process*). Take for the state space $\mathbb{I}$ the set of all positive integers, and let $\lambda_i = i\lambda$ for $i \geqslant 1$. Transition probabilities are given by the Pascal distribution

$$p_{ij}(t) = \binom{j-1}{j-i} e^{-i\lambda t}(1-e^{-\lambda t})^{j-i}, \qquad j \geqslant i \geqslant 1,$$

with the mean

$$\mathbb{E}^i X_t = ie^{\lambda t}.$$

Evidently, one has here the stochastic case with $p_{i\mathbb{I}}(t) = 1$. Moreover,

$$u_{ij} = 1/(j\lambda) \qquad \text{for} \quad j \geqslant i.$$

Example 3. *Quadratic growth*. Take for $\mathbb{I}$ the set of all positive integers, and let $\lambda_i = i^2\lambda$, $i \geqslant 1$. The series in the 0–1 law converges in this case, giving thus the simplest example of a sub-stochastic matrix $\mathbf{P}(t)$. It is known that now

$$L_i(t) = 1 + \sum_{j=i}^{\infty} a_j^{(i)} e^{-j^2\lambda t}, \qquad t \geqslant 0,$$

where

$$a_j^{(i)} = 2\binom{j+i-1}{i-1}\binom{j-1}{i-1}(-1)^{j-i+1} \qquad \text{for} \quad i \geqslant 1,$$

and

$$\mathbb{E}^i(\tau,\ \tau < \infty) = \frac{1}{\lambda}\sum_{j=i}^{\infty} j^{-2} < \infty.$$

§ 7.2.6. Death process. Although a death process is a special case of a birth-and-death process corresponding to downward transitions only ($\lambda_i = 0$ for all i), its special structure deserves separate discussion. The results given here are well known; see references for details.

a) The state space $\mathbb{I}$ is again the set of all non-negative integers, and the intensity matrix $\mathbf{Q} = (q_{ij})$ has the form

$$\begin{aligned} q_i &= q_{i,i-1} = \mu_i > 0 && \text{for } i \geqslant 1, \\ q_0 &= \mu_0 = 0, \\ q_{ij} &= 0 && \text{otherwise,} \end{aligned}$$

so 0 is by definition an absorbing state. Transition probabilities are of interest only for $i \geqslant j$, and $p_{ij}(t) = 0$ for $i < j$. The (B)- and (F)-equations are easily written down, and their common solution (in transformed form) is, for $0 \leqslant j \leqslant i$,

$$\begin{aligned} u_{ij}^{\alpha} &= \frac{1}{\alpha+\mu_j}\,\frac{\mu_{j+1}}{\alpha+\mu_{j+1}} \cdots \frac{\mu_i}{\alpha+\mu_i}, && j < i, \\ u_{jj}^{\alpha} &= \frac{1}{\alpha+\mu_j}, && j = i, \end{aligned}$$

with $u_{ij}^{\alpha} = 0$ for $j > i$. It is clear that all states, except the absorbing 0, are transient, and that

$$\sum_{j=0}^{i} p_{ij}(t) = 1.$$

The first passage relation $u_{ij}^{\alpha} = f_{ij}^{\alpha}\, u_{jj}^{\alpha}$ $(i \neq j)$ yields

$$f_{ij}^{\alpha} = u_{i,j+1}^{\alpha}\, \mu_{j+1} \qquad \text{for} \quad i > j,$$

which corresponds to the convolution of exponentials with means $1/\mu_{j+1}, \ldots, 1/\mu_i$. Clearly, $f_{jj}^{\alpha} = 0$. In particular, time to reach the absorbing state 0 has transform

$$f_{i0}^{\alpha} = u_{i1}^{\alpha}\, \mu_1 = \alpha u_{i0}^{\alpha}.$$

The following examples are classical.

Example 1. $\mu_i = i\mu$, $i \geqslant 0$. One has now

$$p_{ij}(t) = \binom{i}{j} e^{-j\mu t}\left(1 - e^{-\mu t}\right)^{i-j}, \qquad 0 \leqslant j \leqslant i,$$

$$f_{i0}(t) = i\mu e^{-\mu t}\left(1 - e^{-\mu t}\right)^{i-1}, \qquad i \geqslant 0.$$

Example 2. $\mu_i = \mu$ for $i \geqslant 1$, $\mu_0 = 0$. Then,

$$p_{ij}(t) = \frac{(\mu t)^{i-j}}{(i-j)!}\, e^{-\mu t}, \qquad j = 1, 2, \ldots, i,$$

$$p_{i0}(t) = 1 - \sum_{k=0}^{i-1} \frac{(\mu t)^k}{k!}\, e^{-\mu t},$$

and

$$f_{i0}(t) = \frac{(\mu t)^{i-1}}{(i-1)!}\, e^{-\mu t} \mu, \qquad i \geqslant 1.$$

b) *Duality*. Consider now a birth process with transition probabilities $p_{ij}(t)$ and coefficients λ_i $(i \geqslant 0)$. Let ξ be a purely excessive measure with charge $\sigma = \xi(-\mathbf{Q})$ defined in § 7.2. The dual chain has transition probabilities defined by

$$\hat{p}_{ij}(t) = \xi(j) p_{ji}(t)/\xi(i) \qquad \text{for} \quad j \leqslant i$$

(see § 1.3.5). It is clear that $\hat{p}_{ij}(t)$ correspond to a death process, but with an adjoint state Δ, so

$$\hat{p}_{i\Delta}(t) = 1 - \sum_{j=0}^{i} \hat{p}_{ij}(t).$$

The associated intensities are

$$\hat{q}_{i,i-1} = \xi(i-1)\lambda_{i-1}/\xi(i) \qquad \text{for} \quad i \geqslant 1,$$
$$\hat{q}_i = \lambda_i \qquad \text{for all} \quad i > 0.$$

Hence, the matrix $\widehat{\mathbf{Q}}$ is not conservative, and

$$\hat{q}_{i\Delta} = \sigma(i)/\xi(i) \geqslant 0, \qquad i \geqslant 0.$$

Thus, from every state $i \geqslant 0$, the process may jump to the absorbing state Δ; in particular, for state 0 one has only $\hat{q}_0 = \lambda_0$ and $\hat{q}_{0\Delta} = \lambda_0$. The following examples illustrate the role of reference measure ξ.

Example 3. Consider the Poisson process with a parameter λ, and take $\xi(i) = 1$ for all $i \geqslant 0$. Thus $\sigma(i) = 0$ for $i \geqslant 1$, and $\sigma(0) = \lambda$. Hence,

$$\begin{aligned} \hat{\mu}_i &= \lambda \quad \text{for} \quad i \geqslant 1, & \hat{q}_i &= \lambda \quad \text{for} \quad i \geqslant 0, \\ \hat{q}_{i\Delta} &= 0 \quad \text{for} \quad i \geqslant 1, & \hat{q}_{0\Delta} &= \lambda, \end{aligned}$$

with a jump to Δ only from state 0. This is the reverse of the Poisson process.

Example 4. Take $\lambda_i = i\lambda$ for $i \geqslant 0$, and select $\sigma(i) = 1/\lambda_i$ for $i \geqslant 1$. Hence,

$$\hat{\mu}_i = \hat{q}_i = \lambda_i = i\lambda \qquad \text{for} \quad i \geqslant 1, \qquad \hat{q}_{i\Delta} = 0,$$

and $\hat{q}_0 = \lambda_0 = 0$.

Thus, 0 is the absorbing state and one obtains the proper death process (no jumps to Δ) with $\hat{\mu}_i = i\lambda$ for all $i \geqslant 0$. This is Example 1 which is dual to Example 2, § 7.2.5.

§ 7.2.7. Special example. The example considered here exhibits interesting connections with renewal argument and equations for passage distributions; see R. Syski [1986, p. 128].

a) Consider a birth-and-death process and define the auxiliary probability

$$S_{ij}(t) = \sum_{k=j}^{\infty} p_{ik}(t), \qquad i, j \geqslant 0, \quad t \geqslant 0.$$

Evidently, $S_{i0}(t) = 1$ and

$$S_{ij}(0) = \begin{cases} 1 & \text{if} \quad i \geqslant j, \\ 0 & \text{if} \quad i < j. \end{cases}$$

The Chapman–Kolmogorov equation implies that for a fixed j, $S_{\cdot j}(t)$ form the exit law relative to $\mathbf{P}(t)$:

$$S_{ij}(t+h) = \sum_k p_{ik}(t) S_{kj}(h), \qquad h, t \geqslant 0$$

(see § 1.3.4).

The (B)- and (F)-equations for transition probabilities (see Ex. 1, § 1.1.9) imply that

$$\frac{dS_{ij}(t)}{dt} = \begin{cases} \mu_i S_{i-1,j}(t) - (\lambda_i + \mu_i) S_{ij}(t) + \lambda_i S_{i+1,j}(t), & \text{(B)} \\ p_{i,j-1}(t)\lambda_{j-1} - p_{ij}(t)\mu_j. & \text{(F)} \end{cases}$$

However, $S_{ij}(t)$ need not be a monotone function of t. It is of interest to examine situation when $S_{ij}(t)$ behaves like a (complementary) distribution function.

For this purpose consider the partition of the state space

$$H = \{0, 1, \dots, j-1\}, \qquad H^c = \{j, j+1, \dots\},$$

and let $D_i^c(t)$ be the complementary distribution function of the first entrance time T to the set H. Equations for $D_i^c(t)$, $i \geqslant j$, are (see Thm. 6, § 3.1.1):

$$\frac{dD_i^c(t)}{dt} = \mu_i D_{i-1}^c(t) - (\lambda_i + \mu_i) D_i^c(t) + \lambda_i D_{i+1}^c(t), \qquad i > j,$$
$$\frac{dD_j^c(t)}{dt} = -(\lambda_j + \mu_j) D_j^c(t) + \lambda_j D_{j+1}^c(t), \qquad i = j,$$

with $D_i^c(0) = 1$.

Comparison of equations for $S_{ij}(t)$ and $D_i^c(t)$ indicates agreement when $D_{j-1}^c(t) = S_{j-1,j}(t) = 0$. Indeed, according to the cut-off method, all states in H are made absorbing, and therefore $S_{ij}(t) = D_i^c(t)$ for $i \geqslant j$ (fixed j). Actually, $S_{ij}(t)$ is the sum of taboo probabilities in this case.

Passage to the limit leads to the equations for D_i in the Dirichlet problem

$$\mu_i D_{i-1} - (\lambda_i + \mu_i) D_i + \lambda_i D_{i+1} = 0, \qquad i \geqslant j,$$
$$D_i = 1, \qquad i < j.$$

Proceeding as in Theorem 1, § 7.2.1, for $i \geqslant j$ the solution is found to be

$$D_i = \begin{cases} \dfrac{A_i}{1 + A_j} & \text{when} \quad A_j < \infty, \\ 1 & \text{when} \quad A_j = \infty, \end{cases}$$

where

$$A_i = \sum_{k=i}^{\infty} (1/\varrho_k) \qquad \text{for} \quad i \geqslant j$$

and

$$\varrho_i = \frac{\lambda_j}{\mu_j} \cdots \frac{\lambda_i}{\mu_i}, \qquad i \geqslant j.$$

In particular, for the M/M/1 queue this gives for $\mu < \lambda$:

$$D_i = \left(\frac{\mu}{\lambda}\right)^{i-j+1} \qquad \text{for} \quad i \geqslant j.$$

Finally, it follows from definition that

$$p_{ij}(t) = S_{ij}(t) - S_{i,j+1}(t),$$

so if the mean is finite, it is given by

$$\mathbb{E}^i X_t = \sum_{j=1}^{\infty} S_{ij}(t), \qquad i \geqslant 0, \quad t \geqslant 0.$$

Remark. Reversing H and H^c, one obtains the dual version with $S_{ij}(t)$ corresponding to the distribution function $D_i(t)$; see R. Syski [1986, p. 582].

b) The birth-and-death process provides natural interpretation of the first return distribution $D_i^\rho(t)$. With the same sets H and H^c as above, the transform $d_i^{\rho\alpha}$ satisfies the (B)-equation from Theorem 9, § 3.3.1, which reduces now for $i \in H$ to the following equations

$$(\alpha + \lambda_i + \mu_i)d_i^{\rho\alpha} = \lambda_i + \mu_i \qquad \text{for} \quad i = 0, 1, \ldots, j-2,$$
$$(\alpha + \lambda_{j-1} + \mu_{j-1})d_{j-1}^{\rho\alpha} = \mu_{j-1} + \lambda_{j-1}d_j^\alpha \qquad \text{for} \quad i = j-1.$$

Thus, $D_i^\rho(t)$ is simply exponential distribution with parameter $\lambda_i + \mu_i$ for $i < j-1$, but for $i = j-1$ one has

$$D_{j-1}^{\rho c} = \mu_{j-1}(\lambda_{j-1} + \mu_{j-1})^{-1} D_j^c.$$

7.3. SPECIAL CHAINS

Examples selected here have complex structure and lead to rather involved calculations. However, they have interesting properties and received considerable attention in literature. Only passage time aspects will be emphasized here.

§ 7.3.1. "Pole climbing". This is one of a few non-trivial examples where simple explicit expressions are available. The M.C. is characterized by

$$\begin{array}{llll} q_{i,i+1} = \lambda & \text{for all } \ i \geqslant 0, & q_{i0} = \mu & \text{for } \ i \geqslant 1, \\ q_{ij} = 0, & \text{otherwise } (i \neq j), & & \\ q_i = \lambda + \mu & \text{for } \ i \geqslant 1, & q_0 = \lambda. & \end{array}$$

See Fig. 7.3.1.

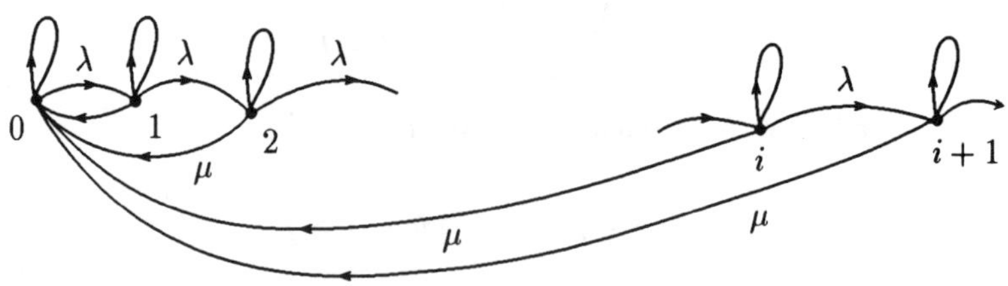

Figure 7.3.1. "Pole climbing".

The (F)-equation

$$\frac{dp_{ij}(t)}{dt} = p_{i,j-1}(t)\lambda - p_{ij}(t)(\lambda + \mu), \qquad j \geqslant 1,$$
$$\frac{dp_{i0}(t)}{dt} = -p_{i0}(t)\lambda + \mu \sum_{k=1}^{\infty} p_{ik}(t), \qquad j = 0,$$

can be solved using the generating function of the transform u_{ij}^α, and the solution is found to be

$$u_{ij}^\alpha = \begin{cases} \dfrac{\mu}{\alpha(\alpha+\mu+\lambda)}\left(\dfrac{\lambda}{\alpha+\mu+\lambda}\right)^j, & j < i, \\ \dfrac{\mu}{\alpha(\alpha+\mu+\lambda)}\left(\dfrac{\lambda}{\alpha+\mu+\lambda}\right)^j + \dfrac{1}{\alpha+\mu+\lambda}\left(\dfrac{\lambda}{\alpha+\mu+\lambda}\right)^{j-i}, & j \geqslant i. \end{cases}$$

Take now $H = \{0\}$. Using passage relations from § 1.1.7, it follows that

$$f_{i0}^\alpha = \frac{\mu}{(\mu+\alpha)} \qquad \text{for} \quad i \geqslant 1, \qquad f_{00}^\alpha = \frac{\lambda}{\lambda+\alpha}\frac{\mu}{\mu+\alpha}.$$

Thus, $D_i(t)$ is exponential, and $D_0^\rho(t)$ is the convolution of two exponentials, as may be expected. Moreover, $D_i = 1$ for all $i \geqslant 0$, in agreement with the fact that the chain is ergodic, with ergodic distribution

$$\varepsilon_j = \frac{\mu}{\mu+\lambda}\left(\frac{\lambda}{\mu+\lambda}\right)^j, \qquad j \geqslant 0.$$

The (F)-equations for taboo probabilities ${}_0p_{ij}(t)$ for $i, j \geqslant 1$ have the form

$$\begin{aligned} \frac{d\,{}_0p_{ij}(t)}{dt} &= {}_0p_{i,j-1}(t)\lambda - {}_0p_{ij}(t)(\lambda+\mu), \qquad && j > i, \\ \frac{d\,{}_0p_{ii}(t)}{dt} &= -{}_0p_{ii}(t)(\lambda+\mu), && j = i, \end{aligned}$$

and ${}_0p_{ij}(t) = 0$ for $j < i$. Using again the generating function, the solution is found to have Poisson form

$${}_0u_{ij}^\alpha = \frac{1}{\alpha+\mu+\lambda}\left(\frac{\lambda}{\alpha+\mu+\lambda}\right)^{j-i}, \qquad j \geqslant i.$$

Hence, again

$$D_i^c(t) = \sum_{j=i}^{\infty} {}_0p_{ij}(t) = e^{-\mu t}$$

and

$$\sum_{j=i}^{\infty} {}_0u_{ij}^\alpha = \frac{1}{(\alpha+\mu)}.$$

On the other hand,

$$n_{0k}^\alpha = \left(\frac{\lambda}{\alpha+\mu+\lambda}\right)^k = u_{0k}^\alpha / u_{00}^\alpha, \qquad k \geqslant 1.$$

Hence,

$$n_0^\alpha = \lambda/(\alpha+\mu), \qquad b_0^\alpha = \alpha\big(n_0^\alpha + 1\big).$$

Consequently,

$$d_i^\alpha = \mathbf{U}^\alpha b_i^\alpha = u_{i0}^\alpha b_0^\alpha = \mu/(\alpha+\mu), \qquad i \geqslant 1,$$

in agreement with earlier result. Note however that $b_0 = 0$, because of ergodicity.

§ 7.3.2. Random walk. This classical example received considerable attention in literature. However, even in the case as simple as this one, powerful analytic methods are needed.

a) The state space $\mathbb{I}$ is now a set of all integers $0, \pm1, \pm2, \ldots$. The intensity matrix $\mathbf{Q}$ is defined by

$$q_{i,i-1} = \mu, \qquad q_{i,i+1} = \lambda, \qquad q_i = \lambda + \mu \qquad \text{for all } i;$$
$$q_{ij} = 0 \qquad \text{otherwise.}$$

In applications X_t represents the position of the random walk at time t with (infinitesimal) transitions restricted to the nearest neighbour ("simple random walk"). As q_i is constant, this is also an example of a compound Poisson process. Furthermore, it is clear that transition probabilities $p_{ij}(t)$ are functions of difference $j - i$ (space homogeneity). Then the (B)- and (F)-equations have the same form, and they are reduced to the single equation

$$\frac{\partial \psi_i}{\partial t} = [\lambda z - (\lambda + \mu) + \mu/z]\psi_i$$

for the generating function ψ_i, defined by

$$\psi_i(t, z) = \sum_j p_{ij}(t) z^j, \qquad 0 < |z| < 1.$$

The solution is found to be

$$\psi_i(t, z) = z^i e^{-t\varphi(z)}, \qquad i = 0, \pm1, \pm2, \ldots,$$

with exponent

$$\varphi(z) = \lambda + \mu - \lambda z - \mu/z.$$

Note that $\psi_i(t, 1) = 1$ indicating that the minimal solution is stochastic. The conditional mean is easily found to be

$$\mathbb{E}^i X_t = i + (\lambda - \mu)t,$$

and it can be verified that the chain X is a process with independent increments.

The explicit expression for transition probabilities is known to be

$$p_{i,i+k}(t) = (\lambda/\mu)^{k/2} e^{-(\lambda+\mu)t} I_k(2t\sqrt{\lambda\mu}), \qquad t \geqslant 0,$$

valid for $j - i = k = 0, \pm1, \pm2, \ldots$, where $I_k(x)$ is the modified Bessel function of the first kind of order k. Note that the above expression is the distribution of the difference of two independent Poisson random variables with means λt and μt. It has been obtained by inversion of the transform $u^{\alpha}_{i,i+k}$ which in turn depends on roots of the "characteristic equation"

$$\lambda z^2 - (\lambda + \mu + \alpha)z + \mu = 0, \qquad \alpha > 0.$$

Discussion of this aspect will be omitted here (Rouché theorem); see references.

It may be of interest, however, to state the limit expressions for $\alpha = 0$:

$$u_{i,i+k} = \begin{cases} (\mu-\lambda)^{-1}(\lambda/\mu)^k & \text{for } k \geqslant 0, \\ (\mu-\lambda)^{-1} & \text{for } k < 0, \end{cases} \quad \text{for } \lambda < \mu;$$

$$u_{i,i+k} = \begin{cases} (\lambda-\mu)^{-1} & \text{for } k \geqslant 0, \\ (\lambda-\mu)^{-1}(\mu/\lambda)^{-k} & \text{for } k < 0, \end{cases} \quad \text{for } \lambda > \mu;$$

$$u_{i,i+k} = \infty \qquad \text{for all } k, \qquad \text{for } \lambda = \mu.$$

Moreover, it is known that the random walk is transient when $\lambda \neq \mu$, and recurrent null when $\lambda = \mu$.

Finally, note that for invariant measures

$$\xi(j-1)\lambda - \xi(j)(\lambda+\mu) + \xi(j+1)\mu = 0 \qquad \text{for all } j,$$

so $\xi(j) = 1$ and $\xi(j) = (\lambda/\mu)^j$ are two independent solutions for $\lambda \neq \mu$.

Dually, for invariant functions

$$\mu h(i-1) - (\lambda+\mu)h(i) + \lambda h(i+1) = 0 \qquad \text{for all } i,$$

the independent solutions are $h(i) = 1$ and $h(i) = (\mu/\lambda)^i$ for $\mu \neq \lambda$.

b) Take now $H = \{n\}$, where n is a fixed state. Equations for the Dirichlet problem now are

$$\mu D_{i-1} - (\lambda+\mu)D_i + \lambda D_{i+1} = 0 \qquad \text{for } i \neq n,$$

with $D_n = 1$. Solving in the usual way, one finds that

$$D_i = \begin{cases} (\mu/\lambda)^{i-n} & \text{for } i > n \\ 1 & \text{for } i < n \end{cases} \quad \text{when } \mu < \lambda;$$

$$D_i = \begin{cases} 1 & \text{for } i > n \\ (\lambda/\mu)^{n-i} & \text{for } i < n \end{cases} \quad \text{when } \mu > \lambda;$$

and

$$D_n^\rho = \begin{cases} 2\lambda/(\lambda+\mu) & \text{when } \mu > \lambda \\ 2\mu/(\lambda+\mu) & \text{when } \mu < \lambda \end{cases} \quad \text{for } i = n.$$

Note that

$$(\lambda+\mu)D_n^\rho = \lambda D_{n+1} + \mu D_{n-1}.$$

Remark. It is possible to obtain explicit expressions for densities $d_i(t)$ and taboo probability ${}_np_{ij}(t)$ in terms of Bessel functions $I_k(x)$.

§ 7.3.3. Compound Poisson process. This is actually a class of Markov chains with a bounded generator, say $q_i = \gamma$ — constant (see Ex. 2, § 1.1.9). It is instructive to illustrate dependence between the minimal chain and its jumps chain R, under taboo set H.

Applying the Poisson randomization to $(\mathbf{R}^a)^n$ from § 3.1.5, one finds

$$\mathbf{S}(t) = \sum_{n=0}^{\infty} (\mathbf{R}^a)^n \frac{(\gamma t)^n}{n!} e^{-\gamma t} = e^{t\mathbf{Q}^a},$$

with $\mathbf{Q}^a = \gamma(\mathbf{R}^a - \mathbf{I})$, whose non-zero components are $\mathbf{Q}_{11} = \gamma(\mathbf{R}_{11} - \mathbf{I}_1)$ and $\mathbf{Q}_{12} = \gamma \mathbf{R}_{12}$.

Hence, non-trivial components are

$$_H\mathbf{P}_{11}(t) = \sum_{n=0}^{\infty} \mathbf{R}_{11}^n \frac{(\gamma t)^n}{n!} e^{-\gamma t} = e^{t\mathbf{Q}_{11}},$$

$$\mathbf{B}_{12}(t) = \sum_{n=1}^{\infty} \left(\sum_{v=0}^{n-1} \mathbf{R}_{11}^v \mathbf{R}_{12} \right) \frac{(\gamma t)^n}{n!} e^{-\gamma t},$$

or, in terms of transforms,

$$_H\mathbf{U}_{11}^\alpha = \left(\alpha \mathbf{I}_1 - \mathbf{Q}_{11}\right)^{-1}, \qquad \mathbf{B}_{12}^\alpha = \gamma\, {_H\mathbf{U}_{11}^\alpha} \mathbf{R}_{12},$$

in agreement with Example 4, § 1.2.6.

It follows that

$$D_{ij}(t) = \sum_{n=1}^{\infty} D_{ij}^*(n) \frac{(\gamma t)^n}{n!} e^{-\gamma t}, \qquad i \in H^c, \quad j \in H,$$

where

$$D_{ij}^*(n) = \sum_{v=0}^{n-1} {_H r_{ij}^{(v+1)}}$$

is the balayage of a jump chain under taboo. This form represents Poisson randomization of a discrete chain (X_n) — see § 3.1.5. Expression of the same form is obtained for $D_{ij}^\rho(t)$ when $i \in H$, $j \in H$.

Finally, observe that

$$d_{ij}^\alpha = {_H u_{ij}^\alpha}(\alpha + \gamma) \qquad \text{for} \quad i \in H^c,\ j \in H,$$

$$n_{ij}^\alpha = (\alpha + \gamma) {_H u_{ij}^{\rho\alpha}} \qquad \text{for} \quad i \in H,\ j \in H.$$

Recall also that the Poisson process as well as random walk are special cases of the compound Poisson process.

§ 7.3.4. Branching process. This classical chain has been already introduced in Example 3, § 1.1.9. As state 0 is absorbing, transition probabilities $p_{ij}(t)$ coincide with the taboo probabilities ${}_0p_{ij}(t)$. Thus, T is the time to extinction defined as the time of the first entrance to the absorbing state 0, and $D_i(t)$ is its distribution function, conditional on initial population of size $i > 0$. Evidently, by branching property,

$$D_i(t) = p_{i0}(t) = \left(p_{10}(t)\right)^i = D_1^i(t),$$

because $D_1(t) = \psi_1(t, 0)$. Hence, the (B)-equation gives equation for $D_1(t)$:

$$\frac{dD_1(t)}{dt} = \beta\left[B(D_1(t)) - D_1(t)\right],$$

with initial condition $D_1(0) = 0$.

Probability of extinction is now $D_i = D_1^i$, and the Dirichlet problem leads to $B(D_1) - D_1 = 0$. By the famous branching theorem, D_1 is the smallest non-negative root of the equation $B(z) = z$, and $D_1 < 1$ iff $m > 1$, or equivalently, $D_1 = 1$ iff $m \leqslant 1$. See references for details.

It is easy to check that

$$\mathbb{E}^i X_t = ie^{\beta(m-1)t}, \qquad t \geqslant 0,$$

and that the Poisson equation yields $D_0^\rho = 1$.

Example 1. Take $\beta = \lambda + \mu$, and define $b(0) = \mu/\beta$, $b(2) = \lambda/\beta$, and $b(k) = 0$ otherwise. It is clear that this choice gives the linear birth-and-death process discussed in Example 3, § 7.2.4. The equation $B(z) = z$ reduces now to

$$(\lambda + \mu)z = \mu + \lambda z^2,$$

and the solution for D_1 is 1 when $\lambda/\mu \leqslant 1$, and μ/λ when $\lambda/\mu > 1$, in agreement with results stated earlier.

Example 2. Take a distribution with a generating function of the form

$$B(z) = 1 - (1 - z)^a + a(1 - z),$$

where $0 < a < 1$. It follows that $m = \infty$, so the chain is sub-stochastic with $\psi_i(t, 1) < 1$.

Example 3. For connection with busy period see § 13.3.3.

CHAPTER III

MEASURE THEORY

This chapter approaches discussion of passage times by methods of measure theory, and examines the role of passage times in transformations of Markov chains.

The basic tool is now a concept of a functional of a M.C., and its various ramifications. Associated martingales play also important role in discussion. Some of results obtained earlier are now re-derived by the new method, and many new results are included. This partial repetition is justified by simplicity of the direct method, and by the fact that measure theoretic methods require large amount of preparatory material.

For this reason Section 8 constitutes rather extensive introduction to functionals and potentials. Main characteristics of integral (IF) and purely discontinuous (DAF) functionals are described, with special attention to *compensators* (*or dual predictable projections*). Numerous applications are mentioned, including local times.

Section 9 treats typical transformations of Markov chains like killing, birthing, stopping, reversing and random time change. These well known operations in theory of general Markov processes are conveniently expressed in terms of functionals, and it is of interest to examine how they influence passage times.

Section 10 discusses several extensions like composite passage times, joint distributions of different passage times, consecutive entrances to a set, and excursions from a set. In particular, connection with Markov renewal processes is indicated.

Topics presented in this chapter have been discussed extensively in advanced literature on Markov processes and martingales. It is believed that presentation here will serve as an introduction to this field, with special emphasis on passage times. Individual references are given in the text.

SECTION 8

FUNCTIONALS AND POTENTIALS

This section is devoted to discussion of functionals and martingales associated with M.C.'s. The notion of a functional is a powerful tool in theoretical considerations of Markov processes, and its use in applications is widespread. Similarly, martingale methods offer many advantages and their use is growing in popularity.

Functionals provide an alternative approach to studies of passage times in M.C.'s. This approach, which is measure-theoretic in character, supplements analytic methods described earlier in Chapter II. It offers new insight into structure of passage times, simplifies presentation and presents neat unifying treatment, but its use requires extensive preliminaries. Situation is similar with use of martingales.

It is therefore desirable to summarize here basic ideas and properties of functionals and martingales in order to illustrate their applicability to passage times. Sub-section 8.1 discusses basic concepts from the above point of view, and Sub-section 8.2 treats extended functionals. Several results obtained in Chapter II are re-derived here using new and simpler method. Advantages of functionals are pointed out in Sub-section 8.3 which re-visits discussion of the last exit.

A local time, which in the case of the usual M.C.'s is a rather simple concept, makes possible by its simplicity a more detailed analysis of behavior of chains. In fact, a local time on a set may be used as the starting point of all considerations in Chapter II, when all material on functionals and martingales is available.

The last sub-section summarises basic properties of balayage, the concept essentially associated with passage times, which now is presented in a new light.

As always, presentation here is based on works dealing with general Markov processes, but specialization to the usual M.C.'s leads to considerable simplification of arguments. For complete treatment see R. M. Blumenthal and R. K. Getoor [1968], K. L. Chung [1982], C. Dellacherie and P. A. Meyer [1982, 1988], J. Doob [1984], D. Williams [1979]. Additional references are given in the text.

8.1. BASIC CONCEPTS

§ 8.1.1. Functions of a chain. Let $X = (X_t,\ 0 \leqslant t < \infty)$ be the usual M.C. on a (complete) probability space $(\Omega, \mathcal{F}, \mathbb{P})$, relative to an increasing family $(\mathcal{F}_t,\ 0 \leqslant t < \infty)$ of natural σ-fields. Taking càdlàg version of X, and assuming that each $\mathcal{F}_t$ contains all $\mathbb{P}$-null sets, it is known that the family $(\mathcal{F}_t)$ is right-continuous, i.e., for every $t \geqslant 0$:

$$\mathcal{F}_t = \mathcal{F}_{t+},$$

where

$$\mathcal{F}_{t-} = \sigma\Big(\bigcup_{s<t} \mathcal{F}_s\Big) \subset \mathcal{F}_t \subset \mathcal{F}_{t+} = \bigcap_{s>t} \mathcal{F}_s.$$

As $\mathcal{F}_t = \sigma(X_s,\ s \leqslant t)$, one can interpret $\mathcal{F}_{t-} = \sigma(X_s,\ s < t)$. Define also

$$\mathcal{G}_t = \sigma(X_s,\ s \geqslant t)$$

for each t, so $\mathcal{G}_0 = \mathcal{F}_\infty$.

The Markov property can also be expressed by saying that for each t, the σ-field $\mathcal{F}_t$ and the σ-field $\mathcal{G}_t$ are *conditionally independent* given X_t, that is,

$$\mathbb{P}(A \cap B \mid X_t) = \mathbb{P}(A \mid X_t)\mathbb{P}(B \mid X_t) \quad \text{a.e.}$$

for every $A \in \mathcal{F}_t$ and $B \in \mathcal{G}_t$. In words, the past $\mathcal{F}_t$ and the future $\mathcal{G}_t$ are independent, given the present X_t.

a) For any $f \in \mathcal{B}$, $f(X_t)$ is a random variable with values $f(X_t)(\omega) = f(i)$ when $X_t(\omega) = i$. Denoting by I_i the indicator of one point set $\{i\}$, so

$$I_i(X_t) = I_{(X_t=i)},$$

one can write

$$f(X_t) = \sum_i f(i) I_i(X_t).$$

The Markov property can also be expressed equivalently in still another form

$$\mathbb{E}\{f(X_{t+s}) \mid \mathcal{F}_s\} = \mathbb{E}\{f(X_{t+s}) \mid X_s\} \quad \text{a.e.}$$

for any $f \in \mathcal{B}$ and $t, s \geqslant 0$. The strong Markov property is expressed by validity of the above relation when times s are replaced by stopping times S (relative to the natural σ-fields).

Hence, one has generalization — and probabilistic interpretation — of analytic expressions stated earlier (in § 1.3.1):

$$\mathbf{P}(t) f(X_s) = \mathbb{E}\{f(X_{t+s}) \mid X_s\} \quad \text{a.e.},$$

with values

$$\mathbf{P}(t) f(i) = \mathbb{E}\{f(X_t) \mid X_0 = i\}, \qquad i \in \mathbb{I}.$$

In particular, for $f = I_i$, the above relations yield transition probabilities

$$p_{X_s, j}(t) = \mathbb{P}(X_{t+s} = j \mid X_s) \quad \text{a.e.},$$

with values $p_{ij}(t)$.

With the help of the shift, one can write

$$\mathbf{P}(t) f(X_s) = \mathbb{E}[f(X_t) \circ \theta_s \mid X_s] = \mathbb{E}^{X_s}[f(X_t)] \quad \text{a.e.}$$

The notation emphasises the fact that $\mathbf{P}(t)f(X_s)$ has been obtained from $\mathbf{P}(t)f(X_0)$ by shift by s. In particular, one may write

$$\mathbb{P}(X_{t+s} = j \mid X_s) = \mathbb{P}(X_t \circ \theta_s = j \mid X_s) = \mathbb{P}^{X_s}(X_t = j) = p_{X_s,j}(t) \quad \text{a.e.}$$

and

$$\mathbb{E}(Y \circ \theta_s \mid \mathcal{F}_s) = \mathbb{E}^{X_s}(Y) \quad \text{a.e.}$$

Notation $\mathbb{E}^i$ and $\mathbb{P}^i$ introduced earlier, for conditioning on $X_0 = i$, is consistent with notation described here.

The strong Markov property applied to random shift leads to the following notation. For any stopping time T:

$$\mathbf{P}(t)f(X_T) = \mathbb{E}[f(X_{T+t}) \mid X_T] = \mathbb{E}[f(X_t) \circ \theta_T \mid X_T] = \mathbb{E}^{X(T)}f(X_t),$$
$$\mathbb{E}(Y \circ \theta_T \mid \mathcal{F}_T) = \mathbb{E}^{X(T)}(Y).$$

Finally, by smoothing properties of conditional expectation, the strong Markov property may be expressed in the form

$$\mathbb{E}[Z \cdot Y \circ \theta_T] = \mathbb{E}[Z\mathbb{E}^{X(T)}(Y)]$$

for any $\mathcal{F}_\infty$-measurable random variable Y and any $\mathcal{F}_T$-measurable random variable Z.

Similarly, for resolvent with $\alpha > 0$,

$$\mathbf{U}^\alpha f(X_s) = \int_0^\infty e^{-\alpha t}\mathbf{P}(t)f(X_s)\,dt = \sum_j u^\alpha_{X_s,j} f(j)$$
$$= \mathbb{E}\left\{\int_0^\infty e^{-\alpha t} f(X_{t+s})\,dt \mid X_s\right\} \quad \text{a.e.},$$

with values

$$\mathbf{U}^\alpha f(i) = \mathbb{E}^i \int_0^\infty e^{-\alpha t} f(X_t)\,dt, \qquad i \in \mathbb{I}.$$

The generalized (weak) infinitesimal generator (also denoted by $\mathbf{Q}$) is defined by

$$\lim_{h\to 0} \mathbb{E}\left\{[f(X_{s+h}) - f(X_s)]/h \mid \mathcal{F}_s\right\} = \mathbf{Q}f(X_s) \quad \text{a.e.},$$

for $f \in \mathcal{B}$, $s \geqslant 0$. This in turn leads to the generalized Kolmogorov equations. In particular, the (F)-equation in the integrated form is

$$\mathbb{E}[f(X_{t+s}) \mid \mathcal{F}_s] - f(X_s) = \mathbb{E}\left[\int_s^{s+t} \mathbf{Q}f(X_u)\,du \mid X_s\right] \quad \text{a.e.}, \qquad \text{(F)}$$

or, in equivalent form,

$$\mathbf{P}(t)f(X_s) - f(X_s) = \int_s^{t+s} \mathbf{P}(u-s)\mathbf{Q}f(X_s)\,du \quad \text{a.e.,} \tag{F}$$

for $f \in \mathcal{B}$ and $t, s \geqslant 0$. For the (B)-equation write $\mathbf{QP}(u-s)$ in the above integral.

Turning now to the probabilistic interpretation of bounded measures, $\mu\mathbf{P}(t)$ defines (for any $\mu \in \mathcal{M}$) a measure $\mathbb{P}^\mu$ by

$$\mu\mathbf{P}(t)(j) = \mathbb{P}^\mu(X_t = j), \qquad j \in \mathbb{I}, \quad t \geqslant 0,$$

and

$$\mu\mathbf{U}^\alpha(j) = \int_0^\infty e^{-\alpha t}\,\mathbb{P}^\mu(X_t = j)\,dt, \qquad j \in \mathbb{I}, \quad \alpha > 0.$$

It may be convenient to write $p_{\mu j}(t)$ and $u^\alpha_{\mu j}$ for values of measures $\mu\mathbf{P}(t)$ and $\mu\mathbf{U}^\alpha$, respectively. In particular, as already noted, with π denoting the initial distribution of X_0, the (absolute) distribution of X_t is $\pi\mathbf{P}(t)$, so

$$\mathbb{P}(X_t = j) = \sum_i \pi_i p_{ij}(t) = p_{\pi j}(t), \qquad t \geqslant 0, \quad j \in \mathbb{I}.$$

More generally, one has, for $\mu \in \mathcal{M}$ and $f \in \mathcal{B}$, the expectation relative to measure $\mathbb{P}^\mu$:

$$\mathbb{E}^\mu f(X_t) = \mu\mathbf{P}(t)f,$$

so the "ordinary" expectation is written as

$$\mathbb{E}f(X_t) = \pi\mathbf{P}(t)f.$$

Note, however, that the identity function $1(i) = i$ is unbounded in general, but the expected values of X_t are

$$\mathbb{E}^i X_t = \mathbf{P}(t)\,1(i), \quad \mathbb{E}X_t = \pi\mathbf{P}(t)\mathbf{1},$$

provided $\mathbf{1}$ is integrable. Similarly, an important example of an unbounded measure is given by the counting measure.

b) More generally, if T and S are two stopping times (relative to the same family of σ-fields), then

$$T_S = S + T \circ \theta_S$$

is also a stopping time, and

$$X_T \circ \theta_S = X(T_S).$$

The stopping time T_S may be regarded as occurrence of event T after event S.

For any stopping time T, relative to the family $(\mathcal{F}_t)$, define the operator $\mathbf{P}^\alpha(T)$ by

$$\mathbf{P}^\alpha(T)f(i) = \mathbb{E}\big(e^{-\alpha T} f(X_T),\ T < \infty \mid X_0 = i\big), \qquad i \in \mathbb{I},$$

for every bounded measurable function f on $\mathbb{I}$ and $\alpha \geqslant 0$. Write $\mathbf{P}(T)$ when $\alpha = 0$.

Furthermore, for any stopping time S,

$$\mathbf{P}^\alpha(T)f(X_S) = \mathbb{E}\{e^{-\alpha T\circ\theta_S} f(X_{S+T\circ\theta_S}),\ T \circ \theta_S < \infty \mid X_S\} \quad \text{a.e.}$$

Of great importance in theoretical considerations is the following product property of operators $\mathbf{P}^\alpha(T)$ which extends semi-group property of operators $\mathbf{P}(t)$.

PRODUCT THEOREM. *For any two stopping times S and T:*

$$\mathbf{P}^\alpha(S)\,\mathbf{P}^\alpha(T) = \mathbf{P}^\alpha(S + T \circ \theta_S), \qquad \alpha \geqslant 0.$$

Proof is obtained by immediate verification from definition, using the strong Markov property and smoothing properties of conditional expectation.

Observe that when S and T are constant, say s and t, then the product theorem reduces to the semi-group property

$$\mathbf{P}^\alpha(s)\mathbf{P}^\alpha(t) = \mathbf{P}^\alpha(s+t),$$

as noted in § 1.1.6. Other special cases are

$$\begin{aligned} \mathbf{P}^\alpha(s)\mathbf{P}^\alpha(T) &= \mathbf{P}^\alpha(s + T \circ \theta_s), \\ \mathbf{P}^\alpha(S)\mathbf{P}^\alpha(t) &= \mathbf{P}^\alpha(S+t) \qquad \text{(strong Markov property)}. \end{aligned}$$

Observe that when $T = T_H$ is the first entrance time to a set H, then T_S is the first entrance time to H after a random time S, as already discussed in § 2.2.2 (in connection with delayed taboo). In this case, the product theorem permits to express the joint distribution of T_S and $X(T_S)$ in terms of the joint distribution of S and $X(S)$, and the joint distribution of T and $X(T)$. Recall definition of $D_{ij}(t)$, and write

$$\mathbb{P}^i\big(X(S) = j,\ S \leqslant t\big) = \widehat{D}_{ij}(t), \qquad \mathbb{P}^i\big(X(T_S) = j,\ T_S \leqslant t\big) = D^S_{ij}(t)$$

for all i and j in $\mathbb{I}$, and t, $0 \leqslant t < \infty$. Denote the corresponding L.S. transforms by d^α_{ij}, $\hat{d}^\alpha_{ij}$, and $d^{S\alpha}_{ij}$, and note that they are kernels of the operators $\mathbf{P}^\alpha(T)$, $\mathbf{P}^\alpha(S)$, and $\mathbf{P}^\alpha(T_S)$, respectively. Hence, the product theorem can be expressed in the form

$$\mathbf{B}^{S\alpha} = \widehat{\mathbf{B}}^\alpha \mathbf{B}^\alpha, \qquad \alpha \geqslant 0. \tag{PT}$$

Recall now special cases obtained earlier

$$\begin{aligned} \mathbf{B}^\alpha &= \mathbf{R}^{a\alpha}\mathbf{B}^\alpha && (\S\ 3.1.5), \\ \mathbf{B}^{s\alpha} &= e^{-s\alpha}\mathbf{P}(s)\mathbf{B}^\alpha && (\S\ 3.2.5\quad \text{for } S = s,\ \text{constant}), \\ \mathbf{B}^{\rho\alpha} &= \mathbf{R}^\alpha\mathbf{B}^\alpha && (\S\ 3.3.5\quad \text{for } S = \rho_i). \end{aligned}$$

Product theorem may be used as the starting point of all discussion of passage times, but the approach through taboo probabilities is more direct. Moreover, the product theorem supports the density argument described in § 2.2.5.

c) The following theorem is crucial in study of stopping times in Markov chains.

DYNKIN THEOREM. *Let $X = (X_t)$ be an usual* M.C. *with a semi-group $\mathbf{P}(t)$, resolvent $\mathbf{U}^\alpha$ and the operator $\mathbf{P}^\alpha(T)$, where T is a stopping time. Then, for any bounded function f on the state space $\mathbb{I}$,*

$$\mathbf{U}^\alpha f(i) = \mathbb{E}^i \int_0^T e^{-\alpha t} f(X_t)\,dt + \mathbf{P}^\alpha(T)\mathbf{U}^\alpha f(i), \qquad i \in \mathbb{I},$$

or, equivalently,

$$\mathbb{E}^i \int_T^\infty e^{-\alpha t} f(X_t)\,dt = \mathbb{E}^i\Big(e^{-\alpha T}\mathbf{U}^\alpha f(X_T)\Big), \qquad i \in \mathbb{I}$$

(the assertion holds also for $T=\infty$ upon defining $f(X_\infty)=0$).

For proof see references; see also discussion in § 8.1.3-d.

As a corollary, one obtain the *Dynkin formula* when $\mathbb{E}^i(T)<\infty$,

$$\mathbb{E}^i f(X_T) - f(i) = \mathbb{E}^i \int_0^T \mathbf{Q}f(X_u)\,du, \qquad i \in \mathbb{I}.$$

Note that the LHS of the above formula may be written as $\mathbf{P}(T)f-f$. In particular, taking $T=\rho_i$, one finds that Dynkin's "characteristic operator" (defined by exits from states) reduces in the case of minimal chain to considerations of the associated jump chain, and hence coincides with the generator $\mathbf{Q}$. Indeed, recalling that the first exit time ρ_i from a state i is exponential with mean $1/q_i$, the characteristic operator becomes

$$\frac{\mathbb{E}^i f(X(\rho_i)) - f(i)}{\mathbb{E}^i(\rho_i)} = q_i\big[\mathbf{R}f(i) - f(i)\big] = \mathbf{Q}f(i)$$

for $f\in\mathcal{B}$, $i\in\mathbb{I}$. Similarly,

$$\mathbf{R}^\alpha f(i) = \mathbb{E}^i\{e^{-\alpha\rho_i} f(X(\rho_i))\}$$

provides further interpretation of the α-chain (see § 1.1.6). Conversely, replacing $\mathbf{R}$ by some one-step transition matrix (for a discrete time chain) and selecting q_i arbitrarily, the above relation may be used to define $\mathbf{Q}$, and hence $\mathbf{P}(t)$.

As already noted, first entrance decomposition theorems (see Section 6) can be obtained from Dynkin theorem by specialization to a stopping time T_S. Indeed, in present notation,

$$\mathbf{U}^\alpha f(i) = {}_H\widehat{\mathbf{U}}^\alpha f(i) + \mathbf{B}^{S\alpha}\mathbf{U}^\alpha f(i), \qquad i\in\mathbb{I}, \quad \alpha>0.$$

For further discussion see § 10.1.5.

§ 8.1.2. Functionals. a) *Definitions.* Consider an usual M.C. $X=(X_t)$, as defined in § 1.1.1, with a discrete state space $\mathbb{I}$ and natural σ-fields $(\mathcal{F}_t)$. The following is the basic

DEFINITION 1. A family $A=(A_t,\ 0\leqslant t<\infty)$ of non-negative random variables on a probability space $(\Omega,\mathcal{F},\mathbb{P})$ is called an *additive functional* (AF) of a M.C. X when:

a) the paths of A are non-decreasing, right-continuous, and $A_0=0$;

b) for each t, A_t is $\mathcal{F}_t$-measurable;

c) for each t and h:

$$A_{t+h} = A_t + A_h \circ \theta_t,$$

where θ_t is the shift operator.

It is assumed that conditions a) and c) hold everywhere (such an AF is said to be perfect). Condition b) means that A is adapted to the increasing family $(\mathcal{F}_t)$. An AF is said to be a raw additive functional (RAF) if it is not adapted. A family of (integrable) random variables (A_t) satisfying conditions a) and b) only is said to be an *increasing process*.

A functional A is said to be *continuous* (CAF) if its paths are continuous; it is said to be *purely discontinuous* (DAF) if its paths are jump functions. A functional (or an increasing process) is said to be *integrable* if $\mathbb{E}A_\infty < \infty$.

For a non-negative function f on $\mathbb{I}$, the *α-potential* of f relative to an AF (or an increasing process) $A = (A_t)$ is defined by (the path integral)

$$\mathbf{U}_A^\alpha f(i) = \mathbb{E}^i \int_0^\infty e^{-\alpha t} f(X_t)\, dA_t = \sum_j u_A^\alpha(i,j)\, f(j), \qquad \alpha \geqslant 0, \quad i \in \mathbb{I},$$

where $u_A^\alpha(i,j)$ is the kernel of the potential operator $\mathbf{U}_A^\alpha$. When $f \equiv 1$, write $\mathbf{U}_A^\alpha \mathbf{1} = u_A^\alpha$. The function u_A^α is called the *α-potential* of a functional (increasing process) A:

$$u_A^\alpha(i) = \mathbb{E}^i \int_0^\infty e^{-\alpha t}\, dA_t, \qquad \alpha \geqslant 0, \quad i \in \mathbb{I}.$$

In particular, when $\alpha = 0$, the *potential* of A is

$$u_A(i) = \mathbb{E}^i A_\infty, \qquad i \in \mathbb{I}.$$

The *support of* A is a subset supp (A) of $\mathbb{I}$ such that A increases at a given instant t iff $X_t \in \operatorname{supp}(A)$. For any A, the function F defined by

$$F_i(t) = \mathbb{E}^i A_t, \qquad t \geqslant 0, \quad i \in \mathbb{I},$$

is called the *characteristic of* A.

The behavior of functionals under random shift is expressed by

$$A_{T_S} - A_T = A_S \circ \theta_T$$

for (finite) stopping time T and S, with $T_S = S + T \circ \theta_S$ (see § 8.1.1).

Of special importance are the following basic types of AF's.

1) An *integral functional* (IF) $A = (A_t)$ is defined by (the path integral)

$$A_t = \int_0^t a(X_s)\, ds, \qquad 0 \leqslant t < \infty,$$

where a is a non-negative bounded function on $\mathbb{I}$, called a *charge*. Note that paths of the charge process $(a(X_t),\ 0 \leqslant t < \infty)$ are right-continuous jump functions, so paths of $A = (A_t)$ are continuous and piece-wise linear. Thus, A is a CAF.

In particular, if a is an indicator of a set $H \subset \mathbb{I}$, then

$$A_t = \int_0^t I_H(X_s)\, ds$$

represents the total time spent in a set H up to instant t (occupation or sojourn time).

2) A purely discontinuous functional (DAF) $A = (A_t)$ is defined by a sum

$$A_t = \sum_{0<s\leqslant t} g(X_{s-}, X_s), \qquad t > 0,$$

with $A_0 = 0$, where $g = g(\cdot\,,\cdot)$ is a non-negative bounded function on $\mathbb{I} \times \mathbb{I}$ such that $g(i,i) = 0$ for all $i \in \mathbb{I}$.

In particular, if $g(i,j) = 1$ for all $i \neq j$, then A_t gives the number of jumps of the chain up to time t (a counting process).

Enumerating consecutive jump times (τ_n) of a chain X, as in § 1.1.3, one can write, with reference to the jump chain (X_n),

$$A_t = \sum_{n=1}^{\infty} g(X_{n-1}, X_n) I_{(0,t]}(\tau_n).$$

The typical example of such a DAF is the counting process (N_t) already introduced in § 1.1.5.

It can be shown that for the usual M.C., any AF is the sum of an IF and an DAF of the above form (see comments in § 8.1.3).

b) *Structure*. Of special interest are considerations which relate time behavior of additive functionals (or, more generally, increasing processes) with sets visited by a Markov process. The notions of a support and of an inverse play basic role in such studies.

1) *The inverse*. Let $A = (A_t)$ be an AF (or an increasing process) relative to the family $(\mathcal{F}_t)$ of natural σ-fields of a M.C. $X = (X_t)$. The *inverse* of A is a family $R = (R_t)$ defined by

$$R_t = \inf\,(s\colon\ A_s > t) \qquad \text{for each} \quad t \geqslant 0,$$

with $R_t = \infty$ if the set in parentheses is empty.

The following properties of R are immediate (see Figs. 8.1–8.5):

i) R is right-continuous, non-decreasing and $R_0 \geqslant 0$;

ii) for all non-negative s and t:

$$(R_s < t) = (A_{t-} > s),$$

iii) for all non-negative t:

$$A_t = \inf\,(s\colon\ R_s > t).$$

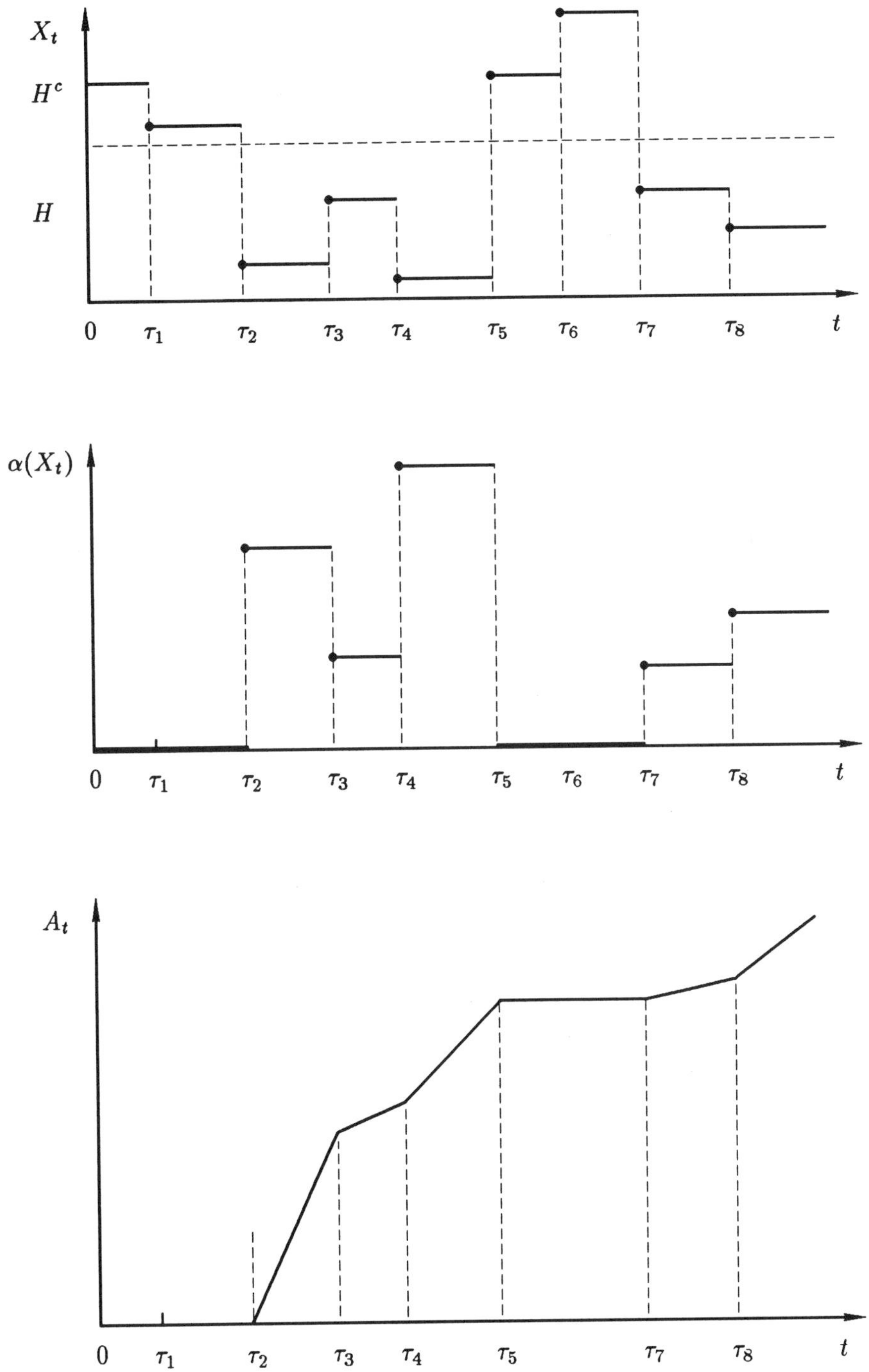

Figure 8.1. Integral functional $A = (A_t)$.

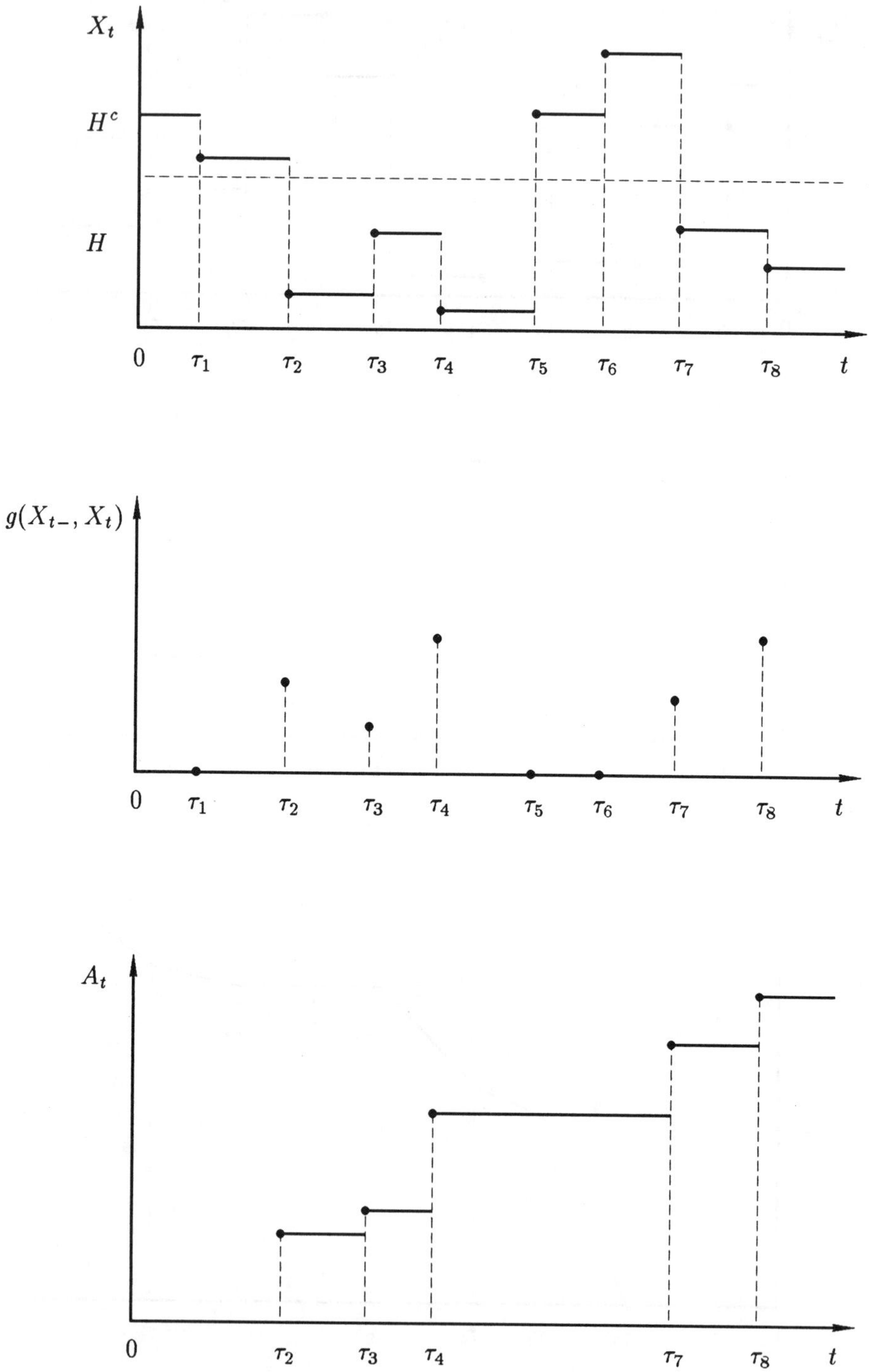

Figure 8.2. Discontinuous functional $A = (A_t)$.

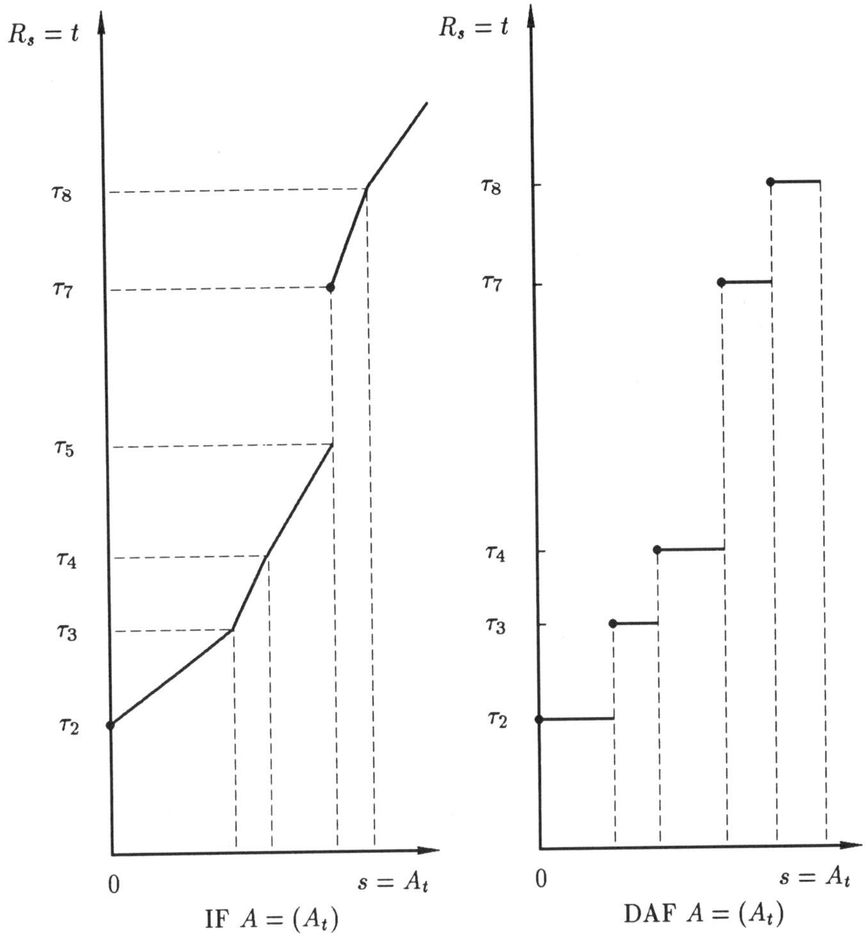

Figure 8.3. The inverse $R = (R_s)$ of an integral and of a discontinuous functional.

Further properties are given by the following basic lemma.

LEMMA 2. *Let A be an* AF *and let R be the inverse of A. Then,*

i) *for each t, R_t is a stopping time relative to the family $(\mathfrak{F}_s)$, i.e.,*

$$\big(R_t \leqslant s\big) \in \mathfrak{F}_s, \qquad s \geqslant 0.$$

Moreover, with $z = h + t - A(R_t)$, one has, for each $h, t \geqslant 0$ and $j \in \mathbb{I}$,

ii) $$R_{t+h} = R_t + R_z \circ \theta_{R_t} \qquad \text{on} \quad \big(R_t < \infty\big),$$

iii) $$\big(X\big(R_z\big) \circ \theta_{R_t} = j\big) = \big(X\big(R_{t+h}\big) = j\big) \qquad \text{on} \quad \big(R_{t+h} < \infty\big).$$

In the case of a CAF, continuity of $t \to A_t$ implies that $s \to R_s$ is strictly increasing on $[0, A_\infty)$ and that $R_s = \max(t\colon\ A_t = s)$. In particular, if $R_s < \infty$,

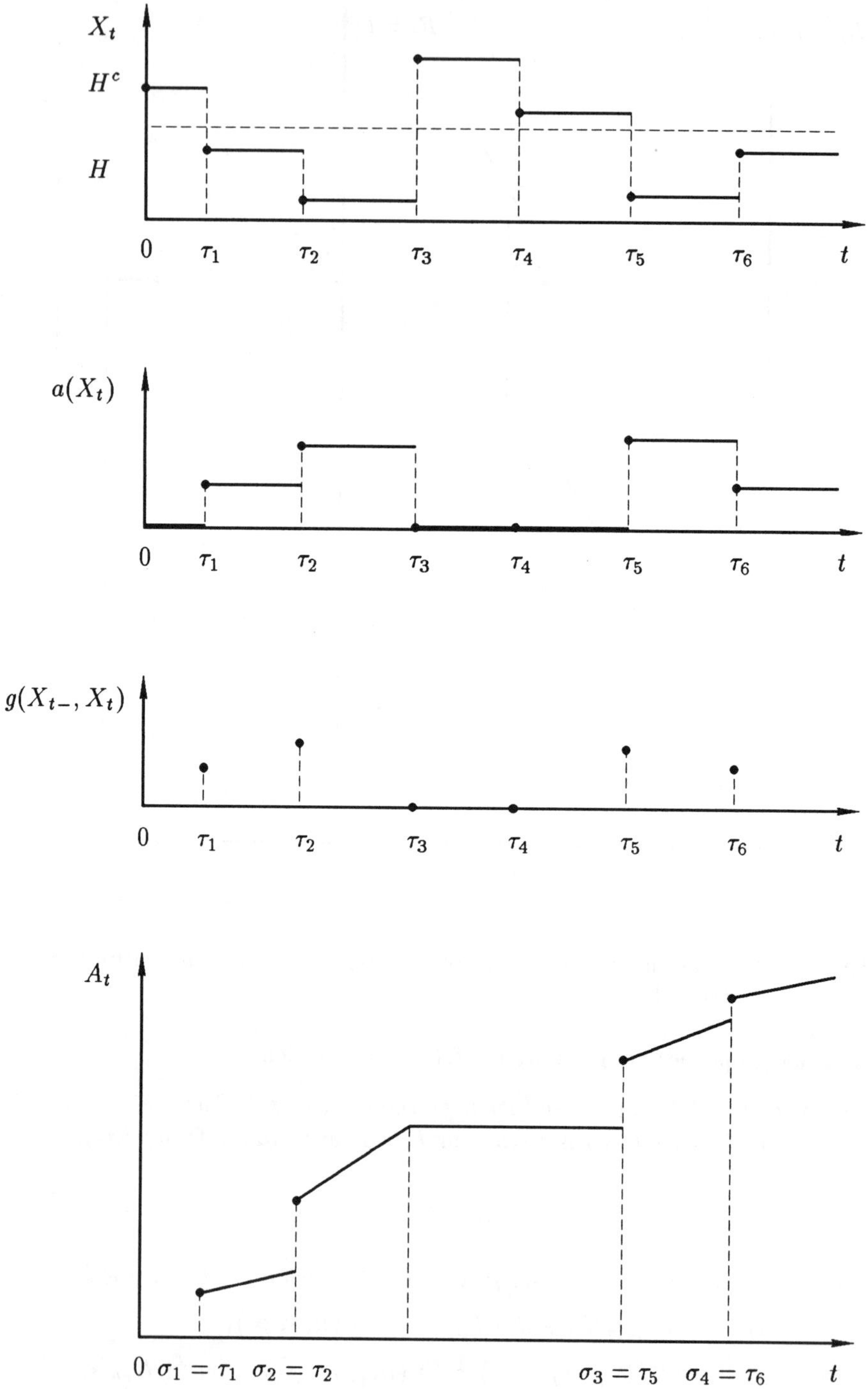

Figure 8.4. General form of an additive functional $A = (A_t)$.

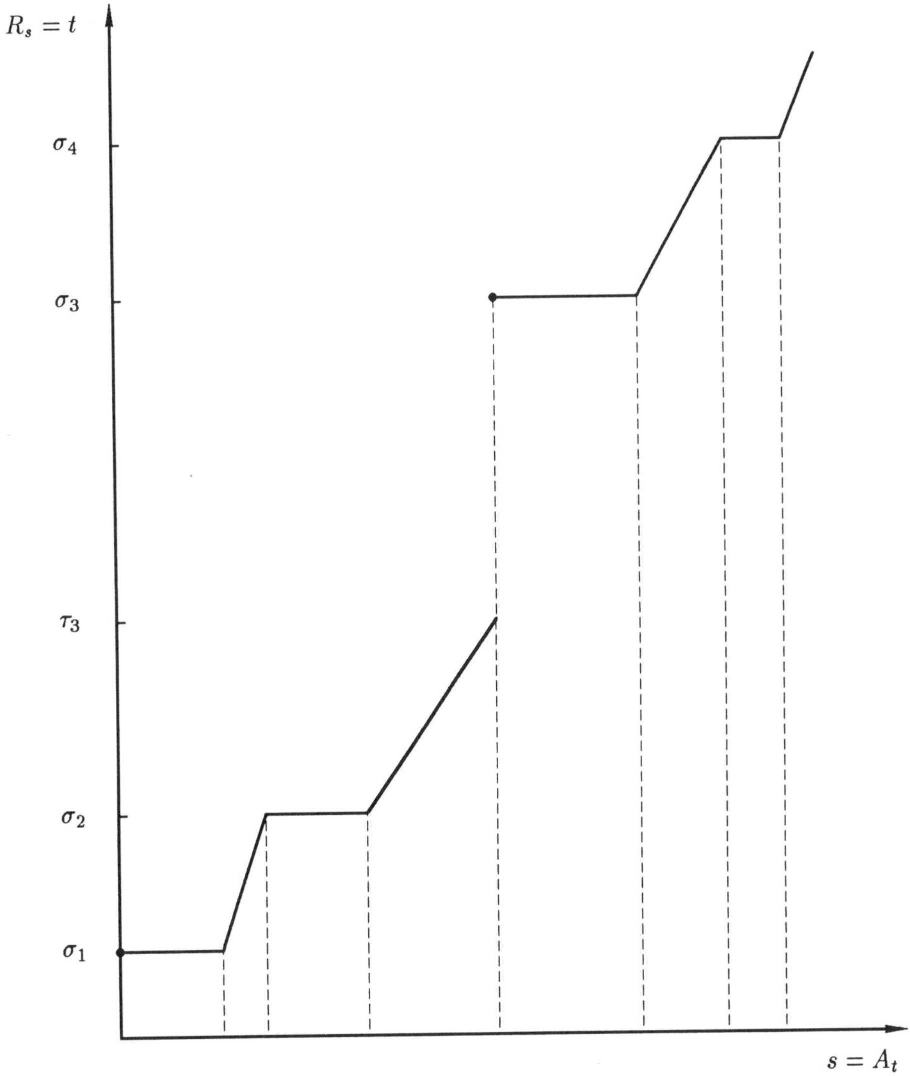

Figure 8.5. The inverse of a general functional.

then

$$A(R_s) = s, \qquad R(A_t) \geqslant t.$$

Consequently, $z = h$ in the above lemma. In the case of a DAF, the inverse is also a right-continuous jump function.

For the proof of Lemma 2, ii)–iii), for CAF, see R. M. Blumenthal and R. K. Getoor [1968, p. 207], C. Dellacherie and P. A. Meyer [1982, p. 121], E. B. Dynkin [1965, p. 321], M. J. Sharpe [1988, p. 305]; J. M. Rolin [1975] has shown that the process R has conditionally independent increments, but they are not necessarily stationary. Parts ii) and iii) for discontinuous AF were investigated by G. Weidenfeld [1980]; see also subsequent work by J. Glover [1983] and J. Mitro [1986].

Recall that the concept of an inverse has been already used in § 2.1.4 in connection with increasing processes (T_s) and (L_t). For other applications see discussion of random time change (§ 9.6.1) and of a local time (§ 8.4.1).

2) *Support.* Let A be an AF, and define R_0 as above

$$R_0 = \inf\big(t\colon\ A_t > 0\big),$$

with $R_0 = \infty$ when the set in parentheses is empty. The *support* of A is a set defined by

$$\operatorname{supp}(A) = \Big(i\colon\ \mathbb{P}^i(R_0 = 0) = 1\Big).$$

Thus, R_0 coincides with the first entrance time to the set $S = \operatorname{supp}(A)$. It can be shown that the potential of I_{S^c} vanishes:

$$\mathbf{U}_A I_{S^c}(i) = \mathbb{E}^i \int_0^\infty I_{S^c}(X_s)\, dA_s = \sum_{j\in S^c} u_A(i,j) = 0, \qquad \text{for } \ i \in \mathbb{I}.$$

An AF (or an increasing process) satisfying the above relation is said to be carried by a set S; this is taken as an equivalent definition of $\operatorname{supp}(A)$.

It is now clear that definitions stated here agree with that introduced in a). Indeed, for a CAF $A = (A_t)$, $\operatorname{supp}(A)$ consists of those states i such that starting from i, $t \to A_t$ begins to increase immediately. For a DAF $A = (A_t)$, $\operatorname{supp}(A)$ consists of those states i such that starting from i, $t \to A_t$ stays constant (till the next jump). Hence, for an IF, $\operatorname{supp}(A)$ consists of these states for which the charge a is strictly positive; for a DAF, $\operatorname{supp}(A)$ consists of those states j for which $g(\cdot\,,j) > 0$. See Figs. 8.1–8.3.

c) *Characteristic.* Let $A = (A_t)$ be an AF and let $F_i(t)$ be its characteristic defined above and assumed finite.

LEMMA 3. *The characteristic satisfies the equation*

$$\mathbf{P}(s)F_i(t) = F_i(t+s) - F_i(s), \qquad s,t \geqslant 0,$$

with $F_i(0) = 0$. *Then, there exists an exit law* $f(\cdot)$ *such that*

$$F_i(t) = \int_0^t f_i(x)\, dx, \qquad t \geqslant 0,$$

and a finite function $F(\infty)$ *on* $\mathbb{I}$ *is purely excessive.*

Proof. It follows from the definition of an AF that

$$F_{X_s}(t) = \mathbb{E}\big(A_t \circ \theta_s \mid X_s\big) = \mathbb{E}\big(A_{t+s} - A_s \mid \mathcal{F}_s\big),$$

so

$$\mathbb{E}^i\big(A_{t+s} - A_s\big) = \mathbb{E}^i F_{X_s}(t) = \mathbf{P}(s)F_i(t), \qquad t,s \geqslant 0.$$

Existence of the exit law

$$\mathbf{P}(s)f_i(t) = f_i(t+s), \qquad t, s \geqslant 0,$$

follows from Theorem 3, § 1.3.4. Finally, with

$$F_i(\infty) = \mathbb{E}^i A_\infty = u_A(i) \qquad \text{for} \quad i \in \mathbb{I},$$

assumed finite, one has

$$\mathbf{P}(s)F_i(\infty) = F_i(\infty) - F_i(s) \longrightarrow 0 \qquad \text{as} \quad s \to \infty,$$

so $F(\infty)$ is purely excessive. □

An example of a function satisfying the equation in Lemma 3 is $\Pi_i(t)$ — see Theorem 6, § 4.1.1.

§ 8.1.3. Special cases. Sample function properties and potentials of functionals offer several advantages in problems concerning passage times for usual M.C.'s. Examples given here provide additional insight into some results from Chapter II.

a) *Continuous functionals.* Consider an IF defined by a charge $a \geqslant 0$:

$$A_t = \int_0^t a(X_s)\, ds, \qquad t \geqslant 0.$$

Evidently,

$$u_A^\alpha = \mathbf{U}^\alpha a, \qquad \mathbf{U}_A^\alpha f(i) = \mathbf{U}^\alpha\big(a(i)f(i)\big), \qquad u_A^\alpha(i,j) = u_{ij}^\alpha a(j),$$

$$F_i(t) = \int_0^t \mathbf{P}(s)a(i)\, ds, \qquad \operatorname{supp}(A) = \big(i\colon\ a(i) > 0\big).$$

In particular, if $a = 1$ identically, so $A_t = t$, and then $\mathbf{U}_A^\alpha f = \mathbf{U}^\alpha f$.

Example 1. Taking for a the escape rate b (defined in § 3.1.3), one has

$$u_A^\alpha(i) = \mathbf{U}^\alpha b_i, \qquad \mathbb{E}^i A_t = \Pi_i(t) \uparrow \Pi_i = \mathbb{E}^i A_\infty = \mathbf{U} b_i,$$

in agreement with the results from § 3.3.1.

Example 2. Taking for a the entrance probability D, one finds by duality that

$$\mathbb{E}^i A_t = \int_0^t \Gamma_i^c(s)\, ds, \qquad u_A^\alpha(i) = \big(1 - \gamma_i^\alpha\big)/\alpha = \mathbf{U}^\alpha D_i,$$

in agreement with Theorem 5, § 4.1.1.

Example 3. Taking for a the probability Π of the last exit from a set H, similar calculations yield

$$\mathbb{E}^i A_t = \int_0^t \left(\Gamma_i^c(s) - \Gamma_i^c\right) ds, \qquad u_A^\alpha(i) = \left(\Gamma_i - \gamma_i^\alpha\right)/\alpha = \mathbf{U}^\alpha \Pi_i,$$

because Γ^c is invariant. Thus,

$$\mathbb{E}^i A_\infty = \mathbb{E}^i(L,\ L < \infty) = \mathbf{U}\Pi_i,$$

in agreement with a formula stated in § 4.1.1.

Example 4. The class of increasing processes defined here has wide applications. Let R be a random time, in particular, a stopping or a co-stopping time. Define, for $0 \leqslant t < \infty$,

$$A_t = \int_0^t I_{(s<R)}\, ds = \min(t, R).$$

Clearly, $A = (A_t)$ is an increasing process which is not adapted, unless R is a stopping time. However, A is not an AF in general, because condition c) in Definition 1, § 8.1.1, need not hold (as, for example, in special cases listed below). However, A has its α-potential and characteristic given by

$$\begin{aligned} u_A^\alpha(i) &= \mathbb{E}^i \int_0^R e^{-\alpha s}\, ds = \left(1 - \mathbb{E}^i e^{-\alpha R}\right)/\alpha \\ &\longrightarrow \mathbb{E}^i\left(A_R\right) = \mathbb{E}^i(R) \leqslant \infty \qquad \text{as} \quad \alpha \to 0, \\ \mathbb{E}^i A_t &= \int_0^t \mathbb{P}^i(s < R)\, ds = t\,\mathbb{P}^i(t < R) + \mathbb{E}^i(R,\ R \leqslant t) \\ &\longrightarrow \mathbb{E}^i A_\infty = \mathbb{E}^i A_R = E^i R \leqslant \infty \qquad \text{as} \quad t \to \infty. \end{aligned}$$

Taking for R times T, T_s, T_ρ, and L, respectively, one finds results given earlier.

b) *Discontinuous functionals.* Consider a DAF $A = (A_t)$ defined in § 8.1.2, in terms of a function $g(i,j)$. Define now a function $g_{\mathbf{Q}}$ on $\mathbb{I}$ by

$$g_{\mathbf{Q}}(i) = \sum_{j \neq i} q_{ij}\, g(i,j), \qquad i \in \mathbb{I},$$

and consider an IF $B = (B_t)$ determined by the charge $g_{\mathbf{Q}}$. Thus,

$$B_t = \int_0^t g_{\mathbf{Q}}(X_s)\, ds, \qquad 0 \leqslant t < \infty,$$

where

$$g_{\mathbf{Q}}(X_s) = \sum_{j \neq X_s} q_{X_s, j}\, g(X_s, j).$$

Connection between functionals A and B is established by the fundamental

THEOREM 1 (Lévy formula). *With (X_t) and $g_{\mathbf{Q}}$ defined above:*

$$\mathbb{E}^i \sum_{0<s\leqslant t} g(X_{s-}, X_s)h(s) = \mathbb{E}^i \int_0^t h(s)g_{\mathbf{Q}}(X_s)\,ds, \qquad t>0, \quad i \in \mathbb{I},$$

for any continuous positive function h.

Observe that for any t, the LHS is the finite sum, as the only non-vanishing terms correspond to the discontinuities of the chain $X = (X_t)$. The Lévy formula may be written in another form

$$\mathbb{E}^i \int_0^t h(s)\,dA_s = \mathbb{E}^i \int_0^t h(s)\,dB_s.$$

It then follows that functionals A and B have the same characteristic and the α-potential

$$\mathbb{E}^i A_t = \mathbb{E}^i B_t = \int_0^t \mathbf{P}(s)g_{\mathbf{Q}}(i)\,ds, \qquad u^\alpha_A(i) = u^\alpha_B(i) = \mathbf{U}^\alpha g_{\mathbf{Q}}(i), \qquad i \in \mathbb{I}.$$

Thus, calculation of these quantities for A may be carried out in terms of those for B. Furthermore, it follows from the Lévy formula that (under integrability condition) differences $A_t - B_t$ form a martingale (relative to the natural σ-fields)

$$\mathbb{E}(A_t - B_t \mid \mathcal{F}_s) = A_s - B_s \quad \text{a.e.}, \qquad t \geqslant s,$$

In the language of modern theory of stochastic processes, these facts are expressed by saying that the continuous functional (B_t) is the "dual predictable projection" or the "compensator" of the discontinuous functional (A_t). The pair consisting of (q_{ij}), with q_{ii} replaced by 0, and of the functional (B_t) is called the Lévy system. In this connection, $g_{\mathbf{Q}}(X_s)$ is called the intensity of DAF (A_t). Note also that in notation of Lemma 3, § 8.1.2, $f_i(0) = g_{\mathbf{Q}}(i)$.

Lévy formula, as stated here, is a special case of the general result established by B. Maisonneuve [1975] for general Markov processes. Direct proof of the classical Lévy formula (h constant) may be obtained by explicit calculation of the conditional expectation $\mathbb{E}^i A_t$ using representation of A_t in terms of the jump chain (see definition of DAF in § 8.1.2). Indeed, Lévy formula describes all jumps of the chain, and establishes connection between the minimal chain and its jump chain, in the same fashion as the Feller relation and the Dynkin's characteristic operator. This leads to the direct analysis of the chain by following its consecutive jumps (essentially the renewal argument). In other words, one returns here to the "density argument" already introduced in § 2.2.5.

The following examples interpret some expressions already mentioned earlier, and illustrate advantages provided by use of the Lévy formula.

The case when the function g factorizes $g(i,j) = g_1(i)\,g_2(j)$, $i \neq j$, leads to several simplifications, and then $g_{\mathbf{Q}}(i) = g_1(i) \bullet (\mathbf{Q} + \mathbf{\Theta})\,g_2(i)$.

The following examples are of special interest. They refer to the number of jumps of a certain type during time interval $(0,t]$. With $E = E_1 \cap E_2$ denoting the appropriate event, the function $g = I_E$ is of the above form with $g_1 = I_{E_1}$ and $g_2 = I_{E_2}$.

For convenience, write $m_t(i)$ and $u^\alpha(i)$ for $\mathbb{E}^i A_t$ and $u^\alpha_A(i)$, respectively, for each of the following DAF.

Example 1. N_t = number of jumps,

$$E = (X_{s-} \neq X_s), \qquad m_t(i) = \int_0^t \mathbf{P}(s) q_i \, ds, \qquad u^\alpha(i) = \mathbf{U}^\alpha q_i.$$

Example 2. N_t^{kj} = number of jumps from state k to state j $(k \neq j)$,

$$E = (X_{s-} = k,\ X_s = j), \qquad m_t(i) = \int_0^t p_{ik}(s) q_{kj} \, ds, \qquad u^\alpha(i) = u^\alpha_{ik} q_{kj}.$$

Example 3. N_t^j = number of jumps to state j,

$$E = (X_{s-} \neq j,\ X_s = j), \qquad m_t(i) = \int_0^t w_{ij}(s) \, ds, \qquad u^\alpha(i) = w^\alpha_{ij}.$$

Example 4. $N_t^{H,j}$ = number of jumps from set H to state j $(j \in H^c)$,

$$E = (X_{s-} \in H,\ X_s = j), \qquad m_t(i) = \int_0^t c_{ij}(s) \, ds, \qquad u^\alpha(i) = c^\alpha_{ij}.$$

Example 5. $N_t^{k,H}$ = number of jumps from state k to set H $(k \in H^c)$,

$$E = (X_{s-} = k,\ X_s \in H), \qquad m_t(i) = \int_0^t p_{ik}(s) \bar{q}_k \, ds, \qquad u^\alpha(i) = u^\alpha_{ik} \bar{q}_k.$$

c) *General form.* Jumps of a M.C. $X = (X_t)$ constitute an important feature of X. It is of interest to investigate how discontinuities of X affect discontinuities of an AF $A = (A_t)$. As sample functions of A are non-decreasing and right-continuous (with left-limits), they have at most a countable number of discontinuities (jumps). Thus, A may be regarded as a random measure for which A_t is a distribution function associating with an interval $(s,t]$ the value $A_t - A_s$. Consequently, A admits a decomposition into a sum of continuous and discontinuous parts. For an usual M.C. (all states stable), the only CAF are the integral functionals, and the only discontinuous functionals are DAF. This is the assertion of the following theorem which gives the general form of an AF.

THEOREM 2. *Any* AF $A = (A_t)$, *for the usual* M.C. $X = (X_t)$, *has the form*

$$A_t = \int_0^t a(X_s) \, ds + \sum_{0 < s \leqslant t} g(X_{s-}, X_s), \qquad t \geqslant 0,$$

for some functions $a \geqslant 0$ and $g \geqslant 0$.

As the proof of Theorem 2 involves results from the general theory of Markov processes, only a few comments can be given here to indicate the role of discontinuities. For complete proof, see A. Benveniste and J. Jacod [1973], C. Dellacherie [1972]; see also § 8.1.4 for additional comments.

Write $(\tau_n) = N$ and $(\sigma_n) = J$ for sets of consecutive discontinuities (jumps) of X and A, respectively, and let $S = \operatorname{supp}(A)$. Clearly, if A is an integral functional then jumps of X correspond to change of slope of A, when $X(\tau_n) \in S$; jumps such that $X(\tau_n) \in S^c$ do not affect A (see Fig. 8.1). If A is a DAF, then discontinuities of A follow those of X, provided again $X(\tau_n) \in S$; otherwise they do not affect A (see Fig. 8.2). Hence, all points t such that $X_t \in S^c$ are continuity points of A, and discontinuity points of A are located in the random set $L = (t\colon X_t \in S)$. As X has intervals of constancy, continuity points of X which lie in L would not affect the value of A, and are therefore continuity points of A. Hence, discontinuity points of A coincide with discontinuity points of X located in L. In other words, (σ_n) is a subset of (τ_n) for which $X(\tau_n) \in S$ (see Fig. 8.4). Recall also that all jumps of X are accounted for by the Lévy formula.

More generally, an AF (A_t) is said to be predictable (or "natural") if (with probability one) $t \to A_t$ and $t \to X_t$ have no common discontinuities. Clearly, a CAF is predictable, whereas a DAF is not. Using concept of a compensator (or a "dual predictable projection") of an increasing process (see § 8.1.4), one finds that the compensator of an AF from Theorem 2 is the sum of the compensator of its DAF component and the continuous component of A. The former, in turn, has been found from Lévy formula to be an integral functional.

According to general results for increasing processes, the compensator of an AF $A = (A_t)$ is continuous if and only if the jump times (σ_n) of A are totally inaccessible (or, equivalently, if A is "quasi-left-continuous"); see § 8.1.4, for further comments. Consequently, all jump times of an AF from Theorem 2 are totally inaccessible. This is, of course, in agreement with the fact that all jumps in the usual M.C. X are totally inaccessible.

Remark. Observe that the set L defined above is a "homogeneous set" in the sense that

$$L \circ \theta_t = L - t.$$

Similarly, the image of the inverse $R = (R_t)$ of an AF $A = (A_t)$, is defined by

$$G = \big(t\colon\ R_s = t \text{ for some } s\big).$$

It can be verified that for a CAF $L = G$, whereas for a DAF $G \subset L$ and $G = J$ is a discrete set.

d) *Dynkin theorem.* The following extension of the classical Dynkin theorem (already noted in § 8.1.1) is of interest.

THEOREM 3. *Let A be an AF of a M.C. X. Then, for any non-negative function f with finite α-potential $\mathbf{U}^\alpha_A f$ and for any stopping time T:*

$$\mathbf{U}^\alpha_A f(i) = \mathbb{E}^i \int_0^T e^{-\alpha t} f(X_t)\, dA_t + \mathbf{P}^\alpha(T)\mathbf{U}^\alpha_A f(i), \qquad i \in \mathbb{I},$$

or, equivalently,

$$\mathbb{E}^i \int_T^\infty e^{-\alpha t} f(X_t)\, dA_t = \mathbf{P}^\alpha(T)\mathbf{U}_A^\alpha f(i), \qquad i \in \mathbb{I}.$$

(The assertion holds also for $T = \infty$ upon defining $f(X_\infty) = 0$).

For the proof see E. B. Dynkin [1965]. Observe that for the degenerate functional $A_t = t$, Theorem 3 reduces to the classical case. Another form of interest is obtained for $f = 1$.

e) *Revuz measure.* There is a connection between potentials of additive functionals and potentials of measures. Theorem 6 here shows that if A is an AF with an α-potential u_A^α, then there exists a measure μ_A, called the Revuz measure, such that the α-potential of A equals the α-potential of μ_A, namely $u_A^\alpha = \mathbf{U}^\alpha \mu_A$. This result was established for general Markov processes by D. Revuz [1970]; see also B. W. Atkinson and J. B. Mitro [1983]. Here adaptation to the usual M.C.'s will be presented.

Recall that for $i \in \mathbb{I}$:

$$u_A^\alpha(i) = \sum_j u_A^\alpha(i,j), \qquad \mathbf{U}^\alpha \mu_A(i) = \sum_j v^\alpha(i,j)\mu_A(j) = \sum_j u_{ij}^\alpha a(j),$$

where $a(j) = \mu_A(j)/\xi(j)$ is the dual of μ_A.

Consider the α-potentials $\mathbf{U}_A^\alpha f$ and $\mathbf{U}^\alpha f$ of a function $f \geqslant 0$, and note that $\mathbf{U}^\alpha f$ corresponds to the degenerate functional $A_t = t$.

LEMMA 4. *Let A_t be finite for all t, and let for $f \geqslant 0$, $\mathbf{U}_A^\alpha f$ and $\mathbf{U}_A^\beta f$ be finite $(\alpha, \beta \geqslant 0)$. Then, for $i \in \mathbb{I}$,*

$$\mathbf{U}_A^\alpha f(i) - \mathbf{U}_A^\beta f(i) = (\beta - \alpha)\mathbf{U}^\alpha \mathbf{U}_A^\beta f(i) = (\beta - \alpha)\mathbf{U}^\beta \mathbf{U}_A^\alpha f(i).$$

The assertion is the extension of the resolvent equation for $\mathbf{U}^\alpha$, and its proof can be obtained in the same manner by direct verification from definitions. From this, one has immediately that the function $\mathbf{U}_A^\alpha f$ is α-excessive.

The Revuz measure μ_A on $\mathbb{I}$, associated with an AF $A = (A_t)$ and taken relative to the reference measure ξ, is defined by

$$\sum_j \mu_A(j) f(j) = \lim_{t \to 0} \frac{1}{t} \mathbb{E}^\xi \int_0^t f(X_s)\, dA_s$$

for a positive bounded function f on $\mathbb{I}$. Here $\mathbb{E}^\xi$ denotes expectation relative to the measure ξ (see § 8.1.1). In particular, $\mu_A(j)$ corresponds to $f(X_s) = I_j(X_s)$. Considering a CAF $B = (B_t)$ with

$$B_t = \int_0^t f(X_s)\, dA_s, \qquad t \geqslant 0,$$

and applying the Abelian theorem for Laplace transforms to the characteristic of B, one obtains:

$$\lim_{t\to 0}\frac{1}{t}\,\mathbb{E}^{\xi}B_t = \lim_{\alpha\to\infty}\sum_i \xi(i)\alpha\mathbf{U}_A^{\alpha} f(i).$$

Hence, one has the following working definition (where $\bullet$ denotes the inner product).

DEFINITION 5. The Revuz measure μ_A associated with an AF A and the reference measure ξ, is given by

$$\lim_{\alpha\to\infty}\xi\bullet\alpha\mathbf{U}_A^{\alpha} f = \mu_A\bullet f \qquad \text{for} \quad f\geqslant 0.$$

The main theorem now follows.

THEOREM 6. *Let A be an AF with finite potential, and let μ_A be its associated Revuz measure. Then*

$$u_A^{\alpha}(i) = \mathbf{U}^{\alpha}\mu_A(i), \qquad i\in\mathbb{I}, \quad \alpha\geqslant 0.$$

Proof (outline). Recall that $u_A^{\alpha} = \mathbf{U}_A^{\alpha}\mathbf{1}$, so Lemma 4 yields

$$\mathbf{U}_A^{\alpha}\mathbf{1} - \mathbf{U}_A^{\alpha+\beta}\mathbf{1} = \beta\mathbf{U}^{\alpha}\,\mathbf{U}_A^{\alpha+\beta}\mathbf{1},$$

and the resolvent equation yields, for $\mathbf{U}^{\alpha}\mu_A$,

$$\mathbf{U}^{\alpha}\mu_A - \mathbf{U}^{\alpha+\beta}\mu_A = \beta\mathbf{U}^{\alpha}\,\mathbf{U}^{\alpha+\beta}\mu_A.$$

Hence, the inner product

$$\xi\bullet\left(\mathbf{U}_A^{\alpha}\mathbf{1} - \mathbf{U}^{\alpha}\mu_A\right) = \lim_{\beta\to\infty}\xi\bullet\mathbf{U}^{\alpha}\left(\beta\mathbf{U}_A^{\alpha+\beta}\mathbf{1} - \beta\mathbf{U}^{\alpha+\beta}\mu_A\right) = 0,$$

because

$$\xi\bullet\beta\mathbf{U}_A^{\beta}\mathbf{1}\longrightarrow\mu_A\bullet\mathbf{1}$$

and, by duality (see § 1.3.5),

$$\xi\bullet\beta\mathbf{U}^{\beta}\mu_A = \xi\bullet\beta\mathbf{U}^{\beta}\hat{\mu}_A = \mu_A\bullet\beta\widehat{\mathbf{U}}^{\beta}\mathbf{1}\longrightarrow\mu_A\bullet\mathbf{1}.$$

It follows that $\mathbf{U}_A^{\alpha}\mathbf{1}(i) = \mathbf{U}^{\alpha}\mu_A(i)$ on $\mathbb{I}$. Extension for $\alpha = 0$ follows by monotone passage to the limit. □

For the usual M.C. it is possible to obtain the explicit expression for the Revuz measure. When μ_A is α-finite, it can be computed from relation

$$\mu_A(j) = \sum_i \xi(i)\mu_A^{i}(j), \qquad j\in\mathbb{I},$$

with the "conditional Revuz measure" defined by

$$\mu^i_A(j) = \lim_{\alpha\to\infty} \alpha \mathbf{U}^\alpha_A I_j(i).$$

THEOREM 7. *Define a measure $\sigma = \xi(-\mathbf{Q})$, where ξ is the reference measure. Then the Revuz measure for the usual* M.C. *is given by $\mu_A = \sigma\mathbf{U}_A$, and the explicit form of the conditional measure is*

$$\mu^i_A(j) = -\sum_k q_{ik} u_A(k,j), \qquad i,j \in \mathbb{I}.$$

Proof (outline). From Lemma 4 with $\beta = 0$:

$$\alpha\mathbf{U}^\alpha_A f(i) = -\alpha\big(\alpha\mathbf{U}^\alpha - \mathbf{I}\big)\mathbf{U}_A f(i).$$

Letting $\alpha \to \infty$:

$$\sum_j \mu^i_A(j)\, f(j) = -\mathbf{Q}\mathbf{U}_A f(i),$$

so $\mu_A \bullet f = \sigma\mathbf{U}_A f$. □

As an example, consider an IF A determined by the charge $a \geqslant 0$. Then,

$$v^\alpha(i,j) = u^\alpha_{ij} a(j), \qquad \mu^i_A(j) = \delta_{ij} a(j),$$

$$\mu_A(j) = \xi(j)\, a(j) = \hat{a}(j), \qquad \mathbf{U}^\alpha \mu_A = \mathbf{U}^\alpha a = u^\alpha_A.$$

When A is a DAF determined by a function g, then consider an IF determined by the charge $g_{\mathbf{Q}}$.

More generally, one has, for $f \geqslant 0$,

$$\mathbf{U}_A f(i) = \sum_j v(i,j)\mu_A(j) f(j), \qquad i \in \mathbb{I}.$$

This extension follows from the observation that $u_A(i,j) = v(i,j)\mu_A(j)$.

As another application, consider again a CAF $B = (B_t)$ defined above in terms of an AF $A = (A_t)$ and a function $f \geqslant 0$. The following result establishes connection between the Revuz measure μ_A of A and the characteristic $G_i(t)$ of B defined by

$$G_i(t) = \mathbb{E}^i \int_0^t f(X_s)\, dA_s, \qquad i \in \mathbb{I}.$$

Clearly, α-potential of B coincides with α-potential of f relative to A:

$$u^\alpha_B(i) = \mathbf{U}^\alpha_A f(i), \qquad i \in \mathbb{I}.$$

COROLLARY 8. *For $i \in \mathbb{I}$ and $t \geqslant 0$:*

$$G_i(t) = \int_0^t g_i(s)\, ds,$$

where

$$g_i(s) = \mathbf{P}(s) g_i(0) \quad \text{and} \quad g_i(0) = \mu^i_A \bullet f.$$

Proof. By Lemma 3, § 8.1.2, there exists an exit law $g_i(s)$, as required. Hence, as $t \to 0$,

$$\frac{1}{t}\,\mathbb{E}^i\big(B_{t+s} - B_s\big) = \mathbf{P}(s)G_i(t)/t \longrightarrow \mathbf{P}(s)g_i(0) = g_i(s), \qquad s \geqslant 0,$$

where

$$g_i(0) = \mu_A^i \bullet f. \qquad \square$$

§ 8.1.4. Martingales. The use of martingales associated with M.C.'s provides additional insight into properties of passage times. It is therefore useful to list here some basic facts on martingales for the purpose of reference.

a) *Supermartingales.* In the following discussion, unless otherwise explicitly stated, the term *supermartingale* will denote a positive, right-continuous supermartingale relative to the natural σ-fields of a M.C. X, say $S = (S_t) = (S_t, \mathcal{F}_t, 0 \leqslant t < \infty)$, where S_t is integrable and $\mathcal{F}_t$-measurable for each t (see Appendix).

As S is positive, $\lim_{t\to\infty} S_t = S_\infty$ a.e. exists, and closes S on the right:

$$\mathbb{E}\big(S_\infty \mid \mathcal{F}_t\big) \leqslant S_t \quad \text{a.e.},$$

and $\lim_{t\to\infty} \mathbb{E}S_t \geqslant \mathbb{E}S_\infty$ (with equality if and only if r.v.'s S_t are uniformly integrable).

A supermartingale S is said to be a *potential supermartingale* when

$$\lim_{t\to\infty} \mathbb{E}S_t = 0.$$

This implies that $S_\infty = 0$ a.e. and that (S_t) is uniformly integrable.

The connection between M.C.'s and supermartingales is established by

THEOREM 1. *Let f be a $\mathbf{P}(t)$-excessive function on $\mathbb{I}$ and $\pi f < \infty$ (where π is the initial distribution). Then, $S = (S_t)$ with $S_t = f(X_t)$, $t \geqslant 0$, is a supermartingale (a martingale, if f is invariant).*

Proof. By direct verification, using Markov property:

$$\mathbb{E}\big(S_{t+h} \mid \mathcal{F}_t\big) = \mathbb{E}\big(S_{t+h} \mid X_t\big) = \mathbf{P}(h)f(X_t) \leqslant f(X_t) = S_t \quad \text{a.e.},$$

and noting that

$$\mathbb{E}f(X_t) = \pi\mathbf{P}(t)f \leqslant \pi f < \infty. \qquad \square$$

Application of the Riesz decomposition I for excessive functions (Thm. 1, § 1.3.3) yields the following decomposition for a supermartingale $S = (S_t)$ from Theorem 1,

$$f(X_t) = p(X_t) + h(X_t), \qquad t \geqslant 0,$$

where $p(X_t)$, $t \geqslant 0$, form a potential supermartingale with $\lim_{s\to\infty} \mathbf{P}(s)p(X_t) = 0$, and $h(X_t)$, $t \geqslant 0$, form a martingale with $h(X_t) = \lim_{s\to\infty} \mathbf{P}(s)f(X_t)$. This decomposition is unique up to equivalence.

Another Riesz decomposition of the supermartingale S from Theorem 1 has the form

$$S_t = V_t + L_t, \qquad t \geqslant 0,$$

where $L_t = \mathbb{E}(S_\infty \mid \mathcal{F}_t)$, $t \geqslant 0$, form a uniformly integrable martingale, and $V_t = \mathbb{E}((S_t - S_\infty) \mid \mathcal{F}_t)$, $t \geqslant 0$, form a supermartingale such that $\lim_{t\to\infty} V_t = 0$. In general, $h(X_t) \geqslant L_t$ and $p(X_t) \leqslant V_t$. When (S_t) is a uniformly integrable family, then both these decompositions coincide.

The next application of Feller Theorem (Thm. 6, § 1.3.3.) for the minimal chain $X = (X_t)$ leads to a further decomposition.

THEOREM 2. *A supermartingale $S = (S_t)$ from Theorem 1 can be represented uniquely (up to equivalence) as the sum*

$$f(X_t) = u(X_t) + w(X_t) + h(X_t), \qquad t \geqslant 0,$$

where $u(X_t)$, $t \geqslant 0$, form a potential of a charge a, and $w(X_t)$, $t \geqslant 0$, form a boundary potential.

Recall that now

$$p(X_t) = u(X_t) + w(X_t), \qquad r(X_t) = h(X_t) + w(X_t).$$

Thus, $u(X_t) = \mathbf{U}a(X_t)$ is a potential supermartingale, but $r(X_t)$ need not be a martingale. Writing assertion of Theorem 2 in the form

$$f(X_t) = u(X_t) + r(X_t),$$

comparison with previously stated Riesz decomposition, indicates that potential component has been restricted, but the martingale component extended.

Suppose that in the Riesz decomposition of an excessive function f, the purely excessive component p is also a potential of a charge $a \geqslant 0$ (so boundary potential w vanishes). Hence, assuming that $\mathbf{U}$ is finite (transient case): $p = \mathbf{U}a$, $a = (-\mathbf{Q})p$. Then, $p(X_t) = \mathbf{U}a(X_t)$, $t \geqslant 0$, defines a potential supermartingale which is also a potential of a charge, and $r(X_t) = h(X_t)$, $t \geqslant 0$, form a martingale. This is the most common situation in applications.

Example. This example provides additional insight concerning the first entrance and the last exit probabilities D and Π. Since D is excessive, $S_t = D(X_t)$, $t \geqslant 0$, form an uniformly integrable supermartingale, with the Riesz decomposition

$$D(X_t) = \Pi(X_t) + \Gamma^c(X_t), \qquad t \geqslant 0,$$

which corresponds to the Poisson representation $D = \Pi + \Gamma^c$, where $\Pi = \mathbf{U}b$ is a purely excessive component which is also a potential of a charge b, and Γ^c is invariant. Hence,

$$\Pi(X_t) = \mathbf{U}b(X_t) \quad \text{and} \quad \Gamma^c(X_t) = \mathbb{E}(S_\infty \mid \mathcal{F}_t), \qquad t \geqslant 0,$$

form a potential supermartingale and a martingale, respectively.

b) *Increasing process.* The following is the summary of some properties of increasing processes associated with an usual M.C. $X = (X_t)$. For the proofs, see references listed in the introduction to this section.

Definition 3. Let $A = (A_t)$ be an integrable increasing process. A process $U = (U_t)$ defined (up to equivalence) by

$$U_t = \mathbb{E}(A_\infty \mid \mathcal{F}_t) - A_t, \qquad t \geqslant 0,$$

is called a *potential generated* by an increasing process.

It is immediately seen that (U_t) is a uniformly integrable potential supermartingale. Moreover, it is shown that the family (U_T), where T runs through all finite stopping times, is uniformly integrable; then, U is said to be "of class (D)". Conversely, the following basic theorem holds.

Theorem 4 (Doob–Meyer decomposition). *Let $U = (U_t)$ be a potential supermartingale of class* (D). *Then, there exists a unique (up to equivalence) integrable and predictable increasing process $A = (A_t)$ such that U is a potential generated by A. Here*

$$U_t = M_t - A_t, \qquad t \geqslant 0,$$

where $M_t = \mathbb{E}(A_\infty \mid \mathcal{F}_t)$ form a uniformly integrable martingale. If U is not of class (D), *then (M_t) must be replaced by a local martingale.*

Proof of Theorem 4 consists essentially of justification of the following limit

$$A_t = \lim_{h \to 0} \frac{1}{h} \int_0^t \Big(U_s - \mathbb{E}\big(U_{s+h} \mid \mathcal{F}_s\big)\Big) ds$$

taken in an appropriate sense; see C. Dellacherie and P. A. Meyer [1982]. For definition of predictable processes, see Appendix.

Definition 5. For an integrable increasing process $A = (A_t)$, its *dual predictable projection* (also called a *compensator* of A) is a predictable increasing process $A^* = (A_t^*)$ defined by

$$\mathbb{E} \int_0^\infty Z_t \, dA_t = \mathbb{E} \int_0^\infty Z_t \, dA_t^*$$

for every non-negative predictable process $Z = (Z_t)$.

Two increasing processes are said to be associated if they have the same compensator. The next theorem establishes connection with martingales.

Theorem 6. *Two increasing processes A' and A'' are associated if and only if the process $A' - A''$ is a martingale; equivalently, they generate the same potential.*

Thus, if A is an integrable increasing process and A^* is its compensator, then $(A_t - A_t^*, \mathcal{F}_t)$ is a martingale, and

$$\mathbb{E}\big(A_{t+s} - A_s \mid \mathcal{F}_s\big) = \mathbb{E}\big(A_{t+s}^* - A_s^* \mid \mathcal{F}_s\big), \qquad t, s \geqslant 0.$$

Note that the increasing process (A_t) is a submartingale, so writing

$$A_t = M_t + A_t^*, \qquad \text{where} \quad M_t = \mathbb{E}\big(A_\infty - A_\infty^* \mid \mathcal{F}_t\big),$$

one has the Doob–Meyer decomposition of a submartingale A; this explains the term "compensator" for A^*.

As an example, consider an integral functional $A = (A_t)$ determined by a charge $a \geqslant 0$, and assume that $\mathbb{E}A_\infty = \pi \mathbf{U}a < \infty$. Then, clearly, the potential generated by an IF is

$$U_t = \mathbb{E}\left(\int_t^\infty a(X_s)\,ds \mid \mathcal{F}_t\right) = \mathbf{U}a(X_t) = u_A(X_t).$$

Similarly, for a DAF determined by a function g, the compensator obtained from Lévy formula is an integral functional determined by charge $g_{\mathbf{Q}}$ (Thm. 1, § 8.1.3). Hence, the common potential generated by A and by B is $U_t = \mathbf{U}g_{\mathbf{Q}}(X_t)$.

Consequently, if $A = (A_t)$ is an AF for the usual M.C. $X = (X_t)$ with the characteristic $F_i(t) = \mathbb{E}^i A_t$, then Lemma 3, § 8.1.2, implies that the following limit, known as the *intensity* of A, exists:

$$\lim_{h\to 0} \mathbb{E}(A_{s+h} - A_s \mid \mathcal{F}_s)/h = \lim_{h\to 0} F_{X_s}(h)/h = f_{X_s}(0) = \Lambda_s \quad \text{a.e.}$$

Hence,

$$\mathbb{E}(A_{s+h} - A_s \mid \mathcal{F}_s) = \mathbb{E}\left(\int_s^{s+h} \Lambda_u\,du \mid \mathcal{F}_s\right) \quad \text{a.e.}$$

and

$$M_t = A_t - \int_0^t \Lambda_s\,ds, \qquad t \geqslant 0,$$

form a martingale. It should be noted that intensity Λ_s is a function of X_s only. Thus, for an integral functional, $\Lambda_s = a(X_s)$, and M_t vanishes, but for a DAF $\Lambda_s = g_{\mathbf{Q}}(X_s)$ and $M_t = A_t - B_t$. For general form of A, see Theorem 2, § 8.1.3.

c) **Q**-*martingales*. There are other martingales associated with a M.C. and determined by functions which need not be excessive nor positive. Recall that the generalized (weak) generator $\mathbf{Q}$ has been already defined in § 8.1.1-a for any bounded function f, and leads to the (F)-equation stated there.

THEOREM 7. *For any bounded function f, the family $(M_t, \mathcal{F}_t,\ 0 \leqslant t < \infty)$, where*

$$M_t = f(X_t) - \int_0^t \mathbf{Q}f(X_u)\,du, \qquad t \geqslant 0,$$

is a martingale with $\mathbb{E}M_t = \pi f$.

Proof. Verification of the martingale property $\mathbb{E}(M_{t+s} \mid \mathcal{F}_s) = M_s$ a.e. follows immediately from the above mentioned (F)-equation. Integrability of M_t follows from the inequality

$$\mathbb{E}^i \int_0^t \left|\mathbf{Q}f(X_u)\right| du \leqslant 2\,\|f\| \int_0^t \mathbf{P}(s)q_i\,ds < \infty \qquad \text{for each} \quad t.$$

Indeed, the last integral represents the average number of jumps up to time t, and this is finite for the usual M.C. Evidently, $\mathbb{E}^i M_t = f(i)$. □

Theorem 7 can be extended to wider class of functions f for which $\pi|f| < \infty$. When f is excessive, then integrability of M_t follows from the assumption that $\pi|f| < \infty$, and one has $(-\mathbf{Q})f = a \geqslant 0$. Thus,

$$M_t = f(X_t) + \int_0^t a(X_u)\,du \geqslant 0,$$

and if f is invariant, then $M_t = f(X_t)$.

By optional stopping theorem, t in Theorem 7 may be replaced by a finite stopping time T, and then (M_0, M_T) is a martingale, so

$$f(i) = \mathbb{E}^i f(X_T) - \mathbb{E}^i \int_0^T \mathbf{Q}f(X_u)\,du.$$

More generally, (M_t^α) defined for a bounded function f by

$$M_t^\alpha = e^{-\alpha t} f(X_t) - \int_0^t e^{-\alpha u}(\mathbf{Q} - \alpha\mathbf{I})f(X_u)\,du \quad \text{a.e.}$$

for $t \geqslant 0$ and $\alpha > 0$, form an uniformly integrable martingale relative to $(\mathcal{F}_t)$, with $\mathbb{E}M_t^\alpha = \pi f$. Replacing t by T (optional sampling theorem) with $\mathbb{E}^i T < \infty$, and letting $\alpha \to 0$, one obtains the Dynkin formula, already stated in § 8.1.1.

Remark. The "martingale problem" refers to determination of a M.C. $X = (X_t)$, for a given operator $\mathbf{Q}$ and a function f, in such a way that (M_t), defined as in Theorem 7, is a martingale.

§ 8.1.5. Multiplicative functionals. Another type of functionals of importance in many considerations is that of multiplicative functionals defined below. These functionals are essential in studies of processes killed at random times, and thus are linked to taboo probabilities. These aspects are treated later in Section 9, but here only basic properties are listed.

a) The following is the basic definition.

DEFINITION 1. A family $M = (M_t,\ 0 \leqslant t < \infty)$ of non-negative random variables on a probability space $(\Omega, \mathcal{F}, \mathbb{P})$ is called a *multiplicative functional* (MF) of a M.C. X, when

a) $0 \leqslant M_t \leqslant 1$;
b) M_t is $\mathcal{F}_t$-measurable for each $t \geqslant 0$;
c) $M_{t+s} = M_t(M_s \circ \theta_t)$ for all $t, s \geqslant 0$.

Condition a) indicates that the functional M is contracting. Condition c) is assumed to hold everywhere (perfect MF). Other terminology from § 8.1.2 applies also to MF's.

Observe that M is right-continuous and non-increasing in t. As $M_0 = M_0^2$, so M_0 can only assume values 0 and 1. A state i is called *permanent* for M if

$$\mathbb{P}^i(M_0 = 1) = 1,$$

and a set of all permanent states will be denoted by $\mathbb{I}_M$. It can be verified that the life time

$$\sigma = \inf\,(t\colon\ M_t = 0)$$

is a terminal stopping time, and that $\mathbb{I}_M$ consists of those states which are irregular for σ:

$$\mathbb{P}^i(\sigma > 0) = 1.$$

Furthermore, the formulae

$$M_t = e^{-A_t}, \qquad A_t = -\log M_t,$$

establish one-to-one correspondence between MF's and AF's.

The following examples are of special interest.

1) Take $M_t = e^{-\alpha t}$ for a fixed $\alpha \geqslant 0$. Clearly $\mathbb{I}_M = \mathbb{I}$.

2) Let T be a terminal time (see Appendix A1); in particular, the first entrance time $T = T_H$.

Define

$$M_t = \begin{cases} 1 & \text{if} \quad t < T, \\ 0 & \text{if} \quad t \geqslant T, \end{cases}$$

or, in another notation,

$$M_t = I_{[0,T)}(t) = I_{(t<T)}.$$

Then, it is easy to verify that $M = (M_t)$ is an MF, and the set $\mathbb{I}_M$ of permanent states coincides with those i for which $\mathbb{P}^i(T > 0) = 1$. Moreover,

$$\mathbb{E}^i \int_0^\infty e^{-\alpha t} M_t\,dt = \left(1 - \mathbb{E}^i e^{-\alpha T}\right)/\alpha, \qquad \alpha > 0;$$

$$\mathbb{E}^i M_t = \mathbb{P}^i(T > t).$$

In particular, when T is the first entrance time to a set H, then $\mathbb{I}_M = H^c$ and $\mathbb{E}^i M_t = D_i^c(t)$.

b) In some considerations, condition a) in Definition 1 is replaced by the weaker condition

$$\text{m)} \qquad \mathbb{E}^i M_t \leqslant 1 \qquad \text{for all} \quad i \quad \text{and} \quad t.$$

An example of an MF which is not contracting, but satisfies condition m), is given by

$$M_t = g(X_t)/g(X_0), \qquad t \geqslant 0,$$

where $g > 0$ is a $\mathbf{P}(t)$-excessive function on $\mathbb{I}$.

Moreover, (M_t) is a positive supermartingale relative to the natural σ-fields:

$$\begin{aligned} \mathbb{E}(M_{t+h} \mid \mathcal{F}_t) &= M_t \mathbb{E}(M_h \circ \theta_t \mid \mathcal{F}_t) = (1/g(X_0))\mathbb{E} g(X_{t+h} \mid X_t) \\ &= \mathbf{P}(h) g(X_t)/g(X_0) \leqslant g(X_t)/g(X_0) = M_t \quad \text{a.e.} \end{aligned}$$

Clearly, the set of permanent states coincides now with the state space $\mathbb{I}$.

Observe, however, that in the present case of non-contracting MF (M_t), the relation $A_t = -\log M_t$ satisfies all conditions in Definition 1, § 8.1.2, except that A_t need not be positive.

8.2. EXTENDED FUNCTIONALS

§ 8.2.1. Definitions. In many applications it is convenient to modify slightly the original definition of an additive functional in the following manner.

Let $M = (M_t)$ be an MF for the usual M.C. $X = (X_t)$, and let $\mathbb{I}_M$ and σ be, respectively, the set of permanent states and the life time of M (see definitions in § 8.1.5).

DEFINITION 1. A family $A = (A_t,\ 0 \leqslant t < \infty)$ of non-negative random variables is called an *additive functional* (AF) of (X, M) when

a) the paths of A are non-decreasing, right-continuous, and $A_0 = 0$;

b) for each t, A_t is $\mathcal{F}_t$-measurable;

c) for each t and h:

$$A_{t+h} = A_t + M_t(A_h \circ \theta_t),$$

where θ_t is the shift operator.

Observe that Definition 1 reduces to that in § 8.1.2 when $M_t = 1$ identically. It is shown later in § 9.1.1, that the MF $M = (M_t)$ generates a semi-group $\widetilde{\mathbf{P}}(t) = (\tilde{p}_{ij}(t))$ of transition probabilities

$$\tilde{p}_{ij}(t) = \mathbb{E}^i(X_t = j, M_t), \qquad i, j \in \mathbb{I}, \quad t \geqslant 0,$$

which is subordinate to the original semi-group $\mathbf{P}(t)$, and corresponds to the killing of the original chain X. Note that

$$\widetilde{\mathbf{P}}(t)\mathbf{1}_i = \mathbb{E}^i M_t = \mathbb{P}^i(\sigma > t),$$

so $\widetilde{\mathbf{P}}(t)$ is sub-stochastic, and $\widetilde{\mathbf{P}}(0)\mathbf{1}_i = 1$ for $i \in \mathbb{I}_M$, and 0 otherwise. Thus, one can restrict $\tilde{p}_{ij}(t)$ to i and j in $\mathbb{I}_M$. Write $\widetilde{\mathbf{U}}^\alpha = (\tilde{u}^\alpha_{ij})$ for the corresponding resolvent.

Let $C = (C_t)$ be an additive functional in the sense of Definition 1 from § 8.1.2, and consider a stochastic integral of the form

$$A_t = \int_0^t M_s\, dC_s, \quad t \geqslant 0.$$

Clearly, $A = (A_t)$ is an increasing process, and $A_t = A_\sigma$ for $t \geqslant \sigma$. The following basic result indicates that this A is an extended AF in the sense of Definition 1.

THEOREM 2. *The process $A = (A_t)$ defined above is an AF satisfying*

$$A_{t+h} = A_t + M_t(A_h \circ \theta_t), \qquad t, h \geqslant 0,$$

and for any non-negative function f, the α-potential of f relative to A is

$$\mathbf{U}_A^\alpha f(i) = \widetilde{\mathbf{U}}_C^\alpha f(i), \qquad i \in \mathbb{I}_M,$$

and the characteristic of A is

$$\mathbb{E}^i A_t = \widetilde{\mathbb{E}}^i C_t, \qquad t \geqslant 0.$$

Proof (outline). From properties of the shift one has

$$\begin{aligned} M_t(A_h \circ \theta_t) &= M_t \int_0^h (M_s \, dC_s) \circ \theta_t = M_t \int_0^h M_s \circ \theta_t \, d(C_s \circ \theta_t) \\ &= \int_0^h M_{s+t} \, dC_{s+t} = \int_t^{t+h} M_u \, dC_u = A_{t+h} - A_t. \end{aligned}$$

Furthermore,

$$\mathbf{U}_A^\alpha f(i) = \mathbb{E}^i \int_0^\infty e^{-\alpha t} M_t \, dC_t = \widetilde{\mathbb{E}}^i \int_0^\infty e^{-\alpha t} \, dC_t = \widetilde{\mathbf{U}}_C^\alpha f(i),$$

$$\mathbb{E}^i A_t = \mathbb{E}^i \int_0^t M_s \, dC_s = \widetilde{\mathbb{E}}^i \int_0^t dC_s = \widetilde{\mathbb{E}}^i C_t,$$

because $\mathbb{E}^i(M_s \, dC_s) = \widetilde{\mathbb{E}}^i \, dC_s$, where $\widetilde{\mathbb{E}}^i$ is the conditional expectation corresponding to $\widetilde{\mathbf{P}}(t)$. □

COROLLARY 3. *Suppose that A_σ is integrable. Then the potential $U = (U_t)$ generated by the extended AF $A = (A_t)$ from Theorem 2, is given by*

$$U_t = M_t u_A(X_t) \quad \text{a.e.}, \qquad t \geqslant 0.$$

Proof (outline).

$$\begin{aligned} U_t &= \mathbb{E}(A_\infty - A_t \mid \mathcal{F}_t) = \mathbb{E}\big(M_t(A_\infty \circ \theta_t) \mid \mathcal{F}_t\big) \\ &= M_t \mathbb{E}(A_\infty \circ \theta_t \mid X_t) = M_t u_A(X_t). \end{aligned}$$ □

The following two results are corollaries of Theorem 2, but because of their importance are stated as theorems.

THEOREM 4. *If C is an IF determined by a charge $a \geqslant 0$, then A is a CAF of the form*

$$A_t = \int_0^t a(X_s)M_s\,ds, \qquad t \geqslant 0,$$

with α-potential and characteristic given by

$$u^\alpha_A(i) = \widetilde{\mathbf{U}}^\alpha a(i), \qquad \mathbb{E}^i A_t = \int_0^t \widetilde{\mathbf{P}}(s)a(i)\,ds, \qquad i \in \mathbb{I}_M.$$

THEOREM 5. *If C is a DAF determined by a function $g \geqslant 0$, then A is a DAF of the form*

$$A_t = \sum_{0<s\leqslant t} g(X_{s-}, X_s)M_s, \qquad t \geqslant 0, \qquad A_0 = 0,$$

with α-potential and characteristic given by

$$u^\alpha_A(i) = \widetilde{\mathbf{U}}^\alpha g_{\mathbf{Q}}(i), \qquad \mathbb{E}^i A_t = \int_0^t \widetilde{\mathbf{P}}(s)g_{\mathbf{Q}}(i)\,ds, \qquad i \in \mathbb{I}_M.$$

Proof of Theorems 4 *and* 5 (outline). Specialisation of Theorem 2 to IF C with $dC_t = a(X_t)\,dt$, leads to Theorem 4. For DAF C, one can use compensator of A defined by

$$B_t = \int_0^t g_{\mathbf{Q}}(X_s)M_s\,ds. \qquad \square$$

The proof of Theorem 5 relies on the extended Lévy formula for M.C.'s which establishes connection between a discontinuous functional A and its compensator B. This provides another illustration of a connection between a jump chain and a minimal chain, under killing determined by a MF M. In view of importance of the Lévy formula, it is stated below. Its proof may be patterned on the proof of the classical Lévy formula stated in § 8.1.3; see also subsequent discussion of compensators in § 8.1.4.

THEOREM 6 (Lévy formula). *For a given MF $M = (M_t)$, the following relation*

$$\mathbb{E}^i \sum_{0<s\leqslant t} Z_s g(X_{s-}, X_s)M_s = \mathbb{E}^i \int_0^t Z_s g_{\mathbf{Q}}(X_s)M_s\,ds, \qquad t > 0, \quad i \in \mathbb{I}_M,$$

holds for all predictable processes $Z = (Z_t) \geqslant 0$.

Observe that for any t, the LHS is the finite sum, as only non-vanishing terms correspond to the discontinuities of the chain $X = (X_t)$. The Lévy formula may be written in another form:

$$\mathbb{E}^i \int_0^t Z_s\,dA_s = \mathbb{E}^i \int_0^t Z_s\,dB_s$$

in agreement with definition stated in § 8.1.4. As noted, $(A_t - B_t, \mathcal{F}_t)$ is a martingale (under integrability condition), and processes A and B have the same potential. The pair $(\mathbf{Q} + \mathbf{\Theta}, B)$ constitutes the Lévy system for the discontinuous functional $A = (A_t)$.

The examples given in the following § 8.2.2–8.2.3 emphasize the role of MF (M_t) in the extended AF (A_t). They should be compared with corresponding examples in § 8.1.3.

§ 8.2.2. Continuous functionals. Several examples of extended CAF of the form

$$A_t = \int_0^t a(X_s) M_s \, ds, \qquad t \geqslant 0,$$

given here, are chosen to provide additional insight into some results obtained in Chapter II. For the purpose of calculations it may be convenient to recall that for $i \in \mathbb{I}_M$:

$$\mathbb{E}^i a(X_s) M_s = \sum_j \tilde{p}_{ij}(s) a(j) = \widetilde{\mathbf{P}}(s) a(i),$$

where $a \geqslant 0$ is bounded.

Example 1. Let T be the first entrance time to a set H, and define a MF $M = (M_t)$ by

$$M_t = I_{(t<T)}, \qquad t \geqslant 0$$

(see § 8.1.4). Thus, $\mathbb{I}_M = H^c$ and $\sigma = T$, and for $i, j \in H^c$:

$$\tilde{p}_{ij}(t) = \mathbb{E}^i(X_t = j,\ M_t) = \mathbb{P}^i(X_t = j,\ t < T \leqslant \infty) = {}_H p_{ij}(t).$$

This means that a M.C. generated by M has the taboo semi-group ${}_H\mathbf{P}(t)$, and has been obtained from the original chain observed prior to the first entrance to the taboo set H. Hence, the use of taboo probabilities is equivalent to killing of the original chain at time T. This basic point of view has been already noted in § 3.1.1 and is further discussed in § 9.1.2.

With a charge a strictly positive on H^c, A defined as above is a CAF by Theorem 4, § 8.2.1, with

$$u_A^\alpha(i) = {}_H\mathbf{U}^\alpha a(i), \qquad \mathbb{E}^i A_t = \int_0^t {}_H\mathbf{P}(s) a(i) \, ds.$$

The following special cases are of interest. Indeed, they may serve as an alternative approach to discussion in Sub-section 3.1.

a) Let $a = 1$, so $A_t = \min(t, T)$ and

$$u_A^\alpha(i) = \sum_{j \in H^c} {}_H u_{ij}^\alpha = \mathbb{E}^i \int_0^T e^{-\alpha t} \, dt, \qquad \mathbb{E}^i A_t = \int_0^t D_i^c(s) \, ds \longrightarrow \mathbb{E}^i T.$$

See Theorems 8 and 10 in § 3.1.1, and also Example 4, § 8.1.3.

b) Let $a = D$, so

$$A_t = \int_0^t D(X_s) I_{(s<T)}\, ds$$

and

$$u_A^\alpha(i) = \sum_{j \in H^c} {}_H u_{ij}^\alpha\, D_j = \mathbb{E}^i \int_0^T e^{-\alpha s} D(X_s)\, ds,$$

$$\mathbb{E}^i A_t = \int_0^t (D_i^c(s) - D_i^c)\, ds \longrightarrow \mathbb{E}^i(T,\ T < \infty) = \mathbb{E}^i A_T.$$

See Theorems 8 and 10 in § 3.1.1, and compare with Example 2, § 8.1.3.

c) Let $a = \bar{q}$, so

$$A_t = \int_0^t \bar{q}(X_s) I_{(s<T)}\, ds, \qquad t \geqslant 0,$$

and

$$u_A^\alpha(i) = \sum_{j \in H^c} {}_H u_{ij}^\alpha\, \bar{q}_j = d_i^\alpha = \mathbb{E}^i \int_0^T e^{-\alpha t} \bar{q}(X_t)\, dt,$$

$$\mathbb{E}^i A_t = D_i(t) \longrightarrow D_i = \mathbb{E}^i A_T.$$

See Theorems 7 and 9 in § 3.1.1.

Example 2. In the analogous manner, consider a CAF $A = (A_t)$ defined by

$$A_t = \int_0^t a(X_s) I_{(s<T_\rho)}\, ds, \qquad t \geqslant 0,$$

where T_ρ is the first return time to a set H. As T_ρ is a terminal time, Theorem 2, § 8.2.1, applies, and

$$u_A^\alpha(i) = {}_H\mathbf{U}^{\rho\alpha} a(i), \qquad \mathbb{E}^i A_t = \int_0^t {}_H\mathbf{P}^\rho(s) a(i)\, ds,$$

with supp $(A) = H^c$. Thus, in comparison with Example 1, the ordinary taboo probabilities are replaced now by the exit taboo probabilities (see § 2.2.4).

Again, the following special cases supplement discussion in Sub-section 3.3.

a) Take $a = 1$, so

$$u_A^\alpha(i) = \sum_j {}_H u_{ij}^{\rho\alpha}, \qquad \mathbb{E}^i A_t = \int_0^t D_i^{\rho c}(s)\, ds \longrightarrow \mathbb{E}^i T_\rho.$$

See Theorems 8 and 10 in § 3.3.1, and also Example 4, § 8.1.3.

b) Take $a = D$, so

$$u_A^\alpha(i) = \sum_j {}_H u_{ij}^{\rho\alpha} D_j = (D_i^\rho - d_i^{\rho\alpha})/\alpha = \mathbb{E}^i \int_0^{T_\rho} e^{-\alpha s} D(X_s)\, ds,$$

$$\mathbb{E}^i A_t = \int_0^t (D_i^{\rho c}(s) - D_i^{\rho c})\, ds \longrightarrow \mathbb{E}^i(T_\rho,\ T_\rho < \infty).$$

See Theorem 8 and Corollary 5 in § 3.3.1, and compare with Example 2, § 8.1.3.

c) Take $a = \bar{q}$, so

$$u_A^\alpha(i) = d_i^{\rho\alpha}, \qquad \mathbb{E}^i A_t = D_i^\rho(t).$$

See Theorems 7 and 9 in § 3.3.1.

Example 3. Consider the first entrance time T_s after an instant s to a set H and define, for fixed $s > 0$,

$$A_t = \int_0^t a(X_v) I_{(v<T_s)}\, dv, \qquad t \geqslant 0.$$

Clearly, (A_t) is an (adapted) increasing process, but it is not an AF (because T_s is not a terminal time); see Example 4, § 8.1.3.

However, its α-potential and characteristic are easily found to be

$$u_A^\alpha(i) = {}_H\mathbf{U}^{s\alpha} a(i), \qquad \mathbb{E}^i A_t = \int_0^t {}_H\mathbf{P}^s(v) a(i)\, dv,$$

where ${}_H\mathbf{P}^s(v)$ is the matrix of delayed taboo probabilities introduced in § 2.2.3, and ${}_H\mathbf{U}^{s\alpha}$ is the matrix of their Laplace transforms (see also § 3.3.5).

Example 4. Consider the last exit time L from a set H, and define

$$M_t = I_{(t<L)}, \qquad t \geqslant 0.$$

It is easy to verify that

$$M_t(M_s \circ \theta_t) = M_{t+s}, \qquad t, s \geqslant 0$$

(see Lemma 2, § 2.1.4). However, M_t is not $\mathcal{F}_t$-measurable, because L is a co-stopping time. Hence, $M = (M_t)$ is only "rough" MF. Its set of permanent points consists of those states i for which

$$\mathbb{P}^i(L > 0) = \Gamma_i^c(0) = D_i = 1,$$

and its mean is

$$\mathbb{E}^i M_t = \Gamma_i^c(t).$$

Consider now

$$A_t = \int_0^t a(X_s)M_s\,ds, \qquad t \geqslant 0.$$

It is easy to verify that

$$A_{t+h} = A_t + M_t(A_h \circ \theta_t),$$

but as A_t is not $\mathcal{F}_t$-measurable, so $A = (A_t)$ is only a "rough" AF.

From discussion in § 4.1.3, one could consider

$$\tilde{p}_{ij}(t) = \mathbb{P}^i(X_t = j,\ t < L) = p_{ij}(t)D_j.$$

However, it is evident that these $\tilde{p}_{ij}(t)$ do not form a semi-group (for a proper form of transition probabilities for a chain killed at L, see § 9.1.6). Nevertheless, direct calculation of expectation yields

$$\mathbb{E}^i a(X_t)M_t = \sum_j p_{ij}(t)D_j a(j),$$

so the α-potential and the characteristic of the rough AF A are:

$$u^\alpha_A(i) = \mathbf{U}^\alpha a(i)D_i, \qquad \mathbb{E}^i A_t = \int_0^t \mathbf{P}(s)a(i)D_i\,ds.$$

Note that the case of $a = 1$ has been treated in Example 4, § 8.1.3.

Example 5. This example well illustrates the role of a MF $M = (M_t)$ in definition of an extended AF. Let R be a terminal stopping time, so

$$M_t = I_{(t<R)}, \qquad t \geqslant 0,$$

becomes a MF.

Consider now an increasing process $A = (A_t)$ defined by

$$A_t = \int_0^t M_s\,ds = \min(t, R), \qquad t \geqslant 0,$$

and already considered in Example 4, § 8.1.3. As noted, A is not an AF in general, because

$$A_{t+h} - A_t = \int_t^{t+h} I_{(s<R)}\,ds, \qquad A_h \circ \theta_t = \int_t^{t+h} I_{(s<R_t)}\,ds,$$

where $R_t = t + R \circ \theta_t$. However, A is an extended AF because $R_t = R$ for $t < R$, so

$$M_t(A_h \circ \theta_t) = M_t \int_t^{t+h} I_{(s<R_t)}\,ds = \int_t^{t+h} M_s\,ds = A_{t+h} - A_t.$$

Thus, restricting attention to $t < R$, the original increasing process becomes an AF $A = (A_t,\ 0 \leqslant t < R)$ with life time R. In particular, with reference to Example 4, § 8.1.3, the processes with restricted life time $(A_t,\ t < T)$ and $(A_t,\ t < T_\rho)$ are AF, whereas $(A_t,\ t < L)$ is a rough AF.

§ 8.2.3. Discontinuous functionals. Several examples of extended DAF of the form

$$A_t = \sum_{0<s\leqslant t} g(X_{s-}, X_s)M_s, \qquad t > 0, \qquad A_0 = 0,$$

given here, provide additional insight into some results obtained in Chapter II. Recall that $g \geqslant 0$ is bounded on $\mathbb{I} \times \mathbb{I}$ and vanishes on the diagonal. Calculations of α-potential and characteristic are carried out using Theorem 5, § 8.2.1.

Example 1. Let T be the first entrance time to a set H, and define a MF $M = (M_t)$ by

$$M_t = I_{(t<T)}, \qquad t \geqslant 0.$$

Then,

$$u_A^\alpha(i) = {}_H\mathbf{U}^\alpha g_{\mathbf{Q}}(i), \qquad \mathbb{E}^i A_t = \int_0^t {}_H\mathbf{P}(s) g_{\mathbf{Q}}(i)\, ds.$$

Compare this with Example 1, § 8.2.2. Suppose that g is strictly positive on $H^c \times H^c$. Then, $\operatorname{supp}(A) = H^c$, and the compensator $B = (B_t)$ of A is given by

$$B_t = \int_0^t g_{\mathbf{Q}}(X_s) I_{(s<T)}\, ds.$$

The following special cases are of interest. Indeed, they provide an alternative approach to the role of the jump chain (see Sub-section 3.1).

a) *Counting process.* Let $g(i,j) = 1$ for $i \neq j$, so $g_{\mathbf{Q}}(i) = q_i$. Hence,

$$A_t = \sum_{0<s\leqslant t} I_{(s<T)} = \sum_{n=0}^{\infty} I_{(\tau_n \leqslant t)} \min(\tau_n, T)$$

represents the number of jumps up to time t before the first entrance to H. Clearly,

$$u_A^\alpha(i) = {}_H\mathbf{U}^\alpha q_i, \qquad \mathbb{E}^i A_\infty = \mathbb{E}^i A_T = {}_H\mathbf{U} q_i, \qquad i \in H^c,$$

and

$$B_t = \int_0^{\min(t,T)} q(X_s)\, ds, \qquad t \geqslant 0.$$

b) *Individual jumps.* Let k and j be two different states with $k \in H^c$, and let

$$g(i,j) = \begin{cases} 1 & \text{when } i = k, \\ 0 & \text{when } i \neq k. \end{cases}$$

Then

$$g_{\mathbf{Q}}(i) = \begin{cases} q_{ij} & \text{when } i = k, \\ 0 & \text{when } i \neq k, \end{cases}$$

and

$$A_t = \sum_{0<s\leqslant t} I_{(X_{s-}=k,\ X_s=j)} I_{(s<T)}, \qquad k \in H^c, \quad k \neq j,$$

is the number of jumps from state k to state j up to time t, before the first entrance to H.

Clearly, the compensator has the form

$$B_t = \int_0^t I_k(X_s) q_{X_s,j} I_{(s<T)}\, ds$$

and, therefore,

$$u_A^\alpha(i) = {}_H u_{ik}^\alpha q_{kj}, \qquad \mathbb{E}^i A_t = \int_0^t {}_H p_{ik}(s) q_{kj}\, ds,$$

in agreement with Lemma 9, § 3.1.2, for $j \in H$.

c) Summing over $k \in H^c$ in the preceding example, one obtains A_t which represents the number of jumps to state j up to time t, before the first entrance to H,

$$A_t = \sum_{0<s\leqslant t} I_{(X_{s-}\in H^c,\ X_s=j)} I_{(s<T)}.$$

Then

$$\mathbb{E}^i A_t = \int_0^t \sum_{k\in H^c} {}_H p_{ik}(s) q_{kj}\, ds = \begin{cases} {}_H w_{ij}(t) & \text{for } j \in H^c, \\ D_{ij}(t) & \text{for } j \in H, \end{cases}$$

and

$$u_A^\alpha(i) = \sum_{k\in H^c} {}_H u_{ik}^\alpha\, q_{kj} = \begin{cases} {}_H w_{ij}^\alpha & \text{for } j \in H^c, \\ d_{ij}^\alpha & \text{for } j \in H. \end{cases}$$

See Theorem 6, § 3.1.2, and § 2.2.4. It is obvious that positive ${}_H w_{ij}(t)$ is at least 1, whereas $D_{ij}(t) \leqslant 1$.

d) *Jumps to* H. Let

$$g(i,j) = I_{H^c}(i) I_H(j), \qquad i \neq j,$$

so

$$g_{\mathbf{Q}}(i) = I_{H^c}(i)\bar{q}_i,$$

and

$$A_t = \sum_{0<s\leqslant t} I_{(X_{s-}\in H^c,\ X_s\in H)} I_{(s<T)}, \qquad t > 0,$$

represents the number of jumps up to time t from H^c to H, before the first entrance to H. Obviously, there is at most one such a jump.

The intensity is

$$\Lambda_s = I_{H^c}(X_s)\bar{q}(X_s)I_{(s<T)} = \bar{q}(X_s)I_{(s<T)},$$

and the compensator has the form

$$B_t = \int_0^t \bar{q}(X_s)I_{(s<T)}\,ds.$$

Compare this with Example 1-c, § 8.2.2. It follows that

$$u_A^\alpha(i) = {}_H\mathbf{U}^\alpha \bar{q}_i = d_i^\alpha, \qquad \mathbb{E}^i A_t = \int_0^t {}_H\mathbf{P}(s)\bar{q}_i\,ds = D_i(t), \qquad i \in H^c,$$

in agreement with Theorem 7, § 3.1.1.

Example 2. Let T_ρ be the first return time to a set H, and define a MF $M = (M_t)$ by

$$M_t = I_{(t<T_\rho)}, \qquad t \geqslant 0.$$

Observe that now summation in definition of A_t is actually over $\rho \leqslant s < T_\rho$, because for $s < \rho$ the M.C. is in the initial state. Then,

$$u_A^\alpha(i) = {}_H\mathbf{U}^{\rho\alpha} g_{\mathbf{Q}}(i), \qquad \mathbb{E}^i A_t = \int_0^t {}_H\mathbf{P}^\rho(s) g_{\mathbf{Q}}(i)\,ds, \qquad i \in \mathbb{I}.$$

Compare this with Example 2, § 8.2.2. The compensator $B = (B_t)$ of A is given by

$$B_t = \int_0^t g_{\mathbf{Q}}(X_s)I_{(s<T_\rho)}\,ds.$$

The present example is similar to the preceding Example 1, but the following special case is of interest.

a) Consider

$$A_t = \sum_{0<s\leqslant t} I_{(X_{s-}\neq j,\ X_s=j)} I_{(s<T_\rho)}, \qquad t > 0,$$

for a fixed state $j \in H^c$, and suppose that the initial state i is in H. Then A_t represents the number of visits to state j in H^c up to time t, before the first return to H, when starting at i in H and leaving H immediately on leaving state i.

It is easy to verify from relations in § 3.4.2 that

$$u_A^\alpha(i) = \sum_{m\neq j} {}_H u_{im}^{\rho\alpha} q_{mj} = k_{ij}^\alpha, \qquad i \in H, \quad j \in H^c,$$

and

$$\mathbb{E}^i A_t = \int_0^t \sum_{m \neq j} {}_H p^\rho_{im}(s)\, q_{mj}\, ds = K_{ij}(t) \longrightarrow K_{ij} = \mathbb{E}^i A_\infty .$$

As noted in § 3.4.2, k^α_{ij} refers to the jump chain. Thus, $K_{ij}(t)$ is the average number of visits to state $j \in H^c$ up to time t, when starting from $i \in H$, before the first return to H. Furthermore, K_{ij} is the average for the total time. It is rather obvious now why this situation refers to the jump chain, confirming distinction between co-balayages K_{ij} and N_{ij} discussed in § 3.4.2. As noted there,

$$K_{ij} = {}_H u^\rho_{ij}\, q_j, \qquad N_{ij} = q_i\, {}_H u^\rho_{ij}.$$

Remark. It is of interest to compare the following expressions (for $i \in H$):

$$u^\alpha_A(i) = \sum_{m \neq j} {}_H u^{\rho\alpha}_{im}\, q_{mj} = {}_H w^{\rho\alpha}_{ij} = k^\alpha_{ij} \qquad \text{for} \quad j \in H^c, \qquad (\S\ 3.4.2)$$

$$d^{\rho\alpha}_{ij} = \sum_{m \neq j} {}_H u^{\rho\alpha}_{im}\, q_{mj} \leqslant 1 \qquad \text{for} \quad j \in H. \qquad (\text{Thm. 6, } \S\ 3.3.2)$$

For $i \in H^c$, one obtains ${}_H w^\alpha_{ij}$ and d^α_{ij} from Example 1-c above. In terms of the average number of visits to state j, before returning to H, it is obvious that positive K_{ij} is at least 1, whereas $D^\rho_{ij} \leqslant 1$.

8.3. LAST EXIT

§ 8.3.1. Theorems. This sub-section is devoted to discussion of the last exit time in terms of functionals, and in particular to re-derivation of various joint distributions from Section 4. The treatment here is the adaptation to the usual M.C. $X = (X_t)$ of results obtained by R. K. Getoor and M. J. Sharpe [1973] for the general Markov processes (see also Lévy formula in § 8.2.1).

Let L_t be the last exit time from a set H before time t. Recall that

$$s = L_t < t \qquad \text{iff} \qquad s \in W \quad \text{and} \quad s < t < T_s,$$

where W is the set of consecutive entrances to H^c (see § 2.1.1).

Let $Z = (Z_t)$ be a non-negative (measurable) process, and let f be a positive bounded function on the state space $\mathbb{I}$. Write $Z(L_t)$ for Z_{L_t}. The problem now consists of expressing the expectation

$$\mathbb{E}^i\big(Z(L_t) f(X_t),\ 0 < L_t < t\big) =: E\,(i,t)$$

in terms of appropriate functionals. Subsequently, selection of special forms of f and Z_t would lead to several important formulas. In the following, if $\mu_i(\,\cdot\,)$ is any entrance law indexed by $i \in \mathbb{I}$, then the inner product $\mu_i(t) \bullet f$ will be written as $\mu(t)f(i)$. The main result may be presented in the following two versions.

THEOREM 1. *There exists a DAF $\tilde{A} = (\tilde{A}_t)$ supported on H^c and a family of entrance laws $\tilde{\mu}(\cdot)$ for ${}_H\mathbf{P}(t)$, such that if $Z = (Z_t) \geqslant 0$ is a (measurable) process and $f \geqslant 0$ is bounded on $\mathbb{I}$, then, for all $i \in \mathbb{I}$,*

$$E(i,t) = \mathbb{E}^i \int_0^t Z_s \tilde{\mu}(t-s) f(X_s)\, d\tilde{A}_s,$$

where the integral is taken over the open interval $(0,t)$.

The predictable version of Theorem 1 is obtained by taking the dual predictable projection of the functional $\tilde{A}$ (see § 8.1.4).

THEOREM 2. *There exists a CAF $A = (A_t)$ supported on H and a family of entrance laws $\mu(\cdot)$ for ${}_H\mathbf{P}(t)$, such that, if $Z = (Z_t) \geqslant 0$ is a predictable process and $f \geqslant 0$ is bounded on $\mathbb{I}$, then, for all $i \in \mathbb{I}$,*

$$E(i,t) = \mathbb{E}^i \int_0^t Z_s \mu(t-s) f(X_s)\, dA_s,$$

where the integral is taken over the open interval $(0,t)$.

The important feature of Theorem 2 is the transformation from a discontinuous functional $\tilde{A}$ supported on H^c to a continuous functional A supported on H, and from the entrance law $\tilde{\mu}(t)$ to an entrance law $\mu(t)$. These effects of duality have been already observed in Sections 4 and 6.

Proofs of the above theorems are outlined in the next § 8.3.2 where also expressions for functionals $\tilde{A}$ and A, and for entrance laws $\tilde{\mu}$ and μ, are obtained.

It suffices to prove the following lemma (corresponding to $Z_s \equiv 1$) which covers both Theorem 1 and Theorem 2, and establishes connections with discussion in Chapter II. The extension to predictable (Z_s) follows as in proof of Lévy formula.

LEMMA 3. *For a bounded $f \geqslant 0$ on $\mathbb{I}$:*

$$\mathbb{E}^i \int_0^t \tilde{\mu}(t-s) f(X_s)\, d\tilde{A}_s = \mathbb{E}^i \int_0^t \mu(t-s) f(X_{s-})\, dA_s, \qquad i \in \mathbb{I}, \quad t > 0.$$

Essential link in the proof is the following relation deduced from properties of the set W and valid for a process (Z_s):

$$\begin{aligned} \mathbb{E}^i \sum_{s \in W} Z_s f(X_t) I_{(s<t<T_s)} &= \mathbb{E}^i \sum_{s \in W} Z_s f(X_t) I_{(s=L_s<t)} \\ &= \mathbb{E}^i \big(Z(L_t) f(X_t),\ 0 < L_t < t\big). \end{aligned} \tag{$*$}$$

§ 8.3.2. Outline. a) Consider first Theorem 1 from the preceding § 8.3.1. Let c be a non-negative bounded function on $\mathbb{I}$, supported by H^c, to be used as a normalizing factor. For each $i \in \mathbb{I}$, the measure $\tilde{\mu}_{i\cdot}(t)$ defined for all $j \in \mathbb{I}$ by

$$\tilde{\mu}_{ij}(t) = \begin{cases} c^{-1}(i)\, {}_Hp_{ij}(t) & \text{for} \quad i \in H^c, \\ 0 & \text{for} \quad i \in H, \end{cases}$$

forms the entrance law $\tilde{\mu}_i(\cdot)$ relative to the taboo semi-group ${}_H\mathbf{P}(t)$, $t \geqslant 0$ (see § 1.3.4). This entrance law is supported on H^c, and satisfies

$$\sum_{j \in H^c} \tilde{\mu}_{ij}(t+s) = \sum_{k \in H^c} \tilde{\mu}_{ik}(s) D_k^c(t), \qquad t \geqslant 0, \quad s > 0.$$

For any bounded function f on $\mathbb{I}$, write for the inner product

$$\tilde{\mu}(t)f(i) = \sum_j \tilde{\mu}_{ij}(t) f(j) = c^{-1}(i)_H\mathbf{P}(t)f(i), \qquad i \in \mathbb{I}$$

(with the understanding that 0/0 is taken as 0). Observe that the entrance law defines a semi-group $\widetilde{\mathbf{P}}(t) = (\tilde{p}_{ij}(t))$, where

$$\tilde{p}_{ij}(t) = \tilde{\mu}_{ij}(t)c(j) = c^{-1}(i)\, {}_Hp_{ij}(t)c(j).$$

To compute the LHS expectation in Lemma 3, § 8.3.1, denote by (δ_s) a purely discontinuous AF with unit steps at point of W (a counting process) and supp $(\delta) = H^c$, and take for the required DAF

$$\tilde{A}_t = \sum_{0<s\leqslant t,\ s \in W} h^{-1}(X_{s-})c(X_s) = \int_0^t h^{-1}(X_{s-})c(X_s)\, d\delta_s,$$

where h^{-1} is a positive bounded function on $\mathbb{I}$, supported on H, to be used as another normalizing factor. Then, using relation $(*)$,

$$\begin{aligned}
&\mathbb{E}^i \sum_{s \in W} h^{-1}(X_{s-}) f(X_t) I_{(s<t<T_s)} \\
&= \mathbb{E}^i \sum_{s \in W} h^{-1}(X_{s-}) \mathbb{E}\big(f(X_{t-s}) I_{(0<t-s<T)} \circ \theta_s \mid X_s\big) \\
&= \mathbb{E}^i \sum_{s \in W} h^{-1}(X_{s-})\, {}_H\mathbf{P}(t-s) f(X_s) = \mathbb{E}^i \int_0^t h^{-1}(X_{s-})\, {}_H\mathbf{P}(t-s) f(X_s)\, d\delta_s \\
&= \mathbb{E}^i \int_0^t c^{-1}(X_s)\, {}_H\mathbf{P}(t-s) f(X_s)\, d\tilde{A}_s = \mathbb{E}^i \int_0^t \tilde{\mu}(t-s) f(X_s)\, d\tilde{A}_s.
\end{aligned}$$

b) To consider next Theorem 2, § 8.3.1, note that the DAF $(\tilde{A}_t)$ may be written as

$$\tilde{A}_t = \sum_{0<s\leqslant t} I_H(X_{s-})\, h^{-1}(X_{s-})\, c(X_s)\, I_{H^c}(X_s),$$

with $X_{s-} \in H$ and $X_s \in H^c$, so

$$g(i,j) = I_H(i)\,h^{-1}(i)\,c(j)\,I_{H^c}(j).$$

Hence, in agreement with § 8.1.3,

$$g_{\mathbf{Q}}(i) = \begin{cases} h^{-1}(i)w(i) \equiv a(i), & i \in H, \\ 0, & i \in H^c, \end{cases}$$

where

$$w(i) = I_H(i) \sum_{j \in H^c} q_{ij} c(j), \qquad i \in H.$$

Consequently, the dual predictable projection (a compensator) of $\tilde{A}$ is a CAF $A = (A_t)$ defined by

$$A_t = \int_0^t a(X_{s-})\,ds, \qquad t > 0.$$

Clearly, supp $(A) = H$.

To put Theorem 2 in the required formulation, another entrance law must be introduced. For each $i \in \mathbb{I}$, the measure $\mu_{i\bullet}(t)$ defined for all $j \in H^c$ by

$$\mu_{ij}(t) = \begin{cases} w^{-1}(i) n_{ij}(t) & \text{for} \quad i \in H, \\ 0 & \text{for} \quad i \in H^c, \end{cases}$$

forms the entrance law $\mu_i(\cdot)$ relative to the taboo semi-group ${}_H\mathbf{P}(t)$, $t \geqslant 0$ (see § 1.3.4). This entrance law is supported on H^c and satisfies

$$\sum_{j \in H^c} \mu_{ij}(t+s) = \sum_{k \in H^c} \mu_{ik}(s) D_k^c(t), \qquad t \geqslant 0, \quad s > 0.$$

Recall that it has been mentioned in § 3.4.1 that $n_{i\bullet}(t)$, for fixed i, form an entrance law; hence so does $\mu_i(t)$. For any bounded function f on $\mathbb{I}$, write for the inner product

$$\begin{aligned} \mu(t)f(i) &= \sum_j \mu_{ij}(t) f(j) = w^{-1}(i) \sum_{k \in H^c} q_{ik}\, {}_H\mathbf{P}(t) f(k) \\ &= w^{-1}(i) \mathbf{Q}\, {}_H\mathbf{P}(t) f(i), \qquad i \in H. \end{aligned}$$

One can now compute the RHS expectation in Lemma 3, § 8.3.1:

$$\begin{aligned} \mathbb{E}^i \int_0^t \mu(t-s) f(X_{s-})\,dA_s &= \mathbb{E}^i \int_0^t w^{-1}(X_{s-}) \mathbf{Q}\, {}_H\mathbf{P}(t-s) f(X_{s-})\,dA_s \\ &= \mathbb{E}^i \int_0^t w^{-1}(X_{s-}) \mathbf{Q}\, {}_H\mathbf{P}(t-s) f(X_{s-}) a(X_{s-})\,ds \\ &= \mathbb{E}^i \int_0^t h^{-1}(X_{s-}) \mathbf{Q}\, {}_H\mathbf{P}(t-s) f(X_{s-})\,ds. \end{aligned}$$

c) To verify that AF's and entrance laws found above have the required form, observe that expressions for expectations on the both sides of Lemma 3, § 8.3.1, are equal. Indeed, their common value stated below, follows from Lemma 4, § 2.2.5, (density argument), so Lemma 3, § 8.3.1, agrees with the relation (*)

$$\mathbb{E}^i \sum_{s\in W} h^{-1}(X_{s-})f(X_t)I_{(s<t\leqslant T_s)} = \mathbb{E}^i\big(h^{-1}(X(L_t-))f(X_t),\ 0<L_t<t\big)$$

$$= \int_0^t \sum_{r\in H}\sum_{k\in H^c} p_{ir}(s)\,h^{-1}(r)\,q_{rk}\sum_j {}_H p_{kj}(t-s)f(j)\,ds.$$

Remark. As noted in § 8.1.4, functionals $\tilde{A}$ and A have the same characteristics and α-potentials:

$$\mathbb{E}^i\tilde{A}_t = \mathbb{E}^i A_t = \int_0^t \sum_{r\in H} p_{ir}(s)a(r)\,ds,$$

$$u^\alpha_{\tilde{A}}(i) = u^\alpha_A(i) = \sum_{r\in H} u^\alpha_{ir}a(r).$$

§ 8.3.3. Normalization. In order to determine the normalizing functions c and h, further requirements must be imposed. Several possibilities may be considered, according to required forms of functionals.

a) To obtain (A_t) representing the total time spent in the set H up to time t, one must have $a(i)=1$ for all $i\in H$, so $w(i)=h(i)$ on H. Thus, as in Sub-section 8.4,

$$A_t = \int_0^t I_H(X_{s-})\,ds, \qquad t>0.$$

The form of the DAF (A_t) is now determined by selecting the function c. Thus, taking $c(i)=D^c_i$ for $i\in H^c$ gives $w(i)=b_i$ for $i\in H$, where b is the escape rate (see § 3.1.3). Thus,

$$\tilde{A}_t = \sum_{0<s\leqslant t,\ s\in W} D^c(X_s)/b(X_{s-}), \qquad t>0.$$

b) The choice of $h(i)=1$ for all $i\in H$ is the most common because it corresponds to several interesting special cases. This yields $a(i)=w(i)$ for all $i\in H$, and leaves c arbitrary. The common α-potential is then $u^\alpha_A(i)=\mathbf{U}^\alpha w\,(i)$. Taking $c=D^c$ as before yields now $u^\alpha_A=\mathbf{U}^\alpha b$.

Taking $c(i)=1-d^1_i$ for $i\in H^c$ (with fixed $\alpha=1$) yields $a(i)=n^1_i$ for $i\in H$ (see § 3.4.1). Hence,

$$u^\alpha_A(i) = \sum_{k\in H} u^\alpha_{ik}n^1_k.$$

Note that this differs from the Laplace transform of $\mathbb{P}^i(0<L_t<t)$ which contains the term n^α_k; see the Chung equation in § 4.2.1. Indeed, one can easily verify that

the α-potential $u^{\alpha}_{A^*}$ of an increasing process

$$A^*_t = \max(0, t - T) = \int_0^t I_{(T \leqslant s)}\, ds, \qquad t \geqslant 0,$$

(where T is the first entrance time to H) coincides with d^{α}_i/α given by Corollary 11, § 4.2.1. See also Theorem 2, § 5.2.1, and compare with Example 4 § 8.1.3.

Taking $c(i) = 1$ identically on H^c gives $a(i) = q_i - \bar{q}_i$ for $i \in H$. To obtain $a(i) = I_H(i)$ for $i \in H$, requires that $w(i) = 1$ for $i \in H$.

§ 8.3.4. Special cases. The following special cases are of interest. Note first that the common value of expressions in Lemma 3, § 8.3.1, is $\mathbb{E}^i(0 < L_t < t,\ f(X_t))$ for $h = 1$. This fact provides further insight into duality formulae used in Sections 4 and 6. Indeed, taking f to be the indicator of state j, one has from § 8.3.2-c that

$$\begin{aligned}\mathbb{P}^i(0 < L_t < t,\ X_t = j) &= \int_0^t \sum_{k \in H^c} c_{ik}(s)\, {}_H p_{kj}(t-s)\, ds \\ &= \int_0^t \sum_{r \in H} p_{ir}(s) n_{rj}(t-s)\, ds, \qquad j \in H^c,\end{aligned}$$

and equality of these expressions follows from definitions of $c_{ik}(s)$ and $n_{rj}(t-s)$ (see Thm. 1, § 4.2.3).

Next, note from § 8.3.2 that, irrespective of function h,

$$\begin{aligned}\tilde{\mu}(t-s) f(X_s) &= c^{-1}(X_s)\, {}_H\mathbf{P}(t-s) f(X_s), \\ \mu(t-s) f(X_{s-}) &= w^{-1}(X_{s-}) \mathbf{Q}\, {}_H\mathbf{P}(t-s)\, f(X_{s-}),\end{aligned}$$

which for $f = 1$ yield, respectively,

$$\begin{aligned}\sum_{j \in H^c} \tilde{\mu}_{X_s, j}(t-s) &= c^{-1}(X_s) D^c_{X_s}(t-s), \\ \sum_{j \in H^c} \mu_{X_{s-}, j}(t-s) &= w^{-1}(X_{s-}) \mathbf{Q}\, D^c_{X_{s-}}(t-s).\end{aligned}$$

a) Taking $Z_s = m(X_s)\varphi(s)$, where $m \geqslant 0$ and $\varphi \geqslant 0$ are non-random functions on $\mathbb{I}$ and on $(0, \infty)$, Theorem 1, § 8.3.1, yields (with $h = 1$)

$$\begin{aligned}\mathbb{E}^i\big(m(X(L_t))\varphi(L_t),\ 0 < L_t < t\big) &= \mathbb{E}^i \int_0^t m(X_s)\varphi(s) D^c_{X_s}(t-s)\, d\delta_s \\ &= \int_0^t \varphi(s) \sum_{k \in H^c} c_{ik}(s) m(k) D^c_k(t-s)\, ds.\end{aligned}$$

b) On the other hand, taking $Z_s = m(X_{s-})\varphi(s-)$, Theorem 2, § 8.3.1, yields (with $h = 1$)

$$\mathbb{E}^i\big(m(X(L_t-))\varphi(L_t-),\ 0 < L_t < t\big) = \mathbb{E}^i \int_0^t m(X_{s-})\varphi(s-)\mathbf{Q}\, D^c_{X_{s-}}(t-s)\, ds$$
$$= \int_0^t \varphi(s-) \sum_{r \in H} p_{ir}(s) m(r) n_r(t-s)\, ds.$$

c) Taking $m = 1$, one obtains from the above:

$$\mathbb{E}^i\big(\varphi(L_t),\ 0 < L_t < t\big) = \int_0^t \varphi(v) \sum_{k \in H^c} c_{ik}(v) D^c_k(t-v)\, dv,$$
$$\mathbb{E}^i\big(\varphi(L_t-),\ 0 < L_t < t\big) = \int_0^t \varphi(v-) \sum_{r \in H} p_{ir}(v) n_r(t-v)\, dv.$$

Selecting $\varphi(v) = 1$ for $v \leqslant s$, $\varphi(v) = 0$ for $v > s$, one has, for $s \leqslant t$,

$$\mathbb{P}^i(0 < L_t \leqslant s) = \int_0^s \gamma_i^t(v)\, dv,$$

which is Theorem 9, § 4.2.1. Similarly, taking m and φ to be appropriate indicators, one obtains the joint distributions from Theorem 3, § 4.2.2.

8.4. LOCAL TIME

A notion of a local time, rather esoteric for general Markov processes, reduces to a time spent in a state in the case of the usual M.C. Nevertheless, it exhibits interesting properties and offers several advantages.

§ 8.4.1. Definition. Let $X = (X_t)$ be the usual M.C. with a transition matrix $\mathbf{P}(t)$, and let H be a fixed set in the state space $\mathbb{I}$.

DEFINITION 1. A local time on H for X is given by the integral functional $V = (V_t)$:

$$V_t = \int_0^t I_H(X_s)\, ds, \qquad t \geqslant 0.$$

In particular, when H consists of a single state j, write $V^j = (V_t^j)$ for a *local time for X at a state j*.

Clearly, the integral represents the total time spent in a set H up to the instant t, and one has

$$V_t = \sum_{j \in H} V_t^j, \qquad t \geqslant 0.$$

Thus, V is the *occupation time* of a set H, and V^j may be regarded as a density of the occupation time with respect to the counting measure. It should be remarked that a local time on a state j is often defined as $c_j V^j$, where c_j is a normalizing constant; here for convenience, $c_j = 1$ for all j.

The α-potential of the integral functional V is (see § 8.1.2)

$$u_V^\alpha(i) = \mathbb{E}^i \int_0^\infty e^{-\alpha t}\, dV_t = \mathbf{U}^\alpha I_H(i) = \sum_{j \in H} u_{ij}^\alpha, \qquad i \in \mathbb{I}.$$

Let $T = T_H$ be the first entrance time to the set H, and recall that the α-balayage matrix $\mathbf{B}^\alpha = (d_{ij}^\alpha)$ is given by

$$d_{ij}^\alpha = \mathbb{E}^i(e^{-\alpha T},\ X_T = j,\ T < \infty), \qquad i \in \mathbb{I}, \quad j \in H.$$

THEOREM 2. *A local time V satisfies, for $i \in \mathbb{I}$, $\alpha > 0$,*

$$u_V^\alpha(i) = \sum_{k \in H} d_{ik}^\alpha\, u_V^\alpha(k)$$

or, in matrix notation,

$$u_V^\alpha = \mathbf{B}^\alpha u_V^\alpha.$$

Proof. The assertion follows immediately from the resolvent form of the decomposition theorem (Thm. 1, § 6.1.1) by noting that $u_V^\alpha = \mathbf{U}^\alpha I_H$ and ${}_H\mathbf{U}^\alpha I_H = 0$. Hence, components of u_V^α are, respectively, $\mathbf{U}_{12}^\alpha \mathbf{1} = \mathbf{B}_{12}^\alpha \mathbf{U}_{21}^\alpha \mathbf{1}$ and $\mathbf{U}_{22}^\alpha \mathbf{1}$. Moreover, by projection theorem (Thm. 2, § 6.4.1), the α-balayage of u_V^α coincides with u_V^α, because the charge $a = I_H$ is supported on H. □

The simplicity of the usual M.C.'s allows the direct definition of the local time at state j, instead of indirect one (analogous to Theorem 2) used for general Markov processes. Indeed, for $H = \{j\}$, Theorem 2 reduces to the passage relation $u_{ij}^\alpha = f_{ij}^\alpha u_{jj}^\alpha$.

§ **8.4.2. Characterization.** The inverse $\tau = (\tau_t)$ of V, defined in § 8.1.2, is given by

$$\tau_t = \inf\left(s\colon\ V_s > t\right),$$

for each $0 \leqslant t < V_\infty$. Here τ_t represents the actual time needed to produce time t spent in a set H. Clearly, τ is an increasing, right-continuous process with $\tau_0 = T$, and τ_t is a stopping time for each t. Moreover, for every t,

$$X(\tau_t) \in H \quad \text{and} \quad V(\tau_t) = t;$$

for further comments on the inverse, see § 9.5.1.

In particular, when H consists of a single state j, the inverse of V^j will be denoted by τ^j. Note that if j is transient, then V^j_∞ is finite, so τ^j becomes infinite after the last exit from j. If j is recurrent, then V^j_∞ is infinite and so is τ^j_∞.

Introduce the following notation for $i \in \mathbb{I}$, $j \in H$ and $s, t \geqslant 0$, $\alpha > 0$:

$$\begin{aligned} G_{ij}(s,t) &= \mathbb{P}^i\big(\tau_s \leqslant t,\ X(\tau_s) = j\big), \\ G_i(s,t) &= \mathbb{P}^i\big(\tau_s \leqslant t\big), \\ g^\alpha_{ij}(s) &= \mathbb{E}^i\big(e^{-\alpha\tau_s},\ X(\tau_s) = j,\ \tau_s < \infty\big), \\ g^\alpha_i(s) &= \mathbb{E}^i\big(e^{-\alpha\tau_s},\ \tau_s < \infty\big) = \sum_{j \in H} g^\alpha_{ij}(s), \end{aligned}$$

and observe that

$$G_{ij}(0,t) = D_{ij}(t) \quad \text{and} \quad g^\alpha_{ij}(0) = d^\alpha_{ij}.$$

Moreover, for $\alpha = 0$,

$$g_{ij}(s) = \mathbb{P}^i\big(X(\tau_s) = j,\ \tau_s < \infty\big).$$

In the above formulas, conditioning is on the event $X_0 = i$.

THEOREM 1. *For $i \in \mathbb{I}$, $j \in H$, $t, s \geqslant 0$, and $\alpha > 0$:*

$$g^\alpha_{ij}(t+s) = \sum_{k \in H} g^\alpha_{ik}(t)\, g^\alpha_{kj}(s).$$

Proof. Recall that $\tau_{t+s} = \tau_t + \tau_s \circ \theta_{\tau_t}$, $X(\tau_t) \in H$. Hence

$$\begin{aligned} g^\alpha_{ij}(t+s) &= \mathbb{E}^i\big(e^{-\alpha\tau_{t+s}},\ X(\tau_{t+s}) = j,\ \tau_{t+s} < \infty\big) \\ &= \mathbb{E}^i\Big[e^{-\alpha\tau_t}\,\mathbb{E}\Big\{\Big(e^{-\alpha\tau_s},\ X(\tau_s) = j,\ \tau_s < \infty\Big) \circ \theta_{\tau_t} \mid X(\tau_t)\Big\},\ \tau_t < \infty\Big] \\ &= \mathbb{E}^i\Big[e^{-\alpha\tau_t},\ g_{X(\tau_t),j}(s),\ \tau_t < \infty\Big] = \sum_{k \in H} g^\alpha_{ik}(t)\, g^\alpha_{kj}(s). \end{aligned}$$

□

Of interest are special cases when either $t = 0$ or $s = 0$. For $t = s = 0$, Theorem 1 reduces to $\mathbf{B}^\alpha = \mathbf{B}^\alpha\mathbf{B}^\alpha$. Summation over all $j \in H$ yields

$$g^\alpha_i(t+s) = \sum_{k \in H} g^\alpha_{ik}(t)\, g^\alpha_k(s).$$

THEOREM 2. *For $i \in \mathbb{I}$, $j \in H$, and $\alpha > 0$:*

$$u^\alpha_{ij} = \int_0^\infty g^\alpha_{ij}(s)\, ds, \qquad u^\alpha_V(i) = \int_0^\infty g^\alpha_i(s)\, ds.$$

Proof. For fixed $j \in H$, one has as before

$$\mathbb{E}^i \int_0^\infty e^{-\alpha t} I_j(X_t)\, dV_t = \mathbb{E}^i \int_0^\infty e^{-\alpha t} I_j(X_t)\, dt = u_{ij}^\alpha.$$

On the other hand, using the random time change argument (as in § 9.6.3) with formal substitution $t = \tau_s$, one has

$$\mathbb{E}^i \int_0^\infty e^{-\alpha t} I_j(X_t)\, dV_t = \mathbb{E}^i \int_0^\infty e^{-\alpha \tau_s} I_j(X_{\tau_s})\, ds$$

$$= \int_0^\infty \mathbb{E}^i\big(e^{-\alpha \tau_s},\ X_{\tau_s} = j,\ \tau_s < \infty\big)\, ds = \int_0^\infty g_{ij}^\alpha(s)\, ds.$$

The second assertion follows by addition over all $j \in H$. □

The characteristics of functionals V and V^j by definition are:

$$\mathbb{E}^i V_t = \sum_{j \in H} \mathbb{E}^i V_t^j, \qquad \text{where} \quad \mathbb{E}^i V_t^j = \int_0^t p_{ij}(s)\, ds.$$

Hence,

$$\mathbb{E}^i V_\infty = \sum_{j \in H} u_{ij}, \qquad \text{where} \quad u_{ij} = \mathbb{E}^i V_\infty^j$$

in agreement with Theorem 2 (for $\alpha = 0$).

The next theorem gives another expressions for characteristics.

THEOREM 3. *For $i \in \mathbb{I}$, $j \in H$ and $t \geqslant 0$:*

$$\mathbb{E}^i V_t = \int_0^\infty \mathbb{P}^i(\tau_s \leqslant t)\, ds, \qquad \mathbb{E}^i V_t^j = \int_0^\infty \mathbb{P}^i(\tau_s^j \leqslant t)\, ds.$$

Proof. For a function $h \geqslant 0$, one has using random change argument ($t = \tau_s$):

$$\mathbb{E}^i \int_0^\infty h(t)\, dV_t = \mathbb{E}^i \int_0^\infty h(\tau_s)\, ds = \int_0^\infty ds \int_0^\infty d_t \mathbb{P}^i(\tau_s \leqslant t)\, h(t)$$

$$= \int_0^\infty h(t)\, d_t \int_0^\infty \mathbb{P}^i(\tau_s \leqslant t)\, ds.$$

Take now $h(t) = e^{-\alpha t}$, so

$$u_V^\alpha(i) = \mathbb{E}^i \int_0^\infty e^{-\alpha t}\, dV_t = \int_0^\infty e^{-\alpha t}\, d_t \int_0^\infty \mathbb{P}^i(\tau_s \leqslant t)\, ds,$$

and the result follows from the Laplace transform properties. Assertion for V_t^j is obtained in the same fashion with

$$u_{V^j}^\alpha(i) = u_{ij}^\alpha. \qquad \square$$

Finally observe that the following result holds.

THEOREM 4. *For $i \in \mathbb{I}$ and $s \geqslant 0$:*

$$\int_0^\infty e^{-\alpha t}\, d_t\, \mathbb{P}^i(V_t \leqslant s) = 1 - g_i^\alpha(s).$$

Proof. From definition:

$$\mathbb{P}^i(V_t \leqslant s) = \mathbb{P}^i(\tau_s \geqslant t),$$

so

$$\int_0^\infty e^{-\alpha t}\, \mathbb{P}^i(V_t \leqslant s)\, dt = \int_0^\infty e^{-\alpha t}\, \mathbb{P}^i(\tau_s \geqslant t)\, dt$$

$$= \frac{1}{\alpha} - \int_0^\infty e^{-\alpha t}\, \mathbb{P}^i(\tau_s < t)\, dt = \frac{1}{\alpha}\left[1 - g_i^\alpha(s)\right]$$

and the result follows from the Laplace transform properties. □

§ 8.4.3. Semi-group. Continuing discussion of the inverse process τ, recall that $X(\tau_t)$ takes values in H only. Thus, restricting attention to states in H, one obtains a semi-group from Theorem 1, § 8.4.2 (for fixed $\alpha > 0$),

$$g_{ij}^\alpha(t+s) = \sum_{k \in H} g_{ik}^\alpha(t) g_{kj}^\alpha(s) \qquad \text{for} \quad i, j \in H,\ s, t \geqslant 0,$$

with $g_{ij}^\alpha(0) = \delta_{ij}$. Define now

$$q_{ij}^\alpha = \lim_{s \to 0} g_{ij}^\alpha(s)/s \geqslant 0 \qquad \text{for} \quad i \neq j,$$
$$q_i^\alpha = \lim_{s \to 0} [1 - g_{ii}^\alpha(s)]/s \geqslant 0$$

with $q_{ii}^\alpha = -q_i^\alpha$. Observe that the matrix (q_{ij}^α) of intensities is not conservative, and write

$$b_i^\alpha = -\sum_{j \in H} q_{ij}^\alpha \geqslant 0, \qquad i \in H,$$

for the intensity of disappearance from H, i.e., $q_{i\Delta}^\alpha = b_i^\alpha$. Under the usual assumptions, Kolmogorov equations may be obtained

$$\frac{dg_{ij}^\alpha(t)}{dt} = \begin{cases} \sum_{k \in H} q_{ik}^\alpha\, g_{kj}^\alpha(t), & \text{(B)} \\ \sum_{k \in H} g_{ik}^\alpha(t)\, q_{kj}^\alpha, & \text{(F)} \end{cases}$$

for $i, j \in H$ and $t \geqslant 0$.

Integrating these equations from 0 to ∞, and noting that $g_{ij}^\alpha(\infty) = 0$, one gets from Theorem 1, § 8.4.2,

$$-\delta_{ij} = \sum_{k \in H} q_{ik}^\alpha\, u_{kj}^\alpha, \qquad i, j \in H, \tag{B}$$

$$1 = \sum_{k \in H} u_{ik}^\alpha\, b_k^\alpha, \qquad i \in H. \tag{F}$$

Observe that for $\alpha = 0$, the situation discussed here reduces to trace chain treated in Sub-section 9.5, with $g_{ij}(s) = \tilde{p}_{ij}(s)$ denoting transition probabilities of the trace chain on H.

The following is the main result obtained by argument similar to that used in Theorem 1, § 9.5.4 (essentially the time change; see also § 9.6.5).

THEOREM 1. *For* $i, j \in H$ *and* $\alpha \geqslant 0$:

$$q_{ij}^{\alpha} = q_{ij} + \sum_{k \in H^c} q_{ik}\, d_{kj}^{\alpha}.$$

Then it follows immediately from Theorem 7, § 3.3.2, that

$$q_{ij}^{\alpha} = (\alpha + q_i)\, d_{ij}^{\rho\alpha} \qquad \text{for} \quad i \neq j,$$
$$q_i^{\alpha} = (\alpha + q_i)(1 - d_{ii}^{\rho\alpha}).$$

It is now clear that

$$b_i^{\alpha} = (\alpha + q_i)(1 - d_i^{\rho\alpha}) = \alpha\,(1 + n_i^{\alpha}), \qquad i \in H,$$

where the second expression follows from § 3.4.1. Hence, b_i^{α} defined here coincides with b_i^{α} introduced in § 5.2.1.

Thus, the above (F)-equation coincides with the Chung equation (Cor. 11, § 4.2.1). Indeed, one has here the transformed Poisson equation $\mathbf{1} = \mathbf{U}^{\alpha} b^{\alpha}$ discussed in Theorem 2, § 5.2.1. As noted there, the limit, as $\alpha \to 0$, yields the Poisson decomposition $\mathbf{1} = \mathbf{U}b + s$ on H. Moreover, discussion in § 4.2.1 yields another representation

$$b_i^{\alpha} = \alpha + b_i + \int_{(0,\infty)} (1 - e^{-\alpha s})\, \mu_i(ds), \qquad i \in H.$$

On the other hand, the (B)-equation stated above reduces to the decomposition theorem for i and j in H (Thm. 1, § 6.1.2).

In summary, in view of the above discussion, expression $g_{ij}^{\alpha}(s)$ may be considered as known for i, j in H. Extension to $i \in H^c$ (and $j \in H$) is then obtained from Theorem 1, § 8.4.2 (with $t = 0$).

§ 8.4.4. Special cases. Explicit expressions are easily obtained for the local time V^j on a state j by considering the special case $H = \{j\}$. As already noted, the α-potential and the characteristic of V^j are:

$$u_{V^j}^{\alpha}(i) = u_{ij}^{\alpha}, \qquad \mathbb{E}^i V_t^j = \int_0^t p_{ij}(s)\, ds.$$

Recall now (from Ex. 1, § 3.3.4, Ex. 1, § 3.4.3) that

$$d_{ij}^{\alpha} = f_{ij}^{\alpha} \qquad \text{for} \quad i \neq j, \qquad d_{jj}^{\rho\alpha} = f_{jj}^{\alpha},$$
$$b_j^{\alpha} = \alpha\,(1 + n_j^{\alpha}) = (\alpha + q_j)(1 - f_{jj}^{\alpha}) = 1/u_{jj}^{\alpha}.$$

As now $X(\tau_s^j) = j$, one can write, for a fixed j,

$$\mathbb{E}^i(e^{-\alpha\tau_s^j},\ \tau_s^j < \infty) = g_i^\alpha(s), \qquad i \in \mathbb{I}.$$

Then, Theorem 1, § 8.4.2, and argument in § 8.4.3, yield

$$g_j^\alpha(t+s) = g_j^\alpha(s)g_j^\alpha(t),$$

with solution

$$g_j^\alpha(s) = e^{-sb_j^\alpha}.$$

Hence, Theorem 1, § 8.4.2, yields the distribution of τ_s^j in the form

$$g_i^\alpha(s) = f_{ij}^\alpha\, e^{-sb_j^\alpha}, \qquad i \neq j,$$

and, by Theorem 2, § 8.4.2, one has

$$\int_0^\infty g_i^\alpha(s)\, ds = f_{ij}^\alpha / b_j^\alpha.$$

Thus, Theorem 2, § 8.4.2, is reduced to the passage relation

$$u_{ij}^\alpha = f_{ij}^\alpha\, u_{jj}^\alpha, \qquad i \neq j,$$

in agreement with Theorem 2, § 8.4.1.

From Theorem 3 and Theorem 4 in § 8.4.2 one obtains

$$\int_0^\infty e^{-\alpha t}\, d_t\mathbb{P}^i(V_t^j \leqslant s) = 1 - f_{ij}^\alpha\, e^{-sb_j^\alpha}, \qquad i \neq j,$$

with f_{ij}^α replaced by 1 when $i = j$. Moreover,

$$\mathbb{E}^j\, V_\infty^j = \int_0^\infty \mathbb{P}^j(\tau_s^j < \infty)\, ds = u_{ij} = 1/b_j.$$

Still further considerations are of great interest. Recall from § 3.4.3 that, for $H = \{j\}$,

$$n_j(s) = \sum_{k\neq j} q_{jk}\left[1 - F_{kj}(s)\right], \qquad 0 \leqslant s < \infty,$$

with $n_j(0) = q_j$ and $\lim_{s\to\infty} n_j(s) = b_j = q_j(1 - f_{jj})$.

It is clear that $n_j(s)$, as a function of s, is non-increasing, continuous and finite on $(0, \infty)$. Define now the measure μ_j on the positive real line by

$$\mu_j(s, \infty] = n_j(s), \qquad 0 \leqslant s < \infty,$$

with

$$\mu_j(0,\infty] = q_j, \qquad \mu_j(\{\infty\}) = b_j$$

(see §§ 3.4.1 and 4.2.1).

Let n_j^α be the Laplace transform of $n_j(s)$. Then, by properties of transforms,

$$\alpha n_j^\alpha = \int_{(0,\infty]} (1 - e^{-\alpha s})\, \mu_j(ds) = b_j + \int_{(0,\infty)} (1 - e^{-\alpha s})\, \mu_j(ds)$$

and therefore, as noted in § 4.2.1, one has

$$b_j^\alpha = \alpha + \alpha\, n_j^\alpha = \alpha + b_j + \int_{(0,\infty)} (1 - e^{-\alpha s})\, \mu_j(ds)$$

in agreement with formula stated in § 6.4.3.

Hence, inversion of the Chung equation $u_{jj}^\alpha\, b_j^\alpha = 1$ yields

$$\int_0^t \left[\hat{n}_j\,(t-s) + b_j + \delta(t-s)\right] dE(s) = 1,$$

where $E(s) = \mathbb{E}^j\, V_s^j$ and $\hat{n}_j(s) = n_j(s) - b_j$.

Finally, note that the mean of measure μ_j restricted to $(0,\infty)$ is

$$\int_{(0,\infty)} x\, \mu_j\,(dx) = \int_{(0,\infty)} (n_j(s) - b_j)\, ds.$$

Remarks. 1. It can be verified that the process $(\tau_s^j, \mathbb{P}^j)$ has independent (non-negative) increments, and is in fact the compound Poisson process with drift. The Poisson jumps have rate $n_j(0) = q_j$, the magnitude of jumps having distribution $\mu_j(dt)/q_j$ over $(0,\infty]$, and τ_s^j increases linearly between jumps with rate 1. Recall that jumps of (τ_s^j) correspond to excursions from a state j.

The process $(\tau_s^j, \mathbb{P}^j)$ is called a sub-ordinator, μ_j being the Lévy measure and b_j^α the exponent associated with the process. Actually, this is the generalized sub-ordinator which allows τ^j to take the value ∞. The atom $\mu_j(\{\infty\}) = b_j$ corresponds to a jump to ∞ by τ^j, and this in turn corresponds to $\sup\,(t\colon V_t^j = j) < \infty$.

For further details see R. M. Blumenthal and R. K. Getoor [1968, p. 227] and K. L. Chung [1970, p. 90]. These results are in agreement with general properties of additive processes; see, for example, B. Maisonneuve [1971], E. Çinlar [1972].

2. The principal difference between the case of an arbitrary set H and the case of H consisting of a single state, lies in the Chung equation

$$\sum_{j\in H} u_{ij}^\alpha\, b_j^\alpha = 1, \qquad i \in H,$$

which reduces to $u_{jj}^\alpha\, b_j^\alpha = 1$. This in turn leads to the exponential solution for $g_j^\alpha(s)$, here, in contrast with the general solution for $g_{ij}^\alpha(s)$ with i and j in H, as indicated in § 6.4.2.

Thus, it should be noted that $g_i^\alpha(s)$ obtained here ($i \neq j$) is different from $g_{ij}^\alpha(s)$ for the general H; indeed for $s = 0$ they are, respectively, f_{ij}^α and d_{ij}^α.

Nevertheless, one has for the general H and for $i \in H^c$,

$$u_V^\alpha(i) = \sum_{j \in H} \mathbb{E}^i \int_0^\infty e^{-\alpha t}\, dV_t^j = \sum_{j \in H} \int_0^\infty \mathbb{E}^i(e^{-\alpha \tau_s^j})\, ds$$

$$= \sum_{j \in H} \int_0^\infty f_{ij}^\alpha\, e^{-s b_j^\alpha}\, ds = \sum_{j \in H} f_{ij}^\alpha\, u_{jj}^\alpha = \sum_{j \in H} u_{ij}^\alpha.$$

3. Validity of the Chung equation for all t (and local time for points at the boundary) has been known as "Chung conjecture"; cf. K. L. Chung [1970, p. 63], H. Kesten [1969], P. Assoud [1971], G. Giroux [1970].

§ 8.4.5. Supermartingale. In closing this sub-section, comments on connection between local times and supermartingales should be made. A local time, as an increasing process, generates a supermartingale. Conversely, a potential supermartingale can be used to define a local time.

Let V^j be a local time for a state j, and write u_j^α for its α-potential, where $u_j^\alpha(i) = u_{ij}^\alpha$. Recall that u_j^α is $\mathbf{P}(t)$-purely excessive for $\alpha \geqslant 0$. Hence, $u_j^\alpha(X_t)$ defines a potential supermartingale (see § 8.1.4). In fact, one has

THEOREM 1. *Define for $\alpha > 0$, $t \geqslant 0$, and fixed j:*

$$u_j^\alpha(X_t) = \mathbb{E}\left(\int_t^\infty e^{-\alpha s}\, dV_s^j \,\middle|\, \mathcal{F}_t\right).$$

Then, $(u_j^\alpha(X_t), \mathcal{F}_t,\ 0 \leqslant t < \infty)$ is a right-continuous, uniformly integrable, potential supermartingale.

Proof. By smoothing properties of conditional expectation

$$\mathbb{E}\big(u_j^\alpha(X_{t+h}) \mid \mathcal{F}_t\big) = u_j^\alpha(X_t) - \mathbb{E}\left(\int_t^{t+h} e^{-\alpha s}\, dV_s^j \,\middle|\, \mathcal{F}_t\right) \leqslant u_j^\alpha(X_t),$$

so $u_j^\alpha(X_t)$ form a supermartingale. This is a potential supermartingale because

$$\mathbb{E}u_j^\alpha(X_t) = \mathbb{E}\int_t^\infty e^{-\alpha s}\, dV_s^j \longrightarrow 0 \qquad \text{as} \quad t \to \infty.$$

As a positive supermartingale always converges, so by Fatou Lemma

$$u_j^\alpha(X_t) \longrightarrow 0 \quad \text{a.e.} \qquad \text{as} \quad t \to \infty.$$

Consequently, $(u_j^\alpha(X_t))$ are uniformly integrable (by Vitali Theorem), and

$$\mathbb{E}u_j^\alpha(X_t) \leqslant \mathbb{E}u_j^\alpha(X_0) = \sum_i \pi_i\, u_{ij} < \infty.$$

Finally, the right-continuity follows from the right-continuity of the chain (X_t). □

Define for $\alpha > 0$, $t \geqslant 0$, and for fixed j:

$$V_t^\alpha = \int_0^t e^{-\alpha s}\, dV_s^j, \qquad M_t^\alpha = \mathbb{E}(V_\infty^\alpha \mid \mathcal{F}_t).$$

Thus, (V_t^α) is an integrable increasing process and (M_t^α) is an uniformly integrable martingale. Hence, $(u_j^\alpha(X_t))$ is a supermartingale generated by the increasing process (V_t^α) (see Def. 3, § 8.1.4) and

$$u_j^\alpha(X_t) = M_t^\alpha - V_t^\alpha, \qquad \alpha > 0, \quad t \geqslant 0.$$

These relations hold also for $\alpha = 0$ when the state j is transient.

THEOREM 2. *The local time V^j, for $\alpha > 0$ and $t \geqslant 0$, satisfies*

$$\mathbb{E}\left(\int_t^\infty e^{-\alpha s}\, dV_s^j \,\Big|\, \mathcal{F}_t\right) = u_{jj}^\alpha\, \mathbb{E}\left(e^{-\alpha T_t} \mid \mathcal{F}_t\right),$$

where $T_t = t + T \circ \theta_t$ and $T = T_j$ is the first entrance time to a state j.

Proof. Recall that V^j increases only on the set $S_j = (t\colon\ T_t = t)$, and $X(T_t) = j$. Hence, by the strong Markov property,

$$\begin{aligned}
\mathbb{E}\left(\int_t^\infty e^{-\alpha s}\, dV_s^j \,\Big|\, \mathcal{F}_t\right) &= \mathbb{E}\left(\int_{T_t}^\infty e^{-\alpha s}\, dV_s^j \,\Big|\, \mathcal{F}_t\right) \\
&= \mathbb{E}\left[e^{-\alpha T_t}\, \mathbb{E}\left(\int_0^\infty e^{-\alpha x}\, dV_x^j \circ \theta_{T_t} \,\Big|\, X(T_t)\right) \,\Big|\, \mathcal{F}_t\right] \\
&= \mathbb{E}\left(e^{-\alpha T_t}\, u_{jj}^\alpha \mid \mathcal{F}_t\right).
\end{aligned}$$
□

Note that Theorem 2 is the extension of the passage relation $u_{ij}^\alpha = f_{ij}^\alpha\, u_{ij}^\alpha$ (see Thm. 2, § 8.4.1) to which it reduces when $t = 0$. Assertion of Theorem 2 is sometime taken as a definition of a local time V^j.

It is also easy to verify that the family $(S_t^\alpha, \mathcal{F}_t, 0 \leqslant t < \infty)$, where

$$S_t^\alpha = \mathbb{E}\left(e^{-\alpha T_t} \mid \mathcal{F}_t\right),$$

is a potential supermartingale.

8.5. BALAYAGE

§ 8.5.1. Operator. Importance of a balayage has been already pointed out earlier. In particular, Sub-section 5.3 treated "reduced functions" aspects, and Sub-section 6.4 indicated connections with projection theorems. These analytic approaches are now supplemented by a measure-theoretic approach to a balayage operator on a M.C.

The presentation here is the specialization of results from general Markov processes to the usual Markov chains considered in this book. For complete discussion see references, in particular, R. M. Blumenthal and R. K. Getoor [1968], K. L. Chung [1982], J. Doob [1984], J. Bliedtner and W. Hansen [1986].

The operator $\mathbf{P}^\alpha(T)$ on measurable functions on the state space $\mathbb{I}$ has been already introduced in § 8.1.1 for any stopping time T, relative to a family $(\mathcal{F}_t)$. When $T = T_H$, the first entrance time to a set H (Def. 1, § 2.1.2), one can write $\mathbf{P}_H^\alpha$ instead of $\mathbf{P}^\alpha(T_H)$; the operator $\mathbf{P}_H^\alpha$ is known as an α-balayage. For a fixed set H, this is the operator $\mathbf{B}^\alpha$ introduced in § 3.1.5, and determined by the kernel (d_{ij}^α). Thus, for any positive (or any bounded) function f on $\mathbb{I}$:

$$\mathbf{B}^\alpha f(i) = \mathbb{E}^i\big(e^{-\alpha T} f(X_T),\ T < \infty\big), \qquad i \in \mathbb{I}, \quad \alpha \geqslant 0.$$

The function $\mathbf{B}^\alpha f$ is called the α-balayage of f on H; note that $\mathbf{B}^\alpha f(i) = f(i)$ for $i \in H$. When $\alpha = 0$, it will be dropped from notation, and $\mathbf{B}f$ is referred simply as the balayage of a function f on the set H. In particular, $\mathbf{B1} = D$.

The family of random variables $\mathbf{B}^\alpha f(X_t)$, $0 \leqslant t < \infty$, defined by

$$\mathbf{B}^\alpha f(X_t) = \mathbb{E}\big(e^{-\alpha T\circ\theta_t} f(X(T_t)),\ T\circ\theta_t < \infty \mid X_t\big) \quad \text{a.e.}$$

(where $T_t = t + T\circ\theta_t$) is referred to as the α-balayage of the family $f(X_t)$, $0 \leqslant t < \infty$.

As noted earlier in Sub-section 6.1, the first entrance decomposition theorem is a special case of the Dynkin theorem (§ 8.1.1) applied to the first entrance time T. Indeed, for any bounded function f on $\mathbb{I}$ one has from Theorem 1, § 6.1.1:

$$\mathbf{U}^\alpha f(i) = {}_H\mathbf{U}^\alpha f(i) + \mathbf{B}^\alpha \mathbf{U}^\alpha f(i), \qquad i \in \mathbb{I},$$

with components given by integrals

$$\begin{aligned}\mathbb{E}^i \int_0^T e^{-\alpha t} f(X_t)\,dt &= \mathbb{E}^i \int_0^\infty e^{-\alpha t} f(X_t) I_{(t<T)}\,dt \\ &= \int_0^\infty e^{-\alpha t} \sum_j \mathbb{P}^i(X_t = j, t < T) f(j)\,dt \\ &= \sum_j {}_H u_{ij}^\alpha f(j) = {}_H\mathbf{U}^\alpha f(i)\end{aligned}$$

and

$$\mathbb{E}^i \int_T^\infty e^{-\alpha t} f(X_t)\,dt = \mathbb{E}^i e^{-\alpha T} \int_0^\infty e^{-\alpha s} f(X_{s+T})\,ds$$
$$= \mathbb{E}^i\big(e^{-\alpha T}\mathbf{U}^\alpha f(X_T)\big) = \mathbf{B}^\alpha \mathbf{U}^\alpha f(i).$$

Similarly, Theorem 1, § 6.1.2, is a special case of the Dynkin theorem applied to the stopping time $T = T_\rho$. Indeed, for any bounded function f on $\mathbb{I}$ one has:

$$\mathbf{U}^\alpha f(i) = {}_H\mathbf{U}^{\rho\alpha} f(i) + \mathbf{B}^{\rho\alpha}\mathbf{U}^\alpha f(i),$$

where

$$\mathbb{E}^i \int_0^\infty e^{-\alpha t} f(X_t) I_{(t<T_\rho)}\,dt = {}_H\mathbf{U}^{\rho\alpha} f(i),$$

$$\mathbb{E}^i \int_T^\infty e^{-\alpha t} f(X_t)\,dt = \mathbb{E}^i\big(e^{-\alpha T_\rho}\mathbf{U}^\alpha f(X_{T_\rho})\big) = \mathbf{B}^{\rho\alpha}\mathbf{U}^\alpha f(i).$$

These integrals may be used to evaluate characteristics of integral functionals at stopping times. Thus, if A is an integral functional determined by a charge $a \geqslant 0$, then

$$\mathbb{E}^i \int_0^T a(X_s)\,ds = {}_H\mathbf{U}a(i), \qquad \mathbb{E}^i \int_0^{T_\rho} a(X_s)\,ds = {}_H\mathbf{U}^\rho a(i),$$

in agreement with discussion of extended functionals in § 8.2.2.

As an another example, let p be a purely excessive function, so $(-\mathbf{Q})p = a \geqslant 0$. Then, the martingale argument (§ 8.1.4) shows that

$$p(i) = \mathbf{B}p(i) + {}_H\mathbf{U}a(i)$$

in agreement with projection theorem (see § 6.4.3). Clearly, for finite T:

$$\mathbf{B}p(i) = \mathbb{E}^i p(X_T).$$

Evidently,

$$p(X_t) = \mathbf{B}p(X_t) + {}_H\mathbf{U}a(X_t), \qquad t \geqslant 0,$$

form a potential supermartingale (§ 8.1.4), and the above representation is another form of the Dynkin theorem (§ 8.1.3). One can easily verify that

$${}_H\mathbf{U}a(X_t) = \mathbb{E}(A_T \circ \theta_t \mid \mathcal{F}_t) \quad \text{a.e.},$$

where (A_t) is an integral functional determined by a charge a.

In particular, when $p = \mathbf{U}a$ (see Thm. 7, § 1.3.3), so $p(X_t) = \mathbf{U}a(X_t)$, one can write

$$\mathbf{B}p(X_t) = \mathbf{U}\bar{a}(X_t) \quad \text{a.e.},$$

where $\bar{a} = \mathbf{C}a$ (Thm. 2, § 6.4.1). Hence, in agreement with Definition 3, § 8.1.4,

$$\mathbf{B}p(X_t) = \mathbb{E}(\bar{A}_\infty \mid \mathcal{F}_t) - \bar{A}_t \quad \text{a.e.},$$

where $(\bar{A}_t)$ is an integral functional determined by a charge $\bar{a}$.

Remark. Similar analysis may be carried out for the last exit decomposition. Theorem 1, § 6.2.1, is a special case of the duality interpretation (see § 6.3.3). For any bounded function f on $\mathbb{I}$ one has:

$$\mathbf{U}^\alpha f(i) = {}_H\mathbf{U}^\alpha f(i) + \mathbf{U}^\alpha \mathbf{C}^\alpha f(i),$$

where, in the partitioned form,

$$\mathbf{C}^\alpha f = \begin{cases} 0, & \text{on} \quad H^c, \\ \mathbf{C}^\alpha_{21} f_1 + f_2 & \text{on} \quad H. \end{cases}$$

Hence, in particular when $p = \mathbf{U}a$, then

$$\mathbf{B}p(X_t) = \mathbf{U}\bar{a}(X_t), \qquad t \geqslant 0,$$

where $\bar{a} = \mathbf{C}a$.

§ 8.5.2. Potentials. Let $A = (A_t)$ be an increasing process, or an AF, with finite α-potential u^α_A. If there exists an increasing process, or an AF, $\bar{A}^\alpha = (\bar{A}^\alpha_t)$ depending on α, whose α-potential is such that

$$u^\alpha_{\bar{A}^\alpha} = \mathbf{B}^\alpha u^\alpha_A$$

where $\mathbf{B}^\alpha$ is the α-balayage operator, then $\bar{A}^\alpha$ is called the *α-balayage of A on H*. Observe that the above equation is of interest only on H^c, because obviously $\mathbf{B}^\alpha u^\alpha_A = u^\alpha_A$ on H.

It has been already noted in § 6.4.1, that the notion of a balayage is related to that of a projection. This is the immediate consequence of the extended Dynkin theorem from § 8.1.3. The result is repeated here for convenience.

THEOREM 1. *Let $A = (A_t)$ be an AF with finite α-potential u^α_A, and let $T = T_H$ be the first entrance time to a set H. Then,*

$$\mathbf{B}^\alpha u^\alpha_A(i) = u^\alpha_A(i) - \mathbb{E}^i \int_0^T e^{-\alpha t}\, dA_t$$

or, equivalently,

$$\mathbf{B}^\alpha u^\alpha_A(i) = \mathbf{E}^i \int_T^\infty e^{-\alpha t}\, dA_t.$$

In the special case when A is an IF determined by a charge a, then

$$u^\alpha_A(i) = \mathbf{U}^\alpha a(i),$$

$${}_H\mathbf{U}^\alpha a(i) = \mathbb{E}^i \int_0^T e^{-\alpha t} a(X_t)\, dt,$$

$$\mathbf{B}^\alpha u^\alpha_A(i) = \mathbb{E}^i\big(e^{-\alpha T} a(X_T),\ T < \infty\big)$$

which yields the projection theorem (Thm. 2, § 6.4.1), where $p^\alpha = u^\alpha_A$. See also § 8.5.1.

THEOREM 2. *Let A be an* AF *with finite α-potential u_A^α. Then, there exists a unique* CAF *$\bar{A}^\alpha$ which is the α-balayage of A on H, such that*

$$\operatorname{supp} \bar{A}^\alpha = H.$$

Proof (outline). The function $\mathbf{B}^\alpha u_A^\alpha$ is (uniformly) α-excessive relative to $\mathbf{P}(t)$, and is also purely excessive (by Theorem 1). Hence, the first assertion follows from an existence theorem for functionals; see R. M. Blumenthal and R. K. Getoor [1968, p. 169]. As $\mathbf{B}^\alpha$ is idempotent,

$$B^\alpha u_{\bar{A}^\alpha}^\alpha = u_{\bar{A}^\alpha}^\alpha,$$

so the support of $\bar{A}^\alpha$ is H. □

The cases of an integral functional and of a purely discontinuos functional can be treated directly. Indeed, one has

THEOREM 3. i) *Let A be an* IF *determined by a charge $a \geqslant 0$ on $\mathbb{I}$. Then its α-balayage $\bar{A}^\alpha$ is an* IF *determined by a charge $\bar{a}^\alpha = \mathbf{C}^\alpha a$, supported on H.*

ii) *Let A be a* DAF *determined by a function $g \geqslant 0$. Then its α-balayage $\bar{A}^\alpha$ is an* IF *determined by the charge $\bar{a}^\alpha = \mathbf{C}^\alpha g_{\mathbf{Q}}$, supported on H.*

Proof. Here $\mathbf{C}^\alpha$ is the α-co-balayage (see § 3.4.4). Define now

$$\bar{A}_t^\alpha = \int_0^t \bar{a}^\alpha(X_s)\, ds, \qquad t \geqslant 0,$$

where $\bar{a}^\alpha(i) = \mathbf{C}^\alpha a(i)$, as found in Theorem 3, § 6.4.1. Hence,

$$u_{\bar{A}^\alpha}^\alpha(i) = \mathbf{U}^\alpha \bar{a}^\alpha(i) = \mathbf{U}^\alpha \mathbf{C}^\alpha a(i) = \mathbf{B}^\alpha \mathbf{U}^\alpha a(i) = \mathbf{B}^\alpha u_A^\alpha(i).$$

Since $\bar{a}^\alpha$ is supported on H, so $\operatorname{supp}(\bar{A}^\alpha) = H$.

Proof for DAF follows the same pattern, with charge a replaced by charge $g_{\mathbf{Q}}$ (defined in § 8.1.2). □

It is instructive to consider some simply examples.

Example 1. Let $a = 1$ identically on $\mathbb{I}$, so $u_A^\alpha = 1/\alpha$ and, therefore, $\mathbf{B}^\alpha u_A^\alpha = d^\alpha/\alpha$. On the other hand, Theorem 3, § 6.4.1, yields the expression for the charge $\bar{a}^\alpha(i) = 1 + n_i^\alpha$ ($i \in H$), and Theorem 1, § 6.2.1, shows that indeed

$$\mathbf{U}^\alpha \bar{a}^\alpha(i) = d_i^\alpha/\alpha, \qquad i \in H^c.$$

Example 2. Let $a = I_{H^c}$, so $\operatorname{supp}(A) = H^c$, and

$$u_A^\alpha(i) = \sum_{j \in H^c} u_{ij}^\alpha.$$

Hence,

$$\mathbf{B}^\alpha u_A^\alpha(i) = \sum_{j \in H^c} \sum_{k \in H} d_{ik}^\alpha\, u_{kj}^\alpha.$$

Again, Theorem 3, § 6.4.1, yields $\bar{a}^\alpha(i) = n_i^\alpha$ for $i \in H$, so

$$\mathbf{U}^\alpha \bar{a}^\alpha(i) = \sum_{j \in H^c} \sum_{k \in H} u_{ik}^\alpha\, n_{kj}^\alpha$$

and equality of both expressions follows from duality (see § 6.3.2).

Example 3. Counting process (N_t) — see Example 1, § 8.1.3. Here $u_A^\alpha(i) = \mathbf{U}^\alpha q_i$, so

$$\mathbf{B}^\alpha u_A^\alpha(i) = \mathbf{U}^\alpha \bar{a}^\alpha(i),$$

where

$$\bar{a}^\alpha(i) = q_i + \sum_{k\in H^c} n_{ik}^\alpha q_k, \qquad i \in H.$$

It is now possible to elaborate on remark made in § 6.4.3. Let $p^\alpha = \mathbf{U}^\alpha a$ be the α-potential of a charge $a \geqslant 0$, and let $\bar{p}^\alpha = \mathbf{B}^\alpha p^\alpha$ be its α-balayage (see Def. 1, § 6.4.1). Using notation introduced in § 8.5.1, the potential supermartingale

$$p^\alpha(X_t) = \mathbf{U}^\alpha a(X_t) = \mathbb{E}\left(\int_0^\infty e^{-\alpha s} a(X_s)\, ds \circ \theta_t \mid X_t\right)$$

has balayage given by

$$\begin{aligned}\bar{p}^\alpha(X_t) &= \mathbf{B}^\alpha p^\alpha(X_t) = \mathbb{E}\big(e^{-\alpha T\circ\theta_t} p^\alpha(X_{T_t}),\ T\circ\theta_t < \infty \mid X_t\big) \\ &= \mathbb{E}\left(e^{-\alpha T\circ\theta_t}\int_0^\infty e^{-\alpha s} a(X_{s+T_t})\, ds,\ T\circ\theta_t < \infty \mid X_t\right) \quad \text{a.e.}, \qquad t \geqslant 0,\end{aligned}$$

where $T_t = t + T\circ\theta_t$. Alternatively,

$$\bar{p}^\alpha(X_t) = \mathbf{B}^\alpha \mathbf{U}^\alpha a(X_t) = \mathbb{E}\left(\int_T^\infty e^{-\alpha s} a(X_s)\, ds \circ \theta_t \mid X_t\right).$$

This again gives the Dynkin formula in the form

$$p^\alpha(X_t) = \bar{p}^\alpha(X_t) + \mathbb{E}\big(A_T^\alpha \circ \theta_t \mid \mathcal{F}_t\big) \quad \text{a.e.}, \qquad t \geqslant 0,$$

where A_t^α is an IF with a charge $a^\alpha(X_t)$.

Proceeding as in Theorem 2, $\bar{p}^\alpha(X_t)$ may be written as a potential of a charge $\bar{a}^\alpha(X_t)$:

$$\bar{p}^\alpha(X_t) = \mathbf{U}^\alpha\, \bar{a}^\alpha(X_t) = \bar{U}_t^\alpha$$

so that $(\bar{U}_t^\alpha)$ is a potential generated by an increasing process

$$\bar{A}_t^\alpha = \int_0^t \bar{a}^\alpha(X_s)\, ds,$$

and

$$\bar{U}_t^\alpha = \mathbb{E}\big(\bar{A}_\infty^\alpha - \bar{A}_t^\alpha \mid \mathcal{F}_t\big) \quad \text{a.e.}$$

Compare these considerations with those in § 8.5.1; the above expressions hold also for $\alpha = 0$ under finiteness assumptions.

SECTION 9

TRANSFORMATIONS

In many situations a M.C. undergoes modification according to a certain rule to produce a modified chain $\widetilde{X}$. It is then required to verify that the modified chain is Markovian, to find its new semi-group and to express properties of the modified chain in terms of those of the original chain. The typical examples are provided by killing, birthing, and reversing of the original chain at some random time R. Random times R which preserve Markov property (or only some form of it) are known as splitting times. Typical examples of such times are the first entrance time T and the last exit time L.

More precisely, with a M.C. $X = (X_t)$ associate a secondary (or transformed) chain $\widetilde{X} = (\widetilde{X}_t)$ is determined by a random time as follows.

a) *Killing.* A chain killed at R is defined by

$$\widetilde{X}_t = \begin{cases} X_t & \text{for} \quad t < R, \\ \Delta & \text{for} \quad t \geqslant R, \end{cases}$$

where Δ is the adjoint absorbing state. R is then called the death time of the chain, and the new chain is also referred to as a pre-R chain.

b) *Birthing.* A chain started at R is defined by

$$\widetilde{X}_t = X(R+t), \qquad t \geqslant 0.$$

R is then called the birth time of the chain, and the new chain is also referred to as a post-R chain.

c) *Reversing.* A chain reversed at R is defined by

$$\widetilde{X}_t = \begin{cases} X(R-t), & 0 < t \leqslant R, \\ \Delta, & t > R, \end{cases}$$

where Δ the adjoint isolated state. R is then called the reversal time of the chain, and the new chain, is also referred to as a reverse chain.

Another type of transformations involve randomisation of time (random time change), and replacement of the intensity matrix $\mathbf{Q}$ by a modified matrix $\widetilde{\mathbf{Q}}$ constructed independently of the chain, with Markov property preserved by construction (perturbation).

Splitting times and random time change received considerable attention in literature and are object of current research. The basic paper in this field is that of P. A. Meyer, R. T. Smythe and J. B. Walsh [1972] and further developments were obtained by A. O. Pittenger and C. T. Shih [1973], R. K. Getoor and M. J. Sharpe

[1973], B. Maisonneuve and P. A. Meyer [1974]; see also the review by P. W. Millar [1977]. For recent work in this direction see series of papers by J. Glover [1981 a, 1981 b, 1983], A. O. Pittenger [1981 a, 1981 b], A. O. Pittenger and M. J. Sharpe [1981]; additional references are given in the text.

This section is based on the above papers, with specialisation to the usual Markov chains. Special structure of such chains allows great simplification of proofs, so emphasis is here on final results and their connection with passage times.

In summary:

i) if R is a stopping time T, then the strong Markov property assures that the post-T chain is Markov and has the same semi-group as the original chain. However, the chain killed at T is Markovian only if T is a terminal time.

ii) if R is a co-stopping time L, then the chain killed at L is Markovian, but with a modified semi-group. However, the post-L chain is Markovian only if L is a co-terminal time.

iii) if R is a terminal time or a co-stopping time, then the reverse chain is Markovian and dual to the original chain.

As a rule, the new transition matrix $\widetilde{\mathbf{P}}(t)$ is different from the original matrix $\mathbf{P}(t)$. For example, killing at the first entrance time T produces taboo probabilities and killing at the last exit time L involves hitting probability (note that T and L are both death and birth times).

Random time change transformation has a rather different character. To simplify its delicate argument, it is convenient to consider first the special case of a "trace chain".

9.1. KILLING

The killing operation is of great importance for this book because taboo probabilities are obtained in this way. The study of killing of a M.C. can be approached in several ways, the most general formulation being through multiplicative functionals.

§ 9.1.1. Definitions. Let $X = (X_t, \mathcal{F}_t,\ 0 \leqslant t < \infty)$ be the usual M.C. with a state space $\mathbb{I}$, standard stochastic transition matrix $\mathbf{P}(t)$ and a conservative intensity matrix $\mathbf{Q}$, taken relative to the family $(\mathcal{F}_t)$ of natural σ-fields on a probability space $(\Omega, \mathcal{F}, \mathbb{P})$. Let $M = (M_t)$ be a contracting right-continuous MF of the usual M.C. X (see § 8.1.5). Recall that

$$0 \leqslant M_t \leqslant 1, \qquad M_{t+h} = M_t \cdot \big(M_h \circ \theta_t\big).$$

Let $\mathbb{I}_M$ be the set of all permanent points of M, i.e., the set of all states i such that

$$\mathbb{P}^i\big(M_0 = 1\big) = 1.$$

Define for each $t \geqslant 0$ an operator $\widetilde{\mathbf{P}}(t)$ on the class of all bounded functions f on $\mathbb{I}$ by

$$\widetilde{\mathbf{P}}(t)f(i) = \mathbb{E}^i\big(f(X_t)M_t\big), \qquad i \in \mathbb{I},$$

(its domain can be extended to $\mathbb{I} \cup \{\Delta\}$ by taking $f(\Delta) = 0$). In particular,

$$\tilde{p}_{ij}(t) = \mathbb{P}^i\big(X_t = j,\ M_t\big) := \mathbb{E}^i\big(M_t,\ X_t = j\big), \qquad i,j \in \mathbb{I}, \quad t \geqslant 0.$$

The following basic theorem indicates that, given a MF, there exist a M.C. $\widetilde{X}$ and a (terminal) stopping time $\widetilde{T}$ such that $\widetilde{X}$ is obtained by killing the original chain X at time $\widetilde{T}$, and that $\widetilde{X}$ has the above $\widetilde{\mathbf{P}}(t)$ as its transition matrix. The proof defines a random probability measure α on Borel sets of $[0,\infty]$ such that, for each $\omega \in \Omega$,

$$\alpha_\omega((s,\infty]) = M_s(\omega) \qquad \text{for} \quad 0 \leqslant s \leqslant \infty,$$

with $\alpha_\omega(\{0\}) = 0$ and $\alpha_\omega([0,\infty]) = M_0(\omega) = 1$. It is then shown that

$$\mathbb{P}^i\big(\widetilde{T} > t\big) = \mathbb{E}^i M_t, \qquad i \in \mathbb{I}_M.$$

Thus, M_t denotes the random probability that the original chain did not terminate before time $\widetilde{T}$. In other words, the instantaneous killing is done at the rate $-dM_t/M_t$ which can be interpreted (in analogy with hazard rate) as the "conditional probability" that X_t is killed in the interval $(t, t+dt)$, given that it is still alive at time t.

THEOREM 1. *Let $M = (M_t)$ be a given contracting* MF *of an usual* M.C. $X = (X_t)$. *Then, there exists a* M.C. $\widetilde{X} = (\widetilde{X}_t)$ *with a state space* $\mathbb{I} \cup \{\Delta\}$ *such that*

$$\mathbb{E}\big(f(\widetilde{X}_t) \mid \widetilde{X}_0 = i\big) = \mathbb{E}^i\big(f(X_t)M_t\big), \qquad i \in \mathbb{I},$$

for all bounded functions f on $\mathbb{I}$ (with $f(\Delta) = 0$). That is, $\widetilde{X}$ has a Markov semi-group $\widetilde{\mathbf{P}}(t) = (\tilde{p}_{ij}(t))$, $t \geqslant 0$, defined above (said to be generated by the MF *M). However, $\widetilde{\mathbf{P}}(t)$ is only sub-stochastic on $\mathbb{I}$, and $\widetilde{\mathbf{P}}(0)$ need not be the identity* **I**. *Moreover, $\widetilde{\mathbf{P}}(t)$, $t \geqslant 0$, is sub-ordinate to the original semi-group $\mathbf{P}(t)$, $t \geqslant 0$, in the sense that, for all bounded positive functions f on $\mathbb{I}$,*

$$\widetilde{\mathbf{P}}(t)f \leqslant \mathbf{P}(t)f.$$

Proof. For existence of $\widetilde{X}$ and verification of its Markov property see R. M. Blumenthal and R. K. Getoor [1968, p. 106], E. B. Dynkin [1965, p. 308] or D. Williams [1979, p. 148]. Here only comments on transition matrix $\widetilde{\mathbf{P}}(t)$ can be given.

It is clear that $\widetilde{\mathbf{P}}(t)$ is a positive linear operator on the class of all bounded functions on $\mathbb{I}$ (see § 1.3.1). The MF yields the semi-group property

$$\begin{aligned}\widetilde{\mathbf{P}}(t+s)f(i) &= \mathbb{E}^i\Big(f(X_{t+s})M_{t+s}\Big) = \mathbb{E}^i\big(f(X_t)\circ\theta_s \cdot M_s(M_t\circ\theta_s)\big)\\ &= \mathbb{E}^i\Big[M_s\mathbb{E}(f(X_t)\circ\theta_s \mid X_s)\Big] = \mathbb{E}^i\Big(M_s\widetilde{\mathbf{P}}(t)f(X_s)\Big)\\ &= \widetilde{\mathbf{P}}(s)\widetilde{\mathbf{P}}(t)f(i).\end{aligned}$$

Moreover,

$$\widetilde{\mathbf{P}}(t)\mathbf{1}_i = \mathbb{E}^i M_t.$$

Hence,

$$\lim_{t\to 0} \widetilde{\mathbf{P}}(t)\mathbf{1}_i = \widetilde{\mathbf{P}}(0)\mathbf{1}_i = \mathbb{E}^i M_0,$$

and this is 1 for $i \in \mathbb{I}_M$ and 0 otherwise. □

Thus,

$$\tilde{p}_{ij}(0) = \begin{cases} \delta_{ij}, & i \in \mathbb{I}_M, \\ 0, & i \in \mathbb{I} \setminus \mathbb{I}_M. \end{cases}$$

Consequently, $\tilde{p}_{ij}(t) = 0$ for all $t \geqslant 0$ when i or j are in $\mathbb{I} \setminus \mathbb{I}_M$. Thus, in the matrix notation, $\widetilde{\mathbf{P}}(t)$ partitioned according to $\mathbb{I}_M$ and $\mathbb{I} \setminus \mathbb{I}_M$, has the only non-zero component $\widetilde{\mathbf{P}}_{11}(t)$, and $\widetilde{\mathbf{P}}(0) = \mathbf{J}_1$.

Hence, one could restrict the state space to $\mathbb{I}_M$ and consider only functions f which vanish on $\mathbb{I} \setminus \mathbb{I}_M$, requiring the domination relation to hold only on $\mathbb{I}_M$. A M.C. satisfying these restrictive conditions is said to be a *sub-chain* of the original chain.

§ **9.1.2. Terminal time.** A different point of view is usually accepted in applications. Let $X = (X_t)$ be the usual M.C. with stochastic transition matrix $\mathbf{P}(t)$. For any stopping time T (relative to the natural σ-fields of X) define the killing operator $\mathcal{R}_T$ (on Ω to Ω) by

$$X_t(\mathcal{R}_T\omega) = \begin{cases} X_t(\omega) & \text{if } t < T, \\ \Delta & \text{if } t \geqslant T. \end{cases}$$

Then the random variables of the killed chain $\widetilde{X} = (\widetilde{X}_t)$ may be written as

$$\widetilde{X}_t = X_t \circ \mathcal{R}_T, \qquad t \geqslant 0.$$

Suppose now that T is a terminal stopping time. Then, as shown in § 8.1.5,

$$M_t = I_{(T>t)}, \qquad t \geqslant 0,$$

define a multiplicative functional $M = (M_t)$. Application of Theorem 1, § 9.1.1, to this MF yields the following important theorem (which can also be proved directly).

THEOREM 1. *The killed chain* $\widetilde{X} = (\widetilde{X}_t)$ *defined above, is the usual* M.C. *with the state space* $\mathbb{I} \cup \{\Delta\}$ *and with the transition probabilities given by*

$$\begin{aligned} \tilde{p}_{ij}(t) &= \mathbb{P}(X_t = j,\ t < T \mid X_0 = i,\ 0 < T), && i, j \in \mathbb{I}, \\ \tilde{p}_{i\Delta}(t) &= \mathbb{P}(T \leqslant t \mid X_0 = i,\ 0 < T), && i \in \mathbb{I}, \\ \tilde{p}_{\Delta j}(t) &= \delta_{\Delta j}. \end{aligned}$$

Moreover, the new matrix $\widetilde{\mathbf{P}}(t) = (\tilde{p}_{ij}(t))$ *is sub-stochastic on* $\mathbb{I}$:

$$\sum_j \tilde{p}_{ij}(t) = \mathbb{P}(T > t \mid X_0 = i,\ 0 < T),$$

and the semi-group $\widetilde{\mathbf{P}}(t)$, $t \geqslant 0$, *is sub-ordinate to the original semi-group* $\mathbf{P}(t)$ *in the sense that*

$$\tilde{p}_{ij}(t) \leqslant p_{ij}(t) \qquad \text{for all } i, j, \text{ and } t.$$

The most typical example of a terminal time is of course the first entrance time $T = T_H$ to a set H. In this case $\tilde{p}_{ij}(t)$ become taboo probabilities ${}_Hp_{ij}(t)$, with the state space $H^c \cup \{\Delta\}$. Thus, the use of taboo probabilities is equivalent to consideration of a M.C. killed at the first entrance to the taboo set (see Ex. 3, § 9.1.4). This point of view is basic for this book, as already pointed out.

§ 9.1.3. Decomposition theorems. The resolvent $\widetilde{\mathbf{U}}^\alpha$ and the generator $\widetilde{\mathbf{Q}}$ for the killed M.C. $\widetilde{X}$, corresponding to a contracting MF $M = (M_t)$, are known only for special forms of M. Killing argument produces decomposition theorems involving $\widetilde{\mathbf{U}}^\alpha$, and this in turn leads to expression for $\widetilde{\mathbf{Q}}$. The following definitions resemble these for the original M.C. (see § 8.1.1). For a function $f \geqslant 0$ on $\mathbb{I}$:

$$\widetilde{\mathbf{U}}^\alpha f(i) = \int_0^\infty e^{-\alpha t}\widetilde{\mathbf{P}}(t) f(i)\, dt = \mathbb{E}^i \int_0^\infty e^{-\alpha t} f(X_t) M_t\, dt, \qquad i \in \mathbb{I}.$$

As $\widetilde{\mathbf{P}}(0)f(i) = f(i)$ for $i \in \mathbb{I}_M$, and vanishes for $i \in \mathbb{I} \setminus \mathbb{I}_M$, it is sufficient to consider only states in $\mathbb{I}_M$, and one has

$$\lim_{\alpha \to \infty} \alpha \widetilde{\mathbf{U}}^\alpha f = \widetilde{\mathbf{P}}(0) f.$$

Hence, the generator $\widetilde{\mathbf{Q}}$ is defined by

$$\widetilde{\mathbf{Q}} f = \lim_{t \to 0} \frac{1}{t}\big(\widetilde{\mathbf{P}}(t) - \widetilde{\mathbf{P}}(0)\big) f = \lim_{\alpha \to \infty} \alpha\big(\alpha \widetilde{\mathbf{U}}^\alpha - \widetilde{\mathbf{P}}(0)\big) f.$$

a) For the purpose of the first decomposition theorem, introduce the operator $\mathbf{K}(t)$, acting on (bounded) positive functions f on $\mathbb{I}$, defined by

$$\mathbf{K}(t) f(i) = \mathbb{E}^i \int_0^t f(X_s)(-dM_s), \qquad t \geqslant 0,$$

with L.S. transform

$$\mathbf{K}^\alpha f(i) = \mathbb{E}^i \int_0^\infty e^{-\alpha s} f(X_s)(-dM_s), \qquad \alpha > 0.$$

Note that

$$\lim_{\alpha \to \infty} \mathbf{K}^\alpha f(i) = \mathbf{K}(0) f(i) = \begin{cases} 0, & i \in \mathbb{I}_M, \\ f(i), & i \in \mathbb{I} \setminus \mathbb{I}_M. \end{cases}$$

The following decomposition theorem provides interpretation of the operator $\mathbf{K}^\alpha$.

THEOREM 1. *Resolvents* $\mathbf{U}^\alpha$ *and* $\widetilde{\mathbf{U}}^\alpha$ *are related by*

$$\mathbf{U}^\alpha f = \widetilde{\mathbf{U}}^\alpha f + \mathbf{K}^\alpha \mathbf{U}^\alpha f, \qquad \alpha > 0,$$

for any bounded non-negative function f.

Proof. Consider the difference of potentials

$$\begin{aligned}
\mathbf{U}^\alpha f(i) - \widetilde{\mathbf{U}}^\alpha f(i) &= \mathbb{E}^i \int_0^\infty e^{-\alpha t} f(X_t)(1 - M_t)\, dt \\
&= \mathbb{E}^i \int_0^\infty e^{-\alpha t} f(X_t)\Big(\int_0^t -dM_s\Big)\, dt \\
&= \mathbb{E}^i \int_0^\infty e^{-\alpha s}(-dM_s) \int_s^\infty e^{-\alpha(t-s)} f(X_t)\, dt \\
&= \mathbb{E}^i \int_0^\infty e^{-\alpha s}(-dM_s)\mathbb{E}\Big(\int_s^\infty e^{-\alpha(t-s)} f(X_t)\, dt \mid X_s\Big) \\
&= \mathbb{E}^i \int_0^\infty e^{-\alpha s}\mathbf{U}^\alpha f(X_s)(-dM_s) = \mathbf{K}^\alpha \mathbf{U}^\alpha f(i). \qquad \square
\end{aligned}$$

Explicitly, the operators $\mathbf{P}(t)$ and $\widetilde{\mathbf{P}}(t)$ are related by the decomposition theorem

$$\mathbf{P}(t) = \widetilde{\mathbf{P}}(t) + \int_{0-}^t d\mathbf{K}(s)\mathbf{P}(t-s), \qquad t \geqslant 0,$$

with $\mathbf{I} = \mathbf{P}(0) = \widetilde{\mathbf{P}}(0) + \mathbf{K}(0)$.

Theorem 1 leads to several interesting consequences, stated here in rather formal way.

COROLLARY 2.

$$\mathbf{K}(t) - \mathbf{K}(0) = \int_0^t \widetilde{\mathbf{P}}(s)(\mathbf{Q} - \widetilde{\mathbf{Q}})\, ds, \qquad t \geqslant 0,$$

$$\mathbf{K}^\alpha - \mathbf{K}(0) = \widetilde{\mathbf{U}}^\alpha(\mathbf{Q} - \widetilde{\mathbf{Q}}), \qquad \alpha > 0.$$

Proof (outline). Rewrite assertion of Theorem 1 in the form

$$\alpha\mathbf{U}^\alpha - \mathbf{I} = \alpha\widetilde{\mathbf{U}}^\alpha - \widetilde{\mathbf{P}}(0) + \mathbf{K}^\alpha \alpha\mathbf{U}^\alpha - \mathbf{K}(0)$$

and use the (F)-equations for both semi-groups. □

It is easy to verify directly that $\mathbf{K}(t)$ satisfies the relation

$$\widetilde{\mathbf{P}}(t)\mathbf{K}(s) = \mathbf{K}(s+t) - \mathbf{K}(t), \qquad t, s \geqslant 0.$$

Hence, by Theorem 3, § 1.3.4, there exists an exit law $k(\cdot)$ relative to $\widetilde{\mathbf{P}}(t)$, such that $\mathbf{K}(t)$ may be represented as an integral. Thus, Corollary 2 shows that $k(t) = \widetilde{\mathbf{P}}(t)(\mathbf{Q} - \widetilde{\mathbf{Q}})$. Thus, with the right-side derivative at $t = 0$, $\mathbf{K}'(0) = \widetilde{\mathbf{P}}(0)(\mathbf{Q} - \widetilde{\mathbf{Q}})$.

COROLLARY 3.
$$\mathbf{Q} - \widetilde{\mathbf{Q}} = \mathbf{K}(0)\mathbf{Q} + \lim_{\alpha\to\infty} \alpha\big(\mathbf{K}^\alpha - \mathbf{K}(0)\big).$$

Proof (outline). Rewrite assertion of Theorem 1 in the form

$$\alpha\big(\alpha\mathbf{U}^\alpha - \mathbf{I}\big) - \alpha\big(\alpha\widetilde{\mathbf{U}}^\alpha - \widetilde{\mathbf{P}}(0)\big) = \mathbf{K}^\alpha\alpha\big(\alpha\mathbf{U}^\alpha - \mathbf{I}\big) + \alpha\big(\mathbf{K}^\alpha - \mathbf{K}(0)\big)$$

and pass to the limit as $\alpha \to \infty$. □

It is immediate that assertion of Theorem 1 can be written in the equivalent form, known as the "second resolvent equation".

COROLLARY 4.

$$\widetilde{\mathbf{P}}(0)\mathbf{U}^\alpha = \widetilde{\mathbf{U}}^\alpha + \widetilde{\mathbf{U}}^\alpha(\mathbf{Q} - \widetilde{\mathbf{Q}})\mathbf{U}^\alpha, \qquad \alpha > 0.$$

Furthermore, assertion of Theorem 1 may be regarded as a Markov renewal equation for the unknown $\mathbf{U}^\alpha f$, when $\widetilde{\mathbf{U}}^\alpha f$ and $\mathbf{K}^\alpha$ are given (see § 10.2.2).

b) Let $M = (M_t)$ be an MF, and let $A = (A_t)$ be a corresponding AF related by

$$M_t = e^{-A_t}, \qquad t \geqslant 0.$$

Recall that the α-potential of A has been defined by

$$\mathbf{U}_A^\alpha f(i) = \mathbb{E}^i \int_0^\infty e^{-\alpha t} f(X_t)\, dA_t, \qquad i \in \mathbb{I},$$

for any bounded function f. Note also that for the killed chain

$$\widetilde{\mathbf{U}}^\alpha\widetilde{\mathbf{P}}(0)f(i) = \begin{cases} \widetilde{\mathbf{U}}^\alpha f(i) \geqslant 0 & \text{for} \quad i \in \mathbb{I}_M, \\ 0 & \text{for} \quad i \in \mathbb{I}\setminus\mathbb{I}_M, \end{cases}$$

for any bounded non-negative function f. Hence,

$$\begin{aligned} \widetilde{\mathbf{U}}^\alpha f(X_s) &= \widetilde{\mathbf{U}}^\alpha\widetilde{\mathbf{P}}(0)f(X_s) \\ &= \frac{1}{M_s}\mathbb{E}\left(\int_s^\infty e^{-\alpha(t-s)}\widetilde{\mathbf{P}}(0)f(X_t)M_t\,dt \mid X_s\right) \quad \text{a.e.} \end{aligned}$$

With the above notation, one has another decomposition theorem.

THEOREM 5. *Resolvents $\mathbf{U}^\alpha$ and $\widetilde{\mathbf{U}}^\alpha$ are related by*

$$\mathbf{U}^\alpha\widetilde{\mathbf{P}}(0)f(i) = \widetilde{\mathbf{U}}^\alpha f(i) + \mathbf{U}_A^\alpha\,\widetilde{\mathbf{U}}^\alpha f(i), \qquad i \in \mathbb{I}, \quad \alpha > 0,$$

for any bounded non-negative function f.

Proof. See R. M. Blumenthal and R. K. Getoor [1968, p. 156], E. B. Dynkin [1965, Thm. 9.5] for details. Here only a formal verification is given, using Markov property,

$$\begin{aligned}
\mathbf{U}_A^\alpha \widetilde{\mathbf{U}}^\alpha f(i) &= \mathbf{U}_A^\alpha \, \widetilde{\mathbf{U}}^\alpha \, \widetilde{\mathbf{P}}(0) f(i) = \mathbb{E}^i \int_0^\infty e^{-\alpha s} \widetilde{\mathbf{U}}^\alpha \, \widetilde{\mathbf{P}}(0) f(X_s) \, dA_s \\
&= \mathbb{E}^i \int_0^\infty (dA_s/M_s) \int_s^\infty e^{-\alpha t} \widetilde{\mathbf{P}}(0) f(X_t) M_t \, dt \\
&= \mathbb{E}^i \int_0^\infty e^{-\alpha t} \widetilde{\mathbf{P}}(0) f(X_t) M_t \, dt \int_0^t dA_s/M_s \\
&= \mathbb{E}^i \int_0^\infty e^{-\alpha t} \widetilde{\mathbf{P}}(0) f(X_t)(1 - M_t) \, dt = \mathbf{U}^\alpha \widetilde{\mathbf{P}}(0) f(i) - \widetilde{\mathbf{U}}^\alpha f(i). \qquad \square
\end{aligned}$$

For the purpose of the next theorem, recall definition of the Revuz measure μ_A^i associated with the AF A, conditional on the initial state i (see § 8.1.3-e).

THEOREM 6. *The generator* $\widetilde{\mathbf{Q}}$ *of the killed chain satisfies the equation*

$$(\mathbf{Q}\widetilde{\mathbf{P}}(0) - \widetilde{\mathbf{Q}}) f(i) = \sum_{j \in \mathbb{I}_M} \mu_A^i(j) f(j), \qquad i \in \mathbb{I},$$

for bounded $f \geqslant 0$.

Proof (outline). Rewrite assertion of Theorem 5 in the form

$$\alpha(\alpha \mathbf{U}^\alpha - \mathbf{I})\widetilde{\mathbf{P}}(0) - \alpha(\alpha \widetilde{\mathbf{U}}^\alpha - \widetilde{\mathbf{P}}(0)) = \alpha \mathbf{U}_A^\alpha (\alpha \widetilde{\mathbf{U}}^\alpha - \widetilde{\mathbf{P}}(0)) + \alpha \mathbf{U}_A^\alpha \widetilde{\mathbf{P}}(0).$$

Passage to the limit as $\alpha \to \infty$ yields

$$(\mathbf{Q}\widetilde{\mathbf{P}}(0) - \widetilde{\mathbf{Q}}) f(i) = \lim_{\alpha \to \infty} \alpha \mathbf{U}_A^\alpha \widetilde{\mathbf{P}}(0) f(i) = \mu_A^i \bullet \widetilde{\mathbf{P}}(0) f$$

where $\bullet$ denotes the inner product. $\square$

From the semi-group property of $\widetilde{\mathbf{P}}(t)$ it follows that

$$\widetilde{\mathbf{U}}^\alpha \widetilde{\mathbf{P}}(0) = \widetilde{\mathbf{P}}(0) \widetilde{\mathbf{U}}^\alpha = \widetilde{\mathbf{U}}^\alpha, \qquad \widetilde{\mathbf{Q}}\widetilde{\mathbf{P}}(0) = \widetilde{\mathbf{P}}(0)\widetilde{\mathbf{Q}} = \widetilde{\mathbf{Q}}.$$

Hence, one can deduce from Theorem 5 that

$$\mathbf{U}_A^\alpha \widetilde{\mathbf{P}}(0) = \mathbf{U}^\alpha (\mathbf{Q} - \widetilde{\mathbf{Q}}) \widetilde{\mathbf{P}}(0)$$

in agreement with Theorem 6. It is then immediate that assertion of Theorem 5 can also be written in the equivalent form of the "second resolvent equation".

COROLLARY 7.

$$\mathbf{U}^\alpha \widetilde{\mathbf{P}}(0) = \widetilde{\mathbf{U}}^\alpha + \mathbf{U}^\alpha (\mathbf{Q} - \widetilde{\mathbf{Q}}) \widetilde{\mathbf{U}}^\alpha, \qquad \alpha > 0.$$

Remark. It is of interest to mention an alternative approach to Theorem 6. Consider an CAF $B = (B_t)$ defined by

$$B_t = \int_0^t \widetilde{\mathbf{P}}(0) f(X_s)\, dA_s, \qquad t \geqslant 0,$$

where as before $A_s = -\log M_s$. Denote by $G_i(t)$ the characteristic of B, and observe that

$$G_i(t) = \mathbb{E}^i B_t = \mathbb{E}^i \int_0^t \widetilde{\mathbf{P}}(0) f(X_s)\, dA_s, \qquad i \in \mathbb{I}, \quad t \geqslant 0,$$

with $G_i(0) = 0$, and that the following relation holds between α-potentials:

$$u_B^\alpha(i) = \mathbb{E}^i \int_0^\infty e^{-\alpha s}\, dB_s = \mathbf{U}_A^\alpha \widetilde{\mathbf{P}}(0) f(i), \qquad i \in \mathbb{I}, \quad \alpha > 0.$$

By the same argument as in Corollary 8, § 8.1.3,

$$\mathbf{P}(s)\, G_i(t) = G_i(t+s) - G_i(s), \qquad t, s \geqslant 0.$$

Hence, there exists an exit law $(g_i(s))$, relative to the semi-group $\mathbf{P}(t)$, $t \geqslant 0$, such that

$$G_i(t) = \int_0^t g_i(s)\, ds,$$

where $g_i(s) = \mathbf{P}(s)\, g_i(0)$ and $g_i(0) = \mu_A^i \bullet \widetilde{\mathbf{P}}(0) f$. Thus, as $t \to 0$,

$$\frac{1}{t} \mathbb{E}^i \int_s^{s+t} \widetilde{\mathbf{P}}(0) f(X_v)\, dA_v \longrightarrow g_i(0),$$

in agreement with Theorem 6.

c) Comparison of Theorem 1 with Theorem 5 indicates duality between these two decomposition theorems. This fact is more visible from comparison of corresponding "second resolvent equations" (Corollaries 4 and 7). It is easy to obtain the following formal relations:

$$\widetilde{\mathbf{P}}(0)\mathbf{U}^\alpha(\mathbf{Q} - \widetilde{\mathbf{Q}})\widetilde{\mathbf{U}}^\alpha = \widetilde{\mathbf{U}}^\alpha(\mathbf{Q} - \widetilde{\mathbf{Q}})\mathbf{U}^\alpha\widetilde{\mathbf{P}}(0),$$
$$\mathbf{U}^\alpha(\mathbf{Q} - \widetilde{\mathbf{Q}})\widetilde{\mathbf{P}}(0) = \mathbf{U}_A^\alpha\widetilde{\mathbf{P}}(0),$$
$$\mathbf{K}^\alpha\mathbf{U}^\alpha\widetilde{\mathbf{P}}(0) = \mathbf{U}_A^\alpha\widetilde{\mathbf{U}}^\alpha.$$

Introducing an α-potential of A relative to the killed semi-group,

$$\widetilde{\mathbf{U}}_A^\alpha f(i) = \mathbb{E}^i \int_0^\infty e^{-\alpha t} f(X_t) M_t\, dA_t, \qquad \alpha > 0,$$

one finds that $\mathbf{K}^\alpha = \widetilde{\mathbf{U}}^\alpha_A$.

Finally, it is shown later in Sub-section 11.1 that the dual $\widehat{\mathbf{K}}^\alpha$ satisfies

$$\mathbf{U}^\alpha = \widetilde{\mathbf{U}}^\alpha + \mathbf{U}^\alpha \widehat{\mathbf{K}}^\alpha \qquad \text{with} \quad \widehat{\mathbf{K}}^\alpha - \widehat{\mathbf{K}}(0) = (\mathbf{Q} - \widetilde{\mathbf{Q}})\widetilde{\mathbf{U}}^\alpha,$$

which again leads to Corollary 7. This fact in conjunction with Theorem 5 yields

$$\mathbf{U}^\alpha_A = \mathbf{U}^\alpha(\mathbf{Q} - \widetilde{\mathbf{Q}})\widetilde{\mathbf{P}}(0).$$

§ 9.1.4. Examples. The following special cases are of great interest. They have been already mentioned earlier and are of basic importance for this book.

Example 1. *Killing at constant rate.* Let $\alpha > 0$ be a constant, and consider the functionals

$$A_t = \alpha t, \qquad M_t = e^{-\alpha t}, \qquad t \geqslant 0,$$

so the rate is α, and $\mathbb{I}_M = \mathbb{I}$. Then, for $i, j \in \mathbb{I}$,

$$\tilde{p}_{ij}(t) = e^{-\alpha t} p_{ij}(t), \qquad \tilde{p}_{i\Delta}(t) = 1 - e^{-\alpha t}, \qquad \tilde{q}_{ij} = q_{ij} - \alpha\delta_{ij},$$

or, in matrix notation,

$$\widetilde{\mathbf{P}}(t) = e^{-\alpha t}\mathbf{P}(t) = \mathbf{P}^\alpha(t), \qquad \widetilde{\mathbf{Q}} = \mathbf{Q} - \alpha \mathbf{I} = \mathbf{Q}^\alpha.$$

Thus, the killed chain corresponds to the α-chain discussed in § 1.1.6. This in turn corresponds to the killing of a M.C. X at a random time independent of X and having exponential distribution with a parameter α:

$$\tilde{p}_{ij}(t) = \mathbb{E}^i(e^{-\alpha t},\ X_t = j) = \mathbb{P}^i(T > t,\ X_t = j).$$

Observe also that both decomposition theorems reduce here to triviality

$$p_{ij}(t) - \tilde{p}_{ij}(t) = (1 - e^{-\alpha t})p_{ij}(t).$$

Example 2. *Killing at state-dependent rate.* Consider an IF $A = (A_t)$ determined by a bounded charge $a \geqslant 0$, and the corresponding contracting MF $M = (M_t)$:

$$A_t = \int_0^t a(X_s)\, ds, \qquad M_t = e^{-A_t}, \qquad t \geqslant 0,$$

so the rate is $-dM_t/M_t = a(X_t)\, dt$. Note that A_t is a piece-wise linear and $A_t < \infty$ for each t, because paths of the usual M.C. are step functions with infinite life time. However, A_∞ may be either finite (so $M_\infty > 0$) or infinite (so $M_\infty = 0$), but the life time of M is always infinite. Thus, $\mathbb{I}_M = \mathbb{I}$. Recall that supp$(A)$ consists of those states i for which $a(i) > 0$.

Theorem 1, § 9.1.1, gives now

$$\tilde{p}_{ij}(t) = \mathbb{E}^i(e^{-A_t},\ X_t = j), \qquad i, j \in \mathbb{I},$$

with

$$\sum_j \tilde{p}_{ij}(t) = \mathbb{E}^i e^{-A_t} = 1 - \tilde{p}_{i\Delta}(t), \qquad t \geqslant 0,$$

representing distribution $\mathbb{P}^i(\widetilde{T} > t)$ of the associated life time $\widetilde{T}$.

As the α-potential of A is $\mathbf{U}_A^\alpha \mathbf{1}(i) = \mathbf{U}^\alpha a(i)$, and $\widetilde{\mathbf{P}}(0) = \mathbf{I}$, Theorem 5, § 9.1.3, yields

$$u_{ij}^\alpha - \tilde{u}_{ij}^\alpha = \sum_k u_{ik}^\alpha\, a(k) \tilde{u}_{kj}^\alpha, \qquad i, j \in \mathbb{I}.$$

Similarly, $\mathbf{K}^\alpha \mathbf{1}(i) = \widetilde{\mathbf{U}}^\alpha a(i)$, so Theorem 1, § 9.1.3, yields

$$u_{ij}^\alpha - \tilde{u}_{ij}^\alpha = \sum_k \tilde{u}_{ik}^\alpha\, a(k) u_{kj}^\alpha, \qquad i, j \in \mathbb{I}.$$

Furthermore, the Revuz measure now is

$$\mu_A^i(j) = \lim_{t \to 0} \frac{1}{t} \int_0^t p_{ij}(s) a(j)\, ds = \delta_{ij} a(j).$$

Hence, one has for generators

$$\tilde{q}_{ij} = q_{ij} - a(j)\delta_{ij},$$

or, in matrix notation,

$$\widetilde{\mathbf{Q}} = \mathbf{Q} - a\mathbf{I}.$$

The (absolute) Revuz measure is obviously

$$\mu_A(j) = \xi(j) a(j), \qquad j \in \mathbb{I}.$$

It is easy to verify directly that

$$u_A^\alpha(i) = \mathbf{U}^\alpha \mu_A(i) = \mathbf{U}^\alpha a(i), \qquad i \in \mathbb{I},$$

in agreement with Theorem 7, § 8.1.3, as already noted.

Example 3. *Taboo.* Let $T = T_H$ be the first entrance time to a set H, and consider the right-continuous MF given by

$$M_t = I_{[0,T)}(t), \qquad t \geqslant 0,$$

with life time T and $\mathbb{I}_M = H^c$ (see § 8.1.5). Then, obviously,

$$\tilde{p}_{ij}(t) = {}_H p_{ij}(t), \qquad 1 - \tilde{p}_{i\Delta}(t) = D_i^c(t) = \mathbb{E}^i M_t \qquad \text{for} \quad i, j \in \mathbb{I},$$

and these probabilities vanish for states in H (see § 1.2.1 and Thm. 1, § 3.1.1). In matrix notation, with reference to the state space $\mathbb{I}$, one can write

$$\widetilde{\mathbf{P}}(t) = {}_H\mathbf{P}(t), \qquad \widetilde{\mathbf{U}}^\alpha = {}_H\mathbf{U}^\alpha, \qquad \widetilde{\mathbf{Q}} = {}_H\mathbf{Q}.$$

Since effectively the state space is $H^c \cup \{\Delta\}$, one could use notation $\mathbf{P}^{\Delta}(t)$ etc. (from § 1.2.4).

In fact, the taboo probabilities refer to the original M.C. prior to the first entrance to the taboo set H, and the life time of the killed chain is the same as the life time of M, namely T. Thus, the use of taboo probabilities is equivalent to consideration of a M.C. killed at the first entrance time T. This point of view is basic for this book and has been already stressed earlier (see § 9.1.1).

Thus, theory developed in this sub-section applies and taboo semi-group provides the most typical example of sub-ordinate semi-groups. However, in the present case (with reference to state space $\mathbb{I}$), $\widetilde{\mathbf{P}}(0) = \mathbf{J}_1$.

Thus, for $i \in \mathbb{I}$,

$$\mathbb{E}^i\big(f(X_s),\ s < T\big) = {}_H\mathbf{P}(s)\, f(i),$$

$$\mathbf{K}^\alpha f(i) = \mathbb{E}^i\big(e^{-\alpha T} f(X_T),\ T < \infty\big) = \mathbf{B}^\alpha f(i),$$

so $\mathbf{K}^\alpha = \mathbf{B}^\alpha$ with $\mathbf{K}(0) = \mathbf{J}_2$.

Consequently, Theorem 1, § 9.1.3, reduces to Theorem 1, § 6.1.1 (Dynkin formula)

$$\mathbf{U}^\alpha = {}_H\mathbf{U}^\alpha + \mathbf{B}^\alpha \mathbf{U}^\alpha.$$

Similarly, Corollaries 2 and 4 from § 9.1.3 reduce, respectively, to relations already stated in § 6.1.1:

$$\mathbf{B}^\alpha - \mathbf{J}_2 = {}_H\mathbf{U}^\alpha\big(\mathbf{Q} - {}_H\mathbf{Q}\big),$$
$$\mathbf{J}_1\mathbf{U}^\alpha = {}_H\mathbf{U}^\alpha + {}_H\mathbf{U}^\alpha\big(\mathbf{Q} - {}_H\mathbf{Q}\big)\mathbf{U}^\alpha.$$

However, Theorem 5, § 9.1.3, is not valid as it stands, because $A_t = -\log M_t$ produces a degenerate functional with infinite potential. See Example 5 below for another interpretation.

Remark. The first entrance decomposition Theorem 3, § 6.1.1, may be written in the matrix form

$$\mathbf{U}^\alpha = {}_H\mathbf{U}^\alpha + {}_H\mathbf{U}^\alpha\, \overline{\mathbf{B}}^\alpha,$$

where $\overline{\mathbf{B}}^\alpha = \mathbf{Q}\mathbf{J}_2\mathbf{U}^\alpha$ is the matrix introduced in § 6.3.1. Thus, one has here

$$\mathbf{J}_1\mathbf{U}^\alpha = {}_H\mathbf{U}^\alpha + {}_H\mathbf{U}^\alpha\big(\mathbf{Q} - {}_H\mathbf{Q}\big)\mathbf{U}^\alpha.$$

Moreover,

$${}_H\mathbf{U}^\alpha\, \overline{\mathbf{B}}^\alpha f(i) = \mathbb{E}^i \int_0^\infty e^{-\alpha t}\, \overline{\mathbf{B}}^\alpha f(X_t) M_t\, dt,$$

and

$$\int_0^t {}_H\mathbf{P}(s)\, d\overline{\mathbf{B}}(t-s) f(i) = \mathbb{E}^i \int_0^t \big(d\overline{\mathbf{B}}(t-s)\, f(X_s),\ s < T\big)$$

for $T \leqslant t$ and $i \in H^c$.

Example 4. *Exit taboo.* Let T_ρ be the first return time to a set H (see § 3.3.1 and recall that T_ρ is a terminal time). Consider the right-continuous MF $M = (M_t)$ given by

$$M_t = I_{(T_\rho > t)}, \qquad t \geqslant 0,$$

with the life time T_ρ and $\mathbb{I}_M = \mathbb{I}$ (see § 8.1.5). Then, obviously,

$$\tilde{p}_{ij}(t) = {}_H p^\rho_{ij}(t), \qquad 1 - \tilde{p}_{i\Delta}(t) = D^{\rho c}_i(t) = \mathbb{E}^i M_t \qquad \text{for} \quad i, j \in \mathbb{I}$$

(see § 2.2.4 and Lemma 1, § 3.3.1). In matrix notation, with reference to the state space $\mathbb{I}$, one can write

$$\widetilde{\mathbf{P}}(t) = {}_H\mathbf{P}^\rho(t), \qquad \widetilde{\mathbf{U}}^\alpha = {}_H\mathbf{U}^{\rho\alpha}, \qquad \widetilde{\mathbf{Q}} = {}_H\mathbf{Q}^\rho,$$

with $\widetilde{\mathbf{P}}(0) = \mathbf{I}$ (see §§ 2.2.4 and 3.3.5).

Thus, for $i \in \mathbb{I}$,

$$\begin{aligned} &\mathbb{E}^i\big(f(X_s),\ s < T_\rho\big) = {}_H\mathbf{P}^\rho(s) f(i), \\ &\mathbf{K}^\alpha f(i) = \mathbb{E}^i\big(e^{-\alpha T_\rho} f(X_T),\ T < \infty\big) = \mathbf{B}^{\rho\alpha} f(i), \end{aligned}$$

so $\mathbf{K}^\alpha = \mathbf{B}^{\rho\alpha}$ with $\mathbf{K}(0) = 0$.

Consequently, Theorem 1, § 9.1.3, reduces to Theorem 1, § 6.1.2,

$$\mathbf{U}^\alpha = {}_H\mathbf{U}^{\rho\alpha} + \mathbf{B}^{\rho\alpha}\mathbf{U}^\alpha.$$

Similarly, Corollaries 2 and 4 from § 9.1.3 reduce, respectively, to relations already stated in § 6.2.1:

$$\begin{aligned} \mathbf{B}^{\rho\alpha} &= {}_H\mathbf{U}^{\rho\alpha}\big(\mathbf{Q} - {}_H\mathbf{Q}^\rho\big), \\ \mathbf{U}^\alpha &= {}_H\mathbf{U}^{\rho\alpha} + {}_H\mathbf{U}^{\rho\alpha}\big(\mathbf{Q} - {}_H\mathbf{Q}^\rho\big)\mathbf{U}^\alpha. \end{aligned}$$

However, Theorem 5, § 9.1.3, is not valid as it stands, because $A_t = -\log M_t$ produces a degenerate functional with infinite potential. See Example 5 below for another interpretation.

Remark. The first return decomposition has the dual form

$$\mathbf{U}^\alpha = {}_H\mathbf{U}^{\rho\alpha} + {}_H\mathbf{U}^{\rho\alpha}\,\overline{\mathbf{B}}^{\rho\alpha},$$

where

$$\overline{\mathbf{B}}^{\rho\alpha} = \big(\mathbf{Q} - {}_H\mathbf{Q}^\rho\big)\mathbf{U}^\alpha.$$

This evidently coincides with Theorem 3, § 6.1.2, with $\overline{\mathbf{B}}^{\rho\alpha} = (h^{\rho\alpha}_{ij})$ defined in § 3.1.2 and § 3.3.2 for $i \in H^c$ and $i \in H$, respectively.

Hence, again

$$\mathbf{U}^\alpha = {}_H\mathbf{U}^{\rho\alpha} + {}_H\mathbf{U}^{\rho\alpha}\big(\mathbf{Q} - {}_H\mathbf{Q}^\rho\big)\mathbf{U}^\alpha.$$

Moreover,

$${}_H\mathbf{U}^{\rho\alpha}\,\overline{\mathbf{B}}^{\rho\alpha} f(i) = \mathbb{E}^i \int_0^\infty e^{-\alpha t}\,\overline{\mathbf{B}}^{\rho\alpha} f(X_t) M_t\, dt,$$

and

$$\int_0^t {}_H\mathbf{P}^\rho(s)\, d\overline{\mathbf{B}}^\rho(t-s)\, f(i) = \mathbb{E}^i \int_0^t \left(d\overline{\mathbf{B}}^\rho(t-s) f(X_s),\ s < T_\rho\right)$$

for $T_\rho \leqslant t$ and $i \in \mathbb{I}$.

Example 5. *Terminal time.* Let T be a terminal stopping time, and consider the right-continuous MF $M = (M_t)$ given by

$$M_t = I_{(T>t)}, \qquad t \geqslant 0,$$

with the life time T and the set $\mathbb{I}_M$ of permanent points (see § 9.1.2). Note that Examples 3 and 4 are special cases.

Evidently, the corresponding AF $A = (A_t) = (-\log M_t)$, is degenerate: $A_t = 0$ for $t < T$ and $A_t = \infty$ for $t \geqslant T$, so its α-potential is infinite. Consequently, Theorem 5, § 9.1.3, does not hold, as it stands. Nevertheless, this decomposition theorem remains valid provided $\mathbf{U}_A^\alpha$ is given another interpretation. Indeed, proceeding formally, write

$$\mathbf{U}^\alpha\, \widetilde{\mathbf{P}}(0) = \widetilde{\mathbf{U}}^\alpha + \mathbf{U}_A^\alpha \widetilde{\mathbf{U}}^\alpha,$$

where $\mathbf{U}_A^\alpha$ is to be determined. Multiplication by $(\alpha\mathbf{I} - \widetilde{\mathbf{Q}})$ from right and the (F)-equation for $\widetilde{\mathbf{U}}^\alpha$ lead to

$$\mathbf{U}_A^\alpha\, \widetilde{\mathbf{P}}(0) = \mathbf{U}^\alpha\, (\mathbf{Q} - \widetilde{\mathbf{Q}})\widetilde{\mathbf{P}}(0).$$

This defines $\mathbf{U}_A^\alpha$, and one has (as in Corollary 7, § 9.1.3)

$$\mathbf{U}^\alpha\, \widetilde{\mathbf{P}}(0) = \widetilde{\mathbf{U}}^\alpha + \mathbf{U}^\alpha\, (\mathbf{Q} - \widetilde{\mathbf{Q}})\widetilde{\mathbf{U}}^\alpha.$$

Such decompositions occur in discussion of perturbation of M.C.'s, and $\mathbf{U}_A^\alpha$ is related to a compensation kernel (see § 11.2.1). In the present case, components of $\mathbf{U}_A^\alpha$ are of interest only for states in $\mathbb{I}_M$, so

$$u_A^\alpha(i,j) = \sum_{k\in\mathbb{I}_M} u_{ik}^\alpha\left(q_{kj} - \tilde{q}_{kj}\right), \qquad i \in \mathbb{I}, \quad j \in \mathbb{I}_M,$$

and $u_A^\alpha(i,j) = 0$ for $j \in \mathbb{I} \setminus \mathbb{I}_M$. Put

$$u_A^\alpha(i) = \sum_j u_A^\alpha(i,j), \qquad i \in \mathbb{I}.$$

Next, proceeding as in Theorem 6, § 9.1.3 (see also Theorem 7, § 8.1.3), one obtains a well-defined measure supported on $\mathbb{I}_M$:

$$\mu_A^i(j) = -\sum_k q_{ik} u_A(k,j) = q_{ij} - \tilde{q}_{ij}$$

and

$$\mu_A(j) = \sum_i \xi(i)\big(q_{ij} - \tilde{q}_{ij}\big), \qquad j \in \mathbb{I}_M.$$

However, this is not a Revuz measure. Indeed, it is easy to see that

$$\mathbf{U}^\alpha \mu_A(i) = \mathbf{U}^\alpha a(i), \qquad i \in \mathbb{I},$$

where $a(j) = \mu_A(j)/\xi(j)$ for $j \in \mathbb{I}_M$, and 0 for $j \in \mathbb{I} \setminus \mathbb{I}_M$. Consequently, u_A^α and $\mathbf{U}^\alpha \mu_A$ are different, contradicting the basic property expressed by Theorem 6, § 8.1.3.

On the other hand, consider an integral functional $A^* = (A_t^*)$ determined by the charge $a \geqslant 0$ defined above. By Example 2 above, the α-potential of A^* and the α-potential of the Revuz measure μ_{A^*} of A^* coincide. It is easy to verify that their common value is

$$u_{A^*}^\alpha(i) = \mathbf{U}^\alpha \mu_{A^*}(i) = \mathbf{U}^\alpha a(i), \qquad i \in \mathbb{I}.$$

Moreover, the Revuz measure μ_{A^*} coincides with the measure μ_A:

$$\mu_{A^*}^i(j) = \delta_{ij} a(j), \qquad \mu_{A^*}(j) = \xi(j)\, a(j) = \mu_A(j), \qquad \mathbf{U}^\alpha \mu_{A^*} = \mathbf{U}^\alpha \mu_A.$$

It is therefore permissible to associate with the degenerate AF (A_t), corresponding to the MF (M_t), the Revuz measure of the integral functional (A_t^*), and refer to it as a quasi-Revuz measure of A.

To illustrate the above considerations, return to Examples 3 and 4.

Example 3 (continued). The analogue of Theorem 5, § 9.1.3, is

$$\mathbf{U}^\alpha \mathbf{J}_1 = {}_H\mathbf{U}^\alpha + \mathbf{U}_A^\alpha {}_H\mathbf{U}^\alpha.$$

This evidently coincides with the last exit distribution, expressed by formula (4) after Theorem 3, § 6.2.1. Consequently,

$$\mathbf{U}_A^\alpha \mathbf{J}_1 = \mathbf{U}^\alpha \left(\mathbf{Q} - {}_H\mathbf{Q}\right) \mathbf{J}_1 = \widetilde{\mathbf{C}}^\alpha \mathbf{J}_1,$$

where $\widetilde{\mathbf{C}}^\alpha = \mathbf{U}^\alpha \mathbf{J}_2 \mathbf{Q}$, as noted in § 6.2.1. Hence, in agreement with Corollary 7, § 9.1.3,

$$\mathbf{U}^\alpha \mathbf{J}_1 = {}_H\mathbf{U}^\alpha + \mathbf{U}^\alpha \left(\mathbf{Q} - {}_H\mathbf{Q}\right) {}_H\mathbf{U}^\alpha = {}_H\mathbf{U}^\alpha + \widetilde{\mathbf{C}}^\alpha {}_H\mathbf{U}^\alpha.$$

Thus,

$$u_A^\alpha(i,j) = c_{ij}^\alpha \qquad \text{for} \quad i \in \mathbb{I},\ j \in H^c.$$

Therefore, for $j \in H^c$,

$$\mu_A^i(j) = \begin{cases} 0, & i \in H^c, \\ q_{ij}, & i \in H, \end{cases}$$

and the quasi-Revuz measure is

$$\mu_A(j) = \sum_{i \in H} \xi(i) q_{ij}, \qquad j \in H^c.$$

Example 4 (continued). The analogue of Theorem 5, § 9.1.3, is

$$\mathbf{U}^\alpha = {}_H\mathbf{U}^{\rho\alpha} + \mathbf{U}^\alpha_A \, {}_H\mathbf{U}^{\rho\alpha}.$$

Consequently,

$$\mathbf{U}^\alpha_A = \mathbf{U}^\alpha \left(\mathbf{Q} - {}_H\mathbf{Q}^\rho\right).$$

Hence, in agreement with Corollary 7, § 9.1.3,

$$\mathbf{U}^\alpha = {}_H\mathbf{U}^{\rho\alpha} + \mathbf{U}^\alpha \left(\mathbf{Q} - {}_H\mathbf{Q}^\rho\right) {}_H\mathbf{U}^{\rho\alpha}.$$

Thus,

$$u^\alpha_A(i,j) = u^\alpha_{ij}\left(\alpha + q_j\right) - \delta_{ij}, \qquad i \in \mathbb{I}, \quad j \in H,$$

and this vanishes for $j \in H^c$. Therefore, for $j \in H$,

$$\mu^i_A(j) = q_{ij} + \delta_{ij} q_j, \qquad i \in \mathbb{I}, \quad j \in H,$$

with $\mu^i_A(H) = \bar{q}_i$ for $i \in \mathbb{I}$. It follows that the quasi-Revuz measure is

$$\mu_A(j) = \sum_{i \neq j} \xi(i) q_{ij}, \qquad j \in H.$$

Note that the above decomposition theorem shoud be distinguished from Theorem 1, § 6.2.2.

§ 9.1.5. Non-contracting MF. Let $M = (M_t)$ be a non-contracting MF which satisfies the following weaker condition

$$\mathbb{E}^i M_t \leqslant 1 \qquad \text{for all } t \text{ and } i, \tag{m}$$

instead of condition $0 \leqslant M_t \leqslant 1$ (see § 8.1.5). Most of the results obtained earlier hold, but the resulting semi-group $\widetilde{\mathbf{P}}(t)$, $t \geqslant 0$, is no longer subordinate to the original semi-group $\mathbf{P}(t)$, $t \geqslant 0$.

Define the following kernel:

$$\tilde{p}_{ij}(t) = \mathbb{E}^i\left(X_t = j,\ M_t\right), \qquad i, j \in \mathbb{I}, \quad t \geqslant 0.$$

Then, $\widetilde{\mathbf{P}}(t) = (\tilde{p}_{ij}(t))$ is a Markov semi-group (said to be generated by the MF M) with

$$1 - \tilde{p}_{i\Delta}(t) = \mathbb{E}^i M_t, \qquad \lim_{t \to 0} \widetilde{\mathbf{P}}(t)\,\mathbf{1}_i = \widetilde{\mathbf{P}}(0)\,\mathbf{1}_i = \mathbb{E}^i M_0.$$

Proof is identical to that in Theorem 1, § 9.1.1; cf. E. B. Dynkin [1965, Thm. 9.1]. Furthermore, under some additional conditions, extension of that theorem to the present case holds, but the semi-group of the killed chain is no longer subordinate to that of the original chain. See E. B. Dynkin [1965, Thm. 10.5].

In particular, a large class of semi-groups generated by MF's satisfying condition (m), is obtained with the help of a MF defined by

$$M_t = g(X_t)/g(X_0), \qquad t \geqslant 0,$$

where g is a strictly positive $\mathbf{P}(t)$-excessive function on $\mathbb{I}$.

THEOREM 1. *$M = (M_t)$ defined by g, as above, is a MF satisfying condition (m), and the corresponding Markov semi-group $\widetilde{\mathbf{P}}(t) = (p_{ij}(t))$ is given by*

$$\tilde{p}_{ij}(t) = p_{ij}(t)\, g(j)/g(i), \qquad t \geqslant 0, \quad i, j \in \mathbb{I},$$
$$\tilde{p}_{i\Delta}(t) = 1 - \mathbb{E}^i M_t,$$

with a resolvent and a generator of the form

$$\tilde{u}^{\alpha}_{ij} = u^{\alpha}_{ij}\, g(j)/g(i) \qquad \text{and} \qquad \tilde{q}_{ij} = q_{ij}\, g(j)/g(i), \qquad i, j \in \mathbb{I},$$

Moreover, (M_t) is a supermartingale relative to the natural σ-fields.

Proof. It is clear that M_t is $\mathcal{F}_t$-measurable for each t. The relation $M_{t+h} = M_t\,(M_h \circ \theta_t)$ is immediate from definition. The fact that g is $\mathbf{P}(t)$-excessive gives condition (m):

$$\mathbb{E}^i M_t = \mathbb{E}^i g(X_t)/g(i) = \mathbf{P}(t) g(i)/g(i) \leqslant 1 \quad \text{and} \quad \mathbb{E} M_t \leqslant 1.$$

Similarly,

$$\mathbb{E}\big(M_{t+h} \mid \mathcal{F}_t\big) = M_t \mathbb{E}\big(M_h \circ \theta_t \mid \mathcal{F}_t\big) = \big(M_t/g(X_t)\big)\mathbb{E}\big(g(X_{t+h}) \mid X_t\big)$$
$$\leqslant M_t g(X_t)/g(X_t) = M_t \quad \text{a.e.} \qquad \square$$

As (M_t) is a positive supermartingale, it converges a.e. to M_∞. Clearly, the set of permanent points coincides with the state space $\mathbb{I}$.

Remark. See § 9.4.1 for applications.

§ 9.1.6. Co-stopping time. Killing operation may be extended to random times which are not necessarily stopping times. P. A. Meyer, R. T. Smythe, and J. B. Walsh [1972] considered a general Markov process killed at a co-stopping time L and showed its Markovian character. The following is the adaptation of their result to the usual M.C. with a semi-group $\mathbf{P}(t)$, $t \geqslant 0$.

The main property of L to be used is that $L \circ \theta_t = (L - t)^+$.

Let $X = (X_t)$ be the usual chain. The chain killed at L is defined in terms of the killing operator $\mathcal{R}_L$ as $\widetilde{X} = (\widetilde{X}_t)$, where $\widetilde{X}_t = X_t \circ \mathcal{R}_L$. Next define a $\mathbf{P}(t)$-excessive function c by

$$c(i) = \mathbb{P}\big(L > 0 \mid X_0 = i\big), \qquad i \in \mathbb{I}.$$

THEOREM 1. *Let L be a co-stopping time for a M.C. $X = (X_t)$ with the natural σ-fields and the state space $\mathbb{I}$, and let c be an excessive function defined above. Then, the chain $\widetilde{X} = (\widetilde{X}_t)$ killed at L*

$$\widetilde{X}_t = \begin{cases} X_t & \text{for} \quad t < L, \\ \Delta & \text{for} \quad t \geqslant L, \end{cases}$$

is Markovian with a state space (sub-set of) $\mathbb{I} \cup \{\Delta\}$, and with transition probabilities given by

$$\tilde{p}_{ij}(t) = \begin{cases} p_{ij}(t) c(j)/c(i) & \textit{if} \quad c(i) > 0, \\ \delta_{\Delta j} & \textit{if} \quad c(i) = 0, \ i = \Delta, \end{cases}$$

$$\tilde{p}_{i\Delta}(t) = 1 - \sum_j \tilde{p}_{ij}(t).$$

Note that the effective life time is L, and that the chain disappears afterwards. The transition matrix $\widetilde{\mathbf{P}}(t) = (\tilde{p}_{ij}(t))$ is different from the original matrix $\mathbf{P}(t)$, and

$$\tilde{p}_{ij}(t) = \mathbb{P}\big(X_t = j,\ t < L \mid X_0 = i,\ 0 < L\big).$$

It is clear that $\widetilde{\mathbf{P}}(t)$, $t \geqslant 0$, is a semi-group on $\mathbb{I} \cup \{\Delta\}$, but it is not dominated by the original semi-group $\mathbf{P}(t)$, $t \geqslant 0$. This is due in part to different initial conditions of both chains.

Furthermore, the initial distribution of the killed chain is

$$\mathbb{P}\big(\widetilde{X}_0 = i\big) = \pi_i c(i) \qquad \text{for} \quad i \in \mathbb{I},$$
$$\mathbb{P}\big(\widetilde{X}_0 = \Delta\big) = 1 - \sum_i \pi_i c(i),$$

and it differs from the initial distribution π of the original chain X. It is clear that the absolute distribution of $\widetilde{X}_t$ satisfies the equation

$$\mathbb{P}\big(\widetilde{X}_t \in \mathbb{I}\big) = \pi \mathbf{P}(t) c.$$

Remark. As the function c is $\mathbf{P}(t)$-excessive, one could consider a multiplicative functional $M = (M_t)$ defined as in § 9.1.5 by $M_t = c(X_t)/c(X_0)$. However, Theorem 1, § 9.1.5, as it stands, does not apply because of different measurability conditions. Consequently, the separate proof is needed. It should be added that recent work on splitting times produces general theorems which embraces both cases; see A. O. Pittenger and M. J. Sharpe [1981].

The most typical example of a co-stopping time is of course the last exit time $L = L_H$ from a set H, as discussed in Sub-section 4.1. It is worthwhile to revisit this case.

Example. Killing at the last exit. Let L be the last exit time from a set H, with the distribution function $\Gamma_i(t)$, found already in § 4.1.1. As L is a co-stopping time, theory described here applies, and

$$c(i) = \mathbb{P}^i(L > 0) = \Gamma_i^c(0) = D_i.$$

Thus, Theorem 1 yields:

$$\tilde{p}_{ij}(t) = p_{ij}(t) D_j / D_i \qquad \text{for} \quad D_i > 0,$$
$$\tilde{p}_{i\Delta}(t) = 1 - \Gamma_i^c(t)/D_i.$$

Indeed, in view of results in §§ 2.1.4 and 4.1.1, one has, for $t < L$,

$$\tilde{p}_{ij}(t) = \mathbb{P}\big(\widetilde{X}_t = j \mid \widetilde{X}_0 = i\big) = \mathbb{P}\big(X_t = j,\ t < L \mid X_0 = i,\ 0 < L\big)$$
$$= \mathbb{P}\big(X_t = j,\ t < L \mid X_0 = i\big)/\mathbb{P}\big(0 < L \mid X_0 = i\big) = p_{ij}(t) D_j / D_i,$$

and

$$\sum_j p_{ij}(t)D_j/D_i = \Gamma_i^c(t)/D_i = \mathbb{P}(t < L \mid X_0 = i,\ 0 < L).$$

Furthermore, the initial distribution is

$$\mathbb{P}(\widetilde{X}_0 = i) = \begin{cases} \pi_i D_i, & i \in \mathbb{I}, \\ \sum_k \pi_k D_k^c, & i = \Delta, \end{cases}$$

and

$$\mathbb{P}(\widetilde{X}_t \in \mathbb{I}) = \sum_i \pi_i \Gamma_i^c(t).$$

Remark. Taking $M_t = I_{(L>t)}$, one obtains only a contracting rough (i.e., non-adapted) MF. Its set of permanent states is determined by the condition

$$\mathbb{P}^i(L > 0) = \Gamma_i^c(0) = D_i = 1.$$

In accordance with Theorem 2, § 4.1.1, one could consider

$$\tilde{p}_{ij}(t) = \mathbb{P}(X_t = j,\ t < L \mid X_0 = i) = p_{ij}(t)D_j,$$
$$\tilde{p}_{i\Delta}(t) = \Gamma_i(t),$$

with $\mathbb{E}^i M_t = \Gamma_i^c(t)$. However, it is evident that these $(\tilde{p}_{ij}(t))$ do not form a semi-group. Thus, the analogue of Theorem 1, § 9.1.1, does not hold.

Therefore, the proper form of the semi-group $\widetilde{\mathbf{P}}(t)$, $t \geqslant 0$, must be obtained from Theorem 1 here, and is given by the above example.

9.2. BIRTHING

Birthing of a M.C. is obtained by shifting the time origin to some random time R, called the birth time, and counting time from R. A new chain is then started at R — it is also referred to as the post-R chain. This operation is the natural extension of the strong Markov property from stopping times to more general random times.

§ 9.2.1. Definitions. Let $X = (X_t, \mathfrak{F}_t,\ 0 \leqslant t < \infty)$ be an original M.C. satisfying the usual conditions, and let R be a random time (non-negative random variable). The new chain $\widetilde{X} = (\widetilde{X}_t, \widetilde{\mathfrak{F}}_t,\ 0 \leqslant t < \infty)$ started at R is defined by

$$\widetilde{X}_t = X(t+R), \qquad \widetilde{\mathfrak{F}}_t = \mathfrak{F}_{t+R}, \qquad t \geqslant 0,$$

or, in terms of the shift operator, $\widetilde{X}_t = X_t \circ \theta_R$. Observe that the post-$R$ chain need not be defined for $t = 0$, and its state space may be a proper subset of $\mathbb{I}$.

§ 9.2.2. Stopping time. Consider first the case when $R = T$ is a stopping time relative to a family $(\mathcal{F}_t)$. Then, the following theorem clearly holds.

THEOREM 1. *For any (finite) stopping time T, the chain started at T with $\widetilde{X}_t = X(t+T)$, $t \geqslant 0$, is again the usual M.C. with the same state space $\mathbb{I}$ and the same transition probabilities as the original M.C., and with the initial distribution*

$$\mathbb{P}(\widetilde{X}_0 = i) = \mathbb{P}(X_T = i), \qquad i \in \mathbb{I}.$$

The assertion is just the strong Markov property (as noted in § 1.1.1). Furthermore, the post-T chain is conditionally independent of events before T, given X_T, i.e.,

$$\mathbb{P}(A \cap B \mid X_T) = \mathbb{P}(A \mid X_T)\mathbb{P}(B \mid X_T) \quad \text{a.e.}$$

for every $A \in \mathcal{F}_T$ and $B \in \sigma(X_s,\ s \geqslant T)$.

In the special case when $T = T_H$ is the first entrance time to a set H, with the hitting distribution $\mathbf{B} = (D_{ij})$, the post-T chain continues with the same transition probabilities as the original chain, but its initial distribution $\pi\mathbf{B}$ is supported by H (here π is the initial distribution of X). Since T is finite, $\mathbb{P}(\widetilde{X}_0 \in H) = 1$. It is of interest to note that if π coincides with the ergodic distribution of X, so the absolute distribution of X, namely π, is invariant and the absolute distribution of $\widetilde{X}$, namely $\pi\mathbf{BP}(t)$, cannot be invariant.

See Example 1, § 5.4.2. Recall also that the post-ρ_i chain has been already used in connection with the minimal chain and exit taboo probabilities (see §§ 1.1.4, 2.2.4, and 2.2.5).

§ 9.2.3. Co-stopping time. For co-stopping times situation is more involved. It was resolved by P. A. Meyer, R. T. Smythe, and J. B. Walsh [1972] who considered a general Markov process started at a co-terminal time L, and showed its Markovian character. Alternative analysis was carried out by A. O. Pittenger and C. T. Shih [1973]; see also A. O. Pittenger [1981 a, 1981 b, 1982], J. Glover [1981], and P. W. Millar [1977]. The following is the summary of these results, adapted to the usual M.C.'s.

Let $X = (X_t, \mathcal{F}_t,\ 0 \leqslant t < \infty)$ be the original M.C. satisfying the usual conditions. For each co-terminal time L assumed finite, define the associated time T_L by

$$T_L = \inf\big(t\colon\ t > 0,\ L \circ \mathcal{R}_t > 0\big),$$

where $\mathcal{R}_t$ is the killing operator (see § 9.1.2). One can verify that T_L is a terminal time. Next, let $\overline{\mathbf{P}}(t) = (\bar{p}_{ij}(t))$ be a transition matrix of a M.C. $\overline{X}$ obtained by killing X at T_L. Hence, by Theorem 1, § 9.1.2,

$$\bar{p}_{ij}(t) = \mathbb{P}^i\big(X_t = j,\ t < T_L\big).$$

Observe that the function g defined by

$$g(i) = \mathbb{P}^i(L = 0) = \mathbb{P}^i\big(T_L = \infty\big), \qquad i \in \mathbb{I},$$

is $\overline{\mathbf{P}}(t)$-invariant.

Next define the following expression

$$\tilde{p}_{ij}(t) = \begin{cases} \bar{p}_{ij}(t)g(j)/g(i) & \text{when } g(i) > 0, \\ \delta_{\Delta j} & \text{when } g(i) = 0. \end{cases}$$

It is clear that $\widetilde{\mathbf{P}}(t) = (\tilde{p}_{ij}(t))$ is a semi-group on $\mathbb{I} \cup \{\Delta\}$ and $\widetilde{\mathbf{P}}(t)\mathbf{1} = \mathbf{1}$ for $g(i) > 0$. The main result is

THEOREM 1. *Let L be a (finite) co-terminal time (relative to the natural σ-fields of X). Then, the chain $\widetilde{X}$ started at L with $\widetilde{X}_t = X(L+t)$, $t > 0$, is again the usual* M.C. *with the state space $\mathbb{I} \cup \{\Delta\}$, and with $\widetilde{\mathbf{P}}(t)$ defined above, as its transition matrix.*

The matrices $\mathbf{P}(t)$ and $\widetilde{\mathbf{P}}(t)$ are different, and $\tilde{p}_{ij}(t)$ may be interpreted as refering to transitions of the original M.C. conditional on $L = 0$:

$$\tilde{p}_{ij}(t) = \mathbb{P}(X_t = j \mid X_0 = i,\ L = 0).$$

Observe that the post-L chain is defined for $t > 0$, although in the case of the usual M.C. definition can be extended to $t \geqslant 0$, with $\mathbb{P}(\widetilde{X}_0 = i) = \mathbb{P}(X_L = i)$. The σ-field $\mathcal{F}_L$ is defined as a collection of all sets $A \in \mathcal{F}_\infty$ such that for every $t \geqslant 0$, there exists $A_t \in \mathcal{F}_t$ such that $A(L \leqslant t) = A_t(L \leqslant t)$. Note that X_L is $\mathcal{F}_L$-measurable.

Furthermore, the post-L chain is conditionally independent of events prior to L, given X_L, i.e.,

$$\mathbb{P}(A \cap B \mid X_L) = \mathbb{P}(A \mid X_L)\mathbb{P}(B \mid X_L) \quad \text{a.e.}$$

for every $A \in \mathcal{F}_L$ and $B \in \sigma(X_s,\ s \geqslant L)$.

§ 9.2.4. Last exit time. In the special case when $L = L_M$ is the last exit time from a set H, with the exit distribution $\mathbf{\Gamma} = (\Gamma_{ij})$, the post-$L$ chain stays in H^c, and $L \circ \mathcal{R}_t = L_{t-}$ so $T_L = T_H$. Hence, $\bar{p}_{ij}(t) = {}_Hp_{ij}(t)$ and $g(i) = D_i^c$. Consequently, one has

THEOREM 1. *Let $L = L_H$ be the last exit time (assumed finite) from a set H. Then, the post-L chain is also a usual* M.C., *having transition probabilities of the form*

$$\tilde{p}_{ij}(t) = {}_Hp_{ij}(t)\, D_j^c / D_i^c, \qquad i, j \in H^c, \quad t > 0,$$

for $D_i^c > 0$, and with the initial distribution

$$\mathbb{P}(\widetilde{X}_0 = i) = \mathbb{P}(X_L = i) = \sum_k \pi_k \Gamma_{ki}, \qquad i \in H^c,$$

where π is the initial distribution of X.

Recall that $\Gamma_i(t)$ is the distribution function of L, with $\Gamma_i(0) = D_i^c$ and $\Gamma_i = 1$ (by assumption). Since D_i^c is ${}_H\mathbf{P}(t)$-invariant on H^c, so

$$\sum_{j \in H^c} \tilde{p}_{ij}(t) = 1, \qquad i \in H^c.$$

Moreover, $\tilde{p}_{ij}(t)$ may be interpreted as refering to the original chain, conditional not to hit the taboo set H:

$$\tilde{p}_{ij}(t) = \mathbb{P}\big(X_t = j \mid X_0 = i,\ L = 0\big), \qquad i, j \in H^c.$$

See Example 2, § 5.4.2, for an example of a post-L chain.

It is of interest to make the following remarks.

1. For those $i \in H^c$ for which $D_i^c = 0$, define (as in § 9.2.3) $\tilde{p}_{ij}(t) = \delta_{\Delta j}$. Thus, the state space in Theorem 1 is actually $H^c \cup \{\Delta\}$.

2. By assumption L is finite, so $s_i = 0$ in the Poisson decomposition (see § 5.2.1). Hence, H is an equilibrium set. Moreover, for $i \in H^c$, $D_i^c = \mathbb{P}^i(L = 0) > 0$ implies that $D_i = \mathbb{P}^i(0 < L < \infty) < 1$.

3. Observe that it is impossible to have H^c finite. Indeed, this would imply that $D_i^c = 0$ for all i (see § 5.2.3) and consequently $L = \infty$ ($s_i = 1$). Thus, H would be visited infinitely often. This however violates Remark 2 above. In particular, it is impossible for a single state which has been visited before time L, to become absorbing after L.

4. Theorem 1 indicates that for fixed i the measure $\tilde{p}_{i\bullet}(t)$ is absolutely continuous with respect to the measure ${}_Hp_{i\bullet}(t)$, with the Radon–Nikodym derivative D_j^c/D_i^c. Define now

$$M_t = D^c(X_t)/D^c(X_0), \qquad t \geqslant 0,$$

where (X_t) is the original M.C. with its natural σ-fields, and X_0 takes values in H^c with $D^c(X_0) > 0$. Then the following theorem holds.

THEOREM 2. *$M = (M_t)$ is a* MF *and a martingale (relative to the natural σ-fields of X) restricted to H^c, with $M_0 = 1$, $\mathbb{E}^i M_t = 1$, and $L = \sup(t\colon\ M_\infty \circ \theta_t = 0)$.*

Proof. It is clear that M_t is $\mathcal{F}_t$-measurable for each t. The relation

$$M_{t+h} = M_t\big(M_h \circ \theta_t\big)$$

is immediate from the definition. The fact that $D^c(\cdot)$ is ${}_H\mathbf{P}(t)$-invariant implies that

$$\mathbb{E}^i M_t = \mathbb{E}^i D^c(X_t)/D_i^c = {}_H\mathbf{P}(t)\, D_i^c/D_i^c = 1.$$

Thus, M in a non-contracting MF satisfying condition (m) from § 8.1.5. Similarly,

$$\begin{aligned} \mathbb{E}(M_{t+h} \mid \mathcal{F}_t) &= M_t \mathbb{E}(M_h \circ \theta_t \mid X_t) = \mathbb{E}\Big(D^c(X_{t+h}) \mid X_t\Big)/D^c(X_0) \\ &= D^c(X_t)/D^c(X_0) = M_t \qquad \text{a.e.,} \end{aligned}$$

so M is a positive martingale. Hence, by martingale convergence theorem $M_t \to M_\infty$ with probability one, and $M_t = \mathbb{E}(M_\infty \mid \mathcal{F}_t)$ as (M_t) is uniformly integrable.

Finally, if $t < L$, then there exists $s > t$ such that $X_s \in H$. Hence, $M_{s-t} \circ \theta_t = 0$, so $M_\infty \circ \theta_t = 0$. □

§ 9.2.5. Density argument. There is an obvious connection between the post-L chain with $L = L_H$ and the argument used in Corollary 6, § 2.2.5. As noted there, the co-stopping time L has been essentially identified with the first exit time ρ_r from the set H. It is therefore of interest to compare results obtained here with those in § 4.1.2.

THEOREM 1. *For $i \in \mathbb{I}$, $r \in H$, $j,k \in H^c$, and $L + s < t$ with $t, s > 0$:*

$$\mathbb{P}^i\big(X(L-) = r,\ X(L) = j,\ 0 < L \leqslant t,\ X(L+s) = k\big) = \int_0^t p_{ir}(\sigma)\, d\sigma\, q_{rj}\, D_j^c\, \tilde{p}_{jk}(s).$$

Proof follows by argument similar to that used in Theorem 6, § 4.1.2, to which it reduces when $s = 0$. Alternatively note that assertion may be written with the post-L chain $\widetilde{X}_0 = j$ and $\widetilde{X}_s = k$.

Hence, in particular, using Theorem 4, § 4.1.2, and Theorem 3, § 3.4.1, respectively, one has

COROLLARY 2. *For $i \in \mathbb{I}$, $k \in H^c$, and $s > 0$:*

a) *for $j \in H^c$ and $t \geqslant 0$:*

$$\mathbb{P}^i\big(X(L) = j,\ 0 \leqslant L \leqslant t,\ X(L+s) = k\big) = \Gamma_{ij}(t)\tilde{p}_{jk}(s);$$

b) *for $r \in H$ and $t > 0$:*

$$\mathbb{P}^i\big(X(L-) = r,\ 0 < L \leqslant t,\ X(L+s) = k\big) = \Pi_{ir}(t)\tilde{n}_{rk}(s),$$

where

$$\tilde{n}_{rk}(s) = n_{rk}(s) D_k^c / b_r, \qquad r \in H, \quad k \in H^c.$$

Observe that for fixed r, $\tilde{n}_{r\bullet}(s)$ forms the normalised entrance law for $\widetilde{\mathbf{P}}(t)$ on H^c:

$$\sum_{k \in H^c} \tilde{n}_{rk}(s)\, \tilde{p}_{kj}(t) = \tilde{n}_{rj}(s+t), \qquad \sum_{k \in H^c} \tilde{n}_{rk}(s) = 1.$$

Compare these formulae with those in § 3.4.1.

In the next theorem the time distance between indicated states is now a random variable $t - L$.

THEOREM 3. *For $i \in \mathbb{I}$, $k \in H^c$:*

a) *for $j \in H^c$ and $t \geqslant 0$:*

$$\mathbb{P}^i\big(X(L) = j,\ 0 \leqslant L \leqslant t,\ X_t = k\big) = \int_{0-}^t d\Gamma_{ij}(\sigma)\, \tilde{p}_{jk}(t - \sigma);$$

b) *for $r \in H$ and $t > 0$:*

$$\mathbb{P}^i\big(X(L-) = r,\ 0 < L \leqslant t,\ X_t = k\big) = \int_0^t d\Pi_{ir}(\sigma)\, \tilde{n}_{rk}(t - \sigma).$$

Observe that Theorem 3 (with $t > 0$ in a)) is exactly the same as Theorem 7, § 4.1.2.

The density argument used in § 4.1.2 is made precise by the conditional probabilities in the following way (see also § 4.2.2).

THEOREM 4. *For* $r \in H$, $j, k \in H^c$:

$$\tilde{p}_{jk}(t-\sigma) = \mathbb{P}(X_t = k \mid L = \sigma,\ X(L) = j),$$
$$\tilde{n}_{rk}(t-\sigma) = \mathbb{P}(X_t = k \mid L \doteq \sigma,\ X(L-) = r).$$

Remark. Corollary 2 and Theorem 3 agree with specialization of general results (expressed in terms of entrance laws) obtained by A. O. Pittenger and C. T. Shih [1973, Theorems 8 and 6].

9.3. STOPPING

The stopping operation is one of the oldest operations considered in stochastic processes, and is of great theoretical and practical importance. Here only brief comments will be made, with emphasis on stopping at the first entrance and at the last exit.

§ 9.3.1. Definitions. Let $X = (X_t, \mathfrak{F}_t,\ 0 \leqslant t < \infty)$ be the usual M.C. with a state space $\mathbb{I}$, a standard stochastic transition matrix $\mathbf{P}(t)$ and a conservative intensity matrix $\mathbf{Q}$, taken relative to a family $(\mathfrak{F}_t)$ of natural σ-fields on a probability space $(\Omega, \mathfrak{F}, \mathbb{P})$. Let T be a stopping time relative to this family $(\mathfrak{F}_t)$, and let $\mathfrak{F}_T$ be the associated "pre-T" σ-field. Recall that the compound r.v. X_T is $\mathfrak{F}_T$-measurable (see § 1.1.1).

Define new r.v.'s by

$$\widetilde{X}_t = \begin{cases} X_t & \text{when} \quad t < T, \\ X_T & \text{when} \quad t \geqslant T, \end{cases}$$

or $\widetilde{X}_t = X_{t\wedge T}$. Similarly define σ-fields

$$\widetilde{\mathfrak{F}}_t = \mathfrak{F}_{t\wedge T}, \qquad 0 \leqslant t < \infty.$$

The family $(\widetilde{\mathfrak{F}}_t)$ is increasing and $(\widetilde{X}_t)$ is adapted to $(\widetilde{\mathfrak{F}}_t)$.

The new chain $\widetilde{X} = (\widetilde{X}_t, \widetilde{\mathfrak{F}}_t,\ 0 \leqslant t < \infty)$ is said to be obtained by *stopping* the original chain X at the stopping time T (this is actually the origin of the name "stopping time"). Stopping should be distinguished from killing (considered in Sub-section 9.1); indeed, for $t \geqslant T$, the stopped chain stays at the value X_T whereas the killed chain disappears.

One may consider the stopping operator a_T such that $X_t \circ a_T = X_{t\wedge T}$. Note also that $\widetilde{\mathfrak{F}}_t$ is actually a class of all sets $A \in \mathfrak{F}_\infty$ such that, for every t,

$$A \cap (T > t) \in \mathfrak{F}_t.$$

It may be added that the σ-field $\mathfrak{F}_{T-}$ of events strictly before the stopping time T is the σ-field generated by $\mathfrak{F}_0$ and all sets of the form $A \cap (t < T)$, for all t and $A \in \mathfrak{F}_t$.

§ 9.3.2. First entrance time. Restrict now attention to the first entrance time $T = T_H$ to a set H. Hence, the stopping makes states in H absorbing ("cut-off method" in § 3.1.6), and the new chain stays in H^c for $t < T$, but X_T takes values in H only.

The main result is that the stopped chain is also a M.C. satisfying the usual conditions, and that its matrix $\widetilde{\mathbf{P}}(t)$ of transition probabilities

$$\tilde{p}_{ij}(t) = \mathbb{P}(\widetilde{X}_t = j \mid \widetilde{X}_0 = i)$$

coincides with the matrix $\mathbf{S}(t)$ defined in § 3.1.5. Thus,

$$\tilde{p}_{ij}(t) = {}_Hp_{ij}(t) + D_{ij}(t), \qquad i, j \in \mathbb{I}.$$

Observe that for $i \in H$, this reduces to δ_{ij}, and for $i \in H^c$ the terms ${}_Hp_{ij}(t)$ and $D_{ij}(t)$ vanish for $j \in H$ and $j \in H^c$, respectively. It has been shown in § 3.1.5 that $\mathbf{S}(t)$, $t \geqslant 0$, form a semi-group with $\mathbf{S}(0) = \mathbf{I}$, $\mathbf{S}(t)\mathbf{1} = \mathbf{1}$. As noted in § 3.1.6, one has here the decomposition

$$\begin{aligned}\mathbb{P}(\widetilde{X}_t = j \mid \widetilde{X}_0 = i) &= \mathbb{P}(X_t = j,\ T > t \mid X_0 = i)\\ &\quad + \mathbb{P}(X_T = j,\ T \leqslant t \mid X_0 = i) = {}_Hp_{ij}(t) + D_{ij}(t),\end{aligned}$$

noting that $\widetilde{X}_0 = X_0$. Indeed, the first term is just killing at T, and the second term represents the "over-kill".

The following theorem summarizes these results. Its proof may be obtained by adaptation of the general result for Markov processes; see, for example, E. B. Dynkin [1965, Thm. 10.2].

THEOREM 1. *Let X be the usual* M.C. *with a state space $\mathbb{I}$ and the semi-group $\mathbf{P}(t)$, $t \geqslant 0$. Then, the chain $\widetilde{X}$ stopped at the first entrance time $T = T_H$, defined above, is also Markovian satisfying the usual conditions, with the same state space $\mathbb{I}$ and the new semi-group $\widetilde{\mathbf{P}}(t)$ given by $\mathbf{S}(t)$, $t \geqslant 0$.*

Recall also that the first entrance probability D is invariant for $\mathbf{S}(t)$; see § 5.1.3.

§ 9.3.3. Last exit time. Stopping of a M.C. at a co-stopping time leads to some complications, so here attention will be restricted to consideration of the last exit time from a set.

Let $X = (X_t,\ 0 \leqslant t < \infty)$ be a M.C. satisfying the usual conditions, with a state space $\mathbb{I}$ and a semi-group $\mathbf{P}(t)$, $t \geqslant 0$. Let $L = L_H$ be the last exit time from a set H (see Section 4), and assume that L is finite. Define new r.v.'s by

$$\widetilde{X}_t = \begin{cases} X_t & \text{for } t < L,\\ X_L & \text{for } t \geqslant L,\end{cases} \qquad \text{or} \qquad \widetilde{X}_t = X_{t\wedge L}.$$

The new chain $\widetilde{X} = (\widetilde{X}_t,\ 0 \leqslant t < \infty)$ is said to be obtained by *stopping* the original chain X at the last exit from the set H. The new chain $\widetilde{X}$ visits all states in $\mathbb{I}$ before L, but after L it stays at X_L (indicating dependence on future). By right-continuity, X_L takes values in H^c, but X_{L-} in H. Nevertheless, as in the case of killing, the new chain $\widetilde{X}$ remains Markovian, but it is no longer time-homogeneous (i.e., its transition probabilities depend both on the length of the interval and its initial point). However, there is some kind of time homogeneity under restriction that the chain is not yet absorbed, as shown in Proposition 1 and its corollary.

Before proceeding further, it may be convenient to recall some distributional properties from § 4.1.2. Thus, for $i \in \mathbb{I}$,

$$\begin{aligned}\Gamma_{ij}(t) &= \mathbb{P}(X_L = j,\ L \leqslant t \mid X_0 = i),\\ \Gamma_{ij}(t) - \Gamma_{ij}(0) &= \mathbb{P}^i(X_L = j,\ 0 < L \leqslant t),\\ p_{ij}(t)D_j &= \mathbb{P}(X_t = j,\ L > t \mid X_0 = i),\\ \sum_j p_{ij}(t)D_j &= \mathbb{P}(L > t \mid X_0 = i),\\ \sum_j (\Gamma_{ij}(t) - \Gamma_{ij}(0)) &= \mathbb{P}^i(0 < L \leqslant t).\end{aligned}$$

Moreover, assumption that $\Gamma_i = \mathbb{P}^i(L < \infty) = 1$ implies that

$$D_i = \mathbb{P}^i(0 < L < \infty), \qquad D_i^c = \mathbb{P}^i(L = 0),$$

but it is impossible to have $D_i = 1$ identically on $\mathbb{I}$ (as noted in § 5.2.4). Recall also that

$$L \circ \theta_s = (L - s)^+, \qquad X_L \circ \theta_s = X_{s+L\circ\theta_s}.$$

PROPOSITION 1. *The chain $\widetilde{X}$ stopped at the last exit time L, as defined above, satisfies for all $i, j \in \mathbb{I}$ and $t \geqslant 0$, the equality*

$$\mathbb{P}(\widetilde{X}_{t+s} = j \mid X_s = i,\ s < L) = p_{ij}(t)\, D_j / D_i + (\Gamma_{ij}(t) - \Gamma_{ij}(0))/D_i,$$

where the RHS *does not depend on $s \geqslant 0$. Addition over all $j \in \mathbb{I}$ yields* 1.

Moreover,

$$\mathbb{P}(\widetilde{X}_{t+s} = j \mid X_s = i,\ L \leqslant s) = \delta_{ij}.$$

Proof. Write

$$\begin{aligned}\mathbb{P}(\widetilde{X}_{t+s} = j \mid X_s = i,\ s < L) &= \mathbb{P}(\widetilde{X}_{t+s} = j,\ t+s < L \mid X_s = i,\ s < L)\\ &\quad + \mathbb{P}(\widetilde{X}_{t+s} = j,\ t+s \geqslant L \mid X_s = i,\ s < L).\end{aligned}$$

The first term on the RHS corresponds to killing at L:

$$\begin{aligned}&\mathbb{P}(X_{t+s} = j,\ t+s < L \mid X_s = i,\ s < L)\\ &\quad = \mathbb{P}(X_t \circ \theta_s = j,\ t < L \circ \theta_s \mid X_0 \circ \theta_s = i)/\mathbb{P}(0 < L \circ \theta_s \mid X_0 \circ \theta_s = i)\\ &\quad = p_{ij}(t)\, D_j / D_i.\end{aligned}$$

The second term gives the "over-kill":

$$\begin{aligned}&\mathbb{P}(X_L = j,\ t+s \geqslant L \mid X_s = i,\ s < L)\\ &\quad = \mathbb{P}(X_L = j,\ s < L \leqslant t+s \mid X_s = i)/\mathbb{P}(s < L \mid X_s = i)\\ &\quad = \mathbb{P}(X_L \circ \theta_s = j,\ 0 < L \circ \theta_s \leqslant t \mid X_0 \circ \theta_s = i)/\mathbb{P}(0 < L \circ \theta_s \mid X_0 \circ \theta_s = i)\\ &\quad = (\Gamma_{ij}(t) - \Gamma_{ij}(0))/D_i.\end{aligned}$$

Hence, the required expression for $\mathbb{P}(\widetilde{X}_{t+s} = j \mid X_s = i,\ s < L)$ follows, independently of s. It is easy to see that addition over all $j \in \mathbb{I}$ yields 1.

On the other hand, one has evidently

$$\mathbb{P}\big(\widetilde{X}_{t+s} = j \mid X_s = i,\ s \geqslant L\big) = \delta_{ij}. \ \square$$

Observe that taking $s = 0$ in Proposition 1, one finds

$$\mathbb{P}\big(\widetilde{X}_t = j,\ t < L \mid X_0 = i,\ 0 < L\big) = p_{ij}(t)\, D_j / D_i,$$

which should be distinguished from the expression $p_{ij}(t) D_j$ stated above.

Conditional probabilities from Proposition 1 satisfy the restricted semi-group property in the following sense.

COROLLARY 2. *For the stopped chain:*

$$\begin{aligned} &p_{ij}(t+s)\, D_j/D_i + \big(\Gamma_{ij}(t+s) - \Gamma_{ij}(0)\big)/D_i \\ &\quad = \sum_k p_{ik}(s)\big(D_k/D_i\big)\big[p_{kj}(t)\, D_j/D_k + \big(\Gamma_{kj}(t) - \Gamma_{kj}(0)\big)/D_k\big] \\ &\qquad + \sum_k \delta_{kj}\big(\Gamma_{ik}(s) - \Gamma_{ik}(0)\big)/D_i. \end{aligned}$$

Proof. This follows immediately from the semi-group property of $(p_{ij}(t))$ and from Theorem 2, § 4.1.2. Probabilistic interpretation is supplied by Proposition 1. Consider the decomposition

$$\begin{aligned} &\mathbb{P}(\widetilde{X}_{t+s} = j \mid \widetilde{X}_0 = i,\ 0 < L) \\ &\quad = \sum_k \mathbb{P}\big(\widetilde{X}_{t+s} = j,\ \widetilde{X}_s = k,\ s < L \mid \widetilde{X}_0 = i,\ 0 < L\big) \\ &\qquad + \sum_k \mathbb{P}\big(\widetilde{X}_{t+s} = j,\ \widetilde{X}_s = k,\ L \leqslant s \mid \widetilde{X}_0 = i,\ 0 < L\big). \end{aligned}$$

The term in the first sum may be written in the form

$$\begin{aligned} &\mathbb{P}\big(\widetilde{X}_{t+s} = j \mid \widetilde{X}_s = k,\ s < L,\ \widetilde{X}_0 = i,\ 0 < L\big) \\ &\qquad \times \mathbb{P}\big(\widetilde{X}_s = k,\ s < L \mid \widetilde{X}_0 = i,\ 0 < L\big) \\ &\quad = \Big[p_{kj}(t)\, D_j/D_k + \big(\Gamma_{kj}(t) - \Gamma_{kj}(0)\big)/D_k\Big] p_{ik}(s) D_k/D_i, \end{aligned}$$

and the second term becomes

$$\begin{aligned} &\mathbb{P}\big(\widetilde{X}_{t+s} = j \mid \widetilde{X}_s = k,\ 0 < L \leqslant s,\ \widetilde{X}_0 = i\big) \\ &\qquad \times \mathbb{P}\big(\widetilde{X}_s = k,\ L \leqslant s \mid \widetilde{X}_0 = i,\ 0 < L\big) = \delta_{kj}\big(\Gamma_{ik}(s) - \Gamma_{ik}(0)\big)/D_i. \end{aligned}$$

Hence, the assertion follows. $\square$

9.4. REVERSAL

A reverse chain is obtained by reversing time in a M.C. at some random time R. Questions of importance concern conditions which must be satisfied by R to preserve Markov property, and the form of the new semi-group. The general theory will be illustrated here on reversal at the first entrance time and at the last exit time. The close relationship between reversal procedure and killing operation and duality becomes apparent.

§ 9.4.1. Definitions. Let $X = (X_t, \mathfrak{F}_t,\ 0 \leqslant t < \infty)$ be a direct (or original) M.C. satisfying the usual conditions. Let R be a random time such that R is finite and $R \circ \theta_t = (R-t)^+$. Recall that this condition is satisfied by terminal stopping times and by co-stopping times. R is called a *random reversal time*.

The general construction of reversal involves two steps: killing at the random time and the actual reversal at the killing time. Consider first a M.C. $\overline{X} = (\overline{X}_t, \overline{\mathfrak{F}}_t)$ obtained by killing the original chain X at time R (see § 9.1.1). Let $E \cup \{\Delta\}$ be the state space of $\overline{X}$, where E is a subset of $\mathbb{I}$ and Δ is the adjoint absorbing state. Write for $\overline{X}$ the stochastic matrix

$$\overline{\mathbf{P}}(t) = \big(\bar{p}_{ij}(t)\big), \qquad i,j \in E \cup \{\Delta\}, \quad t \geqslant 0.$$

Next, the *reverse chain* $\widetilde{X} = (\widetilde{X}_t, \widetilde{\mathfrak{F}}_t,\ 0 \leqslant t < \infty)$ is defined by

$$\widetilde{X}_t = \begin{cases} \overline{X}_{R-t} & \text{for } \ 0 < t \leqslant R, \\ \widetilde{\Delta} & \text{for } \ t > R, \end{cases}$$

where $\widetilde{\Delta}$ is an isolated state adjoint to the state space, and $\widetilde{\mathfrak{F}}_t$ are corresponding natural σ-fields.

Observe that the reverse chain takes values in the state space E, and is killed by assuming value $\widetilde{\Delta}$; it never takes value Δ. Moreover, the reverse chain is not defined at $t = 0$. The sample functions of $\widetilde{X}$ are those of $\overline{X}$ with t reversed in direction; hence they are left-continuous. See Fig. 9.4.1 and Fig. 9.4.2. below.

The fact of great importance is that $\widetilde{X}$ is a time-homogeneous Markov chain, and that its transition semi-group is

$$\widetilde{\mathbf{P}}(t) = \big(\tilde{p}_{ij}(t)\big), \qquad t \geqslant 0,$$

where

$$\tilde{p}_{ij}(t) = \mathbb{P}\big(\widetilde{X}_{t+s} = j \mid \widetilde{X}_s = i\big)$$

is expressed in a simple fashion in relation to $\overline{\mathbf{P}}(t)$. These results are stated in Theorem 1 after some preliminaries.

Let $\xi_i(\cdot)$ be the probability entrance law relative to $\overline{\mathbf{P}}(t)$ (see Def. 1, § 1.3.4), so, for $t \geqslant s$,

$$\mathbb{P}\big(\overline{X}_s = i,\ \overline{X}_t = j\big) = \xi_i(s)\bar{p}_{ij}(t), \qquad i,j \in E \cup \{\Delta\}.$$

Let

$$\xi(i) = \int_0^\infty \xi_i(t)\,dt < \infty.$$

Suppose now that there exists the exit law $f_i(\cdot)$ relative to $\overline{\mathbf{P}}(t)$, with $f_\Delta(t) = 0$, such that

$$\mathbb{P}(0 < R \leqslant t \mid \overline{X}_0 = i) = \int_0^t f_i(s)\,ds = F_i(t), \qquad i \in E,$$

with $F_i(0) = 0,\ F_i(\infty) = 1$; recall that $(\overline{X}_0 = i) = (X_0 = i,\ 0 < R)$.

Write

$$\phi(s+t) = \sum_{k \in E} \xi_k(s) f_k(t)$$

(see Thm. 6, § 1.3.4). It can be shown that the absolute distribution of R is

$$\mathbb{P}(s < R \leqslant s+t) = \int_s^{s+t} \phi(\tau)\,d\tau, \qquad s,t \geqslant 0,$$

with

$$\mathbb{P}(R = 0) = 1 - \int_0^\infty \phi(t)\,dt.$$

The following theorem specializes, to the usual M.C.'s considered here, the known results for general Markov processes obtained by authors mentioned in the introduction to this section.

THEOREM 1. *The reverse chain $\widetilde{X}$ is a time homogeneous* M.C. *with a state space $E \cup \{\tilde{\Delta}\}$, having transition probabilities given by*

$$\tilde{p}_{ij}(t) = \xi(j)\bar{p}_{ji}(t)/\xi(i), \qquad i,j \in E, \quad t \geqslant 0,$$

$$\tilde{p}_{i\tilde{\Delta}}(t) = \int_0^t \xi_i(s)\,ds/\xi(i), \qquad i \in E, \quad t \geqslant 0,$$

and the initial distribution

$$\mathbb{P}(\widetilde{X}_0 = i) = \xi(i) f_i(0) \qquad \text{for} \quad i \in E,$$

$$\mathbb{P}(\widetilde{X}_0 = \tilde{\Delta}) = 1 - \int_0^\infty \phi(s)\,ds.$$

Observe that the joint distribution for $0 < s \leqslant t$ and $i,j \in E$ has the form

$$\mathbb{P}(\widetilde{X}_s = i,\ \widetilde{X}_t = j) = \xi(j)\,\bar{p}_{ji}(t-s) f_i(s),$$

and the absolute distribution for $t > 0$ is

$$\mathbb{P}(\widetilde{X}_t = j) = \xi(j) f_j(t) \qquad \text{for} \quad j \in E,$$

$$\mathbb{P}(\widetilde{X}_t = \widetilde{\Delta}) = 1 - \int_t^\infty \phi(s)\, ds.$$

For the proof of the Markov property see P. A. Meyer [1968], or other references mentioned earlier. Verification of expressions for transition probabilities involves explicit calculation.

Note. The reverse chain may be made right-continuous by considering instead the chain $(\widetilde{X}_{t+})$.

§ 9.4.2. Reversal at *T*. Let $T = T_H$ be the first entrance time to a set H, and let $D_i(t)$ be its distribution (conditional on initial state i). Assume that T is finite, so $D_i = 1$ identically in i. The chain $\overline{X}$ killed at T has been discussed in Example 3, § 9.1.4, so $E = H^c$ and

$$\bar{p}_{ij}(t) = {}_H p_{ij}(t), \qquad i, j \in H^c.$$

The reverse chain $\widetilde{X}$ takes values in H^c (before being killed on entering $\widetilde{\Delta}$). Note that both $\widetilde{X}_T = X_0$ and $\widetilde{X}(0+) = X(T-)$ take values in H^c. See Fig. 9.4.1.

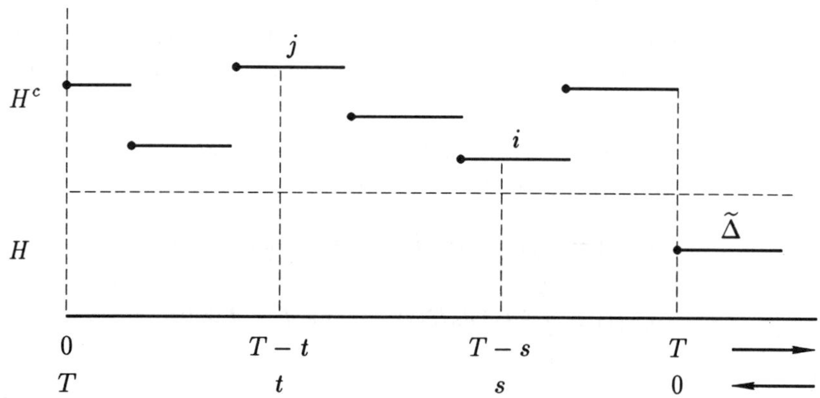

Figure 9.4.1. Reversal at T.

Let π be the initial distribution (supported on H^c) of the original chain X. Write for averages

$${}_H p_{\pi j}(t) = \sum_{i \in H^c} \pi_i \, {}_H p_{ij}(t), \qquad D_\pi(t) = \sum_{i \in H^c} \pi_i D_i(t),$$

and similarly for other functions.

It is clear now that

$$\xi_j(t) = {}_H p_{\pi j}(t), \qquad \xi(j) = {}_H u_{\pi j}, \qquad f_i(t) = d_i(t), \qquad \phi(t) = d_\pi(t),$$

and $\pi\, {}_H\mathbf{U}\mathbf{1} = \mathbb{E}T$. Hence, Theorem 1, § 9.4.1, yields now

$$\tilde{p}_{ij}(t) = {}_H u_{\pi j}\, {}_H p_{ji}(t) / {}_H u_{\pi i}, \qquad i, j \in H^c, \quad t \geqslant 0,$$

$$\tilde{p}_{i\tilde{\Delta}}(t) = \int_0^t {}_H p_{\pi i}(s)\, ds / {}_H u_{\pi i}, \qquad i \in H^c, \quad t \geqslant 0.$$

The absolute distribution is

$$\mathbb{P}(\widetilde{X}_t = j) = {}_H u_{\pi j}\, d_j(t), \qquad \mathbb{P}(\widetilde{X}_t = \widetilde{\Delta}) = D_\pi(t),$$

with the initial distribution (concentrated on H^c)

$$\mathbb{P}(\widetilde{X}_0 = i) = {}_H u_{\pi i}\, \bar{q}_i, \qquad i \in H^c.$$

§ 9.4.3. Reversal at *L*. Let $L = L_H$ be the last exit time from a set H, and let $\Gamma_i(t)$ be its distribution (conditional on the initial state i). Assume that L is finite, so $\Gamma_i = 1$ identically in i. The chain $\overline{X}$ killed at L has been discussed in § 9.1.6, so $E = \mathbb{I}$ and

$$\bar{p}_{ij}(t) = p_{ij}(t)\, D_j / D_i, \qquad i, j \in \mathbb{I}.$$

The reverse chain $\widetilde{X}$ takes values in $\mathbb{I}$ (before being killed on entering $\widetilde{\Delta}$). Note that

$$\widetilde{X}(L) = X_0 \in \mathbb{I}, \qquad \widetilde{X}(0+) = X(L-) \in H, \qquad X(L) \in H^c,$$

see Fig. 9.4.2.

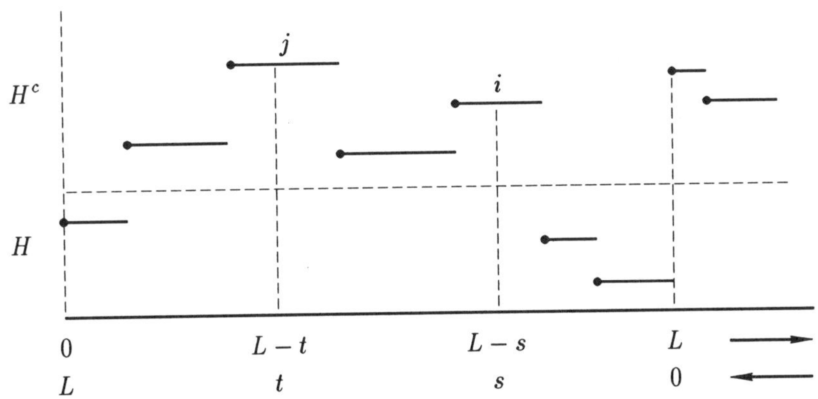

Figure 9.4.2. Reversal at L.

Let π be the initial distribution of the original chain X on $\mathbb{I}$. Write now for averages (as in § 8.1.1): $p_{\pi j}(t)$, $u_{\pi j}$, and

$$\Gamma_\pi(t) = \sum_i \pi_i \Gamma_i(t).$$

By assumption, $\Gamma_i = 1$ and $\Gamma_i(0) = D_i^c > 0$, so $\Gamma_i(t)$ has a jump at $t = 0$, and is differentiable for $t > 0$ with derivative $\gamma_i(t)$, as noted in § 4.1.1. By the argument in § 4.1.2, it follows that $\gamma_i(t)$ determines the exit law for $\mathbf{P}(t)$ such that

$$\int_0^\infty \gamma_i(t)\,dt = D_i, \qquad \gamma_i(0) = b_i, \qquad \gamma_i(t) = \mathbf{P}(t)b_i,$$

where b_i is the escape rate supported on H (see Thm. 4, § 4.1.1).

Consequently,

$$\xi_j(t) = p_{\pi j}(t)D_j, \qquad \xi(j) = u_{\pi j}D_j, \qquad f_i(t) = \gamma_i(t)/D_i, \qquad \phi(t) = \gamma_\pi(t).$$

Note that $f_i(t)$ is the density of L, conditional that $L > 0$:

$$F_i(t) = \frac{\Gamma_i(t) - \Gamma_i(0)}{1 - \Gamma_i(0)}.$$

Theorem 1, § 9.4.1, yields now

$$\tilde{p}_{ij}(t) = u_{\pi j}\, p_{ji}(t)/u_{\pi i}, \qquad i, j \in \mathbb{I}, \quad t \geqslant 0,$$

$$\tilde{p}_{i\tilde{\Delta}}(t) = \int_0^t p_{\pi i}(s)\,ds/u_{\pi i}, \qquad i \in \mathbb{I}, \quad t \geqslant 0.$$

The absolute distribution is

$$\mathbb{P}(\widetilde{X}_t = j) = u_{\pi j}\,\gamma_j(t), \qquad \mathbb{P}(\widetilde{X}_t = \widetilde{\Delta}) = \Gamma_\pi(t),$$

with the initial distribution

$$\mathbb{P}(\widetilde{X}_0 = i) = u_{\pi i} b_i, \qquad \mathbb{P}(\widetilde{X}_0 = \widetilde{\Delta}) = D_\pi^c.$$

Remark. It has been assumed here that $u_{\pi j} < \infty$ for each $j \in \mathbb{I}$ (transient chain). Recall that the measure $\pi\mathbf{U}$ on $\mathbb{I}$ is $\mathbf{P}(t)$-excessive and σ-finite (see § 1.3.5).

§ 9.4.4. Reversal at the life time. When R is the life time of a M.C. $X = (X_t)$, i.e., $R = \inf(t\colon\ t > 0,\ X_t = \Delta)$, then, by definition, X and the killed process $\underline{X}$ coincide. Their common transition probabilities yield

$$F_i(t) = \mathbb{P}^i(0 < R \leqslant t) = \mathbb{P}^i(X_t = \Delta) = 1 - \sum_j p_{ij}(t), \qquad i \in \mathbb{I}, \quad t \geqslant 0.$$

Furthermore, let the absolute distribution be

$$\mathbb{P}(X_t = j) = \xi_j(t), \qquad j \in \mathbb{I}, \quad t \geqslant 0,$$

so the potential measure is

$$\xi(j) = \int_0^\infty \mathbb{P}(X_t = j)\, dt, \qquad j \in \mathbb{I}.$$

Hence, by Theorem 1, § 9.4.1, the transition probabilities of the reverse chain $\widetilde{X} = (\widetilde{X}_t)$ are

$$\tilde{p}_{ij}(t) = \xi(j)\, p_{ji}(t)/\xi(i), \qquad i, j \in \mathbb{I}, \quad t \geqslant 0,$$

$$\tilde{p}_{i\tilde{\Delta}}(t) = \int_0^t \xi_i(s)\, ds/\xi(i), \qquad i \in \mathbb{I}, \quad t \geqslant 0,$$

with the joint distribution

$$\mathbb{P}(\widetilde{X}_s = i,\ \widetilde{X}_t = j) = \xi(j)\, p_{ji}(t - s)\, f_i(s), \qquad i, j \in \mathbb{I},$$

and the absolute distribution

$$\mathbb{P}(\widetilde{X}_t = j) = \xi(j)\, f_j(t), \qquad j \in \mathbb{I}.$$

This case, for general Markov processes, was discussed at length by K. L. Chung and J. B. Walsh [1969]; see also D. Williams [1979, p. 201]. It should be mentioned here that K. L. Chung [1967, p. 265] considered the reversal of a M.C. killed at the first infinity τ (cf. §§ 1.1.3 and 1.1.5), with $\mathbf{P}(t)$ the minimal matrix (necessarily sub-stochastic). Then, $f_i(t) = l_i(t)$, $i \in \mathbb{I}$, as defined in § 12.2.1, and with the initial distribution π:

$$\xi_i(t) = p_{\pi j}(t), \qquad \xi(i) = u_{\pi i}, \qquad i \in \mathbb{I}.$$

Note. An extension of the above procedure is obtained by taking for ξ more general measures. Let μ be a measure on $\mathbb{I}$ and take

$$\xi_j(t) = \mathbb{P}^\mu(X_t = j) = \mu\mathbf{P}(t)(j), \qquad \xi(j) = \mu\mathbf{U}(j)$$

(see § 8.1.1 for notation). Then transition probabilities are again of the above form.

§ 9.4.5. Duality. Let now ξ be an arbitrary strictly positive, σ-finite, $\mathbf{P}(t)$-excessive measure (a reference measure). The dual of a semi-group $\mathbf{P}(t)$, $t \geqslant 0$, relative to ξ has been defined (in § 1.3.5) as a semi-group $\widehat{\mathbf{P}}(t)$, $t \geqslant 0$, such that

$$\hat{p}_{ij}(t) = \xi(j)\, p_{ji}(t)/\xi(i), \qquad i, j \in \mathbb{I}.$$

The matrix $\widehat{\mathbf{P}}(t)$ is sub-stochastic, but can be extended to a stochastic matrix by enlarging the state space in the usual way.

Let $X = (X_t)$ be the usual M.C. with a stochastic transition matrix $\mathbf{P}(t)$. The *dual chain* is now defined as a M.C. $\widehat{X} = (\widehat{X}_t)$ with $\widehat{\mathbf{P}}(t)$ as its transition matrix. Thus, two M.C.'s X and $\widehat{X}$ are said to be in duality relative to ξ if their semi-groups are in duality relative to ξ. Observe that the Markov property is a part of definition, and no reversal is pre-supposed. However, in view of interpretation given above, the term "reverse chain" is also used. For earlier discussion of duality see Sub-section 6.3. Here only some comments will be added. Although the dual chain can be interpreted as the reverse chain, the probabilistic relationship between these two processes may be complicated.

For simplicity of discussion consider the reversal at the terminal time R, especially the life time of the M.C. X. The original chain X is thus killed at R, so without loss of generality one may assume that semi-groups $\mathbf{P}(t)$ and $\overline{\mathbf{P}}(t)$ coincide.

From the observation of sample functions it is clear that the last exit time $\widehat{L}$ and the first entrance time $\widehat{T}$ for the reverse chain $\widehat{X}$ correspond, respectively, to the first entrance time T and the last exit time L for the original chain X (see Fig. 9.4.3), and $T + \widehat{L} = R = L + \widehat{T}$.

As shown in Theorem 2, § 6.3.3, the dual of the first entrance decomposition for the direct chain is the last exit decomposition for the reverse chain. Consider now the dual chain $\widehat{X}$ with balayage $\widehat{\mathbf{B}}^\alpha = (\hat{d}^\alpha_{ij})$, so

$$\hat{d}^\alpha_{ij} = \sum_{k \in H^c} {}_H\hat{u}^\alpha_{ik}\, \hat{q}_{kj} = \text{dual}\,(n^\alpha_{ji}), \qquad i \in H^c, \quad j \in H,$$

and recall that, for example,

$$\widehat{\mathbf{B}} f(i) = \mathbb{E}^i\big(f(\widehat{X}(\widehat{T})),\ \widehat{T} < \infty\big).$$

Then, the dual of $\widehat{\mathbf{B}}^\alpha$ is given by the co-balayage $\mathbf{C}^\alpha$ for the direct chain, namely,

$$\xi(j)\hat{d}^\alpha_{ji}/\xi(i) = \sum_{k \in H^c} q_{ik}\, {}_H u^\alpha_{kj} = n^\alpha_{ij}, \qquad i \in H, \quad j \in H^c.$$

See §§ 3.4.1 and 6.3.3. In the same fashion, say, distribution of the post-L chain could be obtained by reversal of the dual chain at $\widehat{T}$.

It is of interest to make the following observations. Let A be an AF of a direct chain X, having the Revuz measure μ_A (as defined § 8.1.3-e). Recall from § 1.3.5 that the function $\mathbf{U}^\alpha \mu_A$ is the density (with respect of a reference measure) of the measure $\mu_A \widehat{\mathbf{U}}^\alpha$. Hence, Theorem 7, § 8.1.3, gives

$$\xi(j)\, u^\alpha_A(j) = \mu_A\, \widehat{\mathbf{U}}^\alpha(j).$$

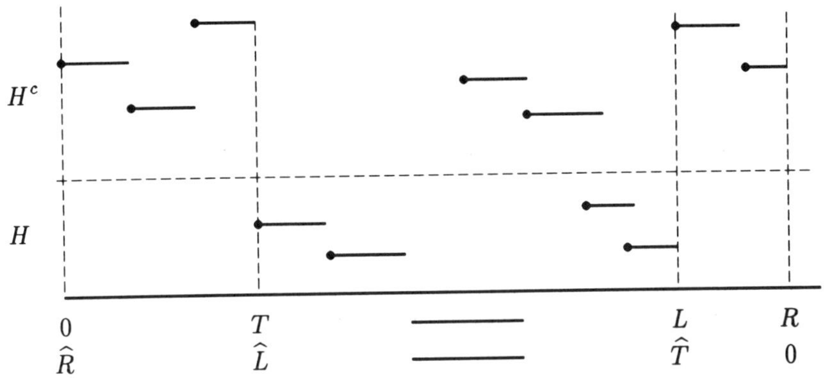

$X(T) \in H$ | right-continuous paths | $X(L) \in H^c$
$X(T-) \in H^c$ | | $X(L-) \in H$

$\hat{X}(\hat{L}) \in H$ | left-continuous paths | $\hat{X}(\hat{T}) \in H^c$
$\hat{X}(\hat{L}+) \in H^c$ | | $\hat{X}(\hat{T}+) \in H$

$$\hat{X}_t = X(R-t), \qquad 0 < t \leqslant R$$

Figure 9.4.3. Duality.

Consider now two integral functionals determined by the same charge $a \geqslant 0$, but taken relative to the direct chain X and the dual chain $\hat{X}$:

$$A_t = \int_0^t a(X_s)\, ds, \qquad \hat{A}_t = \int_0^t a(\hat{X}_s)\, ds, \qquad t \geqslant 0.$$

It is easy to verify that both functionals have the same Revuz measure $\xi(j)a(j)$, $j \in \mathbb{I}$. For further discussion of duality see papers by J. Mitro [1979 a, 1979 b].

Remark. The dual chain may be made right-continuous by considering $\hat{X}_{t+}$. Intuitive interpretation of duality as reversal may be obtained by reversing time at the origin: $\hat{X}_t = X_{-t}$ for $t \geqslant 0$.

9.5. TRACE CHAIN

In several applications it is required to restrict attention to a M.C. when it is in the set H, and disregard its presence in the complementary set H^c. A new chain obtained in this fashion from the original M.C. (with a state space $\mathbb{I}$) by elimination of excursions to H^c is called the *trace chain*. Clearly, the state space of the trace chain is the set H.

Although the trace chain is obtained by a special form of a random time change (discussed in the next Sub-section 9.6), it is instructive to discuss it separately.

This approach will explain better the role of successive entrances to sets H and H^c, of crucial importance for random time change.

§ 9.5.1. Definitions. Let $X = (X_t,\ 0 \leqslant t < \infty)$ be the usual M.C. with a state space $\mathbb{I}$, a stochastic matrix $\mathbf{P}(t)$ and a conservative intensity matrix $\mathbf{Q}$. Consider an integral functional $V = (V_t)$ defined by

$$V_t = \int_0^t I_H(X_s)\, ds, \qquad t \geqslant 0.$$

This is the time spent in the set H (occupation time); see § 8.4.1. Note that support of IF V is H, and that $V_\infty \leqslant \infty$. Recall also from Section 2 that $(t\colon\ X_t \in H)$ is the sojourn time in the set H.

Consider now the inverse $\tau = (\tau_t)$ of V, defined in § 8.1.2 as $\tau_t = \inf\,(s\colon\ V_s > t)$ for each $0 \leqslant t < V_\infty$. The inverse τ_t represents the actual time needed to produce time t spent in the set H.

Clearly, τ is right-continuous and *strictly increasing* on the interval $[0, V_\infty)$, and $(\tau_t < s) = (V_s > t)$ with $\tau_0 = T$ (the first entrance time to H). Moreover, τ_t is a stopping time for each t, and

$$\tau_{t+h} = \tau_t + \tau_h \circ \theta_{\tau_t} \qquad \text{on} \quad \tau_t < \infty,$$

$$X(\tau_h) \circ \theta_{\tau_t} = X(\tau_{t+h}), \qquad h, t \geqslant 0,$$

and

$$V(\tau_t) = t, \qquad \tau(V_t) \geqslant t.$$

The *trace chain* $\widetilde{X} = (\widetilde{X}_t)$ is now defined by $\widetilde{X}_t = X(\tau_t)$, $0 \leqslant t < V_\infty$. Its state space is H and its life time is V_∞ corresponding to $(\tau_t < \infty) = (V_\infty > t)$. It is known that $\widetilde{X}$ is also usual M.C. (see Thm. 1, § 9.6.1). Its transition probabilities are

$$\begin{aligned}\tilde{p}_{ij}(t) &= \mathbb{P}(\widetilde{X}_t = j \mid \widetilde{X}_0 = i)\\ &= \mathbb{P}\big(X(\tau_t) = j,\ \tau_t < \infty \mid X(\tau_0) = i\big), \qquad i, j \in H, \quad t \geqslant 0.\end{aligned}$$

Note that the matrix $\widetilde{\mathbf{P}}(t) = (\tilde{p}_{ij}(t))$ is sub-stochastic:

$$\sum_{j \in H} \tilde{p}_{ij}(t) = \mathbb{P}(\tau_t < \infty \mid X_T = i) = \mathbb{P}(V_\infty > t \mid X_T = i), \qquad i \in H.$$

It may be made stochastic on $H \cup \{\Delta\}$ by adjoining the absorbing state Δ in the usual way.

Note that for $t = 0$, $\tau_0 = T$, so $\widetilde{X}_0 = X_T$ on $T < \infty$, with X_T taking values in H. Hence, the initial distribution of $\widetilde{X}$ is supported on H. Observe also that

$$\begin{aligned}&\mathbb{P}\Big(X(\tau_t) = j,\ \tau_t < \infty \mid X_0 = i\Big)\\ &\quad = \sum_{k \in H} \mathbb{P}\Big(X(\tau_t) = j,\ \tau_t < \infty,\ X(\tau_0) = k \mid X_0 = i\Big)\\ &\quad = \sum_{k \in H} D_{ik}\, \tilde{p}_{kj}(t) = \tilde{p}_{ij}(t)\end{aligned}$$

for i and j in H. Therefore, conditioning on $X_T = i$ may be replaced by conditioning on $X_0 = i$ when $i \in H$.

Furthermore, $X(\tau_t)$ is in fact a compound r.v. with $\tau_t < \infty$; see § 9.6.1 for further comments. Note also that the chain $(\widetilde{X}_t)$ was treated extensively in Subsection 8.4 (on local time) from another point of view. In particular, Theorem 1, § 8.4.2, indicates that $\widetilde{\mathbf{P}}(t)$, $t \geqslant 0$, form a semi-group.

§ 9.5.2. Sample functions. It is of interest to examine more closely the behavior of paths of the trace chain $\widetilde{X}$; see Fig. 9.5.1–9.5.4 below. First, recall random sets for the original chain X (introduced in § 2.1.1), namely the set W_H of consecutive exits from H and the set V_H of consecutive entrances to H. Clearly, V_H is the set of strictly positive left-end points of intervals forming the sojourn set S_H, and W_H is the set of strictly positive left-end points of intervals contiguous to S_H. (For further discussion of consecutive entrances to H and to H^c, see § 9.6.2.)

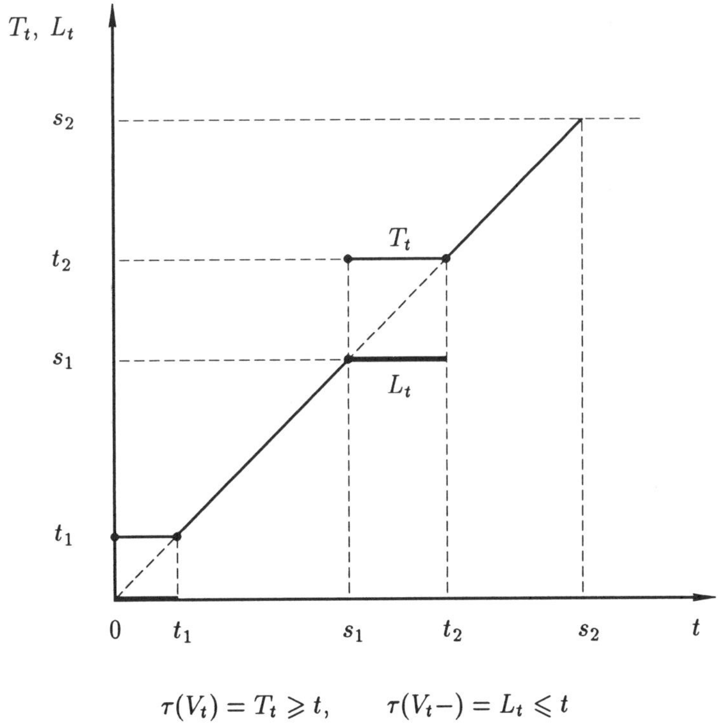

Figure 9.5.1. Comparison.

As noted in §§ 2.1.2 and 2.1.4, W_H coincides with discontinuities of an increasing process (T_t), where T_t is the first entrance time to H after t, and V_H coincides with discontinuities of an increasing process (L_t), where L_t is the last exit time from H before t (see Fig. 2.4 and Fig. 2.8).

Observe that if t is a discontinuity point of the increasing process (τ_t), then for $t > 0$, τ_{t-} lies in W_H, but τ_t lies in V_H (see Fig. 8.3 and Fig. 8.5). Moreover, the

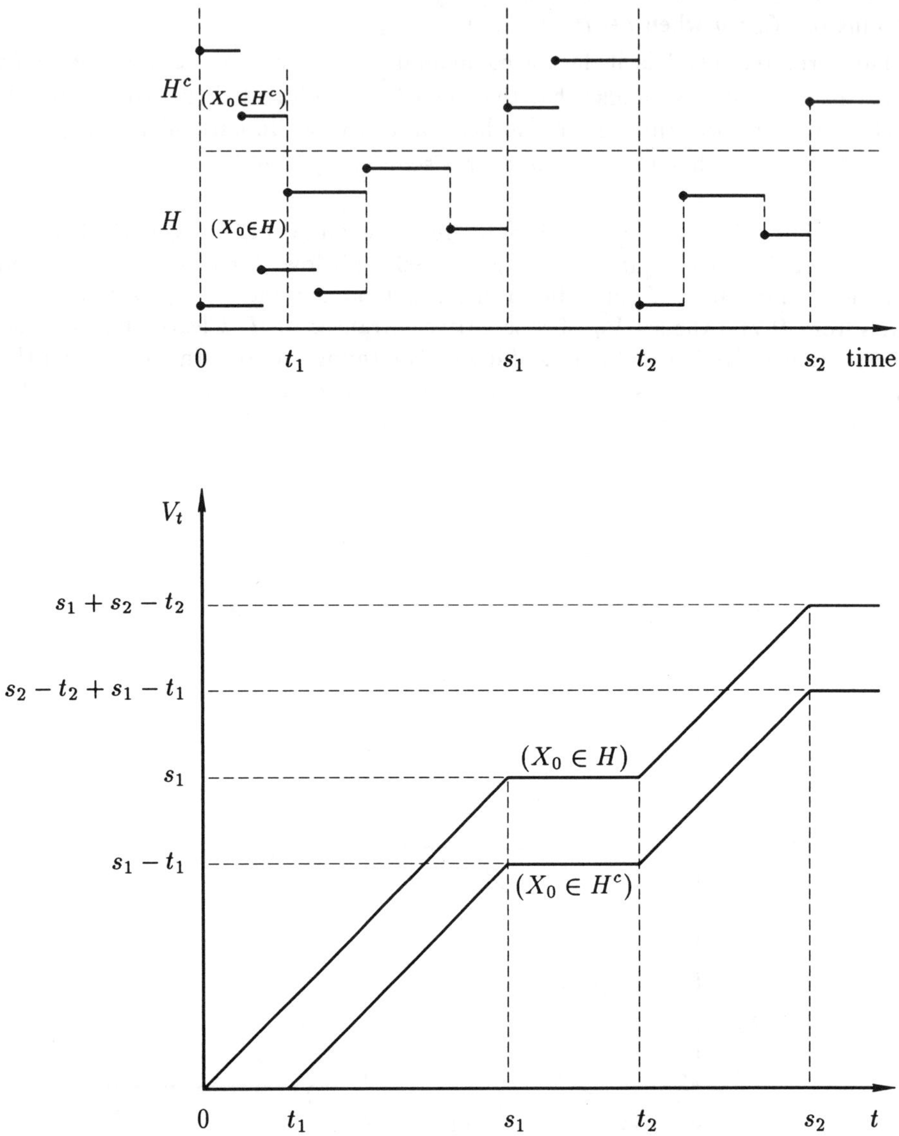

$$V_t = \int_0^t I_H(X_s)\, ds$$

Figure 9.5.2. Occupation time.

intervals contiguous to random intervals $[\tau_{t-}, \tau_t)$ in the range of τ correspond to sojourn times in the set H. Evidently, for $t > 0$,

$$\tau(V_t) = T_t \geqslant t, \qquad \tau(V_{t-}) = L_t \leqslant t.$$

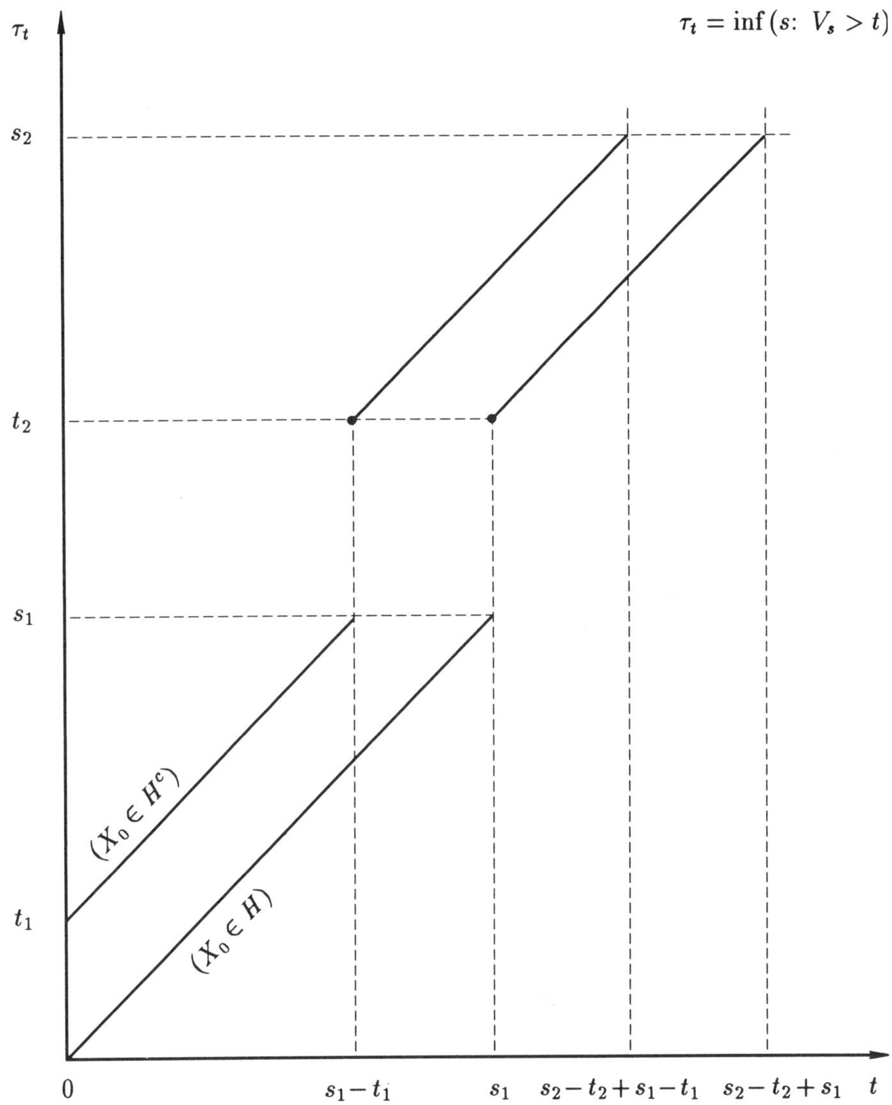

Figure 9.5.3. The inverse of the occupation time.

Although processes (T_t) and (τ_t) are different, there is a close similarity in their behavior with respect to the set H.

It is now clear that sample functions of the process $\widetilde{X}$ (necessarily taking values in the set H) remain the same as those of the original chain X, but with elimination of excursions to H^c. In particular, in the present case durations of visits to each state in H are unchanged.

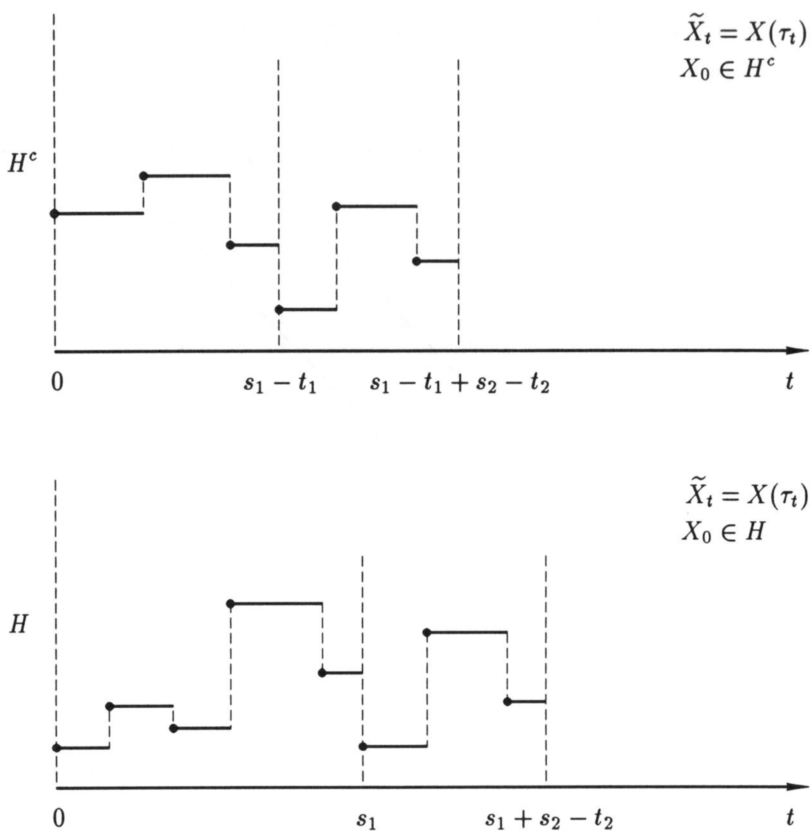

Figure 9.5.4. The trace chain.

§ 9.5.3. Resolvent. Assuming that u_{ij} are finite, one has, for a positive function f on H,

$$\widetilde{\mathbf{U}}f(i) = \mathbb{E}^i \int_0^{V_\infty} f(\widetilde{X}_t)\, dt = \mathbb{E}^i \int_0^\infty f(X_s)\, dV_s$$

$$= \mathbb{E}^i \int_0^\infty f(X_s) I_H(X_s)\, ds = \sum_{j \in H} u_{ij} f(j), \qquad i \in H,$$

by transformation $s = \tau_t$, and by comments at the end of § 9.5.1. Hence,

$$\tilde{u}_{ij} = u_{ij} \qquad \text{for} \quad i, j \in H.$$

Note that this result follows also from Theorem 2, § 8.4.2.

However, for $\alpha > 0$, the resolvent $\widetilde{\mathbf{U}}^\alpha$ can be expressed as a potential of some extended functional (see § 9.6.4).

Calculating the mean in two ways, one has

$$\mathbb{E}^i V_\infty = \mathbb{E}^i \int_0^\infty I_H(X_s)\,ds = \sum_{j\in H} \int_0^\infty p_{ij}(s)\,ds = \sum_{j\in H} u_{ij},$$

$$\mathbb{E}^i V_\infty = \int_0^\infty \mathbb{P}^i(V_\infty > t)\,dt = \sum_{j\in H} \int_0^\infty \tilde{p}_{ij}(t)\,dt = \sum_{j\in H} \tilde{u}_{ij},$$

but $p_{ij}(s)$ and $\tilde{p}_{ij}(s)$ are different.

For discussion of potentials see § 8.4.2.

§ 9.5.4. Generator. To obtain the generator $\widetilde{\mathbf{Q}} = (\tilde{q}_{ij})$ for the trace chain, one must evaluate the limit

$$\begin{aligned}\tilde{q}_{ij} &= \lim_{t\to 0} \tilde{p}_{ij}(t)/t \\ &= \lim_{t\to 0} \mathbb{P}\big(X(\tau_t) = j,\ \tau_t < \infty \mid X(\tau_0) = i\big)/t, \qquad i \neq j, \quad i,j \in H.\end{aligned}$$

Observe that as $\tau_t \to \tau_0$, when $t \to 0$,

$$\widetilde{X}(t) \longrightarrow \widetilde{X}(0) = X_T,$$

so $X(\tau_t)$ traces the original chain (X_s) through H as s approaches T from the right, because by construction the trace chain has the same paths as the original chain.

Recall now the increasing process (T_t), where T_t is the first entrance time to H after t. Lemma 5, § 2.1.2, yields

$$T_{T+t} = T + T_t \circ \theta_T \longrightarrow T, \qquad X(T_{T+t}) \longrightarrow X(T_T) = X_T \qquad \text{as} \quad t \to 0.$$

See graph of T_t in Fig. 9.5.1 and note that graphs of both T_t and τ_t are linear functions when sample functions are in H. Hence, $X(T_{T+t}) = X(\tau_t)$ as t approaches 0.

Therefore, for $i \neq j$ and $i, j \in H$,

$$\begin{aligned}\mathbb{P}\big(X(T_{h+T}) = j \mid X(T) = i\big) &= \mathbb{P}\big(X(T + T_h \circ \theta_T) = j \mid X(T) = i\big) \\ = \mathbb{P}\big(X(T_h) \circ \theta_T = j \mid X_0 \circ \theta_T = i\big) &= \mathbb{P}\big(X(T_h) = j \mid X_0 = i\big)\end{aligned}$$

by the strong Markov property.

Hence, the required limit is equal to

$$\begin{aligned}\tilde{q}_{ij} &= \lim_{h\to 0} \frac{1}{h}\, \mathbb{P}\big(X(T_h) = j \mid X_0 = i\big) = \lim_{h\to 0} \frac{1}{h}\, D^h_{ij} \\ &= \lim_{h\to 0} \frac{1}{h} \sum_k p_{ik}(h) D_{kj} = \sum_k q_{ik}\, D_{kj}\end{aligned}$$

in view of Theorem 5, § 3.2.2. Similar analysis shows that $\tilde{q}_i = -\tilde{q}_{ii}$ is also given by the above expression for $i = j$.

The following theorem summarizes the result.

THEOREM 1. *Let X be the usual* M.C. *with the generator (intensity matrix)* $\mathbf{Q}$ *and the state space* $\mathbb{I}$. *Let* $\widetilde{X}$ *be the trace chain obtained from* X *by restriction of transitions to subset* H *as defined above. Then the generator* $\widetilde{\mathbf{Q}}$ *of the* M.C. $\widetilde{X}$ *is given by*

$$\tilde{q}_{ij} = q_{ij} + \sum_{k \in H^c} q_{ik}\, D_{kj} \qquad \text{for} \quad i, j \in H,$$

or, in matrix notation (subscripts indicating restriction to H*),*

$$\widetilde{\mathbf{Q}}_{22} = \mathbf{Q}_{22} + \mathbf{Q}_{21}\, \mathbf{B}_{12}$$

where $\mathbf{B}_{12} = (D_{ij})$ *is the balayage matrix (first entrance to* H*) for the chain* X.

Theorem 1 connects the trace chain (obtained by the random time change through occupation time V) with properties of the matrix $\mathbf{QB}$ discussed in §§ 3.1.3 and 3.2.5, and provides additional insight into discussion in Section 3. The following comments explore this connection.

By Lemma 4, § 3.1.3, and Theorem 2, § 3.3.3, one has, for i and j in H,

$$\begin{aligned} \tilde{q}_{ij} &= q_i\, D^{\rho}_{ij} = -b_{ij} \geqslant 0 \qquad \text{for} \quad i \neq j, \\ \tilde{q}_i &= q_i (1 - D^{\rho}_{ii}). \end{aligned}$$

The matrix $\widetilde{\mathbf{Q}}_{22}$ is not conservative:

$$\sum_{j \in H} \tilde{q}_{ij} = -\sum_{k \in H^c} q_{ik}\, D^c_k = -b_i \leqslant 0, \qquad i \in H,$$

where $b_i = q_i D^{\rho c}_i$ is the escape rate (Thm. 2, § 3.1.3). Hence, extension to $H \cup \{\Delta\}$ yields $\tilde{q}_{i\Delta} = b_i$.

Next recall that in the matrix $\mathbf{QB}$ the only non-zero component coincides with $\widetilde{\mathbf{Q}}_{22}$ from Theorem 1, and note that the term $\mathbf{Q}_{22}$ corresponds to the original transitions within H and the term $\mathbf{Q}_{21}\mathbf{B}_{12}$ accounts for jumps to H^c and subsequent (first) returns to H. As already noted, consecutive returns to H occur at discontinuity points of τ.

Observe also that Theorem 1 is the special case of Theorem 1, § 8.4.3.

§ 9.5.5. Comments. Concluding discussion of the trace chain, it may be useful to recall connections with some earlier discussions.

1. The generator of the form $\mathbf{Q}_{21}\mathbf{B}_{12} + \mathbf{Q}_{22}$ occurs also in the perturbation method with the balayage $\mathbf{B}$ in the role of the replacement matrix. See Example 3, § 5.4.2, and § 11.2.3.

2. Consider the balayage matrix $\mathbf{B}^{\rho} = \mathbf{RB}$ introduced in § 3.3.5. Using relation between the jump matrix $\mathbf{R}$ and the generator $\mathbf{Q}$ (stated in § 1.1.4), one has

$$\mathbf{QB} = \mathbf{\Theta}(\mathbf{B}^{\rho} - \mathbf{B})$$

and its non-zero component is

$$\widetilde{\mathbf{Q}}_{22} = \mathbf{\Theta}_2 (\mathbf{B}^{\rho}_{22} - \mathbf{I}_2),$$

which is the matrix form of relations stated in § 9.5.4, after Theorem 1.

3. Consider the balayage matrix $\mathbf{B}^t = \mathbf{P}(t)\mathbf{B}$ introduced in § 3.2.5. Then Theorem 2, § 3.2.3, indicates that

$$\mathbf{QB} = \left.\frac{d\mathbf{B}^t}{dt}\right|_{t=0}.$$

4. Using the first entrance decomposition (Cor. 2, § 6.1.1) and the duality relation (Thm. 1, § 6.3.2), under the assumption that $\mathbf{U}$ is finite, one has the interesting relation

$$\widetilde{\mathbf{Q}}_{22}\,\mathbf{U}_{22} = \mathbf{U}_{22}\,\widetilde{\mathbf{Q}}_{22} = -\mathbf{I}_2.$$

5. As already noted, this sub-section should be compared with Sub-section 8.4 on local time.

9.6. RANDOM TIME CHANGE

Transformation by a random time change (or time substitution) is a well-known device in general Markov processes, but only recently has been applied to M.C.'s. Basically such a transformation is a kind of randomisation of time, "a change of clock" in watching development of a process. This is achieved with the help of the inverse of an additive functional of the process, and requires proof that the transformed process is still Markovian. The distributional properties of this new process are of interest, but are known explicitely only in some special cases.

For the general account of a random time change determined by continuous additive functionals see R. M. Blumenthal and R. K. Getoor [1968], and E. B. Dynkin [1965] for strictly increasing CAF's. See also Y. Le Jan [1979], J. Lamperti [1967], C. Dellacherie and P. A. Meyer [1982]. Random time change has been recently studied extensively in series of papers by H. Gzyl [1977, 1980], J. Mitro [1986 a, b], J. Glover [1981, 1983 a, b], H. Kaspi [1988]. Several references will be given later.

The special structure of the usual M.C.'s provides significant simplification of the general theory. Yet the argument is fairly complex although intuitively direct. Presentation here is the improved version of a report by R. Syski [1978].

§ 9.6.1. Definitions. Let $X = (X_t, \mathcal{F}_t,\ 0 \leqslant t < \infty)$ be a M.C. satisfying the usual conditions, with a state space $\mathbb{I}$, a transition matrix $\mathbf{P}(t)$ and a generator $\mathbf{Q}$. Let $A = (A_t,\ 0 \leqslant t < \infty)$ be an integral functional of X defined by

$$A_t = \int_0^t a(X_s)\, ds, \qquad t \geqslant 0,$$

where a is a non-negative bounded function on $\mathbb{I}$, which vanishes only on H^c. Thus, the set $H = (i\colon\ a(i) > 0)$ is the support of A (see § 8.1.2).

Let $\tau = (\tau_t,\ 0 \leqslant t < \infty)$ be the inverse of A defined by $\tau_t = \inf\,(s\colon\ A_s > t)$, see § 8.1.2. (The symbol τ is now used instead of R, in agreement with classical notation.)

Clearly, τ is right-continuous and *strictly increasing* on the interval $[0, A_\infty)$, and $(\tau_t < s) = (A_s > t)$ with $\tau_0 = T$ (the first entrance time to H). Moreover, τ_t is, for each t, a stopping time (relative to the natural σ-fields of X). Recall that, for each $h, t \geqslant 0$, and $j \in \mathbb{I}$,

$$\tau_{t+h} = \tau_t + \tau_h \circ \theta_{\tau_t} \qquad \text{on} \quad \tau_t < \infty,$$

$$\big(X(\tau_h) \circ \theta_{\tau_t} = j\big) = \big(X(\tau_{t+h}) = j\big) \qquad \text{on} \quad \tau_{t+h} < \infty.$$

Now, for each t with $0 \leqslant t < A_\infty$, define

$$\widetilde{X}_t = X(\tau_t), \qquad \widetilde{\mathfrak{F}}_t = \mathfrak{F}_{\tau_t}, \qquad \tilde{\theta}_t = \theta_{\tau_t}.$$

Then, $\widetilde{X} = (\widetilde{X}_t, \widetilde{\mathfrak{F}}_t,\ 0 \leqslant t < A_\infty)$ is called the *time changed process* obtained from X by the random time change τ, corresponding to the IF A.

The question of primary importance is whether the random time change preserves the Markov property. The positive answer is provided by the following theorem which is the special case of the result from general Markov processes. For proof of the general theorem see R. M. Blumenthal and R. K. Getoor [1968, pp. 212, 232–233] and (for strictly increasing A) E. B. Dynkin [1965, Thms. 10.10 and 10.11]; see also D. Williams [1979, p. 150].

THEOREM 1. *The time changed process $\widetilde{X}$ defined above is a* M.C. *satisfying the usual conditions (including the strong Markov property), with the state space H and the life time A_∞.*

The corresponding semi-group $\widetilde{\mathbf{P}}(t) = (\tilde{p}_{ij}(t))$ of transition probabilities is defined for $t \geqslant 0$ by

$$\widetilde{\mathbf{P}}(t)\, f(i) = \mathbb{E}\Big(f(X_{\tau_t}),\ \tau_t < \infty \mid X_{\tau_0} = i\Big), \qquad i \in H,$$

for any bounded function f on H. In particular,

$$\tilde{p}_{ij}(t) = \mathbb{P}\big(X_{\tau_t} = j,\ \tau_t < \infty \mid X_{\tau_0} = i\big), \qquad i, j \in H.$$

Moreover, $\widetilde{\mathbf{P}}(t)$ is sub-stochastic on H.

It follows from Theorem 1 that

$$\begin{aligned}
\widetilde{\mathbf{P}}(t)\, f(\widetilde{X}_s) &= \mathbb{E}\Big(f(\widetilde{X}_{t+s}) \mid \widetilde{X}_s\Big) = \mathbb{E}\Big(f\big[X(\tau_{t+s})\big],\ \tau_{t+s} < \infty \mid X(\tau_s)\Big) \\
&= \mathbb{E}\Big(f\big[X(\tau_t)\big] \circ \theta_{\tau_s},\ \tau_t \circ \theta_{\tau_s} < \infty \mid X(\tau_0) \circ \theta_{\tau_s}\Big) \\
&= \mathbb{E}\Big(f\big[X(\tau_t)\big],\ \tau_t < \infty \mid X(\tau_0)\Big),
\end{aligned}$$

so

$$\widetilde{\mathbf{P}}(t)\, f(i) = \sum_{j \in H} \tilde{p}_{ij}(t)\, f(j), \qquad i \in H,$$

with

$$\tilde{p}_{ij}(t) = \mathbb{P}\big(X(\tau_t) = j,\ \tau_t < \infty \mid X_T = i\big),$$

because $\tau_0 = T$.

Observe, however, that, for $i \in \mathbb{I}$ and $j \in H$,

$$\mathbb{P}(X_{\tau_t} = j,\ \tau_t < \infty \mid X_0 = i) = \sum_{k \in H} \mathbb{P}(X_{\tau_t} = j,\ \tau_t < \infty \mid X_{\tau_0} = k)$$
$$\times\, \mathbb{P}(X_T = k,\ T < \infty \mid X_0 = i) = \sum_{k \in H} D_{ik}\, \tilde{p}_{kj}(t),$$

and this reduces to $\tilde{p}_{ij}(t)$ when $i \in H$. Hence, for $i \in H$, conditioning on $X_T = i$ may be replaced by conditioning on $X_0 = i$. Indeed, $\tilde{p}_{ij}(t)$ is simply the conditional distribution of the compound r.v. $X(\tau_t)$ with $\tau_t < \infty$.

Consider now a matrix $\mathbf{B}^\tau$ defined for each $t > 0$ by

$$\begin{pmatrix} 0 & \mathbf{B}_{12} \\ 0 & \mathbf{I}_2 \end{pmatrix} \begin{pmatrix} 0 & 0 \\ 0 & \widetilde{\mathbf{P}}_{22}(t) \end{pmatrix} = \begin{pmatrix} 0 & \mathbf{B}_{12}\widetilde{\mathbf{P}}_{22}(t) \\ 0 & \widetilde{\mathbf{P}}_{22}(t) \end{pmatrix},$$

where $\widetilde{\mathbf{P}}_{22}(t) = \widetilde{\mathbf{P}}(t)$. For $t = 0$, this reduces to $\mathbf{B}$. It is clear that elements of $\mathbf{B}^\tau$ have the form given above. Thus, $\mathbf{B}^\tau$ is in fact a balayage operator for the stopping time τ_t. (Compare this with analogous expressions for the stopping time T_t in § 3.2.5.)

Furthermore, using relations $\tau_t - \tau_0 = \tau_t \circ \theta_{\tau_0}$, $X_{\tau_t} \circ \theta_{\tau_0} = X_{\tau_t}$ and the strong Markov property, one can deduce in the usual way that, for $t > 0$ and $i \in \mathbb{I}$,

$$\mathbb{P}(X_{\tau_t} = j,\ \tau_0 < \tau_t \leqslant s \mid X_0 = i) = \sum_{k \in H} \int_0^s D_{ik}(s - x)\, dF^t_{kj}(x),$$

which for $i \in H$ reduces to the definition of $F^t_{ij}(s)$. One has also $dF^t_{kj}(x) = dF^t_k(x)\, p_{kj}(x)$. Hence, it follows that, for $i, j \in H$ and $t > 0$,

$$\tilde{p}_{ij}(t) = \int_0^\infty dF^t_i(x)\, p_{ij}(x),$$

where

$$F^t_i(x) = \mathbb{P}(0 < \tau_t \leqslant x \mid X_0 = i).$$

Thus, in agreement with the result stated above, one has $\tilde{p}_{ij}(t) = F^t_{ij}$.

§ 9.6.2. Sample functions. It is important to examine more closely the behavior of sample functions of $\widetilde{X}$; see § 9.5.2 for the special case. Consider now random sets associated with a functional and its inverse. Clearly, the range G of the inverse τ coincides with the random set $\mathcal{I}_r$ of points of right increase of A, whereas random intervals contiguous to $\mathcal{I}_r$ represent magnitudes of discontinuities of τ (see § 8.1.3). Then, the random set

$$W = (\tau_{s-}(\omega)\colon\ \tau_{s-}(\omega) \neq \tau_s(\omega))$$

consists of instants of entrances to H^c, and is at most countable. By right-continuity of paths

$$(u \in W\colon X_u \in H) = \varnothing, \qquad (u \in W\colon X_u \in H^c) = W,$$

whereas $X_t \in H$ for $t \in \mathcal{I}_r$. Hence, W is the set of strictly positive left-end points of the intervals contiguous to the random set $\mathcal{I}_r$ (see the analogous situation in § 2.1.1).

Similarly, the random set

$$V = \big(\tau_s(\omega)\colon\ \tau_{s-}(\omega) \neq \tau_s(\omega)\big)$$

consists of instants of entrances to H.

It is clear that $A_{\tau_s} = s$ for all s, and in particular for s being a discontinuity point of τ:

$$A_{\tau_s} = A_{\tau_{s-}} = s.$$

On the other hand, $\tau_{A_t} = t$ when t is in the range of τ (i.e., $t \in \mathcal{I}_r$), and $\tau_{A_t} = t' > t$, when t is not in the range of τ, and t' is the right-end point of an interval containing t (first entrance to H). See Fig. 9.6.1–9.6.4, and also Fig. 2.5, Fig. 8.4–8.5, and Fig. 9.5.1–9.5.4.

Consequently, Theorem 1, § 9.6.1, may now be interpreted as follows. Under the random time change, the sample functions of the process $\widetilde{X}$ (necessarily taking values in set H) remain the same as far as the transitions are concerned, undergoing only change in duration of states (excursions to H^c being eliminated). In this sense, $\widetilde{X}$ is obtained from process X sampled when it is in H, but with "a change of clock".

Remark. As the life time of the chain $\widetilde{X}$ is A_∞, so $(\tau_t < \infty) = (A_\infty > t)$. Hence, for each $i \in H$, one has

$$\mathbb{P}\big(\widetilde{X}_t \in H \mid \widetilde{X}_0 = i\big) = \sum_{j \in H} \tilde{p}_{ij}(t) = \mathbb{P}^i(\tau_t < \infty) = \mathbb{P}^i(A_\infty > t).$$

The deficiency is interpreted in the usual way by adjoining to H an absorbing state Δ, and letting $\widetilde{X}_t = \Delta$ for all $t \geqslant A_\infty$. Thus, A_∞ is now the first entrance time to Δ:

$$A_\infty = \inf\,(t\colon\ t > 0,\ \widetilde{X}_t = \Delta),$$

and its distribution is given by

$$\tilde{p}_{i\Delta}(t) = 1 - \widetilde{\mathbf{P}}(t)\mathbf{1}_i = \mathbb{P}^i(\tau_t = \infty) = \mathbb{P}^i(A_\infty \leqslant t), \qquad i \in H.$$

Write $\widetilde{\mathbf{P}}^\Delta(t)$ for the stochastic matrix which extends $\widetilde{\mathbf{P}}(t)$ on H to $H^\Delta = H \cup \{\Delta\}$. Extend f on H to f^Δ on H^Δ. Hence,

$$\widetilde{\mathbf{P}}^\Delta(t)\, f^\Delta(i) = \begin{cases} \widetilde{\mathbf{P}}(t)f(i) + \tilde{p}_{i\Delta}(t)f(\Delta), & i \in H, \\ f(\Delta), & i = \Delta. \end{cases}$$

Taking $f(\Delta) = 0$ leads to the convenient simplification $\widetilde{\mathbf{P}}^\Delta(t)f = \widetilde{\mathbf{P}}(t)f$.

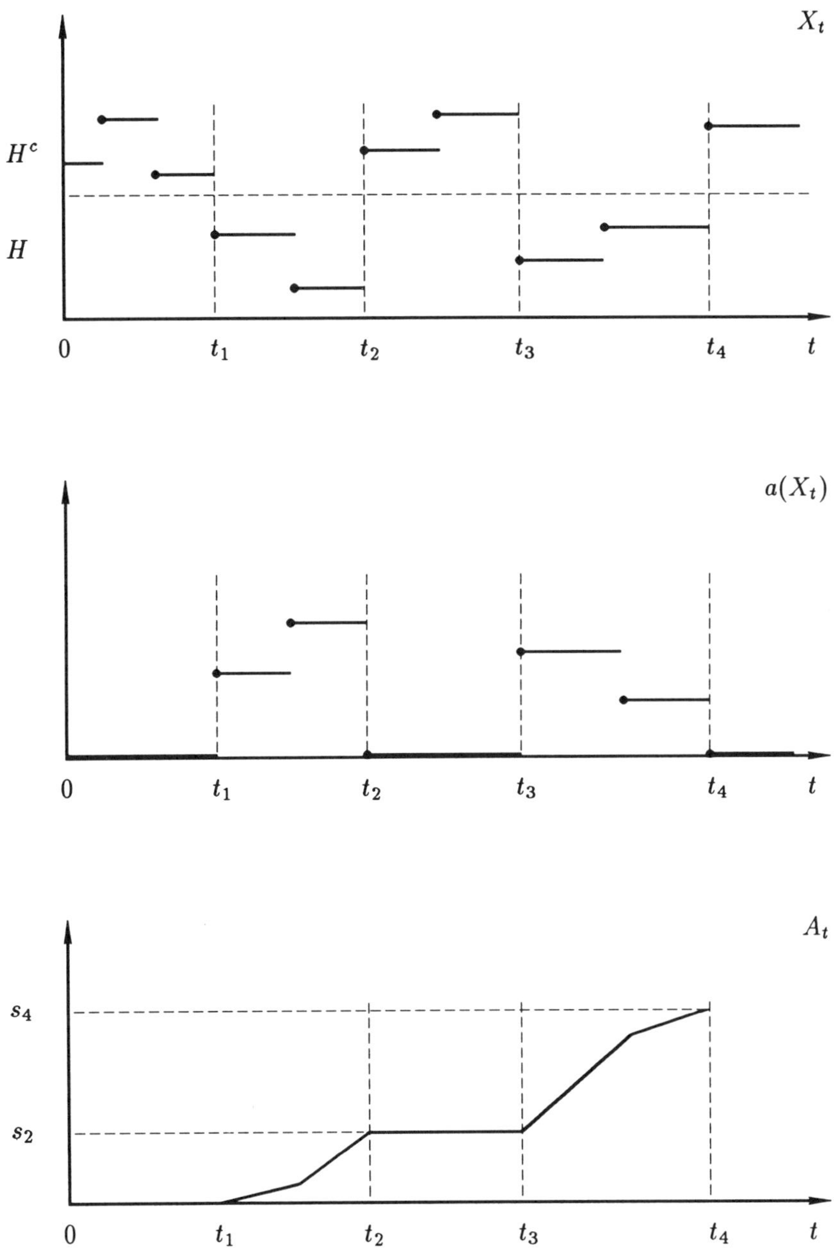

Figure 9.6.1. Integral functional determined by a charge a for a process starting in H^c.

§ 9.6.3. Change of variable. The immediate advantage of processes with changed time is the availability of change of variable formulas in stochastic integrals. Here two very useful formulas will be stated.

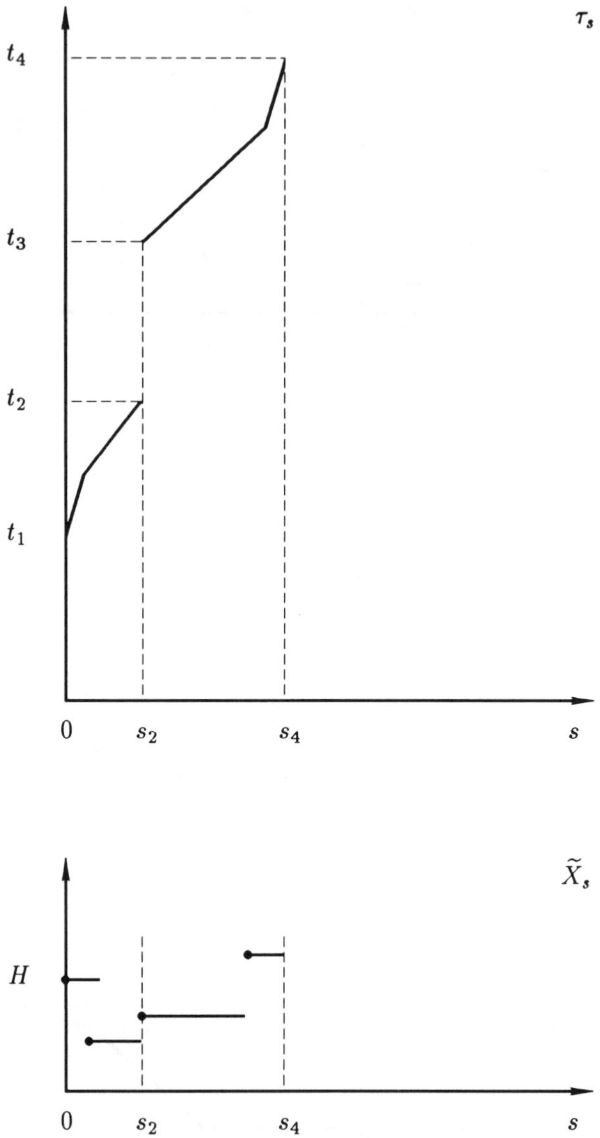

Figure 9.6.2. The inverse and the time changed process starting in H^c.

The first formula is

$$\int_0^\infty f(X_t)\, dA_t = \int_0^{A_\infty} f(\widetilde{X}_s)\, ds \qquad (1)$$

for a positive function f. For the proof see, for example, C. Dellacherie and P. A. Meyer [1982, p. 122]. Formally, the result follows by substitution $t = \tau_s$.

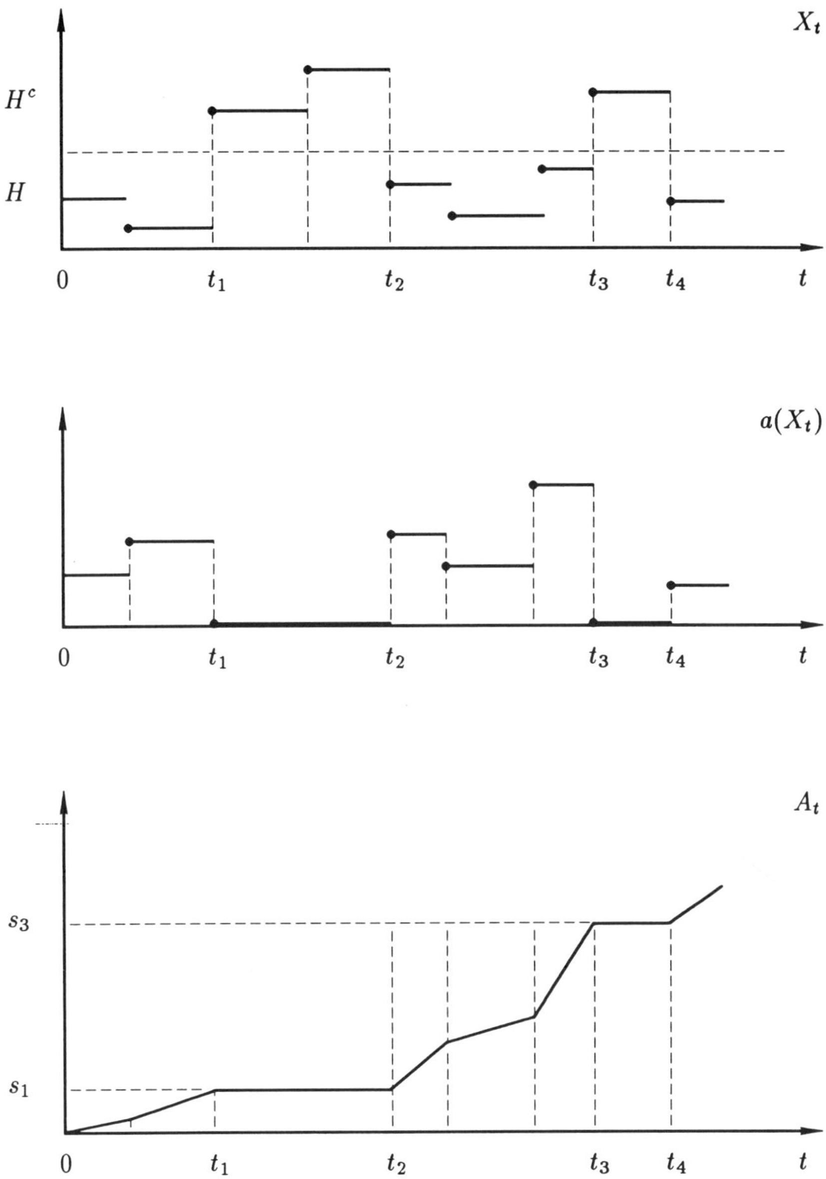

Figure 9.6.3. Integral functional determined by a charge a for a process starting in H.

The second formula is

$$\int_{\tau_0}^{\tau_t} f(X_u)\,du = \int_0^t f(\widetilde{X}_s)\,ds/a(\widetilde{X}_s) + \sum_{0<s\leqslant t} \int_{\tau_{s-}}^{\tau_s} f(X_u)\,du. \tag{2}$$

If the integral on the LHS is taken from 0 to τ_t, the sum on the RHS should be

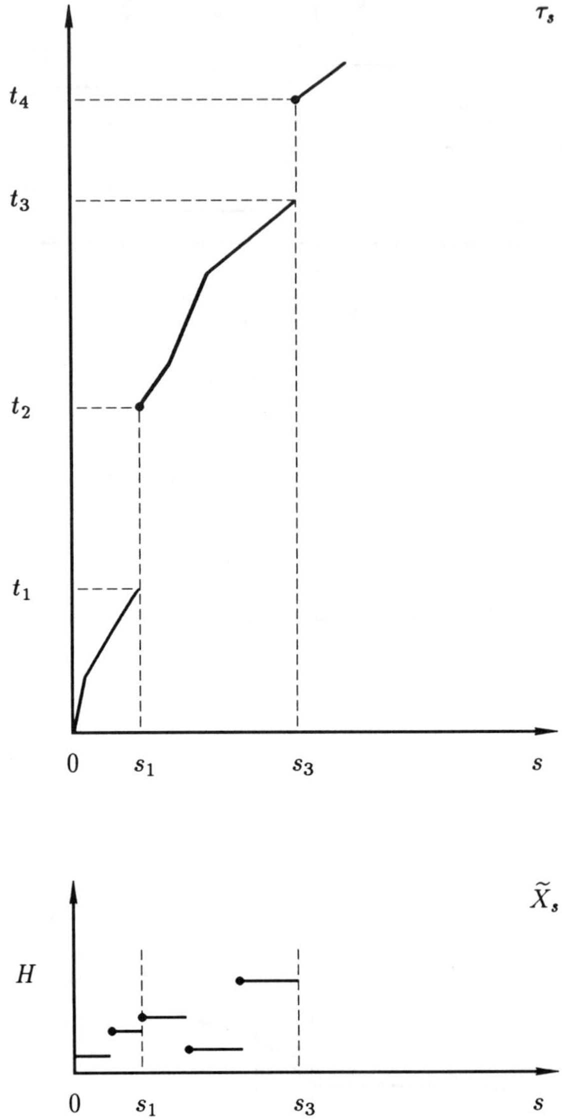

Figure 9.6.4. The inverse and the time changed process starting in H.

taken over $[0, t]$.

We decompose the increasing process $\tau = (\tau_s)$ into a continuous component τ^c and a discontinuous component τ^d:

$$\tau_s = \tau_s^c + \tau_s^d$$

(see § 8.1.3), and observe that for the continuous component (necessarily strictly increasing)

$$d\tau_s^c = ds/a(X_{\tau_s}),$$

whereas the discontinuous component has jumps $\Delta\tau_s = \tau_s - \tau_{s-}$.

To justify (2), apply the change of variable formula (Itô formula) to an increasing process τ and a function F such that

$$F(\tau_t) = \int_0^{\tau_t} f(X_u)\,du,$$

see C. Dellacherie and P. A. Meyer [1982, p. 337] or P. A. Meyer [1976, p. 301]. This yields

$$\int_{\tau_0}^{\tau_t} f(X_u)\,du = \int_{(0,t]} f(\widetilde{X}_{s-})\,d\tau_s + \sum_{0<s\leqslant t}\left(\int_{\tau_{s-}}^{\tau_s} f(X_u)\,du - f(\widetilde{X}_{s-})\Delta\tau_s\right). \tag{3}$$

Note that the term corresponding to F'' in Itô formula vanishes, because the martingale component in representation of τ as a semi-martingale is absent.

If the integral on the LHS of (3) is taken from 0 to τ_t, then on the RHS the integral must be over $[0,t]$ and the sum over $0 \leqslant s \leqslant t$. However, the form stated above with the left-continuous process $(\widetilde{X}_{s-})$ is usual for stochastic integration.

Next observe that using decomposition of τ, one has

$$\int_{(0,t]} f(\widetilde{X}_{s-})\,d\tau_s = \int_0^t f(\widetilde{X}_{s-})\,d\tau_s^c + \sum_{0<s\leqslant t} f(\widetilde{X}_{s-})\Delta\tau_s,$$

where the integral with respect to τ_s^c corresponds to the integration over $u = \tau_s$ in the range of τ.

Integration over u outside the range of τ yields the term

$$\sum_{0<s\leqslant t} \int_{\tau_{s-}}^{\tau_s} f(X_u)\,du.$$

Thus, (3) expresses splitting of integration over u into two regions, namely the range of τ and its complement. Finally, formula (2) is obtained from (3) by the fact that over continuity points $\tau_s = \tau_{s-}$.

§ 9.6.4. Resolvent. The resolvent $\widetilde{\mathbf{U}}^\alpha = (\tilde{u}_{ij}^\alpha)$ of the time changed process $\widetilde{X}$ with the semi-group $\widetilde{\mathbf{P}}(t)$, $t \geqslant 0$, is defined by

$$\widetilde{\mathbf{U}}^\alpha f(i) = \mathbb{E}\left(\int_0^{A_\infty} e^{-\alpha t} f(\widetilde{X}_t)\,dt \mid \widetilde{X}_0 = i\right)$$

$$= \int_0^\infty e^{-\alpha t}\widetilde{\mathbf{P}}(t) f(i)\,dt = \sum_{j\in H} \tilde{u}_{ij}^\alpha\, f(j), \qquad i \in H,$$

for any bounded (or positive) function f on H, which vanishes on Δ. By the random time change $s = \tau_t$, as in (1), § 9.6.3, and by comments at the end of § 9.6.1, this is the same as

$$\widetilde{\mathbf{U}}^\alpha f(i) = \mathbb{E}\left(\int_0^\infty e^{-\alpha A_s} f(X_s)\, dA_s \mid X_0 = i \right), \qquad i \in H,$$

where the integral is actually taken over $[T, \infty)$, because $T = \inf(s\colon A_s > 0)$.

Hence, in particular for $\alpha = 0$ (assuming u_{ij} finite)

$$\widetilde{\mathbf{U}} f(i) = \sum_{j \in H} u_{ij}\, a(j) f(j), \qquad i \in H.$$

Consequently, one obtains an interesting relation

$$\tilde{u}_{ij} = u_{ij}\, a(j), \qquad i, j \in H.$$

The resolvent $\widetilde{\mathbf{U}}^\alpha$ can be simply expressed as a potential of an extended CAF $(\tilde{A}_t^\alpha)$ depending on α, defined by

$$A_t^\alpha = \int_0^t e^{-\alpha A_s}\, dA_s, \qquad t \geqslant 0.$$

It is easy to verify that (see § 8.2.1)

$$A_{t+h}^\alpha - A_t^\alpha = e^{-\alpha A_t} (A_h^\alpha \circ \theta_t), \qquad h, t \geqslant 0,$$

with multiplicative functional

$$M_t = e^{-\alpha A_t}.$$

It is clear that $\operatorname{supp}(A^\alpha) = H$.

Hence, for any bounded (or positive) function f on H,

$$\widetilde{\mathbf{U}}^\alpha f(i) = \mathbb{E}^i \int_0^\infty f(X_s)\, dA_s^\alpha = \mathbf{U}_{A^\alpha} f(i), \qquad i \in H,$$

where the last expression follows from Theorem 2, § 8.2.1.

§ 9.6.5. Generator. In the case of an integral functional A considered here, it is possible to obtain a simple explicit expression for the infinitesimal generator $\widetilde{\mathbf{Q}}$ of the semi-group $\widetilde{\mathbf{P}}(t)$, $t \geqslant 0$, in terms of the generator $\mathbf{Q}$ of the original semi-group $\mathbf{P}(t)$, $t \geqslant 0$, and of the determining function $a(\cdot)$.

The following theorem is the specialization of the general result for Markov processes obtained by H. Gzyl [1977, 1980]. Proof for strictly increasing A (strictly positive a) — hence for a continuous inverse τ — can be found in E. B. Dynkin [1965, Thm. 10.12]; see also D. Williams [1979, p. 151]. It is the presence of discontinuities of τ (corresponding to sojourns in H^c) which makes the analysis difficult, and the Gzyl approach accounts for transitions within H and for excursions to H^c.

THEOREM 1. *Let X be a M.C. with the infinitesimal generator $\mathbf{Q}$ on the state space $\mathbb{I}$, and let $\widetilde{X}$ be a M.C. obtained from X by the random time change τ, corresponding to the integral functional A:*

$$A_t = \int_0^t a(X_s)\, ds, \qquad t \geqslant 0,$$

with supp $A = H$.

Then the infinitesimal generator $\widetilde{\mathbf{Q}} = (\tilde{q}_{ij})$ of the $\widetilde{X}$-chain on H is given by

$$\tilde{q}_{ij} = \frac{1}{a(i)}\left(q_{ij} + \sum_{k \in H^c} q_{ik} D_{kj}\right), \qquad i, j \in H,$$

where $\mathbf{B} = (D_{ij})$ is the balayage matrix of hitting distribution for the first entrance to H (for the chain X).

Simplicity of this result makes it possible to provide a direct argument based on the trace chain $\overline{X}$ induced on H by the original chain X. As noted in Subsection 9.5, the trace chain refers to the consecutive entrances to H of the original chain, but eliminates excursions to H^c. The state space of $\overline{X}$ is H and its generator $\overline{\mathbf{Q}}_{22}$ was shown to be

$$\overline{\mathbf{Q}}_{22} = \mathbf{Q}_{22} + \mathbf{Q}_{21}\mathbf{B}_{12}$$

(see Thm. 1, § 9.5.4).

As a is strictly positive on H, one can apply to the trace chain $\overline{X}$ on H the classical random time change theorem for strictly increasing A (see Dynkin). The required result is then

$$\tilde{q}_{ij} = \bar{q}_{ij}/a(i), \qquad i, j \in H.$$

It is of interest to summarize briefly Gzyl's argument. For a bounded function f, one has from the definition:

$$\widetilde{\mathbf{P}}(t)f(i) = \mathbb{E}^i\big(f(X_{\tau_t}),\ \tau_t < \infty\big), \qquad i \in H.$$

As τ_t is a stopping time, the strong Markov property of the chain (X_t) gives

$$\widetilde{\mathbf{P}}(t)f(i) - f(i) = \mathbb{E}^i \int_0^{\tau_t} \mathbf{Q}f(X_s)\, ds.$$

Hence, Theorem 1 will be proved by evaluating the limit

$$\widetilde{\mathbf{Q}}f(i) = \lim_{t \to 0} \frac{1}{t}\, \mathbb{E}^i \int_0^{\tau_t} \mathbf{Q}f(X_s)\, ds.$$

By the change of variable formula (2) in § 9.6.3, applied to function $\mathbf{Q}f$, one has

$$\int_0^{\tau_t} \mathbf{Q}f(X_s)\, ds = \int_0^t \mathbf{Q}f(\widetilde{X}_s)\, ds/a(\widetilde{X}_s) + \sum_{s \leqslant t} \int_{\tau_{s-}}^{\tau_s} \mathbf{Q}f(X_u)\, du.$$

The first term on the RHS of the above decomposition corresponds to the continuous component of τ (transitions within H) and, when divided by t, leads to $\mathbf{Q}f(i)/a(i)$. The second term on the RHS corresponds to discontinuities of τ (excursions to H^c) and, when divided by t, leads to $\mathbf{Q}(\mathbf{B}-\mathbf{I})f(i)/a(i)$. Proof of this assertion depends essentially on argument related to that used in § 8.3.1. The final form is then

$$\widetilde{\mathbf{Q}}f(i) = \mathbf{Q}\mathbf{B}f(i)/a(i), \qquad i \in H.$$

Remark. One can extend $\widetilde{\mathbf{Q}}$ to $\widetilde{\mathbf{Q}}^{\Delta}$ on the state space $H \cup \{\Delta\}$ by taking

$$\tilde{q}_{i\Delta} = b_i/a(i), \qquad i \in H,$$

where b_i is the escape rate (see § 6.5.4).

§ 9.6.6. Special properties. The following specific features pertinent to random time change are of interest.

a) *Dirichlet problem.* With the stochastic matrix $\widetilde{\mathbf{P}}^{\Delta}(t)$ and its conservative generator $\widetilde{\mathbf{Q}}^{\Delta}$ (both extended to $H \cup \{\Delta\}$), the usual (B)- and (F)-equations hold.

As noted in § 9.6.2, A_∞ is the first entrance time to Δ with distribution

$$D_i^{\Delta}(t) = \mathbb{P}^i(A_\infty \leqslant t) = \tilde{p}_{i\Delta}(t), \qquad i \in H,$$

and $D_{\Delta}^{\Delta}(t) = 1$. Evidently, one has now the (B)-equation

$$\frac{dD_i^{\Delta}(t)}{dt} = \sum_{k \in H} \tilde{q}_{ik} D_k^{\Delta}(t) + \tilde{q}_{i\Delta}, \qquad i \in H,$$

with the initial condition $D_i^{\Delta}(0) = 0$.

It follows that the Dirichlet problem for $D_i^{\Delta} = \mathbb{P}^i(A_\infty < \infty)$ takes the form

$$\begin{aligned} \widetilde{\mathbf{Q}}^{\Delta} D_i^{\Delta} &= 0, \qquad i \in H, \\ D_i^{\Delta} &= 1, \qquad i = \Delta, \end{aligned}$$

(one can replace $\widetilde{\mathbf{Q}}^{\Delta}$ by $\overline{\mathbf{Q}}^{\Delta}$ here).

Denote now by S the first entrance time to the set H^c for the original chain X, and let D_i' be the corresponding first entrance probability. The Dirichlet problem is now

$$\begin{aligned} \mathbf{Q}D_i' &= 0, \qquad i \in H, \\ D_i' &= 1, \qquad i \in H^c. \end{aligned}$$

In general, D_i^{Δ} and D_i' differ, but they coincide when the set H^c is closed.

PROPOSITION 1. *Suppose that in the original chain $\mathbf{Q}_{12} = \mathbf{0}$. Then, the probability D_i^{Δ} that A_∞ is finite, and the probability D_i' that S is finite, are equal for $i \in H$.*

Proof. The result follows from the observation that $\mathbf{B}_{12} = \mathbf{0}$ for the original chain X, so $\overline{\mathbf{Q}}_{22} = \mathbf{Q}_{22}$ and therefore

$$\bar{q}_{i\Delta} = \sum_{k \in H^c} q_{ik}.$$

Hence, both Dirichlet problems coincide. Note, however, that $A_\infty = A_S$ here. □

In the same fashion one can write equations for the joint distribution of the first entrance time to a subset of a state space, and of the position of the first entry, both for the original chain and for the time-changed chain. These equations have form of the equation in Theorem 3, § 3.1.3. The difference between these two distributions is reflected by the difference between matrices $\widetilde{\mathbf{Q}}_{22}$ and $\mathbf{Q}_{22}$. The following is the important special case when this difference disappears.

PROPOSITION 2. *Suppose that the integral functional A is strictly increasing. Then the first entrance distributions to a set in the state space $\mathbb{I}$ for the chain $\widetilde{X}$ and for the chain X are identical.*

Proof. By assumption, charge $a > 0$, so $H = \mathbb{I}$. Consequently, $\mathbf{B}_{12} = \mathbf{0}$ and $\overline{\mathbf{Q}}_{22} = \mathbf{Q}_{22} = \mathbf{Q}$. Hence,

$$\tilde{q}_{ij} = q_{ij}/a(i) \qquad \text{for all } i, j \in \mathbb{I}.$$

Thus quantities depending on jump matrix will be the same for both chains. □

It is therefore clear from the behavior of sample functions of the chain $\widetilde{X}$ and the chain X (both on $\mathbb{I}$) that positions of the first hit for both chains should have the same distribution.

The above result is frequently taken as the characterization of the random time change. R. M. Blumenthal and R. K. Getoor [1968, p. 234] proved that, conversely, if two M.C.'s have the same hitting distributions then they are obtained by the random time change according to some strictly increasing functional A. For further results in this direction see J. Glover [1981, 1983], R. K. Getoor and J. Glover [1983].

b) *Duality*. Another interesting observation is that random time change preserves duality, but with the new duality measure expressed in terms of the Revuz measure.

Let X be the original M.C. with generator $\mathbf{Q}$, and let $\widehat{X}$ be its dual chain relative to a reference measure ξ, with the generator $\widehat{\mathbf{Q}}$ (see § 9.4.5). Next, let $\widetilde{X}$ be the time changed M.C., obtained from the original chain X, relative to the integral functional A, as defined in this sub-section, and let $\widetilde{\mathbf{Q}}$ be the corresponding generator.

Consider now the chain X^* obtained by the time change determined by the same charge a from the dual chain $\widehat{X}$ (see § 9.4.5), and let $\mathbf{Q}^*$ be the corresponding generator. Then the time change preserves duality in the following sense.

THEOREM 3. *The chain X^* (obtained from the dual chain $\widehat{X}$) coincides with the dual of the time changed chain $\widetilde{X}$, with Revuz measure μ_A as a reference measure.*

In terms of generators, the assertion may be expressed symbolicaly as

$$\begin{array}{ccc} & \xi & \\ \mathbf{Q} & \longrightarrow & \widehat{\mathbf{Q}} \\ a \downarrow & & \downarrow a \\ \widetilde{\mathbf{Q}} & \longrightarrow & \mathbf{Q}^* \\ & \mu_A & \end{array}$$

Proof. i) Path $\mathbf{Q} \to \widetilde{\mathbf{Q}} \to \mathbf{Q}^*$. Recall that $\tilde{q}_{ij} = \bar{q}_{ij}/a(i)$ for $i, j \in H$. Recall also that the Revuz measure μ_A of the integral functional A determined by charge a (supported on H) is $\mu_A(j) = \xi(j)a(j)$; see § 8.1.3-e. Hence, the dual of $\widetilde{\mathbf{Q}}$ relative to the reference measure μ_A is given (for $i, j \in H$) by

$$\text{dual } (\tilde{q}_{ij}) = \mu_A(j)\, \tilde{q}_{ji}/\mu_A(i) = \xi(j) \sum_k q_{jk} D_{ki}/\mu_A(i).$$

ii) Path $\mathbf{Q} \to \widehat{\mathbf{Q}} \to \mathbf{Q}^*$. Recall that by duality (relative to ξ) $\hat{q}_{ij} = \xi(j)q_{ji}/\xi(i)$; see § 1.3.5. Applying random time change (determined by the same charge a) to the dual chain $\widehat{X}$, one has in the same fashion (for $i, j \in H$):

$$\bar{\hat{q}}_{ij} = \sum_k \hat{q}_{ik}\, \widehat{D}_{kj}$$

and

$$\tilde{\hat{q}}_{ij} = \bar{\hat{q}}_{ij}/a(i) = \xi(j) \sum_k N_{jk}\, q_{ki}/\mu_A(i),$$

where use has been made of the fact that dual of $\widehat{D}_{kj}$ (relative to ξ) is N_{jk}; see §§ 6.3.3 and 9.4.5.

iii) Finally, recall that Theorem 2, § 3.4.2, asserts that, for $i, j \in H$,

$$\sum_{k \in H^c} q_{jk} D_{ki} = \sum_{k \in H^c} N_{jk} q_{ki}.$$

This can be extended to all k, because $D_{ki} = \delta_{ki}$ and $N_{jk} = \delta_{jk}$ for $k \in H$. Hence, dual $(\tilde{q}_{ij}) = \tilde{\hat{q}}_{ij}$, and their common value is now denoted by q^*_{ij}. □

For further discussion of random time change and duality see J. Mitro [1979 a, 1979 b].

c) *Remark*. Recent investigations indicate that random time change determined by discontinuous functionals need not preserve duality, and even may destroy Markov property. See papers by J. Mitro [1986] and J. Glover [1983] for further discussion. However, when A is an DAF determined by a function g, one can replace it by its compensator which is an integral functional with charge $g_{\mathbf{Q}}$ (see § 8.1.3-b).

§ 9.6.7. Examples. The following simple examples are very instructive. More involved examples will be given in Chapter IV.

Example 1. Take $a = I_H$, so A_t is the occupation time of a set H, discussed extensively in Sub-sections 8.4 and 9.5. The inverse τ_t represents the actual time needed to produce time t spent in the set H.

Evidently, $\tilde{q}_{ij}$ and $\bar{q}_{ij}$ coincide, and one has, for $i, j \in H$,

$$\begin{aligned} \tilde{q}_{ij} &= q_i D^{\rho}_{ij} \qquad \text{for} \quad i \neq j, \\ \tilde{q}_i &= q_i (1 - D^{\rho}_{ii}), \\ \tilde{q}_{i\Delta} &= q_i \, D^{\rho c}_i = b_i. \end{aligned}$$

As already noted, the sample functions retain their shape when in the set H.

Example 2. In the special case when $H = \{j\}$, where j is a fixed state, $D^{\rho}_{jj} = f_{jj} \leqslant 1$, and Example 1 gives

$$\tilde{q}_{j\Delta} = q_j(1 - f_{jj}) = b_j, \qquad u_{jj} b_j = 1.$$

If j is recurrent ($f_{jj} = 1$) for the original chain, then j becomes an absorbing state for the time changed chain. If j is transient ($f_{jj} < 1$) for the original chain, there is a possibility that a chain restricted to j will escape from j. See also discussion of the local time V^j in § 8.4.4.

Example 3. Let X be a birth process with intensities

$$\lambda_i = a(i)\lambda, \qquad i = 0, 1, \dots,$$

where $a(i) > 0$ for all i. Then, $\widetilde{X}$ is the Poisson process with intensity λ, and

$$u_{ij} = 1/\lambda_j, \qquad \tilde{u}_{ij} = 1/\lambda \qquad \text{for} \quad j \geqslant i.$$

For further discussion see W. A. O'N. Waugh [1970], M. F. Neuts and S. I. Resnick [1971].

Example 4. Suppose that the function a is constant, say c, for all $i \in \mathbb{I}$. Therefore, $A_t = ct$ and $\tau_t = t/c$. Hence,

$$\widetilde{X}_t = X(t/c), \qquad 0 \leqslant t < \infty.$$

Evidently,

$$\tilde{p}_{ij}(t) = p_{ij}(t/c) \qquad \text{and} \qquad \tilde{q}_{ij} = q_{ij}/c \qquad \text{for} \quad i, j \in \mathbb{I}.$$

One has here a simple change of scale. The case when c is a random variable was treated by P. D. Feigin [1977].

Example 5. Let X be a branching process with intensities q_{ij} proportional to i, and with $q_i = i\beta$ (see Ex. 3, § 1.1.9). Take $H = \{1, 2, \ldots\}$ and $H^c = \{0\}$, where 0 is the absorbing state. Evidently, $D_{0j} = 0$ for $j \geqslant 1$, so $\bar{q}_{ij} = q_{ij}$.

Take now $a(i) = i > 0$ for all $i \in H$. Hence, for $i, j \in H$,

$$\tilde{q}_{ij} = \beta\, b(j - i + 1) \qquad \text{for } j \geqslant i - 1, \quad i \geqslant 1,$$
$$\tilde{q}_i = \beta \qquad \text{for } i \geqslant 1,$$

but

$$\tilde{q}_{i\Delta} = \begin{cases} \beta\, b(0) & \text{for } i = 1, \\ 0 & \text{for } i > 1. \end{cases}$$

It is clear that $A_t = A_S$ for $t \geqslant S$, where S is the time to absorption at 0 (for the chain X). Therefore, A_S is the life time of the *chain* $\widetilde{X}$, with $\tau_{s-} = S$ and $\tau_s = \infty$, where $s = A_S$.

Moreover, by Remark in § 9.6.2, one has

$$\mathbb{P}^i(A_S \leqslant t) = \tilde{p}_{i\Delta}(t), \qquad i \geqslant 1.$$

Example 6. Consider a M.C. X with the intensity matrix $\mathbf{Q}$ of the form given in Example 3, § 5.4.2. Thus, there are no transitions in H^c, as the chain X can enter a state in H^c only by a jump from H, and then must return to H immediately on completion of its visit to that state. As indicated there, the generator on H has the form $\mathbf{Q}_{22} + \mathbf{Q}_{21}\mathbf{B}_{12}$ which is the same as that for the trace chain (see Thm. 1, § 9.5.4). However, in the present case $\mathbf{B}_{12} = \mathbf{\Theta}_1^{-1}\mathbf{Q}_{12}$, so the generator

$$\overline{\mathbf{Q}}_{22} = \mathbf{Q}_{22} + \mathbf{Q}_{21}\mathbf{\Theta}_1^{-1}\mathbf{Q}_{12}$$

is conservative.

Direct proof that the generator $\overline{\mathbf{Q}}_{22}$ has this form in the case considered here was obtained by P. D. Feigin and E. Rubinstein [1979].

SECTION 10

EXTENSIONS

This section discusses several extensions of supportive character. These include composite passage times, renewal method, joint distributions and excursion theory.

The first return time T_S is a composite time representing the first entrance time to a set H after the last exit time S from H. Evidently, T_S comprises first entrance times discussed in Chapter II, and could be used as a starting point of presentation of passage times. The discussion, however, is based on the product theorem for balayage, and requires familiarity with functionals and martingales. For this reason, it is more convenient to discuss special cases by direct methods.

Consecutive entrances to (or exits from) a set H constitute natural extension of earlier discussions. The analysis is linked with renewal argument and leads to Markov renewal processes. Of special interest are limit theorems and analogies with classical renewal theory. Indeed, one could use this argument as a starting point of presentation as is done in many works, but here renewal properties are treated as a consequence of Markovian structure.

Joint distributions of passage times are treated rigorously with the help of functionals, and this justifies a direct approach through *the density argument*. Finally, excursions from a set form a recent object of studies and clearly extend well known properties of waiting times in a classical renewal theory.

The material presented in this section has been discussed in literature, although properties of composite passage times are often taken for granted. There exists voluminous literature on Markov renewal processes, and the case treated here is a rather obvious specialisation. Excursion theory is growing extensively, but joint distributions received much less attention.

Exact references to literature are given in the text.

10.1. FIRST RETURN TIME

This sub-section generalizes the first return time to a set, discussed already in Sub-section 3.3. The use is made here of delayed taboo probabilities (introduced in § 2.2.2), and the role of the product theorem for stopping times is stressed. Although presentation is analogous to that in Sub-section 3.3, the analysis reveals important structural properties of a M.C. In particular, decomposition theorems, duality, and connection with renewal theory are examined.

§ 10.1.1. Definitions. Let $X = (X_t,\ 0 \leqslant t < \infty)$ be the usual M.C. with a state space $\mathbb{I}$ and let $H \subset \mathbb{I}$ be a taboo set, $H^c = \mathbb{I} \setminus H$. The first entrance time $T = T_H$

to the set H has been already defined in § 2.1.2 and discussed in Sub-section 3.1. Recall that

$$\mathbb{P}^i(X_t = j,\ t < T) = {}_Hp_{ij}(t), \qquad \mathbb{P}^i(X_T = j,\ T \leqslant t) = D_{ij}(t),$$

with the Laplace transform ${}_Hu^\alpha_{ij}$ and the Laplace–Stieltjes transform d^α_{ij}, respectively. Write also $\mathbf{B}(t) = (D_{ij}(t))$ and $\mathbf{B}^\alpha = (d^\alpha_{ij})$ for the balayage matrices.

By definition, T is identically zero when the chain is initially in H. Consider now the following extension. When the chain starts in H, it is allowed to continue in H till the first entrance time S to the set H^c, defined by

$$S = \inf\big(t\colon t > 0,\ X_t \in H^c\big).$$

Afterwards, the chain continues in H^c till subsequent first entrance to H. This takes place at the first entrance time to H after S, denoted by T_S and defined by

$$T_S = \inf\big(t\colon t > S,\ X_t \in H\big).$$

Thus, T_S may be regarded as *the first return time to* H. Observe that one has here a special case of Definition 2, § 2.2.2. Evidently,

$$T_S = S + T \circ \theta_S.$$

Moreover, $X(T_S)$ takes values in H only. The notation

$$\mathbb{P}^i(X(T_S) = j,\ T_S \leqslant t) = D^*_{ij}(t),$$

with $D^*_{ij}(t) = D_{ij}(t)$ when $i \in H^c$, will be used. The Laplace–Stieltjes transform of $D^*_{ij}(t)$ will be denoted by $d^{*\alpha}_{ij}$ with associated balayage matrices $\mathbf{B}^*(t)$ and $\mathbf{B}^{*\alpha}$.

Similarly, write for the time S

$$\mathbb{P}^i(X_t = j,\ t < S) = {}_H\hat{p}_{ij}(t), \qquad \mathbb{P}^i(X_S = j,\ S \leqslant t) = \widehat{D}_{ij}(t),$$

with associated balayage matrices $\widehat{\mathbf{B}}(t)$ and $\widehat{\mathbf{B}}^\alpha = (\hat{d}^\alpha_{ij})$. Note that X_S takes values in H^c.

§ 10.1.2. Delayed taboo. As stated above, the M.C. starting in H proceeds without restriction up to time S, and taboo is imposed only after S. Thus, one has here a special case of a delayed taboo probability, to be denoted by ${}_Hp^*_{ij}(t)$, and defined as in Definition 1, § 2.2.2. For the usual M.C. this definition may be written as

$${}_Hp^*_{ij}(t) = \mathbb{P}^i(S > t,\ X_t = j) + \mathbb{P}^i(S \leqslant t,\ X_u \in H^c,\ S \leqslant u < t,\ X_t = j).$$

When the initial state i is in H^c, then $S = 0$ and the first term on the RHS vanishes, so the delayed taboo probability reduces to the ordinary taboo probability ${}_Hp_{ij}(t)$. On the other hand, when the initial state i is in H, the first term on the RHS vanishes for $j \in H^c$, whereas the second term on the RHS vanishes for $j \in H$.

As it stands, the first term on the right is ${}_H\hat{p}_{ij}(t)$. However, it may be further specified according to the number of jumps within H before time S, with the last jump (out of H) occurring at S. If there is only one jump (as in Sub-section 3.3) this term is simply $\delta_{ij}e^{-q_i t}$, and the delayed taboo probability reduces to ${}_H p^{\rho}_{ij}(t)$, as in § 2.2.4. If no such restriction is imposed, then the first term is the taboo probability ${}_{H^c}p_{ij}(t)$.

Connection between delayed taboo probability and distribution of T_S has been already established by Lemma 3, § 2.2.2, and Theorem 4, § 2.2.2, yields the following basic result.

THEOREM 1. *For $i \in \mathbb{I}$ and $t > 0$:*

$$\mathbb{P}^i(T_S > t) = \sum_j {}_H p^*_{ij}(t),$$

with

$$\mathbb{P}^i(S \leqslant t < T_S) = \sum_{j\in H^c} {}_H p^*_{ij}(t), \qquad \mathbb{P}^i(S > t) = \sum_{j\in H} {}_H p^*_{ij}(t).$$

As both T and S are terminal stopping times, Proposition 5, § 2.2.2, indicates that also T_S is a terminal stopping time. Hence, the delayed taboo probabilities are obtained by killing of the original M.C. at the first entry time T_S to the taboo set H after time S. In other words, killing occurs at the first entrance time T when the chain starts in H^c, and at the first return time T_S when it starts in H.

Consequently, by Theorem 1, § 9.1.2 (see also Thm. 6, § 2.2.2) the transition matrices ${}_H\mathbf{P}^*(t) = ({}_H p^*_{ij}(t))$ form a Markov semi-group.

§ 10.1.3. Equations. As always, the matrix notation is more transparent. The delayed taboo matrix has the partitioned form

$${}_H\mathbf{P}^*(t) = \begin{pmatrix} {}_H\mathbf{P}^*_{11}(t) & \mathbf{0} \\ {}_H\mathbf{P}^*_{21}(t) & {}_H\mathbf{P}^*_{22}(t) \end{pmatrix}, \qquad {}_H\mathbf{P}^*(0) = \mathbf{I},$$

where ${}_H\mathbf{P}^*_{11}(t) = {}_H\mathbf{P}_{11}(t)$ and ${}_H\mathbf{P}^*_{22}(t) = {}_H\widehat{\mathbf{P}}_{22}(t)$. Write also ${}_H\mathbf{U}^{*\alpha} = ({}_H u^{*\alpha}_{ij})$ for the corresponding Laplace transform (resolvent), so

$${}_H\mathbf{U}^{*\alpha} = \begin{pmatrix} {}_H\mathbf{U}^{\alpha}_{11} & \mathbf{0} \\ {}_H\mathbf{U}^{*\alpha}_{21} & {}_H\mathbf{U}^{*\alpha}_{22} \end{pmatrix}, \qquad \alpha > 0.$$

In order to obtain the (B)- and (F)-equations for ${}_H\mathbf{P}^*(t)$, one can proceed in the usual way, taking into account that the matrix ${}_H\mathbf{P}^*(t)$ is only sub-stochastic. The corresponding non-conservative intensity matrix (the generator) ${}_H\mathbf{Q}^*$ has the form

$${}_H\mathbf{Q}^* = \begin{pmatrix} \mathbf{Q}^*_{11} & \mathbf{0} \\ \mathbf{Q}^*_{21} & \mathbf{Q}^*_{22} \end{pmatrix},$$

where components $\mathbf{Q}^*_{11}$ and $\mathbf{Q}^*_{21}$ are the same as components $\mathbf{Q}_{11}$ and $\mathbf{Q}_{21}$ of the original matrix $\mathbf{Q}$, respectively. Indeed, transitions into the set H^c for the killed chain follow those for the original chain. On the other hand, the component $\mathbf{Q}^*_{22}$ reflects the effect of the number of allowed transitions within H up to time S; in the two extreme cases already mentioned this component is $-\mathbf{\Theta}_2$ and $\mathbf{Q}_{22}$, respectively. However, one has always

$$q_i = q_i^* \quad \text{for} \quad i \in H, \qquad q_{ij} \geqslant q_{ij}^* \quad \text{for} \quad i \neq j.$$

In general, it is rather difficult to find a generator of a killed chain (see § 9.1.3), but here the form of ${}_H\mathbf{Q}^*$ is determined by the structure imposed by T_S. For further comments on the relationship between generators $\mathbf{Q}$ and ${}_H\mathbf{Q}^*$, see § 10.1.6.

One has therefore

THEOREM 1. *For $t > 0$:*

$$\frac{d_H\mathbf{P}^*_{ij}(t)}{dt} = \begin{cases} {}_H\mathbf{Q}^*\,{}_H\mathbf{P}^*(t), & \text{(B)} \\ {}_H\mathbf{P}^*(t)\,{}_H\mathbf{Q}^*, & \text{(F)} \end{cases}$$

with ${}_H\mathbf{P}^*(0) = \mathbf{I}$.

Equivalently, in terms of transforms,

$$\alpha\,{}_H\mathbf{U}^{*\alpha} - \mathbf{I} = \begin{cases} {}_H\mathbf{Q}^*\,{}_H\mathbf{U}^{*\alpha}, & \text{(B)} \\ {}_H\mathbf{U}^{*\alpha}\,{}_H\mathbf{Q}^*. & \text{(F)} \end{cases}$$

Consequently, the (B)-equation and Theorem 1, § 10.1.2, yield the equation for the complementary distribution function of T_S.

THEOREM 2. *For $i \in \mathbb{I}$, $t > 0$:*

$$\frac{dD_i^{*c}(t)}{dt} = {}_H\mathbf{Q}^*\,D_i^{*c}(t), \qquad \text{(B)}$$

with $D_i^{*c}(0) = 1$.

Equivalently, with L.S. *transform of $D_i^*(t)$ denoted by $d_i^{*\alpha}$,*

$$\alpha d_i^{*\alpha} = {}_H\mathbf{Q}^* d_i^{*\alpha} - (\mathbf{Q} - {}_H\mathbf{Q}^*)\mathbf{1}_i, \qquad \alpha > 0.$$

Observe that $D_i^*(t)$ has the density $d_i^*(t)$, and that $d_i^*(0) = (\mathbf{Q} - {}_H\mathbf{Q}^*)\mathbf{1}_i$ for $i \in \mathbb{I}$, so

$$d_i^*(0) = \begin{cases} d_i(0) = \mathbf{Q}_{12}\mathbf{1}_i & \text{for} \quad i \in H^c, \\ (\mathbf{Q}_{22} - \mathbf{Q}^*_{22})\mathbf{1}_i & \text{for} \quad i \in H. \end{cases}$$

On the other hand, the (F)-equation yields the explicit expression for $D_i^*(t)$.

THEOREM 3. *For $i \in \mathbb{I}$, $t > 0$:*

$$D_i^*(t) = \int_0^t {}_H\mathbf{P}^*(s) d_i^*(0)\,ds, \qquad \text{(F)}$$

with $D_i^*(0) = 0$.

Equivalently, in terms of transform ($\alpha > 0$):

$$d_i^{*\alpha} = {}_H\mathbf{U}^{*\alpha} d_i^*(0).$$

Proof. Note first that the semi-group property of ${}_H\mathbf{P}^*(t)$ implies that for $t, s \geqslant 0$:

$$D_i^*(t+s) - D_i^*(t) = \sum_k {}_Hp_{ik}^*(t) D_k^*(s).$$

It follows that (D_i^*) is purely excessive on $\mathbb{I}$ relative to ${}_H\mathbf{P}^*(t)$. Hence, there exists a unique exit law $d_i^*(\cdot)$ such that

$$D_i^*(t) = \int_0^t d_i^*(s)\, ds$$

and

$$d_i^*(t+s) = \sum_k {}_Hp_{ik}^*(t) d_k^*(s), \qquad t \geqslant 0, \quad s > 0.$$

Validity of this equation for $s = 0$ yields

$$d_i^*(t) = \sum_k {}_Hp_{ik}^*(t) d_k^*(0).$$ □

Thus, in terms of components,

$$D_i(t) = \int_0^t {}_H\mathbf{P}_{11}(s) d_i(0)\, ds, \qquad i \in H^c,$$

$$D_i^*(t) = \int_0^t {}_H\mathbf{P}_{21}^*(s) d_i(0)\, ds + \int_0^t {}_H\mathbf{P}_{22}^*(s) d_i^*(0)\, ds, \qquad i \in H.$$

Note that the above expression for $i \in H$ corresponds to the decomposition

$$D_i^*(t) = \mathbb{P}^i(S < T_S \leqslant t) + \mathbb{P}^i(S = T_S \leqslant t).$$

§ 10.1.4. Balayage. The product theorem (§ 8.1.1) asserts that

$$\mathbf{P}^\alpha(T_S) = \mathbf{P}^\alpha(S)\mathbf{P}^\alpha(T)$$

or, in notation used here,

$$\mathbf{B}^{*\alpha} = \widehat{\mathbf{B}}^\alpha \mathbf{B}^\alpha. \tag{PT}$$

Matrices $\mathbf{B}^{*\alpha}$ and $\mathbf{B}^\alpha$ have the same usual form, but $\widehat{\mathbf{B}}^\alpha = (\hat{d}_{ij}^\alpha)$ has a different form, namely,

$$\widehat{\mathbf{B}}^\alpha = \begin{pmatrix} \mathbf{I}_1 & \mathbf{0} \\ \widehat{\mathbf{B}}_{21}^\alpha & \widehat{\mathbf{B}}_{22}^\alpha \end{pmatrix},$$

where the component $\widehat{\mathbf{B}}_{22}^\alpha$ reflects the effects of jumps within H before time S, as noted in § 10.1.2. Recall that the α-balayage matrix $\mathbf{B}^{*\alpha} = (d_{ij}^{*\alpha})$ has elements

$$d_{ij}^{*\alpha} = \mathbb{E}^i(e^{-\alpha T_S} I_{\{j\}}(X(T_S)),\ T_S < \infty), \qquad i \in \mathbb{I}.$$

The following results is the immediate consequence of the product theorem.

PROPOSITION 1. *The non-zero components of the α-balayage matrix are*

$$\mathbf{B}^{*\alpha}_{12} = \mathbf{B}^{\alpha}_{12}, \qquad \mathbf{B}^{*\alpha}_{22} = \widehat{\mathbf{B}}^{\alpha}_{21}\,\mathbf{B}^{\alpha}_{12} + \widehat{\mathbf{B}}^{\alpha}_{22}.$$

Proof. The following lemma is needed to establish connection between α-balayage and delayed taboo probabilities.

LEMMA 2. *For $\alpha > 0$:*

$${}_H\mathbf{U}^{*\alpha}_{21} = \widehat{\mathbf{B}}^{\alpha}_{21}\,{}_H\mathbf{U}^{\alpha}_{11}.$$

Proof. For $i \in H$ and $j \in H^c$:

$$\begin{aligned}{}_Hp^*_{ij}(t) &= \mathbb{P}^i(S \leqslant t,\ X_u \in H^c,\ S \leqslant u < t,\ X_t = j) = \mathbb{P}^i(S \leqslant t < T_S,\ X_t = j)\\ &= \sum_{k\in H^c} \mathbb{P}^i(X_S = k,\ S \leqslant t < S + T\circ\theta_S,\ X_t = j)\\ &= \sum_{k\in H^c} \int_0^t \mathbb{P}^i(X_S = k,\ S \in ds,\ t-s < T\circ\theta_S,\ X_t = j)\\ &= \sum_{k\in H^c} \int_0^t \mathbb{P}^i(X_S = k,\ S\in ds)\mathbb{P}(X_t = j,\ t-s < T\circ\theta_S \mid X_S = k,\ S = s)\\ &= \sum_{k\in H^c} \int_0^t \hat{d}_{ik}(s)\,{}_Hp_{kj}(t-s)\,ds.\end{aligned}$$

In the transformed form, the above relation is

$${}_Hu^{*\alpha}_{ij} = \sum_{k\in H^c} \hat{d}^{\alpha}_{ik}\,{}_Hu^{\alpha}_{kj} \qquad \text{for} \quad i \in H,\ j \in H^c. \qquad \square$$

In contrast, the term $\widehat{\mathbf{B}}^{\alpha}_{22}$ involves transitions within set H up to time S (before taboo is imposed). Hence, by the similar argument used earlier, depending on the (F)-equation, one has

LEMMA 3. *For $\alpha > 0$:*

$$\widehat{\mathbf{B}}^{\alpha}_{22} = {}_H\mathbf{U}^{*\alpha}_{22}(\mathbf{Q}_{22} - \mathbf{Q}^*_{22}).$$

The explicit expression for the balayage is now given by the following

THEOREM 4. *For $\alpha > 0$:*

$$\mathbf{B}^{*\alpha} = {}_H\mathbf{U}^{*\alpha}(\mathbf{Q} - {}_H\mathbf{Q}^*),$$

or, equivalently, for $t > 0$:

$$\mathbf{B}^*(t) = \int_0^t {}_H\mathbf{P}^*(s)(\mathbf{Q} - {}_H\mathbf{Q}^*)\,ds,$$

with $\mathbf{B}^*(0) = \mathbf{0}$.

Proof. It is more convenient to work with transforms. The assertion follows by expressing the right-hand side in terms of components, and comparing with the left-hand side. One has immediately that

$$\mathbf{B}_{12}^{\alpha} = {}_H\mathbf{U}_{11}^{\alpha}\mathbf{Q}_{12}.$$

Hence, it suffices to consider only the term $\mathbf{B}_{22}^{*\alpha}$. By Lemma 2,

$$\widehat{\mathbf{B}}_{21}^{\alpha}\mathbf{B}_{12}^{\alpha} = \widehat{\mathbf{B}}_{21}^{\alpha}{}_H\mathbf{U}_{11}^{\alpha}\mathbf{Q}_{12} = {}_H\mathbf{U}_{21}^{*\alpha}\mathbf{Q}_{12}.$$

Thus, using Lemma 3,

$$\mathbf{B}_{22}^{*\alpha} = {}_H\mathbf{U}_{21}^{*\alpha}\mathbf{Q}_{12} + {}_H\mathbf{U}_{22}^{*\alpha}\left(\mathbf{Q}_{22} - \mathbf{Q}_{22}^{*}\right). \qquad \square$$

It is worthwhile to interpret the explicit form of Theorem 4. Only the term $\mathbf{B}_{22}^{*}(t)$ is new. Thus, for $i, j \in H$, one has the following decomposition:

$$D_{ij}^{*}(t) = \mathbb{P}^{i}\left(X(T_S) = j,\ S < T_S \leqslant t\right) + \mathbb{P}^{i}\left(X(T_S) = j,\ S = T_S \leqslant t\right),$$

where

$$\mathbb{P}^{i}\left(X(T_S) = j,\ S < T_S \leqslant t\right) = \int_0^t \sum_{k \in H^c} {}_Hp_{ik}^{*}(s)q_{kj}\, ds,$$

$$\mathbb{P}^{i}\left(X(T_S) = j,\ S = T_S \leqslant t\right) = \int_0^t \sum_{k \in H} {}_Hp_{ik}^{*}(s)\left(q_{kj} - q_{kj}^{*}\right) ds.$$

Observe that the above formulas could be obtained directly from the path decomposition ("density argument"). Clearly, if $\mathbf{Q}_{22} = \mathbf{Q}_{22}^{*}$, the second term vanishes. In any case, summation over all $j \in H$ yields the decomposition of $D_i^{*}(t)$ in agreement with § 10.1.3.

Note that $D_{ij}^{*}(t)$ has a density $d_{ij}^{*}(t)$, and evidently

$$\left.\frac{d\mathbf{B}^{*}(t)}{dt}\right|_{t=0} = \lim_{\alpha \to 0} \alpha\mathbf{B}^{*\alpha} = \mathbf{Q} - {}_H\mathbf{Q}^{*}.$$

COROLLARY 5. *For $t, h \geqslant 0$:*

$$\mathbf{B}^{*}(t+h) - \mathbf{B}^{*}(h) = {}_H\mathbf{P}^{*}(h)\mathbf{B}^{*}(t).$$

Proof. Theorem 4 implies that

$$\mathbf{B}^{*}(t+h) - \mathbf{B}^{*}(h) = \int_0^t {}_H\mathbf{P}^{*}(s+h)\left(\mathbf{Q} - {}_H\mathbf{Q}^{*}\right) ds = {}_H\mathbf{P}^{*}(h)\mathbf{B}^{*}(t),$$

where the last step follows from the semi-group property of ${}_H\mathbf{P}^{*}(t)$. $\square$

From Corollary 5 one has the following result which should be compared with Theorem 4.

COROLLARY 6. *For $t > 0$:*

$$\frac{d\mathbf{B}^*(t)}{dt} = {}_H\mathbf{Q}^*\mathbf{B}^*(t) + (\mathbf{Q} - {}_H\mathbf{Q}^*)$$

or, equivalently, for $\alpha > 0$:

$$\alpha\mathbf{B}^{*\alpha} = {}_H\mathbf{Q}^*\mathbf{B}^{*\alpha} + (\mathbf{Q} - {}_H\mathbf{Q}^*).$$

Finally, consider the limit matrix

$$\mathbf{B}^* = \lim_{t\to\infty} \mathbf{B}^*(t) = \lim_{\alpha\to 0} \mathbf{B}^{*\alpha} = (D^*_{ij}).$$

Then one has immediately

COROLLARY 7.

$$\begin{aligned} \mathbf{B}^* - \mathbf{B}^*(t) &= {}_H\mathbf{P}^*(t)\mathbf{B}^*, \\ -{}_H\mathbf{Q}^*\,\mathbf{B}^* &= \mathbf{Q} - {}_H\mathbf{Q}^*, \\ \mathbf{Q}\,\mathbf{B}^* &= (\mathbf{Q} - {}_H\mathbf{Q}^*)(\mathbf{B}^* - \mathbf{I}). \end{aligned}$$

See also § 10.1.6 for further comments on $\mathbf{Q} - {}_H\mathbf{Q}^*$.

§ 10.1.5. Decomposition theorems. The following decomposition theorem is actually a consequence of the Dynkin theorem applied to the stopping time T_S (see § 8.1.1).

THEOREM 1. *For $\alpha > 0$:*

$$\mathbf{U}^\alpha = {}_H\mathbf{U}^{*\alpha} + \mathbf{B}^{*\alpha}\,\mathbf{U}^\alpha.$$

Proof. Using Theorem 4, § 10.1.4, and the (F)-equation,

$$\mathbf{B}^{*\alpha} = {}_H\mathbf{U}^{*\alpha}(\mathbf{Q} - {}_H\mathbf{Q}^*) = {}_H\mathbf{U}^{*\alpha}\mathbf{Q} - \alpha\,{}_H\mathbf{U}^{*\alpha} + \mathbf{I} = \mathbf{I} + {}_H\mathbf{U}^{*\alpha}(\mathbf{Q} - \alpha\mathbf{I}),$$

so

$$\mathbf{B}^{*\alpha}\mathbf{U}^\alpha = \mathbf{U}^\alpha - {}_H\mathbf{U}^{*\alpha}.$$ □

Thus, in terms of components, one has the following interpretation

$$\begin{aligned} p_{ij}(t) &= {}_Hp^*_{ij}(t) + \int_0^t \sum_{r\in H} d^*_{ir}(s)\,p_{rj}(t-s)\,ds \\ &= \mathbb{P}^i(X_t = j,\ t < T_S) + \mathbb{P}^i(X_t = j,\ T_S \leqslant t) \qquad \text{for} \quad i,j \in \mathbb{I}. \end{aligned}$$

Observe that Theorem 1 is a special case of a decomposition theorem for a killed chain; see Theorem 1, § 9.1.3, with $\mathbf{B}^{*\alpha} = \mathbf{K}^\alpha$. Evidently, Theorem 1 can be re-written as the "second resolvent equation" (Corollary 2).

COROLLARY 2. *For* $\alpha > 0$:

$$\mathbf{U}^\alpha = {}_H\mathbf{U}^{*\alpha} + {}_H\mathbf{U}^{*\alpha}(\mathbf{Q} - {}_H\mathbf{Q}^*)\mathbf{U}^\alpha.$$

The next result exhibits another aspect of structure of the taboo matrix ${}_H\mathbf{U}^{*\alpha}$.

COROLLARY 3. *For* $\alpha > 0$:

$$\mathbf{U}^\alpha - {}_H\mathbf{U}^{*\alpha} = \widehat{\mathbf{B}}^\alpha(\mathbf{U}^\alpha - {}_H\mathbf{U}^\alpha).$$

Proof. The first entrance decomposition theorem (from § 6.1.1),

$$\mathbf{U}^\alpha = {}_H\mathbf{U}^\alpha + \mathbf{B}^\alpha\,\mathbf{U}^\alpha,$$

when multiplied by $\widehat{\mathbf{B}}^\alpha$ from left, yields, in view of the product theorem from § 10.1.4,

$$\widehat{\mathbf{B}}^\alpha\mathbf{U}^\alpha = \widehat{\mathbf{B}}^\alpha{}_H\mathbf{U}^\alpha + \mathbf{B}^{*\alpha}\mathbf{U}^\alpha,$$

and the result follows from Theorem 1. □

The following proposition describes influence of the stopping time S on the structure of the matrix ${}_H\mathbf{U}^{*\alpha}$. Although this is the immediate consequence of Corollary 3, it merits to be stated separately.

PROPOSITION 4. *For* $\alpha > 0$:

$${}_H\mathbf{U}^{*\alpha} = {}_H\widehat{\mathbf{U}}^\alpha + \widehat{\mathbf{B}}^\alpha{}_H\mathbf{U}^\alpha.$$

Proof. One has the first entrance decomposition for S:

$$\mathbf{U}^\alpha = {}_H\widehat{\mathbf{U}}^\alpha + \widehat{\mathbf{B}}^\alpha\mathbf{U}^\alpha.$$

Thus, the assertion follows from Corollary 3. Direct verification (using Lemma 2, § 10.1.4) shows that ${}_H\widehat{\mathbf{U}}^\alpha$ has all components zero, except ${}_H\widehat{\mathbf{U}}^\alpha_{22} = {}_H\mathbf{U}^{*\alpha}_{22}$. □

For interest of clarity, the explicit expressions for balayage $\widehat{D}_{ij}(t)$ are repeated here:

$$\widehat{D}_{ij}(t) = \begin{cases} \mathbb{P}^i(X_S = j,\ S \leqslant t) & \text{for } i \in H,\ j \in H^c, \\ \mathbb{P}^i(X(T_S) = j,\ S = T_S \leqslant t) & \text{for } i, j \in H. \end{cases}$$

§ 10.1.6. Functionals. Additional probabilistic interpretation of the delayed taboo resolvent ${}_H\mathbf{U}^{*\alpha}$ and of the balayage operator $\mathbf{B}^{*\alpha}$ is provided by approach through functionals of a M.C. (see Sub-sections 8.1 and 8.2).

a) As noted in § 9.1.1, killing of a chain is done according to a MF $M = (M_t)$ of the form

$$M_t = I_{(t<T_S)}, \qquad t \geqslant 0.$$

Observe that

$$\mathbb{P}^i(M_0 = 1) = 1, \qquad \mathbb{P}^i(M_\infty = 0) = D_i^{*c} \geqslant 0, \qquad \mathbb{E}^i M_t = D_i^{*c}(t).$$

Hence, the set of all permanent states coincides with the state space $\mathbb{I}$.

The delayed taboo probabilities have been defined by

$${}_H p^*_{ij}(t) = \mathbb{P}^i(X_t = j,\ M_t),$$

or more generally, for any non-negative function f on $\mathbb{I}$,

$${}_H\mathbf{P}^*(t)f(i) = \mathbb{E}^i\big(f(X_t)M_t\big).$$

Consider now an extended AF $V^* = (V_t^*)$ of the form

$$V_t^* = \int_0^t f(X_s)M_s\,ds, \qquad t \geqslant 0,$$

for any non-negative function f on $\mathbb{I}$ (see § 8.2.2). Evidently, the characteristic and the α-potential of V^* are

$$\mathbb{E}^i V_t^* = \int_0^t {}_H\mathbf{P}^*(s)f(i)\,ds, \qquad \mathbb{E}^i \int_0^\infty e^{-\alpha t}\,dV_t^* = {}_H\mathbf{U}^{*\alpha} f(i).$$

In particular, ${}_H u^{*\alpha}_{ij}$ corresponds to $f(i) = \delta_{ij}$ for a fixed state j.

Consider now an extended AF $F^* = (F_t^*)$ of the form

$$F_t^* = \int_0^t f(X_s)(-dM_s), \qquad t \geqslant 0,$$

for any non-negative function f on $\mathbb{I}$. It is immediate to verify that indeed

$$F^*_{t+h} = F^*_t + M_t(F^*_h \circ \theta_t)$$

(see Def. 1, § 8.2.1).

THEOREM 1. *The α-potential of F^* coincides with the α-balayage $\mathbf{B}^{*\alpha}$.*

Proof. From the definition of the killing operation (in § 9.1.1) one has

$$\mathbb{E}^i\big(I_{\{j\}}(X_s)(-dM_s)\big) = \mathbb{P}^i\big(X(T_S) = j,\ T_S \in (s, s+ds)\big) = dD^*_{ij}(s).$$

Hence, the characteristic and the α-potential of F^* are:

$$\mathbb{E}^i F^*_t = \mathbf{B}^*(t)f(i), \qquad \mathbb{E}^i \int_0^\infty e^{-\alpha s}\, dF^*_s = \mathbf{B}^{*\alpha} f(i). \qquad \square$$

It is now clear that operator $\mathbf{B}^{*\alpha}$ coincides with the operator $\mathbf{K}^\alpha$ introduced in § 9.1.3. Thus, Theorem 1, § 10.1.4, applied to any non-negative function f, i.e.,

$$\mathbf{U}^\alpha f = {}_H\mathbf{U}^{*\alpha} f + \mathbf{B}^{*\alpha}\,\mathbf{U}^\alpha f,$$

coincides with Theorem 1, § 9.1.3. Here $\mathbf{U}^\alpha f$ and ${}_H\mathbf{U}^{*\alpha} f$ are α-potentials of $f \geqslant 0$ relative to the original and to the killed chain, respectively. The fact that the taboo semi-group is subordinate to the original semi-group implies that

$$\mathbf{U}^\alpha f \geqslant {}_H\mathbf{U}^{*\alpha} f, \qquad \mathbf{Q} f \geqslant {}_H\mathbf{Q}^* f \qquad \text{for } f \geqslant 0.$$

b) The AF $A = (A_t)$ such that $A_t = -\log M_t$ is degenerate, with α-potential (relative to the killed semi-group)

$$\mathbf{U}^{*\alpha}_A f(i) = \mathbb{E}^i \int_0^\infty e^{-\alpha s} f(X_s) M_s\, dA_s = \mathbb{E}^i \int_0^\infty e^{-\alpha s}\, dF^*_s$$

being infinite. Consequently, the analogue of Theorem 5, § 9.1.3, does not hold as it stands:

$$\mathbf{U}^\alpha f = {}_H\mathbf{U}^{*\alpha} f + \mathbf{U}^{*\alpha}_A\, {}_H\mathbf{U}^{*\alpha} f. \qquad (*)$$

This decomposition theorem remains valid provided $\mathbf{U}^{*\alpha}_A$ is given another interpretation, as already explained in Example 5, § 9.1.4. The argument exhibits an interesting connection between $\mathbf{Q} - {}_H\mathbf{Q}^*$ and the "quasi-Revuz measure" of the functional A.

Indeed, suppose that $(*)$ holds. Then, replacing f by $(\alpha\mathbf{I} - {}_H\mathbf{Q}^*)f$ and using the (F)-equation, one should have

$$\mathbf{U}^{*\alpha}_A = \mathbf{U}^\alpha(\mathbf{Q} - {}_H\mathbf{Q}^*),$$

see remarks following Theorem 6, § 9.1.3. The next theorem indicates that this new interpretation is the correct one.

THEOREM 2. *The quasi-Revuz measure μ_A associated with the functional A is given*

i) *by conditional form, for a fixed $i \in \mathbb{I}$,*

$$\mu^i_A(j) = \begin{cases} q_{ij} - q^*_{ij} \geqslant 0, & j \in H, \\ 0, & j \in H^c; \end{cases}$$

and its total mass is

$$\sum_j \mu^i_A(j) = d^*_i(0);$$

ii) *with reference measure ξ for the chain,*

$$\mu_A = \xi(\mathbf{Q} - {}_H\mathbf{Q}^*),$$

and μ_A is supported on H.

Proof. For a fixed $i \in \mathbb{I}$, consider a well defined measure

$$\mu_A^i(j) = (\mathbf{Q} - {}_H\mathbf{Q}^*)I_{\{j\}}(i) = q_{ij} - q_{ij}^* \geqslant 0,$$

which is non-negative for $j \in H$, but vanishes for $j \in H^c$. Summation over all j yields $d_i^*(0)$, already introduced in § 10.1.3.

Let ξ be a reference measure for the original chain, so $\xi(-\mathbf{Q}) \geqslant 0$. It can be deduced from structure of ${}_H\mathbf{Q}^*$ that ξ is also a reference measure for the killed chain, and that $\sigma^* = \xi(-{}_H\mathbf{Q}^*) \geqslant 0$. Then,

$$\mu_A(j) = \sum_i \xi(i)(q_{ij} - q_{ij}^*)$$

defines a measure on $\mathbb{I}$, supported on H, which however is not a Revuz measure of A.

Nevertheless, consider an integral functional A^* of the original chain, determined by the charge $a \geqslant 0$ such that

$$a(i)\xi(i) = \mu_A(i).$$

This means that the Revuz measure of A^* coincides with μ_A (called now the quasi-Revuz measure of A). Hence, proceeding as in Theorem 7, § 8.1.3, one has, using $\mathbf{U}_A^{*\alpha}$ defined above,

$$\lim_{\alpha\to\infty} \xi\alpha\mathbf{U}_A^{*\alpha} = \sigma\mathbf{U}_A^* = \xi(\mathbf{Q} - {}_H\mathbf{Q}^*) = \mu_A. \qquad \square$$

Observe that by Corollary 7, § 10.1.4, and using notation introduced above, one has

$$\mu_A = \sigma^*\mathbf{B}^*.$$

The next result expresses the basic property of the Revuz measure (see also Thm. 6, § 8.1.3).

THEOREM 3. *The α-potential of the integral functional $A^* = (A_t^*)$ coincides with the α-potential of the measure μ_A.*

Proof. It suffices to recall that

$$u_{A^*}^\alpha(i) = \mathbb{E}^i \int_0^\infty e^{-\alpha s}a(X_s)\,ds = \mathbf{U}^\alpha a(i) = \mathbf{U}^\alpha\mu_A(i), \qquad i \in \mathbb{I}. \qquad \square$$

It should be remarked that with the auxiliary functional A^* one can consider the killing according to the MF $M_t^* = e^{-A_t^*}$, with the killing rate $-dM_t^*/M_t^* = a(X_t)\,dt$. Then theory developed in § 9.1.3 applies.

c) The following result is the immediate consequence of Theorem 2 and Corollary 7, § 10.1.4.

COROLLARY 4. *For* $i \in \mathbb{I}$, $j \in H$:

$$\mu_A^i(j) = -\sum_k q_{ik}^* D_{kj}^* \geqslant 0,$$

$$d_i^*(0) = -\sum_k q_{ik}^* D_k^* \geqslant 0.$$

As a by-product one obtains from Corollary 4 the following form of the Poisson equation.

COROLLARY 5. *The first return probability* D_i^* *satisfies the equation*

$$\sum_j q_{ij} D_j^* = \begin{cases} -a_i & \text{for } i \in H^c, \\ 0 & \text{for } i \in H, \end{cases}$$

where

$$a_i = \sum_{j \in H} q_{ij} D_j^{*c} \geqslant 0 \qquad \text{for } i \in H^c.$$

Observe that this is the reciprocal form of the Poisson equation discussed earlier for which the escape rate has been b_i (for $i \in H$). Note also that the function (D_i^*) is excessive relative to $\mathbf{P}(t)$.

§ 10.1.7. Renewal equation. Define now a renewal matrix

$$\mathbf{W}(t) = \int_0^t \mathbf{P}(s)(\mathbf{Q} - {}_H\mathbf{Q}^*)\, ds, \qquad t \geqslant 0,$$

or, with $\mathbf{W}^\alpha$ denoting the Laplace–Stieltjes transform of $\mathbf{W}(t)$,

$$\mathbf{W}^\alpha = \mathbf{U}^\alpha(\mathbf{Q} - {}_H\mathbf{Q}^*), \qquad \alpha > 0.$$

Then the following matrix renewal equation holds.

THEOREM 1. *For* $\alpha > 0$:

$$\mathbf{W}^\alpha = \mathbf{B}^{*\alpha} + \mathbf{B}^{*\alpha}\mathbf{W}^\alpha.$$

Proof. The assertion follows from Theorem 1, § 10.1.5, and from Theorem 4, § 10.1.4, by multiplication by $\mathbf{Q} - {}_H\mathbf{Q}^*$ from right. □

It is of interest to see the analogy between $\mathbf{B}^{*\alpha}$ as defined in Theorem 4, § 10.1.4, and $\mathbf{W}^\alpha$ defined above; note that $\mathbf{W}^\alpha$ is the same as $\mathbf{U}_A^{*\alpha}$ mentioned in § 10.1.6. Observe that the non-zero components in Theorem 1 are

$$\mathbf{W}_{12}^\alpha = \mathbf{B}_{12}^\alpha + \mathbf{B}_{12}^\alpha \mathbf{W}_{22}^\alpha, \qquad \mathbf{W}_{22}^\alpha = \mathbf{B}_{22}^{*\alpha} + \mathbf{B}_{22}^{*\alpha} \mathbf{W}_{22}^\alpha.$$

Clearly, Theorem 1 is analogous to decomposition theorems in Sub-section 11.1. Hence, by the Riesz decomposition, the formal solution is found to be

$$\mathbf{W}^\alpha = \sum_{n=1}^\infty (\mathbf{B}^{*\alpha})^n + \lim_{n\to\infty} (\mathbf{B}^{*\alpha})^n \mathbf{W}^\alpha,$$

and the solution is unique when the limit on the right vanishes. Then, one has

COROLLARY 2. *The unique solution of the Markov renewal equation in Theorem 1 is*

$$\mathbf{W}^\alpha = \sum_{n=1}^{\infty} (\mathbf{B}^{*\alpha})^n, \qquad \alpha > 0.$$

Consider now the function $w(t)$ defined by

$$w_i(t) = \mathbf{W}(t)\mathbf{1}_i = \int_0^t \mathbf{P}(s) d_i^*(0)\, ds.$$

It follows immediately from Theorem 1 that $w_i(t)$ satisfies the following renewal equation.

PROPOSITION 3. *For* $i \in \mathbb{I}$, $t \geqslant 0$:

$$w_i(t) = D_i^*(t) + \int_0^t \sum_{k\in H} d_{ik}^*(s) w_k(t-s)\, ds.$$

Remark. Let Z be an integral functional determined by the charge $d_i^*(0)$. It is clear that $w_i(t)$ is the characteristic of Z, and its α-potential is

$$u_Z^\alpha(i) = \mathbf{U}^\alpha d_i^*(0).$$

Hence, as $t \to \infty$,

$$w_i(t) \longrightarrow \mathbf{U} d_i^*(0) \quad \text{and} \quad w_i(t)/t \longrightarrow \mathbf{E} d_i^*(0),$$

where $\mathbf{U}$ and $\mathbf{E}$ are potential and limit matrices of $\mathbf{P}(t)$, respectively.

§ 10.1.8. Duality. The decomposition theorem presented here is associated with the last exit time, and is dual to the theorem considered in § 10.1.5 which has been associated with the first entrance time. Derivation here is analytic in form, analogous to that in Sub-sections 6.1 and 6.2.

As a first step, it is necessary to define matrices dual to those used earlier in this sub-section. Recall that the dual of the α-balayage $\mathbf{B}^\alpha$ is the α-co-balayage $\mathbf{C}^\alpha$. Introduce now the dual $\widehat{\mathbf{C}}^\alpha$ of $\widehat{\mathbf{B}}^\alpha$, which is of transposed form

$$\widehat{\mathbf{C}}^\alpha = \begin{pmatrix} \mathbf{I}_1 & \widehat{\mathbf{C}}_{12}^\alpha \\ \mathbf{0} & \widehat{\mathbf{C}}_{22}^\alpha \end{pmatrix}.$$

Then the dual of the product theorem $\mathbf{B}^{*\alpha} = \widehat{\mathbf{B}}^\alpha \mathbf{B}^\alpha$ gives the α-co-balayage

$$\mathbf{C}^{*\alpha} = \mathbf{C}^\alpha \widehat{\mathbf{C}}^\alpha. \qquad \text{(PT)}$$

PROPOSITION 1. *The non-zero components of the α-co-balayage matrix are*

$$\mathbf{C}_{21}^{*\alpha} = \mathbf{C}_{21}^\alpha, \qquad \mathbf{C}_{22}^{*\alpha} = \mathbf{C}_{21}^\alpha \widehat{\mathbf{C}}_{12}^\alpha + \widehat{\mathbf{C}}_{22}^\alpha.$$

Next, denote by ${}_H\widetilde{\mathbf{U}}^{*\alpha}$ the dual of the taboo matrix ${}_H\mathbf{U}^{*\alpha}$, taken to be a semigroup with a generator of the form

$$ {}_H\widetilde{\mathbf{Q}}^* = \begin{pmatrix} \mathbf{Q}_{11} & \mathbf{Q}_{12} \\ \mathbf{0} & \mathbf{Q}^*_{22} \end{pmatrix}. $$

The (B)-equation

$$ \alpha\, {}_H\widetilde{\mathbf{U}}^{*\alpha} - \mathbf{I} = {}_H\widetilde{\mathbf{Q}}^*\, {}_H\widetilde{\mathbf{U}}^{*\alpha} $$

indicates that ${}_H\widetilde{\mathbf{U}}^{*\alpha}$ has the form

$$ {}_H\widetilde{\mathbf{U}}^{*\alpha} = \begin{pmatrix} {}_H\mathbf{U}^{\alpha}_{11} & {}_H\widetilde{\mathbf{U}}^{*\alpha}_{12} \\ \mathbf{0} & {}_H\mathbf{U}^{*\alpha}_{22} \end{pmatrix}, $$

so the only new term is ${}_H\widetilde{\mathbf{U}}^{*\alpha}_{12}$, the dual of ${}_H\mathbf{U}^{*\alpha}_{21}$.

The following results are dual versions of results stated in § 10.1.4, and their proofs, omitted here, follow analogous analytic pattern.

LEMMA 2. *For* $\alpha > 0$:

$$ {}_H\widetilde{\mathbf{U}}^{*\alpha}_{12} = {}_H\mathbf{U}^{\alpha}_{11}\, \widehat{\mathbf{C}}^{\alpha}_{12}. $$

LEMMA 3. *For* $\alpha > 0$:

$$ \widehat{\mathbf{C}}^{\alpha}_{22} = \left(\mathbf{Q}_{22} - \mathbf{Q}^*_{22}\right) {}_H\mathbf{U}^{*\alpha}_{22}. $$

THEOREM 4. *For* $\alpha > 0$:

$$ \mathbf{C}^{*\alpha} = \left(\mathbf{Q} - {}_H\widetilde{\mathbf{Q}}^*\right) {}_H\widetilde{\mathbf{U}}^{*\alpha}. $$

Indeed, here

$$ \mathbf{C}^{\alpha}_{21} = \mathbf{Q}_{21}\, {}_H\mathbf{U}^{\alpha}_{11}, \qquad \mathbf{Q}_{21}\, {}_H\widetilde{\mathbf{U}}^{*\alpha}_{12} = \mathbf{C}^{\alpha}_{21}\, \widehat{\mathbf{C}}^{\alpha}_{12}. $$

Writing $\mathbf{C}^*(t)$ and ${}_H\widetilde{\mathbf{P}}^*(t)$ for time versions of $\mathbf{C}^{*\alpha}$ and ${}_H\widetilde{\mathbf{U}}^{*\alpha}$, respectively, one has the following statements.

COROLLARY 5. *For* $t, h \geqslant 0$:

$$ \mathbf{C}^*(t+h) - \mathbf{C}^*(h) = \mathbf{C}^*(t)\, {}_H\widetilde{\mathbf{P}}^*(h). $$

COROLLARY 6. *For* $\alpha > 0$:

$$ \alpha\mathbf{C}^{*\alpha} = \mathbf{C}^{*\alpha}\, {}_H\widetilde{\mathbf{Q}}^* + \left(\mathbf{Q} - {}_H\widetilde{\mathbf{Q}}^*\right). $$

Now the main decomposition theorem is

THEOREM 7. *For* $\alpha > 0$:

$$ \mathbf{U}^{\alpha} = {}_H\widetilde{\mathbf{U}}^{*\alpha} + \mathbf{U}^{\alpha}\mathbf{C}^{*\alpha}. $$

This theorem, which is dual to Theorem 1, § 10.1.4, follows immediately from Theorem 4 and the (B)-equation. Observe analogy with the dual version mentioned in § 9.1.3-c, with identification $\widehat{\mathbf{K}}^{\alpha} = \mathbf{C}^{*\alpha}$. The corresponding "second resolvent equation" is

COROLLARY 8. *For* $\alpha > 0$:

$$\mathbf{U}^{\alpha} = {}_H\widetilde{\mathbf{U}}^{*\alpha} + \mathbf{U}^{\alpha}(\mathbf{Q} - {}_H\widetilde{\mathbf{Q}}^{*})\,{}_H\widetilde{\mathbf{U}}^{*\alpha}.$$

Next, the last exit decomposition theorem (from § 6.2.1), when multiplied by $\widehat{\mathbf{C}}^{\alpha}$ from right, yields

COROLLARY 9. *For* $\alpha > 0$:

$$\mathbf{U}^{\alpha} - {}_H\widetilde{\mathbf{U}}^{*\alpha} = (\mathbf{U}^{\alpha} - {}_H\mathbf{U}^{\alpha})\widehat{\mathbf{C}}^{\alpha}.$$

Finally, using the last exit decomposition for S,

$$\mathbf{U}^{\alpha} = {}_H\widehat{\mathbf{U}}^{\alpha} + \mathbf{U}^{\alpha}\widehat{\mathbf{C}}^{\alpha},$$

one obtains the dual of Proposition 4, § 10.1.5.

PROPOSITION 10. *For* $\alpha > 0$:

$${}_H\widetilde{\mathbf{U}}^{*\alpha} = {}_H\widehat{\mathbf{U}}^{\alpha} + {}_H\mathbf{U}^{\alpha}\widehat{\mathbf{C}}^{\alpha}.$$

Observe that both these propositions use the same matrix ${}_H\widehat{\mathbf{U}}^{\alpha}$. Indeed, using Lemma 2 above, it is easy to verify that Proposition 10 implies that ${}_H\widehat{\mathbf{U}}^{\alpha}$ has all components equal to zero, except ${}_H\widehat{\mathbf{U}}^{\alpha}_{22}$.

§ 10.1.9. Extreme cases. The selection of the term $\mathbf{Q}^{*}_{22}$ is somewhat arbitrary. It should depend on the number of transitions within the set H before the exit time S. Two extreme cases will be now considered.

Case 1. Assume that only one jump is allowed. Then the M.C. leaves H when the initial state $i \in H$ terminates. This has been the situation treated in Subsection 3.3; see also Example 4, § 9.1.4. Hence, $S = \rho_i$ and $T_S = T_{\rho}$, so the killed chain refers to the exit taboo discussed in § 2.2.4. Consequently,

$${}_H\mathbf{Q}^{*} = {}_H\mathbf{Q}^{\rho}, \qquad {}_H\mathbf{U}^{*\alpha} = {}_H\mathbf{U}^{\rho\alpha},$$

with

$$\mathbf{Q}^{*}_{22} = -\mathbf{\Theta}_2, \qquad {}_H\mathbf{U}^{*\rho}_{22} = (\mathbf{\Theta}^{\alpha}_2)^{-1}.$$

The matrix $\widehat{\mathbf{B}}^{\alpha}$ is related to the α-jump matrix $\mathbf{R}^{\alpha}$ in the fashion

$$\widehat{\mathbf{B}}^{\alpha} = \begin{pmatrix} \mathbf{I}_1 & \mathbf{0} \\ \mathbf{R}^{\alpha}_{21} & \mathbf{R}^{\alpha}_{22} \end{pmatrix},$$

so the product theorem (already mentioned in § 3.3.5) yields

$$\mathbf{B}^{*\alpha} = \mathbf{B}^{\rho\alpha} = \widehat{\mathbf{B}}^{\alpha}\,\mathbf{B}^{\alpha},$$

with components

$$\mathbf{B}^{\rho\alpha}_{12} = \mathbf{B}^{\alpha}_{12}, \qquad \mathbf{B}^{\rho\alpha}_{22} = \mathbf{R}^{\alpha}_{22} + \mathbf{R}^{\alpha}_{21}\,\mathbf{B}^{\alpha}_{12}.$$

The decomposition theorem (Thm. 1, § 10.1.5) reduces now to Theorem 1, § 6.1.2,

$$\mathbf{U}^{\alpha} = {}_H\mathbf{U}^{\rho\alpha} + \mathbf{B}^{\rho\alpha}\mathbf{U}^{\alpha}.$$

Note also that Lemma 2, § 10.1.4, reduces to Theorem 2, § 2.2.4,

$${}_H\mathbf{U}_{21}^{\rho\alpha} = \mathbf{R}_{21}^{\alpha}\,{}_H\mathbf{U}_{11}^{\alpha}.$$

The matrix $\widehat{\mathbf{C}}^{\alpha}$ is related to the dual $\widehat{\mathbf{R}}^{\alpha}$ of $\mathbf{R}^{\alpha}$, defined in § 3.4.4-d, in such a way that

$$\widehat{\mathbf{C}}^{\alpha} = \begin{pmatrix} \mathbf{I}_1 & \widehat{\mathbf{R}}_{12}^{\alpha} \\ \mathbf{0} & \widehat{\mathbf{R}}_{22}^{\alpha} \end{pmatrix}, \qquad \text{where} \qquad \begin{array}{l} \widehat{\mathbf{R}}_{12}^{\alpha} = \boldsymbol{\Theta}_1^{\alpha}\,\mathbf{R}_{12}^{\alpha}(\boldsymbol{\Theta}_2^{\alpha})^{-1}, \\ \widehat{\mathbf{R}}_{22}^{\alpha} = \boldsymbol{\Theta}_2^{\alpha}\,\mathbf{R}_{22}^{\alpha}(\boldsymbol{\Theta}_2^{\alpha})^{-1}, \end{array}$$

so the product theorem (already mentioned in § 3.4.4) yields

$$\mathbf{C}^{*\alpha} = \mathbf{C}^{\rho\alpha} = \mathbf{C}^{\alpha}\widehat{\mathbf{C}}^{\alpha},$$

with components

$$\mathbf{C}_{21}^{\rho\alpha} = \mathbf{C}_{21}^{\alpha}, \qquad \mathbf{C}_{22}^{\rho\alpha} = \widehat{\mathbf{R}}_{22}^{\alpha} + \mathbf{C}_{21}^{\alpha}\,\widehat{\mathbf{R}}_{12}^{\alpha}.$$

The decomposition theorem (Thm. 7, § 10.1.8) reduces now to Theorem 1, § 6.2.2,

$$\mathbf{U}^{\alpha} = {}_H\widetilde{\mathbf{U}}^{\rho\alpha} + \mathbf{U}^{\alpha}\mathbf{C}^{\rho\alpha},$$

where ${}_H\widetilde{\mathbf{U}}^{\rho\alpha} = {}_H\widetilde{\mathbf{U}}^{*\alpha}$ whose the only new component is given by Lemma 2, § 10.1.8,

$${}_H\widetilde{\mathbf{U}}_{12}^{\rho\alpha} = {}_H\mathbf{U}_{11}^{\alpha}\widehat{\mathbf{R}}_{12}^{\alpha}.$$

Case 2. Assume that the number of transitions within H is not fixed. Then, when the chain starts in H, it will continue in H until the first entrance to H^c at time S. Therefore, S is the usual first entrance time to H^c, and consequently, for the killed chain,

$${}_H\mathbf{P}_{22}^{*}(t) = {}_{H^c}\mathbf{P}_{22}(t), \qquad \mathbf{Q}_{22}^{*} = \mathbf{Q}_{22}.$$

Furthermore, $\widehat{\mathbf{B}}_{22} = \mathbf{0}$, so the product theorem reduces to

$$\mathbf{B}_{12}^{*\alpha} = \mathbf{B}_{12}^{\alpha}, \qquad \mathbf{B}_{22}^{*\alpha} = \widehat{\mathbf{B}}_{21}^{\alpha}\,\mathbf{B}_{12}^{\alpha}.$$

Lemma 2, § 10.1.4, gives the explicit expression for the taboo probability

$${}_H\mathbf{U}_{21}^{*\alpha} = \widehat{\mathbf{B}}_{21}^{\alpha}\,{}_H\mathbf{U}_{11}^{\alpha},$$

and

$${}_H\mathbf{U}_{11}^{*\alpha} = {}_H\mathbf{U}_{11}^{\alpha}, \qquad {}_H\mathbf{U}_{22}^{*\alpha} = {}_{H^c}\mathbf{U}_{22}^{\alpha}.$$

By properties of the first entrance times

$$\mathbf{B}_{12}^{\alpha} = {}_H\mathbf{U}_{11}^{\alpha}\,\mathbf{Q}_{12}, \qquad \widehat{\mathbf{B}}_{21}^{\alpha} = {}_{H^c}\mathbf{U}_{22}^{\alpha}\,\mathbf{Q}_{21}.$$

Hence, in agreement with Theorem 4, § 10.1.4,

$$\mathbf{B}_{22}^{*\alpha} = {}_H\mathbf{U}_{21}^{*\alpha}\,\mathbf{Q}_{12},$$

because the only non-zero component of $\mathbf{Q} - {}_H\mathbf{Q}^*$ is $\mathbf{Q}_{12}$. It is worthwhile to state explicitly, for $i, j \in H$,

$$D_{ij}^*(t) = \int_0^t \sum_{k\in H^c} {}_Hp_{ik}^*(s)q_{kj}\,ds = \mathbb{P}^i\big(X(T_S) = j,\ S < T_S \leqslant t\big),$$

with $\mathbb{P}^i(X(T_S) = j,\ S = T_S \leqslant t) = 0$.

Decomposition theorem (Theorem 1, § 10.1.5) holds, so in particular,

$$\mathbf{U}_{21}^{\alpha} = \widehat{\mathbf{B}}_{21}^{\alpha}\,\mathbf{U}_{11}^{\alpha}, \qquad \mathbf{U}_{22}^{\alpha} - {}_H\mathbf{U}_{22}^{*\alpha} = \widehat{\mathbf{B}}_{21}^{\alpha}\,\mathbf{U}_{12}^{\alpha}.$$

(Compare this situation with that in Example 2, § 5.4.2, in which the set H^c was closed.)

Dually, $\widehat{\mathbf{C}}_{22} = \mathbf{0}$, so the product theorem reduces to

$$\mathbf{C}_{21}^{*\alpha} = \mathbf{C}_{21}^{\alpha}, \qquad \mathbf{C}_{22}^{*\alpha} = \mathbf{C}_{21}^{\alpha}\widehat{\mathbf{C}}_{12}^{\alpha}.$$

Lemma 2, § 10.1.8, gives the explicit expression for the dual taboo probability

$${}_H\widetilde{\mathbf{U}}_{12}^{*\alpha} = {}_H\mathbf{U}_{11}^{\alpha}\widehat{\mathbf{C}}_{12}^{\alpha}.$$

By properties of last exit times,

$$\mathbf{C}_{21}^{\alpha} = \mathbf{Q}_{21}\,{}_H\mathbf{U}_{11}^{\alpha}, \qquad \widehat{\mathbf{C}}_{12}^{\alpha} = \mathbf{Q}_{12}\,{}_{H^c}\mathbf{U}_{22}^{\alpha}.$$

Hence, in agreement with Theorem 4, § 10.1.8,

$$\mathbf{C}_{22}^{*\alpha} = \mathbf{Q}_{21}\,{}_H\widetilde{\mathbf{U}}_{12}^{*\alpha},$$

because the only non-zero component of $\mathbf{Q} - {}_H\widetilde{\mathbf{Q}}^*$ is $\mathbf{Q}_{21}$.

Decomposition theorem (Thm. 7, § 10.1.8) holds, so in particular,

$$\mathbf{U}_{12}^{\alpha} = \mathbf{U}_{11}^{\alpha}\widehat{\mathbf{C}}_{12}^{\alpha}, \qquad \mathbf{U}_{22}^{\alpha} - {}_H\widetilde{\mathbf{U}}_{22}^{*\alpha} = \mathbf{U}_{21}^{\alpha}\widehat{\mathbf{C}}_{12}^{\alpha}.$$

(Compare this situation with that in Example 1, § 5.4.2, in which the set H was closed.)

Finally, observe that the non-zero components of the renewal matrix $\mathbf{W}^{\alpha}$, introduced in § 10.1.7, are

$$\mathbf{W}_{12}^{\alpha} = \mathbf{U}_{11}^{\alpha}\mathbf{Q}_{12}, \qquad \mathbf{W}_{22}^{\alpha} = \mathbf{U}_{21}^{\alpha}\mathbf{Q}_{12}.$$

Thus, the renewal equation from Proposition 3, § 10.1.7, holds.

10.2. CONSECUTIVE ENTRANCES

Most of discussion so far has been concerned with individual entrances to (or exits from) a set H. It is, however, important to consider consecutive entrances (or exits), as well as time intervals between them.

The analysis of consecutive entrances to a set H is linked with renewal argument and leads to Markov renewal processes. In fact, a Markov renewal equation for the average number of visits to H has a form of a decomposition theorem, similar to those discussed earlier.

Presentation in this sub-section is based on the approach from Chapter I, combined with the approach through functionals from Section 8. As such, it might be used as the alternative approach to study of passage times. The material is scattered in technical literature, and the clear reference may be found in E. Çinlar [1972, 1975], for Markov renewal processes, and in K. L. Chung [1967]. This sub-section leans heavily on discussion of Markov renewal kernel presented in the previous Sub-section 10.1.

§ 10.2.1. Definitions. Let $X = (X_t,\ 0 \leqslant t < \infty)$ be the usual M.C. with a state space $\mathbb{I}$, and let H be a taboo set, $H^c = \mathbb{I} \setminus H$. Let $T = T_H$ be the first entrance time to a set H (defined in § 2.1.2), and let $S = S_{H^c}$ be the first entrance time to a set H^c (as defined in § 10.1.1). Then,

$$T_S = S + T \circ \theta_S = \inf(t\colon t > S,\ X_t \in H),$$
$$S_T = T + S \circ \theta_T = \inf(t\colon t > T,\ X_t \in H^c)$$

are, respectively, the first entrance time to H (or to H^c) after the first entrance to H^c (or to H). It follows from Proposition 5, § 2.2.2, that T_S and S_T are terminal stopping times (relative to the natural σ-fields of X). Moreover, $X(T_S)$ and $X(S_T)$ take values, respectively, in H only (when $T_S < \infty$) and in H^c only (when $S_T < \infty$). If the chain is initially in H^c, then $T > 0$ and $S_T > T$; if the chain starts from a state in H, then $S > 0$ and $T_S > S$ (see Fig. 10.2.1).

The following notation will be used, in agreement with § 10.1.1:

$$\mathbb{P}^i(X(T_S) = j,\ T_S \leqslant t) = D^*_{ij}(t), \qquad \mathbb{P}^i(X(S_T) = j,\ S_T \leqslant t) = \widehat{D}^*_{ij}(t).$$

As shown in Case 2, § 10.1.9, distribution of T_S satisfies

$$D^{*c}_i(t) = \sum_j {}_H p^*_{ij}(t), \qquad D^*_i(t) = \int_0^t \sum_{j \in H^c} {}_H p^*_{ij}(s)\bar{q}_j\, ds.$$

Dually, distribution of S_T satisfies

$$\widehat{D}^{*c}_i(t) = \sum_j {}_{H^c} p^*_{ij}(t), \qquad \widehat{D}^*_i(t) = \int_0^t \sum_{j \in H} {}_{H^c} p^*_{ij}(s)(q_j - \bar{q}_j)\, ds.$$

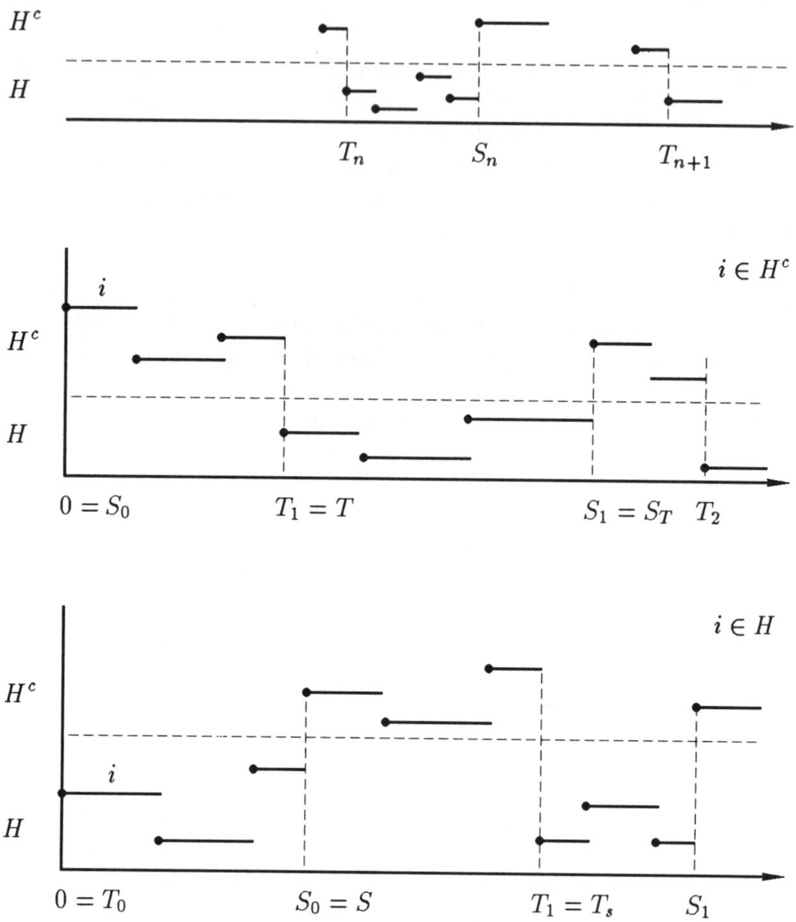

$$T_n = S_{n-1} + T \circ \theta_{S_{n-1}}, \qquad S_n = T_n + S \circ \theta_{T_n},$$
$$T = \inf(t\colon t > 0,\ X_t \in H), \qquad S = \inf(t\colon t > 0,\ X_t \in H^c),$$
$$T_n < S_n < T_{n+1}, \qquad n \geqslant 1,$$
$$T_{n+1} - T_n = T_S \circ \theta_{T_n}, \qquad n \geqslant 1.$$

Figure 10.2.1. Consecutive entrances to a set H.

a) Denote now by T_n, $n \geqslant 1$, the consecutive entrances to a set H, and by S_n, $n \geqslant 1$, the consecutive entrances to a set H^c. Fixing attention on entrances to H, define inductively

$$T_n = \inf\big(t\colon t > S_{n-1},\ X_t \in H\big) = S_{n-1} + T \circ \theta_{S_{n-1}},$$
$$S_n = \inf\big(t\colon t > T_n,\ X_t \in H^c\big) = T_n + S \circ \theta_{T_n},$$

with initial conditions ($X_0 = i$): $S_0 = 0$, $T_1 = T$, $S_1 = S_T$, when $i \in H^c$, and

$T_0 = 0,\ S_0 = S,\ T_1 = T_S,$ when $i \in H$. Thus, for $n \geqslant 1$,

$$T_n < S_n < T_{n+1}.$$

With these definitions, starting at $i \in H^c$, the chain enters H at $T_1 = T > 0$ and then leaves it at $S_1 = S_T$, whereas starting at $i \in H$, the chain leaves H at $S_0 = S > 0$ and then returns to H at $T_1 = T_S$. See Fig. 10.2.1 and also Fig. 9.5.2.

Observe that, apart from the initial segment T_1 (which is different for $i \in H^c$ and for $i \in H$), the consecutive entrances T_n to H are separated by the common time T_S. Indeed, it follows from the above definitions that

$$T_{n+1} - T_n = T_S \circ \theta_{T_n}, \qquad n \geqslant 1.$$

Note also that if $T_m < \infty$ and $T_{m+1} = \infty$ for some m, then $T_n = \infty$ for all $n > m$. Hence $S_m = L$ is the last exit time from H. It will be assumed that

$$\sup_n T_n = \infty.$$

It is important to observe that time instants $(S_n,\ n \geqslant 1)$ coincide with the random set W_H of discontinuities of an increasing process $(T_s,\ s > 0)$, whereas time instants $(T_n,\ n \geqslant 1)$ coincide with random set V_H of discontinuities of an increasing process $(L_s,\ s > 0)$; see definitions in § 2.1.1 and comments in § 9.5.2. Indeed, if (V_t) is the occupation time of a set H, and $\tau = (\tau_t)$ is its inverse (see § 9.5.2), then if t is a discontinuity point of (τ_t), then, for some n,

$$\tau_{t-} = S_n, \qquad \tau_t = T_{n+1}, \qquad V_{S_n} = t.$$

Moreover, the random sets $[T_n, S_n)$ in the range of τ correspond to sojourn times in the set H.

It is easy to verify that T_n and S_n are (for each n) stopping times relative to the natural σ-fields of the chain X. Thus, one can define the following joint distributions (which are natural extensions of the case $n = 1$).

For $i \in \mathbb{I},\ n \geqslant 1$:

$$D_{ij}^{(n)}(t) = \mathbb{P}^i(X_{T_n} = j,\ T_n \leqslant t) \quad \text{for} \quad j \in H \quad (\text{and } 0 \quad \text{for } j \in H^c),$$
$$\widehat{D}_{ij}^{(n)}(t) = \mathbb{P}^i(X_{S_n} = j,\ S_n \leqslant t) \quad \text{for} \quad j \in H^c \quad (\text{and } 0 \quad \text{for } j \in H),$$

with corresponding marginal distributions $D_i^{(n)}(t)$, $D_{ij}^{(n)}$ and $\widehat{D}_i^{(n)}(t)$, $\widehat{D}_{ij}^{(n)}$. In particular, one has, for $i \in H^c$,

$$D_{ij}^{(1)}(t) = D_{ij}(t), \qquad \widehat{D}_{ij}^{(1)}(t) = \widehat{D}_{ij}^*(t),$$

and, for $i \in H$,

$$D_{ij}^{(1)}(t) = D_{ij}^*(t), \qquad \widehat{D}_{ij}^{(0)}(t) = \widehat{D}_{ij}(t).$$

Corresponding notation for L.S. transforms will be $d_{ij}^{(n)\alpha}$ and $\hat{d}_{ij}^{(n)\alpha}$.

b) In the present set-up, the joint distribution $D_{ij}^*(t)$ of T_S and $X(T_S)$ plays an important role. The following proposition connect this distribution with analogous distributions for first entrances to H and to H^c.

PROPOSITION 1. *For $i, j \in H$ and $t \geqslant 0$:*

$$D^*_{ij}(t) = \int_0^t \sum_{k\in H^c} \hat{d}_{ik}(s) D_{kj}(t-s)\, ds$$

with the density

$$d^*_{ij}(t) = \int_0^t \sum_{k\in H^c} \hat{d}_{ik}(s) d_{kj}(t-s)\, ds$$

or, in terms of transforms,

$$d^{*\alpha}_{ij} = \sum_{k\in H^c} \hat{d}^\alpha_{ik}\, d^\alpha_{kj}.$$

For $i \in H^c$, $j \in H$:

$$D^*_{ij}(t) = D_{ij}(t).$$

It is clear that Proposition 1 corresponds to the product theorem for the stopping time T_S. For the proof see Proposition 1, § 10.1.4, and note that situation here corresponds to Case 2 discussed in § 10.1.9. The following interpretation may be useful. Starting in H, the chain remains in H till the time s when it enters H^c at state k. Then, taboo is imposed and subsequently the chain returns to H in the remaining time $t-s$, entering at state j. When starting in H^c, taboo is imposed at the beginning, and clearly T_S coincides with T. Thus, $D^*_{ij}(t)$ for $i \in H$ may be interpreted as the proper probability of the first return to set H.

c) Direct analysis of sample functions of a M.C. X indicates that, for $i \in \mathbb{I}$:

$$\mathbb{P}^i(X_t = j,\ t < T_n) = {}_H p^{(n)}_{ij}(t), \qquad j \in \mathbb{I},$$

$$\mathbb{P}^i(X_t = j,\ T_n \leqslant t < S_n) = \int_0^t \sum_{r\in H} d^{(n)}_{ir}(s)\, {}_{H^c}p_{rj}(t-s)\, ds, \qquad j \in H,$$

$$\mathbb{P}^i(X_t = j,\ S_n \leqslant t < T_{n+1}) = \int_0^t \sum_{k\in H^c} \hat{d}^{(n)}_{ik}(s)\, {}_H p_{kj}(t-s)\, ds, \qquad j \in H^c,$$

$$\mathbb{P}^i(X_t = j,\ T_{n+1} \leqslant t) = \int_0^t \sum_{r\in H} d^{(n+1)}_{ir}(s)\, p_{rj}(t-s)\, ds, \qquad j \in \mathbb{I}.$$

Addition of these terms yields decomposition of $\mathbb{P}^i(X_t = j) = p_{ij}(t)$; cf. Theorem 1, § 10.1.5.

§ 10.2.2. Markov renewal equation. The integral equation of renewal type for the average number of visits to the set H will be now derived.

1) The following are the basic recursive relations.

Lemma 1. *For $i \in \mathbb{I}$, $j \in H$, and $t \geqslant 0$, $D_{ij}^{(n)}(t)$ has the density $d_{ij}^{(n)}(t)$ for $n \geqslant 1$, and satisfies*

$$D_{ij}^{(n+1)}(t) = \int_0^t \sum_{r \in H} d_{ir}^{(1)}(s) D_{rj}^{(n)}(t-s)\, ds, \qquad n \geqslant 1,$$

where

$$d_{ir}^{(1)}(s) = \begin{cases} d_{ir}(s) & \text{for} \quad i \in H^c, \\ d_{ir}^*(s) & \text{for} \quad i \in H. \end{cases}$$

Dually,

$$D_{ij}^{(n+1)}(t) = \int_0^t \sum_{r \in H} d_{ir}^{(n)}(s) D_{rj}^*(t-s)\, ds, \qquad n \geqslant 1.$$

Proof (outline). Using $T_{n+1} = T_n + T_S \circ \theta_{T_n}$ $(n \geqslant 1)$, the dual relation follows from the decomposition

$$D_{ij}^{(n+1)}(t) = \sum_{r \in H} \mathbb{P}^i(X(T_n) = r,\ X(T_{n+1}) = j,\ T_{n+1} \leqslant t),$$

by the strong Markov property and conditioning on $X(T_n)$ and T_n.

For the first relation, two cases should be considered, according to the initial state i in $\mathbb{I}$.

a) Let $i \in H$. It is easy to verify that now

$$T_{n+1} = T_S + T_n \circ \theta_{T_S} \qquad \text{for} \quad n \geqslant 1 \quad \text{with} \quad T_1 = T_S,$$

and the result follows in the same manner by conditioning on $X(T_S)$ and T_S.

b) Let $i \in H^c$. Consideration of the sequence (S_n) yields the analogous relation

$$S_n = S_T + S_{n-1} \circ \theta_{S_T} \qquad \text{for} \quad n \geqslant 1 \quad \text{with} \quad S_0 = 0.$$

Hence, definition of T_{n+1} yields dually

$$T_{n+1} = S_T + T_n \circ \theta_{S_T} \qquad \text{for} \quad n \geqslant 1 \quad \text{with} \quad T_1 = T.$$

Proceeding as before, one finds

$$D_{ij}^{(n+1)}(t) = \int_0^t \sum_{k \in H^c} \hat{d}_{ik}^*(s) D_{kj}^{(n)}(t-s)\, ds.$$

The required form of the assertion follows from the product theorem expressing balayage of S_T in terms of balayages of S and of T (dual form of Proposition 1, § 10.2.1). □

Lemma 1 expresses the main fact that T_S represents life time between consecutive extrances to H. By construction, the first entrance to H occurs at time $T_1 = T > 0$ when $i \in H^c$, but at time $T_1 = T_S > 0$ when $i \in H$.

2) Let N_t be the number of consecutive entrances to the set H during the time interval $(0, t]$, defined by

$$N_t = \sum_{n=1}^{\infty} I_{(T_n \leqslant t)}, \qquad t > 0,$$

with $N_0 = 0$. Evidently, $N = (N_t,\ 0 \leqslant t < \infty)$ is an increasing process, and $N_t = \sup(n\colon\ T_n \leqslant t)$. Hence, for $i \in \mathbb{I}$,

$$\mathbb{P}^i(N_t = 0) = \mathbb{P}^i(T_1 > t) = 1 - D_i^{(1)}(t),$$
$$\mathbb{P}^i(N_t = n) = \mathbb{P}^i(T_n \leqslant t < T_{n+1}) = D_i^{(n)}(t) - D_i^{(n+1)}(t), \qquad n \geqslant 1,$$

with

$$D_i^{(1)}(t) = \begin{cases} D_i(t) & \text{for} \quad i \in H^c, \\ D_i^*(t) & \text{for} \quad i \in H. \end{cases}$$

Clearly, $\mathbb{P}^i(N_t > n) = D_i^{(n+1)}(t)$.

For a stable M.C., the number of jumps within a finite interval cannot be infinite, so $D_i^{(n)}(t) \to 0$ as $n \to \infty$ (for each t). Hence, the mean number of visits to H is

$$\mathbb{E}^i N_t = \sum_{n=1}^{\infty} D_i^{(n)}(t), \qquad 0 \leqslant t < \infty.$$

It will be convenient to write $U_i(t) = \mathbb{E}^i N_t$. The next theorem gives the basic integral equation for the "renewal function" $U_i(t)$.

THEOREM 2 (Markov renewal equation). *For $i \in \mathbb{I}$ and $t \geqslant 0$:*

$$U_i(t) = D_i^{(1)}(t) + \int_0^t \sum_{r \in H} d_{ir}^{(1)}(s) U_r(t-s)\, ds,$$

where

$$D_i^{(1)}(t) = \begin{cases} D_i(t) & \textit{for} \quad i \in H^c, \\ D_i^*(t) & \textit{for} \quad i \in H. \end{cases}$$

Proof. The result follows in the usual way by writing

$$U_i(t) = D_i^{(1)}(t) + \sum_{n=2}^{\infty} D_i^{(n)}(t)$$

and using Lemma 1. □

Denoting the L.S. transform of $U_i(t)$ by u_i^α, Theorem 2 becomes, in the transformed form,

$$u_i^\alpha = d_i^{(1)\alpha} + \sum_{r \in H} d_{ir}^{(1)\alpha}\, u_r^\alpha.$$

Remark. In works on Markov renewal processes, N_t is defined as the number of renewals in the closed interval $[0, t]$, including the jump at $t = 0$. Writing $U_i^0(t) = \mathbb{E}^i N_t$ for this extended N_t, one finds that now

$$U_i^0(t) = \sum_{n=0}^{\infty} D_i^{(n)}(t),$$

with $D_i^{(0)}(t)$ equal to 1 for $i \in H$, and to 0 for $i \in H^c$. Consequently, the Markov renewal equation takes the form

$$U_i^0(t) = 1 + \int_0^t \sum_{r \in H} d_{ir}^{(1)}(s) U_r^0(t-s)\, ds \qquad \text{for} \quad i \in H.$$

and $U_i^0(t) = U_i(t)$ for $i \in H^c$.

The choice of $[0, t]$ is in agreement with interpretation of $U_i^0(t)$ as a potential operator. On the other hand, selection of $(0, t]$ corresponds to regarding $N = (N_t)$ as an increasing process (with $N_0 = 0$), in agreement with § 8.1.2. Moreover, the choice of semi-open interval agrees with the classical form of renewal equation for $U_i(t)$. Thus, for convenience, the form $(0, t]$ is taken here.

3) In the same fashion, consider, for $t \geqslant 0$ and $j \in H$,

$$U_{ij}(t) = \sum_{n=1}^{\infty} D_{ij}^{(n)}(t) = \mathbb{E}^i(N_t,\ X(T_{N_t}) = j) \qquad \text{with} \qquad U_i(t) = \sum_{j \in H} U_{ij}(t).$$

Analogous procedure, using dual form of Lemma 1, yields the dual version of Theorem 2.

THEOREM 3. *For $i \in \mathbb{I}$ and $t \geqslant 0$:*

$$U_i(t) = D_i^{(1)}(t) + \int_0^t \sum_{r \in H} D_r^*(t-s)\, dU_{ir}(s).$$

In terms of the L.S. transforms, Theorem 3 becomes:

$$u_i^\alpha = d_i^{(1)\alpha} + \sum_{r \in H} u_{ir}^\alpha\, d_r^{*\alpha}.$$

Warning. Do not confuse L.S. transform u_{ij}^α of $U_{ij}(t)$ with α-resolvent $\mathbf{U}^\alpha = (u_{ij}^\alpha)$.

§ 10.2.3. Solution. In the Markov situation, the explicit solution of the integral equation for $U_i(t)$ is available.

THEOREM 1. *For $i \in \mathbb{I}$, the unique solution $U_i(t)$ of the Markov renewal equation in Theorem 2, § 10.2.2, is given by*

$$U_i(t) = \int_0^t \sum_{j \in H^c} p_{ij}(s)\bar{q}_j \, ds, \qquad t \geqslant 0,$$

with "renewal density" for jumps from H^c to H of the form

$$u_i(t) = \sum_{j \in H^c} p_{ij}(t)\bar{q}_j, \qquad t \geqslant 0.$$

Proof. a) Define $N = (N_t)$ alternatively as a discontinuous AF,

$$N_t = \sum_{0<s\leqslant t} I_{H^c}(X_{s-})I_H(X_s), \qquad t > 0,$$

with $N_0 = 0$. Proceeding as in § 8.1.3, one finds that

$$g_{\mathbf{Q}}(i) = \sum_{j \neq i} I_{H^c}(i)\, I_H(j)\, q_{ij} = I_{H^c}(i)\bar{q}_i.$$

Then the compensator $N^* = (N_t^*)$ of $N = (N_t)$ is a CAF of the form

$$N_t^* = \int_0^t g_{\mathbf{Q}}(X_s)\, ds, \qquad t \geqslant 0.$$

Hence, N and N^* have the same α-potential and characteristic, so

$$u_N^\alpha(i) = \mathbb{E}^i \int_0^\infty e^{-\alpha t}\, dN_t = \sum_{j \in H^c} u_{ij}^\alpha\, \bar{q}_j, \qquad \alpha > 0,$$

$$\mathbb{E}^i N_t = \int_0^t \mathbf{P}(s)\, g_{\mathbf{Q}}(i)\, ds = \int_0^t \sum_{j \in H^c} p_{ij}(s)\, \bar{q}_j\, ds, \qquad t \geqslant 0$$

(see § 8.1.3). Evidently, $u_N^\alpha(i) = u_i^\alpha$ and $\mathbb{E}^i N_t = U_i(t)$.

b) In the analogous manner, consider the extended DAF of the form

$${}_H N_t = \sum_{0<s\leqslant t} I_{H^c}(X_{s-})\, I_H(X_s)\, I_{(s<T_S)}, \qquad t > 0,$$

with ${}_H N_0 = 0$. This represents the number of jumps from H^c to H before time T_S; there is at most one such jump.

The compensator is now given by an AF of the form

$$ {}_H N_t^* = \int_0^t g_{\mathbf{Q}}(X_s) I_{(s<T_S)}\, ds $$

and its characteristic is (see Ex. 1-d, § 8.2.3)

$$ \mathbb{E}^i\, {}_H N_t^* = \int_0^t \sum_{j \in H^c} {}_H p_{ij}^*(s) \bar{q}_j \, ds. $$

c) To show that $U_i(t)$ has the required form, it suffices to observe that one deals here with Case 2 considered in § 10.1.9. Using notation from § 10.1.7, one finds from comparison of expressions that

$$ \mathbb{E}^i\, N_t = w_i(t), \qquad \mathbb{E}^i\, {}_H N_t^* = D_i^*(t). $$

Hence, the assertion follows from Proposition 3, § 10.1.7.

Alternatively, one can use the decomposition theorem for $p_{ij}(t)$, relative to T_S, stated in terms of components after Theorem 1, § 10.1.5. Multiplication by $\bar{q}_j$, summation over all $j \in H^c$, and integration from 0 to t, yields Proposition 3, § 10.1.7.

d) Uniqueness of the solution follows from comments in §§ 10.1.7 and 10.2.6. □

It is now clear that

$$ U_{ij}(t) = \int_0^t \sum_{k \in H^c} p_{ik}(s) q_{kj}\, ds, \qquad j \in H. $$

Remark. Observe that summation in expression for $U_i(t)$ in Theorem 1 is extended only over the set H^c. It may be of interest to see what happens when summation is extended over the whole state space $\mathbb{I}$. For this purpose consider a DAF of the form

$$ \widetilde{N}_t = \sum_{0<s\leqslant t} I_H(X_{s-}) I_H(X_s) I_{(X_{s-} \neq X_s)}, \qquad t > 0, $$

with $\widetilde{N}_0 = 0$. Thus, $\widetilde{N}_t$ represents the number of jumps within the set H up to time t.

Proceeding as before, one finds

$$ g_{\mathbf{Q}}(i) = I_H(i)\, \bar{q}_i. $$

The compensator is therefore

$$ \widetilde{N}_t^* = \int_0^t I_H(X_s) \bar{q}(X_s)\, ds, $$

and its characteristic is

$$\mathbb{E}^i \widetilde{N}_t = \int_0^t \sum_{k\in H} p_{ik}(s)\bar{q}_k \, ds.$$

Consequently, for $i \in \mathbb{I}$ and $t \geqslant 0$,

$$\int_0^t \sum_k p_{ik}(s)\bar{q}_k \, ds = \mathbb{E}^i N_t + \mathbb{E}^i \widetilde{N}_t.$$

Hence, the above expression represents the sum of the average number of jumps within H and the average number of visits to H up to time t. In other words, jumps within H^c are condensed to a single jump $H^c \to H$.

§ **10.2.4. Comments.** It may be convenient to pause briefly in order to comment on connection with other stochastic processes.

1) *Markov renewal process.* Define now a random sequence $(X_n,\ n = 0, 1, 2, \ldots)$ by

$$X_n = X(T_n), \qquad n = 0, 1, 2, \ldots, \qquad \text{with} \quad T_0 = 0.$$

By construction, r.v.'s X_n take values only in H for $n \geqslant 1$, but X_0 may be either in H or in H^c. The first result here elucidates meaning of the kernel $(D^*_{ij}(t))$ associated with the stopping time T_S.

PROPOSITION 1. *For any $i, j \in H$, and $t \geqslant 0$, $n \geqslant 1$:*

$$\mathbb{P}(X_{n+1} = j,\ T_{n+1} - T_n \leqslant t \mid X_n = i,\ T_n < \infty) = D^*_{ij}(t).$$

This assertion follows from the relation $T_{n+1} - T_n = T_S \circ \theta_{T_n}$, and the strong Markov property. Thus, T_S is the actual time between consecutive entrances to the set H. States i and j are positions in H at the beginning and the end of an interval of duration $T_{n+1} - T_n$. Moreover, the above transition probability does not depend on n.

The usual sample function analysis and the strong Markov property imply that

$$\begin{aligned} &\mathbb{P}(X_{n+1} = j,\ T_{n+1} - T_n \leqslant t \mid X_0, \ldots, X_n; T_0, \ldots, T_n) \\ &= \mathbb{P}(X_{n+1} = j,\ T_{n+1} - T_n \leqslant t \mid X_n) \quad \text{a.e.} \end{aligned}$$

for all $n \geqslant 0$, $j \in H$, and $t \geqslant 0$. Thus, the stochastic process $\big((X_n, T_n),\ n \geqslant 0\big)$ is a Markov renewal process; see E. Çinlar [1972] for details. Its semi-Markov kernel over H is $(D^*_{ij}(t))$ provided $n \geqslant 1$. For $n = 0$, it is $(D^{(1)}_{ij}(t))$ which coincides with $(D^*_{ij}(t))$ if $i \in H$, but equals $(D_{ij}(t))$ if $i \in H^c$. Thus, if the initial state is in H, (X_n, T_n) is a time homogeneous Markov renewal process, but if the initial state is in H^c, one has the "delayed" Markov renewal process with $(D_{ij}(t))$ for $n = 0$ (cf. the classical delayed renewal process).

It then follows that $(X_n,\ n \geqslant 0)$ is a Markov chain with a stochastic matrix (D^*_{ij}) for $n \geqslant 1$, and $(D^{(1)}_{ij})$ for $n = 0$. It is assumed here that all T_n are finite, so $D^*_i = 1$.

The fact that (X_n, T_n) is a Markov renewal process implies the following important result which may be also obtained by the usual sample function analysis ("density argument").

PROPOSITION 2. *For any $n \geqslant 0$ and $t_1, \ldots, t_{n+1}$:*

$$\begin{aligned}&\mathbb{P}(T_1 - T_0 \leqslant t_1,\ T_2 - T_1 \leqslant t_2,\ \ldots,\ T_{n+1} - T_n \leqslant t_{n+1} \mid X_0, \ldots, X_{n+1})\\ &\quad = \mathbb{P}(T_1 - T_0 \leqslant t_1 \mid X_0, X_1)\mathbb{P}(T_2 - T_1 \leqslant t_2 \mid X_1, X_2) \ldots\\ &\quad\quad \times \mathbb{P}(T_{n+1} - T_n \leqslant t_{n+1} \mid X_n, X_{n+1}) \quad \text{a.e.}\end{aligned}$$

Proposition 2 expresses the fact that for each n, the increments $T_{n+1} - T_n$ are *conditionally independent*, given the Markov chain $(X_n,\ n \geqslant 0)$. The distributions of $T_{n+1} - T_n$ depend only on X_n and X_{n+1}.

An useful form of the conditional independence property is the following

COROLLARY 3. *For $i \in \mathbb{I}$, $r, j \in H$, $s' \leqslant s$, and $n > 1$:*

$$\mathbb{P}^i(X_n = r,\ T_n \leqslant s',\ X_{n+1} = j,\ T_{n+1} \leqslant s) = \int_0^{s'} d_{ir}^{(n)}(x) D_{rj}^*(s - x)\, dx.$$

Observe that for $s' = s$ and summation over all $r \in H$, Corollary 3 yields the dual recursive relation for $D_{ij}^{(n+1)}(s)$, stated in § 10.2.2.

One can see from Corollary 3 that

$$\mathbb{P}(T_{n+1} \leqslant s,\ X_{n+1} = j \mid T_n = x,\ X_n = r) = D_{rj}^*(s - x), \qquad s \geqslant x,$$

independently of n.

2) *Counting process.* The expression for the number N_t of visits to the set H during $(0, t]$, already stated in § 10.2.2,

$$N_t = \sum_{n=1}^{\infty} I_{(T_n \leqslant t)}, \qquad t \geqslant 0, \qquad N_0 = 0,$$

indicates that $N = (N_t,\ 0 \leqslant t < \infty)$ is a counting point process. Evidently, N is an increasing process and also a purely discontinuous AF (because T_n are terminal times). Furthermore, as already shown, distributions of each T_n have densities.

The general theory of point processes of the above form provides the explicit expression for the compensator (dual predictable projection) $N^* = (N_t^*)$ of N. It is therefore of interest to show that this expression agrees, in the case considered here, with that already found in § 10.2.3.

For this purpose consider the conditional probabilities

$$\begin{aligned}&K_n(t) = \mathbb{P}(T_n \leqslant t \mid T_1, \ldots, T_{n-1}) \qquad \text{for} \quad n \geqslant 2,\\ &K_1(t) = \mathbb{P}(T_1 \leqslant t),\end{aligned}$$

with densities $k_n(t)$. Note that $K_n(t)$ vanishes for $t < T_{n-1}$. Introduce next the hazard functions

$$G_n(t) = \int_0^t g_n(s)\, ds,$$

with hazard rates

$$g_n(t) = k_n(t)/K_n^c(t) \equiv g_n(t \mid T_1, \ldots, T_{n-1}).$$

Then the basic theorem asserts that

$$N_t^* = \sum_{n=1}^{\infty} G_n(t \wedge T_n);$$

see R. Sh. Liptser and A. N. Shiryajev [1981, p. 590], P. Bremaud [1981, p. 61], M. Jacobsen [1982], E. Slud [1991].

The following lemma refers to the case considered here.

LEMMA 4. *For each n and $t > 0$:*

$$\mathbb{E}\big(G_n(t \wedge T_n) \mid T_1, \ldots, T_{n-1}\big) = K_n(t).$$

Proof.

$$\mathbb{E}\left(\int_0^{t\wedge T_n} g_n(u)\,du \mid T_1, \ldots, T_{n-1}\right) = \int_0^{\infty} dK_n(s) \int_0^{t\wedge s} g_n(u)\,du$$

$$= \int_0^t G_n(s)\,dK_n(s) + (1 - K_n(t))G_n(t)$$

$$= G_n(t)K_n(t) - \int_0^t K_n(s)\,dG_n(s) + G_n(t) - K_n(t)G_n(t)$$

$$= \int_0^t (1 - K_n(s))\,dG_n(s) = K_n(t),$$

using integration by parts and definition of $G_n(t)$. □

THEOREM 5. *For $i \in \mathbb{I}$, $t \geqslant 0$:*

$$\mathbb{E}^i N_t^* = \mathbb{E}^i N_t = U_i(t),$$

and $N_t - N_t^*$, $t \geqslant 0$, form a martingale.

Proof. Clearly,

$$\mathbb{E}^i\, K_n(t) = \mathbb{P}^i(T_n \leqslant t) = D_i^{(n)}(t),$$

so the first assertion follows from Lemma 4.

From results in § 8.1.4, it follows that increasing processes N and N^* have the same α-potential

$$\mathbb{E}^i \int_0^{\infty} e^{-\alpha t}\,dN_t = \mathbb{E}^i \int_0^{\infty} e^{-\alpha t}\,dN_t^* = \sum_{n=1}^{\infty} d_i^{\alpha(n)} = \mathbf{U}^{\alpha}\mathbf{1}_i,$$

and the same characteristic $U_i(t)$. Furthermore, $N - N^*$ is a martingale. □

N^* can be put in an equivalent form which expresses it as a CAF, namely as an integral of the intensity,

$$N_t^* = \int_0^t \Lambda(s)\, ds.$$

Indeed, consider the conditional probabilities

$$F_n(t) = \mathbb{P}(T_n - T_{n-1} \leqslant t \mid T_1, \dots, T_{n-1}) \qquad \text{for} \quad n \geqslant 2,$$
$$F_1(t) = \mathbb{P}(T_1 \leqslant t),$$

with densities $f_n(t)$. Introduce also the hazard rates

$$r_n(t) = f_n(t)/F_n^c(t).$$

Finally, define the intensity by

$$\Lambda(t) = \sum_{n=1}^{\infty} r_n(t - T_{n-1}) I_{(T_{n-1} \leqslant t < T_n)}.$$

PROPOSITION 6. *Both above given expressions for N_t^* coincide.*

Proof. Observe that

$$F_n(t - T_{n-1}) = K_n(t), \qquad r_n(t - T_{n-1}) = g_n(t).$$

Equivalence follows by considering t in intervals $T_{j-1} \leqslant t < T_j$ for each j. □

For further remarks on intensity $\Lambda(t)$ see § 8.1.4.

3) *Normalization.* It is apparent that the time spent in a set H and the number of visits to H represent dual aspects of the same situation. Actually, it is more natural to consider the number of exits from the set H.

Let $\widehat{N} = (\widehat{N}_t)$ be a purely discontinuous AF with jumps at consecutive entrances to the set H^c (described by a random set W introduced in § 2.1.1):

$$\widehat{N}_t = \sum_{0 < s \leqslant t} I_H(X_{s-}) I_{H^c}(X_s), \qquad t > 0,$$

with $\widehat{N}_0 = 0$, and $\operatorname{supp}(\widehat{N}) = H^c$. By argument in § 8.1.3, one has

$$g_{\mathbf{Q}}(i) = (q_i - \bar{q}_i) I_H(i),$$

and the corresponding compensator $N^* = (N_t^*)$ of $\widehat{N}$ is given by an integral functional

$$\widehat{N}_t^* = \int_0^t I_H(X_s)\big(q(X_s) - \bar{q}(X_s)\big)\, ds.$$

Hence,

$$\mathbb{E}^i \widehat{N}_t = \mathbb{E}^i \widehat{N}_t^* = \int_0^t \sum_{j \in H} p_{ij}(s)(q_j - \bar{q}_j)\, ds.$$

Moreover, proceeding in the same fashion as before, one gets

$$\widehat{U}_{ij}(t) = \int_0^t \sum_{k \in H} p_{ik}(s) q_{kj}\, ds.$$

To obtain the proper occupation time $V = (V_t)$ of the set H, defined in § 8.4.1, it is necessary to introduce normalization. The following argument is the same as that in Sub-section 8.3. Consider a DAF $\tilde{A} = (\tilde{A}_t)$ introduced in § 8.3.2 and defined with help of normalizing functions $h > 0$ and $c > 0$ (supported on H and H^c, respectively). Thus, as in § 8.3.2,

$$g_{\mathbf{Q}}(i) = I_H(i) h^{-1}(i) \sum_{j \in H^c} q_{ij}\, c(j), \qquad i \in H.$$

Selecting h and c such that

$$\sum_{j \in H^c} q_{ij} c(j) = h(i), \qquad i \in H,$$

gives $g_{\mathbf{Q}}(i) = I_H(i)$. For pairs h and c satisfying the above condition, see § 8.3.3 (in some cases it is possible to have $h = 1$ identically). Next, the argument in § 8.4.2 shows that the CAF V is the dual predictable projection of the DAF $\tilde{A}$, with supports H and H^c, respectively. Thus, potentials and characteristics of V and of $\tilde{A}$ agree. In particular,

$$\mathbb{E}^i \tilde{A}_t = \mathbb{E}^i V_t = \int_0^t \sum_{j \in H} p_{ij}(s)\, ds.$$

§ 10.2.5. Limit theorems. The number of successive entrances to a set H during all time $(0, \infty)$ is a r.v. N_∞ defined by the limit with probability one:

$$N_\infty = \lim_{t \to \infty} N_t \quad \text{a.e.}$$

Although N_t is finite for each t, N_∞ may be infinite. Some comments are in order here. By definition (§ 10.2.1), the event $(N_\infty < \infty)$ indicates that for some n, $T_n < \infty$, but $T_{n+1} = \infty$. In particular, if the chain starts in H and remains in H all the time ($T_0 = 0$, $S = \infty$, so $T_1 = \infty$), this is counted as no visit. If the chain starts in H^c and enters H at T, and then stays in H ($T_1 = T < \infty$, $S = \infty$, so $T_2 = \infty$), this is counted as one visit. Thus, remaining forever in H does not contribute to the number of visits.

Note that the relation $T_{n+1} = S_n + T \circ \theta_{S_n}$ implies that the event $(N_\infty = n)$ occurs when either $S_n < \infty$ and $T = \infty$, or $S_n = \infty$. In the first case the chain leaves H at time S_n and remains in H^c afterwards; in the second case the chain enters H at time T_n and remains in H afterwards. It is therefore clear that the last exit time L from the set H is

$$L = S_{N_\infty}$$

(see § 4.1.1). However, in the stable chain, finite L can occur only if N_∞ is finite. On the other hand, infinite L occurs whenever N_∞ is infinite. Recall also that the event $(L = \infty)$ is equivalent to the event $(X_n \in H \text{ i.o.})$ for the jump chain (thus, the M.C. does not remain in H forever); see § 5.2.3. Thus,

$$\mathbb{P}^i(L = \infty) = s_i, \qquad i \in \mathbb{I},$$

with s_i introduced in § 5.2.3. In particular, when $N_\infty = 0$, then $L = S$, if the initial state $i \in H$, and $L = 0$ if $i \in H^c$.

A Markov renewal process is said to be recurrent when its basic life time T_S is finite, and is said to be transient if T_S assumes infinite value. This corresponds, respectively, to $D^* = 1$ and $D^* < 1$. Both these cases must be treated separately. For discussion of limit theorems in Markov renewal processes see E. Çinlar [1975].

Recall that the distribution of N_t has been found in § 10.2.2, and the explicit expression for $U_i(t) = \mathbb{E}^i N_t$ has been found in § 10.2.3. Write $U_i(t) \to U_i$ as $t \to \infty$, and note that

$$U_i = \sum_{j \in H^c} u_{ij} \bar{q}_j \leqslant \infty, \qquad i \in \mathbb{I}.$$

Warning. Do not confuse mean U_i with resolvent $\mathbf{U} = (u_{ij})$.

a) *Recurrent case.* The time T_S between two consecutive entrances to the set H is assumed finite, so its (conditional) distribution satisfies $D_r^* = 1$ for all $r \in H$. Proposition 1, § 10.2.1, indicates that $D_i^* = 1$ for all $i \in \mathbb{I}$ is equivalent to $D_i = \widehat{D}_i = 1$ for all $i \in \mathbb{I}$. The recursive relations from Lemma 1, § 10.2.2, imply that also $D_i^{(n)} = 1$ for all $n \geqslant 1$ and all $i \in \mathbb{I}$. On the other hand, recall from § 5.2.3 that $D_i = 1$ for all i implies that the set H is visited infinitely often with probability one ($s_i = 1$ identically). Therefore, $\mathbb{P}^i(N_\infty < \infty) = 0$, as in the classical renewal process. Evidently, then $U_i = \infty$.

The following version of the Markov renewal theorem extends the classical theorem corresponding to H consisting of a single state.

THEOREM 1. *Suppose that the* M.C. X *is ergodic with limit distribution* $(\varepsilon_j) = \varepsilon$, $1/\varepsilon_j = m_{jj}\, q_j$, *where* m_{jj} *is the mean recurrence time of a state* j *(see* § 1.1.8*). Then*

$$\lim_{t \to \infty} U_i(t)/t = \sum_{j \in H} \varepsilon_j (q_j - \bar{q}_j).$$

Proof. The assertion follows from the explicit expression for $U_i(t)$ found earlier. □

It may be of interest to interpret the above result in the manner analogous to that in the classical renewal theorem. Indeed, N_t represents the number of entrances to H and the occupation time V_t represents the time spent in H, both up to time t. Hence, in the limit, the expected rate of visits is given by Theorem 1, and the expected proportion of time spent in H is

$$\mathbb{E}^i\, V_t/t \longrightarrow \sum_{j\in H} \varepsilon_j \qquad \text{as} \quad t\to\infty.$$

The next renewal theorem (analogous to the Blackwell theorem in renewal theory) can be proved in the present case in the same fashion as Theorem 1, using occupation time (V_t) or, equivalently, increasing process (N_t).

THEOREM 2. *Suppose that the usual* M.C. X *is ergodic with the invariant distribution* $\varepsilon = (\varepsilon_j)$. *Then*

$$\lim_{t\to\infty} \left[U_i(t+h) - U_i(t)\right] = mh, \qquad h > 0,$$

where m *is the value of the limit in Theorem* 1.

b) *Transient case.* The time T_S between consecutive entrances to H is assumed to have the (conditional) distribution satisfying $D_r^* < 1$ for all $r \in H$. Proposition 1, § 10.2.1, indicates that $D_i^* < 1$ for all $i \in \mathbb{I}$ whenever at least one of D_i and $\widehat{D}_i$ is less than 1.

The recursive relations from Lemma 1, § 10.2.2, imply then that

$$D_i^{(n+1)} < D_i^{(n)} \qquad \text{for all} \quad n \geqslant 1,$$

so the $\lim_{n\to\infty} D_i^{(n)} = D_i^{(\infty)}$ exists for all $i \in \mathbb{I}$. Recall also that $D_i^{(1)}$ equals D_i when $i \in H^c$, and D_i^* when $i \in H$.

Passage to the limit (as $t \to \infty$) in the distribution of N_t, yields the distribution of N_∞,

$$\mathbb{P}^i(N_\infty = 0) = 1 - D_i^{(1)},$$
$$\mathbb{P}^i(N_\infty = n) = D_i^{(n)} - D_i^{(n+1)}, \qquad n \geqslant 1,$$

in agreement with relation $(N_\infty = n) = (T_n < \infty,\ T_{n+1} = \infty)$. Hence,

$$\mathbb{P}^i(N_\infty < \infty) = 1 - D_i^{(\infty)},$$

so as already noted, $D_i^{(\infty)} = s_i$. Indeed, one has from Lemma 1, § 10.2.2, that

$$D_i^{(\infty)} = \sum_{r\in H} D_{ir} D_r^{(\infty)}, \qquad i \in H^c,$$

in agreement with the relation $\mathbf{B}s = s$, stated in § 5.2.3. Thus, in the transient case, N_∞ may be infinite with positive probability. In particular, if H is totally transient set then N_∞ is finite with probability one.

The alternative forms of the distribution of N_∞ are perhaps more informative.

THEOREM 3. *Under assumption that $D_i < 1$ for $i \in H^c$ and $D_i^* < 1$ for $i \in H$:*

$$\mathbb{P}^i(N_\infty = n) = \sum_{r \in H} D_{ir}^{(n)}(1 - D_r^*)$$
$$= \sum_{r \in H} D_{ir}^{(n)}\Big(\sum_{k \in H^c} \widehat{D}_{rk}\, D_k^c + \widehat{D}_r^c\Big), \qquad n \geqslant 1, \quad i \in \mathbb{I}.$$

Proof. The first expression follows from the dual form of Lemma 1, § 10.2.2,

$$D_i^{(n+1)} = \sum_{r \in H} D_{ir}^{(n)} D_r^*.$$

Next, Proposition 1, § 10.2.1, implies that

$$D_r^* = \sum_{k \in H^c} \widehat{D}_{rk}\, D_k, \qquad r \in H,$$

and the second expression follows by substitution. □

Theorem 3 has an intuitive interpretation. To have exactly n visits to H (during the whole time), the n-th visit occurs at the instant T_n and no more visits can take place. This last event is achieved either by the chain jumping back to H^c and staying there, or by staying in H forever.

Next observe that if H^c is a finite set, then (by Thm. 1, § 10.1.2) T_S would be finite with probability one (assuming S finite), contradicting assumption that $D^* < 1$. Moreover, it has been shown in § 9.2.4 (with reference to the post-L chain) that finite H^c implies that H would be visited i.o. with probability one. Hence, L would be infinite with probability one, so $D = 1$ identically. Thus, H^c must be an infinite set.

PROPOSITION 4. *For $i \in \mathbb{I}$ and $t > 0$:*

$$\mathbb{P}^i(S_{N_\infty} \leqslant t,\ N_\infty < \infty) = \mathbb{P}^i(0 < L \leqslant t).$$

Proof. By the usual decomposition, the LHS is

$$\sum_n \mathbb{P}^i(S_n \leqslant t,\ N_\infty = n) = \sum_n \sum_{j \in H^c} \mathbb{P}^i(S_n \leqslant t,\ X(S_n) = j,\ T_n < \infty,\ T_{n+1} = \infty)$$
$$= \sum_n \sum_{j \in H^c} \widehat{D}_{ij}^{(n)}(t)\, D_j^c = \sum_{j \in H^c} \widehat{U}_{ij}(t)\, D_j^c,$$

where summation in the series for $\widehat{U}_{ij}(t)$ is extended over $n \geqslant 0$ for $i \in H$, and over $n \geqslant 1$ for $i \in H^c$ (this is the consequence of enumeration of S_n introduced in § 10.2.1). Substitution of expression for $\widehat{U}_{ij}(t)$ found in § 10.2.4, yields, with the help of relations from § 4.1.1,

$$\mathbb{P}^i(S_{N_\infty} \leqslant t,\ N_\infty < \infty) = \int_0^t \sum_{k \in H} p_{ik}(s)\, b_k\, ds = \Pi_i(t). \qquad \square$$

Clearly,

$$\mathbb{P}^i(S_{N_\infty} < \infty,\ N_\infty < \infty) = \Pi_i,$$
$$\mathbb{P}^i(S_{N_\infty} = \infty,\ N_\infty = \infty) = s_i,$$

and

$$\mathbb{E}^i(S_{N_\infty},\ S_{N_\infty} < \infty,\ N_\infty < \infty) = \sum_j u_{ij}\,\Pi_j, \qquad i \in \mathbb{I},$$

in agreement with the expression stated in § 4.1.1.

c) *Duality.* In the transient case the means of N_∞ and of $\widehat{N}_\infty$ enjoy interesting properties. As already noted, the average number of visits to the set H is

$$U_i = \sum_{j \in H^c} u_{ij}\bar{q}_j, \qquad i \in \mathbb{I}.$$

Dually, the average number of visits to the set H^c is (from § 10.2.4)

$$\widehat{U}_i = \sum_{j \in H} u_{ij}\,(q_j - \bar{q}_j), \qquad i \in \mathbb{I}.$$

Recall from § 8.1.2 that the average of the total number of jumps in the chain is

$$\sum_j u_{ij}\,q_j \geqslant U_i + \widehat{U}_i.$$

See also examples in § 10.2.7.

LEMMA 5. *The function (U_i) (resp., $(\widehat{U}_i)$) on the state space $\mathbb{I}$ is a potential of a charge supported on H^c (resp., H) and satisfies the Poisson equation relative to the first entrance time S to a set H^c (resp., the first entrance time T to a set H).*

Proof. Consider, for $i \in \mathbb{I}$,

$$\begin{aligned} \mathbf{Q}U_i &= \sum_k q_{ik}U_k = \sum_k q_{ik} \sum_{j \in H^c} u_{kj}\bar{q}_j = \sum_{j \in H^c} \bar{q}_j \sum_k q_{ik}u_{kj} \\ &= -\sum_{j \in H^c} \delta_{ij}\bar{q}_i = \begin{cases} 0, & i \in H, \\ -\bar{q}_i, & i \in H^c. \end{cases} \end{aligned}$$

Analogously,

$$\mathbf{Q}\,\widehat{U}_i = \begin{cases} 0, & i \in H^c, \\ -(q_i - \bar{q}_i), & i \in H. \end{cases}$$

Thus, (U_i) (resp., $(\widehat{U}_i)$) is excessive and a potential of a charge q vanishing on H (resp., $q - \bar{q}$ vanishing on H^c). Hence, it is also purely excessive function by Theorem 5, § 1.3.2. □

The following form of the maximum principle holds (see § 5.3.2).

PROPOSITION 6. *Suppose that (U_i) (resp., $(\widehat{U}_i)$) is bounded on H^c (resp., on H). Then it is bounded on $\mathbb{I}$.*

Proof. Write c_i for values of (U_i) on H^c; by assumption, $c_i \leqslant k$ on H^c. By Lemma 5, (U_i) satisfies the Dirichlet problem (relative to S),

$$\begin{aligned} \mathbf{Q}U_i &= 0, & i &\in H, \\ U_i &= c_i, & i &\in H^c, \end{aligned}$$

with solution $U_i = \widehat{\mathbf{B}} c_i$. Hence

$$U_i \leqslant k\widehat{\mathbf{B}}\,\mathbf{1}_i = k\widehat{D}_i \leqslant k, \qquad i \in \mathbb{I}.$$

For $(\widehat{U}_i)$ one has analogously the Dirichlet problem

$$\begin{aligned} \mathbf{Q}\,\widehat{U}_i &= 0, & i &\in H^c, \\ \widehat{U}_i &= \hat{c}_i, & i &\in H, \end{aligned}$$

with solution $\widehat{U}_i = \mathbf{B}\hat{c}_i$. □

Consider now the last exit time L from the set H and introduce its reciprocal, the last exit time $\hat{L}$ from the set H^c. The following lemma exhibits duality between these times.

LEMMA 7. *For* $i \in \mathbb{I}$:

$$\begin{aligned} \mathbb{P}^i(0 < N_\infty < \infty) &= \mathbf{U}b_i^* = \mathbb{P}^i(0 < \hat{L} < \infty), \\ \mathbb{P}^i(0 < \widehat{N}_\infty < \infty) &= \mathbf{U}b_i = \mathbb{P}^i(0 < L < \infty), \end{aligned}$$

where

$$b_i = \sum_{r \in H^c} q_{ir}\, D_r^c \quad \text{for} \quad i \in H; \qquad b_i = 0 \quad \text{for} \quad i \in H^c,$$
$$b_i^* = \sum_{r \in H} q_{ir}\, D_r^{*c} \quad \text{for} \quad i \in H^c; \qquad b_i^* = 0 \quad \text{for} \quad i \in H.$$

Proof. Using distribution of N_∞ from Theorem 3 and expression for U_{ir}, one has

$$\mathbb{P}^i(0 < N_\infty < \infty) = \sum_{r \in H} U_{ir} D_r^{*c} = \sum_{k \in H^c} u_{ik} b_k^* = \mathbf{U}b_i^*.$$

The expression for $\widehat{N}_\infty$ follows in the analogous fashion (see § 10.2.4-3 and Prop. 4). □

Write now

$$D_i^{(1)} = \begin{cases} D_i, & i \in H^c, \\ D_i^*, & i \in H, \end{cases} \qquad \widetilde{D}_i = \begin{cases} \widehat{D}_i^*, & i \in H^c, \\ \widehat{D}_i, & i \in H. \end{cases}$$

It follows from Lemma 7 that the dual Poisson equations hold (see also Cor. 5, § 10.1.6).

PROPOSITION 8. *The functions* $(D_i^{(1)})$ *and* $(\widetilde{D}_i)$ *satisfy, respectively, the equations*

$$\mathbf{Q}D_i^{(1)} = \begin{cases} 0, & i \in H, \\ -b_i^*, & i \in H^c, \end{cases} \qquad \mathbf{Q}\widetilde{D}_i = \begin{cases} 0, & i \in H^c, \\ -b_i, & i \in H. \end{cases}$$

Several comments are in order. The minimal solutions of Proposition 8 are of the form

$$D_i^{(1)} = \mathbf{U}b_i^*, \qquad \widetilde{D}_i = \mathbf{U}b_i.$$

As $U_i \geqslant D_i^{(1)}$, $\widehat{U}_i \geqslant \widetilde{D}_i$, so $(D_i^{(1)})$ and $(\widetilde{D}_i)$ are also purely excessive functions. Hence, the invariant component in Riesz decomposition from Proposition 8 must vanish; as it represents infinitely many visits, it means that both H and H^c are totally transient. These results have intuitive interpretation. Note also the symmetry involved.

Finally, it is of interest to observe that it is impossible to have $U_i \geqslant 1$ for all i, despite intuitive expectation. If this were true, then the constant 1 would be an invariant function dominated by a purely excessive function (U_i) which is impossible by Theorem 5, § 1.3.2. The same remark applies to $\widehat{U}_i$.

§ 10.2.6. Matrices. Matrix notation provides concise formulation of results obtained earlier in this sub-section.

a) The t-balayage matrices $\mathbf{B}(t)$, $\mathbf{B}^*(t)$, $\widehat{\mathbf{B}}(t)$, $\widehat{\mathbf{B}}^*(t)$, and their L.S. transforms, the α-balayage matrices $\mathbf{B}^\alpha$, $\mathbf{B}^{*\alpha}$, $\widehat{\mathbf{B}}^\alpha$, and $\widehat{\mathbf{B}}^{*\alpha}$, have been already discussed in § 10.1.4. These matrices represent balayage for T, T_S, S, and S_T, respectively. In the present case:

$$\mathbf{B}^\alpha = \begin{pmatrix} 0 & \mathbf{B}_{12}^\alpha \\ 0 & \mathbf{I}_2 \end{pmatrix}, \qquad \mathbf{B}^{*\alpha} = \begin{pmatrix} 0 & \mathbf{B}_{12}^\alpha \\ 0 & \mathbf{B}_{22}^{*\alpha} \end{pmatrix},$$

$$\widehat{\mathbf{B}}^\alpha = \begin{pmatrix} \mathbf{I}_1 & 0 \\ \widehat{\mathbf{B}}_{21}^\alpha & 0 \end{pmatrix}, \qquad \widehat{\mathbf{B}}^{*\alpha} = \begin{pmatrix} \widehat{\mathbf{B}}_{11}^{*\alpha} & 0 \\ \widehat{\mathbf{B}}_{21}^\alpha & 0 \end{pmatrix}.$$

The product theorem yields

$$\mathbf{B}^{*\alpha} = \widehat{\mathbf{B}}^\alpha \mathbf{B}^\alpha \quad \text{with} \quad \mathbf{B}_{12}^{*\alpha} = \mathbf{B}_{12}^\alpha, \quad \mathbf{B}_{22}^{*\alpha} = \widehat{\mathbf{B}}_{21}^\alpha \mathbf{B}_{12}^\alpha,$$

and dually,

$$\widehat{\mathbf{B}}^{*\alpha} = \mathbf{B}^\alpha \widehat{\mathbf{B}}^\alpha \quad \text{with} \quad \widehat{\mathbf{B}}_{21}^{*\alpha} = \widehat{\mathbf{B}}_{21}^\alpha, \quad \widehat{\mathbf{B}}_{11}^{*\alpha} = \mathbf{B}_{12}^\alpha \widehat{\mathbf{B}}_{21}^\alpha.$$

(See Prop. 1, § 10.2.1, and its dual mentioned in Lemma 1, § 10.2.2.)

b) Define now matrices $\mathbf{B}^{(n)}(t) = (D_{ij}^{(n)})$ for each $n \geqslant 1$, with $\mathbf{B}^{(1)}(t) = \mathbf{B}^*(t)$, and note that the L.S. transform of $\mathbf{B}^{(n)}(t)$ is the n-th power of $\mathbf{B}^{*\alpha}$:

$$\mathbf{B}^{(n)\alpha} = (\mathbf{B}^{*\alpha})^n = (d_{ij}^{(n)\alpha})$$

with non-zero components

$$\mathbf{B}_{12}^{(n)\alpha} = \mathbf{B}_{12}^{*\alpha}(\mathbf{B}_{22}^{*\alpha})^{n-1}, \qquad \mathbf{B}_{22}^{(n)\alpha} = (\mathbf{B}_{22}^{*\alpha})^n.$$

(See Lemma 1, § 10.2.2.)

c) Introduce now the matrix $\mathbf{U}(t) = (U_{ij}(t))$ and its L.S. transform $\mathbf{U}^\alpha = (U_{ij}^\alpha)$ defined by

$$\mathbf{U}(t) = \sum_{n=1}^{\infty} \mathbf{B}^{(n)}(t), \qquad t \geqslant 0, \qquad \mathbf{U}^\alpha = \sum_{n=1}^{\infty} \mathbf{B}^{(n)\alpha}, \qquad \alpha > 0.$$

Then the Markov renewal equations from § 10.2.2 lead to the following matrix renewal equation

$$\mathbf{U}(t) = \mathbf{B}^*(t) + \int_0^t d\mathbf{B}^*(s)\mathbf{U}(t-s) = \mathbf{B}^*(t) + \int_0^t d\mathbf{U}(s)\,\mathbf{B}^*(t-s),$$

or, in terms of transforms,

$$\mathbf{U}^\alpha = \mathbf{B}^{*\alpha} + \mathbf{B}^{*\alpha}\mathbf{U}^\alpha = \mathbf{B}^{*\alpha} + \mathbf{U}^\alpha\mathbf{B}^{*\alpha}.$$

This has the form of decomposition theorems of the type discussed in Subsection 11.1. Hence, as in the Riesz decomposition, the formal solution is found to be

$$\mathbf{U}^\alpha = \sum_{n=1}^{\infty} (\mathbf{B}^{*\alpha})^n + \mathbf{W}^\alpha\mathbf{U}^\alpha \qquad \text{with} \quad \mathbf{W}^\alpha = \lim_{n\to\infty} (\mathbf{B}^{*\alpha})^n.$$

The solution is unique when $\mathbf{W}^\alpha$ vanishes (see situation in § 10.1.7).

Observe that the matrix $\mathbf{U}^\alpha$ used here should not be confused with the resolvent matrix $\mathbf{U}^\alpha = (u_{ij}^\alpha)$. Indeed, from § 10.2.3 one has in terms of elements

$$U_{ij}^\alpha = \sum_{k\in H^c} u_{ik}^\alpha q_{kj}, \qquad i \in \mathbb{I}, \quad j \in H.$$

All the above discussion may be repeated for the dual matrix $\widehat{\mathbf{U}}(t)$.

d) Note that the average number $U_i(t)$ of visits to the set H is represented by the function

$$U_i(t) = \mathbf{U}(t)\mathbf{1}_i = \sum_{j\in H} U_{ij}(t), \qquad i \in \mathbb{I}.$$

This observation leads to Markov renewal equations in Theorems 2 and 3, § 10.2.1, assuming uniqueness of solution. Analogous results hold for $\widehat{U}_i(t) = \widehat{\mathbf{U}}(t)\mathbf{1}_i$.

e) Let $\mathbf{C}^\alpha$ be the dual of $\mathbf{B}^\alpha$ (introduced in § 3.4.4) and let $\widehat{\mathbf{C}}^\alpha$ be the dual of $\widehat{\mathbf{B}}^\alpha$ (introduced in § 10.1.8). Analogously, let $\widehat{\mathbf{C}}^{*\alpha}$ and $\mathbf{C}^{*\alpha}$ be duals of $\widehat{\mathbf{B}}^{*\alpha}$ and $\mathbf{B}^{*\alpha}$. Here

$$\mathbf{C}^\alpha = \begin{pmatrix} \mathbf{0} & \mathbf{0} \\ \mathbf{C}_{21}^\alpha & \mathbf{I}_2 \end{pmatrix}, \qquad \widehat{\mathbf{C}}^\alpha = \begin{pmatrix} \mathbf{I}_1 & \widehat{\mathbf{C}}_{12}^\alpha \\ \mathbf{0} & \mathbf{0} \end{pmatrix}$$

(cf. Case 2, § 10.1.9).

The dual form of the product theorem is now

$$\mathbf{C}^{*\alpha} = \mathbf{C}^\alpha\widehat{\mathbf{C}}^\alpha, \qquad \widehat{\mathbf{C}}^{*\alpha} = \widehat{\mathbf{C}}^\alpha\mathbf{C}^\alpha,$$

with non-zero components

$$\mathbf{C}^{*\alpha}_{21} = \mathbf{C}^{\alpha}_{21}, \qquad \mathbf{C}^{*\alpha}_{22} = \mathbf{C}^{\alpha}_{21}\widehat{\mathbf{C}}^{\alpha}_{12},$$
$$\widehat{\mathbf{C}}^{*\alpha}_{11} = \widehat{\mathbf{C}}^{\alpha}_{12}\mathbf{C}^{\alpha}_{21}, \qquad \widehat{\mathbf{C}}^{*\alpha}_{12} = \widehat{\mathbf{C}}^{\alpha}_{12}.$$

See also discussion in § 10.1.8.

f) Consider now the balayage matrix $\mathbf{B}^*$ defined as a limit

$$\mathbf{B}^* = \lim_{t\to\infty} \mathbf{B}^*(t) = \lim_{\alpha\to 0} \alpha\mathbf{B}^{*\alpha},$$

so in the partitioned form

$$\mathbf{B}^* = \begin{pmatrix} 0 & \mathbf{B}_{12} \\ 0 & \mathbf{B}^*_{22} \end{pmatrix}.$$

Clearly,

$$\mathbf{B}^* = \widehat{\mathbf{B}}\mathbf{B} \qquad \text{with} \quad \mathbf{B}^*_{22} = \widehat{\mathbf{B}}_{21}\mathbf{B}_{12},$$

where the limiting matrices $\widehat{\mathbf{B}}$ and $\mathbf{B}$ are defined in the same manner.

Observe that the n-th power for $n \geqslant 1$ is

$$(\mathbf{B}^*)^n = \begin{pmatrix} 0 & \mathbf{B}_{12}(\mathbf{B}^*_{22})^{n-1} \\ 0 & (\mathbf{B}^*_{22})^n \end{pmatrix}$$

and

$$(\mathbf{B}^*)^n \mathbf{1}_i = D_i^{(n)}, \qquad n \geqslant 1, \quad i \in \mathbb{I}.$$

In the transient case, passage to the limit in the Markov renewal equation yields

$$\mathbf{U} = \mathbf{B}^* + \mathbf{B}^*\mathbf{U},$$

where $\mathbf{U} = (U_{ij})$. Thus, in the terms of non-zero components,

$$\mathbf{U}_{12} = \mathbf{B}_{12} + \mathbf{B}_{12}\mathbf{U}_{22}, \qquad \mathbf{U}_{22} = \mathbf{B}^*_{22} + \mathbf{B}^*_{22}\mathbf{U}_{22}.$$

Moreover,

$$\mathbb{P}^i(N_\infty = n) = (\mathbf{B}^*)^n(\mathbf{I} - \mathbf{B}^*)\mathbf{1}_i \qquad \text{for} \quad n \geqslant 0.$$

Addition over all n yields

$$U_i = D_i^{(1)} + \sum_{r\in H} U_{ir} D_r \qquad \text{for} \quad i \in \mathbb{I},$$

which is dual to

$$U_i = D_i^{(1)} + \sum_{r\in H} D^*_{ir} U_r \qquad \text{for} \quad i \in \mathbb{I}.$$

As already shown, (U_i) is a purely excessive function and satisfies the Poisson equation.

Finally, it is easy to verify that the matrix $\mathbf{QB}^*$ has only one non-zero component, namely,

$$(\mathbf{QB}^*)_{12} = \mathbf{Q}_{11}\mathbf{B}_{12} + \mathbf{Q}_{12}\mathbf{B}^*_{22}.$$

As $D^*_i = \mathbf{B}^*\mathbf{1}_i$, one finds immediately that the function (D^*_i) on $\mathbb{I}$ satisfies the (reciprocal) Poisson equation, in agreement with Proposition 8, § 10.2.5.

§ 10.2.7. Classical case. It is of interest to consider the special case when the taboo set H consists of a single state, say j. In this situation one obtains the proper renewal process.

a) Let $H = \{j\}$, $H^c = \{k : k \neq j\}$. Hence, as in Example 1, § 3.1.4,

$$D_{ij}(t) = F_{ij}(t), \qquad d_{ij}^{\alpha} = f_{ij}^{\alpha}, \qquad i \neq j.$$

It is clear that S coincides now with $\rho = \rho_j$, so, by Example 2, § 3.1.4 (with H and H^c interchanged),

$$\widehat{D}_{jk}(t) = (1 - e^{-q_j t}) r_{jk}, \qquad \hat{d}_{jk}^{\alpha} = q_{jk}\,(\alpha + q_j)^{-1} = r_{jk}^{\alpha}, \qquad j \neq k.$$

Consequently, $T_S = T_\rho$, and the product theorem (Prop. 1, § 10.2.1) yields

$$d_{jj}^{*\alpha} = \sum_{k \neq j} \hat{d}_{jk}^{\alpha}\, d_{kj}^{\alpha} = \sum_{k \neq j} r_{jk}^{\alpha}\, f_{kj}^{\alpha}.$$

THEOREM 1. $d_{jj}^{*\alpha} = f_{jj}^{\alpha}$ or $D_{jj}^{*}(t) = F_{jj}(t)$.

Proof. By Theorem 7, § 3.3.2,

$$d_{jj}^{\rho\alpha} = \sum_{k \neq j} r_{jk}^{\alpha}\, d_{kj}^{\alpha},$$

so, by Example 1, § 3.3.4,

$$d_{jj}^{\rho\alpha} = f_{jj}^{\alpha}.$$

□

Observe that one has now the usual passage relations

$$u_{ij}^{\alpha} = d_{ij}^{\alpha}\, u_{jj}^{\alpha}, \qquad i \neq j, \qquad 1 = (\alpha + q_j) u_{jj}^{\alpha}\,(1 - d_{jj}^{*\alpha}).$$

It is important to observe that Proposition 2, § 10.2.4, indicates that increments $T_{n+1} - T_n$ are independent and identically distributed, with distribution $F_{jj}(t)$. Thus, $(T_n,\ n = 0, 1, \ldots)$ is a renewal process.

Noting that $d_i^{\alpha} = f_{ij}^{\alpha}$ and $d_j^{*\alpha} = f_{jj}^{\alpha}$, the Markov renewal equation (Thm. 2, § 10.2.2) reduces to the ordinary renewal equations

$$u_i^{\alpha} = f_{ij}^{\alpha} + f_{ij}^{\alpha} u_j^{\alpha}, \qquad i \neq j,$$
$$u_j^{\alpha} = f_{jj}^{\alpha} + f_{jj}^{\alpha} u_j^{\alpha}.$$

The solution of the second equation is

$$u_j^{\alpha} = f_{jj}^{\alpha}(1 - f_{jj}^{\alpha})^{-1},$$

so substitution in the first equation yields

$$u_i^{\alpha} = f_{ij}^{\alpha}(1 - f_{jj}^{\alpha})^{-1}.$$

Consequently, one has from the above relations.

THEOREM 2.

$$u_{ij}^{\alpha}(\alpha + q_j) = \begin{cases} u_i^{\alpha}, & i \neq j, \\ u_j^{\alpha}/f_{jj}^{\alpha}, & i = j. \end{cases}$$

And, in agreement with recursive relation in Lemma 1, § 10.2.2, one has

THEOREM 3. *For* $n \geqslant 1$:

$$d_{ij}^{\alpha(n)} = f_{ij}^{\alpha}(f_{jj}^{\alpha})^{n-1}, \qquad i \neq j,$$
$$d_{jj}^{\alpha(n)} = (f_{jj}^{\alpha})^{n}.$$

b) As H consists of a single state j, so $\bar{q}_i = q_{ij}$ for $i \neq j$, and $\bar{q}_j = 0$. Hence, Theorem 1, § 10.2.3, yields

$$u_i^{\alpha} = \sum_{k \neq j} u_{ik}^{\alpha}\, \bar{q}_k = u_{ij}^{\alpha}(\alpha + q_j) - \delta_{ij}. \tag{F}$$

Evidently, this expression for u_i^{α} agrees with Theorem 2 above.

Next, concerning the limits, one has in the recurrent case (Thm. 1, § 10.2.5):

$$u_i(t) \to \varepsilon_j q_j = 1/m_{jj} \qquad \text{as} \quad t \to \infty.$$

On the other hand, in the transient case ($f_{jj} < 1$) one has, for $i \neq j$,

$$\mathbb{P}^i(N_\infty = 0) = 1 - f_{ij},$$
$$\mathbb{P}^i(N_\infty = n) = f_{ij}(1 - f_{jj})(f_{jj})^{n-1} \qquad \text{for} \quad n \geqslant 1,$$

and, for $i = j$,

$$\mathbb{P}^j(N_\infty = n) = (1 - f_{jj})(f_{jj})^{n} \qquad \text{for} \quad n \geqslant 0.$$

Obviously, for any i the mean is

$$U_i = f_{ij}/(1 - f_{jj}).$$

c) The special example considered here possesses a feature not shared by other return times T_ρ.

Recall that the exit taboo probabilities ${}_H u_{ik}^{\rho\alpha}$ discussed in § 2.2.4 have for $i \in H$ a component which corresponds to transitions within H. Consequently, the general expression for $d_i^{\rho\alpha}$ obtained from the (F)-equation in Theorem 9, § 3.3.1, has an additional term $\bar{q}_i/(\alpha + q_i)$ corresponding to these transitions.

In the present case with H consisting of a single state j, $\bar{q}_j = 0$ and this additional term vanishes. Hence, Theorem 9, § 3.3.1, yields now

$$d_j^{\rho\alpha} = \sum_{k \neq j} {}_j u_{jk}^{\rho\alpha}\, \bar{q}_k.$$

For the taboo probability ${}_j u_{jk}^{\rho\alpha}$ see Example 1, § 2.2.4, and note that ${}_j u_{jj}^{\rho\alpha} = (\alpha + q_j)^{-1}$.

With reference to discussion in § 10.1.9 (Case 1), one has the following interpretation: $d_j^{\rho\alpha}$ given above is the L.S. transform of $\mathbb{P}^j(\rho < T_\rho \leqslant t)$, whereas the L.S. transform of $\mathbb{P}^j(\rho = T_\rho \leqslant t)$ is

$${}_j u_{jj}^{\rho\alpha}\, \bar{q}_j = \bar{q}_j/(\alpha + q_j) = 0.$$

Furthermore, by concluding remarks in § 10.2.3, the average number $\mathbb{E}^i \widetilde{N}_t$ of jumps within H is obviously zero.

§ 10.2.8. Randomized entrances. Let N_t be the number of consecutive visits to a set H in the time interval $(0,t]$. As in the classical renewal theory, it is of interest to consider the compound r.v.'s T_{N_t} and S_{N_t} which represent instants of consecutive entrances to sets H and H^c, respectively, for a random number of visits to the set H up to time t.

Of special interest are the following two results with rather intuitive interpretation, but whose formal verification is tedious.

THEOREM 1. *For $i \in \mathbb{I}$ and for a fixed instant $t > 0$:*

$$\mathbb{P}^i(S_{N_t} \leqslant t < T_{N_t+1}) = \sum_{j \in H^c} p_{ij}(t) = \mathbb{P}^i(L_t < t) = \mathbb{P}^i(T_t > t),$$
$$\mathbb{P}^i(T_{N_t} \leqslant t < S_{N_t}) = \sum_{j \in H} p_{ij}(t) = \mathbb{P}^i(L_t = t) = \mathbb{P}^i(T_t = t).$$

Proof. Decompose the first assertion in the usual way:

$$\sum_{n=0}^{\infty} \mathbb{P}^i(S_n \leqslant t < T_{n+1},\ N_t = n) = \sum_{n=0}^{\infty} \mathbb{P}^i(S_n \leqslant t < T_{n+1})$$
$$= \sum_{n=0}^{\infty} \left(\widehat{D}_i^{(n)}(t) - D_i^{(n+1)}(t)\right) = \widehat{U}_i(t) - U_i(t) + I_{H^c}(i) = \sum_{j \in H^c} p_{ij}(t).$$

The second equality follows from formulas in § 10.2.1-c; in the third one the summation in $\widehat{U}_i(t)$ is taken over $n \geqslant 0$ for $i \in H$, and over $n \geqslant 1$ for H^c (see definitions in § 10.2.1). The last equality follows from expressions for $U_i(t)$ and $\widehat{U}_i(t)$ found in §§ 10.2.3 and 10.2.4, respectively. Interpretation in terms of T_t and L_t follows from §§ 3.2.1 and 4.2.1, respectively. The second assertion is obtained in the similar fashion. □

The next result refers to the random time spent outside set H.

THEOREM 2. *For $i \in \mathbb{I}$ and $v > t$ (with fixed $t > 0$):*

$$\mathbb{P}^i(0 < T_{N_t+1} - S_{N_t} \leqslant v,\ S_{N_t} < t)$$
$$= \int_0^t \sum_{r \in H} p_{ir}(s)\big(n_r(t-s) - n_r(v)\big)\, ds.$$

Proof. Using density argument (as in § 2.2.5) one has (see Fig. 10.2.2)

$$\mathbb{P}^i(0 < T_{n+1} - S_n \leqslant v,\ S_n < t < T_{n+1})$$
$$= \sum_{r \in H} \int_0^{\infty} d_{ir}^{(n)}(\tau)\, d\tau \sum_{k \in H^c} \int_{\tau}^{t} \hat{d}_{rk}(\sigma - \tau) \sum_{j \in H^c} {}_H p_{kj}(t-\sigma)\, d\tau \int_t^{\sigma+v} d_j(w-t)\, dw$$
$$= \sum_{k \in H^c} D_{ik}^{(n)}(t) D_k(v) - D_i^{(n+1)}(t),$$

where use has been made of Theorem 4, § 3.1.1, Proposition 1, § 10.2.1, and Lemma 1, § 10.2.2.

Next, decomposing the assertion in the usual way, one has

$$\sum_n \mathbb{P}^i(0 < T_{n+1} - S_n \leqslant v,\ S_n < t < T_{n+1})$$
$$= \sum_{j \in H^c} p_{ij}(t) - D_i^c(t) - \sum_{k \in H^c} \widehat{U}_{ik}(t) D_k^c(v)$$
$$= \mathbb{P}^i(0 < L_t < t) - \sum_{r \in H} \int_0^t p_{ir}(s)\, ds\, n_r(v),$$

where use has been made of Proposition 8, § 4.2.1, Theorem 3, § 4.2.3, and expression for $\widehat{U}_{ik}(t)$ from § 10.2.4. The final result follows then from Theorem 9, § 4.2.1. □

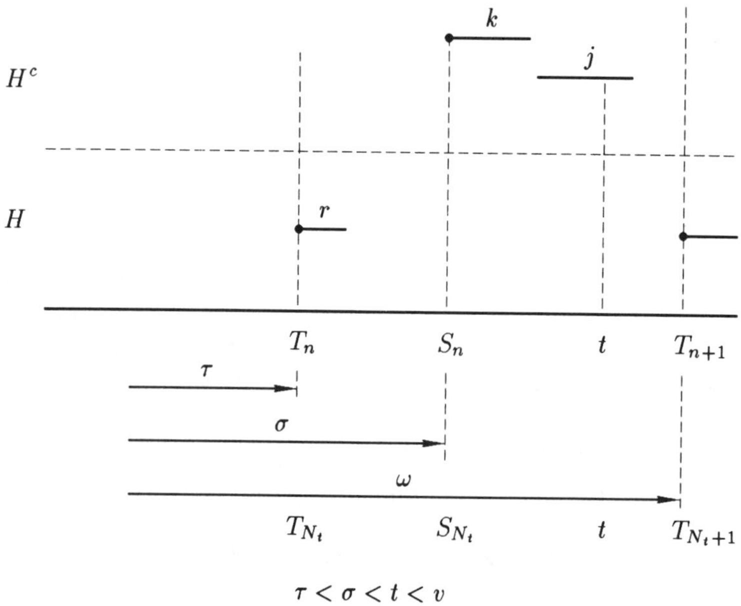

Figure 10.2.2. Illustration of Theorem 2, § 10.2.8.

Remark. As noted earlier, a M.C. (X_t) may be obtained by randomization of its jump chain (X_n). This effect is present again here in the sample function analysis ("density argument") and leads to correspondence between S_{N_t} and L_t, and between T_{N_t+1} and T_t. Direct analysis (not involving renewal argument) of the joint behavior of L_t and T_t (in Sub-section 10.3), as well as of "excursions from set H" (in Sub-section 10.4), leads to corresponding results. Indeed, Theorem 2 agrees with Corollary 4, § 10.3.1, and Remark 1, § 10.4.2. See also analogies with the local time (Sub-section 8.4) and the trace chain (Sub-section 9.5).

10.3. JOINT DISTRIBUTION

In Chapter II the passage times T_t and L_t were discussed separately. In this sub-section joint distribution will be considered. The result is the specialization of a more general formula obtained by H. Gzyl and C. Gzyl [1978] by the method of functionals and the random time change. For the usual M.C., however, the direct method (the density argument) is available, and this sub-section is in fact the continuation of Sub-sections 3.2 and 4.2.

§ 10.3.1. Distributions. Let $X = (X_t)$ be the usual M.C. with a stochastic matrix $\mathbf{P}(t)$, and let H be a fixed subset of the state space $\mathbb{I}$. Recall definitions of the first entrance time T_t and the last exit time L_t for the set H:

$$T_t = \inf(s\colon\ s > t,\ X_s \in H) = t + T \circ \theta_t, \qquad T \circ \theta_t = \inf(s\colon\ s > 0,\ X_{s+t} \in H),$$
$$L_t = \sup(s\colon\ 0 < s \leqslant t,\ X_s \in H) = t - \inf(s\colon\ s > 0,\ X_{t-s} \in H).$$

In the functional approach, one considers the local time $V = (V_t)$ on H and its inverse $\tau = (\tau_t)$, as discussed in Sub-section 8.4. The use is also made of relations

$$T_t = \tau_{V_t}, \qquad L_t = \tau_{V_t -}.$$

See Sub-section 9.5 and Fig. 9.5.1-4.

In the direct method the following lemma is basic. Note that in this lemma as well as in all its consequences, the instant t is fixed and conditioning is on $X_0 = i$.

Clearly

$$L_t < t < T_t,$$

and by the right-continuity of paths $X(L_t)$ takes values in H^c whereas $X(T_t)$ takes values in H (see Fig. 10.3.1).

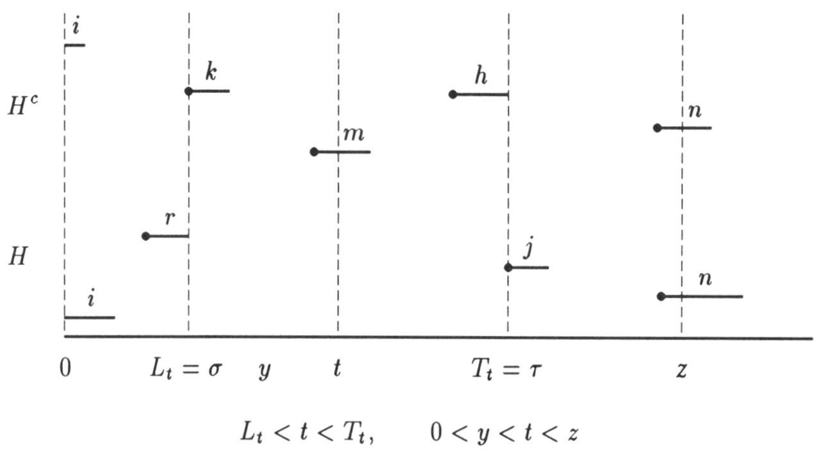

Figure 10.3.1. Joint distributions.

LEMMA 1. *For $i \in \mathbb{I}$, $r, j \in H$, $k, m, h \in H^c$, $n \in \mathbb{I}$, and $0 < y < t < z$:*

$$\mathbb{P}^i\Big(X(L_t-) = r,\ X(L_t) = k,\ 0 < L_t \leqslant y,\ X_t = m,$$
$$X(T_t-) = h,\ X(T_t) = j,\ t < T_t \leqslant z,\ X_z = n\Big)$$
$$= \int_0^y p_{ir}(\sigma) q_{rk}\ {}_H p_{km}(t-\sigma)\, d\sigma \int_t^z {}_H p_{mh}(\tau - t)\, q_{hj}\, p_{jn}(z-\tau)\, d\tau.$$

Proof. The argument is the combination of path arguments used in lemmas of § 2.2.5. □

It is convenient to use now the expression

$$c_{ik}(t) = \sum_{r \in H} p_{ir}(t) q_{rk} \qquad \text{for} \quad i \in \mathbb{I},\ k \in H^c,\ t \geqslant 0,$$

already introduced in § 2.2.6.

The joint distribution of $X(L_t)$, $X(T_t)$, L_t, and T_t is then given by

THEOREM 2. *For $k \in H^c$, $j \in H$, and $0 < y < t < z$:*

$$\mathbb{P}^i\big(X(L_t) = k,\ 0 < L_t \leqslant y,\ X(T_t) = j,\ t < T_t \leqslant z\big)$$
$$= \int_0^y c_{ik}(\sigma)\big(D_{kj}(z-\sigma) - D_{kj}(t-\sigma)\big)\, d\sigma.$$

Proof. The result follows from Lemma 1 by summation over all r, m, h, and n. □

The joint distribution of L_t and T_t is now given by

THEOREM 3. *For $0 < y < t < z$:*

$$\mathbb{P}^i(0 < L_t \leqslant y,\ t < T_t \leqslant z)$$
$$= \int_0^y \sum_{k \in H^c} c_{ik}(\sigma)\big(D_k^c(t-\sigma) - D_k^c(z-\sigma)\big)\, d\sigma.$$

Proof. Using the notation $\gamma_i^t(s)$ from § 4.2.1, the assertion of Theorem 3 can be written as

$$\mathbb{P}^i(0 < L_t \leqslant y < L_z) = \int_0^y \big(\gamma_i^t(\sigma) - \gamma_i^z(\sigma)\big)\, d\sigma$$

in view of the fact that L_s is a non-decreasing function of s, and that events $(T_t > z)$ and $(L_z \leqslant t)$ are equivalent. This observation provides the alternative quick proof of Theorem 3. It is also clear that L_t and T_t are dependent. □

Recall from Theorem 9, § 4.2.2, that

$$\mathbb{P}^i(0 < L_t \leqslant y) = \int_0^y \gamma_i^t(\sigma)\, d\sigma = \Pi_i^t(y), \qquad y < t,$$

with similar expression for L (when $t \to \infty$), and that $L_t = L$ when $L \leqslant t$ (see Lemma 7, § 2.1.4). Letting, respectively, $y \uparrow t$ and $z \to \infty$ in Theorem 3, one finds the marginal distributions

$$\mathbb{P}^i(0 < L_t < t,\ t < T_t \leqslant z) = \Pi_i^t - \Pi_i^z(t), \qquad t < z,$$
$$\mathbb{P}^i(0 < L_t \leqslant y,\ t < T_t < \infty) = \Pi_i^t(y) - \Pi_i(y), \qquad y < t.$$

Observe that the distribution in Theorem 3 has the joint density

$$f_t(y,z) = \sum_{k \in H^c} c_{ik}(y) d_k(z-y) \qquad \text{for} \quad 0 < y < t < z.$$

Integrating $f_t(y,z)$ over the region $s < z - y < v$ yields the expression for

$$\mathbb{P}^i(0 < L_t \leqslant y,\ s < T_t - L_t \leqslant v)$$

for $0 < y < t$, $0 < s < v$, $t - y < v$, in the form equivalent to that given by Corollary 5, § 10.4.2. In particular, in the limit as $y \uparrow t$ and $s \downarrow 0$, one obtains directly

COROLLARY 4. *For* $v > 0$:

$$\begin{aligned}
&\mathbb{P}^i(0 < L_t < t,\ 0 < T_t - L_t \leqslant v) \\
&= \int_{\max(0,t-v)}^{t} \sum_{k \in H^c} c_{ik}(y)\big(D_k^c(t-y) - D_k^c(v)\big)\, dy \\
&= \int_{\max(0,t-v)}^{t} \big(\gamma_i^t(y) - \gamma_i^{v+y}(y)\big)\, dy \\
&= \sum_{r \in H} \int_{\max(0,t-v)}^{t} p_{ir}(\sigma)\big(n_r(t-\sigma) - n_r(v)\big)\, d\sigma.
\end{aligned}$$

Note that for $v \to \infty$, this yields

$$\sum_{r \in H} \int_0^t p_{ir}(\sigma)\big(n_r(t-\sigma) - b_r\big)\, d\sigma = \Pi_i^t - \Pi_i(t),$$

in agreement with Theorem 4, § 4.1.1, and Theorem 10, § 4.2.1.

§ 10.3.2. Transforms. Recall definition of the balayage d_{ij}^{α} and write c_{ij}^{α} for the Laplace transform of $c_{ij}(t)$. Let $\rho > 0$, $\lambda > 0$, and $\alpha > 0$, and write, for $t > 0$,

$$J_{\rho,\lambda}(t) = \mathbb{E}^i\big(e^{-\rho(t-L_t)-\lambda(T_t-t)},\ X(L_t) = k,\ X(T_t) = j\big)$$

and

$$J_{\rho,\lambda}^{\alpha} = \int_0^{\infty} e^{-\alpha t} J_{\rho,\lambda}(t)\, dt.$$

The following is the main result.

THEOREM 1. *For $i \in \mathbb{I}$, $k \in H^c$, $j \in H$, and $\lambda \neq \alpha + \rho$:*

$$J_{\rho,\lambda}^{\alpha} = c_{ik}^{\alpha}(d_{kj}^{\lambda} - d_{kj}^{\rho+\alpha})/(\alpha + \rho - \lambda).$$

Proof. Observe that the distribution in Theorem 2, § 10.3.1, has the joint density

$$c_{ik}(y)d_{kj}(z-y) \qquad \text{for} \quad 0 < y < t < z.$$

Hence,

$$\begin{aligned} J_{\rho,\lambda}(t) &= \int_0^t c_{ik}(y)e^{-\rho(t-y)}\, dy \int_t^{\infty} e^{-\lambda(z-y)-\lambda(y-t)}\, d_{kj}(z-y)\, dz \\ &= \int_0^t c_{ik}(y)e^{-(\rho-\lambda)(t-y)}\, dy \bigg(d_{kj}^{\lambda} - \int_0^{t-y} e^{-\lambda u}\, d_{kj}(u)\, du \bigg). \end{aligned}$$

The result follows by taking transforms of convolution integrals. □

Special cases of Theorem 1 are of interest. Taking $\rho = \lambda$ one has

$$J_{\lambda}^{\alpha} = c_{ik}^{\alpha}(d_{kj}^{\lambda} - d_{kj}^{\lambda+\alpha})/\alpha$$

which corresponds to the transform of $T_t - L_t$ (see Cor. 4, § 10.3.1). In particular, $\rho = \lambda = 0$ gives

$$J_0^{\alpha} = c_{ik}^{\alpha}(D_{kj} - d_{kj}^{\alpha})/\alpha,$$

which is clearly the transform of

$$\begin{aligned} &\mathbb{P}^i(X(L_t) = k,\ 0 < L_t < t,\ X(T_t) = j,\ t < T_t < \infty) \\ &= \int_0^t c_{ik}(\sigma)\big(D_{kj} - D_{kj}(t-\sigma)\big)\, d\sigma. \end{aligned}$$

The above expression has been obtained from Theorem 2, § 10.3.1, with $y \uparrow t$, $z \to \infty$. This result should be compared with Theorem 5, § 3.1.2, Theorem 8, § 4.2.2, and the Chung equation (Cor. 11, § 4.2.1).

10.4. EXCURSIONS

The idea of considering a M.C. only between random times leads to many important applications. Excursion processes for general Markov processes received considerable attention in literature; see, for example, R. K. Getoor [1979], B. Maisonneuve [1975, 1979], B. Maisonneuve and P. A. Meyer [1974], H. Kaspi [1983]. These authors made extensive use of functionals of Markov processes. Because of the simple structure of M.C.'s considered here, direct analysis by "density argument" (as in Sub-section 10.3) is possible. The results obtained here in rather heuristic manner agree with those obtained by the rigorous approach given by R. K. Getoor [1979]; see also R. Syski [1984, 1986].

§ **10.4.1. Basic Lemma.** Let $X = (X_t)$ be a M.C. satisfying the usual conditions, and let H be a fixed subset of a state space $\mathbb{I}$. For a given t, let L_t and T_t be, respectively, the last exit time from H before t and the first entrance time to H after t, as defined in Section 2. As noted in § 2.1.4, the processes (L_t) and (T_t) are right-continuous and inverse of each other. The following definition has been already mentioned in § 2.1.4.

The random interval (L_t, T_t) is called the *excursion interval* (from H) *straddling* t (for $t > 0$). The process $\{X(L_t + s),\ 0 < s < T_t - L_t\}$ defined on the random interval (L_t, T_t) is called the *excursion process straddling* t. By the right-continuity of the M.C. X, the excursion process takes values in H^c only, and $L_t < t < T_t$.

Of primary interest is the distribution of the excursion as well as the distribution of the life time $T_t - L_t$.

The following Lemma is basic. Its main advantage is the fact that required probabilities are expressed in terms of known taboo probabilities. See Fig. 10.4.1 for interpretation.

LEMMA 1. *For* $k, j \in H^c,\ i \in \mathbb{I}$, *and* $0 < \sigma < t,\ 0 < s < v,\ t - \sigma < v$:

$$\begin{aligned}&\mathbb{P}\{X(L_t) = k,\ L_t \in d\sigma,\ X(L_t + s) = j,\ T_t - L_t \in dv \mid X_0 = i\}\\ &\quad = c_{ik}(\sigma)d\sigma\ {}_Hp_{kj}(s)d_j(v - s)\, dv.\end{aligned}$$

Proof (outline). Using the density argument (see § 2.2.5) consider the conditioning

$$\begin{aligned}&\mathbb{P}^i\{X(L_t) = k,\ L_t \in d\sigma,\ X(L_t + s) = j,\ T_t - L_t \in dv\}\\ &\quad = \mathbb{P}^i\{X(L_t) = k,\ L_t \in d\sigma\}\\ &\quad\quad \times \mathbb{P}\{X(L_t + s) = j,\ T_t - L_t \in dv \mid X(L_t) = k,\ L_t = \sigma\}.\end{aligned}$$

By Theorem 8, § 4.2.2, the first density is, for $0 < \sigma < t$,

$$\mathbb{P}^i(X(L_t) = k,\ L_t \in d\sigma) = c_{ik}(\sigma)\, d\sigma\, D_k^c(t - \sigma).$$

The second density is evaluated with the help of the strong Markov property (at the instants of jumps $L_t = \sigma$) and the fact that $T_\tau = T_t$ for any τ in the interval

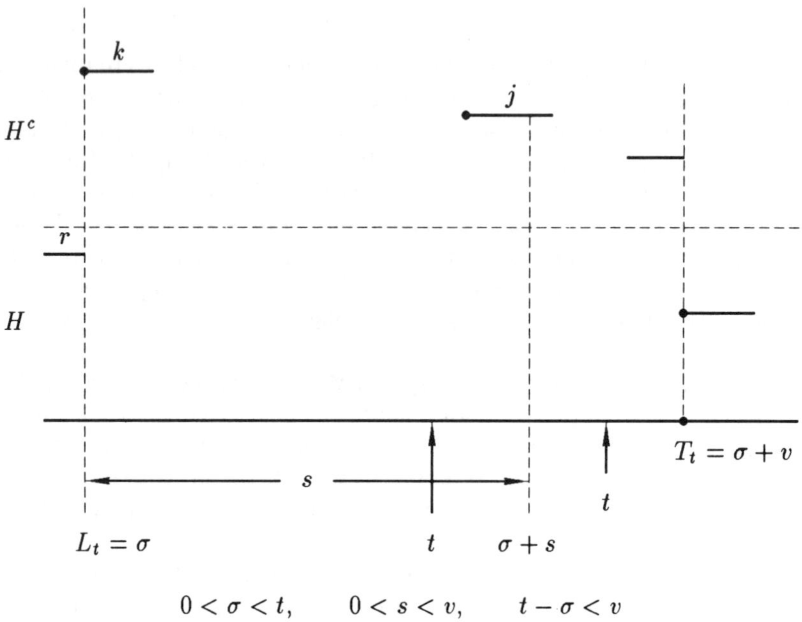

Figure 10.4.1. Excursions.

(L_t, T_t). Indeed, in this random interval the M.C. is necessarily in H^c; recall from § 2.1.4 that events $(T_\sigma > t)$ and $(X_\tau \in H^c,\ \sigma < \tau < t)$ are equivalent. Hence,

$$\begin{aligned}
&\mathbb{P}(X(L_t + s) = j,\ T_t - L_t \in dv \mid X(L_t) = k,\ L_t = \sigma) \\
&\quad = \mathbb{P}(X_{\sigma+s} = j,\ X_\tau \in H^c,\ \sigma < \tau < \sigma + dv \mid X_\sigma = k,\ T_\sigma > t) \\
&\quad = \mathbb{P}(X_{\sigma+s} = j,\ T_\sigma > \sigma + s,\ T_{\sigma+s} \in \sigma + dv \mid X_\sigma = k)/\mathbb{P}(T_\sigma > t \mid X_\sigma = k) \\
&\quad = {}_H p_{kj}(s)\, d_j(v - s)\, ds / D_k^c(t - \sigma),
\end{aligned}$$

because

$$\begin{aligned}
&\mathbb{P}(X_{\sigma+s} = j,\ T_\sigma > \sigma + s,\ T_{\sigma+s} \in \sigma + dv \mid X_\sigma = k) \\
&\quad = \mathbb{P}(X_s = j,\ T > s \mid X_0 = k)\, \mathbb{P}(T \circ \theta_{\sigma+s} \in dv - s \mid X_0 \circ \theta_{\sigma+s} = j) \\
&\quad = {}_H p_{kj}(s)\, d_j(v - s)\, dv,
\end{aligned}$$

and

$$\mathbb{P}(T_\sigma > t \mid X_\sigma = k) = \mathbb{P}(T \circ \theta_\sigma > t - \sigma \mid X_\sigma = k) = D_k^c(t - \sigma).$$

The required probability density is then the product of the first and the second density. □

§ **10.4.2. Theorems.** The following basic result is the direct consequence of Lemma 1, § 10.4.1, and yields the marginal distributions of interest. Observe that here t and s are fixed, and σ and v are variables of time distributions.

THEOREM 1. *For $k,j \in H^c$, $i \in \mathbb{I}$, and $0 < \sigma < t$, $0 < s < v$, $t - \sigma < v$:*

$$\mathbb{P}^i\big(X(L_t) = k,\ 0 < L_t \leqslant \sigma,\ X(L_t + s) = j,\ s < T_t - L_t \leqslant v\big)$$

$$= \begin{cases} \displaystyle\int_{\max(t-v,0)}^{\sigma} c_{ik}(x)\,dx\ {}_Hp_{kj}(s) D_j(v-s) \\ \displaystyle - \int_{\max(t-v,0)}^{t-s} c_{ik}(x)\,dx\ {}_Hp_{kj}(s) D_j(t-x-s) & \text{for}\quad s < t,\ t-s < \sigma < t, \\ \displaystyle\int_{\max(t-v,0)}^{\sigma} c_{ik}(x)\,dx\ {}_Hp_{kj}(s) \\ \times\big[D_j(v-s) - D_j(t-x-s)\big] & \text{for}\quad s < t,\ 0 < \sigma \leqslant t - s, \\ \displaystyle\int_0^{\sigma} c_{ik}(x)\,dx\ {}_Hp_{kj}(s) D_j(v-s) & \text{for}\quad s \geqslant t. \end{cases}$$

Recalling definitions of $c_{ik}(x)$ and $n_{rj}(x)$ stated in § 2.2.6, one has

COROLLARY 2. *For $j \in H^c$, $i \in \mathbb{I}$, and $0 < \sigma < t$, $0 < s < v$, $t - \sigma < v$:*

$$\mathbb{P}^i(0 < L_t \leqslant \sigma,\ X(L_t + s) = j,\ s < T_t - L_t \leqslant v)$$

$$= \begin{cases} \displaystyle\int_{\max(0,t-v)}^{\sigma} \sum_{r \in H} p_{ir}(x)\,dx\, n_{rj}(s) D_j(v-s) \\ \displaystyle - \int_{\max(0,t-v)}^{\min(\sigma,t-s)} \sum_{r \in H} p_{ir}(x)\,dx\, n_{rj}(s) D_j(t-x-s) & \text{for}\quad s < t, \\ \displaystyle\int_0^{\sigma} \sum_{r \in H} p_{ir}(x)\,dx\, n_{rj}(s) D_j(v-s) & \text{for}\quad s \geqslant t. \end{cases}$$

Next, letting $v \to \infty$, $\sigma \uparrow t$ and summing over all j, respectively, one obtains the following marginal distributions.

COROLLARY 3. *For $j \in H^c$, $i \in \mathbb{I}$, and $0 < \sigma < t$, $0 < s < \infty$, $t - \sigma < \infty$,*

$$\mathbb{P}^i(0 < L_t \leqslant \sigma,\ X(L_t + s) = j,\ s < T_t - L_t < \infty)$$

$$= \begin{cases} \displaystyle\int_0^{\sigma} \sum_{r \in H} p_{ir}(x)\,dx\, n_{rj}(s) D_j \\ \displaystyle - \int_0^{\min(\sigma,t-s)} \sum_{r \in H} p_{ir}(x)\,dx\ n_{rj}(s) D_j(t-x-s) & \text{for}\quad s < t, \\ \displaystyle\int_0^{\sigma} \sum_{r \in H} p_{ir}(x)\,dx\, n_{rj}(s) D_j & \text{for}\quad s \geqslant t. \end{cases}$$

COROLLARY 4. *For $j \in H^c$, $i \in \mathbb{I}$, and $0 < s < v$:*

$$\mathbb{P}^i(0 < L_t < t,\ X(L_t + s) = j,\ s < T_t - L_t \leqslant v)$$

$$= \begin{cases} \int_{\max(0,t-v)}^{t} \sum_{r \in H} p_{ir}(x)\, dx\, n_{rj}(s) D_j(v-s) \\ - \int_{\max(0,t-v)}^{t-s} \sum_{r \in H} p_{ir}(x)\, dx\, n_{rj}(s) D_j(t-x-s) & \text{for } s < t, \\ \int_0^t \sum_{r \in H} p_{ir}(x)\, dx\, n_{rj}(s) D_j(v-s) & \text{for } s \geqslant t. \end{cases}$$

COROLLARY 5. *For $i \in \mathbb{I}$ and $0 < \sigma < t$, $0 < s < v$, $t - \sigma < v$:*

$$\mathbb{P}^i(0 < L_t \leqslant \sigma,\ s < T_t - L_t \leqslant v)$$

$$= \begin{cases} \int_{\max(0,t-v)}^{\sigma} \sum_{r \in H} p_{ir}(x)\, dx\big[n_r(s) - n_r(v)\big] \\ - \int_{\max(0,t-v)}^{\min(\sigma,t-s)} \sum_{r \in H} p_{ir}(x)\, dx\big[n_r(s) - n_r(t-x)\big] & \text{for } s < t, \\ \int_0^{\sigma} \sum_{r \in H} p_{ir}(x)\, dx\big[n_r(s) - n_r(v)\big] & \text{for } s \geqslant t. \end{cases}$$

In Corollary 5 use has been made from the fact that for fixed r, $(n_{rj}(\cdot))$ form an entrance law relative to the taboo semi-group, as noted in § 3.4.1. Thus,

$$\sum_{j \in H^c} n_{rj}(s) D_j(u-s) = n_r(s) - n_r(u), \qquad u \geqslant s.$$

Alternative expressions for Corollary 5 may be obtained using density $\gamma_i^t(\sigma)$ from § 4.2.1 or the function $\varphi_r(y)$ from § 3.4.2.

Remarks. 1. It should be noted that the marginal distribution in Corollary 5 depends on s. Letting $s \downarrow 0$, one obtains the joint distribution

$$\mathbb{P}^i(0 < L_t \leqslant \sigma,\ 0 < T_t - L_t \leqslant v)$$
$$= \sum_{r \in H} \int_{\max(0,t-v)}^{\sigma} p_{ir}(x)\big[n_r(t-x) - n_r(v)\big]\, dx,$$

and this agrees with Corollary 4, § 10.3.1, when $\sigma \uparrow t$.

2. It is of interest to extend Corollary 2 to the boundary cases of $L_t = 0$ and $L_t = t$, using results from § 4.2.1 and § 4.2.3.

3. From the proof of Theorem 1, one has immediately for $k, j \in H^c$ and $0 < \sigma < t$, $0 < s < v$, $t - \sigma < v$:

$$\mathbb{P}(X(L_t + s) = j,\ T_t - L_t > v \mid L_t = \sigma,\ X(L_t) = k)$$
$$= {}_H p_{kj}(s) D_j^c(v-s) / D_k^c(t-\sigma).$$

Letting $t \to \infty$ and $v \to \infty$, one obtains Theorem 1, § 9.2.4, for the post-L chain. Moreover

$$\mathbb{P}(T_t - L_t > v \mid L_t = \sigma,\ X(L_t) = k) = D^c(v)/D^c_k(t-\sigma).$$

Compare this result with Theorem 9, § 4.2.2. See also discussion in § 9.2.5.

4. It is of interest to recall connection with argument based on randomization of consecutive entrances to a set (see Remark in § 10.2.8). Indeed, the distribution of $T_t - L_t$ straddling t is reminiscent of an age distribution in the renewal theory. See also discussion of the trace chain in Sub-section 9.5.

§ **10.4.3. Examples.** The following examples occur frequently in applications. Specific examples will be discussed in Chapter IV.

Example 1. *Excursion from a fixed state* r. Take $H = \{r\}$, so by Example 1, § 3.1.4, $D_i(t) = F_{ir}(t)$ for $i \neq r$, and therefore

$$n_r(t) = \sum_{k \neq r} q_{rk} F^c_{kr}(t)$$

(see Ex. 1, § 3.4.3). Hence, Corollary 5, § 10.4.2, yields

$$\mathbb{P}^i(0 < L_t < t,\ 0 < T_t - L_t \leqslant v) = \int_{\max(0,t-v)}^{t} p_{ir}(x)\,dx\,\big[n_r(t-x) - n_r(v)\big].$$

For connection with the local time see Remark 1, § 8.4.4.

Example 2. Take now $H^c = \{k\}$ for a fixed state k. By Example 2, § 3.1.4, $D^c_k(t) = e^{-q_k t}$, so

$$n_r(t) = q_{rk} e^{-q_k t}, \qquad r \neq k$$

(see Ex. 2, § 3.4.3). Hence, Corollary 5, § 10.4.2, yields

$$\mathbb{P}^i(0 < L_t < t,\ 0 < T_t - L_t \leqslant v) = \int_{\max(0,t-v)}^{t} w_{ik}(x)\,dx\,\big[e^{-q_k(t-x)} - e^{-q_k v}\big],$$

where $w_{ik}(x)$ has been defined in § 1.1.2. Letting $v \to \infty$, one has

$$\mathbb{P}^i(0 < L_t < t) = \int_0^t w_{ik}(x)\,dx\, e^{-q_k(t-x)} = p_{ik}(t) - \delta_{ik} e^{-q_k t}.$$

Observe that Remark 3, § 10.4.2, yields

$$\mathbb{P}(T_t - L_t > v \mid L_t = \sigma,\ X(L_t) = k) = e^{-q_k(v-t+\sigma)},$$

as expected.

CHAPTER IV

APPLICATIONS

This chapter is devoted to applications of passage times to various Markovian structures. There are two kinds of such applications. In the first one, examination of a particular structure will reveal presence of passage times either in explicit or implicit form. It is then of interest to investigate the role passage times exert on that structure. The first two sections of this chapter are devoted to such a task. In the second kind of applications, one deals with a detailed study of a particular chain leading to explicit expressions characterizing that chain. The last section in this chapter treats such problems.

Applications have been selected from huge literature on the subject. However, emphasis has been here on the role of passage times. Section 11 treats general structures, and Section 12 presents brief introduction to more general chains. Material in Section 13 contains new calculations. Individual references can be found in the text.

SECTION 11

STRUCTURES

This section is devoted to special topics concerning structure of Markov semi-groups. Emphasis is placed, as always, on the role of passage times.

Of special interest is a more general look at *decomposition theorems*, according to an arbitrary kernel **K** (Sub-section 11.1) This includes as particular cases decompositions according to the first entrance time and to the last exist time, where **K** is, respectively, the balayage **B** and the co-balayage **C**. Killing operation is an another example.

Related problems concerning structure involve *perturbation* and *revival* operations (Sub-section 11.2). In perturbation the original semi-group is modified according to the external mechanism described by a replacement matrix **R**. Several applications of this method are presented. Revival is similar in character, and concerns the situation when a chain is "revived" or re-started on entering an absorbing set.

Aggregation refers to the modification of a semi-group obtained by grouping states (Sub-section 11.3). This operation destroys Markov property in general, and it is of interest to know when the Markov property is preserved. This sub-section discusses also some less known operations, valid only for special chains.

The remaining two sub-sections dealing with *capacity* and *energy* are of a different character. Capacity (Sub-section 11.4) is a set function, generalizing measure, well known in Functional Analysis (Choquet capacity), and it is of interest to see its connections with balayage concept.

Concept of energy gain widespread interest in recent years. It generalises a similar concept from classical Potential Theory, but lack of symmetry leads to additional features. One may hope that energy will become a useful characterisation of Markovian systems.

As always, presentation here is based on works dealing with general Markov processes, but specialization to the usual M.C. chains leads to considerable simplification. Individual references are given in the text. It should be mentioned, however, that Sub-sections 11.1, 11.2, and 11.3 have been based on papers by J. Keilson, M. Neuts, H. Cai, and R. Syski. Capacity concept is extensively discussed in literature. Sub-section on energy has been based on work of P. Meyer, K. L. Chung, and J. Glover.

11.1. DECOMPOSITION THEOREMS

In study of Markovian systems it is of interest to have representation theorems which relate structure of two associated semi-groups. Typical examples are provided by the well-known "first entrance" and the "last exit" decompositions (Section 6), and by various transformations of processes, like killing (Section 9). This sub-section presents comments on a general form of such decomposition theorems, with emphasis on perturbation of semi-groups.

§ 11.1.1. Kernel. Let $\mathbf{P}(t)$ be a standard stochastic transition matrix of an usual M.C. $X = (X_t)$; recall that $\mathbf{P}(0) = \mathbf{I}$ and $\mathbf{P}(t)\mathbf{1} = \mathbf{1}$ for all $t \geqslant 0$. Let $\widetilde{\mathbf{P}}(t)$ be a sub-stochastic matrix corresponding to some other Markov chain $\widetilde{X}$; here

$$\widetilde{\mathbf{P}}(0) \leqslant \mathbf{I} \quad \text{and} \quad \widetilde{\mathbf{P}}(t)\mathbf{1} \leqslant \mathbf{1} \qquad \text{for all} \quad t \geqslant 0.$$

Denote by $\mathbf{Q}$ and $\widetilde{\mathbf{Q}}$ the corresponding generators, and write $\mathbf{U}^\alpha$ and $\widetilde{\mathbf{U}}^\alpha$ for resolvents. Note that $\widetilde{\mathbf{Q}}$ need not be conservative.

The operators $\mathbf{P}(t)$ and $\widetilde{\mathbf{P}}(t)$ are said to be related by a decomposition theorem when

$$\mathbf{P}(t) = \widetilde{\mathbf{P}}(t) + (\mathbf{K} * \mathbf{P})(t), \qquad t \geqslant 0,$$

where $\mathbf{K}(t)$ is a positive operator, defined by a kernel $(k_{ij}(t))$, and $*$ denotes convolution:

$$(\mathbf{K} * \mathbf{P})(t) = \int_{0-}^{t} d\mathbf{K}(s)\, \mathbf{P}(t-s), \qquad t \geqslant 0,$$

with integral taken over the closed interval $[0,t]$.

In applications, $\mathbf{K}(t)$ is given and $\widetilde{\mathbf{P}}(t)$ is the primary (original, basic etc.) semi-group, whereas $\mathbf{P}(t)$ is the secondary (transformed, perturbed etc.) semi-group. The above defining decomposition formula is then the Markov renewal equation for $\mathbf{P}(t)$.

It is more convenient to write the decomposition formula in terms of resolvents:

$$\mathbf{U}^\alpha = \widetilde{\mathbf{U}}^\alpha + \mathbf{K}^\alpha\, \mathbf{U}^\alpha, \qquad \alpha > 0,$$

where

$$\mathbf{K}^\alpha = \int_{0-}^{\infty} e^{-\alpha t}\, d\mathbf{K}(t), \qquad \lim_{\alpha\to\infty} \mathbf{K}^\alpha = \mathbf{K}(0).$$

As an illustration, recall the following two examples (more examples will be given later in this sub-section).

a) The (B)-equation in the integrated form

$$p_{ij}(t) = \delta_{ij} e^{-q_i t} + \sum_k \int_0^t r_{ik}(s) p_{kj}(t-s)\, ds, \qquad i,j \in \mathbb{I},$$

where $\widetilde{\mathbf{P}}(t)$ is the diagonal matrix of exponential terms, and $\mathbf{K}(t)$ with $\mathbf{K}^\alpha = (r^\alpha_{ij})$ is the α-jump matrix. See Example 1, § 11.1.3, for further discussion.

b) The first entrance decomposition

$$\mathbf{U}^\alpha = {}_H\mathbf{U}^\alpha + \mathbf{B}^\alpha\,\mathbf{U}^\alpha$$

with $\widetilde{\mathbf{U}}^\alpha = {}_H\mathbf{U}^\alpha$, $\mathbf{K}^\alpha = \mathbf{B}^\alpha$ and $\widetilde{\mathbf{P}}(0) = \mathbf{J}_1$, $\mathbf{K}(0) = \mathbf{J}_2$, $\widetilde{\mathbf{Q}} = {}_H\mathbf{Q}$. See Example 2, § 11.1.3, for further discussion.

Returning to the general case, the defining formula yields

$$\widetilde{\mathbf{P}}(0) + \mathbf{K}(0) = \mathbf{P}(0) = \mathbf{I}, \qquad \widetilde{\mathbf{P}}(t)\mathbf{1} + \mathbf{K}(t)\mathbf{1} = \mathbf{P}(t)\mathbf{1} = \mathbf{1}.$$

By assumption, the secondary semi-group $\mathbf{P}(t)$, $t \geqslant 0$, has all the usual properties. On the other hand, the primary semi-group $\widetilde{\mathbf{P}}(t)$, $t \geqslant 0$, exhibits some peculiar properties. From the Chapman–Kolmogorov equation it follows that

$$\widetilde{\mathbf{P}}(0)\widetilde{\mathbf{P}}(t) = \widetilde{\mathbf{P}}(t)\widetilde{\mathbf{P}}(0) = \widetilde{\mathbf{P}}(t), \qquad t \geqslant 0,$$

so $\widetilde{\mathbf{P}}(0)$ is idempotent,

$$\widetilde{\mathbf{P}}(0)\widetilde{\mathbf{P}}(0) = \widetilde{\mathbf{P}}(0).$$

Consequently,

$$\mathbf{K}(0)\widetilde{\mathbf{P}}(t) = \widetilde{\mathbf{P}}(t)\mathbf{K}(0) = \mathbf{0}, \qquad t \geqslant 0.$$

With $\widetilde{\mathbf{P}}(t) \to \widetilde{\mathbf{P}}(0)$ as $t \to 0$, the generator $\widetilde{\mathbf{Q}}$ is now defined by

$$\widetilde{\mathbf{Q}} = \lim_{t\to 0} \big(\widetilde{\mathbf{P}}(t) - \widetilde{\mathbf{P}}(0)\big)/t$$

(pointwise limit), and the Kolmogorov equations take the form

$$\alpha\widetilde{\mathbf{U}}^\alpha - \widetilde{\mathbf{P}}(0) = \begin{cases} \widetilde{\mathbf{Q}}\widetilde{\mathbf{U}}^\alpha, & \text{(B)} \\ \widetilde{\mathbf{U}}^\alpha\widetilde{\mathbf{Q}}. & \text{(F)} \end{cases}$$

It follows that

$$\widetilde{\mathbf{Q}}\widetilde{\mathbf{P}}(0) = \widetilde{\mathbf{P}}(0)\widetilde{\mathbf{Q}} = \widetilde{\mathbf{Q}}, \qquad \widetilde{\mathbf{Q}}\mathbf{K}(0) = \mathbf{K}(0)\widetilde{\mathbf{Q}} = \mathbf{0},$$

and that $\widetilde{\mathbf{Q}}$ is not conservative, in general.

The following results follow from the fact that both $\mathbf{P}(t)$ and $\widetilde{\mathbf{P}}(t)$, $t \geqslant 0$, are semi-groups, and from definition of operator $\mathbf{K}(t)$.

LEMMA 1. *For* $t, s \geqslant 0$:

$$\widetilde{\mathbf{P}}(t)\mathbf{K}(s) = \mathbf{K}(s+t) - \mathbf{K}(t).$$

In particular, one has

$$\widetilde{\mathbf{P}}(0)\mathbf{K}(s) = \mathbf{K}(s) - \mathbf{K}(0), \qquad s \geqslant 0,$$

or in terms of L.S. *transforms,*

$$\widetilde{\mathbf{P}}(0)\mathbf{K}^\alpha = \mathbf{K}^\alpha - \mathbf{K}(0), \qquad \alpha > 0.$$

LEMMA 2.

$$\tilde{\mathbf{Q}}\mathbf{K}(s) = \mathbf{K}'(s) - \mathbf{K}'(0), \qquad s \geqslant 0,$$

or

$$\tilde{\mathbf{Q}}\mathbf{K}^\alpha = \alpha\big(\mathbf{K}^\alpha - \mathbf{K}(0)\big) - \mathbf{K}'(0), \qquad \alpha > 0,$$

where ′ denotes the differentiation.

THEOREM 3.

$$\mathbf{K}(t) - \mathbf{K}(0) = \int_0^t \tilde{\mathbf{P}}(s)(\mathbf{Q} - \tilde{\mathbf{Q}})\, ds, \qquad t \geqslant 0,$$

or

$$\mathbf{K}^\alpha - \mathbf{K}(0) = \tilde{\mathbf{U}}^\alpha(\mathbf{Q} - \tilde{\mathbf{Q}}), \qquad \alpha > 0.$$

It is immediate that the defining relation can be written in the equivalent form, known as the "second resolvent equation", which expresses $\mathbf{U}^\alpha$ directly in terms of the primary process; see E. Hille and R. S. Phillips [1957].

COROLLARY 4.

$$\tilde{\mathbf{P}}(0)\mathbf{U}^\alpha = \tilde{\mathbf{U}}^\alpha + \tilde{\mathbf{U}}^\alpha(\mathbf{Q} - \tilde{\mathbf{Q}})\mathbf{U}^\alpha, \qquad \alpha > 0.$$

Furthermore, Theorem 3 yields the following relations between generators.

COROLLARY 5.

$$\mathbf{Q} - \tilde{\mathbf{Q}} = \lim_{\alpha \to \infty} \alpha\big(\mathbf{K}^\alpha \alpha \mathbf{U}^\alpha - \mathbf{K}(0)\big)$$

and

$$\tilde{\mathbf{P}}(0)(\mathbf{Q} - \tilde{\mathbf{Q}}) = \mathbf{K}'(0).$$

§ 11.1.2. Other kernels. Several other forms of the decomposition theorem are of interest. The following result indicates the connection with the (B)-equation for the secondary process.

Define the compensation kernel $\mathbf{C}(t) = (c_{ij}(t))$ by

$$\mathbf{C}(t) = (\mathbf{Q} - \tilde{\mathbf{Q}})\mathbf{P}(t), \qquad t \geqslant 0,$$

or, in the Laplace transform,

$$\mathbf{C}^\alpha = (\mathbf{Q} - \tilde{\mathbf{Q}})\mathbf{U}^\alpha, \qquad \alpha > 0.$$

Note that $\mathbf{C}(0) = \mathbf{Q} - \tilde{\mathbf{Q}}$, and that $\mathbf{C}(t)$ has some negative terms, with

$$\mathbf{C}(t)\mathbf{1} = -\tilde{\mathbf{Q}}\mathbf{1}.$$

Consider now the (B)-equation for the secondary semi-group, and write it in the form

$$\alpha\mathbf{U}^\alpha - \mathbf{I} = \mathbf{Q}\mathbf{U}^\alpha = (\mathbf{Q} - \tilde{\mathbf{Q}})\mathbf{U}^\alpha + \tilde{\mathbf{Q}}\mathbf{U}^\alpha = \mathbf{C}^\alpha + \tilde{\mathbf{Q}}\mathbf{U}^\alpha. \qquad \text{(B)}$$

This is a (transform of) non-homogeneous linear equation with the "forcing term" $\mathbf{C}^\alpha$. As $\tilde{\mathbf{U}}^\alpha$, for the primary process, satisfies a homogeneous linear equation

$$\alpha\tilde{\mathbf{U}}^\alpha - \tilde{\mathbf{P}}(0) = \tilde{\mathbf{Q}}\tilde{\mathbf{U}}^\alpha, \qquad \text{(B)}$$

one can conclude that the solution $\mathbf{U}^\alpha$ satisfies also the defining Markov renewal equation.

THEOREM 1. *Let* $\mathbf{U}^\alpha$ *and* $\widetilde{\mathbf{U}}^\alpha$ *satisfy the above* (B)*-equations, respectively. Then*

$$\widetilde{\mathbf{P}}(0)\mathbf{U}^\alpha = \widetilde{\mathbf{U}}^\alpha + \widetilde{\mathbf{U}}^\alpha\mathbf{C}^\alpha.$$

Proof. Note that the assertion is the same as that in Corollary 4, § 11.1.1. Substitution into the (B)-equation for $\mathbf{U}^\alpha$ verifies that this is indeed the solution. □

The next result indicates that a solution of the Markov renewal equation has essentially a form of the Riesz decomposition. The operator $\mathbf{G}^\alpha$ defined here is the analogue of a renewal function.

THEOREM 2. *For a bounded function* f, *the equation*

$$\mathbf{U}^\alpha f = \widetilde{\mathbf{U}}^\alpha f + \mathbf{K}^\alpha\mathbf{U}^\alpha f$$

has a solution $\mathbf{U}^\alpha f$ *iff the* (*strong*) *limit* $\mathbf{G}^\alpha\widetilde{\mathbf{U}}^\alpha f$ *exists, where*

$$\mathbf{G}^\alpha = \sum_{n=0}^{\infty}(\mathbf{K}^\alpha)^n.$$

All solutions are of the form

$$\mathbf{U}^\alpha f = \mathbf{G}^\alpha\widetilde{\mathbf{U}}^\alpha f + \mathbf{W}^\alpha\mathbf{U}^\alpha f,$$

where

$$\mathbf{W}^\alpha\mathbf{U}^\alpha f = \lim_{n\to\infty}(\mathbf{K}^\alpha)^n\mathbf{U}^\alpha f.$$

The solution is unique when $\mathbf{W}^\alpha\mathbf{U}^\alpha f = 0$.

For complete proof see E. Çinlar [1975].

§ 11.1.3. Examples. The following examples have been already mentioned earlier, but it is of interest to see how they fit into decomposition structure. Several new applications will be presented later in the next sub-section.

Example 1. This is the continuation of example a) from § 11.1.1. This example establishes connection between the usual M.C. and its associated jump chain. With matrices

$$\mathbf{U}^\alpha = (u_{ij}^\alpha), \quad \mathbf{Q} = (q_{ij}) \qquad \text{and} \qquad \widetilde{\mathbf{U}}^\alpha = (\boldsymbol{\Theta}^\alpha)^{-1}, \quad \widetilde{\mathbf{Q}} = -\boldsymbol{\Theta},$$

the relation

$$\mathbf{U}^\alpha = \mathbf{G}^\alpha\widetilde{\mathbf{U}}^\alpha$$

is just the Feller relation from § 1.1.6, confirming that the minimal solution of Kolmogorov equations is related to the structure of the jump chain. Here $\mathbf{G}^\alpha$ is the (Green) potential of the α-jump matrix $\mathbf{R}^\alpha = \mathbf{K}^\alpha$, and the diagonal matrices $\boldsymbol{\Theta}^\alpha$ and $\boldsymbol{\Theta}$ have been introduced in §§ 1.1.6 and 1.1.4. In fact, the formula

$$\mathbf{K}^\alpha = \widetilde{\mathbf{U}}^\alpha(\mathbf{Q} - \widetilde{\mathbf{Q}})$$

from Theorem 3, § 11.1.1, with $\mathbf{K}(0) = \mathbf{0}$, is the same as the formula

$$\mathbf{R}^\alpha - \mathbf{I} = (\boldsymbol{\Theta}^\alpha)^{-1}\mathbf{Q}^\alpha$$

from § 1.1.6 for the α-jump chain. In this sense, the Feller relation is equivalent to the Dynkin characteristic operator, as noted in § 8.1.1.

Furthermore, this example indicates that the Feller relation for a minimal chain corresponds to the Riesz decomposition with component $\mathbf{W}^\alpha = \mathbf{R}^{\alpha\infty}$ vanishing. See Theorem 2, § 11.1.2, and comments on Theorems 6 and 7 in § 1.3.3.

Example 2. This is the continuation of example b) from § 11.1.1. This example establishes connection between the usual semi-group $\mathbf{P}(t)$, $t \geqslant 0$, and its taboo semi-group ${}_H\mathbf{P}(t)$, $t \geqslant 0$, expressed by the decomposition theorem

$$\mathbf{U}^\alpha = {}_H\mathbf{U}^\alpha + \mathbf{B}^\alpha \mathbf{U}^\alpha.$$

Here

$$\widetilde{\mathbf{U}}^\alpha = {}_H\mathbf{U}^\alpha, \qquad \mathbf{K}^\alpha = \mathbf{B}^\alpha, \qquad \widetilde{\mathbf{Q}} = {}_H\mathbf{Q}, \qquad \widetilde{\mathbf{P}}(0) = \mathbf{J}_1, \qquad \mathbf{K}(0) = \mathbf{J}_2.$$

As might be expected, results from § 11.1.1 agree with those stated in § 3.1.5.

Thus, Lemma 1, § 11.1.1, becomes:

$${}_H\mathbf{P}(t)\,\mathbf{B}(s) = \mathbf{B}(t+s) - \mathbf{B}(t) \qquad \text{with} \quad \mathbf{B}^\alpha \mathbf{J}_2 = \mathbf{J}_2,$$

whereas Lemma 2, § 11.1.1, yields for the only non-zero component

$$(\mathbf{Q}_{11} - \alpha \mathbf{I}_1)\mathbf{B}^\alpha_{12} = -\mathbf{Q}_{12},$$

because $\mathbf{B}'_{12}(0) = \mathbf{Q}_{12}$.

Furthermore, Theorem 3, § 11.1.1, reduces to relation $\mathbf{B}^\alpha_{12} = {}_H\mathbf{U}^\alpha_{11}\mathbf{Q}_{12}$ already used in § 3.1.5, and Corollary 4, § 11.1.1, coincides with formula (4) in § 6.1.1.

The compensation kernel

$$\mathbf{C}^\alpha = (\mathbf{Q} - {}_H\mathbf{Q})\mathbf{U}^\alpha$$

from § 11.1.2 has the components

$$\begin{aligned} \mathbf{C}^\alpha_{11} &= \alpha \mathbf{U}^\alpha_{11} - \mathbf{I}_1 - \mathbf{Q}_{11}\mathbf{U}^\alpha_{11}, \qquad & \mathbf{C}^\alpha_{12} &= \alpha \mathbf{U}^\alpha_{12} - \mathbf{Q}_{11}\mathbf{U}^\alpha_{12}, \\ \mathbf{C}^\alpha_{21} &= \alpha \mathbf{U}^\alpha_{21}, & \mathbf{C}^\alpha_{22} &= \alpha \mathbf{U}^\alpha_{22} - \mathbf{I}_2, \end{aligned}$$

and Theorem 1, § 11.1.2, yields the first entrance decomposition (restricted to states in H^c).

On the other hand, $(\mathbf{B}^\alpha)^n = \mathbf{B}^\alpha$ for $n \geqslant 1$, so $\mathbf{B}^\alpha {}_H\mathbf{U}^\alpha = \mathbf{0}$, and therefore, with reference to Theorem 2, § 11.1.2,

$$\mathbf{G}^\alpha {}_H\mathbf{U}^\alpha = {}_H\mathbf{U}^\alpha, \qquad \mathbf{W}^\alpha = \mathbf{B}^\alpha.$$

Hence, the solution of the Markov renewal equation reduces to the decomposition itself

$$\mathbf{U}^\alpha = \mathbf{G}^\alpha {}_H\mathbf{U}^\alpha + \mathbf{W}^\alpha \mathbf{U}^\alpha.$$

Example 3. With the β-jump chain (from § 1.1.6) in the role of the primary chain, i.e., with

$$\widetilde{\mathbf{P}}(t) = e^{-\beta t}\mathbf{P}(t), \qquad \widetilde{\mathbf{Q}} = \mathbf{Q} - \beta \mathbf{I}, \qquad \widetilde{\mathbf{U}}^\alpha = \mathbf{U}^{\alpha+\beta},$$

one obtains the resolvent equation for the secondary chain (see § 1.1.1)

$$\mathbf{U}^\alpha = \mathbf{U}^{\alpha+\beta} + \beta \mathbf{U}^{\alpha+\beta}\mathbf{U}^\alpha,$$

with

$$\mathbf{K}^\alpha = \beta \mathbf{U}^{\alpha+\beta}, \qquad \mathbf{K}(0) = \mathbf{0}.$$

Example 4. This example indicates that the decomposition theorem for the first return time T_S (Thm. 1, § 10.1.5) is a special case of the decomposition theorem discussed here, with the following identification:

$$\widetilde{\mathbf{P}}(t) = {}_H\mathbf{P}^*(t), \qquad \widetilde{\mathbf{P}}(0) = \mathbf{I}, \qquad \widetilde{\mathbf{U}}^\alpha = {}_H\mathbf{U}^{*\alpha}, \qquad \widetilde{\mathbf{Q}} = {}_H\mathbf{Q}^*,$$
$$\mathbf{K}(t) = \mathbf{B}^*(t), \qquad \mathbf{K}(0) = \mathbf{0}, \qquad \mathbf{K}^\alpha = \mathbf{B}^{*\alpha}.$$

By assumption, $\mathbf{P}(t)$, $t \geqslant 0$, is the usual semi-group, whereas ${}_H\mathbf{P}^*(t)$, $t \geqslant 0$, has been shown to be a semi-group of delayed taboo probabilities, corresponding to killing at the terminal time T_S.

Thus, Corollary 5, § 10.1.4, is a special case of Lemma 1, § 11.1.1. Also Theorem 4 and Corollary 6 in § 10.1.4 are special cases of Theorem 3 and Lemma 2 in § 11.1.1, respectively. The expression for $\mathbf{Q} - {}_H\mathbf{Q}^*$ already found in § 10.1.4 agrees with Corollary 5, § 11.1.1. Finally, Corollary 2, § 10.1.4, corresponds to the second resolvent equation in Corollary 4, § 11.1.1. As already noted, it admits interpretation in terms of a compensation kernel $\mathbf{C}^\alpha$ which in this case is a positive operator.

§ 11.1.4. Killing. It is of interest to show that the killing transformation, discussed in § 9.1.3, admits interpretation in terms of the decomposition theorem. Let $X = (X_t)$ be the usual M.C. with a stochastic matrix $\mathbf{P}(t)$ and a generator $\mathbf{Q}$, and let $M = (M_t)$ be a MF associated with X. The process $\widetilde{X} = (\widetilde{X}_t)$, obtained from X by killing it according to M, has transition probabilities determined by

$$\widetilde{\mathbf{P}}(t)f(i) = \mathbb{E}^i f(X_t)M_t, \qquad t \geqslant 0.$$

As shown earlier, $\widetilde{\mathbf{P}}(t)$, $t \geqslant 0$, is a semi-group with a generator $\widetilde{\mathbf{Q}}$. Assume now for simplicity that $\widetilde{\mathbf{P}}(0) = \mathbf{I}$. Note that in terminology of this sub-section, the killed process and its semi-group are regarded as "primary", whereas the original process and its semi-group are "secondary".

The following theorem provides interpretation of the decomposition theorem and of the operator $\mathbf{K}(t)$. For proof see Theorem 1, § 9.1.3.

THEOREM 1. *Let $X = (X_t)$ be an usual* M.C. *with the resolvent $\mathbf{U}^\alpha$, and let $M = (M_t)$ be a* MF *associated with X. Then, for any bounded function f,*

$$\mathbf{K}(t)f(i) = \mathbb{E}^i \int_0^t f(X_s)(-dM_s), \qquad t \geqslant 0,$$

$$\mathbf{K}^\alpha \mathbf{U}^\alpha f(i) = \mathbb{E}^i \int_0^\infty e^{-\alpha s}\mathbf{U}^\alpha f(X_s)(-dM_s), \qquad \alpha > 0.$$

It is now easy to verify directly that $\mathbf{K}(t)$ defined as in Theorem 1, satisfies Lemma 1, § 11.1.1. Introducing a potential relative to the killed semi-group,

$$\widetilde{\mathbf{U}}_A^\alpha f(i) = \mathbb{E}^i \int_0^\infty e^{-\alpha t} f(X_t)M_t\, dA_t, \qquad \alpha > 0,$$

where $A_t = -\log M_t$ is the corresponding AF, one has $\mathbf{K}^\alpha = \widetilde{\mathbf{U}}^\alpha_A$. Hence, the decomposition formula may be written here as

$$\mathbf{U}^\alpha = \widetilde{\mathbf{U}}^\alpha + \widetilde{\mathbf{U}}^\alpha_A \mathbf{U}^\alpha.$$

Note that this is a dual to formula

$$\mathbf{U}^\alpha = \widetilde{\mathbf{U}}^\alpha + \mathbf{U}^\alpha_A \widetilde{\mathbf{U}}^\alpha$$

from Theorem 5, § 9.1.3.

Finally, Corollary 5, § 11.1.1, yields the following relation between generators.

THEOREM 2. *With* $\mathbf{K}(0) = \mathbf{0}$:

$$\mathbf{Q} - \widetilde{\mathbf{Q}} = \lim_{\alpha \to \infty} \alpha \mathbf{K}^\alpha = \mathbf{K}'(0).$$

The following special cases are of interest.

a) $M_t = e^{-\beta t}$. This yields Example 3, § 11.1.3, with $\mathbf{K}^\alpha f(i) = \beta \mathbf{U}^{\alpha+\beta} f(i)$.

b) $M_t = I_{(T>t)}$, where T is the first entrance time to a set H. This yields Example 2, § 11.1.3, with $\widetilde{\mathbf{U}}^\alpha = {}_H\mathbf{U}^\alpha$ and $\mathbf{K}^\alpha = \mathbf{B}^\alpha$, because

$$\mathbf{K}^\alpha f(i) = \mathbb{E}^i \int_0^\infty e^{-\alpha t} f(X_t)(-dM_t) = \mathbb{E}^i \left(e^{-\alpha T} f(X_T),\ T < \infty\right) = \mathbf{B}^\alpha f(i),$$

although $\widetilde{\mathbf{P}}(0) \neq \mathbf{I}$.

c) $M_t = e^{-A_t}$, where $A_t = \int_0^t a(X_s)\, ds$ with $a \geqslant 0$. Thus,

$$-dM_s = M_s a(X_s)\, ds,$$

and

$$\mathbf{K}^\alpha f(i) = \mathbb{E}^i \int_0^\infty e^{-\alpha t} f(X_t) M_t a(X_t)\, dt = \widetilde{\mathbf{U}}^\alpha f(i) a(i) = \widetilde{\mathbf{U}}^\alpha_A f(i),$$

so $\mathbf{K}^\alpha = \widetilde{\mathbf{U}}^\alpha(a\mathbf{I})$.

The decomposition formula takes now the form

$$\mathbf{U}^\alpha f(i) = \widetilde{\mathbf{U}}^\alpha f(i) + \widetilde{\mathbf{U}}^\alpha\big(a(i)\mathbf{U}^\alpha f(i)\big).$$

Compare this (with f being an indicator) with the formula in Example 2, § 9.1.4, in which the last term would be $\mathbf{U}^\alpha a(i) \widetilde{\mathbf{U}}^\alpha f(i)$. Recall that here $\mathbf{Q} - \widetilde{\mathbf{Q}} = a\mathbf{I}$.

§ 11.1.5. Duality. The decomposition theorem considered so far may be thought as being associated with the first entrance decomposition. The dual theorem, associated with the last exit decomposition, takes the form (using the same notation)

$$\mathbf{U}^\alpha = \widetilde{\mathbf{U}}^\alpha + \mathbf{U}^\alpha \mathbf{K}^\alpha,$$

or

$$\mathbf{P}(t) = \widetilde{\mathbf{P}}(t) + \int_{0-}^{t} \mathbf{P}(t-s)\, d\mathbf{K}(s).$$

Properties of this system are dual to that from § 11.1.1. In particular, the (B)- and (F)-equations are interchanged, and the Markov renewal equation is now for measures.

Thus, the analogues of results from § 11.1.1 are

LEMMA 1.

$$\mathbf{K}(s)\widetilde{\mathbf{P}}(t) = \mathbf{K}(s+t) - \mathbf{K}(t).$$

LEMMA 2.

$$\mathbf{K}(s)\widetilde{\mathbf{Q}} = \mathbf{K}'(s) - \mathbf{K}'(0).$$

THEOREM 3.

$$\mathbf{K}^\alpha - \mathbf{K}(0) = (\mathbf{Q} - \widetilde{\mathbf{Q}})\widetilde{\mathbf{U}}^\alpha,$$

using the (B)*-equation for both semi-groups.*

COROLLARY 4.

$$\mathbf{U}^\alpha \widetilde{\mathbf{P}}(0) = \widetilde{\mathbf{U}}^\alpha + \mathbf{U}^\alpha(\mathbf{Q} - \widetilde{\mathbf{Q}})\widetilde{\mathbf{U}}^\alpha,$$

second resolvent equation.

The compensation kernel is now

$$\mathbf{C}^\alpha = \mathbf{U}^\alpha(\mathbf{Q} - \widetilde{\mathbf{Q}}).$$

The same argument as that in § 11.1.2, but applied to the (F)-equation for the secondary semi-group, yields the following analogue of Theorem 1, § 11.1.2,

$$\mathbf{U}^\alpha \widetilde{\mathbf{P}}(0) = \widetilde{\mathbf{U}}^\alpha + \mathbf{C}^\alpha \widetilde{\mathbf{U}}^\alpha.$$

The solution of the Markov renewal equation (for measures) is

$$\mathbf{U}^\alpha = \widetilde{\mathbf{U}}\mathbf{G}^\alpha + \mathbf{U}^\alpha \mathbf{W}^\alpha.$$

The duals of examples from § 11.1.3 are as follows.

Example 1. This is the (F)-equation in the integrated form (see §§ 1.1.2 and 1.1.6), with $\mathbf{K}^\alpha = \widehat{\mathbf{R}}^\alpha$ and $\mathbf{G}^\alpha$ replaced by $\widehat{\mathbf{G}}^\alpha$ from § 1.1.6.

Example 2. This is the last exit decomposition from Theorem 1, § 6.2.1,

$$\mathbf{U}^\alpha = {}_H\mathbf{U}^\alpha + \mathbf{U}^\alpha\mathbf{C}^\alpha$$

with $\mathbf{K}^\alpha = \mathbf{C}^\alpha$, where $\mathbf{C}^\alpha$ is here the α-co-balayage matrix (see § 3.4.4). Observe that Corollary 4 above agrees with formula (4) in § 6.2.1.

Example 3. This remains unchanged, because in the resolvent equation $\mathbf{U}^{\alpha+\beta}\mathbf{U}^\alpha = \mathbf{U}^\alpha\mathbf{U}^{\alpha+\beta}$.

Finally, with reference to killing discussed in § 11.1.4, comparison with Theorem 5, § 9.1.3, indicates that compensation kernel (in the dual decomposition) agrees with the α-potential of the associated AF:

$$\mathbf{U}^\alpha_A = \mathbf{C}^\alpha = \mathbf{U}^\alpha(\mathbf{Q} - \widetilde{\mathbf{Q}}).$$

The case when A is an integral functional has been already noted in example c) in § 11.1.4.

11.2. PERTURBATION AND REVIVAL

Important class of transformations of M.C.'s is obtained by perturbing a given chain according to an external mechanism. The new chain obtained in this manner is called a perturbed chain, and it is of interest to express its properties in terms of those for the original chain. In the revival process, the original chain is killed on entrance to absorbing states and is subsequently "revived" by an external mechanism.

These transformations are related to decomposition theorems from Sub-section 11.1. However, in agreement with customary notation, the symbols X_t, $\mathbf{P}(t)$, $\mathbf{Q}$, $\mathbf{U}^\alpha$ etc. refer now to the original (primary) chain, and the symbols X_t^*, $\mathbf{P}^*(t)$, $\mathbf{Q}^*$, $\mathbf{U}^{*\alpha}$ etc. will be used for the perturbed (secondary) chain.

§ 11.2.1. Perturbation. Consider the usual M.C. $X = (X_t)$ on a state space $\mathbb{I}$, with the standard stochastic matrix $\mathbf{P}(t)$ and the conservative generator $\mathbf{Q}$. Perturbation described here is a transformation of the original chain X into another chain $X^* = (X_t^*)$ with a state space reduced to a subset $H \subset \mathbb{I}$, and the generator modified according to some external mechanism.

1. For the perturbed chain X^* transitions from H to H^c are not permitted. Any such transition in the original chain is replaced by the forced return to H, according to the *replacement matrix* $\mathbf{R}$. Transitions within H consist of those of the original chain and those induced by replacement. For convenience, the perturbed chain X^* will be considered on the state space $\mathbb{I}$, with appropriately extended transition matrix $\mathbf{P}^*(t)$ and the generator $\mathbf{Q}^*$ (it will be assumed that transitions from H^c to H remain unaffected).

Hence, the matrices are in the partitioned form:

$$\mathbf{Q} = \begin{pmatrix} \mathbf{Q}_{11} & \mathbf{Q}_{12} \\ \mathbf{Q}_{21} & \mathbf{Q}_{22} \end{pmatrix}, \qquad \mathbf{Q}^* = \begin{pmatrix} \mathbf{Q}_{11} & \mathbf{Q}_{12} \\ 0 & \mathbf{Q}_{22}^* \end{pmatrix}, \qquad \mathbf{R} = \begin{pmatrix} 0 & \mathbf{R}_{12} \\ 0 & \mathbf{I}_2 \end{pmatrix}.$$

The perturbation is defined by the change of the generator

$$\mathbf{Q} \longrightarrow \mathbf{Q}^* = \mathbf{J}_1\mathbf{Q} + \mathbf{J}_2\mathbf{Q}\mathbf{R},$$

with the non-trivial component

$$\mathbf{Q}^*_{22} = \mathbf{Q}_{22} + \mathbf{Q}_{21}\mathbf{R}_{12}.$$

Here $\mathbf{R}_{12} = (r_{ij})$ is a rectangular matrix whose rows are probability distributions, describing transformation of a state $i \in H^c$ into a state $j \in H$ (do not confuse $\mathbf{R}$ with jump matrix, also denoted by $\mathbf{R}$). Thus $\mathbf{R}_{12}\mathbf{1} = \mathbf{1}$ implies that $\mathbf{Q}^*\mathbf{1} = 0$. Observe that here $\mathbf{P}(0) = \mathbf{P}^*(0) = \mathbf{I}$. The perturbed chain X^* is Markov (by construction) with the transition matrix $\mathbf{P}^*(t)$ satisfying both the (B)- and (F)-equations, and $\mathbf{P}^*(t)\mathbf{1} = \mathbf{1}$. Note that $(\mathbf{Q}^* - \mathbf{Q})\mathbf{R} = \mathbf{0}$.

It is now required to determine properties of the matrix $\mathbf{P}^*(t)$ in terms of $\mathbf{P}(t)$. Actually, it is the component $\mathbf{P}^*_{22}(t)$, describing the behavior on H, which is of interest. The procedure depends essentially on a compensation kernel introduced by J. Keilson [1965] which "compensates" for effects of forced returns to H. Indeed, states in H may be called "permissible" and those in H^c "forbidden", so the replacement matrix $\mathbf{R}$ has natural interpretation. Such perturbation models were considered by R. Syski in the series of papers [1977, 1978, 1982, 1986], where complete discussion can be found. Here only a brief summary will be given, with emphasis on passage times and decomposition theorems.

2. It will be convenient to carry out discussion in terms of resolvents $\mathbf{U}^\alpha$ and $\mathbf{U}^{*\alpha}$. The compensation kernel $\mathbf{C}^\alpha$ has been defined in § 11.1.2 as

$$\mathbf{C}^\alpha = \big(\mathbf{Q}^* - \mathbf{Q}\big)\mathbf{U}^{*\alpha}, \qquad \alpha > 0.$$

Note that $\mathbf{C}^\alpha\mathbf{1} = 0$. Then, the (B)-equation for $\mathbf{U}^{*\alpha}$ yields (as in Thm. 1, § 11.1.2) the "second resolvent equation"

$$\mathbf{U}^{*\alpha} = \mathbf{U}^\alpha + \mathbf{U}^\alpha\mathbf{C}^\alpha, \qquad \alpha > 0.$$

In most applications, the original chain is taken as transient ($\mathbf{U}$ finite), and the secondary chain is required to be ergodic. Hence, under suitable regularity conditions, the limit matrix $\mathbf{E}^*$ for the perturbed chain is expressed in terms of $\mathbf{U}$ and $\mathbf{C}$ as

$$\mathbf{E}^* = \mathbf{U}\mathbf{C},$$

where $\mathbf{C} = \lim \alpha\mathbf{C}^\alpha$ as $\alpha \to 0$.

As the set H is closed for the perturbed chain, it is clear that with reference to the first entrance time T^* to H (for the chain X^*)

$$\mathbf{U}^{*\alpha}_{11} = {}_H\mathbf{U}^{*\alpha}_{11}, \qquad \mathbf{U}^{*\alpha}_{12} = \mathbf{B}^{*\alpha}_{12}\,\mathbf{U}^{*\alpha}_{22},$$

where $\mathbf{B}^{*\alpha}_{12} = {}_H\mathbf{U}^{*\alpha}_{11}\mathbf{Q}_{12}$ is the α-balayage for the chain X^*. Observe that this is in agreement with the post-T^* process on $\mathbb{I}$ discussed in § 9.2.2. Indeed, starting in H^c (forbidden states), the chain X^* enters H eventually with initial distribution $\mathbf{B}^{*\alpha}_{12}$.

Since H is closed, the chain stays there and develops with transition probabilities from $\mathbf{U}^{*\alpha}_{22}$.

As matrices $\mathbf{Q}^*$ and $\mathbf{Q}$ have the same components $\mathbf{Q}_{11}$ and $\mathbf{Q}_{12}$, it is clear that the first entrance times T^* and T to the set H for both chains X^* and X, respectively, should be the same. This fact is confirmed by Theorem 2 below. It is, however, useful to have some computations before this result.

Evaluation of components of the compensation kernel $\mathbf{C}^\alpha$ yields

$$\begin{aligned} \mathbf{C}^\alpha_{11} &= \mathbf{0}, & \mathbf{C}^\alpha_{12} &= \mathbf{0}, \\ \mathbf{C}^\alpha_{21} &= -\mathbf{Q}_{21}\mathbf{U}^{*\alpha}_{11}, & \mathbf{C}^\alpha_{22} &= -\mathbf{Q}_{21}\mathbf{U}^{*\alpha}_{12} + \mathbf{Q}_{21}\mathbf{R}_{12}\mathbf{U}^{*\alpha}_{22}. \end{aligned}$$

Similarly, for the kernel $\mathbf{K}^\alpha = \mathbf{U}^\alpha(\mathbf{Q}^* - \mathbf{Q})$ one finds that

$$\begin{aligned} \mathbf{K}^\alpha_{11} &= -\mathbf{U}^\alpha_{12}\mathbf{Q}_{21}, & \mathbf{K}^\alpha_{12} &= \mathbf{U}^\alpha_{12}\mathbf{Q}_{21}\mathbf{R}_{12}, \\ \mathbf{K}^\alpha_{21} &= -\mathbf{U}^\alpha_{22}\mathbf{Q}_{21}, & \mathbf{K}^\alpha_{22} &= \mathbf{U}^\alpha_{22}\mathbf{Q}_{21}\mathbf{R}_{12}. \end{aligned}$$

Note that $\mathbf{C}^\alpha$ and $\mathbf{K}^\alpha$ have some components negative. It is easy to verify that, indeed,

$$\mathbf{U}^\alpha\mathbf{C}^\alpha = \mathbf{K}^\alpha\mathbf{U}^{*\alpha}.$$

Hence, the explicit form of the decomposition theorem (i.e., the second resolvent equation) can be written as follows.

LEMMA 1. *The components of the matrix* $\mathbf{U}^{*\alpha}$ *satisfy*:

$$\begin{aligned} \mathbf{U}^{*\alpha}_{11} &= \mathbf{U}^\alpha_{11} - \mathbf{U}^\alpha_{12}\mathbf{Q}_{21}\mathbf{U}^{*\alpha}_{11}, & \mathbf{U}^{*\alpha}_{12} &= \mathbf{U}^\alpha_{12} - \mathbf{U}^\alpha_{12}\mathbf{Q}_{21}\mathbf{U}^{*\alpha}_{12} + \mathbf{U}^\alpha_{12}\mathbf{Q}_{21}\mathbf{R}_{12}\mathbf{U}^{*\alpha}_{22}, \\ \mathbf{0} &= \mathbf{U}^\alpha_{21} - \mathbf{U}^\alpha_{22}\mathbf{Q}_{21}\mathbf{U}^{*\alpha}_{11}, & \mathbf{U}^{*\alpha}_{22} &= \mathbf{U}^\alpha_{22} - \mathbf{U}^\alpha_{22}\mathbf{Q}_{21}\mathbf{U}^{*\alpha}_{12} + \mathbf{U}^\alpha_{22}\mathbf{Q}_{21}\mathbf{R}_{12}\mathbf{U}^{*\alpha}_{22}. \end{aligned}$$

THEOREM 2. *Let* ${}_H\mathbf{U}^\alpha$ *and* $\mathbf{B}^\alpha$ *be the taboo and balayage matrices for the original chain* X. *Then*

$$\begin{aligned} \mathbf{U}^{*\alpha}_{11} &= {}_H\mathbf{U}^\alpha_{11}, \qquad \mathbf{U}^{*\alpha}_{12} = \mathbf{B}^\alpha_{12}\mathbf{U}^{*\alpha}_{22}, \\ \mathbf{U}^{*\alpha}_{22} &= \mathbf{U}^\alpha_{22} + \mathbf{U}^\alpha_{22}\mathbf{Q}_{21}(\mathbf{R}_{12} - \mathbf{B}^\alpha_{12})\mathbf{U}^{*\alpha}_{22}. \end{aligned}$$

Proof. Using duality formulas from § 6.3.1, valid for the (B)- and (F)-equations, it is seen immediately that $\mathbf{U}^{*\alpha}_{11} = {}_H\mathbf{U}^\alpha_{11}$ satisfies the equation for $\mathbf{U}^{*\alpha}_{11}$ in Lemma 1. Furthermore,

$$\mathbf{U}^{*\alpha}_{12} = \mathbf{B}^{*\alpha}_{12}\mathbf{U}^{*\alpha}_{22} = {}_H\mathbf{U}^{*\alpha}_{11}\mathbf{Q}_{12}\mathbf{U}^{*\alpha}_{22} = \mathbf{U}^{*\alpha}_{11}\mathbf{Q}_{12}\mathbf{U}^{*\alpha}_{22} = {}_H\mathbf{U}^\alpha_{11}\mathbf{Q}_{12}\mathbf{U}^{*\alpha}_{22} = \mathbf{B}^\alpha_{12}\mathbf{U}^{*\alpha}_{22}.$$

Finally, the last expression follows from the expression for $\mathbf{U}^{*\alpha}_{22}$ in Lemma 1. □

3. The second resolvent equation stated above is suitable for operations on functions. For a function f on $\mathbb{I}$, $\mathbf{C}^\alpha f$ is called a compensation function, and one has the representation

$$\mathbf{U}^{*\alpha}f = \mathbf{U}^\alpha f + \mathbf{U}^\alpha\mathbf{C}^\alpha f, \qquad \alpha > 0,$$

where $\mathbf{C}^\alpha = (\mathbf{Q}^* - \mathbf{Q})\mathbf{U}^{*\alpha}$.

The dual of the second resolvent equation has the form corresponding to that in § 11.1.5:

$$\mathbf{U}^{*\alpha} = \mathbf{U}^{\alpha} + \mathbf{C}^{\alpha}\mathbf{U}^{\alpha}, \qquad \alpha > 0,$$

where now the dual is

$$\mathbf{C}^{\alpha} = \mathbf{U}^{*\alpha}(\mathbf{Q}^{*} - \mathbf{Q}).$$

This form is suitable for measures, and $\mu\mathbf{C}^{\alpha}$ is called a compensation measure. One has the decomposition

$$\mu\mathbf{U}^{*\alpha} = \mu\mathbf{U}^{\alpha} + \mu\mathbf{C}^{\alpha}\mathbf{U}^{\alpha}, \qquad \alpha > 0.$$

Note. It has been assumed here that both matrices $\mathbf{P}(t)$ and $\mathbf{P}^{*}(t)$ are stochastic,

$$\alpha\mathbf{U}^{\alpha}\mathbf{1} = \mathbf{1}, \qquad \alpha\mathbf{U}^{*\alpha}\mathbf{1} = \mathbf{1}, \qquad \alpha > 0.$$

Consequently, for both forms of $\mathbf{C}^{\alpha}$, one has $\mathbf{C}^{\alpha}\mathbf{1} = 0$.

Suppose that only the primary matrix $\mathbf{P}(t)$ is stochastic. Then, for the direct form of $\mathbf{C}^{\alpha}$ one would only have $\mathbf{C}^{\alpha}\mathbf{1} \leqslant 0$, and therefore, $\alpha\mathbf{U}^{*\alpha}\mathbf{1} \leqslant \mathbf{1}$. On the other hand, for the dual form of $\mathbf{C}^{\alpha}$ one has always $\mathbf{C}^{\alpha}\mathbf{1} = 0$, and consequently, $\alpha\mathbf{U}^{*\alpha}\mathbf{1} = \alpha\mathbf{U}^{\alpha}\mathbf{1} = \mathbf{1}$. Thus, $\mathbf{P}^{*}(t)$ must be stochastic.

§ 11.2.2. Revival. Consider the usual M.C. $X = (X_t)$ on the state space $\mathbb{I}$ with the stochastic transition matrix $\mathbf{P}(t)$ and the conservative generator $\mathbf{Q}$. Let H be a fixed set. Assume that H^c is a closed set and that H^c can be reached from any state in H. Thus, $\mathbf{Q}$ and $\mathbf{U}^{\alpha}$ have the zero component:

$$\mathbf{Q}_{12} = \mathbf{0}, \qquad \mathbf{U}^{\alpha}_{12} = \mathbf{0}.$$

Consequently, X is the post-S chain, where S denotes the first entrance time to set H^c. Thus, $\mathbf{U}^{\alpha}_{22}$ refers to transitions on H under taboo H^c, $\mathbf{U}^{\alpha}_{21}$ represents the first entrance to H^c and subsequent transitions in H^c, whereas $\mathbf{U}^{\alpha}_{11}$ corresponds to transitions in the closed set H^c. (Observe that here the usual role of H and H^c has been reversed, as H^c is the taboo set, and elements of $\mathbf{P}_{22}(t)$ are taboo probabilities $({}_{H^c}p_{ij}(t))$.) See Example 2 in § 5.4.2 for the last exit interpretation.

1. Consider now the same perturbation mechanism described in § 11.2.1. Each time the original chain X enters the closed set H^c, it is immediately returned to the set H, according to the replacement matrix $\mathbf{R}_{12}$. Consequently, the decomposition theorem (second resolvent equation) holds. The non-zero components of kernels $\mathbf{C}^{\alpha}$ and $\mathbf{K}^{\alpha}$ are now:

$$\mathbf{C}^{\alpha}_{21} = -\mathbf{Q}_{21}\mathbf{U}^{*\alpha}_{11}, \qquad \mathbf{C}^{\alpha}_{22} = \mathbf{Q}_{21}\mathbf{R}_{12}\mathbf{U}^{*\alpha}_{22},$$
$$\mathbf{K}^{\alpha}_{21} = -\mathbf{U}^{\alpha}_{22}\mathbf{Q}_{21}, \qquad \mathbf{K}^{\alpha}_{22} = \mathbf{U}^{\alpha}_{22}\mathbf{Q}_{21}\mathbf{R}_{12}.$$

According to Lemma 1 and Theorem 2 from § 11.2.1, the generator $\mathbf{Q}^{*}$ and the resolvent $\mathbf{U}^{*\alpha}$ for the perturbed chain $X^{*} = (X^{*}_t)$ have only two non-zero components

$$\mathbf{Q}^{*}_{11} = \mathbf{Q}_{11}, \qquad \mathbf{U}^{*\alpha}_{11} = \mathbf{U}^{\alpha}_{11},$$
$$\mathbf{Q}^{*}_{22} = \mathbf{Q}_{22} + \mathbf{Q}_{21}\mathbf{R}_{12}, \qquad \mathbf{U}^{*\alpha}_{22} = \mathbf{U}^{\alpha}_{22} + \mathbf{U}^{\alpha}_{22}\mathbf{Q}_{21}\mathbf{R}_{12}\mathbf{U}^{*\alpha}_{22}.$$

Hence, there are here two separate sub-chains, one on H and one on H^c. Of primary interest is the chain on H described by $\mathbf{U}_{22}^{*\alpha}$, the chain on H^c (described by $\mathbf{U}_{11}^{*\alpha}$) is an auxiliary construction and may be disregarded.

Indeed, the original chain X, when considered on H only, may be regarded as killed by entering the closed set H^c. The perturbation mechanism "revives" the chain by returning it to H. The perturbed chain, therefore, stays effectively in H. In the special case when H^c consists of absorbing states, the chain is "restarted" on entering an absorbing state. See M. Neuts [1975], R. Syski [1982, 1987], and also § 11.2.3.

2. A general method of restarting a given Markov process at the end of its life time was developed by N. Ikeda et al. [1966]. Simplification and clarification were given by P. A. Meyer [1975], and discussion of the Dirichlet problem was presented by K. Siegrist [1983]. Further extensions were obtained by H. Cai [1987]. In these studies the restarting is obtained by modifying the process in such a way that the resulting "revival process" is still Markovian. The modification is achieved with the help of an "instantaneous distribution".

Adaptation of the general revival method to the usual M.C.'s indicates that the replacement matrix plays the role of the "instantaneous distribution". The following is the brief outline of the argument; for details see R. Syski [1990].

Consider the original chain X restricted to the state space H, with the sub-stochastic matrix $\mathbf{P}_{22}(t)$ and with the life time S. The chain is killed on the first entry to the set H^c at the time S. The revival mechanism returns the chain to H where it stays till the next "death" and the subsequent revival, and the process repeats itself in this manner. The time between revivals is clearly S.

Write $S_1 = S$ and define inductively

$$S_m = S_{m-1} + S \circ \theta_{S_{m-1}}, \qquad m \geqslant 2.$$

Thus, S_m is the instant of the m-th revival with $\lim_{m\to\infty} S_m = \infty$.

The revival chain $X^* = (X_t^*)$ is now defined by

$$X_t^* = \begin{cases} X(t), & 0 \leqslant t < S_1, \\ X(t - S_m), & S_m \leqslant t < S_{m+1}, \quad m \geqslant 1. \end{cases}$$

Thus, the revival chain coincides with the original chain before the killing at leaving set H, and between successive revival instants S_m and S_{m+1} it is the replica of the original chain killed at S.

Analysis of this new chain leads then to another proof of the decomposition theorem for $\mathbf{U}_{22}^{*\alpha}$, with

$$p_{ij}^*(t) = \mathbb{P}^i\big(X_t^* = j,\ t < S\big) + \mathbb{P}^i\big(X_t^* = j,\ S \leqslant t\big), \qquad i, j \in H.$$

Thus, the perturbation argument and the revival method lead to the same result. Revival method provides interpretation of the matrix $\mathbf{K}_{22}^{\alpha}$, whereas the compensation argument stresses the term $\mathbf{C}_{22}^{\alpha}$.

§ 11.2.3. Special cases. The following special cases received considerable attention in literature.

1. *Time change.* Of special interest is the situation when there are no transitions on the set H^c. This means that the original chain X can enter a state in H^c only by a jump from H, and then must return to H at the end of its stay in that state in H^c. Hence, the component $\mathbf{Q}_{11}$ in the intensity matrix $\mathbf{Q}$ has the diagonal form

$$\mathbf{Q}_{11} = -\mathbf{\Theta}_1 = -\left(\delta_{ij} q_j\right).$$

As shown in Example 3, § 5.4.2, the balayage matrix $\mathbf{B}$ has its component $\mathbf{B}_{12}$ equal to the component $\mathbf{R}_{12}$ of the jump matrix, and $\mathbf{R}_{11} = \mathbf{0}$. Hence, $\mathbf{D} = \mathbf{B1} = \mathbf{1}$ here. (The replacement matrix $\mathbf{R}$ should not be confused with the jump matrix $\mathbf{R}$.)

Make now the replacement matrix $\mathbf{R}$ to coincide with the balayage matrix $\mathbf{B}$ for the original chain (so $\mathbf{R}_{12}$ in both interpretations coincide, and $\mathbf{R}_{12} = \mathbf{B}_{12}$). The component $\mathbf{Q}^*_{22}$ of the intensity matrix $\mathbf{Q}^*$ for the perturbed chain X^* has now the form

$$\mathbf{Q}^*_{22} = \mathbf{Q}_{22} + \mathbf{Q}_{21}\mathbf{B}_{12} = \overline{\mathbf{Q}}_{22}$$

in agreement with formulas from § 3.2.5. Note that $\mathbf{Q}^*_{22}$ is the only non-zero component of the matrix $\mathbf{QB}$. The matrix $\overline{\mathbf{Q}}_{22}$ refers to the trace chain, and has been used in connection with the random time change (see § 9.5.4). Direct proof that the generator $\mathbf{Q}^*_{22}$ has the form given here was obtained by P. D. Feigin and E. Rubinstein [1979].

From Theorem 2, § 11.2.1, it follows that

$$\mathbf{B}^{*\alpha}_{12} = \mathbf{B}^{\alpha}_{12} = \left(\alpha\mathbf{I}_1 + \mathbf{\Theta}_1\right)^{-1}\mathbf{Q}_{12} = \mathbf{R}^{\alpha}_{12}$$

(where $\mathbf{R}^{\alpha}_{12}$ refers to the α-chain associated with the original $\mathbf{Q}$), and

$$\mathbf{U}^{*\alpha}_{11} = {}_H\mathbf{U}^{\alpha}_{11} = \left(\alpha\mathbf{I}_1 + \mathbf{\Theta}_1\right)^{-1}, \qquad \mathbf{U}^{*\alpha}_{12} = \mathbf{B}^{\alpha}_{12}\,\mathbf{U}^{*\alpha}_{22}, \qquad \mathbf{U}^{*\alpha}_{22} = \overline{\mathbf{U}}^{\alpha}_{22},$$

where $\overline{\mathbf{U}}^{\alpha}_{22}$ refers to the trace chain from § 9.5.4. Thus, the matrix $\mathbf{U}^{*\alpha}$ is completely determined.

Recall, however, that the component $\mathbf{U}^{*\alpha}_{22}$ satisfies separate the (B)- and (F)-equations. On the other hand, the above perturbation interpretation indicates that it has been obtained by the random time change (with AF determined by a charge $a = I_H$). Observe finally, that if H^c is a closed set (as in revival), then all its states must be absorbing.

2. *Dynkin condition.* Suppose that the original matrix $\mathbf{P}(t)$ has the property

$$\mathbf{RP}(t)\mathbf{R} = \mathbf{P}(t)\mathbf{R}, \tag{D}$$

where $\mathbf{R}$ is the replacement matrix. This condition is equivalent to the condition imposed by E. B. Dynkin [1965, § 10.25] on a certain transformation of the state space (see also § 11.3.1).

It can be verified that the modified matrix $\mathbf{P}^*(t)$ is such that

$$\mathbf{RP}^*(t) = \mathbf{P}(t)\mathbf{R} \quad \text{and} \quad \mathbf{RP}^*(t) = \mathbf{P}^*(t)\mathbf{R}.$$

This yields the special form of the solution on H:

$$\mathbf{P}^*_{22}(t) = \mathbf{P}_{22}(t) + \mathbf{P}_{21}(t)\mathbf{R}_{12}.$$

Moreover,

$$\mathbf{P}^*_{11}(t)\mathbf{R}_{12} + \mathbf{P}^*_{12}(t) = \mathbf{P}_{11}(t)\mathbf{R}_{12} + \mathbf{P}_{12}(t),$$

and

$$\mathbf{QR} = \mathbf{RQ}^* = \mathbf{Q}^*\mathbf{R}.$$

In particular, when $\mathbf{R} = \mathbf{B}$, as in the Case 1 above, then condition (D) takes the form

$$\mathbf{BB}^t = \mathbf{B}^t, \qquad \text{(D)}$$

where $\mathbf{B}^t = \mathbf{P}(t)\mathbf{B}$ is the matrix considered in § 3.2.5.

For full discussion of solutions under condition (D) see R. Syski [1977], E. Rubinstein [1978], P. D. Feigin and E. Rubinstein [1979], where other equivalent forms of condition (D) are discussed.

3. PH-*distributions*. These "phase-type distributions", introduced by M. Neuts [1975], are simply absolute distributions of time to absorption. They can be regarded, however, as special case of perturbation, as shown by R. Syski [1982].

For this purpose suppose that the set H^c is closed for the original chain X with transition matrix $\mathbf{P}(t)$ and the generator $\mathbf{Q}$. Note that $\mathbf{Q}_{12} = \mathbf{0}$ and $\mathbf{P}_{12}(t) = \mathbf{0}$, so one has the situation prevailing in § 11.2.2. As before, let S be the first entrance time to the set H^c. Thus, $\mathbf{P}_{22}(t)$ is the matrix of taboo probabilities (with taboo set H^c) and $\mathbf{P}_{21}(t)$ is the corresponding matrix of transitions from H to and in H^c. Note that if H^c were consisting of absorbing states, then $\mathbf{P}_{21}(t)$ would coincide with the balayage matrix (relative to S).

Let now $\widetilde{D}_i(\cdot)$ be the distribution function of S, conditional on the initial state i. The absolute distribution

$$F(t) = \sum_i \pi_i \widetilde{D}_i(t), \qquad t \geqslant 0,$$

where (π_i) is an initial distribution of X_0 is called the PH-distribution. Write

$$W = \mathbb{E}(S,\ S < \infty).$$

Further specializations are now imposed. Assume that $\mathbf{P}(t)$ is ergodic on the closed set H^c (and necessarily transient on H), and that the perturbed chain X^*, obtained by the perturbation procedure described earlier, is ergodic on H. Thus, $\mathbf{P}^*(t)$ and $\mathbf{Q}^*$ have only two non-zero components; namely both $\mathbf{Q}_{11}$ and $\mathbf{Q}^*_{22}$ are conservative, $\mathbf{P}_{11}(t) = \mathbf{P}^*_{11}(t)$, and $\mathbf{P}^*_{22}(t)$ is ergodic with the limit matrix $\mathbf{E}^*_{22}$. Because of the requirement of ergodicity, one should work not with the potential $\mathbf{U}^\alpha$, but rather with the ergodic potential $\mathbf{Z}^\alpha$ defined by

$$\mathbf{Z}^\alpha = \mathbf{U}^\alpha - (1/\alpha)\mathbf{E}, \qquad \alpha > 0,$$

where $\mathbf{E} = \lim_{\alpha\to 0} \alpha\mathbf{U}^\alpha$ is the limit matrix. Then, under proper regularity conditions,

$$\mathbf{Z} = \lim_{\alpha\to 0} \mathbf{Z}^\alpha$$

is finite. Note that $\mathbf{Z}_{22} = \mathbf{U}_{22} < \infty$. Similarly, write $\mathbf{E}^* = \lim_{\alpha \to 0} \alpha \mathbf{U}^{*\alpha}$ for the limit matrix of the perturbed chain.

As the primary object of interest is the ergodic distribution ε^* on H (given by the identical rows of $\mathbf{E}^*_{22}$), it is necessary to work with the dual form of the decomposition theorem, associated with the (F)-equation, as described in § 11.1.5. Thus, the compensation kernel $\mathbf{C}^\alpha$ is defined by

$$\mathbf{C}^\alpha = \mathbf{U}^{*\alpha}(\mathbf{Q}^* - \mathbf{Q}),$$

and

$$\mathbf{C} = \lim_{\alpha \to 0} \alpha \mathbf{C}^\alpha = \mathbf{E}^*(\mathbf{Q}^* - \mathbf{Q}),$$

with

$$\mathbf{C}_{11} = \mathbf{C}_{12} = \mathbf{0}, \qquad \mathbf{C}_{21} = -\mathbf{E}^*_{22}\mathbf{Q}_{21}, \qquad \mathbf{C}_{22} = \mathbf{E}^*_{22}\mathbf{Q}_{21}\mathbf{R}_{12}.$$

It is evident that rows of $\mathbf{C}$ do not depend on i, say $c_{ij} = c_j$ for every $i \in H$, $j \in \mathbb{I}$. The row vector $c = (c_j)$ is called the compensation measure (its total mass is 0).

The second resolvent equation takes now the form (see §§ 11.1.5 and 11.2.1)

$$\mathbf{U}^{*\alpha} = \mathbf{U}^\alpha + \mathbf{C}^\alpha \mathbf{U}^\alpha.$$

From this, one obtains, with the help of the (F)-equation for $\mathbf{U}^\alpha$,

$$\mathbf{E}^*\mathbf{E} - \mathbf{E} = \lim_{\alpha \to 0} \mathbf{C}^\alpha \mathbf{E},$$

and then the passage to the limit (using $\mathbf{Z}^\alpha$) gives the basic relation

$$\mathbf{E}^* = \mathbf{E}^*\mathbf{E} + \mathbf{C}\mathbf{Z}.$$

For transitions within H this relation reduces to $\mathbf{E}^*_{22} = \mathbf{C}_{22}\mathbf{U}_{22}$. Thus, the solution of $\varepsilon^* \mathbf{Q}^*_{22} = 0$ is explicitly given by

$$\varepsilon^*_j = \sum_{k \in H} c_k u_{kj}, \qquad j \in H,$$

where (c_k) is the compensation measure and (u_{kj}) is the potential for transient $\mathbf{P}_{22}(t)$. The main feature of the compensation method is the fact that ergodic distribution of the modified chain is expressed by the potential of the original transient chain on H. It can also be verified that D_i is 1 identically on $\mathbb{I}$.

Finally, the requirement is imposed that the replacement matrix $\mathbf{R}_{12}$ has all rows identical and of the form

$$r_j = \pi_j \Big/ \sum_{k \in H} \pi_k, \qquad j \in H.$$

For the PH-distribution with this choice of $\mathbf{R}$, it is shown that the compensation kernel $\mathbf{C}$ has identical rows for $i \in H$ given by

$$c_j = \pi_j / W, \qquad j \in H,$$

and $\mathbf{0}$ for $i \in H^c$. Hence the ergodic distribution on H has the form

$$\varepsilon^*_j = (1/W) \sum_{i \in H} \pi_i u_{ij}, \qquad j \in H.$$

See R. Syski [1982] for details.

§ 11.2.4. Comments. Further insight into structure of a revival chain is provided by the following interpretation of results obtained earlier in this sub-section. The approach used here is a modification of that used by H. Cai [1987] for general Markov processes.

1) Let $X = (X_t)$ be an usual M.C. with a semi-group $\mathbf{P}(t)$, $t \geqslant 0$, and a generator $\mathbf{Q}$. Let S be the first entrance time to the set H^c for the original chain. It is clear that the perturbation mechanism, described in § 11.2.1, corresponds to the killing carried out according to the MF of the form

$$M_t = I_{(S>t)}, \qquad t \geqslant 0,$$

with the set $\mathbb{I}_M$ of permanent states taken as the set H. Denoting by $\widetilde{\mathbf{P}}(t)$, $t \geqslant 0$, the semi-group corresponding to the killed chain $\widetilde{X}$, so for any bounded function f,

$$\widetilde{\mathbf{P}}(t)f(i) = \mathbb{E}^i f(X_t)M_t, \qquad i \in \mathbb{I}_M = H,$$

one has the decomposition

$$\mathbf{P}(t)f = \widetilde{\mathbf{P}}(t)f + (\mathbf{K} * \mathbf{P})(t)f,$$

where $\widetilde{\mathbf{P}}(t)$ is the taboo matrix (with the taboo set H^c) and $\mathbf{K}(t)$ is the balayage matrix for the first entrance time S (see Ex. 2, § 11.1.3, with H and H^c interchanged).

The killed chain $\widetilde{X}$ is revived as described in § 11.2.2, according to the replacement matrix $\mathbf{R}$. Thus, in fact, one has here a stopped chain (also denoted by $\widetilde{X}$), obtained by stopping the original chain on its first entrance to the set H^c consisting of absorbing states for $\widetilde{X}$. Consequently, analysis of revival in § 11.2.2 is applicable to the stopped chain $\widetilde{X}$.

Let $X^* = (X_t^*)$ be the revival chain obtained from $\widetilde{X}$, and write $\mathbf{P}^*(t)$, $t \geqslant 0$, and $\mathbf{Q}^*$ for its semi-group and its generator, respectively. Thus,

$$\mathbf{P}^*(t)f(i) = \mathbb{E}^i f(X_t^*), \qquad i \in \mathbb{I},$$

and the decomposition theorem for the revival chain has the following form:

$$\mathbf{P}^*(t)f = \widetilde{\mathbf{P}}(t)f + (\widetilde{\mathbf{K}} * \mathbf{P}^*)(t)f, \qquad i \in \mathbb{I},$$

where $\widetilde{\mathbf{P}}(t)$ is the transition matrix of the stopped chain, and $\widetilde{\mathbf{K}}(t)$ is the kernel such that

$$\widetilde{\mathbf{K}}(t)f(i) = \mathbb{E}^i \int_0^t f(X_s^*)(-dM_s).$$

Observe that $\widetilde{\mathbf{P}}(0) = \mathbf{I}$ and $\widetilde{\mathbf{K}}(0) = \mathbf{0}$ here.

The argument in § 11.2.2 indicates that the above decomposition corresponds to

$$\mathbb{E}^i\Big(f(X_t^*),\ t < S\Big) = \mathbb{E}^i f(X_t^*)M_t, \qquad t \geqslant 0,$$

$$\mathbb{E}^i\Big(f(X_t^*),\ S \leqslant t\Big) = \mathbb{E}^i \int_0^t \mathbf{P}^*(t-s)f(X_s^*)(-dM_s), \qquad t \geqslant 0.$$

In terms of resolvents, the decomposition theorem takes the form

$$\mathbf{U}^{*\alpha} f = \widetilde{\mathbf{U}}^{\alpha} f + \widetilde{\mathbf{K}}^{\alpha} \mathbf{U}^{*\alpha} f.$$

Thus, explicitly, in the partitioned form

$$\begin{aligned}
\widetilde{\mathbf{U}}_{11}^{\alpha} &= (1/\alpha)\mathbf{I}_1, & \widetilde{\mathbf{U}}_{12}^{\alpha} &= \mathbf{0},\\
\alpha\widetilde{\mathbf{U}}_{21}^{\alpha} &= \widetilde{\mathbf{U}}_{22}^{\alpha}\mathbf{Q}_{21}, & \widetilde{\mathbf{U}}_{22}^{\alpha} &= {}_{H^c}\mathbf{U}_{22}^{\alpha},\\
\widetilde{\mathbf{K}}_{11}^{\alpha} &= \widetilde{\mathbf{K}}_{12}^{\alpha} = \mathbf{0}, & &\\
\widetilde{\mathbf{K}}_{21}^{\alpha} &= -\widetilde{\mathbf{U}}_{22}^{\alpha}\mathbf{Q}_{21}, & \widetilde{\mathbf{K}}_{22}^{\alpha} &= \widetilde{\mathbf{U}}_{22}^{\alpha}\mathbf{Q}_{21}\mathbf{R}_{12},
\end{aligned}$$

and

$$\mathbf{U}_{22}^{*\alpha} = \widetilde{\mathbf{U}}_{22}^{\alpha} + \widetilde{\mathbf{U}}_{22}^{\alpha}\mathbf{Q}_{21}\mathbf{R}_{12}\mathbf{U}_{22}^{*\alpha}.$$

Recall that the generator $\widetilde{\mathbf{Q}}$ has only non-zero components $\mathbf{Q}_{21}$ and $\mathbf{Q}_{22}$, whereas the components of the generator $\mathbf{Q}^*$ are zero, except $\mathbf{Q}_{22}^* = \mathbf{Q}_{22} + \mathbf{Q}_{21}\mathbf{R}_{12}$. Note that

$$\mathbf{Q}^* = \widetilde{\mathbf{Q}}\mathbf{R}.$$

Furthermore, Corollary 5, § 11.1.1, indicates that

$$\mathbf{Q}^* - \widetilde{\mathbf{Q}} = \widetilde{\mathbf{K}}'(0),$$

where $'$ denotes the differentiation.

2) Consider an increasing process $B = (B_t)$ defined by

$$B_t = M_0 - M_t = I_{(0<S\leqslant t)}, \qquad t \geqslant 0,$$

with characteristic

$$F_i(t) = \mathbb{E}^i B_t = \mathbb{P}^i(0 < S \leqslant t),$$

and $F_i(\infty) = \mathbb{P}^i(0 < S < \infty) = 1$. Obviously, the non-trivial situation prevails only for $i \in H = \mathbb{I}_M$. Observe that (B_t) is an AF in the extended sense (see § 8.2.1), i.e.,

$$B_{t+h} = B_t + M_t(B_h \circ \theta_t), \qquad t, h \geqslant 0,$$

and it is clear that the potential generated by B coincides with M:

$$\mathbb{E}(B_\infty \mid \mathcal{F}_t) - B_t = M_t, \qquad t \geqslant 0.$$

Define now the intensity of B by

$$\lim_{t\to 0} F_i(t)/t = a(i) \qquad \text{for} \quad i \in H,$$

with $a(i) = 0$ for $i \in H^c$. As $F_i(t)$ is a distribution function of the first entrance time S, it follows from Theorem 7, § 3.1.1 (with H and H^c interchanged), that the following lemma holds.

LEMMA 1. *For $i \in H$:*

$$a(i) = \sum_{j \in H^c} q_{ij}.$$

Applying the definition of the Revuz measure from § 8.1.2-e to B and the chain X^*, one has

THEOREM 2. *The Revuz measure (conditional form) associated with the extended* AF (B_t) *is given by*

$$\mu_B^i(j) = \sum_{k \in H^c} q_{ik}\, r_{kj}, \qquad i, j \in H,$$

and the total measure carried by the set H is $\mu_B^i(H) = a(i)$.

As B is an extended DAF with a single discontinuity at S, determined by the function $g(i,j) = I_H(i) I_{H^c}(j)$, $i \neq j$, the compensator $\mathbf{B}^*$ is an integral functional determined by the function

$$g_{\mathbf{Q}}(i) = I_H(i) a(i).$$

THEOREM 3. *The compensator of B is an extended integral functional $B^* = (B_t^*)$ given by*

$$B_t^* = \int_0^t a(X_s^*) M_s \, ds, \qquad t \geqslant 0,$$

with

$$\mathbb{E}^i B_t^* = F_i(t), \qquad i \in H,$$

and

$$B_{t+h}^* = B_t^* + M_t (B_h^* \circ \theta_t), \qquad t, h \geqslant 0.$$

Recall that by construction, both B and B^* have the same intensity $a(X_s^*) M_s$, generate the same potential M, and $B_t - B_t^*$, $t \geqslant 0$, form a martingale. See also analogous argument in Lemma 3, § 8.1.2.

3) Assume now that H^c consists of a single absorbing state, say Δ, and that returns from Δ are made always to the same state from which Δ has been reached, i.e., $i \to \Delta \to i$. This choice of replacement leads to a special form of the revival chain. Restricting attention to X^* on H, it can be shown that $\mathbf{Q}_{22}^* - \widetilde{\mathbf{Q}}_{22} = a\mathbf{I}_2$, and that the unique solution of the Markov renewal equation for $\mathbf{P}_{22}^*(t) f$ is given by

$$\mathbf{P}_{22}^*(t) f(i) = \mathbb{E}^i f(X_t) N_t, \qquad t \geqslant 0, \quad i \in H,$$

where $N = (N_t)$ with $N_t = M_t e^{A_t}$, $\mathbb{E}^i N_t = 1$, and (A_t), an IF, determined by a, is a non-contracting MF.

For details see H. Cai [1987].

11.3. AGGREGATION

Let $X = (X_t,\ 0 \leqslant t < \infty)$ be an usual M.C. on a state space $\mathbb{I}$. For a function f, the new process $(f(X_t),\ 0 \leqslant t < \infty)$ is not in general Markovian. Preservation of the Markov property received considerable attention in literature, in particular, by M. Rosenblatt [1962, 1970], E. B. Dynkin [1965], and in recent work by J. Glover [1990].

Of special interest is the case of aggregation, or grouping, when several states are combined to form a single state. Such a transformation can be represented by a function f which is constant on each group. A Markov property is then preserved if the transition matrix $\mathbf{P}(t)$ satisfies certain structural conditions. Aggregation transformation is of importance in several applications; see, for example, J. M. Ferrandiz [1986], M. Hofri [1988], R. G. Schehrer [1988]. Presentation here follows E. B. Dynkin [1965]; see also R. Syski [1977].

§ 11.3.1. Dynkin condition. Let $H_1, H_2, \ldots,$ be a partition of a state space $\mathbb{I}$ of an usual M.C. $X = (X_t)$ with a transition matrix $\mathbf{P}(t)$. Let $\mathbb{I}^*$ be a countable state space, taken to be the index set of the partition (H_a) of $\mathbb{I}$. Define a measurable mapping γ from $\mathbb{I}$ into $\mathbb{I}^*$ by

$$\gamma(i) = a, \qquad \gamma^{-1}(\{a\}) = H_a \qquad \text{for} \quad i \in H_a,\ a \in \mathbb{I}^*,$$

in a way that the following condition holds:

for any i, i' such that $\gamma(i) = \gamma(i')$, and all b

$$\sum_{j \in H_b} p_{ij}(t) = \sum_{j \in H_b} p_{i'j}(t), \tag{D}$$

where $j \in \gamma^{-1}(b)$.

Next, define the new chain $X^* = (X_t^*)$ with a state space $\mathbb{I}^*$,

$$X_t^* = \gamma(X_t), \qquad 0 \leqslant t < \infty.$$

It was shown by Dynkin that the new chain X^* is Markovian, with transition matrix $\mathbf{P}^*(t) = (p^*_{ab}(t))$ defined by

$$p^*_{ab}(t) = \sum_{j \in H_b} p_{ij}(t), \qquad i \in H_a.$$

Evidently, $\mathbf{P}^*(0) = \mathbf{I}$, and condition (D) indicates that the intensity matrix $\mathbf{Q}^* = (q^*_{ab})$ has the form

$$q^*_{ab} = \sum_{j \in H_b} q_{ij}, \qquad a \neq b; \qquad q^*_a = -\sum_{j \in H_a} q_{ij}, \qquad i \in H_a,$$

and $\mathbf{Q}^*\mathbf{1} = 0$.

Example 1. Consider the birth-and-death process on the state space $\mathbb{I} = \{0, 1, \ldots\}$ with coefficients λ_i and μ_i, and $q_i = \lambda_i + \mu_i$ for $i \geqslant 1$, $q_0 = \lambda_0$. Let $H_0 = \{0, 2, 4, \ldots\}$ and $H_1 = \{1, 3, 5, \ldots\}$. Take $\mathbb{I}^* = \{0, 1\}$ and define γ by

$$\gamma(i) = \begin{cases} 0 & \text{when} \quad i = 0, 2, 4, \ldots, \\ 1 & \text{when} \quad i = 1, 3, 5, \ldots. \end{cases}$$

Then, condition (D) requires that coefficients be constant in the following sense:

$$\begin{aligned} \alpha &= \lambda_i + \mu_i \qquad \text{for all} \quad i = 1, 3, 5, \ldots; \\ \beta &= \lambda_i + \mu_i \qquad \text{for all} \quad i = 2, 4, 6, \ldots \quad \text{and} \quad \beta = \lambda_0. \end{aligned}$$

Hence,

$$\begin{aligned} q_0^* &= -\alpha, \qquad & q_{01}^* &= \alpha, \\ q_{10}^* &= \beta, \qquad & q_1^* &= -\beta. \end{aligned}$$

Example 2. For application to perturbation see comments on condition (D) in § 11.2.3.

Example 3. In the special case when transformation γ is one-to-one, the condition (D) is automatically satisfied. This is the case when the state space $\mathbb{I}$ is mapped into itself (with $\mathbb{I}$ and $\mathbb{I}^*$ identified). In particular, if the state is finite, the transformation γ represents permutation of states.

Writing now

$$\gamma(i) = j \quad \text{and} \quad i = \gamma^{-1}(j), \qquad i, j \in \mathbb{I},$$

it follows immediately that

$$q_{ij}^* = q_{\gamma^{-1}(i), \gamma^{-1}(j)}, \qquad q_i^* = q_{\gamma^{-1}(i)}.$$

Hence,

$$p_{ij}^*(t) = p_{\gamma^{-1}(i), \gamma^{-1}(j)}(t), \qquad t \geqslant 0, \quad i, j \in \mathbb{I}.$$

For alternative interpretation of this case see § 11.3.2 below.

§ 11.3.2. Similar matrices. The type of transformation discussed here is related both to perturbation and grouping. Let $X = (X_t)$ be an usual M.C. with a transition matrix $\mathbf{P}(t)$ and an intensity matrix $\mathbf{Q}$. Let $\mathbf{A} = (a_{ij})$ be a stochastic matrix with $\mathbf{A1} = \mathbf{1}$ and $\det \mathbf{A} \neq 0$.

Suppose that a matrix $\mathbf{P}^*(t)$ satisfies the following relation:

$$\mathbf{AP}^*(t) = \mathbf{P}(t)\mathbf{A} \qquad \text{for all} \quad t \geqslant 0. \tag{1}$$

It is immediate to verify that

$$\begin{aligned} \mathbf{P}^*(t)\mathbf{P}^*(s) &= \mathbf{P}^*(t+s), \qquad t, s \geqslant 0, \\ \mathbf{P}^*(0) &= \mathbf{I}, \\ \mathbf{P}^*(t)\mathbf{1} &= \mathbf{1}, \end{aligned}$$

and that the generator $\mathbf{Q}^*$ of $\mathbf{P}^*(t)$, $t \geqslant 0$, satisfies the relation

$$\mathbf{A}\mathbf{Q}^* = \mathbf{Q}\mathbf{A}, \tag{2}$$

with $\mathbf{Q}^*\mathbf{1} = 0$, $\mathbf{Q}\mathbf{1} = 0$. Moreover, the (B)- and (F)-equations for $\mathbf{P}^*(t)$ follow immediately from the corresponding equations for $\mathbf{P}(t)$.

Matrices $\mathbf{P}^*(t)$ and $\mathbf{P}(t)$ satisfying relation (1), and matrices $\mathbf{Q}^*$ and $\mathbf{Q}$ satisfying relation (2), are said to be similar. It is known that (finite) similar matrices have the same trace and the determinant.

However, it is clear that in general $\mathbf{P}^*(t)$ need not be a transition matrix, because some of its elements may be negative. Similarly, $\mathbf{Q}^*$ may have negative off-diagonal elements. It is of interest to find conditions on a matrix $\mathbf{A}$ which would assure that $\mathbf{P}^*(t)$ is a proper transition matrix or, equivalently, that $\mathbf{Q}^*$ has all off-diagonal elements non-negative. The fact that the non-singular matrix $\mathbf{A}$ is required to be stochastic introduces additional constraints.

The following is a partial result. Let $\mathbb{I}$ be a set of all non-negative integers, and let γ be a one-to-one mapping of $\mathbb{I}$ onto itself. Take for $\mathbf{A} = (a_{ij})$ the incidence matrix defined by

$$a_{ij} = \begin{cases} 1 & \text{when} \quad j = \gamma(i),\ i = \gamma^{-1}(j), \\ 0 & \text{when} \quad j \neq \gamma(i),\ i \neq \gamma^{-1}(j), \end{cases}$$

for $i, j = 0, 1, 2, \ldots$ Thus, all rows and columns of $\mathbf{A}$ consist of zeros, except of one element which is 1. In particular, when $\mathbb{I}$ is finite, then $\mathbf{A}$ is a permutation matrix. Clearly $\mathbf{A}\mathbf{1} = \mathbf{1}$ and $\det \mathbf{A} = \pm 1$.

Denoting by $\mathbf{A}'$ the transpose of $\mathbf{A}$, one has here

$$\mathbf{A}'\mathbf{A} = \mathbf{I}, \qquad \text{so} \qquad \mathbf{A}^{-1} = \mathbf{A}',$$

indicating that $\mathbf{A}$ is an orthogonal matrix.

It is now easy to verify that the matrix $\mathbf{Q}^* = \mathbf{A}^{-1}\mathbf{Q}\mathbf{A}$ has elements

$$q^*_{ij} = q_{\gamma^{-1}(i),\gamma^{-1}(j)} \qquad \text{for all} \quad i, j \in \mathbb{I},$$

with $q^*_i = q_{\gamma^{-1}(i)}$.

Thus, elements of $\mathbf{Q}^*$ are obtained from those of $\mathbf{Q}$ by permutation.

Example 1. Consider the three-state M.C. with the generator $\mathbf{Q}$ and the matrix $\mathbf{A}$ of the form

$$\mathbf{Q} = \begin{pmatrix} -\lambda & \lambda & 0 \\ \mu & -(\lambda+\mu) & \lambda \\ 0 & \mu & -\mu \end{pmatrix}, \qquad \mathbf{A} = \begin{pmatrix} 0 & 0 & 1 \\ 1 & 0 & 0 \\ 0 & 1 & 0 \end{pmatrix}, \qquad \det \mathbf{A} = 1.$$

Then, an easy calculation shows that

$$\mathbf{Q}^* = \mathbf{A}^{-1}\mathbf{Q}\mathbf{A} = \begin{pmatrix} -(\lambda+\mu) & \lambda & \mu \\ \mu & -\mu & 0 \\ \lambda & 0 & -\lambda \end{pmatrix}.$$

Example 2. Consider the two-state M.C. with the generator $\mathbf{Q}$ and the matrix $\mathbf{A}$ of the following general form

$$\mathbf{Q} = \begin{pmatrix} -\lambda & \lambda \\ \mu & -\mu \end{pmatrix}, \qquad \mathbf{A} = \begin{pmatrix} a & 1-a \\ 1-b & b \end{pmatrix},$$

with $0 < a < 1$, $0 < b < 1$, and $a + b - 1 \neq 0$. Then, an easy calculation shows that the matrix $\mathbf{Q}^* = \mathbf{A}^{-1}\mathbf{Q}\mathbf{A}$ has elements given by

$$\lambda^* = \lambda b + \mu(1-a) > 0, \qquad \mu^* = \mu a + \lambda(1-b) > 0.$$

Thus, in this case, $\mathbf{Q}^*$ is a proper generator for any a and b, and $\lambda^* + \mu^* = \lambda + \mu$. Note that the orthogonal $\mathbf{A}$ is obtained for $a = b = 0$ and trivially for $a = b = 1$.

Example 3. Consider the matrix $\mathbf{Q}$ from Example 1 and take for $\mathbf{A}$ any matrix with $\mathbf{A1} = \mathbf{1}$. Then, it can be verified that the matrix $\mathbf{Q}^*$ will have off-diagonal elements negative in general. With a specified choice of $\mathbf{A}$, $\mathbf{Q}^*$ may be a proper generator, but coefficients of $\mathbf{A}$ would depend on λ and μ.

§ 11.3.3. Reynolds identity. The following problem concerning connections with algebraic identities received some attention in literature. Consider a resolvent operator $\mathbf{U}^\alpha$, $\alpha > 0$, of a Markov semi-group $\mathbf{P}(t)$, $t \geqslant 0$, regarded as an operator on a space of all bounded functions on $\mathbb{I}$. It is required to find $\mathbf{U}^\alpha$ which satisfies the Reynolds identity

$$\alpha\mathbf{U}^\alpha fg = \alpha\mathbf{U}^\alpha f \cdot \alpha\mathbf{U}^\alpha g + \alpha\mathbf{U}^\alpha\big(\alpha\mathbf{U}^\alpha f - f\big)\big(\alpha\mathbf{U}^\alpha g - g\big), \tag{R}$$

for any functions f and g on $\mathbb{I}$, and for each $\alpha > 0$. Here fg and $\mathbf{U}^\alpha f \cdot \mathbf{U}^\alpha g$ denote the ordinary product of two functions.

Evidently, (R) is trivially satisfied by the unit matrix $\alpha\mathbf{U}^\alpha = \mathbf{I}$. Condition (R) is also satisfied by the conditional expectation operator $\mathbb{E}(\cdot \mid \mathcal{B})$ in place of $\alpha\mathbf{U}^\alpha$, because this operator is a projection such that:

$$\alpha\mathbf{U}^\alpha\big(f \cdot \alpha\mathbf{U}^\alpha g\big) = \alpha\mathbf{U}^\alpha f \cdot \alpha\mathbf{U}^\alpha g.$$

However, there are Reynolds operators which are not projections. It was shown by G. C. Rota [1961, 1964] and also by M. M. Rao [1981], that Reynolds identity is satisfied by a Markov semi-group $\mathbf{P}(t)$, $t \geqslant 0$, which is also a homomorphism,

$$\mathbf{P}(t)fg = \mathbf{P}(t)f \cdot \mathbf{P}(t)g. \tag{H}$$

This result will now be verified for a usual M.C.

THEOREM 1. *Suppose that a usual Markovian semi-group satisfies condition* (H). *Then its resolvent satisfies Reynolds identity* (R).

Proof. Kolmogorov equations imply that

$$\frac{d\,\mathbf{P}(t)\mathbf{U}^\alpha}{dt} = \mathbf{P}(t)\big(\alpha\mathbf{U}^\alpha - \mathbf{I}\big).$$

Applying property (H) to functions $\alpha\mathbf{U}^\alpha f$ and $\alpha\mathbf{U}^\alpha g$, one has

$$\left(\frac{d\mathbf{P}(t)}{dt}\right)(\alpha\mathbf{U}^\alpha f\cdot\alpha\mathbf{U}^\alpha g) = \mathbf{P}(t)\alpha(\alpha\mathbf{U}^\alpha f - f)\cdot\mathbf{P}(t)\alpha\mathbf{U}^\alpha g + \mathbf{P}(t)\alpha\mathbf{U}^\alpha f\cdot\mathbf{P}(t)\alpha(\alpha\mathbf{U}^\alpha g - g).$$

Next, recalling that the Laplace transform of $d\mathbf{P}(t)/dt$ is $\alpha\mathbf{U}^\alpha - \mathbf{I}$, one finds that

$$\int_0^\infty e^{-\alpha t}\mathbf{P}(t)(\alpha\mathbf{U}^\alpha f\cdot\alpha\mathbf{U}^\alpha g)\,dt = \alpha\mathbf{U}^\alpha f\cdot\alpha\mathbf{U}^\alpha g + \alpha\int_0^\infty e^{-\alpha t}\Big[(\alpha\mathbf{U}^\alpha f - f)\cdot\alpha\mathbf{U}^\alpha g + \alpha\mathbf{U}^\alpha f\cdot(\alpha\mathbf{U}^\alpha g - g)\Big]\,dt.$$

Hence,

$$\alpha\mathbf{U}^\alpha(\alpha\mathbf{U}^\alpha f\cdot\alpha\mathbf{U}^\alpha g) = \alpha\mathbf{U}^\alpha f\cdot\alpha\mathbf{U}^\alpha g + \alpha\mathbf{U}^\alpha\Big[(\alpha\mathbf{U}^\alpha f - f)\cdot\alpha\mathbf{U}^\alpha g + \alpha\mathbf{U}^\alpha f\cdot(\alpha\mathbf{U}^\alpha g - g)\Big]$$

which is equivalent to (R). □

Formal differentiation of relation (H) indicates that the generator should be a differential operator

$$\mathbf{Q}fg = \mathbf{Q}f\cdot g + f\cdot\mathbf{Q}g.$$

The problem is now reduced to search of conditions which would assure validity of relation (H). The classical result from ergodic theory asserts that (H) holds if and only if the semi-group $\mathbf{P}(t)$, $t \geqslant 0$, is determined by a semi-group τ_t, $t \geqslant 0$, of measure-preserving transformations on the state space $\mathbb{I}$, namely,

$$\mathbf{P}(t)f(i) = f(\tau_t(i)), \qquad i \in \mathbb{I};$$

see P. R. Halmos [1956, p. 45]. The semi-group property of $\mathbf{P}(t)$, $t \geqslant 0$, is verified immediately,

$$\mathbf{P}(s)\mathbf{P}(t)f(i) = \mathbf{P}(s)f(\tau_t(i)) = f(\tau_{t+s}(i)) = \mathbf{P}(s+t)f(i),$$

and condition (H) holds, because its both sides reduce to $f(\tau_t(i))\cdot g(\tau_t(i))$.

The fact that the generator $\mathbf{Q}$ is a differential operator requires the extension of the state space and of the notion of differentiation. Indeed, the derivative $\mathbf{Q}f$ is not defined when $\mathbb{I}$ is discrete. For simplicity, assume that $\mathbb{I}$ is the set of all integers, and imbed it into the real line. Differentiation is then taken in the sense of Schwartz distributions; see J. Horvath [1966, p. 325]. Thus, extending a function f on $\mathbb{I}$ to all real x,

$$f(x) = \sum_i f(i)\delta(x-i),$$

its distributional derivative has the form

$$f'(x) = \sum_i f(i)\delta'(x-i),$$

where δ is the Dirac delta function and δ' is its distributional derivative. Then, in the sense of Schwartz distributions:

$$\mathbf{Q}f(x) = \lim_{t \to 0} (\mathbf{P}(t)f(x) - f(x))/t = f'(x)\tau_0'(x),$$

where τ_0' is the generator of the semi-group τ_t, $t \geqslant 0$.

Remarks.

i) When τ_t is the shift operator, then, for a continuous state space,

$$\mathbf{P}(t)f(x) = f(x+t), \qquad \mathbf{Q}f(x) = f'(x).$$

Here $\mathbf{P}(t)$ corresponds to a special form of a process with independent increments, and $\mathbf{Q}f$ is the ordinary differentiation.

ii) In a discrete M.C. with semi-group $\mathbf{P}^n$, $n \geqslant 0$, and with the shift θ_n, one has

$$\mathbf{P}^n f(i) = f(i+n), \qquad \mathbf{Q}f(i) = f(i+1) - f(i).$$

However, this $\mathbf{Q}$ is not a differential operator, and the corresponding resolvent $\mathbf{R}(y) = (y\mathbf{I} - \mathbf{P})^{-1}$ does not satisfy the Reynolds identity (but only its slightly modified version). The relation (R) is true when $\mathbf{Q}$ is defined as Schwartz distribution.

iii) With an ergodic M.C., $\alpha\mathbf{U}^\alpha f \to \mathbf{E}f = f$ when $\alpha \to 0$, because all invariant functions are constant. Then, the Reynolds identity reduces trivially to

$$\mathbf{E}fg = \mathbf{E}f \cdot \mathbf{E}g.$$

iv) Taking $\mathbf{B}^\alpha$ instead of $\alpha\mathbf{U}^\alpha$, one finds that the Reynolds identity reduces to

$$\mathbf{B}^\alpha fg = \mathbf{B}^\alpha f \cdot \mathbf{B}^\alpha g,$$

whose the non-zero component (in partitioned form) is

$$\mathbf{B}_{12}^\alpha f_2 g_2 = \mathbf{B}_{12}^\alpha f_2 \cdot \mathbf{B}_{12}^\alpha g_2.$$

11.4. CAPACITY

The notion of a balayage already discussed extensively (see Sub-section 5.3) leads further to a notion of a capacity of a set, which is of importance in several applications. This sub-section summarizes basic properties of capacity.

§ 11.4.1. Definitions. Consider the usual M.C. $X = (X_t)$, and suppose that H is an equilibrium set as defined in § 5.2.4. Recall that Theorem 1, § 5.2.4, asserts that for an equilibrium set H

$$\mathbf{B1} = D = \mathbf{U}b = \Pi,$$

where b is the escape rate supported on H, and $\mathbf{B}$ is the balayage matrix.

Let ξ be a reference measure and let $v(i,j)$ be a kernel (introduced in § 1.3.5) such that

$$u_{ij} = v(i,j)\xi(j), \qquad i,j \in \mathbb{I},$$

assuming $u_{ij} < \infty$. The dual of a function b is a measure $\hat{b} = \beta$ defined by

$$\beta(j) = \xi(j)b_j.$$

The following result is then immediate.

THEOREM 1. *For an equilibrium set H, there exists a measure β supported on H and such that*

$$D = \mathbf{U}\beta,$$

where β is the dual of b.

Proof.

$$D_i = \mathbf{U}b_i = \sum_{j\in H} v(i,j)\xi(j)b_j = \sum_{j\in H} v(i,j)\beta(j) = \mathbf{U}\beta(i),$$

using definition of a potential of a measure (from § 1.3.5). □

Thus, the first entrance probability D is a potential of a measure β; recall that $D = \Pi$ is a potential of an integral functional with charge b (see Ex. 1, § 8.1.3). In view of the following definition, the measure β is often called the *capacitary measure*.

DEFINITION 2. A *capacity* of an equilibrium set H is the total capacitary measure of H:

$$C(H) = \sum_i \beta(i) = \beta(H).$$

This expression can be re-written by duality as an inner product

$$C(H) = \beta \bullet \mathbf{1} = \xi \bullet b.$$

PROPOSITION 3. *For an equilibrium set H:*

$$C(H) = \sigma \bullet D,$$

where $\sigma = \xi(-\mathbf{Q}) > 0$.

Proof. Recall from § 5.2.4 that for an equilibrium set H, the function D is necessarily purely excessive. Thus,

$$\xi \bullet b = \xi \bullet (-\mathbf{Q})D = \xi(-\mathbf{Q}) \bullet D = \sigma \bullet D. \qquad \square$$

The important special case is obtained when $\xi = \pi\mathbf{U}$, where π is an initial distribution. Then ξ is purely excessive and $\sigma = \pi$ with $\pi \bullet \mathbf{1} = 1$, and $\xi \bullet \mathbf{1} = \infty$. Hence, for an equilibrium set H:

$$C(H) = \pi \bullet D.$$

This expression is often used as the definition of capacity.

Example 1. Suppose that $H = \{k\}$, where k is a fixed transient state; see Example 3, § 5.4.1. Then, taking $\xi(i) = u_{\pi i}$ for all i, one has

$$C(H) = u_{\pi k}/u_{kk}.$$

§ 11.4.2. Set function. On occasion when several sets H are considered simultaneously, write T_H for the first entrance time to H, and D^H, b^H, s^H etc. for the associated quantities. When H is empty, put $T_H = \infty$ and $D^H = 0$.

The following lemma is easily verified; for the proof see R. M. Blumenthal and R. K. Getoor [1968, p. 53].

LEMMA 0.

i) $H_1 \subset H_2$ *implies* $T_{H_1} \geqslant T_{H_2}$,

ii) $T_{H_1 \cup H_2} = \inf(T_{H_1}, T_{H_2})$,

iii) $T_{H_1 \cap H_2} \geqslant \sup(T_{H_1}, T_{H_2})$,

iv) $H = \cup_n H_n$ *implies* $T_H = \inf T_{H_n}$.

It is evident that for an arbitrary H one has

$$T_H = \inf_{i \in H} T_i,$$

where T_i is the first entrance time to a state i. In the next proposition it is assumed that $C(H) < \infty$ for equilibrium sets H.

PROPOSITION 1. *For equilibrium sets:*

a) $C(H) \leqslant C(H')$, *when* $H \subset H'$ (*monotonicity*),

b) $C(H_1 \cup H_2) + C(H_1 \cap H_2) \leqslant C(H_1) + C(H_2)$ (*strong sub-additivity*).

Proof. It will be convenient to use the form of $C(H)$ from Proposition 3, § 11.4.1.

a) $H \subset H'$ implies $D^H \leqslant D^{H'}$ by Lemma 0, so $\sigma \bullet D^H \leqslant \sigma \bullet D^{H'}$.

b) By Lemma 0,

$$\begin{aligned}(T_{H_1 \cup H_2} < \infty) - (T_{H_1} < \infty) &= (T_{H_1 \cup H_2} < \infty, T_{H_1} = \infty) \\ &\subset (T_{H_2} < \infty, T_{H_1 \cap H_2} = \infty) = (T_{H_2} < \infty) - (T_{H_1 \cap H_2} < \infty).\end{aligned}$$

Hence,

$$D^{H_1 \cup H_2} + D^{H_1 \cap H_2} \leqslant D^{H_1} + D^{H_2},$$

and the assertion follows by averaging with respect to σ. $\square$

Similarly, write L_H for the last exit time from a set H, and recall that s^H is the probability that $L_H = \infty$.

PROPOSITION 2. *Finite unions and finite intersections of equilibrium sets are equilibrium sets.*

Proof. It suffices to consider two sets. Observe that Lemma 0 implies that a subset of an equilibrium set is also an equilibrium set.

a) Let $H = H_1 \cup H_2$, with $s^{H_1} = s^{H_2} = 0$. By Lemma 0,

$$L_H = \sup(L_{H_1}, L_{H_2}),$$

so

$$(L_H = \infty) = (L_{H_1} = \infty) \cup (L_{H_2} = \infty).$$

Hence,

$$s_i^H = \mathbb{P}^i\Big((L_{H_1} = \infty) \cup (L_{H_2} = \infty)\Big) \leqslant s_i^{H_1} + s_i^{H_2} = 0, \qquad i \in \mathbb{I}.$$

b) Let $H = H_1 \cap H_2$, with $s^{H_1} = s^{H_2} = 0$. By Lemma 0,

$$L_H \leqslant \inf(L_{H_1}, L_{H_2}),$$

so

$$s^H \leqslant \inf\left(s^{H_1}, s^{H_2}\right) = 0. \qquad \square$$

An alternative proof may be obtained by recalling that s^H is the probability of infinitely many visits to a set H by the jump chain (see Remark in § 4.1.2). Observe also that by Proposition 1,

$$C(H_1 \cup H_2) \leqslant C(H_1) + C(H_2) < \infty,$$
$$C(H_1 \cap H_2) \leqslant \inf\left(C(H_1), C(H_2)\right).$$

However, countable unions of equilibrium sets need not be equilibrium sets. Indeed, consider an increasing sequence (H_n) of equilibrium sets whose union is H. As D^{H_n} is a monotone function of H_n, so the following limit exists, but by Fatou lemma only

$$D^H = \lim_n D^{H_n} = \lim_n \mathbf{U}\beta^{H_n} \geqslant \mathbf{U}\beta^H.$$

Moreover, by Lemma 0,

$$s^H \geqslant \sup_n s^{H_n} = 0,$$

only. Hence, in agreement with the Poisson decomposition,

$$D^H = \mathbf{U}\beta^H + s^H.$$

Let $\mathcal{K}$ be a family of all equilibrium sets $K \subset \mathbb{I}$ with $C(K) < \infty$. Propositions 1 and 2 indicate that the set function C is positive, monotone and strongly subadditive on $\mathcal{K}$, and that $\mathcal{K}$ is closed under finite unions and finite intersections. The following result shows that under certain condition the class $\mathcal{K}$ may also be closed under countable unions.

PROPOSITION 3. *Let (H_n) be an increasing sequence of equilibrium sets converging to a set H, with $\sup_n C(H_n) < \infty$, and suppose that the corresponding measures converge weakly:*

$$\beta^{H_n} \xrightarrow{w} \beta^H .$$

Then H is an equilibrium set, and

$$C(H) = \sup_n C(H_n).$$

Proof. Proceed as above to evaluate D^H. However, the weak convergence means that

$$\lim_n \mathbf{U}\beta^{H_n} = \mathbf{U}\beta^H ,$$

so $s^H = 0$, and H is an equilibrium set.

By Fatou Lemma,

$$C(H) = \beta^H \bullet \mathbf{1} \leqslant \lim_n \beta^{H_n} \bullet \mathbf{1} = \lim_n C(H_n) < \infty,$$

but $C(H_n) \leqslant C(H)$, and therefore,

$$C(H) = \lim_n C(H_n).$$

□

DEFINITION 4. For any set $H \subset \mathbb{I}$ define

$$C(H) = \sup \big(C(K)\colon K \subset H,\ K \in \mathcal{K}\big),$$

where $\mathcal{K}$ is the class of all equilibrium sets satisfying the conditions in Proposition 3, with $C(K) < \infty$.

This definition (and Proposition 3) indicate that notion of capacity may be extended to all subsets of $\mathbb{I}$ (all subsets are here capacitable). Indeed, conditions imposed here are those of the classical theorem on construction of the Choquet capacity; see P. A. Meyer [1966, p. 42], C. Dellacherie and P. A. Meyer [1978, p. 55], J. Doob [1984, p. 753].

Alternatively, one may start with finite subsets of $\mathbb{I}$, necessarily equilibrium sets for transient chains (see Ex. 2, § 5.4.1). In fact, with discrete topology, the state space $\mathbb{I}$ is locally compact, and the only compact sets are finite sets. The classical construction for topological spaces extends the capacity to all capacitable subsets of $\mathbb{I}$.

§ 11.4.3. Duality. Consider now the dual M.C. $(\widehat{X}_t)$ relative to the reference measure ξ, and let $\widehat{\mathbf{U}}, \widehat{\mathbf{Q}}$, and $\widehat{\mathbf{B}}$ be the associated potential, intensity and balayage matrices, respectively, as introduced in § 6.3.3 and § 9.4.5.

Let H be an equilibrium set for the dual chain, and write $\hat{b} = (-\widehat{\mathbf{Q}})\widehat{D}$ for the corresponding escape rate (a function). Then, as before,

$$\widehat{D} = \widehat{\mathbf{B}}\mathbf{1} = \widehat{\mathbf{U}}\hat{b} = \widehat{\mathbf{U}}\beta,$$

where $\widehat{\beta}$ is a measure defined by $\widehat{\beta}(j) = \xi(j)\hat{b}_j$.

Define now the *co-capacity* of the equilibrium set H as

$$\widehat{C}(H) = \widehat{\beta} \bullet \mathbf{1} = \xi \bullet \hat{b},$$

so again,

$$\widehat{C}(H) = \hat{\sigma} \bullet \widehat{D},$$

where $\hat{\sigma} = \xi(-\widehat{\mathbf{Q}})$.

Recall now from § 6.3.3 that dual $(\widehat{D}) = \xi\mathbf{C}$, where $\mathbf{C}$ is the co-balayage matrix (for the direct chain). Hence,

$$\text{dual } (\hat{b}) = \xi\mathbf{C}(-\mathbf{Q}) = \xi(-\mathbf{Q})\mathbf{B},$$

where the second equation follows from Theorem 1, § 6.3.2.

The following result is of interest.

THEOREM 1. *For a set H which is an equilibrium set for a* M.C. *and its dual, the capacity and the co-capacity coincide*

$$C(H) = \widehat{C}(H).$$

Proof. The inner product is preserved by duality (see § 1.3.5), so

$$\widehat{C}(H) = \xi \bullet \hat{b} = \text{ dual } \hat{b} \bullet \text{dual } \xi = \xi(-\mathbf{Q})\mathbf{B1} = \sigma \bullet D = C(H). \qquad \square$$

11.5. ENERGY

The notion of energy for Markov processes received considerable attention recently, and its connection with passage times is clearly seen in the case of Markov chains. The energy can be introduced either analytically in terms of excessive functions, or in measure theoretic way in terms of additive functionals. Moreover, there is an essential distinction between transient and ergodic chains. The approach used here is based on functionals and combines previous approaches proposed earlier. See references for details.

§ 11.5.1. Definitions. Let $X = (X_t, \mathcal{F}_t)$ be the usual M.C., assumed transient, so its potential $\mathbf{U} = (u_{ij})$ is finite, and let $A = (A_t, \mathcal{F}_t)$ be a predictable and integrable AF (see § 8.1.3-c). Let p be a finite purely excessive function, relative to the semi-group $\mathbf{P}(t)$, $t \geqslant 0$, of X. Recall that r.v.'s

$$p(X_t) = \mathbb{E}(A_\infty \mid \mathcal{F}_t) - A_t, \qquad t \geqslant 0,$$

form a potential super-martingale generated by A, and necessarily of class (D).

LEMMA 1. *Define a function g on $\mathbb{I}$ by*

$$g(i) = \frac{1}{2}\,\mathbb{E}^i A_\infty^2, \qquad i \in \mathbb{I},$$

where $A = (A_t)$ is a predictable integrable AF. *Then g is purely excessive, and there exists a function $c \geqslant 0$ such that g is a potential of a charge c,*

$$g = \mathbf{U}c.$$

Proof. The following argument relays on the fact that A is an AF. From definition of g:

$$g(X_t) = \frac{1}{2}\,\mathbb{E}(A_\infty^2 \circ \theta_t \mid \mathcal{F}_t) = \frac{1}{2}\,\mathbb{E}((A_\infty - A_t)^2 \mid \mathcal{F}_t),$$

because $A_\infty^2 \circ \theta_t = (A_\infty \circ \theta_t)^2 = (A_\infty - A_t)^2$. Furthermore,

$$\mathbf{P}(t)g(i) = \mathbb{E}^i g(X_t) = \frac{1}{2}\,\mathbb{E}^i(A_\infty - A_t)^2 \leqslant \frac{1}{2}\,\mathbb{E}^i A_\infty^2 - \frac{1}{2}\,\mathbb{E}^i A_t^2.$$

Hence,

$$\mathbf{P}(t)g(i) \leqslant g(i) \quad \text{and} \quad \mathbf{P}(t)g(i) \longrightarrow 0, \qquad \text{as} \quad t \to \infty,$$

so g is purely excessive.

Consequently, by Theorem 4, § 1.3.4, there exists an exit law $g(t) = (g_i(t))$ such that its Laplace transform g^α satisfies

$$g - \alpha \mathbf{U}^\alpha g = g^\alpha, \qquad \alpha > 0,$$

with $g^\alpha \to g$ as $\alpha \to 0$.

Using the (F)-equation, one can write

$$g^\alpha = \mathbf{U}^\alpha(-\mathbf{Q})g.$$

Hence, in the limit,

$$g = \mathbf{U}c, \qquad \text{where} \quad c = (-\mathbf{Q})g \geqslant 0. \qquad \square$$

Let ξ be a σ-finite reference measure, and consider measure $\hat{c}$, dual of a function c,

$$\hat{c}(i) = \xi(i)c(i), \qquad i \in \mathbb{I},$$

and introduce a measure $\sigma = \xi(-\mathbf{Q}) \geqslant 0$. It is easy to see that the total mass of $\hat{c}$ equals the average of function g taken with respect to measure σ

$$\hat{c} \bullet \mathbf{1} = c \bullet \xi = \sigma \bullet g.$$

This leads to the following definition used by A. O. Pop-Stojanovic [1981].

DEFINITION 2. Let p be a purely excessive function, and let $A = (A_t)$ be a predictable integrable AF. The energy of a potential supermartingale $(p(X_t), t \geqslant 0)$ generated by A (or energy of A, or of p) is the (finite) number

$$\mathrm{en} = \hat{c} \bullet \mathbf{1} = \sigma \bullet g.$$

The following special cases are of particular interest.

i) Let $p = \mathbf{U}a$ be a potential of a charge $a \geqslant 0$, with $(-\mathbf{Q})p = a$, and let $\hat{p}$ and $\hat{a}$ be their respective duals, relative to the reference measure ξ. Put $c(i) = a(i)p(i)$ for all i. Then Definition 2 corresponds to the classical definition of energy as an integral of a potential with respect to a charge (measure).

DEFINITION 3. The energy of a potential p (or of a charge a) is the inner product of a potential and the dual of its charge (or, of a charge and the dual of its potential)

$$\mathrm{en} = p \bullet \hat{a} = \hat{p} \bullet a.$$

Note that in the special case when $\xi = \pi\mathbf{U}$, where π is the initial distribution, the energy is

$$\mathrm{en} = \sum_i \pi(i) \sum_j u_{ij} a(j) p(j) = \pi \bullet \mathbf{U}(ap).$$

ii) Let A be an integral functional determined by a charge $a \geqslant 0$, with

$$\mathbb{E}^i A_\infty = \mathbf{U}a(i) = p(i).$$

Take as before, $c(i) = a(i)p(i)$, but select ξ of the form $\xi = \pi\mathbf{U}$, where π is the initial distribution. Then $\sigma = \pi$, and Definition 2 reduces to

DEFINITION 4. The energy of a M.C. $X = (X_t)$ is defined for a given charge a (or equivalently, for a given purely excessive function p which is a potential of a charge a), as the energy of a potential supermartingale $p(X_t) = \mathbf{U}a(X_t)$, $t \geqslant 0$, generated by the integral functional A

$$\mathrm{en} = \frac{1}{2}\, \mathbb{E} A_\infty^2 .$$

Further comments on these two special cases are given later in this sub-section.

Furthermore, one has the following extension of Definition 2. Let A and B be two predictable integrable AF's, and let $(p(X_t))$ and $(q(X_t))$ be the corresponding potential supermartingales generated by A and B, respectively, where p and q are purely excessive functions. Then the function h defined by

$$h(i) = \frac{1}{2}\, \mathbb{E}^i A_\infty B_\infty, \qquad i \in \mathbb{I},$$

is necessarily purely excessive. The mutual energy of A and B (or p and q) is defined as the average of h with respect to a measure $\sigma = \xi(-\mathbf{Q}) \geqslant 0$

$$(A, B) = \sigma \bullet h.$$

§ 11.5.2. Functions. Let $\mathcal{P}$ be a set of all potentials p whose charges satisfy the integrability condition

$$\xi \bullet a < \infty,$$

which indicates that the measure $\hat{a}$ is finite ($\hat{a}\bullet\mathbf{1} < \infty$). The condition is equivalent to $\sigma\bullet p < \infty$, where $\sigma = \xi(-\mathbf{Q}) \geqslant 0$. In the special case of $\xi = \pi\mathbf{U}$, this is $\pi\bullet p < \infty$.

It is easy to show that any bounded potential in $\mathcal{P}$ has finite energy. Moreover, any potential with finite support (i.e., whose charge has finite support) belongs to $\mathcal{P}$ and has finite energy.

PROPOSITION 1. *If $p \in \mathcal{P}$ is a potential with support in a set H, then energy depends only on the values of p on H.*

Proof. Write

$$a = \begin{pmatrix} 0 \\ a_2 \end{pmatrix},$$

so

$$p = \mathbf{U}a = \begin{pmatrix} \mathbf{U}_{12}a_2 \\ \mathbf{U}_{22}a_2 \end{pmatrix} = \begin{pmatrix} p_1 \\ p_2 \end{pmatrix},$$

and en $= \hat{p}_2 \bullet a_2$.

Observe that ${}_H\mathbf{U}a = 0$, so by the projection theorem (Thm. 2, § 6.4.1):

$$p = \mathbf{U}a = \mathbf{B}\mathbf{U}a = \mathbf{B}p.$$

Consequently,

$$a = (-\mathbf{Q})p = (-\mathbf{Q})\mathbf{B}p,$$

where the matrix $(-\mathbf{Q})\mathbf{B}$ is given in § 3.2.5. □

PROPOSITION 2. *For an equilibrium set H:*

$$\text{en}\,(D) = C(H).$$

Proof. Here $D = \mathbf{U}b$, where b is the escape rate, and $\beta = \hat{b}$ is the capacitary measure supported on H (see § 11.4.1). Hence,

$$\text{en}\,(D) = \beta \bullet D = \beta \bullet \mathbf{1} = C(H) \qquad \square$$

The fact that capacity of an equilibrium set equals energy of its potential is well known in the classical Potential Theory. However, the classical result that energy of D coincides with infimum of energy of potentials in $\mathcal{P}$ with $\hat{a} \bullet \mathbf{1} = C(H)$, holds only for reversible chains (i.e., self dual with $\mathbf{U} = \widehat{\mathbf{U}}$). Recently, several attempts have been made to extend this result to non-symmetric kernels (see § 11.5.4).

It is of interest to calculate energy of various potentials. For example, the balayage potential of p has been defined as $\bar{p} = \mathbf{B}p$, and for $p = \mathbf{U}a$, also $\bar{p} = \mathbf{U}\bar{a}$ with $\bar{a} = \mathbf{C}a$ (see §§ 5.3.3 and 6.3.3). Then it is easy to see that for a fixed charge $a \geqslant 0$, energy for p and for $\bar{p}$ are $\hat{p}\bullet a$ and $\hat{p}\bullet\bar{a}$, respectively. If a has the support H, then these energies coincide. Similarly, energy of taboo potentials ${}_H\mathbf{U}a$ depends only on values of a on H^c.

Next, let $a \geqslant 0$ and $b \geqslant 0$ be two charges, and let $p = \mathbf{U}a$ and $q = \mathbf{U}b$ be their respective potentials (relative to the same $\mathbf{U}$). Then definition stated in § 11.5.1 leads to the following definition.

DEFINITION 3. The *mutual energy* of two potentials p and q (or of two charges a and b), denoted by (p,q) or by (a,b), is defined by

$$(p,q) = \frac{1}{2}(p \bullet \hat{b} + q \bullet \hat{a}) = \frac{1}{2}(\hat{p} \bullet b + a \bullet \hat{q}) = (a,b).$$

Equivalent definition for measures $\hat{a}$ and $\hat{b}$ is

$$(\hat{a},\hat{b}) = \frac{1}{2}(\hat{a} \bullet \mathbf{U}\hat{b} + \hat{b} \bullet \mathbf{U}\hat{a}),$$

where $\mathbf{U}\hat{a}$ and $\mathbf{U}\hat{b}$ are potentials of measures $\hat{a}$ and $\hat{b}$, respectively (see § 1.3.5).

In the special case when the reference measure is $\xi = \pi\mathbf{U}$, where π is the initial distribution, the mutual energy is

$$(p,q) = \frac{1}{2}\pi \bullet \mathbf{U}(aq + bp).$$

Observe that the mutual energy is the extension of the notion of energy

$$\text{en}\,(p) = (p,p) = (a,a) = \text{en}\,(a).$$

Recall that if a and b are two charges such that $\mathbf{U}a = \mathbf{U}b$, then necessarily $a = b$, so there is one-to-one correspondence between (positive) charges and their potentials. Moreover, the set of positive charges (potentials) is a positive cone. Thus, if p and q are potentials with charges a and b, respectively, then $\alpha a + \beta b$ is the charge of a potential $\alpha p + \beta q$ (for $\alpha, \beta \geqslant 0$).

The following theorem indicates that $\mathcal{P}$ is a convex cone.

THEOREM 4. *The mutual energy behaves like a scalar product on* $\mathcal{P}$, i.e.,

i) $(p,q) = (q,p)$,

ii) $(\alpha p, q) = \alpha(p,q)$ *for all* $\alpha \geqslant 0$,

iii) $(p+q, r) = (p,r) + (q,r)$,

iv) $(p,p) = 0$ *implies* $p = 0$.

Proof. Note that $(p,q) \geqslant 0$ by definition, because p and q are non-negative. Properties i) and ii) are obvious. Property iii) follows from duality, and property iv) from the fact that $\text{en}\,(p) = 0$ implies that $a = 0$, and this is equivalent to $p = 0$. □

One could now consider the "energy norm" of a potential (or a charge), whose square is defined by

$$||p||^2 = (p,p) = \text{en}\,(p) \geqslant 0.$$

However, to prove the Schwartz inequality it is necessary to enlarge $\mathcal{P}$ by inclusion of signed potentials and charges. Indeed, one would consider the energy of a potential $pt + q$, but restriction to $t \geqslant 0$ prevents one from using this classical argument.

Consequently, it must be verified that energy for signed potentials remains positive. There are constructions to achieve this aim (based on Proposition 1) — see, for example, L. L. Helms [1969] and J. G. Kemeny [1966], but they will not be discussed here. The following theorem is stated without proof.

THEOREM 5 (Schwartz inequality). *Let p and q be potentials of finite energy in $\mathcal{P}$. Then*

$$(p,q)^2 \leqslant \text{en}\,(p)\,\text{en}\,(q).$$

It then follows that $||p||$ is a proper norm, and one can consider convergence in energy of potentials in $\mathcal{P}$. Hence, $p_n \to p$ in energy, when $||p_n - p|| \to 0$ as $n \to \infty$.

§ 11.5.3. Increasing process. Another approach to the notion of energy, due to P. A. Meyer [1966], is through the concept of an increasing process (in particular, an AF). Definition of energy is then of the same form as Definition 4, § 11.5.1. More precisely, let $U = (U_t, \mathcal{F}_t)$ be a potential super-martingale of class (D). Then Doob–Meyer decomposition theorem asserts that U is generated by an integrable predictable increasing process $A = (A_t, \mathcal{F}_t)$; see § 8.1.4-b. In this case one has

DEFINITION 1. If the potential supermartingale does not belong to class (D), then U is said to have infinite energy. If U belongs to class (D), the energy of U is the (possibly infinite) number

$$\text{en}\,(U) = \frac{1}{2}\,\mathbb{E}A_\infty^2,$$

where (A_t) is the predictable increasing process which generates (U_t). The integration by parts yields the following "energy formula", see P. A. Meyer [1966, p. 141], C. Dellacherie and P. A. Meyer [1982, p. 156].

THEOREM 2. *Let U be the potential generated by the predictable integrable increasing process A. Then*

$$\text{en}\,(U) = \frac{1}{2}\mathbb{E}\int_0^\infty (U_t + U_{t-})\,dA_t.$$

The following special case is of interest; see R. Syski [1979]. Let $p = \mathbf{U}a$ be a purely excessive function which is a potential of a charge $a \geqslant 0$ (here $\mathbf{U}$ denotes the potential operator). For A take the integral functional determined by a (see § 8.1.2) and assume that $\pi \bullet p < \infty$. Consider

$$U_t = p(X_t) = \mathbf{U}a(X_t), \qquad A_t = \int_0^t a(X_s)\,ds.$$

DEFINITION 3. The energy of a M.C. $X = (X_t)$, denoted by en$\,(X)$, is defined for a given charge a (or equivalently, for a given purely excessive function p which is a potential of a charge a), as the energy en$\,(U)$ of a potential supermartingale $U = (U_t)$ associated with the M.C. X by $U_t = \mathbf{U}a(X_t)$.

Thus,

$$\text{en}\,(X) = \frac{1}{2}\,\mathbb{E}A_\infty^2 = \mathbb{E}\int_0^\infty p(X_t)\,a(X_t)\,dt,$$

where $\mathbb{E}A_\infty = \mathbb{E}U_0 = \pi \bullet p$.

THEOREM 4. *Let $a \geqslant 0$ be a charge, and let $p = \mathbf{U}a$ be a purely excessive function which is a potential of a charge, such that $\pi \bullet p < \infty$. Then the energy of the usual M.C. X is given by*

$$\operatorname{en}(X) = \pi \bullet \mathbf{U}(ap).$$

Proof.

$$\operatorname{en}(X) = \mathbb{E}\int_0^\infty a(X_t)\mathbf{U}a(X_t)\,dt = \mathbb{E}\int_0^\infty a(X_t)\,\mathbb{E}\Big(\int_0^\infty a(X_{t+s})\,ds \mid \mathcal{F}_t\Big)\,dt$$

$$= \int_0^\infty\int_0^\infty \mathbb{E}a(X_t)a(X_{t+s})\,ds\,dt = \sum_i \pi(i)\sum_k \sum_j u_{ik}u_{kj}a(k)a(j)$$

$$= \sum_i \pi(i)\mathbf{U}\big[a(i)\,\mathbf{U}a(i)\big] = \pi \bullet \mathbf{U}(a\mathbf{U}a) = \pi \bullet \mathbf{U}(ap). \qquad \square$$

Evidently, it follows from Theorem 4 that one has here the special form of energy, corresponding to a reference measure $\xi = \pi\mathbf{U}$, as noted in § 11.5.1.

In the similar fashion one can introduce the notion of a mutual energy.

DEFINITION 5. Let $U = (U_t)$ and $V = (V_t)$ be two potential supermartingales of class (D). The *mutual energy* of U and V is the number

$$(U, V) = \frac{1}{2}\,\mathbb{E}A_\infty B_\infty,$$

where $A = (A_t)$ and $B = (B_t)$ denote predictable integrable increasing processes which generate U and V, respectively.

The analogue of Theorem 2, stated here for continuous processes, is the following theorem.

THEOREM 6. *Let A and B be two continuous integrable increasing processes which generate potential supermartingales U and V, respectively. Then,*

$$(U, V) = \frac{1}{2}\mathbb{E}\int_0^\infty U_t\,dB_t + \frac{1}{2}\mathbb{E}\int_0^\infty V_t\,dA_t.$$

Specialization to the usual M.C. leads to the following definition.

DEFINITION 7. The mutual energy of a M.C. $X = (X_t)$, relative to given charges a and b (or, equivalently, relative to given purely excessive functions p and q such that $p = \mathbf{U}a$ and $q = \mathbf{U}b$), is defined as the mutual energy (U, V) of the potential supermartingales $U = (U_t)$ and $V = (V_t)$, associated with the M.C. X by

$$U_t = p(X_t) = \mathbf{U}a(X_t), \qquad V_t = q(X_t) = \mathbf{U}b(X_t),$$

and generated by integrable increasing processes $A = (A_t)$ and $B = (B_t)$, respectively, where

$$A_t = \int_0^t a(X_s)\,ds, \qquad B_t = \int_0^t b(X_s)\,ds.$$

Analogously to Theorem 4, one has now

THEOREM 8. *Let a and b be two charges, let $p = \mathbf{U}a$ and $q = \mathbf{U}b$ be their potentials such that $\pi \bullet p < \infty$ and $\pi \bullet q < \infty$. Then, the mutual energy of a M.C. X is given by*

$$(U, V) = \frac{1}{2}\pi \bullet \mathbf{U}(aq + bp).$$

Clearly, Theorem 8 corresponds to the mutual energy (p, q) relative to the reference measure $\xi = \pi\mathbf{U}$, as noted in § 11.5.1. It can be verified directly that $(p, q) = (U, V)$ behaves like a scalar product on the space $\mathcal{P}$ of potentials. More generally, Definition 5 yields immediately the Schwartz inequality

$$\left(\mathbb{E}A_\infty B_\infty\right)^2 \leqslant \mathbb{E}A_\infty^2\, \mathbb{E}B_\infty^2.$$

Observe that in the case when $\bar{p} = \mathbf{B}p$ is the balayage of p, the energies of $U_t = p(X_t)$ and $\overline{U}_t = \bar{p}(X_t)$ are, respectively,

$$\operatorname{en}(U) = \frac{1}{2}\,\mathbb{E}A_\infty^2, \qquad \operatorname{en}(\overline{U}) = \frac{1}{2}\,\mathbb{E}\bar{A}_\infty^2,$$

and the mutual energy is

$$(U, \overline{U}) = \frac{1}{2}\,\mathbb{E}A_\infty \bar{A}_\infty,$$

where A and $\bar{A}$ are IF determined by charges a and $\bar{a} = \mathbf{C}a$. Moreover, one has here the obvious decomposition $p = \bar{p} + (p - \bar{p})$ (see projection theorem for balayage in § 6.4.3). However, this need not be the orthogonal projection with respect to energy inner product.

Finally, note that positivity of energy of a difference $A - B$ follows from the relation

$$\operatorname{en}(A - B) = \frac{1}{2}\,\mathbb{E}(A_\infty - B_\infty)^2.$$

§ 11.5.4. Symmetrization. The following approach, proposed by K. L. Chung and M. M. Rao [1980] — see also K. L. Chung [1980] and J. Glover [1983] — aims to retain some symmetrization properties of the classical case.

Consider a transient M.C. and let H be an equilibrium set. Essential use is made of distributions of the last exit time L from the set H:

$$\Pi_{ij} = \mathbb{P}^i\left(X(L-) = j,\ 0 < L < \infty\right), \qquad i \in \mathbb{I}, \quad j \in H,$$
$$\Pi_i = \sum_{j \in H} \Pi_{ij} = \mathbb{P}^i(0 < L < \infty).$$

Recall that, for the equilibrium set, $\mathbb{P}^i(L = \infty) = 0$, and

$$\Pi_{ij} = u_{ij} b_j = v(i,j)\beta(j), \qquad \Pi_i = \mathbf{U}b_i = D_i,$$

where b is the escape rate, and β is the equilibrium measure (the dual of b).

From now on, attention will be restricted to the set H, and all measures and functions will be considered only on H. In particular, write for the matrix

$\mathbf{\Pi} = (\Pi_{ij})$, where $i, j \in H$. As $\Pi_i = D_i = 1$ for $i \in H$, it follows that the matrix $\mathbf{\Pi}$ is stochastic. It may be therefore regarded as a one-step transition matrix of some discrete parameter M.C., with the state space H.

Suppose that this chain has an invariant probability measure η:

$$\eta\mathbf{\Pi} = \eta \quad \text{and} \quad \eta \bullet \mathbf{1} = 1.$$

Let h be *a* function on H, defined as the Radon–Nikodym derivative of measure η with respect to the equilibrium measure β,

$$\eta(j) = h(j)\beta(j), \qquad j \in H.$$

It follows from expression for Π_{ij}, that

$$h(j) = \sum_{i \in H} \eta(i)v(i,j), \qquad j \in H.$$

Next, define the kernel $\mathbf{U}_h = (v_h(i,j))$ by

$$v_h(i,j) = v(i,j)/h(j), \qquad i,j \in H.$$

The potential of a measure μ (on H), relative to $\mathbf{U}_h$, is a function (on H) defined as $\mathbf{U}_h\mu$:

$$\mathbf{U}_h\mu(i) = \sum_{j \in H} v_h(i,j)\mu(j), \qquad i \in H.$$

The essential symmetrization is achieved by the following

LEMMA 1. *There exist a function h and a probability measure η satisfying*

$$\sum_{i \in H} \eta(i)v_h(i,j) = 1, \qquad \sum_{j \in H} v_h(i,j)\eta(j) = 1.$$

Proof. Both relations follow immediately from definitions of η and β, respectively. □

In the present approach energies of measures on H are considered. One has

DEFINITION 2. The mutual energy of two measures μ_1 and μ_2 (relative to h) is defined as

$$(\mu_1, \mu_2)_h = \frac{1}{2}\left(\mu_1 \bullet \mathbf{U}_h\mu_2 + \mu_2 \bullet \mathbf{U}_h\mu_1\right),$$

and the energy of a measure μ as

$$\text{en}\,(\mu) = (\mu, \mu)_h = \mu \bullet \mathbf{U}_h\mu = \sum\sum_{i,j \in H} \mu(i)v_h(i,j)\mu(j).$$

Observe that the mutual energy is symmetric:

$$(\mu_1, \mu_2)_h = (\mu_2, \mu_1)_h$$

and bilinear. It will be assumed now that $(\mu_1, \mu_2)_h$ is finite.

THEOREM 3. *For a measure* μ *on* H:

$$(\mu, \eta)_h = \mu \bullet \mathbf{1},$$

and, in particular,

$$\mathrm{en}\,(\eta) = 1, \qquad (\beta, \eta)_h = C(H),$$

where C *denotes capacity.*

Proof. Using Lemma 1, one has

$$\mu \bullet \mathbf{U}_h\,\eta = \sum_{i\in H} \mu(i)\mathbf{U}_h\,\eta(i) = \sum_{i\in H} \mu(i) = \mu \bullet \mathbf{1},$$
$$\eta \bullet \mathbf{U}_h\,\mu = \sum_{i\in H} \eta(i)\mathbf{U}_h\,\mu(i) = \sum_{j\in H} \mu(j) = \mu \bullet \mathbf{1},$$

and the special cases follow from the observation that η is the probability measure, and that $C(H) = \beta \bullet \mathbf{1}$, respectively. □

Definition of the mutual energy can be extended to signed finite measures on H by considering decomposition

$$\mu = \mu^+ - \mu^-, \qquad |\mu| = \mu^+ + \mu^- < \infty,$$

with requirement that

$$\left(|\mu_1|, |\mu_2|\right)_h < \infty.$$

Similarly, the energy $\mathrm{en}\,(\mu)$ of a signed measure μ is defined provided $\mathrm{en}\,(|\mu|) < \infty$. The class of such signed measures will be denoted by $\mathcal{E}$. For $\mu \in \mathcal{E}$, define

$$\phi(\mu) = \mathrm{en}\,(\mu) - 2\,\mu \bullet \mathbf{1},$$

and note that $\phi(\eta) = -1$.

PROPOSITION 4. *If* $\mu \in \mathcal{E}$, *then* $\mu + \eta \in \mathcal{E}$ *and*

$$\phi(\mu + \eta) = \mathrm{en}\,(\mu) - 1.$$

Proof.

$$\begin{aligned}\mathrm{en}\,(|\mu + \eta|) &\leqslant \mathrm{en}\,(|\mu| + \eta) = \left(|\mu| + \eta,\ |\mu| + \eta\right)_h \\ &= \mathrm{en}\,(|\mu|) + 2\left(|\mu|, \eta\right)_h + \mathrm{en}\,(\eta) = \mathrm{en}\,(|\mu|) + 2|\mu| \bullet \mathbf{1} + 1 < \infty,\end{aligned}$$

so $\mu + \eta \in \mathcal{E}$. Similarly,

$$\mathrm{en}\,(\mu + \eta) = \mathrm{en}\,(\mu) + 2(\mu, \eta) + \mathrm{en}\,(\eta) = \mathrm{en}\,(\mu) + 2\mu \bullet \mathbf{1} + 1,$$

and the result follows. □

As in the previous approaches, the main difficulty is whether the energy remains positive. The following partial result yields the minimization corresponding to the classical case (see § 11.5.2).

COROLLARY 5. i) *Suppose that* $\mathrm{en}(\mu) \geqslant 0$ *for every* $\mu \in \mathcal{E}$. *Then*

$$\mathrm{en}(\eta) \leqslant \mathrm{en}(\mu) \qquad \text{for every} \quad \mu \in \mathcal{E}.$$

ii) *If in addition* $\mathrm{en}(\mu) = 0$ *implies* $\mu = 0$ *for every* $\mu \in \mathcal{E}$, *then* η *is the unique probability measure among all probability measures in* $\mathcal{E}$ *with finite energy for which the inequality in* i) *holds.*

Proof. If μ is a probability measure such that $\mu - \eta \in \mathcal{E}$, then

$$\phi(\mu) = \phi(\mu - \eta + \eta) = \mathrm{en}(\mu - \eta) - 1 \quad \text{and} \quad \phi(\mu) = \mathrm{en}(\mu) - 2,$$

so

$$\mathrm{en}(\mu) = \mathrm{en}(\mu - \eta) + \mathrm{en}(\eta) = \mathrm{en}(\mu - \eta) + 1.$$

Hence $\mathrm{en}(\mu) \geqslant \mathrm{en}(\eta)$. Next, if $\mathrm{en}(\mu - \eta) = 0$, so $\mathrm{en}(\mu) = 1$, then by assumption $\mu = \eta$. □

As it may be expected, positivity of energy should be assured when energy is defined through functionals. Recall that if A is an AF, then its potential is $\mathbf{U}_A \mathbf{1} = \mathbf{U}\mu_A$, where μ_A is the associated Revuz measure. More generally, for $f \geqslant 0$ which vanishes outside H, the potential of f is (see § 8.1.3-e)

$$\mathbf{U}_A f(i) = \mathbb{E}^i \int_0^\infty f(X_t)\, dA_t = \sum_{j \in H} v(i,j) \mu_A(j) f(j).$$

Taking $f(i) = h^{-1}(i)$ for $i \in H$, one has

$$\mathbf{U}_A h^{-1}(i) = \mathbb{E}^i B_\infty = \mathbf{U}_h\, \mu_A(i),$$

where $B = (B_t)$ is an AF defined by $dB_t = dA_t / h(X_t)$.

Write $\mathbf{U}_h \mu_A = p$, and note that process $p(X_t)$, $t \geqslant 0$, is right-continuous. By the "energy formula" (noted in § 11.5.3), one has

$$\mathbb{E}^i B_\infty^2 = \mathbb{E}^i \int_0^\infty (p(X_{t-}) + p(X_t))\, dB_t.$$

If B is continuous, then

$$\frac{1}{2}\, \mathbb{E}^i B_\infty^2 = \mathbb{E}^i \int_0^\infty p(X_t)\, dB_t = \mathbb{E}^i \int_0^\infty \frac{p(X_t)}{h(X_t)}\, dA_t$$

$$= \sum_{j \in H} v(i,j)\, \frac{p(j)}{h(j)}\, \mu_A(j) = \sum_{j \in H} v_h(i,j) p(j) \mu_A(j).$$

Integration with respect to the measure η yields (using Lemma 1)

$$\frac{1}{2}\, \mathbb{E}^i B_\infty^2 = \frac{1}{2} \sum_{i \in H} \eta(i)\, \mathbb{E}^i B_\infty^2 = \sum_{j \in H} p(j) \mu_A(j) = \mu_A \bullet \mathbf{U}_h\, \mu_A = \mathrm{en}(\mu_A).$$

Writing $A_t = A_t^+ - A_t^-$ and $\mu_A = \mu_A^+ - \mu_A^-$, the above relations hold for difference of two additive functionals. Hence, positivity of energy is achieved, and the minimization result (Cor. 5-i) holds for such measures. This, however, is not sufficient for uniqueness.

Remark. Observe that

$$v_h(i,j)\xi(j) = u_{ij}/h(j),$$

so the interpretation in terms of a random time change determined by a CAF

$$C_t = \int_0^t h^{-1}(X_s)\,ds$$

is possible (see § 9.6.1).

§ 11.5.5. Ergodic chain. Contrary to previous discussion, it will be assumed here that the M.C. is ergodic. Let $\mathbf{P}(t)$ be the transition matrix, and let $\mathbf{E}$ be its limit matrix. Let ε be the ergodic distribution — take ε as the initial distribution and also as the reference measure for duality. Let

$$\mathbf{Z}(t) = \mathbf{P}(t) - \mathbf{E}, \qquad t \geqslant 0,$$

as defined in § 11.2.3, with Laplace transform

$$\mathbf{Z}^\alpha = \mathbf{U}^\alpha - (1/\alpha)\mathbf{E}, \qquad \alpha > 0,$$

with the limit $\mathbf{Z}$.

$\mathbf{Z}^\alpha$ is called α-ergodic potential operator, and for a function $a \geqslant 0$, $\mathbf{Z}^\alpha a$ is α-ergodic potential of a charge a. When $\alpha = 0$, $\mathbf{Z}a$ is called the ergodic potential of a. Evidently, $\varepsilon\mathbf{Z}(t) = 0$, and $\mathbf{Z}(t)\mathbf{1} = 0$.

Consider the charge process $(a(X_t),\ 0 \leqslant t < \infty)$, and note that

$$\mathbb{E}a(X_t) = \varepsilon \bullet a, \qquad \mathbb{E}\big[a(X_s)a(X_{s+t})\big] = \hat{a} \bullet \mathbf{P}(t)a,$$

where $\hat{a}$ is a measure — the dual of a relative to ε. Write

$$\phi(t) = \operatorname{cov}\big(a(X_s), a(X_{s+t})\big), \qquad \phi(0) = \operatorname{var} a(X_s).$$

The direct calculation yields

LEMMA 1.

$$\phi(t) = \hat{a} \bullet \mathbf{Z}(t)a,$$

with

$$\phi(0) = \hat{a} \bullet (\mathbf{I} - \mathbf{E})a \geqslant 0, \qquad \phi(\infty) = 0.$$

Similarly, writing ϕ^α for the Laplace transform of $\phi(t)$, one has

LEMMA 2.

$$\phi^\alpha = \hat{a} \bullet \mathbf{Z}^\alpha a, \qquad \alpha \geqslant 0,$$

where the series converges absolutely.

In particular, for $\alpha = 0$,

$$\int_0^\infty \phi(t)\,dt = \hat{a} \bullet \mathbf{Z}a.$$

This leads to definition of energy for the ergodic chain. See R. Syski [1978] for details, as here only brief comments can be made.

DEFINITION 3. The scalar product $\hat{a} \bullet \mathbf{Z}a$ defines the *energy* of the ergodic potential $\mathbf{Z}a$, with respect to the measure $\hat{a}$ (dual of the charge a).

As an example, suppose that $\sigma = \mu(-\mathbf{Q}) \geqslant 0$ for some excessive measure μ, and take $a = \hat{\sigma}$. Then,

$$\phi(t) = \sigma\mathbf{Z}(t)\hat{\sigma}, \qquad \text{energy} = \sigma\mathbf{Z}\hat{\sigma}.$$

Recall that if the charge a were excessive relative to $\mathbf{P}(t)$, then, for an ergodic chain, a would be constant, and therefore, $\mathbf{Z}(t)a = 0$. Hence, energy would vanish.

Assuming that $a \geqslant 0$ is not excessive, $\mathbf{Z}(t)a$ is signed. Yet, energy $\hat{a} \bullet \mathbf{Z}a \geqslant 0$.

SECTION 12

BOUNDARY THEORY

This short section contains brief introduction to boundary theory and associated decomposition theorems. As the basic assumptions are now dropped, the material presented here gives a glimpse of what lies beyond structures considered so far.

Sub-section 12.1 discusses again the Dirichlet problem and the Poisson equation, but now using martingale method. This approach is then extended to various boundary problems (still however in the framework of the usual Markov chains), emphasizing connections with passage times.

Sub-section 12.2 drops the basic assumption that the chain be minimal, and presents discussion of the "first infinity". This sub-section is included in order to stress the analogy with the Dirichlet problem discussed earlier, although the situation is quite different. Presentation follows work of K. L. Chung [1967].

Sub-section 12.3 presents a very brief introduction to boundary theory of Markov chains. The aim is to point out formal analogies with passage times on a state space, and to stress the role of decomposition theorems. Brief comments on Martin boundary are also included. This sub-section is based on work of K. L. Chung who developed the boundary theory for Markov chains, and on contributions by D. Williams and many others. The individual references are given in the text.

12.1. BOUNDARY PROBLEMS

Some of the results concerning passage times which have been discussed in Chapter II in analytic terms are more visible when expressed in terms of associated stochastic processes. This is especially true about boundary problems, of which the Dirichlet problem and the Poisson equation (from Section 5) are primary examples. In this sub-section, these two problems will be re-visited — together with other problems — in terms of martingales. This approach offers several advantages. The name in the title of this sub-section comes from the analogy with boundary problems in partial differential equations, studied by methods of probabilistic Potential Theory.

§ 12.1.1. Dirichlet problem. This is the most typical problem and its importance has been already pointed out — see Theorem 1, § 5.1.1, for the statement. It is more convenient to use the formulation presented in § 5.1.3, namely,

$$\mathbf{Q}^a f = 0 \quad \text{on} \quad \mathbb{I} \quad \text{with} \quad f = \varphi \quad \text{on} \quad H,$$

where φ is a given boundary function on H. $\mathbf{Q}^a$ is the generator of the Markov semi-group $\mathbf{S}(t)$, $t \geqslant 0$, of the stopped process

$$\widetilde{X}_t = X_{t\wedge T}, \qquad t \geqslant 0,$$

where T is the first entrance time to the set H (see §§ 3.1.1 and 9.3.1). Hence, f is invariant relative to $\mathbf{S}(t)$ on $\mathbb{I}$, and by assumption f is bounded. Consequently, $f(\widetilde{X}_t)$, $t \geqslant 0$, form a non-negative bounded martingale, relative to natural σ-fields $(\mathfrak{F}_t)$. Hence, random variables $f(\widetilde{X}_t)$ are uniformly integrable, and this martingale converges a.e. to an integrable random variable, say Y_∞, which closes it on the right:

$$f(\widetilde{X}_t) \to Y_\infty \quad \text{a.e.},$$

$$\mathbb{E}(Y_\infty \mid \widehat{\mathfrak{F}}_t) = \mathbb{E}(Y_\infty \mid \widetilde{X}_t) = f(\widetilde{X}_t) \quad \text{a.e.}, \qquad t \geqslant 0.$$

Moreover, Y_∞ is invariant, i.e., $Y_\infty \circ \theta_t = Y_\infty$. This allows to determine f:

$$\mathbb{E}(Y_\infty \mid \widetilde{X}_t) = \mathbb{E}(Y_\infty \circ \theta_t \mid \widetilde{X}_0 \circ \theta_t) = \mathbb{E}(Y_\infty \mid \widetilde{X}_0),$$

so

$$f(i) = \mathbb{E}(Y_\infty \mid \widetilde{X}_0 = i), \qquad i \in \mathbb{I}.$$

Next, one can write

$$Y_\infty = \varphi(X_T) I_{(T<\infty)} + \lim_{t\to\infty} f(X_t) I_{(T=\infty)}.$$

Hence,

$$f(i) = p(i) + h(i), \qquad i \in H^c,$$

where

$$p(i) = \mathbb{E}(\varphi(X_T),\ T < \infty \mid X_0 = i) = \mathbf{B}\varphi(i),$$
$$h(i) = \mathbb{E}\Big(\lim_{t\to\infty} f(X_t),\ T = \infty \mid X_0 = i\Big) = {}_H\mathbf{P}(\infty) f(i),$$

and $f(i) = \varphi(i)$ for $i \in H$. It is clear that the above expressions for components p and h coincide with these given in § 5.1.1 (see also §§ 5.1.3 and 5.3.1).

§ 12.1.2. Poisson equation. This is essentially the Riesz decomposition for supermartingales. Recall the statement from § 5.2.2:

$$\mathbf{Q}f = \begin{cases} 0 & \text{on } H^c, \\ -a & \text{on } H, \end{cases}$$

where $a \geqslant 0$ and $f = \varphi$ on H. By assumption, f is a positive bounded function; moreover, by Theorem 7, § 1.3.3, f is excessive on $\mathbb{I}$. Hence, by Theorem 1, § 8.1.4, $S_t = f(X_t)$, $t \geqslant 0$, form a positive bounded supermartingale which is convergent and uniformly integrable:

$$f(X_t) \xrightarrow{\text{a.e.}} S_\infty, \qquad \lim_{t\to\infty} \mathbb{E}f(X_t) = \mathbb{E}S_\infty.$$

The Riesz decomposition (stated in § 8.1.4-a) gives

$$f(X_t) = \mathbb{E}(S_\infty \mid \mathfrak{F}_t) + p(X_t) \quad \text{a.e.}$$

On the other hand, by Theorem 7, § 8.1.4,

$$M_t = f(X_t) + \int_0^t (-\mathbf{Q})f(X_s)\,ds = f(X_t) + \int_0^t a(X_s)\,ds, \qquad t \geqslant 0,$$

form a martingale with $\mathbb{E}M_t = \pi f < \infty$. Let (A_t) be an increasing process (an integral functional) determined by a charge a (see § 8.1.4), so

$$f(X_t) = M_t - A_t.$$

The potential generated by A_t is purely excessive and also a potential of a charge (see example in § 8.1.4-b), so

$$p(X_t) = \mathbf{U}a(X_t) = \mathbb{E}(A_\infty - A_t \mid \mathfrak{F}_t) \quad \text{a.e.}$$

Hence,

$$f(X_t) = \mathbb{E}(A_\infty + S_\infty \mid \mathfrak{F}_t) - A_t,$$

so

$$M_t = \mathbb{E}(A_\infty + S_\infty \mid \mathfrak{F}_t).$$

§ 12.1.3. Basic problem. Let $\mathbf{P}(t)$ be a transition matrix of the usual Markov chain X, with a conservative intensity matrix $\mathbf{Q}$ and a resolvent $\mathbf{U}^\alpha$. Let $T = T_H$ be the first entrance time to a set H. It will be assumed in this discussion that T is finite ($D \equiv 1$).

The boundary problem relative to a generator $\mathbf{Q}$ consists of finding a (non-negative) bounded function f on the state space $\mathbb{I}$ such that

$$\begin{aligned} \mathbf{Q}f(i) &= -g(i), \qquad && i \in H^c, \\ f(i) &= \varphi(i), && i \in H, \end{aligned}$$

where g and φ are given (non-negative) bounded functions. Here g is called a charge and vanishes on H, whereas φ is called a boundary function and vanishes on H^c. The boundary problem may be expressed by saying that the function f is $\mathbf{P}(t)$-excessive on H^c with charge g, and equal to the boundary function φ on the boundary H. Clearly, this reduces to the Dirichlet problem when g vanishes identically.

THEOREM 1. *Suppose that T is finite with probability one. Then the boundary problem formulated above has the unique solution given by*

$$f(i) = u(i) + z(i), \qquad i \in H^c,$$

where

$$u(i) = \mathbb{E}^i\varphi(X_T), \qquad z(i) = \mathbb{E}^i A_T,$$

with the increasing process (A_t) determined by the charge g.

Proof. Consider a M.C. $\widetilde{X} = (\widetilde{X}_t)$ stopped at time T, so $\mathbf{Q}^a f = -g$ on $\mathbb{I}$. Since f is excessive on $\mathbb{I}$, $f(\widetilde{X}_t)$, $t \geqslant 0$, is a positive bounded supermartingale, hence uniformly integrable and convergent. $T < \infty$ a.e., so

$$\lim_{t\to\infty} f(\widetilde{X}_t) = \varphi(X_T) \quad \text{a.e.},$$

and the limit is invariant under shift. Moreover, $U_t = \mathbb{E}(\varphi(X_T) \mid \widetilde{\mathfrak{F}}_t)$, $t \geqslant 0$, form a martingale, and the equality

$$\mathbb{E}(\varphi(X_T) \mid \widetilde{X}_t) = \mathbb{E}(\varphi(X_T) \mid \widetilde{X}_0)$$

implies that

$$U_t = u(\widetilde{X}_t),$$

where

$$u(i) = \mathbb{E}(\varphi(X_T) \mid \widetilde{X}_0 = i), \qquad i \in H^c.$$

By the Riesz decomposition theorem (from § 8.1.4-a)

$$f(X_t) = U_t + Z_t, \qquad t \geqslant 0,$$

where Z_t, $t \geqslant 0$, is a potential supermartingale. From the fact that

$$\mathbb{E}\Big((Z_{t+h} - Z_t)/h \mid \widetilde{\mathfrak{F}}_t\Big) = \mathbb{E}\Big((f(\widetilde{X}_{t+h}) - f(\widetilde{X}_t))/h \mid \widetilde{X}_t\Big)$$
$$\longrightarrow \mathbf{Q}^a f(\widetilde{X}_t) = -g(\widetilde{X}_t) \quad \text{a.e.}$$

(see § 8.1.4-b), it follows that Z_t is generated by an increasing process of the form

$$A_t = \int_0^t g(\widetilde{X}_s)\, ds.$$

Observe, however, that, for $t < T$,

$$f(\widetilde{X}_t) = f(X_t), \qquad U_t = u(X_t),$$
$$Z_t = f(X_t) - u(X_t), \qquad A_t = \int_0^t g(X_s)\, ds,$$

whereas, for $t \geqslant T$,

$$f(\widetilde{X}_t) = \varphi(X_T), \qquad U_t = \varphi(X_T),$$
$$Z_t = 0, \qquad A_t = A_T = A_\infty.$$

Hence,

$$Z_t = \mathbb{E}(A_T - A_t \mid \mathfrak{F}_t) = \mathbb{E}\left(\int_t^T g(X_s)\, ds \mid \mathfrak{F}_t\right), \qquad t < T,$$

so $Z_t = z(X_t)$ with $z(i) = \mathbb{E}^i A_T$ for $i \in H^c$. □

Clearly, when $g \equiv 0$, one obtains the unique solution of the Dirichlet problem

$$u(i) = \mathbb{E}^i \varphi(X_T) = \mathbf{B}\varphi(i)$$

which corresponds to the martingale $u(\widetilde{X}_t) = \mathbb{E}(\varphi(X_T) \mid \widetilde{\mathfrak{F}}_t)$, in agreement with § 12.1.1.

Proceeding in the same fashion as in the proof of Theorem 1, § 5.1.1, one has

COROLLARY 2. *The boundary problem formulated above has the unique solution (for finite T) given by*

$$f(i) = u(i) + z(i), \qquad i \in H^c,$$

where

$$u(i) = \mathbf{B}_{12}\,\varphi(i), \qquad z(i) = {}_H\mathbf{U}_{11}\,g(i).$$

§ 12.1.4. Extension. The more general form of the boundary problem is

$$\mathbf{Q}f(i) - a(i)\,f(i) = -g(i), \qquad i \in H^c,$$
$$f(i) = \varphi(i), \qquad i \in H,$$

where a is a given non-negative bounded function, and other functions are as in § 12.1.3. The solution for f can be obtained by reducing the problem to the previous one by considering the new generator

$$\widehat{\mathbf{Q}} = \mathbf{Q} - a\mathbf{I},$$

corresponding to a process killed at state dependent rate a, as discussed in Example 2, § 9.1.4, but before entrance to set H.

Introduce now the MF defined by

$$M_t = \begin{cases} e^{-A_t}, & t < T, \\ M_T, & t \geqslant T, \end{cases}$$

where

$$A_t = \int_0^t a(X_s)\,ds.$$

THEOREM 1. *The extended boundary problem formulated above has the unique solution (under assumption that T is finite) given by*

$$f(i) = u(i) + z(i), \qquad i \in H^c,$$

where

$$u(i) = \mathbb{E}^i M_T \varphi(X_T), \qquad z(i) = \mathbb{E}^i \int_0^t M_t g(X_t)\,dt.$$

Proof follows by writing $\widehat{\mathbf{Q}}f = -g$, and applying Theorem 1, § 12.1.3, to an extended AF

$$B_t = \int_0^t M_s g(X_s)\,ds,$$

and using the fact that

$$\widehat{\mathbb{E}}^i f(X_t) = \mathbb{E}^i f(X_t) M_t.$$

In the special case when the function a is constant, say α, one obtains as a corollary

$$u(i) = \mathbf{B}^{\alpha}_{12}\,\varphi(i), \qquad z(i) = {}_H\mathbf{U}^{\alpha}_{11}\,g(i).$$

12.2. FIRST INFINITY

In this sub-section the basic assumption that a M.C. is minimal is now dropped. Instead discussion will be concerned with relationship between a general M.C. and its minimal chain, both having the same generator. As already noted, sample functions have now discontinuities of the second kind, and the minimal chain corresponds to the original chain being killed at the instant of reaching the "first infinity". Thus, formally, the minimal transition probabilities can be regarded as taboo probabilities, with the fictitious taboo state ∞. This sub-section explores this analogy; for full details see K. L. Chung [1967].

§ 12.2.1. Decompositions. The first infinity τ for a M.C. $X = (X_t)$ has been already defined in § 1.1.3 as the life time of the chain. To keep the distinction between the chain itself and its minimal chain, the notation $\mathbf{P}(t) = (p_{ij}(t))$, $\mathbf{U}^\alpha = (u_{ij}^\alpha)$ will refer to the minimal chain, whereas the corresponding notation for the original chain will be indicated by a star: $\mathbf{P}^*(t) = (p_{ij}^*(t))$, $\mathbf{U}^{*\alpha} = (u_{ij}^{*\alpha})$. Without loss of generality, $\mathbf{P}^*(t)$ will be assumed stochastic, but it is essential that $\mathbf{P}(t)$ is only sub-stochastic, $\mathbf{P}(t)\mathbf{1} \leqslant \mathbf{1}$. On the other hand, the infinitesimal generator $\mathbf{Q}$ for the minimal chain, as well as for the original chain, is assumed to be conservative, $\mathbf{Q1} = 0$. This means that there are no jumps to ∞, which is now approached in a finite time by "creeping" along successive jumps with a limit point.

Denoting by $L_i(t)$ the distribution function of τ conditional on initial state i, one has from § 1.1.5, for the complementary distribution function,

$$L_i^c(t) = \sum_j p_{ij}(t) =: p_{i\mathbb{I}}(t), \qquad i \in \mathbb{I}, \quad t \geqslant 0,$$

or, in terms of transforms,

$$l_i^\alpha + \alpha u_{i\mathbb{I}}^\alpha = 1, \qquad \alpha \geqslant 0,$$

where

$$l_i^\alpha = \int_0^\infty e^{-\alpha t}\, dL_i(t), \qquad u_{i\mathbb{I}}^\alpha = \int_0^\infty e^{-\alpha t} p_{i\mathbb{I}}(t)\, dt.$$

Evidently,

$$L_i = \mathbb{P}^i(\tau < \infty), \qquad L_i^c = \mathbb{P}^i(\tau = \infty).$$

It is of interest to observe the analogy between $L_i^c(t)$ and $D_i^c(t)$, and recall from § 1.1.5 that $p_{ij}(t)$ plays the role of a taboo probability:

$$p_{ij}(t) = \mathbb{P}^i\left(X_t = j,\ \tau > t\right), \qquad i, j \in \mathbb{I}.$$

This point of view will be substantiated in the sequel, although important differences will be indicated.

As $p_{ij}(t)$ satisfy the (B)-equation, it follows that so does $L_i^c(t)$, and also $L_i(t)$ because $\mathbf{Q}$ is conservative.

THEOREM 1. *For* $i \in \mathbb{I}$, $t \geqslant 0$:

$$\frac{dL_i(t)}{dt} = \sum_k q_{ik} L_k(t)$$

with $L_i(0) = 0$.

COROLLARY 2. *For* $i \in \mathbb{I}$:

$$\sum_k q_{ik} L_k = 0,$$

where L_k *may be replaced by* L_i^c.

Hence, both (L_i) and (L_i^c) are excessive functions on $\mathbb{I}$. From the semi-group property of $\mathbf{P}(t)$ it follows

THEOREM 3. *For* $t, s \geqslant 0$ *and* $i \in \mathbb{I}$:

$$L_i^c(t+s) = \sum_k p_{ik}(t) L_k^c(s),$$

$$L_i(t+s) - L_i(t) = \sum_k p_{ik}(t) L_k(s).$$

The second equation in Theorem 3 implies that $\mathbf{P}(t)L_i \leqslant L_i$ and that (L_i) is purely excessive relative to $\mathbf{P}(t)$. Consequently (by Thm. 4, § 1.3.4), there exists a unique exit law $l(\cdot)$, necessarily bounded, such that

$$\sum_k p_{ik}(t) l_k(s) = l_i(s+t), \qquad s > 0, \quad t \geqslant 0,$$

$$L_i(t) = \int_0^t l_i(s)\, ds \underset{t \to \infty}{\longrightarrow} \int_0^\infty l_i(s)\, ds = L_i,$$

and Theorem 1 indicates that

$$l_i(t) = \sum_k q_{ik} L_k(t)$$

with $l_i(0) = 0$. Thus, $L_i(t)$ is absolutely continuous with density $l_i(t)$.

On the other hand, the first equation in Theorem 3 indicates that (L_i^c) is invariant relative to $\mathbf{P}(t)$ on $\mathbb{I}$, i.e., $\mathbf{P}(t)L_i^c = L_i^c$. Since the constant $\mathbf{1}$ is now $\mathbf{P}(t)$-excessive, the obvious relation

$$L_i + L_i^c = 1$$

is the Riesz decomposition of $\mathbf{1}$.

Observe that the relation defining the exit law $l(\cdot)$ is stated for $s > 0$. If it were valid for $s = 0$, then $l_i(0) = 0$ would imply that $l_i(t) = 0$, hence $L_i(t) = 0$, identically in t. Thus, $\mathbf{P}(t)$ would be a stochastic matrix. The explanation of this fact is provided in the next § 12.2.2.

§ 12.2.2. Minimal chain. Recall from § 1.1.2 that the minimal $\mathbf{P}(t)$ satisfies both the (B)- and (F)-equations.

THEOREM 1. *If* $\sum_j p_{ij}(t)q_j < \infty$, *then* $\sum_j p_{ij}(t) = 1$.

Proof. Finiteness of the series in assumption justifies interchange of summation, so addition over all j in the (F)-equation yields

$$\frac{dL_i^c(t)}{dt} = \sum_j \Big(\sum_k p_{ik}(t)q_{kj} \Big) = \sum_k p_{ik}(t) \sum_j q_{kj} = 0,$$

because $\mathbf{Q}$ is conservative. □

Note that Theorem 1 gives in effect the sufficient condition for validity of the associativity property

$$\mathbf{P}(t)\mathbf{Q} \cdot \mathbf{1} = \mathbf{P}(t) \cdot \mathbf{Q}\mathbf{1}.$$

Thus, the exit law argument leading to the (F)-equation under associativity condition (as, for example, in § 3.1.1) corresponds to the stochastic case $\mathbf{P}(t)\mathbf{1} = \mathbf{1}$, and reduces to triviality.

For the next theorem, recall the construction of the minimal $\mathbf{P}(t)$ discussed in § 1.1.5. Using the recursive relation (F), one obtains

THEOREM 2. *For* $i \in \mathbb{I}$ *and* $t \geqslant 0$:

$$L_i(t) = \lim_{n \to \infty} \int_0^t \sum_j p_{ij}^{(n)}(s) q_j \, ds.$$

Note that Theorem 2 relates $L_i(t)$ to the (F)-equation, and is in this sense the analogue of Theorem 7, § 3.1.1, for $D_i(t)$.

The important problem of finding necessary and sufficient conditions for the minimal solution to be stochastic was solved through pioneering work of W. Feller, P. Lévy, J. Doob, K. L. Chung, and others. Potential theory offers interpretation of these conditions and stresses analogies with the first passage methods, as already mentioned. The main results are stated here without proofs; see K. L. Chung [1967] for complete discussion.

THEOREM 3. *The first infinity* τ *is infinite with probability one iff*

$$\sum_n \frac{1}{q(X_n)} = \infty \quad \text{a.e.}$$

Here $q(X_n)$ is a r.v. obtained from q_i and the jump chain (X_n). Recall that τ is a random series of independent r.v.'s $\tau_{n+1} - \tau_n$ with exponential distributions, so

$$\mathbb{E}\big(\tau_{n+1} - \tau_n \mid X_n\big) = 1/q\big(X_n\big).$$

COROLLARY 4.

i) *A sufficient condition for* $L_i = 1$ *is* $\sum_j u_{ij} < \infty$.

ii) *A sufficient condition for* $L_i = 0$ *is that all* q_i *are bounded.*

Consider now the jump chain (X_n) restricted to H^c with ${}_H\mathbf{R}$ and ${}_H\mathbf{G}$ introduced in § 1.2.5. Since the constant $\mathbf{1}$ is ${}_H\mathbf{R}$-superharmonic, its Riesz decomposition is

$$1 = u_i + r_i, \qquad i \in H^c,$$

where

$$u_i = {}_H\mathbf{G}c_i \qquad \text{with} \quad c_i = \sum_j r_{ij} \leqslant 1, \qquad \text{and} \qquad r_i = {}_H\mathbf{R}^\infty \mathbf{1}_i.$$

The Dirichlet problem for D shows that (see § 5.1.2)

$$u_i = {}_H\mathbf{U}\bar{q}_i = D_i, \qquad r_i = D_i^c.$$

As already noted

$$s_i = \mathbb{P}^i(X_n \in H \text{ i.o.}),$$

and it follows from argument in § 5.2.3-c that $D_i^c > 0$ implies $s_i < 1$.

The complementary set H^c is called the sojourn set for the jump chain when $s_i < 1$ for some $i \in H^c$. The following theorem of Feller gives the necessary and sufficient condition for the minimal matrix $\mathbf{P}(t)$ to be stochastic. For proof see W. Feller [1940] and K. L. Chung [1967]. The formulation here makes use of the Feller relation for the taboo probabilities ${}_Hg_{ij} = {}_Hu_{ij}\, q_j$ (see § 1.2.5).

THEOREM 5. $L_i > 0$ *if and only if there exists a sojourn set* H^c *for the jump matrix* $\mathbf{R}$ *and an* i *satisfying*

$$r_i = D_i^c > 0$$

and

$${}_Hu_i = \sum_{j \in H^c} {}_Hu_{ij} < \infty.$$

Thus, in words, for a minimal matrix $\mathbf{P}(t)$ to be stochastic, it is necessary and sufficient that the average time spent in H^c, without leaving H^c, would be infinite (${}_Hu_i = \infty$) for every set H^c in which it is possible to survive ($r_i > 0$). It is necessary, but not sufficient, that the mean time spent in the entire state space $\mathbb{I}$ is infinite (see Thm. 1 above). See also §§ 1.1.3 and 1.1.5 for interpretation in terms of sample functions. It can also be shown that when the jump matrix $\mathbf{R}$ is recurrent, then the minimal matrix $\mathbf{P}(t)$ is stochastic.

§ 12.2.3. Comments. It is of interest to examine more closely Riesz decompositions, following discussion in § 12.2.1, and to comment on analogy with the first entrance. Resolvent formulation is more covenient for this purpose.

a) Observe that the (B)-equation for $\mathbf{U}^\alpha$ gives the Riesz decomposition of the constant function $\mathbf{1}$ which is excessive ($\alpha\mathbf{U}^\alpha\mathbf{1} \leqslant \mathbf{1}$),

$$\mathbf{1} = \alpha\mathbf{U}^\alpha\mathbf{1} + (-\mathbf{Q})\mathbf{U}^\alpha\mathbf{1}. \tag{B}$$

This agrees with decomposition already stated in § 12.2.1, because

$$\alpha\mathbf{U}^\alpha\mathbf{1}_i = \alpha u^\alpha_{i\mathbb{I}}, \qquad (-\mathbf{Q})\mathbf{U}^\alpha\mathbf{1}_i = l^\alpha_i, \qquad i \in \mathbb{I}.$$

Passage to the limit as $\alpha \to 0$ yields, respectively, the invariant component (L^c_i) and the purely excessive component (L_i).

b) The constant function $\mathbf{1}$ is only $\mathbf{R}^\alpha$-superharmonic (but $\mathbf{R}$-harmonic), and its Riesz decomposition relative to $\mathbf{R}^\alpha$ is (see § 1.3.3)

$$\mathbf{1} = \mathbf{G}^\alpha c^\alpha + w^\alpha, \qquad \alpha > 0,$$

where $w^\alpha_i = \mathbf{R}^{\alpha\infty}\mathbf{1}_i$ and $c^\alpha_i = \alpha(\alpha + q_i)^{-1}$, so by the Feller relation (§ 1.1.6)

$$\mathbf{G}^\alpha c^\alpha_i = \alpha u^\alpha_{i\mathbb{I}}.$$

Thus, as can be seen directly,

$$w^\alpha_i = l^\alpha_i .$$

Letting $\alpha \to 0$, one obtains again decomposition of unity into invariant component (L^c_i) and a purely excessive component $(L_i) = (w_i)$ which is not a potential of a charge.

This is in agreement with Feller decomposition theorem from § 1.3.3, applied to the function $f = 1$, where (in notation used there) $u = 0$ and $p = w$, $r = h + w = 1$.

c) Note that w^α is defined in terms of the α-chain, whereas l^α refers to the minimal chain. From argument in proof of Theorem 6, § 1.3.3, where $\mathbf{W}^\alpha\mathbf{1} = w^\alpha$, one has using the (B)-equation

$$l^\alpha = (-\mathbf{Q})\mathbf{U}^\alpha\mathbf{1} = \mathbf{U}^\alpha \cdot (-\mathbf{Q})\mathbf{1} + w^\alpha = w^\alpha .$$

Since for the minimal semi-group one has always $\mathbf{Q}\mathbf{U}^\alpha = \mathbf{U}^\alpha\mathbf{Q}$, the associativity $\mathbf{U}^\alpha\mathbf{Q} \cdot \mathbf{1} = \mathbf{U}^\alpha \cdot \mathbf{Q}\mathbf{1}$ would imply that $w^\alpha = 0$. Hence, $\mathbf{P}(t)$ would be stochastic (cf. Thm. 1, § 12.2.2).

Indeed, in the (F)-equation $l^\alpha = \alpha\mathbf{U}^\alpha\mathbf{1}$ the term w^α is missing. Its presence yields the augmented (F)-equation, as already noted in § 1.3.3-iv (see also the example of a birth process in § 7.2.5).

d) It is now possible to comment on the analogy between decompositions of unity in terms of the first entrance probability and in terms of the first infinity probability. As shown, (D_i) is purely excessive (and (D^c_i) invariant) relative to the sub-stochastic ${}_H\mathbf{P}(t)$, whereas (L_i) is purely excessive (and (L^c_i) invariant) relative to the sub-stochastic $\mathbf{P}(t)$. The important difference lies in a component which is a potential of a charge. Indeed, relative to the jump chain on H^c, (D_i) is a potential of a charge $(\bar{q}_i)$, whereas relative to the jump chain on $\mathbb{I}$ the potential component vanishes, because the charge is now $(-\mathbf{Q})\mathbf{1} = 0$, and both (L_i) and (L^c_i) are harmonic.

It is here where the analogy breaks down. Indeed, the Dirichlet problem for (L_i) should be refered to the extended state space $\mathbb{I} \cup \{\infty\}$, with $\mathbb{I}$ in the role of H^c and the boundary point ∞ in the role of H. However, $\mathbf{Q}\mathbf{1} = 0$ so $\mathbb{I}$ does not lead to ∞ for the jump chain. Thus, the Dirichlet problem reduces trivially to Corollary 2, § 12.2.1, with $L_\infty = 1$.

Although, formally, the Dirichlet problem leads to triviality, it nevertheless announces presence of a much deeper problem. In fact, one must consider the approach to the boundary ("creeping") by a proper compactification of the state space. It is here where the boundary theory enters, and provides proper interpretation of the Dirichlet problem, as well as the explicit expression for L_i. For further discussion, see the following Sub-section 12.3.

12.3. BOUNDARY

This sub-section is devoted to brief comments on the boundary theory for Markov chains with intention to stress analogies with passage times on a state space. For complete presentation see references: K. L. Chung [1967, 1970], W. Feller [1957, 1960], T. Watanabe [1960], and also G. E. Denzel et al. [1967], D. Williams [1979], where further references can be found.

§ 12.3.1. Decompositions. Let $\mathbb{I}$ be a denumerable state space and assume that the boundary consists of a single point which will be denoted by ∞. As before, $\mathbf{P}(t) = (p_{ij}(t))$ is the minimal transition matrix, assumed to be sub-stochastic, and $\mathbf{P}^*(t) = (p^*_{ij}(t))$ is a non-minimal transition matrix, taken to be stochastic.

The boundary theory establishes the relation between the transition probabilities of the M.C. and the minimal transition probabilities, in the following form.

THEOREM 1. *For $i, j \in \mathbb{I}$, $t \geqslant 0$:*

$$p^*_{ij}(t) = p_{ij}(t) + \int_0^t l_i(s)\xi_j(t-s)\,ds,$$

where $l_i(\cdot)$ is the density of the first infinity τ, and $\xi_j(\cdot)$ is the distribution of the entrances to the state space ("returns from infinity") such that

$$\sum_j \xi_j(t) = 1 \qquad \text{for all} \quad t \geqslant 0.$$

In the terms of Laplace transform, one has

$$u^{*\alpha}_{ij} = u^\alpha_{ij} + l^\alpha_i \xi^\alpha_j, \qquad \alpha > 0.$$

The analogy with the first entrance decomposition from Theorem 1, § 5.1.1, is now apparent. As noted earlier, the minimal transition probabilities play the role of taboo probabilities, with the boundary as a taboo set. Moreover, $l(\cdot)$ is the exit law relative to the minimal $\mathbf{P}(t)$, and $\xi(\cdot)$ is the normalized entrance law relative to $\mathbf{P}^*(t)$. As $\mathbf{P}^*(t)$ is stochastic,

$$\sum_j p^*_{ij}(t) = p_{i\mathbb{I}}(t) + L_i(t) = 1, \qquad i \in \mathbb{I}, \quad t \geqslant 0.$$

The crucial fact is that both semi-groups $\mathbf{P}(t)$ and $\mathbf{P}^*(t)$ have the same conservative generator $\mathbf{Q} = (q_{ij})$ with $\mathbf{Q1} = 0$. Indeed, one can interpret

$$p_{i\infty}(t) = L_i(t), \qquad i \in \mathbb{I},$$

with $q_{i\infty} = 0$.

It is shown that the following form of the strong Markov property holds:

$$\mathbb{P}(X_{t+\tau} = j \mid \mathcal{F}_{\tau-}) = \mathbb{P}(X_{t+\tau} = j \mid X_{\tau-}) = \xi_j(t) I_\infty(X_{\tau-}), \qquad t \geqslant 0, \quad j \in \mathbb{I},$$

where

$$\mathcal{F}_{\tau-} = \sigma\Big(\bigcup_n \mathcal{F}_{\tau_n}\Big)$$

is properly included in $\mathcal{F}_\tau$ (see §§ 8.1.1 and 1.1.3). The M.C. does not assume the value ∞ at τ, but only approaches ∞, and then "branches" to a state j, so

$$X_{\tau-} = \infty, \qquad X_\tau = j \in \mathbb{I}$$

(and ∞ is called the "branch point").

The value of the above conditional probability may be written as

$$p^*_{\infty j}(t) = \xi_j(t), \qquad t \geqslant 0,$$

with convention that the state space is extended to $\mathbb{I} \cup \{\infty\}$.

In particular, $\xi_j(0)$ is the distribution of the "immediate returns" (or "branching distribution") to the state space $\mathbb{I}$ (formally, $q_{\infty j} = \infty$). Thus, the boundary point ∞ may be regarded as an instantaneous state.

COROLLARY 2. *For $i, j \in \mathbb{I}$ and $\alpha > 0$:*

$$u^{*\alpha}_{ij} = u^\alpha_{ij} + l^\alpha_i \frac{\sum_k \xi_k(0) u^\alpha_{kj}}{1 - \sum_k \xi_k(0) l^\alpha_k}.$$

Proof. As $\xi_j(t)$ is the normalized entrance law relative to $\mathbf{P}^*(t)$, one has in particular

$$\sum_k \xi_k(0) u^{*\alpha}_{kj} = \xi^\alpha_j, \qquad \alpha > 0, \quad j \in \mathbb{I}.$$

Hence, the decomposition may be written in the form

$$u^{*\alpha}_{ij} = u^\alpha_{ij} + l^\alpha_i \sum_k \xi_k(0) u^{*\alpha}_{kj},$$

and the result follows by substituting this equation into itself. □

Note that Corollary 2 considers the consecutive passages through ∞ (renewal argument); cf. Sub-section 10.2.

Formally, one could write the Dirichlet problem relative to the generator $\mathbf{Q}$ of the transition matrix $\mathbf{P}^*(t)$ as

$$\begin{aligned} \mathbf{Q}f(i) &= 0, & i &\in \mathbb{I}, \\ f(i) &= \varphi(i), & i &= \infty. \end{aligned}$$

Writing the associated balayage operator in the form of the vector $(L_i,)$ $i \in \mathbb{I}$, with $L_\infty = 1$, the minimal solution of the Dirichlet problem is now

$$f(i) = L_i \varphi(i), \qquad i \in \mathbb{I}.$$

§ 12.3.2. Further decomposition. Of primary importance is to express the transition matrix $\mathbf{P}^*(t)$ solely in terms of the minimal matrix $\mathbf{P}(t)$, that is in terms of the intensity matrix $\mathbf{Q}$. This problem received considerable attention in literature, especially in work of K. L. Chung and D. Williams. The following decomposition theorem is a special case of a general result from the boundary theory, and is given here for the purpose of illustration. For details see K. L. Chung [1967, 1970], D. Williams [1979]; illustrative examples may be found in H. F. Bushar [1976].

THEOREM 1. *Suppose that all states in* $\mathbb{I}$ *are transient relative to* $\mathbf{P}^*(t)$. *Then there exists a normalized entrance law* $\eta(\cdot)$ *relative to the minimal* $\mathbf{P}(t)$ *such that*

$$u_{ij}^{*\alpha} = u_{ij}^{\alpha} + l_i^{\alpha} E^{\alpha} \eta_j^{\alpha}, \qquad i, j \in \mathbb{I}, \quad \alpha > 0,$$

where E^{α} *is the* L.S. *transform of the distribution function* $E(t)$, $t \geqslant 0$, *which is absolutely continuous, except at a jump at the origin* $t = 0$.

Proof (outline). It can be verified that, for a transient state j,

$$\xi_j = \int_0^{\infty} \xi_j(t)\, dt < \infty.$$

Since $\xi(\cdot)$ is the entrance law relative to $\mathbf{P}^*(t)$, it follows that the measure ξ is purely excessive for $\mathbf{P}^*(t)$, hence, also purely excessive relative to the minimal $\mathbf{P}(t)$. Therefore, by Theorem 4, § 1.3.4, there exists an entrance law $\eta(\cdot)$ relative to $\mathbf{P}(t)$ such that

$$\xi = \int_0^{\infty} \eta(t)\, dt = \eta$$

with the Laplace transform

$$\eta^{\alpha} = \xi(\mathbf{I} - \alpha \mathbf{U}^{\alpha}).$$

Next, define relative to the minimal $\mathbf{P}(t)$,

$$\sigma^{\alpha} = \sum_i \eta_i l_i^{\alpha}$$

as in § 1.3.4.

Then it follows that Theorem 1, § 12.3.1, can be put into the form

$$\xi(\mathbf{I} - \alpha \mathbf{U}^{\alpha}) = (1 + \alpha \sigma^{\alpha}) \xi^{\alpha}, \qquad \alpha > 0,$$

so

$$\xi_j^{\alpha} = \eta_j^{\alpha} (1 + \alpha \sigma^{\alpha})^{-1}.$$

Consequently,

$$l_i^{\alpha} \xi_j^{\alpha} = l_i^{\alpha} E^{\alpha} \eta_j^{\alpha},$$

where

$$E^\alpha = \left(1 + \alpha\sigma^\alpha\right)^{-1}.$$

This yields the required decomposition. Here E^α is the L.S. transform of a distribution function $E(t)$ such that $E(\infty) = 1$, and

$$E(0) = \begin{cases} \left(1 + \sigma(0)\right)^{-1} > 0, & \text{if } \sigma(0) < \infty, \\ 0, & \text{if } \sigma(0) = \infty. \end{cases} \qquad \square$$

Inverting relation defining E^α, one obtains the Chung equation

$$\int_{0-}^{t} [1 + \sigma(t - s)]\, dE(s) = 1$$

which received considerable attention in literature. See K. L. Chung [1970], H. Kesten [1969], and § 4.2.1 for the passage time analogue.

§ 12.3.3. Martin boundary. The following brief remark on the approach to the boundary may serve as a preview of deep results from the theory of Martin boundary. For the general Markov processes see P. A. Meyer [1968], and for a jump chain with matrix **R** or equivalently for a minimal chain with resolvent **U**, see K. L. Chung [1967], D. Williams [1979]. See also J. G. Kemeny et al. [1966], J. Neveu [1964], D. Revuz [1975], R. Syski [1973].

Let $(\widetilde{X}_t)$ be a M.C. reversed at the first infinity,

$$\widetilde{X}_t = X(\tau - t) \qquad \text{for} \quad 0 < t \leqslant \tau,$$

with resolvent $(\tilde{u}_{ij}^\alpha)$, as discussed in § 9.4.4. Assuming transient situation, let $u_{\pi j}$, $j \in \mathbb{I}$, be a reference measure (see § 1.3.5).

Define the Martin kernel

$$K(i, j) = u_{ij}/u_{\pi j} = \tilde{u}_{ji}/u_{\pi i} \qquad \text{for} \quad i, j \in \mathbb{I},$$

where the second expression follows from duality.

The sequence (j_n) in $\mathbb{I}$ tending to ∞ is said to converge to the boundary point ∞, say, if and only if for each i, the sequence $K(i, j_n)$ converges to $K(i, \infty)$. This definition is justified by the fact that it is shown (using martingale convergence theorem) that

$$\lim_{t \to 0} K\left(i, \widetilde{X}_t\right) = K\left(i, X_{\tau-}\right) \quad \text{a.e.} \qquad \text{for every} \quad i \in \mathbb{I}$$

which is equivalent to

$$\widetilde{X}_t \longrightarrow X_{\tau-} \quad \text{a.e.} \qquad \text{as} \quad t \to 0.$$

Here $X_{\tau-}$ takes values in the boundary (∞, in the present case).

Then, with a single boundary point, one has the representation

$$L_i = K(i, \infty)\mu(\infty), \qquad i \in \mathbb{I},$$

where μ is a "harmonic measure" on the boundary.

SECTION 13

EXAMPLES

This section presents several advanced examples which illustrate material from Chapters III and IV. Although most of Markov chains used here are classical, examples themselves are new because they are concerned with aspects discussed in this book.

Sub-section 13.1 provides explicit examples of integral functionals, emphasizing further their important role. Also discontinuous functionals related to point processes are mentioned, including a brief comment on Palm measure. Finally, examples on Revuz measure are included.

Sub-section 13.2 illustrates various aspects of the notion of energy. After simple examples, association with balayage is discussed. This is followed by example of energy changes under transformations like random time change and perturbation. The sub-section closes with a rather meager example of optimisation.

Sub-section 13.3 presents the martingale approach to study of Markov chains and their passage times. Martingale problem is illustrated and its connection with the earlier discussion is examined. Of special interest are applications to the branching process and to the random time change.

The last Sub-section 13.4 treats a rather peculiar structure, called *an escape chain*, which exhibits curious properties. Its energy is also discussed.

This section has a particular feature, namely that the same chain (say, branching process) or the same technique (say, random time change), may be discussed under different headings, depending on a point of view.

13.1. FUNCTIONALS

The importance of functionals of Markov chains has been already pointed out in this book. It may be convenient to illustrate their properties on a few further examples. As noted, the only continuous functionals are the integral functionals, and the discontinuous ones are in fact point processes. Also some calculations of Revuz measure will be mentioned.

§ 13.1.1. Integral functionals. The functional $A = (A_t)$, where

$$A_t = \int_0^t a(X_s)\, ds, \qquad t \geqslant 0,$$

with charge $a \geqslant 0$, received already considerable attention. See, for example, the local time when $a = I_H$, and the case when $a = b = (-\mathbf{Q})D$. Recall that the characteristic and α-potential of A are, respectively,

$$\mathbb{E}^i A_t = \int_0^t \mathbf{P}(s)a(i)\,ds, \qquad u_A^\alpha(i) = \mathbf{U}^\alpha a(i), \qquad i \in \mathbb{I},$$

and note that $a(i)$ is the value of the derivative of the characteristic at $t = 0$.

The following three examples exhibit interesting properties.

Example 1. *Birth-and-death process.* This process has been already discussed in Example 1, § 1.1.9, and in Sub-section 7.2.

a) Consider first the case when the charge a has the form

$$a(i) = \gamma^i, \qquad 0 < \gamma < 1, \quad i = 0, 1, \ldots.$$

Then,

$$\mathbb{E}^i A_t = \int_0^t \psi_i(s,\gamma)\,ds, \qquad u_A^\alpha(i) = \int_0^\infty e^{-\alpha t}\psi_i(t,\gamma)\,dt,$$

with the generating function

$$\psi_i(t,\gamma) = \sum_{j=0}^\infty p_{ij}(t)\gamma^j.$$

Assuming that the chain is transient, the potential of (A_t) is then

$$u_A(i) = \int_0^\infty \psi_i(t,\gamma)\,dt = \mathbf{U}\gamma^i = \mathbf{U}a(i) = p(i) < \infty, \qquad i = 0, 1, \ldots.$$

b) It is of interest to verify that the same function $f(i) = \gamma^i$ is also purely excessive for the chain. Indeed, the expression for charge, found from $a = (-\mathbf{Q})f$, is, according to § 7.2.3,

$$a(i) = \begin{cases} \lambda_0(1-\gamma), & i = 0, \\ (\lambda_i\gamma - \mu_i)\gamma^{i-1}(1-\gamma), & i \geqslant 1. \end{cases}$$

Positivity of $a(i)$ follows from the fact that, for $i \geqslant 1$,

$$\lambda_i/\mu_i > \gamma^{-1} > 1,$$

and this confirms that the chain is transient. Indeed, using

$$\rho_n = \frac{\lambda_0 \ldots \lambda_{n-1}}{\mu_1 \ldots \mu_n} \quad \text{for} \quad n \geqslant 1, \qquad \rho_0 = 1,$$

one has

$$\sum_{n=0}^\infty (\lambda_n\rho_n)^{-1} < \lambda_0^{-1}\sum_{n=0}^\infty \gamma^n < \infty,$$

and this is sufficient for transiency; see criteria in S. Karlin and J. McGregor [1957], J. W. Cohen [1969]. Consequently, f is excessive, and

$$\mathbf{P}(t)f(i) = \mathbf{P}(t)\gamma^i = \psi_i(t,\gamma) \longrightarrow 0, \qquad \text{as} \quad t \to \infty,$$

indicates that it is also purely excessive.

Example 2. *Modified Reynolds identity.* Consider a compound Poisson process with resolvent of the form

$$\mathbf{U}^\alpha = (1+\alpha)^{-1}\sum_{n=0}^{\infty}(1+\alpha)^{-n}\mathbf{R}^n, \qquad \alpha > 0,$$

where $\mathbf{R}$ is the jump matrix (see Ex. 2, § 1.1.9). Assume now that $\mathbf{R}$ has the property

$$\mathbf{R}^n f(i) = f(i+n) \qquad \text{for each } n \text{ and for any } f \geqslant 0.$$

Hence, for non-negative functions f and g,

$$\mathbf{R}^n fg = \mathbf{R}^n f \cdot \mathbf{R}^n g.$$

Clearly, the generator is $\mathbf{Q} = \mathbf{R} - \mathbf{I}$, so $\mathbf{Q}f(i) = f(i+1) - f(i)$.

Rather tedious verification shows that the resolvent $\mathbf{U}^\alpha$ satisfies the modified Reynolds identity:

$$\alpha\mathbf{U}^\alpha fg = \alpha\mathbf{U}^\alpha f \cdot \alpha\mathbf{U}^\alpha g + (1+\alpha)\alpha\mathbf{U}^\alpha(\alpha\mathbf{U}^\alpha f - f)(\alpha U^\alpha g - g)$$

for positive functions f and g.

Finally, taking $a(i) = \gamma^i$ with $0 < \gamma < 1$, it follows that the α-potential of the integral functional A is

$$\mathbf{U}^\alpha a(i) = \gamma^i(1+\alpha-\gamma)^{-1},$$

so $\alpha\mathbf{U}^\alpha a \leqslant a$ indicating that (γ^i) is itself purely excessive.

Example 3. *Heat equation.* Let $f = f(\cdot,\cdot)$ be a function of two variables $i \in \mathbb{I}$ and $t \geqslant 0$, assumed differentiable with respect to t. Consider the equation for f of the following form:

$$\frac{\partial f(i,t)}{\partial t} = \mathbf{Q}f(i,t), \qquad i \in \mathbb{I}, \quad t \geqslant 0,$$

with the initial condition $f(i,0) = a(i) \geqslant 0$. Denoting by f^α the Laplace transform of $f(\cdot,t)$, the above equation takes the form:

$$\mathbf{Q}^\alpha f^\alpha = -a, \qquad \alpha > 0,$$

and its solution is $f^\alpha = \mathbf{U}^\alpha a$. Hence,

$$f(i,t) = \mathbf{P}(t)a(i) = \mathbb{E}^i a(X_t).$$

See Lemma 3, § 8.1.2, and compare also with § 8.3.3-b.

§ 13.1.2. Point processes. The discontinuous functional $A = (A_t)$, where

$$A_t = \sum_{0<s\leqslant t} g(X_{s-}, X_s), \qquad t > 0,$$

with intensity $g \geqslant 0$, has been already mentioned frequently. See, for example, §§ 8.1.2, 8.1.3, and discussion of consecutive entrance in Sub-section 10.2. Recall that the characteristic and the α-potential of A coincide with those of its dual predictable projection (compensator) $B = (B_t)$, which is an integral functional with charge

$$g_{\mathbf{Q}}(i) = \sum_{j\neq i} q_{ij}\, g(i,j), \qquad i \in \mathbb{I}.$$

Thus,

$$\mathbb{E}^i A_t = \int_0^t \mathbf{P}(s) g_{\mathbf{Q}}(i)\, ds, \qquad u_A^\alpha(i) = \mathbf{U}^\alpha g_{\mathbf{Q}}(i), \qquad i \in \mathbb{I}.$$

The following examples, referring to point processes, are of interest.

Example 1. *Single jump.* It is instructive to begin with the simplest situation. Let T be the first entrance time to a set H, and consider an increasing integrable process $A = (A_t)$ defined by

$$A_t = I_{(T\leqslant t)}, \qquad t \geqslant 0.$$

Clearly, A has a single discontinuity at T, and $A_\infty = A_T = 1$, although A is only an extended AF (as defined in § 8.2.3). Nevertheless, there exists a continuous increasing process $B = (B_t)$ such that A and B have the same characteristic and potential,

$$\mathbb{E}^i A_t = D_i(t) = \int_0^t {}_H\mathbf{P}(s)\bar{q}_i\, ds, \qquad u_A^\alpha(i) = d_i^\alpha = {}_H\mathbf{U}^\alpha \bar{q}_i, \qquad i \in \mathbb{I}.$$

Moreover, $(A_t - B_t)$ form a martingale, relative to the same σ-fields $(\mathcal{F}_t)$, so the potentials generated by A and by B coincide, i.e.,

$$\mathbb{E}(A_\infty - A_t \mid \mathcal{F}_t) = \mathbb{E}(B_\infty - B_t \mid \mathcal{F}_t) \quad \text{a.e.}$$

Note that T is taken to be finite.

Taking the natural σ-fields to be those generated by A, it can be shown that

$$\mathcal{F}_t = \sigma(t \wedge T), \qquad t \geqslant 0.$$

It is known that the explicit expression for B can then be written in the form

$$B_t = G(t \wedge T), \qquad t \geqslant 0,$$

where G is the hazard function corresponding to distribution of T. See, e.g., P. Bremaud [1981], P. A. Meyer [1966], E. Slud [1992].

Example 2. *Counting process.* Consider the counting process $N = N_t$ of the form

$$N_t = \sum_{n=1}^{\infty} I_{(\tau_n \leqslant t)}, \qquad t > 0,$$

where (τ_n) is a sequence of consecutive jumps of a M.C. $X = (X_t)$. As $\tau_n \leqslant \tau_{n+1}$, it follows that (N_t) is an AF see § 8.1.3-b. Passing to its compensator $N^* = (N_t^*)$, one has

$$\mathbb{E}^i N_t = \int_0^t \mathbf{P}(s) q_i \, ds, \qquad u_N^\alpha(i) = \mathbf{U}^\alpha q_i, \qquad i \in \mathbb{I}.$$

Moreover, $N_t - N_t^*$, $t \geqslant 0$, form a martingale relative to the natural σ-fields $(\mathcal{F}_t)$ of the M.C. Indeed, the compensator has the form

$$N_t^* = \int_0^t q(X_s) \, ds, \qquad t \geqslant 0.$$

The general theory of point processes provides explicit expression for compensators, but it can be shown that it coincides here with the above expression. The analysis is similar to that in § 10.2.4-2; see also references quoted there.

In particular, for a birth process, recalling that intervals $\tau_{n+1} - \tau_n$ have exponential density with mean $1/\lambda_n$, the compensator is

$$N_t^* = \int_0^t \lambda(N_s) \, ds = \sum_{n=0}^{\infty} \lambda_n \int_0^t I_{(\tau_n \leqslant s < \tau_{n+1})} \, ds.$$

Evidently, for the Poisson process (N_t), one finds the well-known expression for the martingale

$$M_t = N_t - \lambda t, \qquad t \geqslant 0,$$

which is a special form of the Doob–Meyer decomposition.

Example 3. *Palm measure.* Suppose that a M.C. $X = (X_t)$ is stationary with invariant distribution $\pi \mathbf{P}(t) = \pi$. Let N_t^j represent the number of jumps up to time t to state $j \in H$ from a state $i \in H$. Using characteristic from § 8.1.4-b, the expected number of visits is

$$\mathbb{E}\, N_t^j I_H(X_t) = \int_0^t \sum_{i \in H,\ i \neq j} \mathbb{P}(X_s = i) q_{ij} \, ds = t \sum_{i \in H,\ i \neq j} \pi(i) q_{ij}, \qquad j \in H.$$

Denoting by N_t the total number of jumps to set H, from $i \in H$, one finds in the same way (summation over all $j \in H$) that

$$\mathbb{E}\, N_t I_H(X_t) = \int_0^t \sum_{i \in H} \mathbb{P}(X_s = i) \bar{q}_i \, ds = t \sum_{i \in H} \pi(i) \bar{q}_i.$$

Assuming that the above expression is finite, the measure $\mu(\cdot)$ defined as the ratio of these two expressions,

$$\mu(j) = \sum_{i \neq j,\, i \in H} \pi(i) q_{ij} \Big/ \sum_{i \in H} \pi(i) \bar{q}(i), \qquad j \in H,$$

is known as a Palm measure for a M.C. It can be shown that this definition is equivalent to the original definition due to C. Palm; see F. Baccelli and P. Bremaud [1987].

Remark. Let $\widehat{D}_i(t)$ be the distribution of the first entrance time S to the set H^c with density $\hat{d}_i(t)$, $i \in H$. The expression

$$D_0(t) = \sum_{i \in H} \mu(i) \widehat{D}_i(t)$$

is known as the "sojourn time distribution" on H. Define also the "entrance time density"

$$d_e(t) = \sum_{i \in H} \Big(\pi(i) \Big/ \sum_{k \in H} \pi(k) \Big) \hat{d}_i(t).$$

J. Keilson [1974, 1979] has shown that for the ergodic chain

$$d_e(t) = D_0^c(t)/m_0,$$

where m_0 is the mean of distribution $D_0(t)$.

§ 13.1.3. Revuz measure. By Revuz theorem (Thm. 6, § 8.1.3-e) for any functional A, there exists a measure μ_A such that the α-potential of A equals the α-potential of μ_A:

$$u_A^\alpha = \mathbf{U}^\alpha \mu_A.$$

Then, writing $a_A(i) = \mu_A(i)/\xi(i)$ for the dual of Revuz measure, one has

$$u_A^\alpha(i) = \mathbf{U}^\alpha a_A(i).$$

For the integral functional determined by the charge a, indeed, $\mu_A(i) = a(i)\xi(i)$, whereas for the discontinuous functional determined by the intensity g, $\mu_A(i) = g_{\mathbf{Q}}(i)\xi(i)$. Here $\xi(\cdot)$ is the reference measure.

Example 1. Consider an integral functional A determined by the charge a. Then, as noted above, $\mu_A(i) = a(i)\xi(i)$. Thus, taking $p = \mathbf{U}a = u_A$, the energy of A, assumed finite, is

$$\text{en} = p \bullet \hat{a} = \sum_i u_A(i) \mu_A(i).$$

i) In the degenerate situation, $A_t = t$, so $u_A^\alpha(i) = 1/\alpha$ and $\mu_A(i) = \xi(i)$.

ii) Let X be a Poisson process with rate λ. Select for the charge a any probability distribution on non-negative integers with a finite mean m. Using the fact that $\mu_A = \sigma \mathbf{U}_A$, where $\sigma = \xi(-\mathbf{Q})$, one finds that

$$\sigma(j) = \begin{cases} \lambda(\xi(j) - \xi(j-1)) \geqslant 0 & \text{for } j \geqslant 1, \\ \lambda \xi(0) & \text{for } j = 0, \end{cases}$$

so ξ is an increasing σ-finite measure. Note that $\sigma \bullet \mathbf{1} \neq \xi \bullet (-\mathbf{Q})\mathbf{1} = 0$. As for the Poisson process $u_{ij} = 1/\lambda$ for $j \geqslant i$, and $u_{ij} = 0$ for $j < i$, the potential of A is

$$u_A(i) = \mathbb{E}^i A_\infty = \sum_{j=1}^{\infty} a(j)/\lambda \leqslant 1/\lambda,$$

confirming that $\mu_A(j) = a(j)\xi(j)$. The energy is now

$$\text{en} = \sum_i u_A(i) a(i) \xi(i).$$

In particular, taking $\xi(j) = j$, one has $\mu_A(j) = ja(j)$, with total mass $\mu_A \bullet \mathbf{1} = m$, so en $\leqslant m/\lambda$.

Example 2. Consider a discontinuous functional determined by the intensity $g(\cdot,\cdot) \geqslant 0$. Passing to a compensator, one has $\mu_A(j) = g_{\mathbf{Q}}(j)\xi(j)$ and $u_A^\alpha(i) = \mathbf{U}^\alpha g_{\mathbf{Q}}(i)$. Hence, the expression for energy is the same as in Example 1.

i) Let A_t be a number of jumps in the interval $(0, t]$ of a M.C. $X = (X_t)$. Then, as shown in § 8.1.3-b, $g_{\mathbf{Q}}(i) = q_i$, and therefore,

$$u_A^\alpha(i) = \mathbf{U}^\alpha q_i = \mathbf{U}^\alpha \mu_A(i),$$

with $\mu_A(j) = \xi(j) q_j$.

ii) Taking in case 1) above, $X = (X_t)$ to be a Poisson process with rate $\lambda = q_i$ for all i, the Revuz measure is σ-finite, and both $\mathbf{U}\mu_A = u_A$ are infinite.

Example 3. Let T be the first entrance time to a set H, and consider a degenerate AF $A = (A_t)$ of the form

$$A_t = -\log M_t, \qquad t \geqslant 0,$$

with the MF $M_t = I_{(T>t)}$. Clearly, $u_A^\alpha = \infty$, and the Revuz theorem does not hold. It can be saved, however, by passing to the quasi-Revuz measure, as explained in Example 5, § 9.1.4.

With reference to the taboo matrices, consider an operator

$$\mathbf{U}_A^\alpha = \mathbf{U}^\alpha(\mathbf{Q} - {}_H\mathbf{Q})\mathbf{J}_1, \qquad \alpha \geqslant 0,$$

with kernel

$$u_A^\alpha(i,j) = \begin{cases} \sum_{k \in H} u_{ik}^\alpha q_{kj}, & i \in \mathbb{I}, \quad j \in H^c, \\ 0, & i \in \mathbb{I}, \quad j \in H. \end{cases}$$

Next, define the measure $\mu_A = \sigma \mathbf{U}_A$, supported on H^c, where $\sigma = \xi(-\mathbf{Q})$. Hence,

$$\mu_A(j) = \sum_{i \in H} \xi(i) q_{ij}, \qquad j \in H^c.$$

Note that the measure μ_A does not satisfy the Revuz theorem, although it has analogous structure.

Consider now an integral functional $A^* = (A_t^*)$ determined by the charge

$$a(j) = \mu_A(j)/\xi(j) = \sum_{i\in H} \hat{q}_{ji}, \qquad j \in H^c,$$

where $\hat{q}$ is the dual of q.

Evidently, the Revuz measure μ_A^* for A^* coincides with the measure μ_A, and the Revuz theorem holds:

$$u^\alpha_{A^*}(i) = \mathbf{U}^\alpha \mu_{A^*}(i) = \mathbf{U}^\alpha a(i), \qquad i \in \mathbb{I}.$$

This illustrates the fact that it is permissible to associate with a degenerate functional A the Revuz measure of an integral functional A^*, and to call it a quasi-Revuz measure for A.

13.2. ENERGY

Concept of energy received much attention in recent years, so it may be of interest to have several examples of calculations. Basically, energy is a quadratic form, as already explained in Sub-section 11.5. Of special interest is a comparison of energy under transformation of a M.C., like random time change or perturbation. Also energy of a balayage has interesting properties. These topics are illustrated here on examples, preceded by a few simple calculations. Some comments on optimisation are also included.

§ 13.2.1. Simple cases. The main formula for energy in the transient chain is, according to Definition 3, § 11.5.1,

$$\text{en} = \sum_i p(i)\xi(i)a(i),$$

where p is a potential of a charge a, and ξ is a reference measure;

$$p = \mathbf{U}a, \qquad a = (-\mathbf{Q})p.$$

Thus, in calculations, one can take either a charge a or potential p, as independent variables. In selecting ξ it must be verified that it is an excessive measure, so $\sigma = \xi(-\mathbf{Q}) \geqslant 0$. It may be convenient to write en (p) or en (a), according whether p or a are taken as initial quantity.

Dually, for a measure μ on $\mathbb{I}$ with potential $\mathbf{U}\mu$, energy of μ is defined by

$$\text{en}\,(\mu) = \mu \bullet \mathbf{U}\mu = \sum_i \sum_j \mu(i)v(i,j)\mu(j).$$

This reduces to the first formula above when μ is the Revuz measure $\mu_A(i) = a(i)\xi(i)$ of the integral functional A with charge a. Furthermore, taking $\xi = \pi\mathbf{U}$, where π is the initial distribution, the energy is given by $\frac{1}{2}A^2_\infty$ (see Def. 4, § 11.5.1).

Finally, for an ergodic chain, energy is given by a different expression $\hat{a}\mathbf{Z}a$, as stated in § 11.5.5.

Example 1. *One-point.* Suppose that, for a fixed state k and a positive constant c,

$$a(i) = \begin{cases} c > 0, & i = k, \\ 0, & i \neq k, \end{cases}$$

and that the initial distribution is also located on k ($\pi_k = 1$). Then, for a transient chain,

$$p(i) = u_{ik}, \qquad \xi(i) = u_{ki}.$$

Therefore,

$$\text{en}\,(a) = c(u_{kk})^2.$$

Example 2. M/M/1 *chain.* Assume that $\lambda > \mu$ (transient chain). It has been already shown in Example 1, § 13.1.1, that the measure

$$\xi(i) = (\lambda/\mu)^i, \qquad i \geqslant 0,$$

is excessive and σ-finite. Select now

$$a(i) = (\mu/\lambda)^i, \qquad i \geqslant 0.$$

Hence, the potential p is given by

$$p(i) = \big(\lambda + i(\lambda - \mu)\big)a(i)/(\lambda - \mu)^2, \qquad i \geqslant 0,$$

(see Ex. 1, § 7.2.4). Thus the dual of charge a is

$$\hat{a}(i) = a(i)\xi(i) = 1 \qquad \text{for all} \quad i \geqslant 0.$$

Therefore,

$$\text{en}\,(a) = \sum_{i=0}^{\infty} p(i) = \lambda(\lambda + \mu)/(\lambda - \mu)^3.$$

As another possibility, select $a(0) = 1$ and $a(i) = 0$ for all $i \geqslant 1$. Then,

$$p(i) = u_{i0} = (\mu/\lambda)^i(\lambda - \mu)^{-1}.$$

Hence, with arbitrary ξ,

$$\text{en}\,(a) = \xi(0)/(\lambda - \mu).$$

Example 3. *Poisson process.* Taking

$$a(i) = (1 - \gamma)\gamma^i, \qquad i \geqslant 0, \quad 0 < \gamma < 1,$$

one has using expression for u_{ij}:

$$p(i) = \gamma^i/\lambda, \qquad i \geqslant 0$$

(see Ex. 1, § 7.2.5); here λ is the rate of the process.

Next, take as a reference measure (with $\pi_0 = 1$)

$$\xi(i) = u_{0i} = 1/\lambda.$$

Consequently,

$$\text{en}\,(a) = \lambda^{-2}(1 + \gamma)^{-1}.$$

For another application to the Poisson process see Example 1, § 13.1.3.

Example 4. *Two states*. This is a birth-and-death process with two states 0 and 1, already discussed in Example 1, § 7.1.1. The chain is ergodic, with ergodic distribution ε and the α-potential ($\alpha > 0$) already found there.

Easy calculation shows that the ergodic potential $\mathbf{Z} = (z_{ij})$ is now

$$z_{00} = \varepsilon_i w, \qquad z_{01} = -\varepsilon_1 w,$$
$$z_{10} = -\varepsilon_0 w, \qquad z_{11} = \varepsilon_0 w,$$

where $1/w = \lambda + \mu$ (see § 1.1.8 for definition of $\mathbf{Z}$).

Consequently, with measure ε taken as a reference measure, the ergodic energy is, according to § 11.5.5,

$$\text{en}\,(a) = \hat{a}\mathbf{Z}a = \varepsilon_0\varepsilon_1 w\big(a(0) - a(1)\big)^2.$$

§ 13.2.2. Balayage. The notion of a balayage operator $\mathbf{B}$ has been already discussed extensively earlier; see Sub-section 8.5. The balayage $\bar{p}$ of a potential $p = \mathbf{U}a$ of a charge $a \geqslant 0$, is by definition $\bar{p} = \mathbf{B}p$. Recall that p is purely excessive, and so is $\bar{p}$. Furthermore, in view of duality,

$$\bar{p} = \mathbf{BU}a = \mathbf{UC}a = \mathbf{U}\bar{a} \qquad \text{with} \quad \bar{a} = (-\mathbf{Q})\bar{p} = \mathbf{C}a,$$

where $\mathbf{C}$ is the co-balayage operator (see § 8.5.1).

Hence, relative to the same reference measure ξ, energies of p and of $\bar{p}$ are

$$\text{en}\,(p) = p\,\hat{a}, \qquad \text{en}\,(\bar{p}) = \bar{p}\,\hat{\bar{a}}.$$

Using the balayage matrix in partitioned form, the components of $\bar{p}$ are $\bar{p}_1 = \mathbf{B}_{12}p_2$ and $\bar{p}_2 = p_2$. Hence, components of $\bar{a}$ are $\bar{a}_1 = 0$ and $\bar{a}_2 = \mathbf{C}_{21}a_1 + a_2$. Thus,

$$\text{en}\,(\bar{p}) = \sum_{i \in H} p(i)\bar{a}(i)\xi(i).$$

Observe that if the charge a is supported on H (so $a_1 = 0$), then en (p) = en $(\bar{p})$, as already shown in Proposition 1, § 11.5.2.

Next, for a M.C. $X = (X_t)$, the balayage $\bar{p}(X_t)$ of a potential supermartingale $p(X_t)$ is $\bar{p}(X_t) = \mathbf{B}p(X_t)$, $t \geqslant 0$, as shown in Sub-section 8.5. With $A = (A_t)$ and $\bar{A} = (\bar{A}_t)$ representing the integral functionals determined by charges a and $\bar{a}$, respectively, it has been shown that

$$p(X_t) = \mathbf{U}a(X_t) = \mathbb{E}\big(A_\infty - A_t \mid \mathcal{F}_t\big),$$
$$\bar{p}(X_t) = \mathbf{U}\bar{a}(X_t) = \mathbb{E}\big(\bar{A}_\infty - \bar{A}_t \mid \mathcal{F}_t\big) \quad \text{a.e.}$$

For finite T one has $\bar{A}_\infty = A_T$, and (see § 8.5.1)

$$p(X_t) - \bar{p}\,(X_t) = \mathbb{E}\big(A_T \circ \theta_t \mid \mathcal{F}_t\big) \quad \text{a.e.}$$

Taking as a reference measure $u_{\pi i}$, it follows that

$$\text{en}\,(p(X_t)) = \frac{1}{2}\mathbb{E}\,A_\infty^2 = \hat{a}\,p, \qquad \text{en}\,(\bar{p}\,(X_t)) = \frac{1}{2}\mathbb{E}\,\bar{A}_\infty^2 = \hat{\bar{a}}\,\bar{p},$$

with mutual energy

$$\big(p(X_t),\, \bar{p}(X_t)\big) = \frac{1}{2}\mathbb{E}\, A_\infty \bar{A}_\infty .$$

One has the obvious decomposition (projection theorem for balayage; see § 8.5.2)

$$p = \bar{p} + (p - \bar{p}).$$

However, this need not be an orthogonal projection with respect to energy inner product. Indeed, the mutual energy $(p - \bar{p}, \bar{p})$ need not vanish.

Finally, recall from § 8.5.2 that taking for p the potential u_A, it follows that $\bar{A}$ is the balayage of A on H, with $\bar{p} = u_{\bar{A}} = \mathbf{B}u_A$.

Example 1. M/M/1 *chain.* Assume $\lambda > \mu$ and take $H = \{0\}$, so T is the busy period. As shown in Example 1, § 7.2.4, $D_i = (\mu/\lambda)^i$ and $b_0 = \lambda - \mu$.

i) Take $a(i) = D_i$ and $\xi(i) = 1/D_i$. Then en (p) has been already found in Example 2, § 13.2.1,

$$\text{en}\,(p) = \lambda(\lambda + \mu) b_0^{-3}.$$

On the other hand, one finds now for the balayage:

$$\bar{p}(i) = \lambda D_i b_0^{-2}, \qquad i \geqslant 0,$$

$$\bar{a}(i) = \begin{cases} 0 & \text{for } \; i \geqslant 1, \\ \lambda b_0^{-1} & \text{for } \; i = 0. \end{cases}$$

Hence,

$$\text{en}\,(\bar{p}) = \lambda^2 b_0^{-3}$$

and

$$\text{en}\,(p) - \text{en}\,(\bar{p}) = \lambda \mu b_0^{-3}.$$

Note also that $\bar{A}$ is proportional to occupation time of state 0, necessarily finite,

$$\bar{A}_t = \bar{a}\,(0) \int_0^t I_0\,(X_s)\, ds$$

(see § 8.4.1).

ii) In the second case mentioned in Example 2, § 13.2.1, $a(0) = 1$ and $a(i) = 0$ for $i \geqslant 1$, with arbitrary reference measure ξ. One can find immediately that $\bar{p} = p$ and $\bar{a} = a$. Hence en $(\bar{p})$ = en (p).

Example 2. *Compound Poisson process.* This chain has been already discussed in Example 2, § 1.1.9, and in § 7.3.3. As shown, the generator has the form $\mathbf{Q} = \lambda(\mathbf{R} - \mathbf{I})$, where $\mathbf{R}$ is jump matrix. Recall also that $\mathbf{QB}$ has only non-zero component of the form $\overline{\mathbf{Q}}_{22} = \mathbf{Q}_{22} + \mathbf{Q}_{21}\mathbf{R}_{12}$. Consider now a measure μ on $\mathbb{I}$. Its charge is $\sigma = \mu(-\mathbf{Q})$, so the potential of σ is $\mu = \sigma\mathbf{U}$; see § 6.4.2. According to Theorem 3, § 6.4.2, the balayage measure is

$$\bar{\mu} = \mu\mathbf{C} = \sigma\mathbf{UC} = \sigma\mathbf{BU} = \bar{\sigma}\mathbf{U},$$

where

$$\bar{\sigma} = \sigma \mathbf{B}.$$

In the present case, with $\mu = (\mu_1, \mu_2)$,

$$\bar{\sigma} = (\mathbf{0}, \mu_2 \overline{\mathbf{Q}}_{22}) \quad \text{and} \quad \bar{\mu} = (\bar{\sigma}_2 \mathbf{U}_{21}, \bar{\sigma}_2 \mathbf{U}_{22}).$$

Hence,

$$\text{en}\,(\mu) = \mu \bullet \mathbf{U}\mu,$$
$$\text{en}\,(\bar{\mu}) = \bar{\mu} \bullet \mathbf{U}\bar{\mu} = \sum_i \sum_j \bar{\mu}(i) v(i,j) \bar{\mu}(j),$$

using definition from § 13.2.1.

§ 13.2.3. Random time change. The mechanism of a random time change has been already described in Sub-section 9.6. Recall that the transformation has been determined by an integral functional A with a charge $a \geqslant 0$ supported on a set H, and the new chain (assumed transient) has the generator $\widetilde{\mathbf{Q}} = (\tilde{q}_{ij})$ and potential $\widetilde{\mathbf{U}} = (\tilde{u}_{ij})$ on H given by

$$\tilde{q}_{ij} = \bar{q}_{ij}/a(i), \qquad \tilde{u}_{ij} = u_{ij} a(j), \qquad i,j \in H,$$

where $\bar{q}_{ij}$ has been found in § 9.6.5. Let ξ be a reference measure for the original chain on $\mathbb{I}$. As shown in Theorem 3, § 9.6.6, the random time change preserves duality with the Revuz measure of A as the new reference measure (with the same a).

Let $\mathbf{Q}^* = (q^*_{ij})$ be the generator of a time changed dual chain or, equivalently, of the dual of the time changed chain. Then one has from proof of Theorem 3, § 9.6.6, that

$$q^*_{ij} = \hat{\bar{q}}_{ij}/a(i), \qquad i,j \in H,$$

where $\hat{\bar{q}}_{ij}$ is the dual of $\bar{q}_{ij}$ relative to ξ.

It then follows that the corresponding potential $\mathbf{U}^* = (u^*_{ij})$ is defined by

$$u^*_{ij} = \hat{u}_{ij}\, a(j), \qquad i,j \in H,$$

where $\hat{u}_{ij}$ is the dual of u_{ij} relative to ξ.

Consider now a charge $c \geqslant 0$ on H, and write for the corresponding potential $p^* = \mathbf{U}^* c$. Then the energy, relative to the Revuz measure μ_A, can be expressed in terms of the original chain in the following manner:

$$\text{en}\,(p^*) = \sum_{i \in H} c(i) \mu_A(i)\, p^*(i) = (ac) \bullet \widehat{ac}\mathbf{U},$$

where $\widehat{ac}$ is the dual (relative to ξ) of the function ac.

In particular, taking $c = 1$, one gets $\text{en}\,(p^*) = a \bullet \mu_A \mathbf{U}$. The natural choice $c = a$ yields the quadratic expression in a^2.

Remark. It may be worthwhile to mention that energy of a chain and energy of its dual (relative to ξ) coincide by definition (see Def. 3, § 11.5.1).

Example 1. M/M/1 *chain.* Take for reference measure the invariant measure

$$\xi(i) = (\lambda/\mu)^i, \qquad i \geqslant 0,$$

assuming $\lambda > \mu$ (transient) (see Ex. 1, § 13.1.1, and Ex. 1, § 1.1.9). It is clear that the chain is self-dual:

$$q_{ij} = \hat{q}_{ij}.$$

Hence, for strictly positive $a > 0$ on the whole state space $\mathbb{I}$, one finds

$$q^*_{ij} = \tilde{q}_{ij}, \qquad u^*_{ij} = \tilde{u}_{ij}, \qquad \text{for all} \quad i, j.$$

Thus, with $c = 1$, it follows that $\mathrm{en}\,(p^*) = \mathrm{en}\,(p)$.

Example 2. *Poisson process.* Let the original process X be Poisson with rate λ. Take $a(i) > 0$ for all $i \geqslant 0$, so the random time changed process on $\mathbb{I}$ is a birth process with intensity

$$\tilde{\lambda}_i = \lambda/a(i), \qquad i \geqslant 0.$$

With strictly positive reference measure ξ, the dual process, relative to the Revuz measure $\mu_A(i) = a(i)\xi(i) > 0$, is the death process with coefficients

$$\mu^*_i = \hat{\mu}/a(i), \qquad i \geqslant 1,$$

where $\hat{\mu}$ is the dual of λ, relative to ξ.

It is now clear that

$$u_{ij} = \begin{cases} 1/\lambda & \text{for} \quad j \geqslant i, \\ 0 & \text{for} \quad j < i, \end{cases}$$

$$\tilde{u}_{ij} = \begin{cases} a(j)/\lambda & \text{for} \quad j \geqslant i, \\ 0 & \text{for} \quad j < i, \end{cases}$$

but

$$\hat{u}_{ij} = \begin{cases} \xi(j)(\xi(i)\lambda)^{-1} & \text{for} \quad j \geqslant j, \\ 0 & \text{for} \quad i < j, \end{cases}$$

$$u^*_{ij} = \begin{cases} \mu_A(j)(\xi(i)\lambda)^{-1} & \text{for} \quad i \geqslant j, \\ 0 & \text{for} \quad i < j. \end{cases}$$

Hence,

$$p(i) = (1/\lambda) \sum_{j=i}^{\infty} a(j), \qquad i \geqslant 0,$$

$$p^*(i) = (\xi(i)\lambda)^{-1} \sum_{j=0}^{i} \mu_A(j)a(j), \qquad i \geqslant 0.$$

Therefore, energies are

$$\mathrm{en}\,(p) = (1/\lambda) \sum_{i=0}^{\infty} a(i)\xi(i) \sum_{j=i}^{\infty} a(j),$$

$$\mathrm{en}\,(p^*) = (1/\lambda) \sum_{j=0}^{\infty} a^2(j)\xi(j) \sum_{i=j}^{\infty} a^2(i).$$

Example 3. *Branching process.* This chain has been already discussed in Example 3, § 1.1.9, and in § 7.3.4. Using notation introduced there, take now

$$H = \{1, 2, \ldots\}, \qquad H^c = \{0\},$$

and assume that $m < 1$ (so extinction occurs certainly). Take next the purely excessive function $p(i) = i$, so the charge is, as required,

$$a(i) = \begin{cases} i\beta(1-m) & \text{for } i \geqslant 1, \\ 0 & \text{for } i = 0. \end{cases}$$

It is immediate that now

$$u^*_{ij} = \mu_A(j) u_{ij} / \xi(i), \qquad i, j \geqslant 1,$$

where $\mu_A(j) = a(j)\xi(j)$ is the Revuz measure.
Selecting $c = 1$, one has

$$p^*(i) = \sum_{j=1}^{\infty} u^*_{ij}, \qquad i \geqslant 1,$$

and therefore, the energy relative to the Revuz measure is

$$\operatorname{en}(p^*) = \beta(1-m) \sum_j \xi(i) j^2.$$

To make it more definite, take now $\xi(j) = u_{\pi j}$, where π is the initial distribution. The branching property implies that

$$\mathbb{E}^i X_t^2 = i\mathbb{E}^1 X_t^2 + i(i-1)(\mathbb{E}^1 X_t)^2.$$

Rather tedious calculations, using the (B)-equation for the generating function of the branching process, show that

$$\sum_j u_{1j} j^2 = \mathbb{E}^1 \int_0^{\infty} X_t^2 \, dt = \frac{\sigma^2 + (1-m)(2-m)}{2\beta(1-m)^2} < \infty,$$

where σ^2 is the variance of progeny distribution. Therefore,

$$\sum_j u_{ij} j^2 = \frac{1}{2\beta(1-m)^2} \left(i^2 + i \, \frac{\sigma^2 + (1-m)^2}{1-m} \right).$$

Consequently, the above expression for energy becomes now

$$\operatorname{en}(p^*) = \beta(1-m) \sum_i \pi_i \sum_j u_{ij} j^2 < \infty.$$

Alternatively, the same value of energy is obtained from Theorem 4, § 11.5.3, according to which

$$\operatorname{en}(X) = \pi \bullet \mathbf{U}(ap).$$

Anticipating slightly, it may be of interest to compare results obtained here with discussion of branching process and the random time change in subsequent §§ 13.3.3 and 13.3.4, respectively.

Indeed, by Definition 3, § 11.5.3, one has also

$$\operatorname{en}(X) = \frac{1}{2}\mathbb{E}\, A_\infty^2,$$

where A is an integral functional determined by charge $a \geqslant 0$. In the present case,

$$A_t = \beta(1-m)B_t, \qquad t \geqslant 0,$$

and B_∞ is finite when $m < 1$, as shown in § 13.3.3.

Furthermore, the intensity matrix $\mathbf{Q}$ and the balayage matrix $\mathbf{B}$ have in the present case partitioned form

$$\mathbf{Q} = \begin{pmatrix} 0 & 0 \\ \mathbf{Q}_{21} & \mathbf{Q}_{22} \end{pmatrix}, \qquad \mathbf{B} = \begin{pmatrix} 0 & 0 \\ 0 & \mathbf{I}_2 \end{pmatrix},$$

because H^c consists of a single absorbing state 0.

Thus, the balayage potential is $\bar{p} = \mathbf{B}p = p$, so Theorem 3, § 13.3.4, holds. Evidently, charges a and $\bar{a}$ coincide, their common value on H being $\mathbf{Q}_{22}p_2$, and Lemma 1, § 13.3.4, shows that $\tilde{a} = 1$ on H. Discussion in § 13.3.5 applies here, too. See also Example 5, § 9.6.7, for comments on the random time change for branching process.

§ 13.2.4. Perturbation. The mechanism of perturbation has been described in § 11.2.1. As noted, the components of the perturbed generator $\mathbf{Q}^*$ which differ from those in the original generator $\mathbf{Q}$, are

$$\mathbf{Q}_{21}^* = \mathbf{0}, \qquad \mathbf{Q}_{22}^* = \mathbf{Q}_{22} + \mathbf{Q}_{21}\mathbf{R}_{12},$$

where $\mathbf{R}_{12}$ is a perturbation matrix.

In general, reference measures ξ and ξ^* for the original and for the perturbed chain are different, making comparison of energies rather awkward. However, the perturbed chain is of interest only on the set H. Thus, one can restrict attention to H, and it is possible under certain conditions to take for ξ^* a measure which vanishes on H^c and equals ξ on H (see Proposition 1). Next, taking the same charge $a \geqslant 0$, consider potentials $p = \mathbf{U}a$ and $p^* = \mathbf{U}^*a$, and note that components of p^* are $p_1^* = p_1$ and $p_2^* = \mathbf{U}_{22}^* a_2$ (assuming that both chains are transient).

Assuming in addition that $a_1 = 0$, one has for energy

$$\operatorname{en}(p) = a_2 \xi_2 \mathbf{U}_{22} a_2, \qquad \operatorname{en}(p^*) = a_2 \xi_2 \mathbf{U}_{22}^* a_2.$$

PROPOSITION 1. *Let $\xi = (\xi_1, \xi_2)$ be a reference measure for the original chain such that $\xi_2 \mathbf{Q}_{22}^* \leqslant 0$. Then the measure $\xi^* = (\mathbf{0}, \xi_2)$ is a reference measure for the perturbed chain on H.*

Proof. Writing $\mathbf{Q}^{*a} = \mathbf{J}_2\mathbf{Q}\mathbf{R}$ for the matrix obtained from $\mathbf{Q}^*$ by letting $\mathbf{Q}_{11} = \mathbf{Q}_{12} = \mathbf{0}$, and noting that $\xi^* = \xi\mathbf{J}_2$, one has

$$\xi(-\mathbf{Q}^{*a}) = \xi^*(-\mathbf{Q})\mathbf{R} = -(\mathbf{0},\, \xi_2\mathbf{Q}^*_{22}) \geqslant 0. \qquad \square$$

Observe that here $\xi^*(-\mathbf{Q}) = \sigma_0$ is a signed charge, yet $\sigma_0\mathbf{R} \geqslant 0$ (see similar situation in Ex. 2, § 13.2.2).

In the special case when the replacement matrix $\mathbf{R}$ equals the balayage matrix $\mathbf{B}$, as in the trace chain (in § 9.5.4), Proposition 1 holds always, as shown in the following

COROLLARY 2. *Condition in Proposition* 1 *is satisfied when* $\mathbf{R} = \mathbf{B}$.

Proof. Observe that $\mathbf{J}_2\mathbf{Q}\mathbf{B} = \mathbf{Q}\mathbf{B}$, so by properties of $\mathbf{Q}\mathbf{B}$ (see § 3.2.5)

$$\xi^*(-\mathbf{Q})\mathbf{B} = \xi(-\mathbf{Q})\mathbf{B} \geqslant 0. \qquad \square$$

Observe also that Theorem 2, § 11.2.1, indicates that $\mathbf{R} = \mathbf{B}$ implies $\mathbf{U}_{22} = \mathbf{U}^*_{22}$, so energies coincide. See also discussion of the random time change in § 11.2.3.

The following observation is rather useful.

PROPOSITION 3. *Consider two reference measures for the original and for the perturbed chain*:

$$\xi = \pi\mathbf{U}, \qquad \xi^* = \pi\mathbf{U}^*,$$

relative to the same initial distribution π.

Then components of reference measures satisfy

$$-(\xi_1 - \xi_1^*)\mathbf{Q}_{11} = \xi_2\mathbf{Q}_{21} > 0,$$
$$(\xi_1 - \xi_1^*)\mathbf{Q}_{12} + (\xi_2 - \xi_2^*)\mathbf{Q}_{22} = \xi_2^*\mathbf{Q}_{21}\mathbf{R}_{12} > 0.$$

Proof. The result follows from comparison of expressions

$$\pi = \xi(-\mathbf{Q}) = \xi^*(-\mathbf{Q}^*). \qquad \square$$

In particular, choice $\xi_1^* = 0$ and $\xi_2^* = \xi_2$ is permitted, in agreement with Proposition 1.

Example 1. *Simple random walk*. The state space is now the set of all integers, and the birth-and-death coefficients are, for $i = 0, \pm 1, \pm 2, \ldots$,

$$q_{i,i+1} = \lambda, \qquad q_{i,i-1} = \mu, \qquad q_i = \lambda + \mu, \qquad \text{and} \qquad q_{ij} = 0 \qquad \text{otherwise.}$$

Assume transient chain with $\lambda > \mu$. Then, as shown in § 7.3.2,

$$u_{ij} = \begin{cases} (\mu/\lambda)^{i-j}(\lambda-\mu)^{-1}, & j \leqslant i, \\ (\lambda-\mu)^{-1}, & j > i. \end{cases}$$

Select now

$$H = \{0, 1, \ldots\} \qquad \text{and} \qquad H^c = \{\ldots, -2, -1\},$$

and define the replacement matrix $\mathbf{R} = (r_{ij})$ by

$$r_{ij} = \begin{cases} 1 & \text{for} \quad i < 0,\ j = 0, \\ 0 & \text{for} \quad i < 0,\ j > 0. \end{cases}$$

From the relation $\mathbf{Q}_{22}^* = \mathbf{Q}_{22} + \mathbf{Q}_{21}\mathbf{R}_{12}$ it is clear that the perturbed chain is the M/M/1 queue on $H = \{0, 1, \ldots\}$, necessarily transient. Observe also that, according to § 7.3.2, $D_i = D_{i0} = 1$ for $i < 0$. It then follows that in this case $\mathbf{R}_{12} = \mathbf{B}_{12}$. Hence, by Theorem 2, § 11.2.1, $u_{ij}^* = u_{ij}$ for $i, j \geqslant 0$.

Consider now the charge supported on H,

$$a(i) = \begin{cases} (1-\gamma)\gamma^i, & i \geqslant 0, \quad 0 < \gamma < 1, \\ 0, & i < 0. \end{cases}$$

Then the potentials

$$p(i) = \sum_{j=0}^{\infty} u_{ij} a(j) \qquad \text{and} \qquad p^*(i) = \sum_{j=0}^{\infty} u_{ij}^* a(j)$$

coincide on H, with common value $(\lambda - \mu)^{-1}$ for all $i \geqslant 0$.

Next, note that

$$\xi(i) = (\lambda/\mu)^i \qquad \text{for} \quad i = 0, \pm 1, \pm 2, \ldots,$$

is an invariant measure for the random walk, and its component ξ_2, restricted to H, is invariant for the M/M/1 queue.

Consequently, energies coincide, and

$$\text{en}\,(p) = \text{en}\,(p^*) = (1-\gamma)(\lambda - \mu)^{-1}(1 - \gamma\lambda/\mu)^{-1},$$

provided $\gamma < \mu/\lambda < 1$.

Example 2. *Birth-and-death process.* Consider a transient birth-and-death process with coefficients λ_i and μ_i, as discussed in Example 1, § 1.1.9, and in Subsection 7.2. Take now, for a fixed state $s \geqslant 1$,

$$H^c = \{0, 1, \ldots, s-1\}, \qquad H = \{s, s+1, \ldots\}.$$

An easy calculation shows that, for $i < s$,

$$D_{ij} = \begin{cases} 0 & \text{for} \quad j > s, \\ {}_H u_{i,s-1} \lambda_{s-1} & \text{for} \quad j = s, \end{cases}$$

so

$$D_i = D_{is} = 1, \qquad i \in H^c.$$

(See Remark in § 7.2.7.)

Take again $\mathbf{R}_{12} = \mathbf{B}_{12}$ (as in the trace chain), so evidently, the perturbed chain on H is the ordinary birth-and-death process with the same coefficients λ_i and μ_i for $i \geqslant s+1$, but with $q_s = \lambda_s$, $q_{s,s+1} = \lambda_s$ for $i = s$.

As before, with a charge a vanishing on H^c, and with the same reference measure ξ, one has $\text{en}\,(p) = \text{en}\,(p^*)$.

Example 3. *Two-point replacement.* Consider a system discussed in § 11.2.3-1 with $\mathbf{Q}_{11} = (-\delta_{ij} q_j)$, corresponding to no transitions in the set H^c ("time change"); other components of $\mathbf{Q}$ being arbitrary. As noted, the balayage matrix is, according to Theorem 4, § 3.1.2, of the form

$$\mathbf{B}_{12} = {}_H\mathbf{U}_{11}\mathbf{U}_{11}\mathbf{Q}_{12}.$$

Obviously,

$${}_H\mathbf{U}_{11} = (\delta_{ij}/q_j),$$

so

$$D_i = 1 \qquad \text{for} \quad i \in H^c.$$

Next, suppose that forbidden transitions from H to H^c are replaced by the forced returns to two fixed states n and m in H. The replacement matrix $\mathbf{R}_{12} = (r_{ij})$ has the form, for $i \in H^c$ and $j \in H$,

$$r_{ij} = \begin{cases} \alpha & \text{for} \quad j = n, \\ \beta & \text{for} \quad j = m, \qquad \text{where} \quad \alpha + \beta = 1. \\ 0 & \text{otherwise,} \end{cases}$$

An easy calculation shows that the matrix $\mathbf{Q}^*_{22} = \mathbf{Q}_{22} + \mathbf{Q}_{21}\mathbf{R}_{12}$ has components, for i and j in H,

$$q^*_{ij} = \begin{cases} q_{ij} & \text{for} \quad j \neq n, m, \\ q_{in} + (q_i - \bar{q}_i)\alpha & \text{for} \quad j = n, \\ q_{im} + (q_i - \bar{q}_i)\beta & \text{for} \quad j = m. \end{cases}$$

Take now a charge a which vanishes on H^c. The corresponding potentials $p = \mathbf{U}a$ and $p^* = \mathbf{U}^* a$ have components on H given by

$$p_2 = \mathbf{U}_{22}\, a_2, \qquad p^*_2 = \mathbf{U}^*_{22}\, a_2,$$

where relation between $\mathbf{U}_{22}$ and $\mathbf{U}^*_{22}$ is given in Theorem 2, § 11.2.1.

With reference measures $\xi = (\xi_1, \xi_2)$ and $\xi^* = (\mathbf{0}, \xi_2)$ for the original and for the perturbed chain, respectively, one finds energies

$$\text{en}\,(p) = a_2 \xi_2 p_2, \qquad \text{en}\,(p^*) = a_2 \xi_2 p^*_2.$$

§ 13.2.5. Optimisation. Using reference measure $\xi(i) = u_{\pi i}$, energy $\text{en}\,(a) = \hat{a}p$ can be written as a homogeneous (non-symmetric) quadratic form, in infinitely many variables $a(i)$,

$$\text{en}\,(a) = \sum_i \sum_j a(i) e_{ij} a(j),$$

where

$$e_{ij} = u_{\pi i} u_{ij}, \qquad i, j \in \mathbb{I},$$

and $e_{ij} \neq e_{ji}$.

Optimisation of such a form under constraints is a rather delicate matter. Here some brief informal comments will be made in the case when the state space $\mathbb{I}$ is the set of all non-negative integers, and when there is only one linear constraint

$$\sum_{i=0}^{\infty} \alpha_i a(i) = m, \qquad \alpha_i \geqslant 0,$$

where α_i and m are known.

Write a for the column vector with components $a(i)$, and denote by a' its transpose (the row vector). Similarly, write α and α' for vectors with components α_i. Let $\mathbf{E} = (e_{ij})$ and $\mathbf{E}' = (e_{ji})$ be a matrix of coefficients and its transpose, respectively.

Then, in matrix notation, the quadratic form becomes

$$\text{en}\,(a) = a'\mathbf{E}a,$$

and the constraint equation

$$m = a'\alpha = \alpha' a.$$

THEOREM 1. *Extremum value of energy* en (a) *occurs at* $a(i)$ *such that*

$$(1/\alpha_i) \sum_{k=0}^{\infty} (e_{ik} + e_{ki}) a(k) = 2\text{en}\,(a)/m = \text{const}$$

for each $i = 0, 1, \ldots$.

Proof. For optimisation, consider the expression F defined by

$$F(a) = \text{en}\,(a) - \Lambda(a'\alpha - m),$$

where Λ is a Lagrange multiplier. Then,

$$\frac{\partial F}{\partial a(i)} = \frac{\partial \text{en}(a)}{\partial a(i)} - \Lambda \alpha_i = 0, \qquad i = 0, 1, \ldots.$$

This system of equations and the single constraint equation determine all $a(i)$ and Λ.

It is easy to see that

$$\frac{\partial \text{en}\,(a)}{\partial a(i)} = \sum_{k=0}^{\infty} (e_{ik} + e_{ki}) a(k), \qquad i = 0, 1, \ldots.$$

Introducing a symmetric matrix $\mathbf{S} = (s_{ij})$, where $s_{ij} = e_{ij} + e_{ji}$, one can consider a column vector $\mathbf{S}a$ with components

$$\frac{\partial \text{en}\,(a)}{\partial a(i)} = \mathbf{S}a(i), \qquad i = 0, 1, \ldots.$$

Consequently, condition for extremum is now

$$\mathbf{S}a = \Lambda\alpha.$$

Assuming existence of the inverse $\mathbf{S}^{-1}$, the formal solution is

$$a = \Lambda\mathbf{S}^{-1}\alpha.$$

Note that $\mathbf{S}^{-1}$ is also a symmetric matrix.

To determine Λ, the constraint equation yields

$$\Lambda m = a'\Lambda\alpha = a'\mathbf{S}a = a'(\mathbf{E} + \mathbf{E}')a = 2\mathrm{en}\,(a),$$

using the fact that $a'\mathbf{E}a = a'\mathbf{E}'a$. Indeed, in terms of components,

$$\sum_{i=0}^{\infty} a(i)\,\frac{\partial \mathrm{en}\,(a)}{\partial a(i)} = 2\mathrm{en}\,(a).$$

Observe also that in the same fashion

$$m = \alpha' a = \alpha'\Lambda\mathbf{S}^{-1}\alpha,$$

so

$$\Lambda = m/\alpha'\mathbf{S}^{-1}\alpha. \qquad \square$$

Theorem 1 provides the necessary condition for existence of an extremum. In order to determine whether there is a maximum or a minimum, observe that the second partial derivatives are

$$\frac{\partial^2 \mathrm{en}\,(a)}{\partial a(i)\partial a(j)} = s_{ij} \geqslant 0 \qquad \text{for all} \quad i, j.$$

The minimum will occur when all Hessian determinants obtained from the matrix $\mathbf{S}$:

$$\mathbf{H}_n = \begin{vmatrix} s_{00} & s_{01} & \cdots & s_{0n} \\ s_{10} & s_{11} & \cdots & s_{1n} \\ \cdots & \cdots & \cdots & \cdots \\ s_{n0} & s_{n1} & \cdots & s_{nn} \end{vmatrix}, \qquad n = 0, 1, \ldots,$$

are strictly positive for all n. Equivalently, when the (symmetric) quadratic form $x'\mathbf{S}x$ is positive definite (i.e., strictly positive for all positive vectors x). This is the case in examples discussed here.

Remark. Observe that, taking $\alpha_i = u_{\pi i}$, one has here a calculus of variation problem of finding a charge $a \geqslant 0$ which minimises energy

$$\mathrm{en}\,(a) = \frac{1}{2}\mathbb{E}\,A_{\infty}^2,$$

subject to constraint equation

$$\mathbb{E}\,A_{\infty} = \pi\mathbf{U}a = m,$$

where (A_t) is an integral functional determined by a charge $a \geqslant 0$.

Example 1. *Birth process.* Take $\pi_0 = 1$, so $u_{\pi i} = u_{0i}$. Recalling that $u_{ij} = 1/\lambda_j$ for $j \geqslant i$, and $u_{ij} = 0$ for $j < i$, one finds

$$e_{ij} = \begin{cases} (\lambda_i\lambda_j)^{-1} & \text{for} \quad j \geqslant i, \\ 0 & \text{for} \quad j < i. \end{cases}$$

Thus,

$$s_{ij} = \begin{cases} (\lambda_i\lambda_j)^{-1} & \text{for} \quad j \neq i, \\ 2/\lambda_i^2 & \text{for} \quad j = i. \end{cases}$$

The energy is now

$$\text{en}\,(a) = \sum_{i=0}^{\infty} \frac{a(i)}{\lambda_i} \sum_{j=i}^{\infty} \frac{a(j)}{\lambda_j}.$$

Hence,

$$\frac{\partial \text{en}\,(a)}{\partial a(i)} = (1/\lambda_i)\big(A + a(i)/\lambda_i\big) = \Lambda\alpha_i,$$

$$2\text{en}\,(a) = \sum_{i=0}^{\infty} a(i)\frac{\partial \text{en}\,(a)}{\partial a(i)} = A^2 + S^2 = \Lambda m,$$

with

$$A = \sum_{k=0}^{\infty} a(k)/\lambda_k, \qquad S^2 = \sum_{k=0}^{\infty} \big(a(k)/\lambda_k\big)^2,$$

assumed finite.

Note that here

$$\frac{\partial^2 \text{en}\,(a)}{\partial a^2(i)} = 2e_{ii} > 0,$$

$$\frac{\partial^2 \text{en}\,(a)}{\partial a(i)\partial a(j)} = e_{ij} + e_{ji} > 0, \qquad i \neq j.$$

It is of interest to consider the inverse procedure. Select an arbitrary probability distribution say $a(i)$, on non-negative integers. Then, calculate A, S^2, and en (a), then find $\Lambda = 2\text{en}\,(a)/m$, and finally determine α_i from

$$\alpha_i = m\big(A + a(i)/\lambda_i\big)\big(2\text{en}\,(a)\lambda_i\big)^{-1}.$$

Example 2. *Poisson process.* Specializing the previous example to $\lambda_i = \lambda$ for all i, one finds

$$\Lambda\alpha_i = \big(a + a(i)\big)\lambda^2, \qquad 2\text{en}\,(a) = (a^2 + s^2)/\lambda^2 = \Lambda m,$$

where

$$a = \sum_i a(i), \qquad s^2 = \sum_i a^2(i).$$

Selecting now $a(i) = (1-\rho)\rho^i$ with $0 < \rho < 1$, one has

$$A = 1/\lambda, \qquad S^2 = (1-\rho)(1+\rho)^{-1}\lambda^{-2},$$

$$\text{en}\,(a) = (1+\rho)^{-1}\lambda^{-2},$$

and so

$$A^2 + S^2 = 2\text{en}\,(a).$$

Hence, there will be minimum if all α_i are of the form

$$\alpha_i = (1+\rho)m\big(1+a(i)\big)/2, \qquad i = 0, 1, \ldots.$$

13.3. Q-MARTINGALES

Usefulness of martingales is clearly demonstrated by increasing number of applications to diverse fields. Here several simple examples are presented, illustrating the role of passage times. An interesting connection with the random time change is also mentioned. For further applications see books by P. Bremaud [1981], P. Hall and C. C. Heyde [1980], and S. Karlin and M. H. Taylor [1975]; see also S. N. Ethier and T. G. Kurtz [1986] for discussion of a martingale problem.

§ 13.3.1. Martingale problem. Consider a $\mathbf{Q}$-martingale for a fixed function f on the state space $\mathbb{I}$, as introduced already in § 8.1.4-c. For purpose of examples, assume that f is a bounded excessive function. Hence, the charge

$$a = (-\mathbf{Q})f = (-\mathbf{Q})p \geqslant 0,$$

where $f = p + h$ is the Riesz decomposition of f. The $\mathbf{Q}$-martingale $M = (M_t)$ relative to the natural σ-fields $(\mathcal{F}_t)$ of the usual M.C. (X_t) takes the form

$$M_t = f(X_t) + \int_0^t a(X_s)\,ds, \qquad t \geqslant 0,$$

with finite $\mathbb{E}M_t = \pi f$. This is in fact the Doob–Meyer decomposition of a supermartingale (S_t) in terms of a martingale (M_t) and an increasing process (A_t), where

$$S_t = f(X_t) \geqslant 0, \qquad A_t = \int_0^t a(X_s)\,ds \geqslant 0.$$

Recall that the martingale property follows from the extended Kolmogorov equation

$$\mathbb{E}\big(f(X_{t+h}) \mid \mathcal{F}_t\big) - f(X_t) = -\int_0^h \mathbf{P}(s)a(X_t)\,ds \quad \text{a.e.}$$

(see § 8.1.1), from which the intensity is

$$\lim_{h\to 0} \mathbb{E}\big(f(X_{t+h}) - f(X_t) \mid \mathcal{F}_t\big)/h = -a(X_t) \quad \text{a.e.}$$

Observe that, for each $t \geqslant 0$,

$$\mathbb{E}^i M_t = \mathbf{P}(t)f(i) + \int_0^t \mathbf{P}(s)a(i)\,ds = f(i), \qquad i \in \mathbb{I}.$$

Hence, as $t \to \infty$,

$$\mathbb{E}^i S_t = \mathbf{P}(t)f(i) \longrightarrow h(i), \qquad \text{invariant,}$$

$$\mathbb{E}^i A_t = \int_0^t \mathbf{P}(s)a(i)\,ds \longrightarrow \mathbf{U}a(i), \qquad \text{purely excessive,}$$

so $p = \mathbf{U}a$ is also a potential of a charge.

As a positive supermartingale always converges, the limit $S_t \to S_\infty$, as $t \to \infty$, exists a.e. and is integrable. Boundedness of f implies that (S_t) is uniformly integrable, so

$$\mathbb{E}S_t \longrightarrow \mathbb{E}S_\infty,$$

and

$$\mathbb{E}(S_\infty \mid \mathcal{F}_t) = h(X_t)$$

form an uniformly integrable martingale, and the potential generated by an integrable increasing process (A_t) is now

$$U_t = \mathbb{E}(A_\infty - A_t \mid \mathcal{F}_t) = \mathbf{U}a(X_t) = p(X_t).$$

Observe that now

$$M_t = \mathbb{E}(M_\infty \mid \mathcal{F}_t),$$

so (M_t) is also an uniformly integrable martingale.

Hence, summarizing, one has the Riesz decomposition

$$S_t = h(X_t) + p(X_t) = M_t - A_t.$$

Remark. Definition of a **Q**-martingale M can be extended to excessive functions f such that $\pi f < \infty$.

§ 13.3.2. Passage times. As the first application, consider probability D_i of the first entrance to the set H, conditioned on the initial state i. Recall that D_i is an excessive function on $\mathbb{I}$, so using the **Q**-martingale with $f = D$ and $a = (-\mathbf{Q})D = b$, one obtains the uniformly integrable martingale (M_t), relative to the natural σ-fields $(\mathcal{F}_t)$, which converges a.e. to an integrable M_∞,

$$M_t = D(X_t) + \int_0^t b(X_s)\,ds = \mathbb{E}(M_\infty \mid \mathcal{F}_t), \qquad t \geqslant 0.$$

Moreover, for $i \in \mathbb{I}$,

$$\mathbb{E}^i M_t = \mathbb{E}^i D(X_t) + \mathbb{E}^i \int_0^t b(X_s)\, ds = D_i \qquad \text{for each} \quad t \geqslant 0,$$

and, as $t \to \infty$,

$$\mathbb{E}^i D_i(X_t) = \mathbf{P}(t) D_i \longrightarrow s_i,$$

$$\mathbb{E}^i \int_0^t b(X_s)\, ds = \int_0^t \mathbf{P}(s) b_i\, ds = \Pi_i(t) \longrightarrow \Pi_i = \mathbf{U} b_i.$$

Noting that $(-\mathbf{Q})D = b$ is actually the Poisson equation, the above argument indicates that its unique solution leads to the Riesz decomposition

$$D_i = \mathbf{U} b_i + s_i, \qquad i \in \mathbb{I},$$

as already indicated earlier.

Note also that $D(X_t)$ form a positive uniformly integrable supermartingale, so

$$\lim_{t \to \infty} \mathbb{E}^i D(X_t) = \mathbb{E}^i \big(\lim_{t \to \infty} D(X_t) \big) = s_i = \mathbb{P}^i (L = \infty).$$

Moreover,

$$\mathbb{E} M_t = \pi \bullet D = C(H)$$

is a capacity of the set H.

Example 1. M/M/1 *queue*. Suppose that X_t represents the number of customers at time t in the queueing system M/M/1. Take $H = \{0\}$, so $T = T_H$ represents the busy period. Assuming transient chain with $\lambda/\mu > 1$, one has

$$D_i = (\mu/\lambda)^i = u_{i0} b_0 \qquad \text{with} \quad b_0 = \lambda - \mu > 0, \ s_i = 0,$$

as shown in Example 1, § 7.2.4. Thus,

$$M_t = (\mu/\lambda)^{X_t} + (\lambda - \mu) \int_0^t I_{\{0\}}(X_s)\, ds = D(X_s) + A_t$$

form an uniformly integrable martingale. Note that the integral in the above expression represents time spent in state 0 up to instant t.

Hence, as $t \to \infty$,

$$\mathbb{E}^i D(X_t) = \sum_j p_{ij}(t) (\mu/\lambda)^i = \mathbb{P}^i (L > t) \longrightarrow s_i = 0,$$

$$\mathbb{E}^i A_t = (\lambda - \mu) \int_0^t p_{i0}(s)\, ds \longrightarrow (\lambda - \mu) u_{i0} = D_i,$$

$$\mathbb{E}^i M_t = D_i, \qquad \mathbb{E} M_t = \sum_i \pi_i (\mu/\lambda)^i.$$

Example 2. *Occupation time.* Consider the time spent in a set H up to the instant t, as discussed in § 8.4.1. Thus, with $a = I_H$,

$$A_t = \int_0^t I_H(X_s)\, ds, \qquad t \geqslant 0.$$

Evidently, the potential $p = \mathbf{U}a$ is infinite when the chain is recurrent. Hence, assuming transient chain and also $\pi \bullet p < \infty$, one has integrable A_∞, and

$$p(i) = \sum_j u_{ij} < \infty, \qquad i \in \mathbb{I}$$

(in particular, for finite H).

As p is purely excessive, $p(X_t)$ form a potential supermartingale with $p(X_t) \to 0$ a.e., as $t \to \infty$. Hence, the $\mathbf{Q}$-martingale is now

$$M_t = p(X_t) + A_t, \qquad t \geqslant 0,$$

with $M_\infty = A_\infty$ a.e., so $M_t = \mathbb{E}(M_\infty \mid \mathcal{F}_t)$ and $\mathbb{E}M_t = \pi \bullet p$ for each t.

Furthermore, for $i \in \mathbb{I}$,

$$\mathbb{E}^i M_t = p(i) \qquad \text{for each} \quad t,$$

and, as $t \to \infty$,

$$\mathbb{E}^i p(X_t) \longrightarrow 0, \qquad \mathbb{E}^i A_t \longrightarrow \mathbb{E}^i A_\infty = p(i).$$

Observe that in this example $f = p$ and $h = 0$.

Example 3. *Waiting time.* An interesting application of passage times is provided by a waiting time in Markovian queues. As the basic ideas are deeply rooted in Queueing Theory, here only the classical system M/M/s is mentioned to illustrate the method. For complete discussion see Chapter 11 in the book by R. Syski [1986] where examples and references are given.

A waiting time of a customer is a time interval between the instant of arrival of this customer and the instant at which his service starts. Evidently, the waiting time is influenced by the queue discipline, by service terminations and by arrivals occurring during waiting (who complete for free servers).

Let X_t represent the number of customers in the system at time t after the arrival of a *test customer*, who may influence the waiting of this test customer (the test customer being excluded). The chain $X = (X_t,\ 0 \leqslant t < \infty)$, called the waiting process, follows (as a post-process) the queueing process (representing the total number of customers), but with omission of transitions irrelevant to the test customer. The waiting process starts at the random instant of arrival and stops when the test customer terminates his waiting. Extraction of the waiting process from a queueing process represents another transformation of Markov chains; see R. Syski [1965].

In the M/M/s system, the test customer arriving when the number of busy servers is less than s, does not wait. The waiting occurs only if all the servers are busy when the test customer arrives. In this formulation the waiting time of the test customer corresponds to the first entrance time T to the absorbing taboo set $H = \{0, 1, \ldots, s-1\}$ for the waiting process (X_t). This observation provides the basic link between passage times and waiting times in Markovian queues, and goes back historically to early days of Queueing Theory; see R. Syski [1965].

Denoting by $W(t)$ the distribution of the waiting time, one has (assuming the waiting process in equilibrium) that

$$W^c(t) = \sum_{i=s}^{\infty} \pi_i D_i^c(t), \qquad t \geqslant 0,$$

where π_i is the initial distribution of the waiting process (the probability that the test customer will meet on his arrival i customers who may influence his waiting). Clearly, (π_i) depends on the queue discipline and on the original queueing process. Note also that the distribution $W(t)$ has a jump at $t = 0$, corresponding to probability of not waiting.

It should be pointed out that in the M/M/s system with the strict order service, the waiting process is the death process starting at $i \geqslant s$, with coefficients $s\mu$ (see Ex. 2, § 7.2.6), so the density $d_i(t)$ has gamma form. On the other hand, the queueing process is the birth-and-death process whose distribution for $i \geqslant s$ has geometric form, and coincides with π_i. Thus,

$$W^c(t) = W^c(0)e^{-\mu(1-\rho)t}, \qquad t \geqslant 0,$$

where $W^c(0)$ is given by the Erlang delay formula. For details see R. Syski [1986], and P. Bremaud [1981] for martingale interpretation.

§ 13.3.3. Branching process. Consider a branching process already discussed in Example 3, § 1.1.9, and in § 7.3.4, and assume that mean of progeny distribution is $m < 1$. Hence, by the branching theorem, probability of extinction, conditional on initial state i, is $D_i = 1$, identically for all $i \in \mathbb{I}$.

Take now the unbounded function $f(i) = i$, and note that

$$\mathbf{P}(t)f(i) = \mathbb{E}^i X_t \longrightarrow 0, \qquad \text{as} \quad t \to \infty,$$

so f is purely excessive, and

$$a(i) = (-\mathbf{Q})f(i) = i\beta(1-m) > 0 \qquad \text{for} \quad i \geqslant 1.$$

Then, under condition $\pi \bullet f < \infty$,

$$M_t = X_t + \beta(1-m)B_t, \qquad t \geqslant 0,$$

where

$$B_t = \int_0^t X_s \, ds$$

form a positive martingale, converging a.e. to an integrable M_∞. Note that (X_t) is a potential supermartingale with $\mathbb{E}X_t \to 0$, so $X_t \to 0$ a.e., as $t \to \infty$.

Moreover,

$$\mathbb{E}^i M_t = i,$$

$$\mathbb{E}M_t = \sum_i i\pi_i \qquad \text{for all} \quad t,$$

$$\beta(1-m)\mathbb{E}^i B_t = i(1 - e^{-\beta(1-m)t}) \longrightarrow i \qquad \text{as} \quad t \to \infty.$$

It is of great interest to see that the distribution of the r.v. B_∞ can be obtained explicitly, as shown in the following theorem which treats the general case.

THEOREM 1. *The random variable*

$$B_\infty = \int_0^\infty X_s \, ds$$

has the same conditional distribution as the busy period (started at i) in the M/M/1 queue with group input (with proper identification of parameters).

Proof. i) Let $\widetilde{X} = (\widetilde{X}_t,\ 0 \leqslant t < B_\infty)$ be chain obtained from (X_t) by the random time change determined by the integral functional (B_t) with charge $\tilde{a}(i) = i$ (see Sub-section 9.6). Note that the state space of $\widetilde{X}$ is $H = \operatorname{supp} B = \{1, 2, \ldots\}$ and that $\widetilde{X}$ has the life time B_∞. It is known that the transition matrix $\widetilde{\mathbf{P}}(t) = (\tilde{p}_{ij}(t))$ is sub-stochastic, and

$$\mathbb{P}^i(B_\infty > t) = \sum_{j=1}^\infty \tilde{p}_{ij}(t) =: \tilde{p}_i(t), \qquad t \geqslant 0.$$

It follows from § 9.6.5 that the generator $\widetilde{\mathbf{Q}} = (\tilde{q}_{ij})$ for the chain $\widetilde{X}$ has the form

$$\tilde{q}_{ij} = q_{ij}/i \qquad \text{for} \quad i, j \geqslant 1,$$

because 0 is an absorbing state, and $\tilde{a}(i) = i$ (see also Ex. 5, § 9.6.7). Thus, explicitly,

$$\tilde{q}_{ij} = \begin{cases} \beta b(j-i+1), & j \geqslant i-1, \quad i \geqslant 2, \\ \beta b(j), & j \geqslant 1, \qquad i = 1, \\ 0, & j < i-1, \quad i \geqslant 3, \end{cases}$$

$$\tilde{q}_i = \beta, \qquad i \geqslant 1,$$

with

$$\sum_{j=1}^\infty \tilde{q}_{ij} = \begin{cases} -\beta b(0), & i = 1, \\ 0, & i > 1. \end{cases}$$

ii) Let $Y = (Y_t,\ 0 \leqslant t < \infty)$ be a M/M/1 queue with the group input, where Y_t denotes a number of customers in the system at time t. The generator (intensity matrix) $\mathbf{Q} = (q_{ij})$ is defined by

$$\begin{aligned} q_{i,i+k} &= \lambda p(k), \qquad i = 0, 1, \ldots, \qquad k = 1, 2, \ldots, \\ q_{i,i-1} &= \mu, \qquad i = 1, 2, \ldots, \\ q_i &= \lambda + \mu \qquad \text{for} \quad i \geqslant 1, \qquad q_0 = \lambda, \end{aligned}$$

where p is a given distribution of the group size,

$$\sum_{i=1}^{\infty} p(i) = 1.$$

The busy period for the chain Y is the first entrance time T to the state 0, with complementary distribution $D_i^c(t)$ for $t \geqslant 0,\ i \geqslant 1$.

iii) The (B)-equation for $\tilde{p}_i(t)$ has the form

$$\frac{d\tilde{p}_i(t)}{dt} = \begin{cases} -\beta\tilde{p}_1(t) + \beta \sum\limits_{k=2}^{\infty} b(k)\, \tilde{p}_k(t), & i = 1, \\ \beta b(0)\, \tilde{p}_{i-1}(t) - \beta\tilde{p}_i(t) + \beta \sum\limits_{k=i+1}^{\infty} b(k-i+1)\, \tilde{p}_k(t), & i \geqslant 2. \end{cases}$$

Similarly, writing the (B)-equation for $D_i^c(t)$ (for the Y-chain)

$$\frac{dD_i^c(t)}{dt} = \begin{cases} -(\lambda+\mu) D_1^c(t) + \lambda \sum\limits_{k=2}^{\infty} p(k-1) D_k^c(t), & i = 1, \\ \mu D_{i-1}^c(t) - (\lambda+\mu) D_i^c(t) + \lambda \sum\limits_{k=i+1}^{\infty} p(k-i) D_k^c(t), & i \geqslant 2. \end{cases}$$

With the identification

$$\begin{aligned} \beta &= \lambda + \mu, \\ \lambda p(i-1) &= \beta b(i), \qquad i = 2, 3, \ldots, \\ \mu &= \beta b(0), \qquad i = 0, \end{aligned}$$

the above two equations are the same. Hence $D_i^c(t) = \tilde{p}_i(t)$. □

From the above theorem one has

$$\mathbb{P}^i(B_\infty < \infty) = D_i, \qquad i = 1, 2, \ldots,$$

where D_i refers to the Y-chain. Furthermore, it is easy to see that identification stated in the theorem yields the following relation between mean m and the mean w of the group size

$$\beta m = \lambda(w+1).$$

Writing $\varrho = \lambda w/\mu$, it follows from the Y-chain properties that $D_i = 1$ iff $\varrho \leqslant 1$. Thus, $m < 1$ implies that B_∞ is finite, and it can be verified that

$$\mathbb{E}^i B_\infty = \mathbb{E}^i T < \infty.$$

This is, of course, in agreement with the fact that B_∞ is integrable, because M_∞ is.

Finally, observe that $B_t \leqslant B_\infty$ implies that (B_t) is uniformly integrable. Also, the potential supermartingale (X_t) is always uniformly integrable. Hence, (M_t) is uniformly integrable, so

$$M_t = \mathbb{E}(M_\infty \mid \mathcal{F}_t) \quad \text{a.e.,} \qquad t \geqslant 0.$$

Example 1. Consider a linear birth-and-death process, discussed in Example 1, § 7.3.4, so

$$b(0) = \mu/\beta, \qquad b(2) = \lambda/\beta \qquad \text{with} \quad \beta = \lambda + \mu,$$

and $p(1) = 1$. Thus, B_∞ has the distribution of the busy period in the simple M/M/1 queue. Hence, for $m < 1$ one has

$$\mathbb{E}^i B_\infty = i(\mu - \lambda)^{-1} \qquad \text{for} \quad \lambda < \mu,$$

because $\beta(1 - m) = \mu - \lambda$.

§ 13.3.4. Random time change. Of special interest is the application of a random time change to **Q**-martingales. As before, consider a CAF

$$A_t = \int_0^t a(X_s)\, ds, \qquad t \geqslant 0,$$

where $a = (-\mathbf{Q})f \geqslant 0$ for a bounded excessive function f on $\mathbb{I}$, and take for its support

$$H = \{i\colon\ a(i) > 0\}.$$

Next, consider a time changed M.C. $\widetilde{X} = (\widetilde{X}_t, \widetilde{\mathcal{F}}_t,\ 0 \leqslant t < A_\infty)$ on H, where $\widetilde{X}_t = X(\tau_t)$ and $\tau_t = \inf\{s\colon\ A_s > t\}$, as discussed in Sub-section 9.6. Write $\widetilde{\mathbf{P}}(t)$ for the new transition matrix. As shown in § 9.6.5, the new generator $\widetilde{\mathbf{Q}} = (\tilde{q}_{ij})$ restricted to H, is given by

$$\tilde{q}_{ij} = \bar{q}_{ij}/a(i) \qquad \text{for} \quad i, j \in H.$$

Here $\bar{q}_{ij}$ are components of $\overline{\mathbf{Q}} = \mathbf{QB}$, restricted to H, where $\mathbf{B} = (D_{ij})$ is the balayage.

Define now on H

$$\tilde{a} = (-\widetilde{\mathbf{Q}})f, \qquad \bar{a} = (-\overline{\mathbf{Q}})f,$$

for the same function f.

LEMMA 1. *Let $f \geqslant 0$ be $\mathbf{P}(t)$-excessive on $\mathbb{I}$. Then f is also $\widetilde{\mathbf{P}}(t)$-excessive on H, and*

$$\tilde{a}(i) = \bar{a}(i)/a(i) \qquad \textit{for} \quad i \in H.$$

Proof. Recall that the balayage of an excessive function f is $\bar{f} = \mathbf{B}f$, and is also excessive. Hence,

$$\tilde{a}(i) = (-\overline{\mathbf{Q}})f(i)/a(i) = (-\mathbf{Q})\mathbf{B}f(i)/a(i) = (-\mathbf{Q})\bar{f}(i)/a(i) = \bar{a}(i)/a(i) \geqslant 0. \quad \square$$

Consider now an increasing process $(\tilde{A}_t) = \tilde{A}$ with

$$\tilde{A}_t = \int_0^t \tilde{a}(\widetilde{X}_s)\,ds, \qquad t \geqslant 0,$$

and note that its potential is given by

$$\mathbb{E}^i \tilde{A}_\infty = \widetilde{\mathbf{U}}\tilde{a}(i) = \mathbf{U}\bar{a}(i).$$

Forming now a $\widetilde{\mathbf{Q}}$-martingale, as in § 13.3.1, corresponding to the transformed chain $(\widetilde{X}_t)$, one has immediately

PROPOSITION 2. *The process* $\widetilde{M} = (\widetilde{M}_t, \widetilde{\mathfrak{F}}_t)$, *where*

$$\widetilde{M}_t = f(\widetilde{X}_t) + \tilde{A}_t, \qquad 0 \leqslant t < A_\infty,$$

is an uniformly integrable martingale with $\mathbb{E}\widetilde{M}_t = \pi f$.

Consider now another approach. Reversing the order, form first the $\mathbf{Q}$-martingale determined by the charge $a \geqslant 0$, as described in § 13.3.1,

$$M_t = f(X_t) + A_t, \qquad t \geqslant 0.$$

Then the random time change, determined by the same charge a, transforms it to $M_t^* = M(\tau_t)$, so

$$M_t^* = f(\widetilde{X}_t) + t, \qquad 0 \leqslant t < A_\infty,$$

because $A(\tau_t) = t$.

It may be of interest to compare $(\widetilde{M}_t)$ with (M_t^*). The following table illustrates the problem.

$$X_t \begin{cases} \xrightarrow{\tau} & \widetilde{X}_t \xrightarrow{\tilde{Q}} \widetilde{M}_t \\ \xrightarrow{Q} & M_t \xrightarrow{\tau} M_t^*. \end{cases}$$

THEOREM 3. $(M_t^*, \widetilde{\mathfrak{F}}_t)$ *need not be a martingale, unless* $\mathbf{B}f = f$.

Proof. Using the extended Kolmogorov equation (as in § 13.3.1) for the transformed chain $(\widetilde{X}_t)$, one has, assuming that $\widetilde{\mathbf{P}}(s)$ is a stochastic matrix,

$$\mathbb{E}\big(f(\widetilde{X}_{t+h}) \mid \widetilde{\mathfrak{F}}_t\big) - f(\widetilde{X}_t) = -\int_0^h \widetilde{\mathbf{P}}(s)\tilde{a}(\widetilde{X}_t)\,ds.$$

Hence, in view of definition of M_t^*,

$$\mathbb{E}(M_{t+h}^* \mid \widetilde{\mathfrak{F}}_t) = M_t^* + h - \int_0^h \widetilde{\mathbf{P}}(s)\tilde{a}(\widetilde{X}_t)\,ds.$$

However, when $\bar{f} = \mathbf{B}f = f$, then by Lemma 1, $\bar{a} = a$, so $\tilde{a} = 1$ on H, and therefore the equalities $M_t^* = \widetilde{M}_t = t$ hold. This is, however, a special case, as in general $\bar{f} \leqslant f$. □

Observe that M_t^* and $\widetilde{M}_t$ will coincide when $a(i) > 0$ for all $i \in \mathbb{I}$, so A_t is strictly increasing on the whole state space $\mathbb{I}$. Indeed, then $\bar{a} = a$, so again $\tilde{a} = 1$.

Also, if $f = \mathbf{B}g$, where g is $\mathbf{P}(t)$-excessive function, then $\mathbf{B}f = \mathbf{B}g = f$. Hence, by Theorem 3, $M_t^* = \widetilde{M}_t$. This is the case when $f = \mathbf{B1} = D$.

Finally, when $\widetilde{\mathbf{P}}(s)$ is only sub-stochastic, one should use extension $\widetilde{\mathbf{P}}^{\Delta}(s)$ to the state space $H \cup \{\Delta\}$ with a suitable definition of $\tilde{a}(\Delta)$, as in § 9.6.2.

§ 13.3.5. Application. This example extends Theorem 1, § 13.3.3, using the random time change. Let $(\widetilde{X}_t)$ be a process obtained from the usual M.C. (X_t) with the help of the random time change determined by the integral functional (A_t) with the charge $a \geqslant 0$, and $\operatorname{supp}(A) = H$.

Consider now an auxiliary chain (X_t^*) on the same state space $\mathbb{I}$, with a stochastic matrix $\mathbf{P}^*(t)$ and a generator $\mathbf{Q}^*$. Assume that in the partitioned form the component for transitions on H is the same as the corresponding component in the generator $\widetilde{\mathbf{Q}}$, i.e.,

$$\mathbf{Q}_{22}^* = \widetilde{\mathbf{Q}}_{22},$$

the remaining components of $\mathbf{Q}^*$ being arbitrary (with $\mathbf{Q}^*\mathbf{1} = 0$).

Let S be the first entrance time to the complementary set H^c for the chain (X_t^*). Then the following theorem holds.

THEOREM 1. *The life time A_∞ of the time changed chain $(\widetilde{X}_t)$ and the first entrance time S to the set H^c for the secondary chain (X_t^*) have the same distribution*

$$\mathbb{P}^i(A_\infty \leqslant t) = \mathbb{P}^i(S \leqslant t), \qquad i \in H, \quad t \geqslant 0.$$

Proof. Identify the auxiliary state Δ for $(\widetilde{X}_t)$ with H^c for (X_t^*). By the cut-off method, the corresponding taboo matrix has all components zero, except the component on H, which is clearly $\widetilde{\mathbf{Q}}_{22}$.

Writing the (B)-equation for each distribution, one can see that these equations have the same coefficients and the same form ($i \in H$). □

13.4. ESCAPE CHAIN

This sub-section describes properties of a modified chain obtained from a transient chain by a perturbation determined by the structure of the original chain itself.

§ 13.4.1. Construction. Let $X = (X_t)$ be the usual M.C. with a transition matrix $\mathbf{P}(t)$, an intensity matrix $\mathbf{Q}$ and a resolvent matrix $\mathbf{U}^\alpha$. Assume that X is transient, so the potential $U < \infty$. Let H be a fixed set in the state space $\mathbb{I}$, and let T and L denote, respectively, the first entrance and the last exit times for H. Let $\mathbf{\Pi} = (\Pi_{ij})$, where

$$\Pi_{ij} = \mathbb{P}^i(X(L-) = j,\ 0 < L < \infty), \qquad i \in \mathbb{I}, \quad j \in H,$$

with $\Pi_{ij} = 0$ for $j \in H^c$. Recall the Poisson representation of the first entrance probability D:

$$D_i = \mathbf{U}b_i + s_i, \qquad i \in \mathbb{I},$$

with $D_i = 1$ for $i \in H$, where

$$\mathbf{U}b_i = \Pi_i = \mathbb{P}^i(0 < L < \infty), \qquad s_i = \mathbb{P}^i(L = \infty),$$

and

$$b_i = \mathbf{Q}D_i^c = q_i D_i^{\rho c} \geqslant 0$$

is the escape rate supported on H.

Assume now that H is an equilibrium set, so $s = 0$ identically on $\mathbb{I}$. Consequently, D cannot be identically 1. In particular, H^c cannot be a finite set. Recall that

$$\mathbf{Q\Pi} = (-\delta_{ij} b_j)$$

by Theorem 5, § 4.1.5, and that $\mathbf{\Pi 1}$ has components $\Pi_i = D_i$ on H^c and $\Pi_i = 1$ on H.

Introduce now the *replacement matrix* $\mathbf{R}$ of the partitioned form

$$\mathbf{R} = \begin{pmatrix} \mathbf{R}_{11} & \mathbf{\Pi}_{12} \\ \mathbf{0} & \mathbf{\Pi}_{22} \end{pmatrix},$$

where $\mathbf{R}_{11} = (\delta_{ij} D_j^c)$. This choice of $\mathbf{R}_{11}$ has been made in order to obtain a stochastic matrix $\mathbf{R}$, with $\mathbf{R1} = \mathbf{1}$. See § 11.2.1 for the role of replacement matrix and for description of perturbation mechanism.

Next, define a new intensity matrix $\mathbf{Q}^* = (q_{ij}^*)$ by

$$\mathbf{Q}^* = \mathbf{QR},$$

or, in the partitioned form,

$$\mathbf{Q}^* = \begin{pmatrix} \mathbf{Q}_{11}^* & \mathbf{0} \\ \mathbf{Q}_{21}^* & \mathbf{Q}_{22}^* \end{pmatrix},$$

where

$$\mathbf{Q}_{11}^* = (q_{ij} D_j^c), \qquad \mathbf{Q}_{12}^* = \mathbf{0},$$
$$\mathbf{Q}_{21}^* = (q_{ij} D_j^c), \qquad \mathbf{Q}_{22}^* = (-\delta_{ij} b_j).$$

These expressions follow from observation that $\mathbf{Q}^* = \mathbf{Q}(\mathbf{RJ}_1) + \mathbf{Q\Pi}$. Using the Poisson equation for D_i, one has evidently that

$$\mathbf{Q}^*\mathbf{1} = 0 \quad \text{and} \quad \mathbf{Q}_{11}^* \mathbf{1}_1 = 0,$$

and

$$q_i^* = \begin{cases} q_i & \text{for } i \in H^c, \\ b_i & \text{for } i \in H. \end{cases}$$

Thus, $\mathbf{Q}^*$ is in fact a conservative intensity matrix, for which a set H^c is a closed set, and the component $\mathbf{Q}_{22}^*$ is a diagonal matrix.

Let $X^* = (X_t^*,\ 0 \leqslant t < \infty)$ be a "perturbed" M.C. corresponding to the generator $\mathbf{Q}^*$, with a standard transition matrix $\mathbf{P}^*(t)$ satisfying the (B)- and (F)-equations. It may be assumed that $\mathbf{P}^*(t)$ is stochastic, $\mathbf{P}^*(t)\mathbf{1} = \mathbf{1}$. In partitioned form

$$\mathbf{P}^*(t) = \begin{pmatrix} \mathbf{P}_{11}^*(t) & \mathbf{0} \\ \mathbf{P}_{21}^*(t) & \mathbf{P}_{22}^*(t) \end{pmatrix}.$$

The generator $\mathbf{Q}^*$ is uniquely defined by the generator $\mathbf{Q}$ (note that the replacement matrix $\mathbf{R}$ also depends on $\mathbf{Q}$). Thus, with each transient chain X, there is associated a modified chain X^* of such peculiar form. Indeed, the modified chain X^* stays essentially in the closed set H^c. However, no transitions within H are possible, and the chain leaves H on the first exit from each state in H. Thus, X^* may be called an *escape chain*, as it really escapes from H.

It is clear that Kolmogorov equations for $\mathbf{P}^*(t)$ reduce to separate equations on H^c and on H. Taking, say, the (B)-equation, one has (in resolvent form)

$$\begin{aligned} \alpha\mathbf{U}_{11}^{*\alpha} - \mathbf{I}_1 &= \mathbf{Q}_{11}^*\mathbf{U}_{11}^{*\alpha}, \\ \alpha\mathbf{U}_{22}^{*\alpha} - \mathbf{I}_2 &= \mathbf{Q}_{22}^*\mathbf{U}_{22}^{*\alpha}, \\ \alpha\mathbf{U}_{21}^{*\alpha} &= \mathbf{Q}_{21}^*\mathbf{U}_{11}^{*\alpha} + \mathbf{Q}_{22}^*\mathbf{U}_{21}^{*\alpha}, \qquad \alpha > 0. \end{aligned}$$

Hence,

$$u_{ij}^{*\alpha} = \begin{cases} \delta_{ij}(\alpha + b_j)^{-1}, & i, j \in H, \\ (\alpha + b_i)^{-1} \sum\limits_{k \in H^c} q_{ik}^* u_{kj}^*, & i \in H, \quad j \in H^c, \quad \alpha > 0, \end{cases}$$

but $u_{ij}^{*\alpha}$ for $i, j \in H^c$ must be found independently.

Note that the matrix $\mathbf{P}^*(t)$ is analogous to the matrix ${}_H\mathbf{P}^\rho(t)$ for the exit taboo probabilities, discussed in § 2.2.4. Indeed, $\mathbf{P}_{11}^*(t)$ corresponds trivially to ${}_H\mathbf{P}_{11}^*(t)$, because H^c is a closed set. On H, the matrix $\mathbf{P}_{22}^*(t)$ is diagonal with exponential terms having $q_i^* = b_i$. This fact may be interpreted by saying that the chain on H is killed at the first entrance time to H^c (which coincides with the first exit time ρ from a state in H), with distribution

$$\mathbb{P}^i(\rho > t) = e^{-b_i t} = \sum_{j \in H} p_{ij}^*(t), \qquad i \in H.$$

Alternatively, one can use the argument from Example 2, § 5.4.2. Combining the above equation for $\mathbf{U}_{21}^{*\alpha}$ with the (F)-equation for $\mathbf{U}_{22}^{*\alpha}$ one finds that

$$\mathbf{U}_{21}^{*\alpha} = \mathbf{U}_{22}^{*\alpha}\mathbf{Q}_{21}^*\mathbf{U}_{11}^{*\alpha}, \qquad \alpha > 0.$$

Hence, recalling the expression for the co-balayage $\mathbf{C}_{21}^{*\alpha}$ (see § 3.4.4), one obtains the following useful relation:

$$\mathbf{U}_{21}^{*\alpha} = \mathbf{U}_{22}^{*\alpha}\mathbf{C}_{21}^{*\alpha}, \qquad \alpha > 0,$$

or explicitly,

$$u_{ij}^{*\alpha} = (\alpha + b_i)^{-1} n_{ij}^{*\alpha}, \qquad i \in H, \quad j \in H^c, \quad \alpha > 0,$$

where $n_{ij}^{*\alpha}$ agrees with expression in § 3.4.1.

Indeed, it is obvious that the chain X^* may enter set H^c only at the last exit time L^* from H. Thus, in the last exit decomposition

$$\mathbf{U}^{*\alpha} = {}_H\mathbf{U}^{*\alpha} + \mathbf{U}^{*\alpha}\mathbf{C}^{*\alpha}, \qquad \alpha > 0,$$

the term ${}_H\mathbf{U}^{*\alpha}$ has the only non-zero component $\mathbf{U}_{11}^{*\alpha}$, and the term $\mathbf{U}^{*\alpha}\mathbf{C}^{*\alpha}$ has two non-zero components, namely $\mathbf{U}_{21}^{*\alpha}$ and $\mathbf{U}_{22}^{*\alpha}$.

Furthermore, Kolmogorov equations for $\mathbf{U}^{*\alpha}$ and $\mathbf{U}^*$ yield the following "second resolvent equation" stated here in the direct and dual form ($\alpha > 0$):

$$\mathbf{U}^{*\alpha} = \mathbf{U}^{\alpha} + \mathbf{U}^{\alpha}(\mathbf{Q}^* - \mathbf{Q})\mathbf{U}^{*\alpha}, \qquad \text{(direct)}$$

$$\mathbf{U}^{*\alpha} = \mathbf{U}^{\alpha} + \mathbf{U}^{*\alpha}(\mathbf{Q}^* - \mathbf{Q})\mathbf{U}^{\alpha} \qquad \text{(dual)}$$

(see § 11.2.1) with $\mathbf{Q}^* - \mathbf{Q} = \mathbf{Q}(\mathbf{R} - \mathbf{I})$. Observe that the dual equation implies that when $\alpha\mathbf{U}^{\alpha}\mathbf{1} = \mathbf{1}$, then also $\alpha\mathbf{U}^{*\alpha}\mathbf{1} = \mathbf{1}$.

The main problem of the perturbation method is to express properties of the escape chain with matrix $\mathbf{P}^*(t)$ in terms of properties of the original chain with matrix $\mathbf{P}(t)$.

§ 13.4.2. Properties. Let T^* be the first entrance time to H for the escape chain X^*, and let D^* be the corresponding first entrance probability. As H^c is a closed set, it is clear that the Dirichlet problem yields

$$D_i^* = \begin{cases} 0 & \text{for} \quad i \in H^c, \\ 1 & \text{for} \quad i \in H. \end{cases}$$

Obviously, as shown above, all states in H are transient. The following lemma strengthens this fact.

LEMMA 1. *In the Poisson decomposition*

$$D_i^* = \mathbf{U}^* b_i^* + s_i^*, \qquad i \in \mathbb{I},$$

$b_i^* = b_i$ and $s_i^* = 0$ for all i. Thus, the set H is an equilibrium set for the escape chain X^*.

Proof.

$$b_i^* = \sum_j q_{ij}^* D_j^{*c} = \sum_{j \in H^c} q_{ij}^* = \sum_{j \in H^c} q_{ij} D_j^c = b_i,$$

and this vanishes on H^c. Hence,

$$D_i^* = \begin{cases} \sum_{j \in H} u_{ij}^* b_j + s_i^*, & i \in H^c, \\ \sum_{j \in H} (\delta_{ij}/b_i) b_j + s_i^*, & i \in H, \end{cases}$$

so $s_i^* = 0$ for all i. □

Using the fact that $\alpha \mathbf{U}^{*\alpha}\mathbf{1} = \mathbf{1}$, the expressions for $u_{ij}^{*\alpha}$ (for $i \in H$, $j \in H^c$) from § 13.4.1 yield

$$\alpha \sum_{j \in H^c} u_{ij}^{*\alpha} = (\alpha + b_i)^{-1} b_i = (\alpha + b_i)^{-1} \alpha n_i^{*\alpha},$$

with

$$\alpha n_i^{*\alpha} = b_i^* = b_i,$$

in agreement with expressions in § 3.4.1.

The escape chain X^* restricted to the closed set H^c is described by the generator $\mathbf{Q}_{11}^*$. The following lemma indicates that such a chain cannot be ergodic on H^c (i.e., it cannot have non-zero finite invariant measure).

LEMMA 2. *Let μ_1 be a finite measure on H^c such that $\mu_1 \mathbf{Q}_{11}^* = 0$. Then μ_1 vanishes identically.*

Proof. i) Suppose that

$$\sum_{i \in H^c} \mu(i) q_{ij}^* = 0, \qquad j \in H^c,$$

has a non-zero finite solution μ. Then also

$$\sum_{i \in H^c} \mu(i) q_{ij} = 0, \qquad j \in H^c.$$

Extending μ to all $\mathbb{I}$ by letting $\mu(i) = 0$ for $i \in H$, one obtains a non-zero finite μ on $\mathbb{I}$ satisfying $\mu \mathbf{Q} = 0$. This is impossible because the original chain X is transient.

ii) Alternatively, adding over all $j \in H^c$, one has, using the Dirichlet problem for the first entrance probability D,

$$0 = \sum_{j \in H^c} \sum_{i \in H^c} \mu(i) q_{ij}^* = \sum_{i \in H^c} \mu(i) \sum_{j \in H^c} q_{ij} D_j^c = - \sum_{i \in H^c} \mu(i) \sum_{j \in H} q_{ij} = -\mu \bullet \bar{q},$$

where $\bar{q}$ (given in § 1.2.1) does not vanish identically. □

Of course, there may exist σ-finite measures satisfying $\mu_1 \mathbf{Q}_{11}^* = 0$. Note also that $\mu \mathbf{Q}^* = 0 = \mu \mathbf{Q} \mathbf{R}$.

Remark. As the escape chain leaves H never to return to it, it is clear that $T^* = \infty$, and therefore (as in Ex. 2, § 5.4.2),

$$\alpha \mathbf{C}_{21}^{*\alpha} \mathbf{1}_1 = b_2, \text{ constant}, \qquad \text{for} \quad \alpha > 0.$$

On the other hand, the fact that the escape chain is transient on H^c, implies that in the limit, as $\alpha \to 0$,

$$\mathbf{C}_{21}^{*\alpha} \longrightarrow \mathbf{C}_{21}^* = \mathbf{Q}_{21}^* \mathbf{U}_{11}^*,$$

and

$$\mathbf{U}^*_{21} = \mathbf{U}^*_{22}\mathbf{C}^*_{21},$$

or explicitly (see § 3.4.1),

$$u^*_{ij} = b_i^{-1} N^*_{ij}, \qquad i \in H, \quad j \in H^c,$$

with $N^*_i = \infty$. Let ξ be a reference measure for $\mathbf{P}(t)$. Clearly, ξ is also excessive for $\mathbf{P}^*(t)$,

$$\xi\mathbf{Q}^* = \xi\mathbf{QR} \leqslant 0.$$

THEOREM 3. *Assume that ξ is the reference measure for both $\mathbf{P}(t)$ and $\mathbf{P}^*(t)$. Then the set H has the same capacity relative to the original chain X and the to escape chain X^*.*

Proof. Recall that the capacity of H is the total mass of the equilibrium measure (dual of the equilibrium charge $b = b^*$):

$$C^*(H) = \sum_{i \in H} b_i^* \xi(i) = \sum_{i \in H} b_i \xi(i) = C(H). \qquad \square$$

§ 13.4.3. Energy. It is of interest to apply the method from § 11.5.4 to the present case. Treating the stochastic matrix $\mathbf{\Pi}_{22} = (\Pi_{ij})$ as a one-step transition matrix of some discrete parameter M.C., the invariant distribution on H is given by $\eta\mathbf{\Pi}_{22} = \eta$. With $h(j) = \eta(j)/\beta(j)$ and $v_h(i,j) = u_{ij}/\hat{h}(j)$, for i, j in H and $\hat{h}(j)$ being the dual of $h(j)$, the energy of a measure μ_2, supported by H, is given by

$$\text{en}\,(\mu_2) = \mu_2 \bullet \mathbf{U}_h \mu_2 = \sum_{i \in H} \mu(i) \sum_{j \in H} v_h(i,j)\mu(j).$$

Note that en $(\eta) = 1$.

Apply now the same argument to the modified chain X^* on H with matrix $\mathbf{U}^*_{22}$. By Lemma 1, § 13.4.2,

$$\Pi^*_{ij} = u^*_{ij} b_j = \delta_{ij}, \qquad i, j \in H,$$

so η is any distribution.

Using again the same reference measure ξ, one finds

$$h^*(j) = \eta(j)/\beta(j), \qquad v^*_{h^*}(i,j) = \delta_{ij}/\eta(j), \qquad i, j \in H.$$

Hence the energy of any measure μ_2 supported by H is given by

$$\text{en}\,(\mu_2) = \mu_2 \bullet \mathbf{U}^*_{h^*} \mu_2 = \sum_{j \in H} \mu^2(j)/\eta(j).$$

And again, en $(\eta) = 1$.

APPENDIX

This Appendix, provided for reader's convenience, lists several definitions and theorems used implicitly in the text.

Stopping times are presented in A1. Basic properties of stochastic processes are summarized in A2, including brief comments on Markov processes, martingales, projections and conditioning. Analytic aspects are presented in A3, by adaptation of classical results.

The material is standard, and is easily accessible in literature. For example, treatment of general stochastic processes can be found in the following books: K. L. Chung [1982], C. Dellacherie and P. A. Meyer [1978], D. Williams [1979]; see also a review article in Encyclopedia of Statistical Sciences by R. Syski. For analytic aspects see, for example, P. Billingsley [1986], E. Hewitt and K. Stromberg [1965].

A1. STOPPING TIMES

§ **A1. 1.** Consider a complete probability space $(\Omega, \mathcal{F}, \mathbb{P})$ and a right-continuous increasing family $(\mathcal{F}_t,\ 0 \leqslant t < \infty)$ of sub-σ-fields of $\mathcal{F}$, such that $\mathcal{F}_0$ contains all $\mathbb{P}$-null sets of $\mathcal{F}$, and

$$\mathcal{F}_\infty = \sigma\Big(\bigcup_t \mathcal{F}_t\Big) \subseteq \mathcal{F}.$$

A stopping time T with respect to a family $(\mathcal{F}_t)$ is a positive random variable T on $(\Omega, \mathcal{F})$, satisfying

$$(T \leqslant t) \in \mathcal{F}_t \qquad \text{for every} \quad t.$$

T may assume infinite values, and $(T = \infty) \in \mathcal{F}_\infty$. A stopping time is also called an optional time or a random time.

Events in $\mathcal{F}_t$ are said to occur prior to time t. A σ-field of events occurring prior to T, denoted by $\mathcal{F}_T$ and called a pre-T σ-field, is defined as a collection of all events A in $\mathcal{F}_\infty$ such that

$$A \cap (T \leqslant t) \in \mathcal{F}_t \qquad \text{for every} \quad t.$$

A σ-field $\mathcal{F}_{T-}$ of all events occurring strictly before T is the σ-field generated by all the sets of the form $A \cap (t < T)$, where $A \in \mathcal{F}_t$ and $0 \leqslant t < \infty$. Evidently, $\mathcal{F}_{T-} \subseteq \mathcal{F}_T$ and $\mathcal{F}_S \subseteq \mathcal{F}_T$ for any two stopping times $S \leqslant T$.

§ **A1. 2.** A stopping time T is predictable if there exists an increasing sequence (T_n) of stopping times such that

$$\lim T_n = T \qquad \text{and} \qquad T_n < T \qquad \text{for all} \quad n \text{ on } (T > 0).$$

The sequence (T_n) is said to foretell or to announce T.

A stopping time T is totally inaccessible if for every predictable stopping time S

$$\mathbb{P}(T = S < \infty) = 0,$$

or, alternatively, if the graphs of T and S are disjoint, i.e.,

$$[\![T]\!] \cap [\![S]\!] = \varnothing \qquad \text{up to modification.}$$

A stopping time T is accessible if there exists a sequence (T_n) of predictable stopping times such that

$$\mathbb{P}\Big(\bigcup_n (T_n = T < \infty)\Big) = 1,$$

or, alternatively, if the graph of T is contained in the union of graphs of T_n, i.e.,

$$[\![T]\!] \subset \bigcup_n [\![T_n]\!] \qquad \text{up to modification.}$$

Every predictable stopping time is accessible. Conversely, accessible stopping times are predictable when the family $(\mathcal{F}_t)$ is quasi-left continuous (q.l.c.), i.e., when

$$\mathcal{F}_T = \mathcal{F}_{T-}$$

for all predictable stopping times.

§ **A1.3.** For two stopping times S and T (relative to the same families of σ-fields), the random variable

$$T_S = S + T \circ \theta_S$$

defined as a mapping $w \to S(w) + T(\theta_S(w))$, where θ_S is a random shift, is also a stopping time. T_S is the first time where the "event T" occurs after the "event S" has occurred. In particular, when S is a constant, say s, then

$$T_s = s + T \circ \theta_s.$$

A stopping time T is called a terminal time, provided $T_s = T$ when $T > s$ for all $s \geqslant 0$.

A2. PROCESSES

§ **A2.1. Conditional expectation.** Let $(\Omega, \mathcal{F}, \mathbb{P})$ be a probability space and let $\mathcal{B}$ be a sub-σ-field of $\mathcal{F}$. Let X be an integrable real-valued random variable on $(\Omega, \mathcal{F}, \mathbb{P})$, $\mathbb{E}|X| < \infty$.

A conditional expectation (c.exp.) of X given $\mathcal{B}$ is a $\mathcal{B}$-measurable random variable $\mathbb{E}(X \mid \mathcal{B})$ (unique up to equivalence) such that

$$\int_B \mathbb{E}(X \mid \mathcal{B})\, d\mathbb{P}_{\mathcal{B}} = \int_B X\, d\mathbb{P}$$

for every $B \in \mathcal{B}$, where $\mathbb{P}_{\mathcal{B}}$ is the restriction of $\mathbb{P}$ to $\mathcal{B}$.

Existence of c.exp. follows from the Radon–Nikodym theorem. C.exp. operator $\mathbb{E}(\,\cdot \mid \mathcal{B})$ is a linear operator mapping of an $\mathcal{F}$-measurable function to a $\mathcal{B}$-measurable function, with the following smoothing properties:

i) $\mathbb{E}(XY \mid \mathcal{B}) = Y\mathbb{E}(X \mid \mathcal{B})$ a.e. for any $\mathcal{B}$-measurable Y and integrable XY;

ii) $\mathbb{E}(\mathbb{E}(X \mid \mathcal{B}_1) \mid \mathcal{B}_2) = \mathbb{E}(\mathbb{E}(X \mid \mathcal{B}_2) \mid \mathcal{B}_1)$
$= \mathbb{E}(X \mid \mathcal{B}_1)$ a.e. whenever $\mathcal{B}_1 \subset \mathcal{B}_2$.

Note that $\mathbb{E}(\,\cdot \mid \mathcal{B})$ is a projection operator, preserving constants, and

$$\mathbb{E}(\mathbb{E}(X \mid \mathcal{B})) = \mathbb{E}(X).$$

The c.exp. of $X = I_A$ (for $A \in \mathcal{F}$) given $\mathcal{B}$, denoted by $\mathbb{P}(A \mid \mathcal{B})$, is called the conditional probability (c.pr.) of A given $\mathcal{B}$.

When $\mathcal{B} = \sigma(X_s,\ s \leqslant t)$ is a σ-field generated by random variables X_s, $s \leqslant t$, then the c.exp. of X given $\mathcal{B}$, denoted by $\mathbb{E}(X \mid X_s,\ s \leqslant t)$, is called the c.exp. of X given X_s, $s \leqslant t$.

§ **A2.2. Markov process.** Let $X = (X_t,\ 0 \leqslant t < \infty)$ be a family of random variables defined on a (complete) pr. space $(\Omega, \mathcal{F}, \mathbb{P})$ and taking values in a state space $(\mathsf{E}, \mathcal{E})$ where E is a metric space and $\mathcal{E}$ its Borel field. Let $(\mathcal{F}_t,\ 0 \leqslant t < \infty)$ be a family of increasing sub-σ-fields of $\mathcal{F}$ satisfying the usual conditions (right-continuity, augmentation), and such that (X_t) is adapted to $(\mathcal{F}_t)$, that is X_t is $\mathcal{F}_t$-measurable for each t. In particular $(\mathcal{F}_t)$ may be the natural σ-fields of the process X.

The process $X = (X_t)$ is said to be a Markov process with respect to the family $(\mathcal{F}_t)$ if it is has the Markov property: the σ-fields $\mathcal{F}_t$ and $\sigma(X_s,\ s \geqslant t)$ are for each t conditionally independent given X_t, i.e.,

$$\mathbb{P}(A \cap B \mid X_t) = \mathbb{P}(A \mid X_t)\,\mathbb{P}(B \mid X_t) \qquad \text{for} \quad A \in \mathcal{F}_t \quad \text{and} \quad B \in \sigma(X_s,\ s \geqslant t),$$

or, equivalently, if for every bounded Borel function f on $(\mathsf{E}, \mathcal{E})$ and for every pair $s \leqslant t$

$$\mathbb{E}(f(X_t) \mid \mathcal{F}_s) = \mathbb{E}(f(X_t) \mid X_s) \quad \text{a.e.}$$

A Markov process X is said to have the strong Markov property (or to be strongly Markov) when for each stopping time T and for a bounded Borel function f one has for each t

$$\mathbb{E}(f(X_{t+T}) \mid \mathcal{F}_T) = \mathbb{E}(f(X_{t+T}) \mid X_T) \quad \text{a.e.}$$

A Hunt process is a Markov process which is strong Markov, q.l.c. and càdlàg (i.e., its paths are right-continuous with left limits). These properties are shown to hold for a Feller process (defined in analytic terms).

It may be assumed that there exists a family of transition functions for a Markov process X, taken to be time-homogeneous,

$$\mathbf{P}_t(x,B) = \mathbb{P}(X_t \in B \mid X_0 = x), \qquad x \in \mathsf{E}, \quad B \in \mathcal{E},$$

which is a measure on $\mathcal{E}$ for each fixed x, and a measurable function on E for each fixed B. The kernel $\mathbf{P}_t(\cdot,\cdot)$ on $\mathsf{E} \times \mathcal{E}$ defines for each t the operator $\mathbf{P}_t$ acting on the space of bounded measurable function on E, and dually, on the space of bounded measures on $\mathcal{E}$ by

$$\mathbf{P}_t f(x) = \int_{\mathsf{E}} \mathbf{P}_t(x,dy)\, f(y) = \mathbb{E}^x f(X_t),$$

$$\mu\mathbf{P}_t(B) = \int_{\mathsf{E}} \mu(dx)\,\mathbf{P}_t(x,B) = \mathbb{P}^\mu(X_t \in B)$$

for $x \in \mathsf{E}$, $B \in \mathcal{E}$. Markov property implies that operators $(\mathbf{P}_t)$ form a contraction semi-group:

$$\mathbf{P}_{t+s} = \mathbf{P}_t\mathbf{P}_s, \qquad t,s \geqslant 0 \qquad \text{(Chapman–Kolmogorov equation)}.$$

For a stopping time T and for $\alpha \geqslant 0$, define the operator $\mathbf{P}_T^\alpha$ by

$$\mathbf{P}_T^\alpha f(x) = \mathbb{E}^x\big(e^{-\alpha T} f(X_T),\ T < \infty\big), \qquad x \in \mathsf{E},$$

for any bounded (or positive finite) measurable function f on E. When $\alpha = 0$, write $\mathbf{P}_T$ instead of $\mathbf{P}_T^0$.

The "product theorem" asserts that, for any two stopping times T and S,

$$\mathbf{P}_S^\alpha\,\mathbf{P}_T^\alpha = \mathbf{P}_{T_S}^\alpha, \qquad \alpha \geqslant 0.$$

Note that when both S and T are constants or if only S is constant, then the product theorem reduces to the semi-group property and the strong Markov property, respectively.

When $T = T_H$ is the first entrance time to a set H, then $\mathbf{P}_T^\alpha$, written simply as $\mathbf{P}_H^\alpha$, is known as an α-balayage on H.

The measure $P_H^\alpha(x,B) = \mathbf{P}_H^\alpha I_B(x)$ defined for any Borel set B is called the α-hitting distribution of H (starting at x) or an α-harmonic measure of H (relative to x). Thus, for $\alpha = 0$,

$$P_H(x,B) = \mathbb{P}^x\big(X(T_H) \in B,\ T_H < \infty\big).$$

The right-continuity of paths and the 0–1 law imply that $P_H^\alpha(x,\cdot)$ is concentrated on a set $H \cup H^r$, known as a fine closure of H. Here H^r is the set of regular points of H, namely those points x such that

$$\mathbb{P}^x(T_H = 0) = 1.$$

Finally, one has corresponding taboo probabilities defined by

$${}_H P_t(x,B) = \mathbb{P}^x\big(X_t \in B,\ t > T_H\big).$$

§ **A2.3. Martingales.** Let $(\Omega, \mathcal{F}, \mathbb{P})$ be a (complete) probability space, let $(\mathcal{F}_t)$ be an increasing family of sub-σ-fields of $\mathcal{F}$ satisfying the usual conditions, and let (X_t) be a family of random variables such that X_t is a measurable with respect to $\mathcal{F}_t$ for each t.

A process $X = (X_t, \mathcal{F}_t,\ 0 \leqslant t < \infty)$ is said to be a super-martingale (super-mtg) with respect to a family $(\mathcal{F}_t)$ if each X_t is integrable, and if

$$\mathbb{E}(X_t \mid \mathcal{F}_s) \leqslant X_s \quad \text{a.e.} \qquad \text{for every pair} \quad s \leqslant t.$$

A process (X_t) is said to be a martingale (mtg) if the strict equality holds in the above defining relation.

Note that for a super-mtg $\mathbb{E}X_t \leqslant \mathbb{E}X_s$ for $s \leqslant t$, with an equality for a mtg. A super-mtg $X = (X_t, \mathcal{F}_t,\ 0 \leqslant t < \infty)$ such that the function $t \to \mathbb{E}X_t$ is right-continuous for all t, has a càdlàg modification.

THEOREM (Doob martingale convergence theorem). *Let $X = (X_t, \mathcal{F}_t,\ 0 \leqslant t < \infty)$ be a right-continuous super-mtg.*

a) *Suppose that $\sup_t \mathbb{E}X_t^- < \infty$ (or equivalently, $\sup_t \mathbb{E}|X_t| < \infty$). Then $\lim_{t\to\infty} X_t = X_\infty$ exists a.e., and is integrable.*

b) *This condition holds, in particular, when $X_t \geqslant 0$ for all t; then*

$$(X_t, \mathcal{F}_t,\ 0 \leqslant t \leqslant \infty)$$

is a super-mtg.

c) *If X_t are uniformly integrable, then the condition in a) holds, and*

$$(X_t, \mathcal{F}_t,\ 0 \leqslant t \leqslant \infty)$$

is a super-mtg, and $X_t \to X_\infty$ in L_1 norm.

d) *If (X_t) is an uniformly integrable mtg, then $(X_t,\ 0 \leqslant t \leqslant \infty)$ is a mtg, and $X_t = \mathbb{E}(X_\infty \mid \mathcal{F}_t)$ a.e.*

THEOREM (Doob stopping theorem). *A right-continuous super-mtg*

$$(X_t,\ \mathcal{F}_t,\ 0 \leqslant t < \infty)$$

is a strong super-mtg, i.e., for every pair of bounded stopping times such that $S \leqslant T$, one has

$$\mathbb{E}(X_T \mid \mathcal{F}_S) \leqslant X_S \quad \text{a.e.}$$

(with equality if X is a mtg).

A super-mtg $(X_t, \mathcal{F}_t,\ 0 \leqslant t < \infty)$ is said to be of class (D) if the family (X_T), where T ranges through all finite stopping times relative to $(\mathcal{F}_t)$, is uniformly integrable. A positive right-continuous super-mtg X is called a potential if $\lim \mathbb{E}X_t = 0$, as $t \to \infty$; hence a potential is always uniformly integrable.

THEOREM (Riesz decompositions). *A positive, right-continuous super-mtg*

$$(X_t, \mathcal{F}_t,\ 0 \leqslant t < \infty)$$

has unique (up to equivalence) decompositions:

a) *$X_t = V_t + L_t$, where $L_t = \mathbb{E}(X_\infty \mid \mathcal{F}_t)$ form a uniformly integrable mtg, and V_t form a positive super-mtg such that $V_t \to 0$ a.e.,*

b) *$X_t = P_t + M_t$, where M_t form a mtg, and P_t form a potential.*

In general, $M_t \geqslant L_t$ and $P_t \leqslant V_t$. Both decompositions coincide when (X_t) is uniformly integrable.

§ A2.4. Stochastic processes. Let $X = (X_t, \mathcal{F}_t,\ 0 \leqslant t < \infty)$ be a stochastic process on a (complete) probability space $(\Omega, \mathcal{F}, \mathbb{P})$ and taking values in a state space $(\mathsf{E}, \mathcal{E})$, where E is a metric space and $\mathcal{E}$ its Borel field, and $(\mathcal{F}_t)$ is an increasing family of sub-σ-fields of $\mathcal{F}$, satisfying the usual conditions (right-continuity, augmentation); moreover, (X_t) is adapted to $(\mathcal{F}_t)$. Regarding the process X as a mapping

$$(t, w) \longrightarrow X(t, w)$$

from a product space $\mathbf{R}_+ \times \Omega$ into a state space E, the process X is measurable if this mapping is measurable relative to the product σ-field $\mathcal{B} \otimes \mathcal{F}$ (where $\mathcal{B}$ is the Borel field on the positive real line $\mathbf{R}_+$). A process X with values in $(\mathsf{E}, \mathcal{E})$ is said to be progressively measurable (or progressive) if for every $t \in \mathbf{R}_+$, the mapping

$$(s, w) \longrightarrow X(s, w)$$

of $[0, t] \times \Omega$ into E is measurable with respect to the σ-field $\mathcal{B}_t \otimes \mathcal{F}_t$, where $\mathcal{B}_t$ is a Borel field on $[0, t]$. A progressive process is adapted and measurable. A process X, adapted to $(\mathcal{F}_t)$ and taking values in a metric space $(\mathsf{E}, \mathcal{E})$ and whose paths are right (or left) continuous, is always progressively measurable.

For two stopping times S and T such that $S \leqslant T$, a stochastic interval relative to the family $(\mathcal{F}_t)$, is a subset of $\mathbf{R}_+ \times \Omega$ defined by

$$[\![S, T[\![= \{(t, w) \colon\ S(w) \leqslant t < T(w)\}.$$

The σ-field generated by intervals $[\![S, T[\![$, when S and T range over all stopping times such that $S \leqslant T$, is called the optional (or well-measurable) σ-field, denoted by $\mathcal{O}$.

The σ-field generated by intervals $[\![S, T[\![$, when S and T range over all predictable stopping times such that $S \leqslant T$, is called the predictable σ-field, denoted by $\mathcal{P}$.

A process X is said to be optional (respectively, predictable) if the function

$$(t, w) \longrightarrow X(t, w)$$

on $\mathbf{R}_+ \times \Omega$ to E is measurable relative to the optional (respectively, predictable) σ-field.

THEOREM. *The following relation holds*

$$\mathcal{P} \subset \mathcal{O} \subset \mathcal{M},$$

where $\mathcal{M}$ is the σ-field of progressive sets.

A process $X = (X_t, \mathcal{F}_t,\ 0 \leqslant t < \infty)$ is said to be q.l.c. if for any increasing sequence (T_n) of stopping times relative to $(\mathcal{F}_t)$ with limit T:

$$\lim_n X(T_n) = X(T) \quad \text{a.e.} \qquad \text{on} \quad (T < \infty).$$

This implies that for the càdlàg process (i.e., right-continuous with left limits)

$$X(T-) = X(T) \quad \text{a.e.,}$$

and that the process X has no fixed times of discontinuity.

§ **A2.5. Projections.** Let $X = (X_t)$ be a positive or a bounded measurable process. The optional projection of X is defined to be the optional process ${}^oX$ such that for every optional time T

$$\mathbb{E}(X_T,\ T < \infty) = \mathbb{E}({}^oX_T,\ T < \infty).$$

The predictable projection of X is defined to be the predictable process ${}^pX$ such that, for every predictable time T,

$$\mathbb{E}(X_T,\ T < \infty) = \mathbb{E}({}^pX_T,\ T < \infty).$$

Both projections are unique up to evanescence. It follows that for every positive measurable process X and for an increasing process $A = (A_t)$:

$$\mathbb{E}\int_{0-}^{\infty} X_t\, dA_t = \mathbb{E}\int_{0-}^{\infty} {}^oX_t\, dA_t \qquad \text{for optional} \quad A,$$

$$\mathbb{E}\int_{0-}^{\infty} X_t\, dA_t = \mathbb{E}\int_{0-}^{\infty} {}^pX_t\, dA_t \qquad \text{for predictable} \quad A.$$

Note that every increasing process is optional.

A $\mathbb{P}$-measure (or a stochastic measure) is a bounded measure μ on the product σ-field $\mathcal{B}\otimes\mathcal{F}$ in the product space $\mathbf{R}_+ \times \Omega$, unique up to evanescent sets. Associate with each μ an integrable increasing process (unique up to evanescence) $A = (A_t)$ such that

$$\mu(X) = \mathbb{E}\int_{0-}^{\infty} X_s\, dA_s$$

for any bounded measurable process X.

THEOREM. *For an increasing integrable process A to be predictable (respectively, optional) it is necessary and sufficient that*

$$\mu(X) = \mu({}^pX) \qquad \big(\textit{respectively,} \quad \mu(X) = \mu({}^oX)\big)$$

for every bounded measurable process X. Then μ is also called predictable (respectively, optional).

DEFINITION. The predictable (respectively, optional) projection of a stochastic measure μ is a stochastic measure μ^p (respectively, μ^o) defined for every measurable process X by

$$\mu^p(X) = \mu({}^pX) \qquad \big(\text{respectively, } \mu^o(X) = \mu({}^oX)\big).$$

DEFINITION. The dual predictable (respectively, optional) projection of an increasing process A is the predictable (respectively, optional) process A^p (respectively, A^o) defined by

$$\mathbb{E}\int_{0-}^{\infty} X_s\, dA_s^p = \mathbb{E}\int_{0-}^{\infty} {}^pX_s\, dA_s,$$

respectively,

$$\mathbb{E}\int_{0-}^{\infty} X_s\, dA_s^o = \mathbb{E}\int_{0-}^{\infty} {}^oX_s\, dA_s.$$

Dual predictable projection of A is called a (predictable) compensator of A. It is shown that a compensator A^p is continuous if and only if A is q.l.c. (or, equivalently, if and only if the jumps of A are totally inaccessible).

THEOREM. *Two integrable increasing processes A and B have the same dual predictable projection if and only if*

$$\mathbb{E}(A_\infty - A_t \mid \mathcal{F}_t) = \mathbb{E}(B_\infty - B_t \mid \mathcal{F}_t) \quad \text{a.e.}, \qquad t \geqslant 0,$$

i.e., when $(A_t - B_t, \mathcal{F}_t)$ is a martingale. Note that A and B generate the same potential.

Let (X_t) be a càdlàg adapted process, and let T be a stopping time. Then (X_t) charges T if $\mathbb{P}(X_T \neq X_{T-},\ T < \infty) > 0$, and (X_t) has a jump at T if $X_T \neq X_{T-}$ a.c. on $(T < \infty)$.

THEOREM. *Let $X = (X_t)$ be a càdlàg adapted process. Then the following three conditions are equivalent:*

i) *all jumps of X are totally inaccessible;*

ii) *X does not charge any predictable stopping time T;*

iii) *X is q.l.c.*

A3. INTERCHANGES

The following notation is used here: $\mathbf{P}(t) = (p_{ij}(t))$ is a stochastic transition matrix, $\mathbf{Q} = (q_{ij})$ is a conservative generator. On the state space $\mathbb{I}$, all (signed) measures are finite and all functions are bounded (with possible extension to σ-finite measures and positive finite functions).

§ A3.1.

THEOREM (Helly–Bray). *With limit along t,*

$$\lim_t \sum_k \mu(k,t) f(k) = \sum_k \lim_t \mu(k,t) f(k) \tag{B}$$

for any bounded function f on $\mathbb{I}$, provided

$$\lim_t \mu(k,t) = \mu(k) \qquad \text{and} \qquad \lim_t \mu(\mathbb{I},t) = \mu(\mathbb{I}).$$

THEOREM (dominated convergence theorem). *With limit along* t,

$$\lim_t \sum_k \mu(k) f(k,t) = \sum_k \mu(k) \lim_t f(k,t) \tag{F}$$

for any finite (signed) measure μ, *provided there is a function* g *such that*

$$|f(k,t)| \leqslant g(k) \qquad \text{and} \qquad \sum_k |\mu(k)| g(k) < \infty.$$

In particular, taking

$$\mu(k,t) = \sum_i \pi(i) p_{ik}(t)$$

for a probability measure π, one has

$$\lim(\pi \mathbf{P}(t) \bullet f) = \pi(\lim \mathbf{P}(t)) \bullet f,$$

and, taking

$$f(k,t) = \sum_j p_{kj}(t) f(j)$$

for a bounded function f, one has

$$\lim(\pi \bullet \mathbf{P}(t) f) = \pi \bullet (\lim \mathbf{P}(t)) f.$$

§ A3.2.

THEOREM (Fubini). *Define for* $i, j \in I$ *and* $t \geqslant 0$:

$$c(i,j) = \sigma(i) p_{ij}(t) a(j),$$

where σ *and* a *are, respectively, a finite (signed) measure and a bounded function. Then, for the equality*

$$\sum_i \sum_j c(i,j) = \sum_j \sum_i c(i,j)$$

to be true, it is sufficient that at least one of the following conditions holds:

a) $c(i,j) \geqslant 0$,

b) $\sum_i \sum_j |c(i,j)| < \infty$,

c) $\sum_j \sum_i |c(i,j)| < \infty$.

Thus, if b) or c) holds, then

$$\sigma \bullet \mathbf{P}(t) a = \sigma \mathbf{P}(t) \bullet a \qquad \text{(associativity)}.$$

Hence, writing

$$\sigma = \mu(-\mathbf{Q}), \qquad a = (-\mathbf{Q}) f$$

for a measure μ and a function f, one has

$$\mu \bullet (-\mathbf{Q})\mathbf{P}(t)a = \sigma\mathbf{P}(t) \bullet a \qquad \text{and} \qquad \sigma\mathbf{P}(t)(-\mathbf{Q}) \bullet f = \sigma \bullet \mathbf{P}(t)a.$$

Thus, under validity of the (B)- and (F)-equations, the following relations hold:

$$\mathbf{QP}(t) \cdot a = \mathbf{Q} \cdot \mathbf{P}(t)a, \qquad \mu\mathbf{Q} \cdot \mathbf{P}(t) = \mu \cdot \mathbf{QP}(t), \tag{B}$$

$$\mathbf{P}(t) \cdot \mathbf{Q}f = \mathbf{P}(t)\mathbf{Q} \cdot f, \qquad \mu\mathbf{P}(t) \cdot \mathbf{Q} = \mu \cdot \mathbf{P}(t)\mathbf{Q}. \tag{F}$$

For example, taking (i fixed)

$$c(k,j) = q_{ik}\, p_{kj}(t)a(j), \qquad c(k,j) = p_{ik}(t)q_{kj}f(j),$$

in the (B)- and (F)-equations for bounded functions, the b) condition yields, respectively,

$$q_i < \infty, \qquad \sum_k p_{ik}(t)q_k < \infty.$$

In the same fashion, taking

$$c(i,j) = \mu(i)q_{ij}f(j),$$

validity of b) and c) yields

$$\mu(-\mathbf{Q})f = \mu \bullet a = \sigma \bullet f.$$

Observe that for $f = 1$, one has $a = 0$, so $\sigma \bullet \mathbf{1} = 0$. Also, condition b) yields, for a bounded f,

$$\sum_i \mu(i)q_i < \infty.$$

Similar analysis applies (for $\alpha > 0$) to

$$\mathbf{QU}^\alpha \cdot a = \mathbf{Q} \cdot \mathbf{U}^\alpha a, \tag{B}$$

$$\mathbf{U}^\alpha\mathbf{Q} \cdot f = \mathbf{U}^\alpha \cdot \mathbf{Q}f, \tag{F}$$

where $\mathbf{U}^\alpha$ is the Laplace transform of $\mathbf{P}(t)$.

Thus, taking, for a fixed i and for $\alpha > 0$,

$$c(k,j) = q_{ik}\, u^\alpha_{kj}\, a(j), \qquad c(k,j) = u^\alpha_{ik}\, q_{kj}\, f(j)$$

in the (B)- and (F)-equations, the b) condition yields, respectively,

$$q_i/\alpha < \infty, \qquad \sum_k u^\alpha_{ik}\, q_k < \infty.$$

REFERENCES

This list contains only works which have been used in preparation of the book. In addition to authors explicitely mentioned in the text, some publications of relevant interest are also included.

ABATE, J. AND WHITT, W. Aproximations for the M/M/1 busy-period distribution, in: *Queueing Theory and its Applications, Liber Amicorum for J. W. Cohen,* O. J. Boxma, R. Syski (Eds.), 1988, pp. 149–192.

ALDOUS, D. *Probability Approximations via the Poisson Clumping Heuristic,* Springer-Verlag, Berlin, 1989.

ARCHINARD, E. Taboo probabilities in the entrance boundary theory of Markov chains, *Z. Wahrsch. verw. Gebiete,* (1974) **29,** 165–179.

ASMUSSEN, S. *Applied Probability and Queues,* John Wiley, N.Y., 1987.

ASSOUAD, P. Demonstration dela "conjecture de Chung" par Carleson, *Lecture Notes in Math.,* Vol. 191, Springer-Verlag, Berlin, 1971, pp. 17–20.

ATKINSON, B. W. AND MITRO, J. B. Applications of Revuz and Palm type measures for additive functionals in weak duality, in: *Seminar on Stoch. Proc. 1982,* Birkhauser, Boston, 1983, pp. 23–50.

BACCELLI, F. AND BREMAUD, P. *Palm Probabilities and Stationary Queues,* Springer-Verlag, Berlin, 1987.

BARLOW, M. T. Construction of a martingale with given absolute value, *Ann. Prob.,* (1981) **9,** 314–320.

BENVENISTE, A. AND JACOD, J. Systemes de Lévy des processus de Markov, *Invest. Math.,* (1973) **21,** 183–198.

BILLINGSLEY, P. *Probability and Measure,* 2nd ed., John Wiley, N.Y., 1986.

BLAKE, L. H. The preservation of regularity of conditional probabilities. Ph. D. Thesis, Dept. Math., Univ. Maryland, 1969.

BLAKE, L. H. A generalization of martingales and two consequent convergence theorems, *Pacific J. Math.,* (1970) **35,** 279–283.

BLAKE, L. H. Canonical extensions of measures and the extension of regularity of conditional probabilities, *Pacific J. Math.,* (1972) **41,** 25–31.

BLAKE, L. H. Simple extensions of measures and the preservation of regularity of conditional probabilities, *Pacific J. Math.,* (1973) **46,** 355–359.

BLAKE, L. H. Every amart is a martingale in the limit, *J. London Math. Soc.* 2, (1978) **18,** 381–384.

BLAKE, L. H. Tempered processes and Riesz decomposition for some martingales in the limit, *Glasgow Math. J.*, (1981) **22,** 9–17.

BLAKE, L. H. Generalized martingales, generalized Markov chains and generalized harmonic functions, *Prob. Th. Rel. Fields,* (1988) **79,** 405–430.

BLIEDTNER, J. AND HANSEN, W. *Potential Theory,* Springer-Verlag, Berlin, 1986.

BLUMENTHAL, R. M. AND GETOOR, R. K. *Markov Processes and Potential Theory,* Academic Press, N.Y., 1968.

BOEL, R., VARAIYA, P., AND WONG, E. Martingales on jump processes, *SIAM J. Control,* (1975) **13**: I. Representation results, pp. 999–1021, II. Applications, pp. 1022–1061.

BOURGIN, R. D. AND COGBURN, R. On determining absorption probabilities for Markov chains in random environments, *Adv. Appl. Prob.*, (1981) **13,** 369–387.

BREMAUD, P. *Point Processes and Queues,* Springer-Verlag, Berlin, 1981.

BROWN, T. A martingale approach to the Poisson convergence of simple point processes, *Ann. Prob.*, (1978) **6,** 615–628.

BURMAN, D. Y. AND SMITH, D. R. A light-traffic theorem for multi-server queues, *Math. Oper. Res.*, (1983) **8,** 15–25.

BUSHAR, H. F. Continuous parameter Markov chains, classification of boundary points and use of local time. Ph. D. Thesis, Dept. Math., Univ. of Maryland, 1976.

CAI, H. On reviving Markov processes and applications. Ph. D. Thesis, Dept. Math., Univ. of Maryland, 1987.

CHAMBERLAIN, M. W. The excessive functions of a continuous time Markov chain, *Ann. Prob.*, (1974) **2,** 1075–1089.

CHAMBERLAIN, M. W. Last exit times from the boundary of a continuous time Markov chain, *Ann. Prob.*, (1977) **5,** 152–156.

CHUNG, K. L. Markov processes with infinities, in: *Markov Processes and Potential Theory,* J. Chover (Ed.), John Wiley, N.Y., 1967, pp. 23–32.

CHUNG, K. L. *Markov Chains with Stationary Transition Probabilities,* 2nd ed., Springer-Verlag, Berlin, 1967.

CHUNG, K. L. Boundary behavior of Markov chains and its contributions to general processes, in: *Actes Congres Intern. Math.*, Nice, 1970, Vol. 2, 499–505.

CHUNG, K. L. *Lectures on Boundary Theory for Markov Chains,* (Annals of Math. Studies, No. 65) Princeton Univ. Press, 1970.

CHUNG, K. L. Note on last exit decomposition, in: *Seminaire de Probabilities VIII,* (*Lecture Notes in Math.*, Vol. 381), Springer-Verlag, Berlin, 1974, pp. 20–21.

CHUNG, K. L. *Lectures from Markov Processes to Brownian Motion*, Springer-Verlag, Berlin, 1982.

CHUNG, K. L. AND DOOB, J. L. Fields, optionality and measurability, *Amer. J. Math.*, (1965) **87**, 397–424.

CHUNG, K. L. AND RAO, M. Equilibrium and energy, *Prob. and Math. Stat.*, (1980) **1**, 99–108.

CHUNG, K. L. AND WALSH, J. B. To reverse a Markov process, *Acta Math.*, (1969) **123**, 225–251.

ÇINLAR, E.Markov additive processes I., *Z. Wahrsch. verw. Gebiete*, (1972) **24**, 85–93.

ÇINLAR, E.Markov additive processes II., *Z. Wahrsch. verw. Gebiete*, (1972) **24**, 95–121.

ÇINLAR, E.Markov renewal theory: a survey, *Management Science*, (1975) **21**, 724–752.

ÇINLAR, E.Lévy systems of Markov additive processes, *Z. Wahrsch. verw. Gebiete*, (1975) **31**, 175–185.

ÇINLAR, E.On increassing continuous process, *Stoch. Proc. Appl.*, (1979) **9**, 147–154.

ÇINLAR, E. AND KASPI, H.Regenerative systems and Markov additive processes, in: *Seminar on Stoch. Proc. 1982*, Birkhauser, Boston, 1983, pp. 123–137.

COHEN, J. W. *On Regenerative Processes in Queueing Theory*, Springer-Verlag, Berlin, 1976.

COHEN, J. W. On up- and down-crossings, *J. Appl. Prob.*, (1977) **14**, 405–410.

COHEN, J. W. *The Single Server Queue*, 2nd ed., North-Holland, Amsterdam, 1982.

COHEN, J. W. Who introduced the birth-and-death technique?, in: *Proc. 11th ITC*, Kyoto, 1985.

COHEN, J. W. AND BOXMA, O. J. *Boundary Value Problems in Queueing System Analysis*, North-Holland, Amsterdam, 1983.

CONNOLLY, B. W. AND CHAN, J. Generalized birth and death queueing processes: recent results, *Adv. Appl. Prob.*, (1977) **9**, 125–140.

COX, J. TH. Entrance laws for Markov chains, *Ann. Prob.*, (1977) **5**, 533–549.

DALEY, D. J. Queueing output processes, *Adv. Appl. Prob.*, (1976) **8**, 395–415.

DAVIS, M. H. A. Piecewise-deterministic Markov processes, *J. R. Stat. Soc.* B, (1984), 353–388.

DELLACHERIE, C. *Capacities et Processus Stochastiques*, Springer-Verlag, Berlin, 1972.

DELLACHERIE, C. AND MEYER, P. A. *Probabilities and Potentials,* North-Holland, Amsterdam, Vol. A–1978, Vol. B–1982, Vol. C–1988.

DENZEL, G. E., KEMENY, J. G., AND SNELL, J. L. Excessive functions for a class of continuous time Markov chains, in: *Markov Processes and Potential Theory,* J. Chover (Ed.), John Wiley, N.Y., 1967, pp. 61–88.

DOOB, J. L. Semimartingales and subharmonic functions, *Trans. Amer. Math. Soc.,* (1954) **77,** 85–121.

DOOB, J. L. Probability methods applied to the first boundary value problem, in: *Proc. 3rd. Berkeley Symp. Math. Stat. Prob.,* Univ. California Press, 1956, pp. 49–80.

DOOB, J. L. Discrete potential theory and boundaries, *J. Math. Mech.,* (1959) **8,** 433–458. (Correction, p. 993.)

DOOB, J. L. Compactification of the discrete state space of a Markov process, *Z. Wahrsch. verw. Gebiete,* (1968) **10,** 236–251.

DOOB, J. L. State spaces for Markov chains, *Trans. Amer. Math. Soc.,* (1970) **149,** 279–305.

DOOB, J. L. The structure of a Markov chain, in: *Proc. 6th Berkeley Symp. Math. Stat. Prob.,* Vol. 3, Univ. California Press, 1972, pp. 131–141.

DOOB, J. L. *Classical Potential Theory and its Probabilistic Counterpart,* Springer-Verlag, Berlin, 1984.

VAN DOORN, E. *Stochastic Monotonicity and Queueing Applications of Birth-Death Processes,* Springer-Verlag, Berlin, 1980.

DURRETT, R. *Brownian Motion and Martingales in Analysis,* Wadsworth, California, 1984.

DYNKIN, E. B. *Theory of Markov Processes,* Prentice–Hall, N.Y., 1961.

DYNKIN, E. B. *Markov Processes,* Vols. 1 and 2, Springer-Verlag, Berlin, 1965.

DYNKIN, E. B. Excessive functions and exit spaces of a Markov process. *Probability Theory, SIAM translations,* (1970) **15,** No 1.

DYNKIN, E. B. AND YUSHKEVICH, A. A. *Markov Processes* (*Theorems and Problems*), Plenum Press, N.Y., 1969.

ERICKSON, R. V. Stationary measures for a class of storage models driven by a Markov chain, *Ann. Math. Stat.,* (1972) **43,** 997–1007.

ETHIER, S. N. AND KURTZ, T. G. *Markov processes* (*Characterization and Convergence*), John Wiley, N.Y., 1986.

EVERITT, D. E. AND DOWNS, T. The output of multi-server queueing systems with Poisson input, in: *Proc. 9th ITC,* 1979, paper 415.

FEINSILVER, P. J. *Special Functions, Probability Semigroups, and Hamiltonian Flows,* (*Lecture Notes in Math.,* Vol. 696), Springer-Verlag, Berlin, 1978.

FEIGIN, P. D. *Conditional Distributions of Successive Birth Times and a Characterization of the Poisson Process,* Technion, Haifa, Mimeograph series, No. 192, 1977.

FEIGIN, P. D. AND RUBINSTEIN, E. Equivalent descriptions of perturbed Markov processes, *Stoch. Proc. Appl.*, (1979) **9,** 261–272.

FELLER, W. On the integro-differential equations of purely discontinuous Markov processes, *Trans. Amer. Math. Soc.*, (1940) **48,** 488–575. (Errata ibid. (1954) **58,** p. 478.)

FELLER, W. Boundaries induced by non-negative matrices, *Trans. Amer. Math. Soc.*, (1956) **83,** 19–54.

FELLER, W. On boundaries and lateral conditions for the Kolmogorov differential equations, *Ann. Math.*, (1957) **65,** 527–570.

FELLER, W. Some new connections between probability and classical analysis, in: *Proc. ICM 1958,* Cambridge Univ. Press, 1960, pp. 69–86.

FELLER, W. *An Introduction to Probability Theory and its Applications,* John Wiley, N.Y., Vol. 1–1968, Vol. 2–1966.

FERRANDIZ, J. M. AND LAZAR, A. A. Geometric aggregation, expansion and partial observations for Markov processes. Elect. Eng. Dept. and Center for Comm. Research, Columbia Univ., N.Y., Preprint, Sept. 1986.

FRANKEN, P., KÖNIG, D., ARNDT, U., AND SCHMIDT, V. *Queues and Point Processes,* John Wiley, N.Y., 1982.

FREEDMAN, D. *Markov Chains,* Holden Day, California, 1971.

GETOOR, R. K. Some remarks on a paper of Kingman, *Adv. Appl. Prob.*, (1974) **6,** 757–767.

GETOOR, R. K. Excursions of a Markov process, *Ann. Prob.*, (1979) **7,** 244–266.

GETOOR, R. K. *Markov Processes: Ray Processes and Right Processes,* (*Lecture Notes in Math.*, Vol. 440), Springer-Verlag, Berlin, 1975.

GETOOR, R. K. AND GLOVER, J. Markov processes with identical excessive measure, *Math. Z.*, (1983) **184,** 287–300.

GETOOR, R. K. AND SHARPE, M. J. Last exit decompositions and distributions, *Indiana Univ. Math. J.*, (1973) **23,** 377–404.

GETOOR, R. K. AND SHARPE, M. J. Last exit times and additive functionals, *Ann. Prob.*, (1973) **1,** 550–569.

GIROUX, G. Sur les lois d'entree et de sortie d'une chaine de Markov, *Ann. Inst. H. Poincare,* (1970) **6B,** 345–362.

GLOVER, J. Energy and the maximum principle for nonsymmetric Hunt process, *Theor. Probab. Appl.*, (1981) **26,** 757–768.

GLOVER, J. Raw time changes of Markov processes, *Ann. Prob.*, (1981) **9,** 90–102.

GLOVER, J. Applications of raw time changes to Markov processes, *Ann. Prob.*, (1981) **9**, 1019–1029.

GLOVER, J. Intrinsically homogeneous sets, splitting times, and the big shift, *Z. Wahrsch. verw. Gebiete,* (1981) **56**, 132–144.

GLOVER, J. Identifying Markov processes up to time change, in: *Seminar on Stoch. Proc. 1982,* Birkhauser, Boston, 1983, pp. 171–194.

GLOVER, J. Topics in energy and potential theory, in: *Seminar on Stoch. Proc. 1982,* Birkhauser, Boston, 1983, pp. 195–202.

GLOVER, J. Discontinuous time changes of semi-regenerative processes and balayage theorems, *Z. Wahrsch. verw. Gebiete,* (1983) **65**, 145–160.

GLOVER, J. Symmetry groups of Markov processes and the diagonal principle, *Ann. Prob.*, (1991) **19**, 562–586.

GLOVER, J. AND MITRO, J. Symmetries and functions of Markov processes. *Ann. Prob.*, (1990) **18**, 655–668.

GRAVES, S. C. AND KEILSON, J. The compensation method applied to a one-product production/inventory problem, *Math. Oper. Research,* (1981) **6**, 246–262.

GRIEGO, R. J. Local time as a derivative of occupation times, *Illinois J. Math.*, (1967) **11**, 54–63.

GZYL, H. Lévy systems for time-changed processes, *Ann. Prob.*, (1977) **5**, 565–570.

GZYL, H. On generators of subordinate semi-groups, *Ann. Prob.*, (1978) **6**, 975–983.

GZYL, H. Infinitesimal generators of time changed processes, *Ann. Prob.*, (1980) **8**, 716–726.

GZYL, C. AND GZYL, H. On entrance-exit distributions of Markov processes, *J. Appl. Prob.*, (1978) **15**, 78–86.

HALMOS, P. R. *Lectures on Ergodic Theory,* Math. Soc. Japan, 1956.

HALL, P. AND HEYDE, C. C. *Martingale Limit Theory and its Applications,* Academic Press, N.Y., 1980.

HELMS, L. L. *Introduction to Potential Theory,* John Wiley, N.Y., 1969.

HEWITT, E. AND STROMBERG, K. *Real and Abstract Analysis,* Springer-Verlag, Berlin, 1965.

HILLE, E. AND PHILLIPS, R. S. *Functional Analysis and Semi-Groups,* (Colloq. Publ.), Amer. Math. Soc., 1957.

HOFRI, M. Reduced state-space heuristic for symmetric Markovian queueing systems, in: *Queueing Theory and its Applications, Liber Amicorum for*

J. W. Cohen, O. J. Boxma, R. Syski (Eds.), North-Holland, Amsterdam, 1988, pp. 342–361.

HORVATH, J. *Topological Vector Spaces and Distributions*, Addison-Wesley, Mass., 1966.

HUNT, G. A. Markov processes and potentials, *Illinois J. Math.*, I–(1957) **1,** 49–93; II–(1957) **1,** 316–369; III–(1958) **2,** 151–213.

IKEDA, N., NAGASAWA, M., AND WATANABE, S. A construction of Markov process by piecing out, *Proc. Japan Acad.*, (1966) **42,** 370–375.

ISAACSON, D. L. AND MADSEN, R. W. *Markov Chains Theory and Applications*, John Wiley, N.Y., 1976.

JACOBSEN, M. Splitting times for Markov processes and a generalized Markov property for diffusions, *Z. Wahrsch. verw. Gebiete*, (1974) **30,** 27–43.

JACOBSEN, M. *Statistical Analysis of Counting Processes*, (*Lecture Notes in Stat.*, Vol. 12), Springer-Verlag, Berlin, 1982.

JACOBSEN, M. AND PITMAN, J. W. Birth, death, and conditioning of Markov chains, *Ann. Prob.*, (1977) **5,** 430–450.

JACOD, J. Multivariate point processes: predictable projections, Radon–Nikodym derivatives, representation of martingales, *Z. Wahrsch. verw. Gebiete*, (1975) **31,** 235–253.

JENSEN, A. *A Distribution Model*, Munksgaard, Copenhagen, 1954.

JENSEN, A. AND KENDALL, D. Denumerable Markov processes with bounded generators: a routine for calculating $p_{ij}(\infty)$, *J. Appl. Prob.*, (1971) **8,** 423–427.

KALIKOV, S. An entrance law which reaches equilibrium, *Ann. Prob.*, (1977) **5,** 467–469.

KALLENBERG, O. *Random Measures*, Academic Press, N.Y., 1976.

KALLENBERG, O. Splitting at backward times in regenerative sets, *Ann. Prob.*, (1981) **9,** 781–799.

KARLIN, S. AND MCGREGOR, J. The classification of birth-and-death processes, *Trans. Amer. Math. Soc.*, (1957) **86,** 366–400.

KARLIN, S. AND TAYLOR, H. M. *A First Course in Stochastic Processes*, 1975; *A Second Course in Stochastic Processes*, 1981; Academic Press, N.Y.

KARR, A. F. Markov chains and processes with a prescribed invariant measure, *Stoch. Proc. Appl.*, (1978) **7,** 277–290.

KARR, A. F. *Point Processes and their Statistical Inference*, Marcel Dekker, N.Y., 1986.

KARR, A. F. AND PITTENGER, A. O. The inverse balayage problem for Markov chains, *Stoch. Proc. Appl.*, (1978) **7,** 165–178.

KARR, A. F. AND PITTENGER, A. O. The inverse balayage problem for Markov chains, part II, *Stoch. Proc. Appl.*, (1979) **9,** 35–53.

KARR, A. F. AND PITTENGER, A. O. Structural properties of randomized times, *Prob. Th. Rel. Fields,* (1986) **72,** 395–415.

KASPI, H. Excursions of Markov processes via Markov additive processes, *Z. Wahrsch. verw. Gebiete,* (1983) **64,** 251–268.

KASPI, H. On invariant measures and dual excursions of Markov processes, *Z. Wahrsch. verw. Gebiete,* (1984) **66,** 185–204.

KASPI, H. Excursion laws of Markov processes in classical duality, *Ann. Prob.*, (1985) **13,** 492–518.

KASPI, H. Random time changes for processes with random birth and death, *Ann. Prob.*, (1988) **16,** 586–599.

KASPI, H. AND MAISONNEUVE, B. Regenerative systems on the real line, *Ann. Prob.*, (1988) **16,** 1306–1332.

KASPI, H. AND MAISONNEUVE, B. Predictable local times and exit systems. Preprint, Technion.

KASPI, H. AND PERRY, D. Inventory systems for perishable commodities with renewal input and Poisson output, *Adv. Appl. Prob.*, (1984) **16,** 402–421.

KASPI, H. AND RUBINOVITCH, M. Regenerative sets and their applications to Markov storage systems, in: *Queueing Theory and its Applications, Liber Amicorum for J. W. Cohen,* O. J. Boxma, R. Syski (Eds.), North-Holland, Amsterdam, 1988, pp. 413–427.

KEILSON, J. *Green's Function Methods in Probability Theory,* Griffin, London, 1965.

KEILSON, J. The role of Green's functions in congestion theory, in: *Congestion Theory,* W. L. Smith and W. E. Wilkinson (Eds.), Univ. N. Carolina Press, 1965, pp. 43–71.

KEILSON, J. Sojourn times, exit times and jitter in multivariate Markov processes, *Adv. Appl. Prob.*, (1974) **6,** 747–756.

KEILSON, J. *Markov Chains Models — Rarity and Exponentiality,* Springer-Verlag, Berlin, 1979.

KEILSON, J. AND SYSKI, R. Compensation measures in the theory of Markov chains, *Stoch. Proc. Appl.*, (1974) **2,** 59–72.

KEILSON, J. AND WISHART, D. M. G. Boundary problems for additive processes defined on a finite Markov chain, *Proc. Cambridge Phil. Soc.*, (1965) **61,** 173–190.

KELLY, F. P. *Reversibility and Stochastic Networks,* John Wiley, N.Y., 1979.

KELLY, F. P. Invariant measures and the Q-matrix, in: *Probability, Statistics and Analysis,* J. F. C. Kingman and G. H. E. Reuter (Eds.), Cambridge Univ. Press, 1983.

KELLY, J. L. *General Topology,* van Nostrand, N.Y., 1955.

KEMENY, J. G., SNELL, J. L. AND KNAPP, A. W. *Denumerable Markov Chains,* van Nostrand, N.Y., 1966.

KEMPERMAN, J. H. B. *The Passage Problem for a Stationary Markov Chain,* Univ. Chicago Press, 1961.

KESTEN, H. Hitting probabilities for single points for processes with stationary independent increments, *Mem. Amer. Math. Soc.,* (1969) **93.**

KINGMAN, J. F. C. Regenerative phenomena and the characterization of Markov transition probabilities, in: *Proc. 6th Berkeley Symp. Math. Stat. Prob.,* Vol. 3, Univ. California Press, 1972, pp. 241–262.

KINGMAN, J. F. C. *Regenerative Phenomena,* John Wiley, N.Y., 1972.

KURTZ, T. G. A note on sequences of continuous parameter Markov chains, *Ann. Math. Stat.,* (1969) **40,** 1078–1082.

KUSSMAUL, A. V. *Stochastic Integration and Generalized Martingales,* Pitman, London, 1978.

LAMB, CH. W. Decomposition and construction of Markov chains, *Z. Wahrsch. verw. Gebiete,* (1971) **19,** 213–224.

LAMPERTI, J. On random time substitution and the Feller property, in: *Markov Processes and Potential Theory,* J. Chover (Ed.), John Wiley, N.Y., 1967, pp. 87–102.

LAMPERTI, J. *Stochastic Processes (A Survey of the Mathematical Theory),* Springer-Verlag, Berlin, 1977.

LAMPERTI, J. AND SNELL, J. L. Martin boundaries for certain Markov chains, *J. Math. Soc. Japan,* (1963) **15,** 113–128.

LE GALL, P. *Les Systemes avec ou sans Attente et les Processus Stochastiques,* Dunode, Paris, 1962.

LE JAN, Y. Martingales et changements de temps, in: *Seminaire de Prob. XIII,* (*Lecture Notes in Math.,* Vol. 721), Springer-Verlag, Berlin, 1979, pp. 385–399.

LENZ, N. Extinction probability, regularity and asymptotic growth of Markovian population, *J. Appl. Prob.,* (1981) **18,** 1–18.

LEVIATAN, T. Perturbations of Markov processes, *J. Functional Analysis,* (1972) **10,** 309–325.

LIPTSER, R. S. AND SHIRYAEV, A. N. *Statistics of Random Processes,* Springer-Verlag, 1977. (Polish Edition: PAN, Warsaw, 1981.)

LOÉVE, M. *Probability Theory,* van Nostrand, N.Y., 1963.

MAISONNEUVE, B. Ensembles regeneratifs, temps locaux et subordinateurs, *Lecture Notes Math.*, Vol. 191, Springer-Verlag, Berlin, 1971, pp. 147–169.

MAISONNEUVE, B. Exit systems, *Ann. Prob.*, (1975) **3,** 399–411.

MAISONNEUVE, B. On the structure of a certain excursion of a Markov process, *Z. Wahrsch. verw. Gebiete,* (1979) **47,** 61–67.

MAISONNEUVE, B. AND MEYER, P. A. Ensembles aleatoires Markoviens homogenes, *Lecture Notes in Math.*, Vol. 381, Springer-Verlag, Berlin, 1974, pp. 172–261.

MEYER, P. A. *Probability and Potentials,* Blaisdell Publ. Co., Mass., 1966.

MEYER, P. A. *Processus de Markov: la Frontiere de Martin,* (*Lecture Notes in Math.*, Vol. 77), Springer-Verlag, Berlin, 1968.

MEYER, P. A. Renaissance, recollements, melanges, relentissement de processus de Markov, *Ann. Inst. Fourier, Grenoble,* (1975) **25,** 465–497.

MEYER, P. A., SMYTHE, R. T. AND WALSH, J. B. Birth and death of Markov process, in: *Proc. 6th Berkeley Symp. Math. Stat. Prob.*, Vol. 3, Univ. California Press, 1972,pp. 295–305.

MILLAR, P. W. Random times and decomposition theorems, in: *Symp. Pure Math.*, Vol. 31, J. L. Doob (Ed.), Amer. Math. Soc., 1977, pp. 91–103.

MILLAR, P. W. A path decomposition for Markov processes, *Ann. Prob.*, (1978) **6,** 345–348.

MILNE, C. Transient behavior of the interrupted Poisson process, *J. R. Stat. Soc.* B, (1982) **44,** 398–405.

MITRO, J. B. Dual Markov processes: construction of a useful auxiliary process, *Z. Wahrsch. verw. Gebiete,* (1979) **47,** 139–156.

MITRO, J. B. Dual Markov processes: applications of a useful auxiliary process, *Z. Wahrsch. verw. Gebiete,* (1979) **48,** 97–114.

MITRO, J. B. A discontinuous time change for natural additive functionals which preserves duality, *Prob. Th. Rel. Fields,* (1986) **72,** 463–470.

MITRO, J. B. A note on discontinuous time changes of Markov processes, *Stoch. Proc. Appl.*, (1986) **23,** 343–348.

NEUTS, M. F. Probability distributions of phase type, in: *Liber Amicorum Prof. H. Florin,* Univ. Louvain, 1975, pp. 173–205.

NEUTS, M. F. Computational uses of the method of phases in the theory of queues, *Comput. Math. Appl.*, (1975) **1,** 151–166.

NEUTS, M. F. *Matrix-Geometric Solutions in Stochastic Models* (*an Algorithmic Approach*), The Johns Hopkins Univ. Press, Baltimore, 1981.

NEUTS, M. F. The caudal characteristic curve of queues, *Adv. Appl. Prob.*, (1986) **18,** 221–254.

NEUTS, M. F. AND RESNICK, S. I. On the times of birth in a linear birth process, *J. Austral. Math. Soc.*, (1971) **12,** 473–475.

NEVEU, J. Entrance, exit and fictitious states for Markov chains, in: *Proc. Aarhus Colloq. Combin. Prob.*, 1962, pp. 64–68.

NEVEU, J. Chaines de Markov et Theorie du Potentiel, *Ann. Fac. Sciences, Clermont-Ferrand,* (1964) **3,** 37–89.

NEVEU, J. *Mathematical Foundations of the Calculus of Probability,* Holden-Day, California, 1965.

PARTHASARATHY, P. R. A transient solution to an M/M/1 queue: a simple approach, *Adv. Appl. Prob.*, (1987) **19,** 997–998.

PEGDEN, C. D. AND ROSENSHINE, M. Some new results for the M/M/1 queue, *Management Science,* (1982) **28,** 821–828.

PEIRAM, L. Accuracy of measurements on Markovian service systems, in: *Proc. 9th ITC,* Torremolinos, 1979, paper 236.

PITMAN, J. W. Lévy systems and path decompositions, in: *Seminar Stoch. Proc. 1981,* Birkhauser, Boston, 1981, pp. 79–110.

PITMAN, J. W. AND SPEED, T. P. A note on random times, *Stoch. Proc. Appl.*, (1973) **1,** 369–374.

PITTENGER, A. O. Boundary decomposition of locally Hunt processes, in: *Advances in Prob. and Related Topics,* P. Ney (Ed.), Marcel Dekker, N.Y., 1970, pp. 117–160.

PITTENGER, A. O. Last exit times and the Q-matrices of Markov chains, *Z. Wahrsch. verw. Gebiete,* (1971) **20,** 143–162.

PITTENGER, A. O. Regular birth and death times, in: *Seminar Stoch. Proc. 1981,* Birkhauser, Boston, 1981, pp. 111–134.

PITTENGER, A. O. Regular birth times for Markov processes, *Ann. Prob.*, (1981) **9,** 769–780.

PITTENGER, A. O. Time changes of Markov chains, *Stoch. Proc. Appl.*, (1982) **13,** 189–199.

PITTENGER, A. O. Hitting times of sequences, *Stoch. Proc. Appl.*, (1987) **24,** 225–240.

PITTENGER, A. O. AND SHARPE, M. J. Regular birth and death times, *Z. Wahrsch. verw. Gebiete,* (1981) **58,** 231–246.

PITTENGER, A. O. AND SHIH, C. T. Coterminal families and the strong Markov property, *Trans. Amer. Math. Soc.*, (1973) **182,** 1–42.

POP-STOJANOVIC, Z. R. AND RAO, M. Some results on energy, in: *Seminar Stoch. Proc. 1981,* Birkhauser, Boston, 1981, pp. 135–150.

POP-STOJANOVIC, Z. R. AND RAO, M. On convergence in energy (abstract), in: *14th Stoch. Proc. Appl. Conference,* Göteborg, 1984.

PRABHU, N. U. *Queues and Inventories*, John Wiley, N.Y., 1965.

PRABHU, N. U. *Stochastic Storage Processes*, Springer-Verlag, Berlin, 1980.

PROTTER, P. AND SHARPE, M. J. Martingales with given absolute value, *Ann. Prob.*, (1979) **7**, 1056–1058.

RAO, M. M. *Foundations of Stochastic Analysis*, Academic Press, N.Y., 1981.

REVUZ, D. Mesures associees aux fonctionnelles additive de Markov, I. *Trans. Amer. Math. Soc.*, (1970) **148**, 501–531; II. *Z. Wahrsch. verw. Gebiete*, (1970) **16**, 336–344.

REVUZ, D. *Markov Chains*, North-Holland, Amsterdam, 1975.

ROLIN, J. M. The inverse of the continuous additive functional, *Pacific J. Math.*, (1975) **58**, 585–604.

ROLSKI, T. *Stationary Random Processes Associated with Point Processes*, (*Lecture Notes in Stat.*, Vol. 5), Springer-Verlag, Berlin, 1981.

ROSENBLATT, M. *Random Processes*, Oxford Univ. Press, 1962.

ROSENBLATT, M. *Markov Processes. Structure and Asymptotic Behavior*, Springer-Verlag, Berlin, 1970.

ROSENKRANTZ, W. A. Some martingales associated with queueing and storage processes, *Z. Wahrsch. verw. Gebiete*, (1981) **58**, 205–222.

ROSENKRANTZ, W. A. Calculation of the Laplace transform of the length of the busy period for M/G/1 queue via martingales, *Ann. Prob.*, (1983) **11**, 817–818.

ROTA, G. C. Une theorie unifiee des martingales et des moyenues ergodique, *C. R. Acad. Sciences, Paris*, (1961) **252**, 2064–2066.

RUBINSTEIN, E. Perturbation model in Markov chains. Thesis, Technion, Haifa, 1978.

SAATY, T. L. Some stochastic processes with absorbing barriers, *J. R. Stat. Soc.* B, (1961) **23**, 319–334.

SCHASSBERGER, R. On the work load process in a general preemptive resume priority queue, *J. Appl. Prob.*, (1972) **9**, 588–603.

SCHEHRER, R. G. On generalized state equations, cyclic processes and cut-off priority systems with hysteresis, in: *Proc. 12th ITC*, Torino, 1988, paper 2.4B.3.1.

SCHWEITZER, P. J. A survey of aggregation-disaggregation in large Markov chains. Working paper No. QM-89-12, Grad. School Business Adm., Univ. Rochester, Rochester, N.Y., 1989.

SHARPE, M. *General Theory of Markov Processes*, Academic Press, N.Y., 1988.

SIEGRIST, K. Harmonic functions and the Dirichlet problem for revival Markov processes, *Ann. Prob.*, (1983) **11**, 624–634.

SILVERSTEIN, M. L. *Symmetric Markov Processes, (Lecture Notes in Math.,* Vol. 426), Springer-Verlag, Berlin, 1974.

SLUD, E. V. *Martingale Methods in Statistics,* John Wiley, (in press).

SMYTHE, R. T. Approximation of local time at regular boundary points of Markov chains, *Z. Wahrsch. verw. Gebiete,* (1971) **20,** 134–142.

SPITZER, F. *Principles of Random Walk,* van Nostrand, N.Y., 1964.

SYSKI, R. Markovian queues, in: *Congestion Theory,* W. L. Smith and W. E. Wilkinson (Eds.), Univ. N. Carolina Press, 1965, pp. 170–227.

SYSKI, R. Potential theory for Markov chains, in: *Probabilistic Methods in Appl. Math.,* Vol. 3, A. T. Bharucha-Reid (Ed.), Academic Press, N.Y., 1973, pp. 213–276.

SYSKI, R. Queues and potentials, in: *Proc. XX TIMS Meeting,* Vol. 2, E. Shlifer (Ed.), Jerusalem Acad. Press, 1975, pp. 547–554.

SYSKI, R. Perturbation models, *Stoch. Proc. Appl.,* (1977) **5,** 93–130.

SYSKI, R. Ergodic potential, *Stoch. Proc. Appl.,* (1978) **7,** 311–336.

SYSKI, R. Random time change in Markov chains. Techn. Report, Univ. Maryland, 1978, (unpublished).

SYSKI, R. Energy of Markov chains, *Adv. Appl. Prob.,* (1979) **11,** 542–575.

SYSKI, R. Phase-type distributions and perturbation model, *Applicationes Mathematicae* (Warsaw), (1982) **XVII,** 3, 377–399.

SYSKI, R. Markov chains in Geology (abstract), *Stoch. Proc. Appl.,* (1984) **17,** 17.

SYSKI, R. Markovian queues in teletraffic theory, in: *Proc. 3rd Int. ITC Seminar on Fundamentals of Teletraffic Theory,* Inst. Probl. Information Transmission, Moscow, 1984, pp. 430–440.

SYSKI, R. *Introduction to Congestion Theory in Telephone Systems,* 2nd ed., North-Holland, Amsterdam, 1986.

SYSKI, R. Markovian functionals in teletraffic analysis, in: *Teletraffic Analysis and Computer Performance Evaluation,* O. J. Boxma, J. W. Cohen and H. C. Tijms (Eds.), North-Holland, Amsterdam, 1986, pp. 303–317.

SYSKI, R. Note on a random walk with absorption. Techn. Report No. 420, Math. Dept., Univ. of Utrecht, 1986.

SYSKI, R. Markov processes and teletraffic, in: *Queueing Theory and its Applications, Liber Amicorum for J. W. Cohen,* O. J. Boxma and R. Syski (Eds.), North-Holland, Amsterdam, 1988, pp. 95–126.

SYSKI, R. Stochastic processes, in: *Encycl. Statist. Sciences,* Vol. 8, S. Kotz and N. L. Johnson (Eds.), John Wiley, N.Y., 1988, pp. 836–851.

SYSKI, R. Further comments on the solution of the M/M/1 queue, *Adv. Appl. Prob.*, (1988) **20,** 693.

SYSKI, R. A note on revival models for Markov chains, *Amer. J. Math. Mgt. Sciences,* (1990) **10,** 159–183.

TAKACS, L. *Introduction to the Theory of Queues,* Oxford Univ. Press, 1962.

TAKACS, L. A combinatorial method in the theory of Markov chains, *J. Math. Anal. Appl.*, (1964) **9,** 153–161.

TAKACS, L. Occupation time problems in the theory of queues, *Oper. Res.*, (1964) **12,** 753–767.

TAKACS, L. Application of ballot theorems in the theory of queues, in: *Congestion Theory,* W. L. Smith and W. E. Wilkinson (Eds.), Univ. N. Carolina Press, 1965, pp. 337–398.

TAKACS, L. *Combinatorial Methods in the Theory of Stochastic Processes,* John Wiley, N.Y., 1967.

TAKACS, L. On a linear transformation in the theory of probability, *Acta Scient. Mathem.*, (1972) **33,** 15–24.

TAKACS, L. On fluctuations of sums of random variables, *Studies in Probability and Ergodic Theory,* (1978) **2,** 45–93.

TAKACS, L. Fluctuation problems for Bernoulli trials, *SIAM Rev.*, (1979) **21,** 222–228.

TAKACS, L. The problem of coincidences, *Archive for History of Exact Sciences,* (1980) **21,** 229–244.

TAKACS, L. Random flights on regular polytopes, *SIAM J. Alg. Disc. Math.*, (1981) **2,** 153–171.

TZELNIC, P. The length of path for finite Markov chains and its application to modelling program behavior and interleaved memory systems, in: *Appl. Prob. Computer Sciences: the Interface,* Vol. 2, R. L. Disney and T. J. Ott (Eds.), Birkhauser, Boston, 1982, pp. 375–405.

WATANABE, T. On the theory of Martin boundaries induced by countable Markov processes, *Mem. Coll. Sci. Univ. Kyoto,* (1960) **A 33,** 39–108.

WAUGH, W. A. O'N. Transformation of a birth process into a Poisson process, *J. R. Stat. Soc.* B, (1970) **32,** 418–431.

WEIDENFELD, G. Changements de temps de processus de Markov, *Z. Wahrsch. verw. Gebiete,* (1980) **53,** 123–146.

WHITT, W. The continuity of queues, *Adv. Appl. Prob.*, (1974) **6,** 175–183.

WHITT, W. Some useful functions for functional limit theorems, *Math. Oper. Research,* (1980) **5,** 67–81.

WHITT, W. Comparing counting processes and queues, *Adv. Appl. Prob.*, (1981) **13,** 207–220.

WILLIAMS, D. Fictitious states, coupled laws and local times, *Z. Wahrsch. verw. Gebiete*, (1969) **11**, 288–310.

WILLIAMS, D. The Q-matrix problem, *Lecture Notes in Math.*, Vol. 511, Springer-Verlag, Berlin, 1976, pp. 216–234.

WILLIAMS, D. The Q-matrix problem, II: Kolmogorov backward equations, *Lecture Notes in Math.*, Vol. 511, Springer-Verlag, Berlin, 1976, pp. 505–520.

WILLIAMS, D. Some Q-matrix problems, in: *Symposium on Pure Math.*, Vol. 3, J. L. Doob (Ed.), Amer. Math. Soc., ,1977, pp. 165–169.

WILLIAMS, D. The Q-matrix problem, III: The Lévy kernel problem for chains, *Lecture Notes in Math.*, Vol. 649, Springer-Verlag, Berlin, 1978, pp. 310–332.

WILLIAMS, D. *Diffusions, Markov Processes and Martingales.* Vol. 1, John Wiley, N.Y., 1979; Vol. 2, *Ito Calculus* (jointly with L. C. G. Rogers), 1987.

INDEX OF NAMES

INDEX OF SYMBOLS

This is the list of main symbols used consistently throughout the book. For further notation see also pages 62–65 (§ 2.1.5) *and page* 78 (§ 2.2.6).

INDEX OF SUBJECTS

Entries are listed under the main subject, and cross-referenced. For example, "energy of a potential supermartingale" is under "energy", but "potential supermartingale" under "potential", whereas the entry "supermartingale" carries cross-reference to "potentials".

Entries with prefixes are listed under corresponding entries without them. For example, "co-balayage" under "balayage", "α-chain" under "chain"; similarly with "pre-" and "post-".

As a rule, page numbers refer to the definition or to the first appearance of a subject; successive appearances are listed when alternative aspects of a subject are presented.

S

T

U

V

W